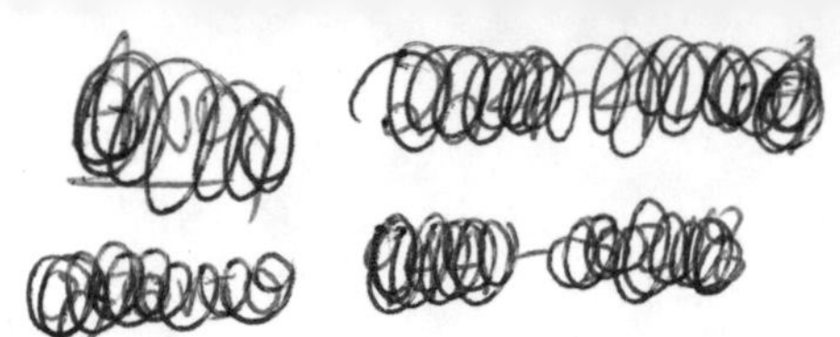

The Earth

FOURTH EDITION

The Earth

An Introduction to Physical Geology

Edward J. Tarbuck
Frederick K. Lutgens

ILLINOIS CENTRAL COLLEGE

Macmillan Publishing Company
New York

Maxwell Macmillan Canada
Toronto

Maxwell Macmillan International
New York Oxford Singapore Sydney

Cover photo © 1992 Kerrick James. Temple of the Sun (sandstone monolith), Cathedral Valley. Capitol Reef National Park, Utah.

Editor: Robert A. McConnin
Production Editor: Rex Davidson
Art Coordinator: Lorraine Woost
Photo Researcher: Eloise Marion
Text Designer: Cynthia Brunk
Cover Designer: Cathleen Norz
Production Buyer: Pamela D. Bennett
Illustrations by Dennis Tasa, Tasa Graphic Arts, Inc.

This book was set in Garamond by York Graphic Services, Inc. and was printed and bound by R. R. Donnelley & Sons Company. The cover was printed by Lehigh Press, Inc.

Printed in the United States of America

Macmillan Publishing Company
866 Third Avenue
New York, New York 10022

Macmillan Publishing Company is part of the Maxwell Communication Group of Companies.

Maxwell Macmillan Canada, Inc.
1200 Eglinton Avenue East, Suite 200
Don Mills, Ontario M3C 3N1

Library of Congress Cataloging-in-Publication Data
Tarbuck, Edward J.
The earth : an introduction to physical geology / Edward J. Tarbuck, Frederick K. Lutgens.—4th ed.
p. cm.
Includes index.
ISBN 0-02-419012-8
1. Physical geology. I. Lutgens, Frederick K.
II. Title.
QE28.2.T37 1993
550—dc20 92-18209
CIP

Printing: 2 3 4 5 6 7 8 9
Year: 3 4 5 6 7

To our students, who have often been our best teachers.

Preface

In recent years, media reports have made us increasingly aware of the geological forces at work in our physical environment. News stories graphically portray the violent force of a volcanic eruption, the devastation created by a strong earthquake, and the large numbers left homeless by mudflows and flooding. Such events, and many others as well, are destructive to life and property, and we must be better able to understand and deal with them. However, our natural environment has an even greater importance, for the earth is our home. The earth not only provides the mineral resources so basic to modern society, but it is also the source of most of the ingredients necessary to support life. Therefore, as many members of society as possible should acquire a basic understanding of how the earth works.

With this in mind, we have written a text to help people increase their understanding of our physical environment. We hope this knowledge will encourage some to actively participate in the preservation of the environment, while others may be sufficiently stimulated to pursue a career in the earth sciences. Equally important, however, is our belief that a basic understanding of the earth will greatly enhance appreciation of our planet and thereby enrich the reader's life.

The Fourth Edition of *The Earth: An Introduction to Physical Geology,* like its predecessors, is intended for both science majors and non-majors taking their first course in geology. We have attempted to write a text that is not only informative and timely, but one that is highly usable as well. The language is straightforward and written to be understood by a student with little or no college-level science experience. We have deliberately refrained from using excessive jargon, and when new terms are introduced, they are placed in boldface and defined. A list of key terms with page references is found at the end of each chapter, and a glossary is included at the conclusion of the text for easy reference to important terms. Further, review questions conclude each chapter to help the student prepare for exams and quizzes. Useful information on metric conversions, the periodic table of the elements, common minerals, and topographic maps is also provided in the appendices.

In the first three editions of this text special attention was given to the quality of photographs and artwork because geology is a highly visual science. This emphasis has been maintained in the Fourth Edition. More than 150 new color photographs appear in this revision. The photographs and images were carefully selected to add realism to the subject and to heighten the readers interest. Moreover, the already excellent art program of the earlier editions has been strengthened in the Fourth Edition. Because we believe that carefully planned and executed line art will significantly aid student understanding by making difficult concepts less abstract, more than 125 new and redrawn figures appear in the Fourth Edition. Once again, the text has benefited greatly from the talents and imaginative production of Dennis Tasa of Tasa Graphic Arts, Inc.

The Fourth Edition of *The Earth: An Introduction to Physical Geology* represents a thorough revision. Extensive rewriting has made many discussions more timely and more readable. It should be emphasized, however, that the main focus of the Fourth Edition remains the same as in the first three editions—to foster a basic understanding of physical geology. As much as possible, we have attempted to provide the reader with a sense of the observational techniques and reasoning processes that constitute the discipline of geology. As with other sciences, geology is much more than a mere collection of facts. At its heart are the various methods of probing the earth aimed at uncovering its secrets. These methods involve the collection of the necessary data used to test hypotheses about the nature of the forces that shape our changing planet. In addition to gaining a better understanding of these natural processes, this activity often leads to a re-evaluation of ideas formulated at a time when less information was available. An excellent example of the way geological "truths" are uncovered and reworked is found in Chapter 18. Here we trace the historical formation and subsequent rejection of the hypothesis that continents drift about the face of the earth and then examine the data that led to the "rebirth" of this idea as part of a more encompassing theory known as plate tectonics.

The organization of the text remains intentionally traditional. Following the overview of geology in the introductory chapter, we turn to a discussion of earth materials and the related processes of volcanism and weathering. Next, a discussion of a most basic topic, geologic time, is followed by an examination of the geological work of gravity, water, wind, and ice in modifying and sculpturing landscapes. After this look at external processes, we examine the earth's internal structure and the processes that deform rocks and give rise to mountains. Finally, the text concludes with chapters on resources and the solar system. This particular organization was selected largely to accommodate the study of minerals and rocks in the laboratory, which usually comes early in the course. Realizing that some instructors may prefer to structure their courses differently, we made each chapter self-contained so that chapters may be taught in a different sequence. Thus, the instructor who wishes to discuss earthquakes, plate tectonics, and mountain building prior to dealing with erosional processes may do so without difficulty. We also chose to introduce plate tectonics in the first chapter so that this important theory could be incorporated in appropriate places throughout the text. Although plate tectonics is an integral part of this book, we have not included it at the expense of other topics. While it is true that plate tectonics is fascinating and of utmost importance in understanding the dynamics of the earth, other topics are equally worthwhile and interesting to the beginning student. It should also be noted that a separate chapter on environmental problems has not been included. Instead, we have incorporated these topics into the text at appropriate places. For example, discussions of the pollution of wells and land subsidence associated with groundwater withdrawal are treated in the chapter on groundwater (Chapter 11), while shoreline erosion problems are taken up in the chapter on shorelines (Chapter 14).

A comparison of the Fourth Edition with earlier versions will reveal that a major change to the text is the addition of 52 special interest boxes. This feature was added in order that applications, case studies, and interesting facts and related science principles could be touched upon without significantly disrupting the flow of the basic text discussions. We believe that the boxes provide the reader with a diversity of information that will make the study of geology more interesting and meaningful.

In addition to the new special interest boxes, discussions of many basic topics have been strengthened in the Fourth Edition. Expanded and updated coverage of igneous rocks, the origin of magma, earthquake prediction, plate tectonics and earthquakes, and Andean-type orogenesis will be seen. Moreover, sections on the Ice Age, shoreline erosion problems, the driving mechanisms of plate tectonics, and the geology of Venus and Neptune have also been revised. A new discussion on soil erosion has been added to Chapter 5.

Acknowledgments

As with any project of this scope, the contributions of others were very important and too numerous for us to give proper credit to each person involved. The credit for the content must go to our professors, colleagues, and students, who challenged us to search for a deeper understanding. We wish to express our thanks to the many individuals, institutions, and government agencies that provided information, photographs, and illustrations for use in this text. A special debt of gratitude goes to those colleagues who prepared in-depth prerevision reviews of the Third Edition of *The Earth*. Their critical comments and thoughtful input helped guide our revision and strengthen the Fourth Edition.

We first wish to thank the reviewers of previous editions: Anne Erdmann, University of Minnesota; David T. King, Jr., Auburn University; George S. Koch, Jr., University of Georgia; Carleton B. Moore, Arizona State University; Anne Pasch, University of Alaska-Anchorage; Vernon P. Scott, Oklahoma State University; Chris Suczek, Western Washington University; and J. Marion Wampler, Georgia Institute of Technology. We thank the reviewers of this edition: Andrew P. Barth, Indiana U./Purdue U. at Indianapolis; Vincent S. Cronin, University of Wisconsin–Milwaukee; Gene L. LaBerge, University of Wisconsin–Oshkosh; David T. King, Jr., Auburn University; George S. Koch, Jr., University of Georgia; Richard L. Mauger, East Carolina University; Judith B. Moody, Moody & Associates, Athens, OH; Carleton B. Moore, Arizona State University; Mary J. Richardson, Texas A & M University; Charles J. Ritter, University of Dayton; Marian M. Smith, Western Michigan University; William A. Smith, Western Michigan University; and Charles W. Yarrison, Kutztown University of Pennsylvania.

Our thanks also go to our senior editor Bob McConnin. His professional and efficient style makes a big difference to us. We must also express appreciation to our production editor Rex Davidson and to the rest of the Macmillan production team. As always, they skillfully transformed our manuscript into a finished product. They are true professionals with whom we feel fortunate to be associated.

Brief Contents

Contents

8

Geologic Time

9

Mass Wasting

10

Running Water

11

Groundwater

12

Glaciers and Glaciation

13

Deserts and Winds

14

Shorelines

22
Planetary Geology

Appendices

The Earth

1
An Introduction to Geology

Opposite: This lava fountain and river of lava were produced by a recent eruption of Hawaii's Kilauea Volcano. (Douglas Peebles Photography). (Top photo by J. D. Griggs, U.S. Geological Survey)

The spectacular eruption of a volcano, the terror brought by an earthquake, the magnificent scenery of a mountain valley, and the destruction created by a landslide are all subjects for the geologist (Figure 1.1). The study of geology deals with many fascinating and practical questions about our physical environment. What forces produce mountains? Will there soon be another great earthquake in San Francisco? What was the Ice Age like? Will there be another? What created this cave and the stone icicles hanging from its ceilings? Should we look for water here? Is strip mining practical in this area? Will oil be found if a well is drilled at that location? What if the landfill is located in the old quarry?

The subject of this text is **geology**, a word that literally means "the study of the earth." To under-

FIGURE 1.1
Snow covers Mesa Arch in Utah's Canyonlands National Park as sunset light strikes the LaSal Mountains in the background. Geologists study the processes that created this scene. (Photo by Jack W. Dykinga)

stand the earth is not an easy task because our planet is not an unchanging mass of rock but rather a dynamic body with a long and complex history.

The science of geology is traditionally divided into two broad areas–physical and historical. **Physical geology**, which is the primary focus of this book, examines the materials composing the earth and seeks to understand the many processes that operate beneath and upon its surface. The aim of **historical geology**, on the other hand, is to understand the origin of the earth and its development through time. Thus, it strives to establish an orderly chronological arrangement of the multitude of physical and biological changes that have occurred in the geologic past. The study of physical geology logically precedes the study of earth history because we must first understand how the earth works before we attempt to unravel its past.

SOME HISTORICAL NOTES ABOUT GEOLOGY

The nature of our earth—its materials and processes—has been a focus of study for centuries. Writings about such topics as fossils, gems, earthquakes, and volcanoes date back to the early Greeks, more than 2300 years ago. Certainly the most influential Greek philosopher was Aristotle. Unfortunately, Aristotle's explanations about the natural world were not based on keen observations and experiments. Instead they were arbitrary pronouncements. He believed that rocks were created under the "influence" of the stars and that earthquakes occurred when air crowded into the ground, was heated by central fires, and escaped explosively. When confronted with a fossil fish, he explained that, "a great many fishes live in the earth motionless and are found when excavations are made." Although Aristotle's explanations may have been adequate for his day, they unfortunately continued to be expounded for many centuries, thus thwarting the acceptance of more up-to-date accounts. Frank D. Adams states in *The Birth and Development of the Geological Sciences* (New York: Dover, 1938) that "throughout the Middle Ages Aristotle was regarded as the head and chief of all philosophers; one whose opinion on any subject was authoritative and final."

Catastrophism

During the seventeenth and eighteenth centuries the doctrine of **catastrophism** strongly influenced the formulation of explanations about the dynamics of the earth. Briefly stated, catastrophists believed that the earth's landscape had been shaped primarily by great catastrophes. Features such as mountains and canyons, which today we know take great periods of time to form, were explained as having been produced by sudden and often worldwide disasters produced by unknowable causes that no longer operate. This philosophy was an attempt to fit the rates of earth processes to the then-current ideas on the age of the earth. In the mid-seventeenth century, James Ussher, Anglican Archbishop of Armagh, Primate of all Ireland, published a major work that had immediate and profound influence. A respected scholar of the Bible, Ussher constructed a chronology of human and earth history in which he determined that the earth was only a few thousands of years old, having been created in 4004 B.C. Ussher's treatise earned widespread acceptance among scientific and religious leaders alike, and his chronology was soon printed in the margins of the Bible itself.

The relationship between catastrophism and the age of the earth has been summarized as follows:

> That the earth had been through tremendous adventures and had seen mighty changes during its obscure past was plainly evident to every inquiring eye; but to concentrate these changes into a few brief millenniums required a tailor-made philosophy, a philosophy whose basis was sudden and violent change.*

The Birth of Modern Geology

The late eighteenth century is generally regarded as the beginning of modern geology, for it was during this time that James Hutton (Figure 1.2), a Scottish physician and gentleman farmer, published his *Theory of the Earth* in which he put forth a principle that came to be known as the doctrine of **uniformitarianism.** Uniformitarianism is a basic part of modern geology. It simply states that the physical, chemical, and biological laws that operate today have also operated in the geologic past. That is to say that the forces and processes that we observe presently shaping our planet have been at work for a very long time. Thus, to understand ancient rocks, we must first understand present-day processes and their results. This idea is commonly stated by saying "the present is the key to the past."

Prior to Hutton's *Theory of the Earth,* no one had effectively demonstrated that geological processes occur over extremely long periods of time. However, Hutton persuasively argued that forces which appear

*H. E. Brown, V. E. Monnett, and J. W. Stovall, *Introduction to Geology* (New York: Blaisdell, 1958).

FIGURE 1.2
James Hutton, the 18th century Scottish geologist who is often called the "father of modern geology." (Photo courtesy of the British Museum)

small could, over long spans of time, produce effects that were just as great as those resulting from sudden catastrophic events. Unlike his predecessors, Hutton carefully cited verifiable observations to support his ideas. For example, when he argued that mountains are sculptured and ultimately destroyed by weathering and the work of running water, and that their wastes are carried to the oceans by processes that can be observed, Hutton said, "We have a chain of facts which clearly demonstrates . . . that the materials of the wasted mountains have traveled through the rivers"; and further, "There is not one step in all this progress . . . that is not to be actually perceived." He then went on to summarize this thought by asking a question and immediately providing the answer: "What more can we require? Nothing but time."

Since Hutton's literary style was cumbersome and difficult, his work was not widely read nor easily understood. However, that began to change in 1802, when Hutton's friend and colleague, John Playfair, published *Illustrations of the Huttonian Theory,* a volume in which he presented Hutton's ideas in a much clearer and attractive form. The following well-known passage from Playfair's work, which is a restatement of Hutton's basic principle, illustrates this style:

> Amid all the revolutions of the Globe, the economy of nature has been uniform and her laws are the only things which have resisted the general movement. The rivers and the rocks, the seas and the continents have been changed in all their parts; but the laws which direct those changes, and the rules to which they are subject, have remained invariably the same.

Although Playfair's book gave impetus to Hutton's ideas and aided the cause of modern geology, it is the English geologist Sir Charles Lyell (Figure 1.3) who is given the most credit for advancing the basic principles of modern geology. Between 1830 and 1872 he produced eleven editions of his great work, *Principles of Geology.* As was customary, Lyell's book had a rather lengthy subtitle that outlined the main theme of the work: *Being an Attempt to Explain the Former Changes of the Earth's Surface, by Reference to Causes now in Operation.* In the text, he painstakingly illustrated the concept of the uniformity of nature through time. He was able to show more convincingly than his predecessors that the geologic processes which are observed today can be assumed to have operated in the past. Although the doctrine of uniformitarianism did not originate with Lyell, a fact that he openly acknowledged, he is the person who was most successful in interpreting and publicizing it for society at large.

Today the basic tenets of uniformitarianism are just as viable as in Lyell's day. Indeed, we realize more strongly than ever that the present gives us insight into the past and that the physical, chemical,

FIGURE 1.3
Charles Lyell. Lyell's book, *Principles of Geology,* did much to advance modern geology. (Courtesy of the Institute of Geological Sciences, London)

and biological laws that govern geological processes remain unchanging through time. However, we also understand that the doctrine should not be taken too literally. To say that geological processes in the past were the same as those occurring today is not to suggest that they always had the same relative importance and operated at precisely the same rate. Although the same processes have prevailed through time, their rates have undoubtedly varied.*

The acceptance of uniformitarianism meant the acceptance of a very long history for the earth, for although processes vary in their intensity, they still take a very long time to create or destroy major features of the landscape.

For example, rocks containing fossils of organisms that lived in the sea more than 15 million years ago are now part of mountains that stand 3000 meters (9800 feet) above sea level. This means that the mountains were uplifted 3000 meters in about 15 million years, which works out to a rate of only 0.2 millimeter per year! Rates of erosion (the processes that wear away land) can be equally slow (Figure 1.4). Estimates indicate that the North American continent is being lowered at a rate of just 3 centimeters per 1000 years. Thus, as you can see, tens of millions of years are required for nature to build mountains and wear them down again. But even these time spans are relatively short on the time scale of earth history, for the rock record contains evidence that

*It should be pointed out that during the earth's formative period, when our planet was very different from today, some processes were at work that are no longer operating.

FIGURE 1.4
Weathering and erosion have gradually carved these striking rock masses in Arizona's Monument Valley. Geologic processes often act so slowly that changes may not be visible during an entire human lifetime. (Photo by Michael Collier)

BOX 1.1

Geology: An Environmental Science

Environment refers to everything that surrounds and influences an organism. Some of these conditions are biological and social, others are abiotic or nonliving. The factors in this latter category are collectively referred to as our *physical environment.* The physical environment encompasses water, air, soil, and rock, as well as conditions such as temperature, humidity, and sunlight. Geological phenomena and processes are among the most basic in the physical environment. In this sense, all geology is environmental geology. However, the term *environmental geology* is usually reserved for the part of geology that focuses on the relationships between people and the physical environment. It involves the application of geology to understanding and solving problems that arise from these interactions.

Although the primary focus of this book is to gain an understanding of basic geologic principles, many important aspects of environmental geology will be explored along the way. Some involve hazardous earth processes (Figure 1.A), including such phenomena as volcanoes, earthquakes, landslides, floods, and coastal hazards. Each of these processes is responsible for significant losses of life and property each year. Of course, geologic hazards are nothing more than natural processes. They become hazards only when people try to live in places where these processes occur.

Resources represent another important focus of environmental geology. They include water and soil, a great variety of metallic and nonmetallic minerals, and energy. These substances are the very foundation of modern society. Geology must deal not only with the formation and occurrence of these vital substances but also with maintaining adequate supplies, and with the environmental impact of their extraction and use (Figure 1.B).

Key factors influencing environmental issues of all kinds are the rapid growth in world population and the desire of people to have a better standard of living. It took until the year 1800 for world population to reach 1 billion. By the year 2000, an estimated 7 billion people will inhabit the planet. At the beginning of the 1990s, there was a net gain of about 90 million people each year. The results were an ever-increasing demand for resources and a greater stress on the planet. A rapidly growing population also means that more and more people are living in hostile environments plagued by geologic hazards.

The geologic processes that have formed and modified the earth have operated through long spans of time. When humans become involved, they can dramatically influence the magnitude and frequency of these processes. For example, river flooding is a natural process, but the magnitude and frequency of flooding can be changed significantly by human activities. Unfortunately, natural systems do not always adjust to artificial changes in ways that we can anticipate. Thus, an alteration to the environment that was intended to benefit society often has the opposite effect.

Knowledge about our planet and how it works is necessary to our survival and well being. The earth is the only suitable habitat we have, and its resources are limited.

FIGURE 1.A
This is one of thousands of structures that was damaged or destroyed by an earthquake that struck Mexico City in September, 1985. An estimated 7000 lives were lost. (Photo by James L. Beck)

FIGURE 1.B
As world population grows, demand for mineral and energy resources climbs. Geology must deal with the search for additional supplies of traditional and alternative resources as well as the environmental impact of their extraction and use. (Courtesy of Shell Oil Company)

shows the earth has experienced many cycles of mountain building and erosion. Concerning the ever-changing nature of the earth through great expanses of geologic time James Hutton made a statement that was to become his most famous. In concluding his classic 1788 paper published in the *Transactions of the Royal Society of Edinburgh,* he stated, "The results, therefore, of our present enquiry is, that we find no vestige of a beginning–no prospect of an end." A quote from William L. Stokes sums up the significance of Hutton's basic concept:

> In the sense that uniformitarianism implies the operation of timeless, changeless laws or principles, we can say that nothing in our incomplete but extensive knowledge disagrees with it.*

In the chapters that follow, we shall be examining the materials that compose our planet and the processes that modify it. It will be important to remember that although many features of our physical landscape may seem to be unchanging in terms of the tens of years over which we might observe them, they are nevertheless changing, but on time scales of hundreds, thousands, or even many millions of years.

THE NATURE OF SCIENTIFIC INQUIRY

As members of a modern society, we are constantly reminded of the significant benefits derived from scientific investigations. What exactly is the nature of this inquiry?

All science is based on the assumption that the natural world behaves in a consistent and predictable manner. This implies that the physical laws which govern the smallest atomic particles also operate in the largest, most distant galaxies. Evidence for the existence of these underlying patterns can be found in the physical world as well as the biological world. For example, the same biochemical processes and the same genetic codes that are found in bacterial cells are also found in human cells. The overall goal of science is to discover the underlying patterns in the natural world and then to use this knowledge to make predictions about what should or should not be expected to happen given certain facts or circumstances.

The development of new scientific knowledge involves some basic, logical processes that are universally accepted. To determine what is occurring in the natural world, scientists collect scientific *facts* through observation and measurement (Figure 1.5). These data are essential to science and serve as the springboard for the development of scientific theories and laws.

Once a set of scientific facts (or principles) that describe a natural phenomenon are gathered, investigators try to explain how or why things happen in the manner observed. They can do this by constructing a tentative (or untested) explanation, which we call a scientific **hypothesis.** Often several hypotheses are advanced to explain the same factual evidence. For example, there are currently five major hypotheses that have been proposed to explain the origin of the earth's moon. Until recently, the most widely held hypothesis argued that the moon and the earth formed simultaneously from the same cloud of nebular dust and gases. A newer hypothesis suggests that a Mars-sized body impacted the earth, thereby ejecting a huge quantity of material into earth orbit that eventually accumulated into the moon. Both hypotheses will be submitted to rigorous testing that will undoubtedly result in the modification or rejection

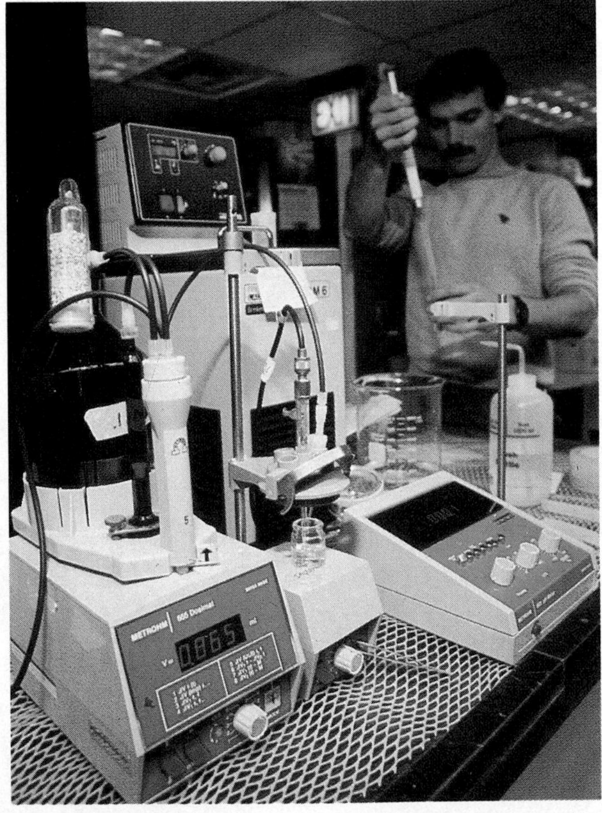

FIGURE 1.5
Scientist analyzing sea-floor samples collected by the drilling ship *JOIDES Resolution.* (Courtesy of Ocean Drilling Program)

**Essentials of Earth History* (Englewood Cliffs, New Jersey: Prentice-Hall, 1966), p. 34.

of one, or perhaps both, of these proposals. The history of science is littered with discarded hypotheses. One of the best known is the idea that the earth was at the center of the universe, a proposal that was supported by the apparent daily motion of the sun, moon, and stars around the earth.

When a hypothesis has survived extensive scrutiny and when competing hypotheses have been eliminated, a hypothesis may be elevated to the status of a scientific **theory**. A scientific theory is a well-tested and widely accepted view that scientists agree best explains certain observable facts. However, scientific theories, like scientific hypotheses, are accepted only provisionally. It is always possible that a theory which has withstood previous testing may eventually be disproven. As theories survive more testing, they are regarded with higher levels of confidence. Theories that have withstood extensive testing, as for example, the theory of plate tectonics or the theory of evolution, are held with a very high degree of confidence.

Some concepts in science are formulated into scientific laws. A scientific **law** is a generalization about the behavior of nature from which there has been no known deviation after numerous observations or experiments. Scientific laws are generally narrower in scope than theories and can usually be expressed mathematically. Examples include Newton's laws of motion and the laws of thermodynamics. Scientific laws describe what happens in nature, but they do not explain how or why things happen this way. Also, unlike hypotheses, which are inventions of the mind to explain scientific facts, scientific laws are based on observations and measurements and thus are rarely overthrown or seriously modified. Nevertheless, even scientific laws are not necessarily "perpetual truths." As the mathematician Jacob Bronowski so ably stated, "Science is a great many things, but in the end they all return to this: Science is the acceptance of what works and the rejection of what does not."

The processes just described, in which scientists gather facts through observations and formulate scientific hypotheses, theories, and laws, is called the *scientific method*. Contrary to popular belief, the scientific method is not a standard recipe that scientists apply in a routine manner to unravel the secrets of our natural world. Neither is this process a haphazard one. Some scientific knowledge is gained through the following steps: (1) the collection of scientific facts through observation and measurement (Figure 1.6); (2) the development of a working hypothesis to explain these facts; (3) construction of experiments to test the hypothesis; and (4) the acceptance, modification, or rejection of the hypothesis based on extensive testing. Other scientific discoveries represent purely theoretical ideas, which stood up to extensive examination. Still other scientific advancements have been made when a totally unexpected happening occurred during an experiment. These so-called serendipitous discoveries are more than pure luck; for as Louis Pasteur said, "In the field of observation, chance favors only the prepared mind." Since scientific knowledge is acquired through several avenues, it might be best to describe the nature of scientific inquiry as the *methods* of science, rather than *the* scientific method.

FIGURE 1.6
Temperature probe of active lava flow, Kilauea Caldera. (Photo by Norman Banks, U.S. Geological Survey)

GEOLOGIC TIME AND THE GEOLOGIC TIME SCALE

Although Hutton, Playfair, Lyell, and others recognized that geologic time is exceedingly long, they had no methods to accurately determine the age of the earth. However, with the discovery of radioactivity near the turn of the twentieth century and the continuing refinement of radiometric dating methods that were first attempted in 1905, geologists are now able to assign fairly accurate, specific dates to events in earth history.* Current estimates put the age of the earth at about 4.6 billion years.

*A more complete discussion of this topic is found in Chapter 8.

During the nineteenth century, long before the advent of radiometric dating, a geologic time scale was developed using principles of relative dating. **Relative dating** means that events are placed in their proper sequence or order without knowing their absolute age in years. This is done by applying principles such as the **law of superposition**, which states that in an undeformed sequence of sedimentary rocks or lava flows, each layer is older than the one above it and younger than the one below it (Figure 1.7). Today such a proposal appears to be elementary, but 300 years ago, it amounted to a major breakthrough in scientific reasoning by establishing a rational basis for relative time measurements. However, since no precise rate of deposition can be determined for most rock layers, the actual length of

FIGURE 1.7
Relative ages of these rock layers exposed in northern Arizona's Vermillion Cliffs can be determined by applying the law of superposition. The rock layers in the background overlie the colorful Chinle Formation in the foreground. Therefore, the Chinle Formation is older. (Photo by Jack W. Dykinga)

geologic time represented by any given layer is unknown.

Fossils, the remains or traces of prehistoric life, were also essential to the development of the geologic time scale (Figure 1.8). A fundamental principle in geology, one that was laboriously worked out over decades by collecting fossils from countless rock layers around the world, is known as the **principle of faunal succession.*** This principle states that fossil organisms succeed one another in a definite and determinable order, and therefore any time period can be recognized by its fossil content. Once established, this principle allowed geologists to identify rocks of the same age in widely separated places and to build the geologic time scale as shown in Figure 1.9.

Notice that units having the same designation do not necessarily extend for the same number of years. For example, the Cambrian period lasted some 65 million years, whereas the Silurian period spanned just 30 million years. As we will emphasize again in Chapter 8, this situation exits because the basis for establishing the time scale was not the regular rhythm of a clock, but the changing character of life forms through time. Specific dates were added long after the time scale was established. A glance at Figure 1.9 also reveals that the Phanerozoic eon is divided into many more units than earlier eons even though it encompasses only about 13 percent of earth history. The meager fossil record for these earlier eons is the primary reason for the lack of detail on this portion of the time scale. Without abundant fossils, geologists lose their primary tool for subdividing geologic time.

*Since this principle is equally applicable to both plants and animals, the term *biotic succession* is preferred by some.

ORIGIN OF THE EARTH

The earth is one of nine planets that along with several dozen moons and numerous smaller bodies revolve around the sun. The orderly nature of our solar system led most astronomers to conclude that its members formed at essentially the same time and from the same primordial material. This proposal, known as the **nebular hypothesis**, suggests that the bodies of our solar system formed from an enormous cloud composed mostly of hydrogen and helium with only a small percentage of the heavier elements.

About 5 billion years ago, and for reasons that are not yet fully understood, this huge cloud of minute rocky fragments and gases began to contract under

A.

B.

FIGURE 1.8
Fossils are important tools for the geologist. In addition to being very important in relative dating, fossils can be useful environmental indicators. **A.** Natural cast of a trilobite. This diverse group of marine organisms was prominent during the Paleozoic era. **B.** This extinct coiled cephalopod, like its modern descendents, was a highly developed marine organism.

GEOLOGIC TIME SCALE

Time Units of the Geologic Time Scale				Development of Plants and Animals
Eon	**Era**	**Period**	**Epoch**	
Phanerozoic	Cenozoic	Quarternary	Holocene	Humans develop
			0.01	
			Pleistocene	
			1.6	
		Tertiary	Pliocene	"Age of Mammals"
			5.3	
			Miocene	
			23.7	
			Oligocene	
			36.6	
			Eocene	
			57.8	
			Paleocene	
			66.4	Extinction of dinosaurs and many other species
	Mesozoic	Cretaceous	"Age of Reptiles"	First flowering plants
		144		
		Jurassic		First birds
		208		
		Triassic		Dinosaurs dominant
		245		
	Paleozoic	Permian	"Age of Amphibians"	Extinction of trilobites and many other marine animals
		286		
		Carboniferous: Pennsylvanian		First reptiles Large coal swamps
		320		
		Carboniferous: Mississippian		Amphibians abundant
		360		
		Devonian	"Age of Fishes"	First insect fossils Fishes dominant
		408		
		Silurian		First land plants
		438		
		Ordovician	"Age of Invertebrates"	First fishes Trilobites dominant
		505		
		Cambrian		First organisms with shells
		570		
Proterozoic		Collectively called Precambrian, comprises about 87% of the geologic time scale		First multicelled organisms
	2500			
Archean				First one-celled organisms
	3800			Age of oldest rocks
Hadean				Origin of the earth
	4600			

FIGURE 1.9
The geologic time scale. Numbers on the time scale represent time in millions of years before the present. These dates were added long after the time scale had been established using relative dating techniques. The Precambrian accounts for more than 85 percent of geologic time. (Data from Geological Society of America)

BOX 1.2

The Magnitude of Geologic Time

The concept of geologic time is new to many nongeologists. People are accustomed to dealing with increments of time that are measured in hours, days, weeks, and years. Our history books often examine events over spans of centuries, but even a century is difficult to appreciate fully. For most of us, someone or something that is 90 years old is *very old,* and a 1000-year-old artifact is *ancient.*

By contrast, those who study geology must routinely deal with vast time periods—millions or billions (thousands of millions) of years. When viewed in the context of the earth's 4.6-billion-year history, a geologic event that occurred 100 million years ago may be characterized as "recent" by a geologist, and a rock sample that has been dated at 10 million years may be called "young."

An appreciation for the magnitude of geologic time is important in the study of geology because many processes are so gradual that vast spans of time are needed before significant changes occur. Hendrick Van Loon tries to convey this idea in the following passage:

> High up in the North in the land called Svithjod, there stands a rock. It is a hundred miles high and a hundred miles wide. Once every thousand years a little bird comes to this rock to sharpen its beak. When the rock has thus been worn away, then a single day of eternity will have gone by.*

Although the preceding quotation is an exaggeration, it makes the point that geologic time is vast and that earth features, which seem to be everlasting and unchanging to us and in fact to generations of people, are indeed slowly changing. Thus, over millions of years, mountains rise and are eroded to hills, and rivers excavate deep canyons.

How long is 4.6 billion years? If you were to begin counting at the rate of one number per second and continued 24 hours a day, 7 days a week, and never stopped, it would take about two lifetimes (150 years) to reach 4.6 billion! Another interesting basis for comparison is as follows:

> Compress for example, the entire 4.5 billion years of geologic time into a single year. On that scale, the oldest rocks we know date from about mid-March. Living things first appeared in the sea in May. Land plants and animals emerged in late November and the widespread swamps that formed the Pennsylvanian coal deposits flourished for about four days in early December. Dinosaurs became dominant in mid-December, but disappeared on the 26th, at about the time the Rocky Mountains were first uplifted. Manlike creatures appeared sometime during the evening of December 31st, and the most recent continental ice sheets began to recede from the Great Lakes area and from northern Europe about 1 minute and 15 seconds before midnight on the 31st. Rome ruled the Western world for 5 seconds from 11:59:45 to 11:59:50. Columbus discovered America 3 seconds before midnight, and the science of geology was born with the writings of James Hutton just slightly more than one second before the end of our eventful year of years.†

The foregoing is just one of many analogies that have been conceived in an attempt to convey the magnitude of geologic time. Although helpful, all of them, no matter how cleaver, only begin to help us comprehend the vast expanse of earth history.

*Hendrick Van Loon (1951), *The Story of Mankind,* ed. Black and Gold (New York: Liveright Publishers, 1962), p. 2.
†Don L. Eicher, *Geologic Time,* 2nd ed. (Englewood Cliffs, New Jersey: Prentice-Hall, 1978), pp. 18–19. Reprinted by permission.

its own gravitational influence (Figure 1.10). The contracting material is assumed to have had some component of rotational motion, and so, like a spinning ice skater pulling in her arms, rotated faster and faster as it contracted. This rotation in turn caused the nebular cloud to assume a flattened disklike shape. Within the rotating disk, relatively small eddylike contractions formed the nuclei from which the planets would eventually develop. The greatest concentration of material was pulled toward the center of this rotating mass and gravitationally heated, forming the hot *protosun.*

In a relatively short time after the formation of the protosun, the temperature within the rotating disk dropped significantly. This decrease in temperature caused substances with high melting points to condense into small particles, perhaps the size of sand grains. Materials such as iron and nickel solidified first. Next to condense were the elements of which rocky substances are composed. As these fragments collided, they joined into larger objects that in a few tens of millions of years accreted into the planets. In the same manner, but on a lesser scale, the processes of condensation and accretion acted to form the moons and other small bodies of the solar system.

As the *protoplanets* (planets in the making) accumulated more and more debris, the solar system began to clear. The removal of debris allowed sunlight to heat the surfaces of the newly formed planets. The resulting high surface temperatures of the

FIGURE 1.10
Nebular hypothesis. **A.** A huge rotating cloud of dust and gases (nebula) begins to contract. **B.** Most of the material is gravitationally swept toward the center, producing the sun. However, due to rotational motion some dust and gases remain orbiting the central body as a flattened disk. **C.** The planets begin to acrete from the material that is orbiting within the flattened disk. **D.** In time most of the remaining debris was either collected into the nine planets and their moons or swept out into space by the solar wind.

inner planets, coupled with the comparatively weak gravitational fields of these bodies, meant that the earth and its neighbors, Mercury, Venus, and Mars, were unable to retain appreciable amounts of the lighter components of the primordial cloud. These materials, which included hydrogen, helium, ammonia, methane, and water, vaporized from their surfaces and were eventually whisked from the inner solar system by the solar winds. At distances beyond Mars, temperatures are quite low. Consequently the large outer planets, Jupiter, Saturn, Uranus, and Neptune, accumulated huge amounts of hydrogen and other light materials from the primordial cloud. The accumulation of these gaseous substances is thought to account for the comparatively large sizes and low densities of the outer planets.

Shortly after the earth formed, the decay of radioactive elements, coupled with heat released by colliding particles, produced at least some melting of the interior. Melting, in turn, is thought to have allowed the heavier elements, principally iron and nickel, to sink, while the lighter rocky components floated upward. This segregation of material, which began early in the earth's history, is believed to still be occurring, but on a much smaller scale. As a result of this chemical differentiation, the earth's interior is not homogeneous. Rather, it consists of shells or spheres composed of materials having different properties. The principal divisions of the earth include: (1) the **inner core**, a solid iron-rich zone having a radius of 1216 kilometers (756 miles); (2) the **outer core**, a molten metallic layer some 2270 kilometers (1410 miles) thick; (3) the **mantle**, a solid rocky layer having a maximum thickness of 2885 kilometers (1789 miles); and (4) the **crust**, a relatively lighter outer skin that ranges from 5 to 40 kilometers (3 to 25 miles) thick (Figure 1.11).

A very important zone exists within the mantle

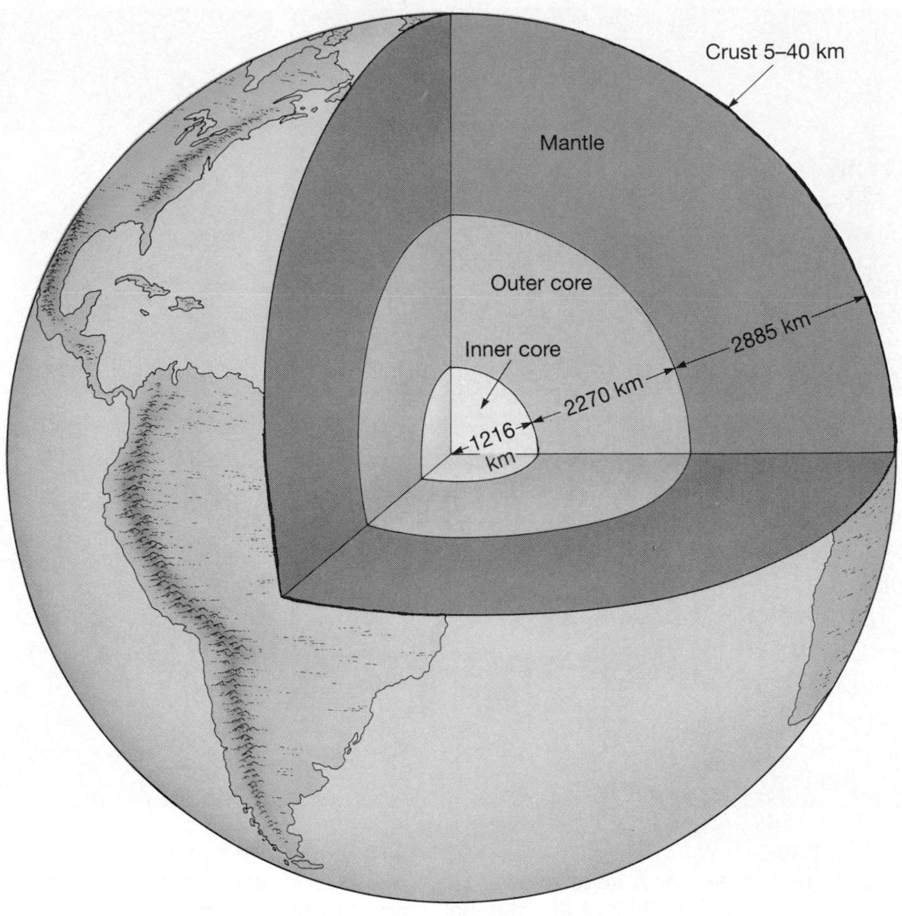

FIGURE 1.11
View of the earth's layered structure. The inner core, outer core, and mantle are drawn to scale, but the thickness of the crust is exaggerated by about five times.

and deserves special mention. This region, called the **asthenosphere**, is located between the depths of 100 and 350 kilometers and may extend down to 700 kilometers. The asthenosphere is a hot, weak zone that is capable of gradual flow. Situated above the asthenosphere, geologists recognize a zone called the **lithosphere** ("sphere of rock"), which includes the crust and uppermost mantle (Figure 1.12). In contrast to the asthenosphere upon which it rests, the lithosphere can be considered to be cool and rigid.

An important consequence of the period of chemical differentiation is that large quantities of gaseous materials were allowed to escape from the earth's interior, as happens today during volcanic eruptions.

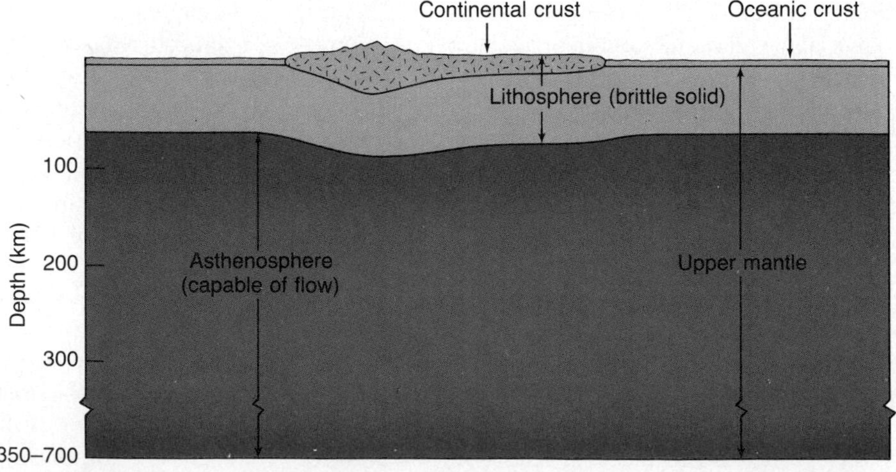

FIGURE 1.12
Schematic drawing of the respective positions of the asthenosphere and lithosphere.

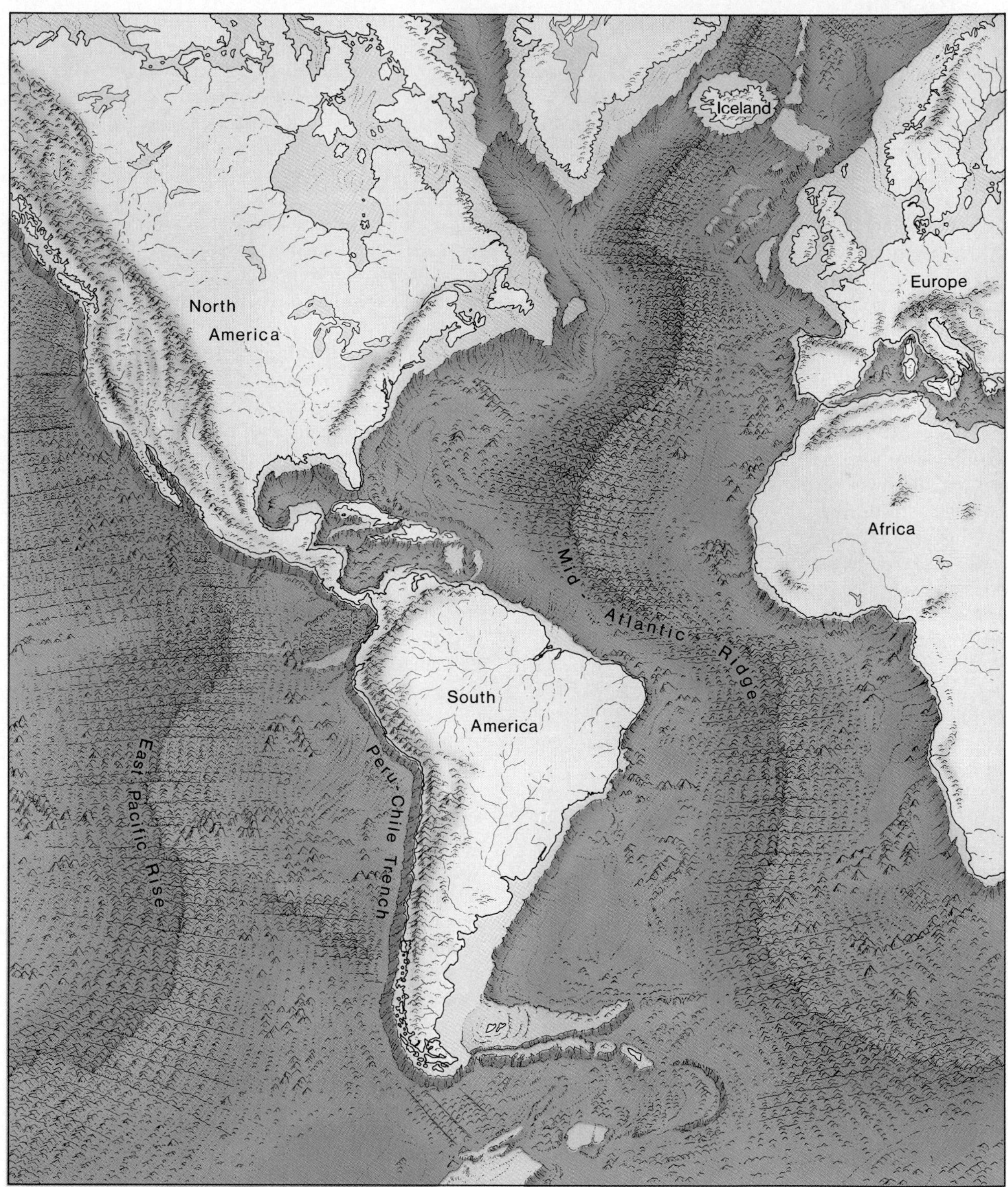

FIGURE 1.15
Major physical features of the continents and ocean basins. Clearly the diversity of features on the ocean floor is as varied as on the continents.

we can obtain clues to the dynamic processes which have shaped our planet.

The two principal divisions of the earth's surface are the continents and the ocean basins. It is important to realize that the present shoreline is not the boundary between these distinct regions. Rather, along most coasts a gently sloping platform of continental material, called the **continental shelf**, extends seaward from the shore. A glance at Figure 1.15 shows that there can be considerable variation in the extent of the continental shelf from one region to another. For example, the shelf is broad along the East and Gulf coasts of the United States, but relatively narrow along the Pacific margin of the continent. The extent of the continental shelf also varies greatly from one time to another. For instance, during the most recent ice age, when more of the world's water was stored on land in the form of glacial ice, the level of the sea was about 150 meters (500 feet) lower than it is today. Consequently, during this period, more of the earth's surface was dry land. The boundary between the continents and the deep-ocean basins is perhaps best placed about halfway down the **continental slopes**, which are steep dropoffs that lead from the edge of the continental shelves to the deep-ocean basins. Using this as the dividing line, we find that about 60 percent of the earth's surface is represented by the ocean basins, whereas the remaining 40 percent exists as continental masses.

The most obvious difference between the continents and the ocean basins is their relative levels. The average elevation of the continents above sea level is about 840 meters (2750 feet), whereas the average depth of the oceans is about 3800 meters (12,500 feet). Thus, the continents stand on the average about 4.6 kilometers above the level of the ocean floor. The elevations of these crustal layers are largely reflections of their densities. The continental blocks are composed of material which has properties similar to those of granite, a common rock with a density about 2.7 times that of water. The crust of ocean basins, on the other hand, is thought to have a composition similar to that of basalt, a rock that is about 3 times denser than water. This difference alone cannot account for the elevated positions of the continents. However, the rocky material located below 100 kilometers is weak and capable of flow. Thus, the rigid outer layer can be thought of as floating on this weak layer, much like an ice cube floats on water. The continental blocks, which consist of thick slabs of less dense rock, float higher than the thinner, more dense oceanic materials.

Within these two diverse provinces, great variations in elevation exist. The most prominent features of the continents are linear mountain belts (see Figure 1.15). Although the distribution of mountains appears to be random, this is not the case. When the youngest mountains are considered, we find they are located principally in two zones. The circum-Pacific belt includes the mountains of the western Americas and continues into the western Pacific in the form of volcanic island arcs. Island arcs are active mountainous regions composed largely of deformed volcanic rocks. Included in this group are the Aleutian Islands, Japan, the Philippines, and New Guinea. The other major mountain belt extends eastward from the Alps through Iran and the Himalayas, and then dips southward into Indonesia. Careful examination of mountainous terrains reveals that most are places where thick sequences of rocks have been squeezed and highly deformed, as if placed in a gigantic vise.

Older mountains are also found on the continents. Examples include the Appalachians in the eastern United States and the Urals in Russia. Their once lofty peaks are now worn low, the result of millions of years of erosion. Still older are the stable continental interiors. Within these stable interiors are areas known as **shields**, extensive and relatively flat expanses composed largely of crystalline material. Radiometric dating of the shields has revealed that they are truly ancient regions. The ages of some samples exceed 3.8 billion years. Even these oldest known rocks exhibit evidence of enormous forces that have folded and deformed them.

At one time the ocean floor was thought to be a rather nondescript region with only an occasional volcanic structure emerging from the otherwise flat, sediment-mantled depths. This perception of the ocean floor was incorrect. Indeed, the ocean basins are now known to contain the most prominent mountain ranges on earth, the **oceanic ridge system**. In Figure 1.15 the Mid-Atlantic Ridge and the East Pacific Rise are parts of this system. This broad elevated feature forms a continuous belt that winds for nearly 65,000 kilometers (40,000 miles) around the globe in a manner similar to the seam of a baseball. Rather than consisting of highly deformed rock, such as most of the mountains found on the continents, the oceanic ridge system consists of layer upon layer of once molten rock which has been fractured and uplifted.

The ocean floor also contains extremely deep grooves that are occasionally more than 11,000 meters (36,000 feet) deep. Although these deep-ocean **trenches** are relatively narrow and represent only a small portion of the ocean floor, they are nevertheless very significant features. Some trenches are located adjacent to young mountains which flank the continents. For example, in Figure 1.15 the Peru-

Chile trench off the west coast of South America parallels the Andes Mountains. Other trenches parallel linear island chains called volcanic island arcs.

What is the connection, if any, between the young, active mountain belts and the oceanic trenches? What is the significance of the enormous ridge system that extends through all the world's oceans? What forces crumple rocks to produce majestic mountain ranges? These questions must be answered as we attempt to discover the dynamic processes which shape our planet.

THE DYNAMIC EARTH

The earth is a dynamic planet. If we could go back in time a billion years or more, we would find a planet with a surface that was dramatically different from what it is today. Such prominent features as the Grand Canyon, the Rocky Mountains, and even the much older Appalachian Mountains did not exist. Moreover, we would find that the continents had different shapes and were located in different positions from those we find them in today. On the other hand, a billion years ago the moon's surface was almost the same as we now find it. In fact, if viewed telescopically from earth, perhaps only a few craters would be missing. Thus, when compared to the earth, the moon is a lifeless body wandering through space and time.

The processes that alter the earth's surface can be divided into two categories. Those forces which wear away the land include weathering and erosion. Unlike the moon, where weathering and erosion progress at infinitesimally slow rates, these processes are continually altering the landscape of the earth. In fact, these destructive forces would have long ago leveled the continents had it not been for opposing constructional processes. Included among the constructional processes are volcanism and mountain building, which increase the average elevation of the land in opposition to gravity. As we shall see, these forces depend upon the earth's internal heat for their source of energy.

Within the last few decades, a great deal has been learned about the workings of our dynamic planet. In fact, many have called this period a revolution in our knowledge about the earth which has been unequalled at any other time. This revolution began in the early part of the twentieth century with the radical proposal that the continents had drifted about the face of the earth. Because this idea contradicted the established view that the continents and ocean basins are permanent and stationary features on the face of the earth, it was received with great skepticism. More than 50 years passed before enough data were gathered to transform this relatively simple hypothesis into a working theory which wove together the basic processes known to operate on the earth. The theory that finally emerged, called **plate tectonics,*** provided geologists with a comprehensive model of the earth's internal workings.

According to the plate tectonics model, the earth's rigid outer shell, the lithosphere, is broken into several individual pieces called **plates** (Figure 1.16). Further, it is known that these rigid plates are slowly, but nevertheless continually, moving. This motion is believed to be driven by a thermal engine, the result of an unequal distribution of heat within the earth. As hot material gradually moves up from deep within the earth and spreads laterally, the plates are set in motion. Ultimately, this movement of the earth's lithospheric plates generates earthquakes, volcanic activity, and the deformation of large masses of rock into mountains.

Because each plate moves as a distinct unit, all interactions among plates occur along their boundaries (Figure 1.17). The first approximations of plate boundaries were made on the basis of earthquake and volcanic activity. Later work indicated the existence of three distinct types of plate boundaries, which are differentiated by the movement they exhibit. These are:

1. **Divergent boundaries**—zones where plates move apart, leaving a gap between them.
2. **Convergent boundaries**—zones where plates move together, causing one to go beneath the other, as happens when oceanic crust is involved; or where plates collide, which occurs when the leading edges are made of continental crust.
3. **Transform boundaries**—zones where plates slide past each other, scraping and deforming as they pass.

Each plate is bounded by a combination of these zones (see Figure 1.16). Movement along one boundary requires that adjustments be made at the others.

Plate spreading (divergence) occurs at the oceanic ridges. As the plates separate, the gap created is immediately filled with molten rock that wells up from the hot asthenosphere (Figure 1.18). This material slowly cools to produce a new sliver of sea floor. Successive separations and fillings continue to add new oceanic lithosphere between the diverging

*Tectonics is the study of large-scale deformation of the earth's lithosphere that results in the formation of major structural features such as those associated with mountains.

plates. This mechanism, which has produced the floor of the Atlantic Ocean during the past 160 million years, is appropriately called **sea-floor spreading**. The typical rate of sea-floor spreading is estimated to be 5 centimeters (2 inches) per year, although it varies considerably from one location to another. This seemingly slow rate of movement is nevertheless rapid enough to have generated all existing ocean basins within the last 5 percent of geologic time.

Although the earth's rigid outer layer is constantly being generated at the oceanic ridges, the total surface area of the earth remains constant. Therefore, lithosphere must be destroyed at the same rate that it is created. The zone of plate convergence is the site of this destruction. As two plates move together, the leading edge of one of the slabs is bent downward, allowing it to slide beneath the other. Whenever continental and oceanic lithosphere collide, it is always the denser oceanic material that plunges into the weak asthenosphere below (Figure 1.18).

The regions where oceanic lithosphere is being consumed are called **subduction zones**. Here, as the solid plates move downward, they enter high pressure and high-temperature environments. Some of the subducted material melts and migrates upward into the overriding plate. Occasionally this molten rock may reach the surface, where it gives rise to explosive volcanic eruptions such as those of Mount St. Helens (Figure 1.19).

Other boundaries, represented by transform faults, are located where plates slip past each other without producing or destroying crust. These faults form in the direction of plate movement and were first discovered in association with offsets in the oceanic ridges (Figure 1.18). Although most transform faults are located within the ocean basins, a few slice through the continents. The San Andreas fault of California is a famous example. Along this fault the Pacific plate is moving toward the northwest, past the North American plate. The movement along this boundary does not go unnoticed. As these plates pass, strain builds in the rocks on opposite sides of the fault and is occasionally released in the form of a great earthquake of the type that devastated San Francisco in 1906.

It has only recently been realized that the interaction of plates along their boundaries initiates most of our planet's volcanism, earthquakes, and mountain building. Further, these boundaries do not remain constant through time. For example, a divergent boundary which runs through eastern Africa appears to have developed in the relatively recent past. If spreading continues there, Africa will split into two continents separated by a new ocean basin. At other locations continents are presently moving toward

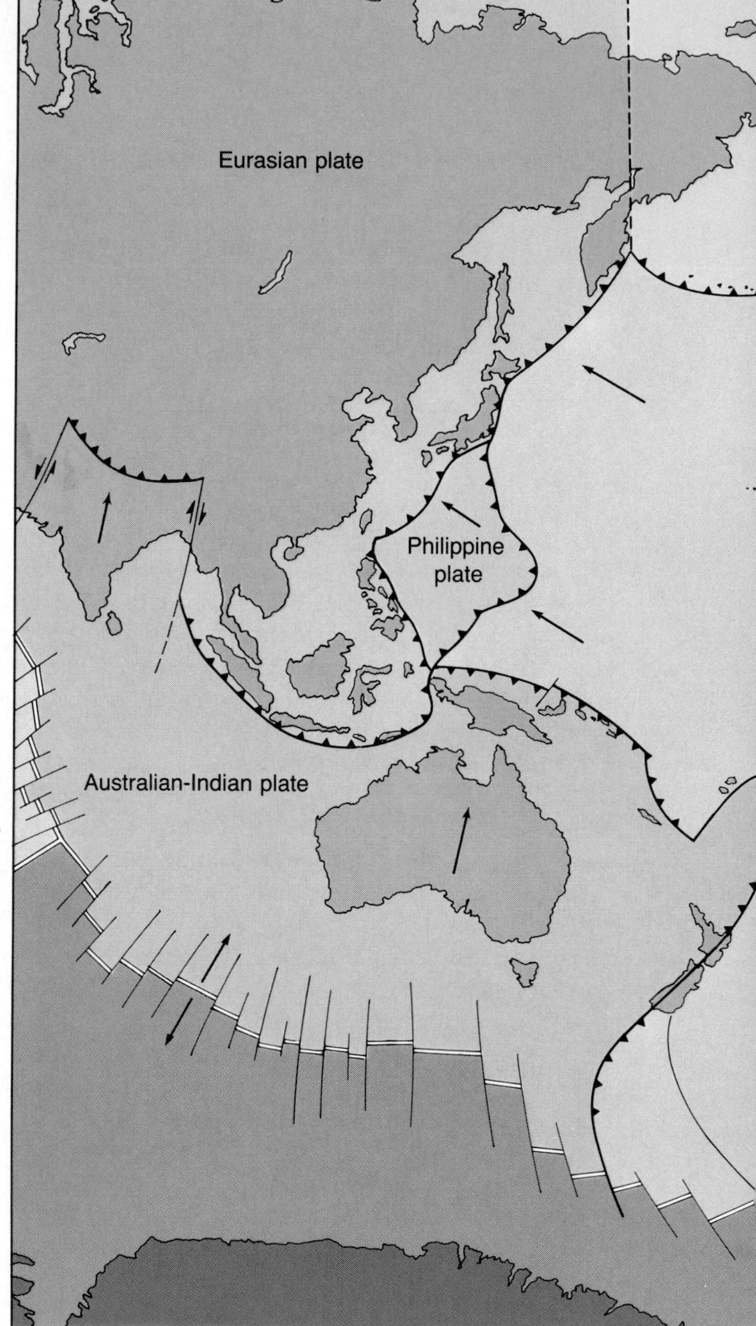

each other and may eventually join into a "supercontinent." When continents collide, the thick accumulations of rocks and sediments along their margins are gradually thrust into majestic mountain ranges.

As long as the temperatures deep within the earth remain significantly higher than those near the surface, the material within the earth will continue to move. This internal flow, in turn, will keep the rigid outer shell of the earth in motion. Thus, while this internal heat engine is operating, the positions and

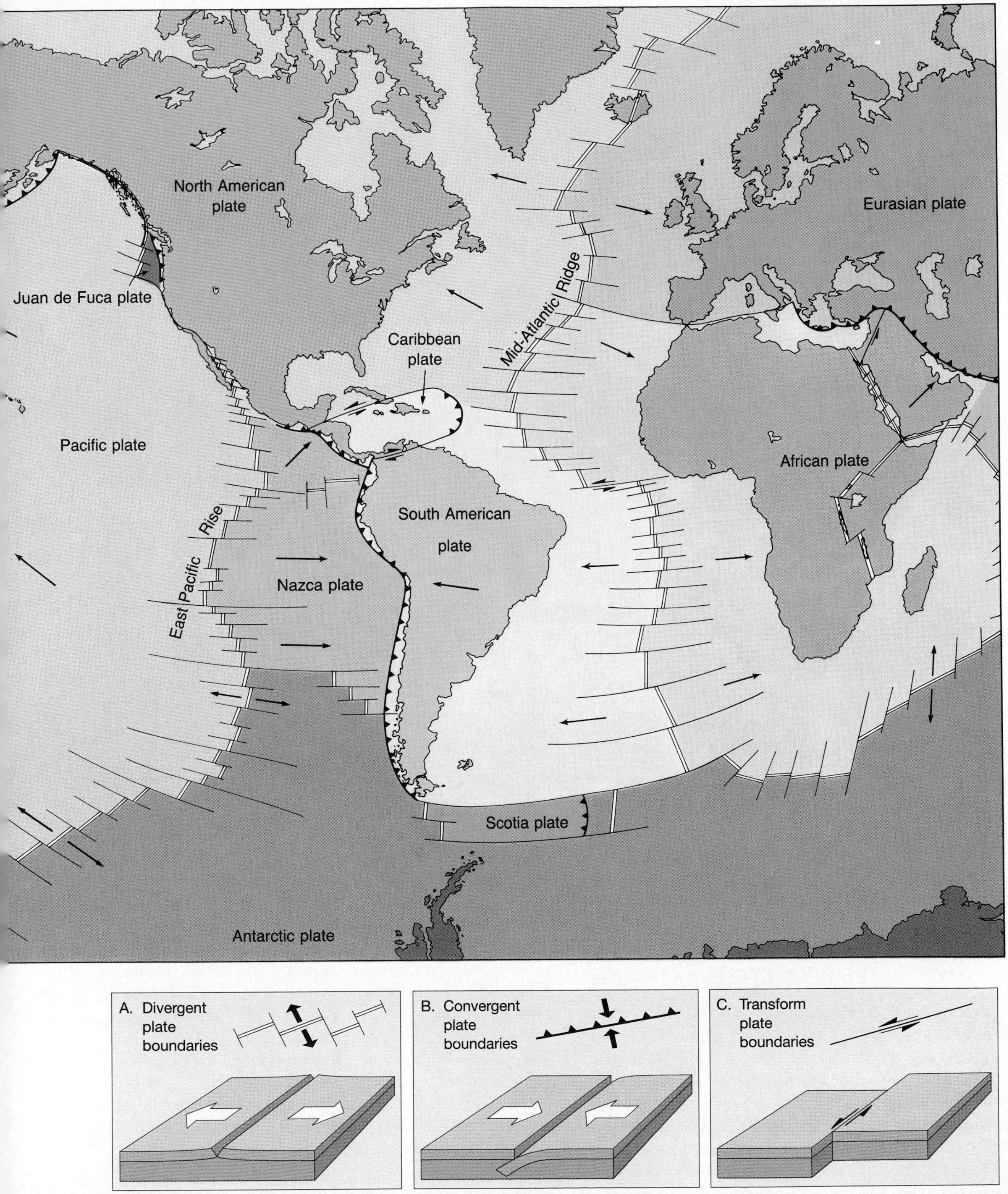

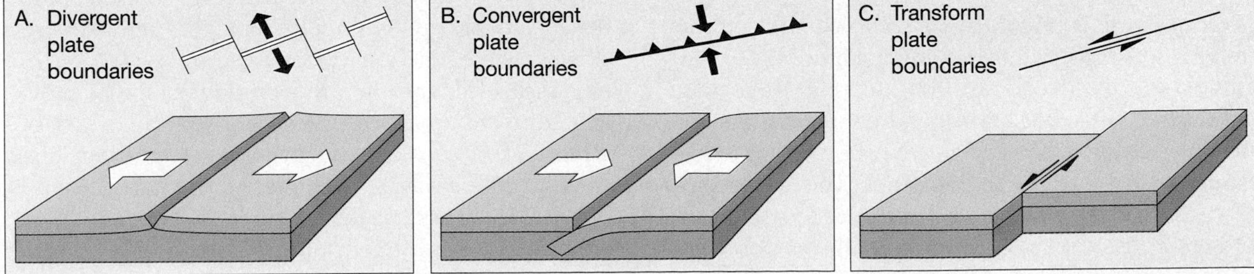

FIGURE 1.16
Mosaic of rigid plates that constitute the earth's outer shell. **A.** Divergent boundary. **B.** Convergent boundary. **C.** Transform fault boundary. (After W. B. Hamilton, U.S. Geological Survey)

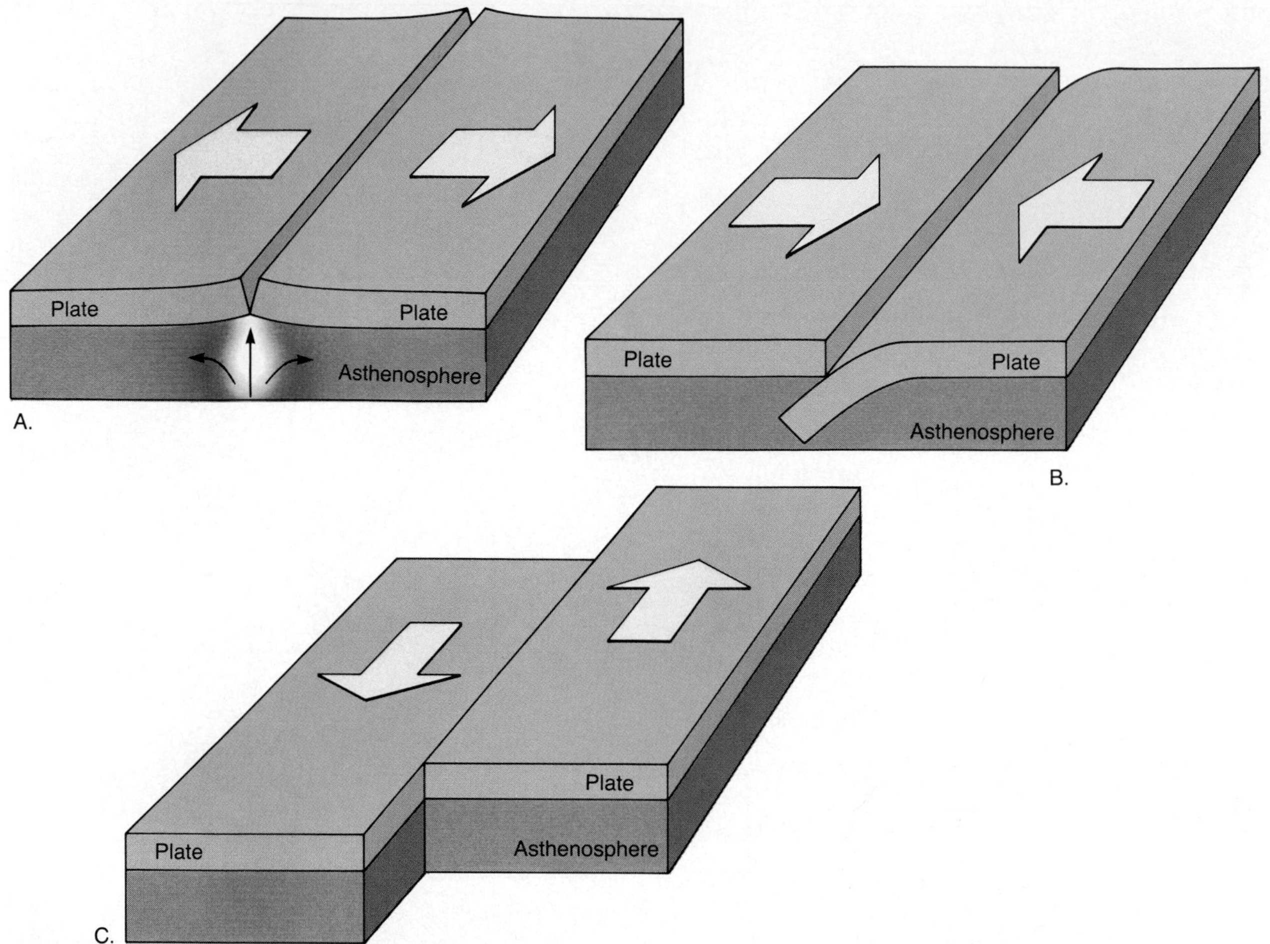

FIGURE 1.17
Schematic drawing showing relative motions along plate boundaries. **A.** Divergent boundary. **B.** Convergent boundary. **C.** Transform boundary.

shapes of the continents and ocean basins will change, and the earth will remain a dynamic planet.

In the remaining chapters we will examine in more detail the workings of our dynamic planet in light of the plate tectonics theory.

THE ROCK CYCLE

The **rock cycle** is one means of viewing many of the interrelationships within geology (Figure 1.20). By studying the rock cycle we may ascertain the origin of the three basic rock types and gain some insight into the role of various geologic processes in transforming one rock type into another. The concept of the rock cycle, which may be considered as a basic outline of physical geology, was initially proposed by James Hutton.

The first rock type, **igneous rock**, originates when molten material called **magma** cools and solidifies. This process, called **crystallization**, may occur either beneath the earth's surface or, following a volcanic eruption, at the surface. Initially, or shortly after forming, the earth's outer shell is believed to have been molten. As this molten material gradually cooled and crystallized, it generated a primitive crust that consisted entirely of igneous rocks.

If igneous rocks are exposed at the surface of the earth, they will undergo **weathering**, in which the day-in-and-day-out influences of the atmosphere slowly disintegrate and decompose rocks. The materials that result are often moved downslope by gravity before being picked up and transported by any of a number of erosional agents—running water, glaciers, wind, or waves. Eventually these particles and dissolved substances, called **sediment**, are deposited. Although most sediment ultimately comes to rest in the ocean, other sites of deposition include river floodplains, desert basins, swamps, and dunes. Next the sediments undergo **lithification**, a term meaning

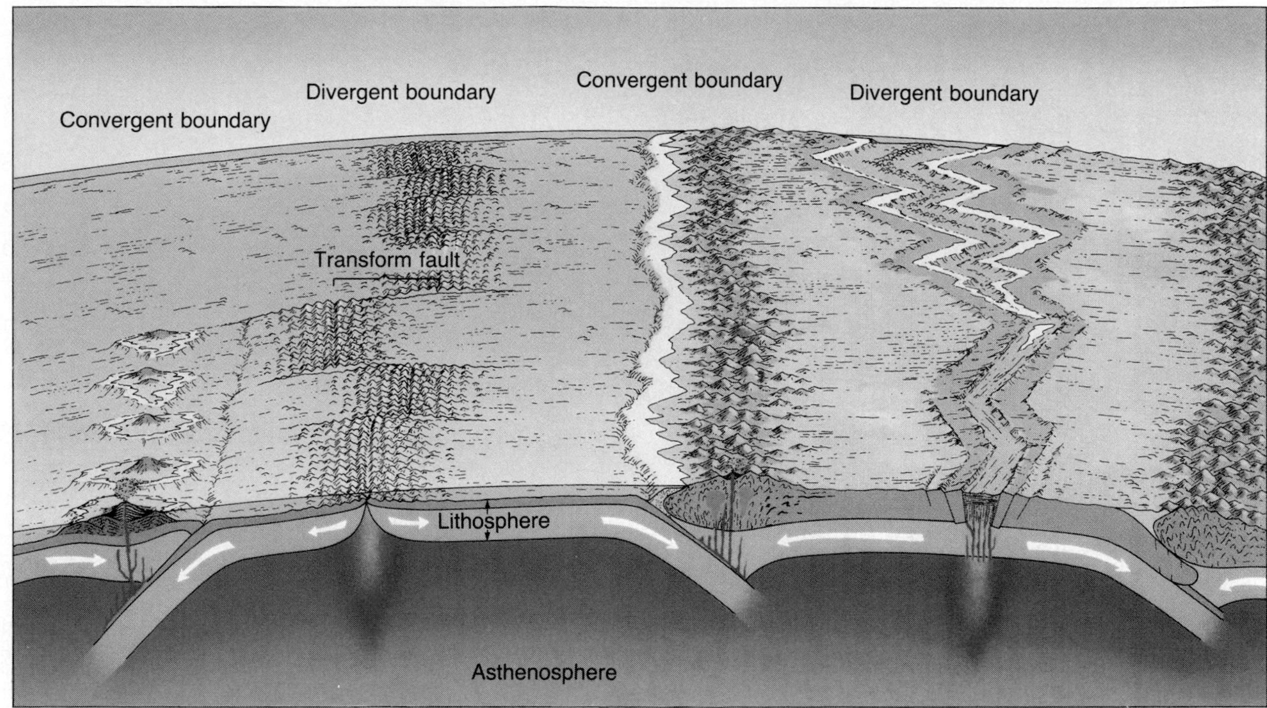

FIGURE 1.18
View of the earth showing the relationship between divergent and convergent plate boundaries.

"conversion into rock." Sediment is usually lithified when compacted by the weight of overlying layers or when cemented as percolating water fills the pores with mineral matter. If the resulting **sedimentary rock** is buried deep within the earth and involved in the dynamics of mountain building, or intruded by a mass of magma it will be subjected to great pressures and heat. The sedimentary rock will react to the

FIGURE 1.19
In May 1980, Washington State's Mount St. Helens erupted explosively, sending huge quantities of volcanic ash and debris into the atmosphere. It is one of the active volcanoes that are part of the Cascade Range. Along with Crater Lake, Mount Rainier, and the other peaks in the range, Mount St. Helens owes its existence to the melting of a subducting plate. (Photo by R. Hoblitt, U.S. Geological Survey)

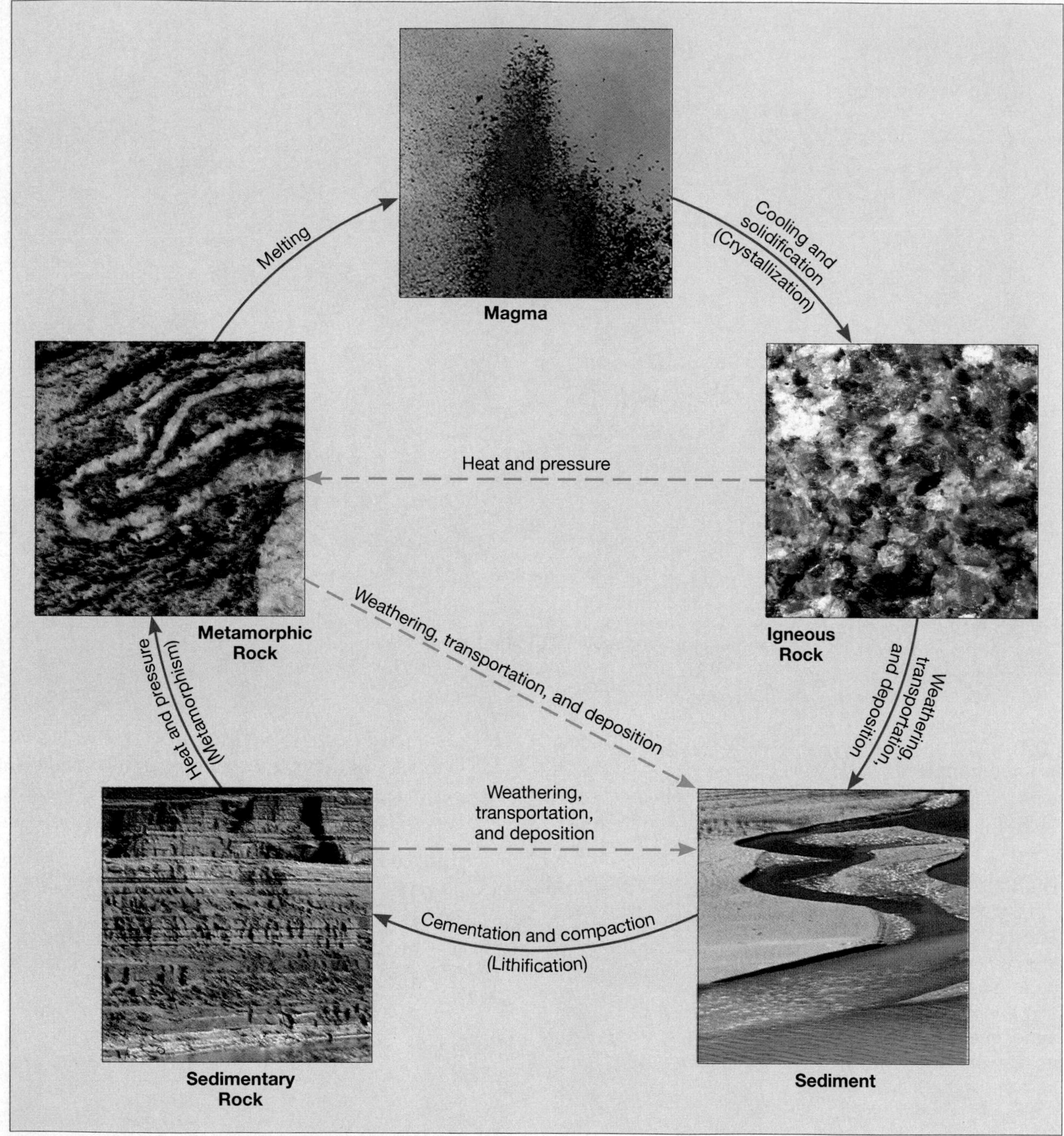

FIGURE 1.20
The rock cycle. Originally proposed by James Hutton, the rock cycle illustrates the role of the various geologic processes which act to transform one rock type into another. [Photos by J. D. Griggs, U.S.G.S. (top), E. J. Tarbuck (top right, bottom right), and Stephen Trimble (top left, bottom left)]

changing environment and turn into the third rock type, **metamorphic rock.** When metamorphic rock is subjected to additional pressure changes or to still higher temperatures, it will melt, creating magma, which will eventually solidify as igneous rock.

The full cycle just described does not always take place. "Shortcuts" in the cycle are indicated by dashed lines in Figure 1.20. Igneous rock, for example, rather than being exposed to weathering and erosion at the earth's surface, may be subjected to the strong compressional forces and high temperatures associated with mountain building and change to metamorphic rock. On the other hand, metamorphic and sedimentary rocks, as well as sediment, may be exposed at the surface and turned into new raw materials for sedimentary rock.

When the rock cycle was first proposed by James Hutton, very little was actually known about the processes by which one rock was transformed into another; only evidence for the transformation existed. In fact, it was not until very recently with the development of the theory of plate tectonics that a complete picture of the rock cycle became clear.

Figure 1.21 illustrates the rock cycle in terms of the plate tectonics model. According to this model, weathered material from elevated landmasses is transported to the continental margins where it is deposited in layers that collectively are thousands of meters thick. Once lithified, these sediments create a thick wedge of sedimentary rocks flanking the continents.

Eventually the relatively quiescent activity of sedimentation along a continental margin may be interrupted if the region becomes a convergent plate boundary. When this occurs, the oceanic lithosphere adjacent to the continent begins to inch downward into the asthenosphere beneath the continent. Along active continental margins such as this, convergence deforms the margin's sedimentary rocks and transforms them into linear belts of metamorphic rocks. Further, as the oceanic plate descends, some of the overlying sediments that were not crumpled into mountains are carried downward into the hot asthenosphere where they too undergo metamorphism. Eventually some of this metamorphic material will be transported to depths where the conditions are conducive to melting. This newly formed magma will then migrate upward through the overlying lithoshere to produce igneous rocks. Some will crystallize prior to reaching the surface and the remainder will erupt and solidify at the surface. When igneous rocks are exposed at the surface they are immediately attacked by the processes of weathering. Thus, the rock cycle is ready to begin anew.

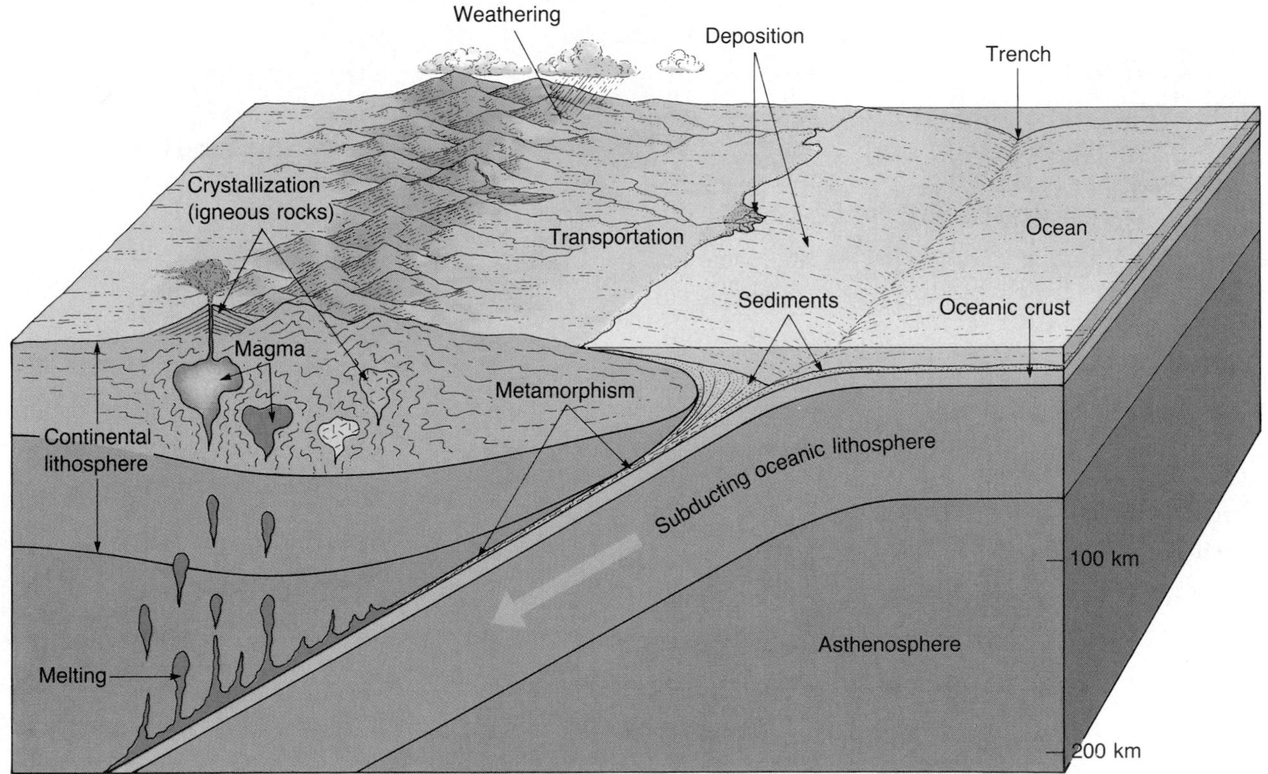

FIGURE 1.21
The rock cycle as it relates to the plate tectonics model.

REVIEW QUESTIONS

1. Geology is traditionally divided into two broad areas. Name and describe these two subdivisions.
2. Briefly describe Aristotle's influence on the science of geology.
3. How did the proponents of catastrophism perceive the age of the earth?
4. Describe the doctrine of uniformitarianism. How did the advocates of this idea view the age of the earth?
5. Briefly describe the contributions of Hutton, Playfair, and Lyell.
6. How is a scientific hypothesis different from a scientific theory?
7. How old is the earth currently thought to be?
8. The geologic time scale was established without the aid of radiometric dating. What principles were used to develop the time scale?
9. Why are the Archean and Proterozoic eons not divided into as many subdivisions as the Phanerozoic eon?
10. Contrast the asthenosphere and the lithosphere.
11. List and briefly describe the four "spheres" that constitute our environment.
12. The present shoreline is not the boundary between the continents and the ocean basin. Explain.
13. With which type of plate boundary is each of the following associated: subduction zone, San Andreas fault, sea-floor spreading, and Mount St. Helens?
14. Using the rock cycle, explain the statement "one rock is the raw material for another."

KEY TERMS

asthenosphere (p. 14)
atmosphere (p. 16)
biosphere (p. 16)
catastrophism (p. 3)
continental shelf (p. 18)
continental slope (p. 18)
convergent boundary (p. 19)
crust (p. 13)
crystallization (p. 22)
divergent boundary (p. 19)
faunal succession, principle of (p. 10)
fossil (p. 10)
geology (p. 2)
historical geology (p. 3)
hydrosphere (p. 15)
hypothesis (p. 7)
igneous rock (p. 22)
inner core (p. 13)
law (p. 8)
lithification (p. 22)
lithosphere (p. 14)
magma (p. 22)
mantle (p. 13)
metamorphic rock (p. 24)
nebular hypothesis (p. 10)
oceanic ridge system (p. 18)
outer core (p. 13)
physical geology (p. 3)
plate (p. 19)
plate tectonics (p. 19)
relative dating (p. 9)
rock cycle (p. 22)
sea-floor spreading (p. 20)
sediment (p. 22)
sedimentary rock (p. 23)
shield (p. 18)
subduction zone (p. 20)
superposition, law of (p. 9)
theory (p. 8)
transform boundary (p. 19)
trench (p. 18)
uniformitarianism (p. 3)
weathering (p. 22)

2

Matter and Minerals

Opposite: Cuprite crystal, Bisbee, Arizona. (Photo by Peter Kresan).
(Top photo E. J. Tarbuck)

The earth's crust and oceans are the source of a wide variety of useful and essential minerals. In fact, practically every manufactured product contains materials obtained from minerals. Most people are familiar with the common uses of many basic metals, including aluminum in beverage cans, copper in electrical wiring, and gold and silver in jewelry. But some people are not aware that pencil lead contains the greasy-feeling mineral graphite and that baby powder comes from a metamorphic rock made of the mineral talc. Moreover, many do not know that drill bits impregnated with diamonds are used to cut through rock layers in search of oil or that the common mineral quartz is the main ingredient in glass. As the material requirements of modern society grow, the need to locate additional supplies of useful minerals also grows, and becomes more challenging as well.

In addition to the economic uses of rocks and minerals, all of the processes studied by geologists are in some way dependent upon the properties of these basic earth materials. Events such as volcanic eruptions, mountain building, weathering and erosion, and even earthquakes involve rocks and minerals. Consequently, a basic knowledge of earth materials is essential to the understanding of all geologic phenomena.

ROCKS VERSUS MINERALS

Many people consider rocks to be rather nondescript objects that are hard and often dirty. Minerals are considered by many to be dietary supplements, or possibly rare ores or precious gems that are mined for their economic value (Figure 2.1). However, these common perceptions are far from the actual situation.

A **rock** can be defined simply as an aggregate of one or more minerals. Here, the term *aggregate* implies that the minerals are found together as a *mixture* in which the properties of the individual minerals are retained. Although most rocks are composed of more than one mineral, certain minerals are commonly found by themselves in large, impure quantities. In these instances they are considered to be a rock. A common example is the mineral calcite, which frequently is the dominant constituent in large rock units, where it is given the name *limestone.*

By contrast, **minerals** are defined as naturally occurring inorganic solids which possess an orderly internal structure and a definite chemical composition. Thus, for any earth material to be considered a mineral, it must exhibit the following characteristics:

1. It must be naturally occurring.
2. It must be inorganic.
3. It must be a solid.
4. It must possess an orderly internal structure; that is, its atoms must be arranged in a definite pattern.
5. It must have a definite chemical composition that may vary within specified limits.

When the term *mineral* is used by geologists, only those substances that meet these precise conditions are considered minerals. Consequently, synthetic diamonds, although chemically the same as natural diamonds, are not considered minerals. Further, oil and natural gas, which are not inorganic solids, and coal, which is produced from decayed inorganic matter, are also not classified as minerals by geologists. It should be noted, however, that miners, who generally classify minerals as anything of commercial value that is extracted from the earth, will refer to coal and petroleum as "mineral resources."

The last two criteria, which refer to the orderly internal structure and definite chemical composition of minerals, are not so obvious and will be discussed in more detail in the following section. Nevertheless, based on the latter characteristics, some naturally occurring inorganic solids are not classified as minerals. For example, the gemstone opal is considered a *mineraloid* because it lacks the orderly internal structure required of minerals. Further, the chemical composition of many minerals actually varies over a rather wide range. As we shall see, these chemical

FIGURE 2.1
Uncut diamond in rock matrix. (Courtesy of Smithsonian Institution)

variations result because certain elements readily substitute for others in crystalline structures.

This chapter deals primarily with the nature of minerals. However, keep in mind that rocks are simply aggregates of minerals. Thus, the properties of rocks are determined largely by the chemical composition and internal structure of those minerals which compose them.

THE COMPOSITION OF MATTER

Minerals, like all matter, are made of **elements**. At present, over 100 elements are known, a dozen and a half of which have been produced only in the laboratory (see Periodic Table in Appendix B). Some minerals such as gold and sulfur are made entirely from one element, but most are a combination of two or more elements joined to form a chemically stable **compound**. In order to better understand how elements combine to form compounds, we must first consider the **atom**, the smallest part of matter that still retains the characteristics of an element, because it is this extremely small particle that does the combining.

ATOMIC STRUCTURE

By the first part of this century, a series of important experiments had shown that atoms are composed of even smaller particles. A simplified model of the structure of an atom is shown in Figure 2.2. Each atom has a central region, called the **nucleus**, which contains very dense positively charged **protons** and equally dense neutral particles called **neutrons**. Orbiting the nucleus are negatively charged particles known as **electrons**. Unlike the orderly orbiting of the planets around the sun, electrons move so rapidly that their positions cannot be pinpointed. Hence, a more realistic picture of the positions of electrons can be obtained by envisioning a cloud of negatively charged electrons surrounding the nucleus. It is also known that individual electrons are located at given distances from the nucleus in regions called **energy-level shells**. As we shall see, an important fact about these shells is that each can hold only a specific number of electrons.

The number of protons found in the nucleus determines the **atomic number** and name of the element. For example, all atoms with six protons are carbon atoms, all those with eight protons are oxygen atoms, and so forth. Since uncombined atoms have the same number of electrons as protons, the atomic number also equals the number of electrons surrounding the nucleus. Moreover, since neutrons have no charge, the positive charge of the protons is exactly balanced by the negative charge of the electrons. Consequently, uncombined atoms as a whole are electrically neutral and have no overall electrical charge. Elements can be considered to be a large collection of electrically neutral atoms, all having the same atomic number.

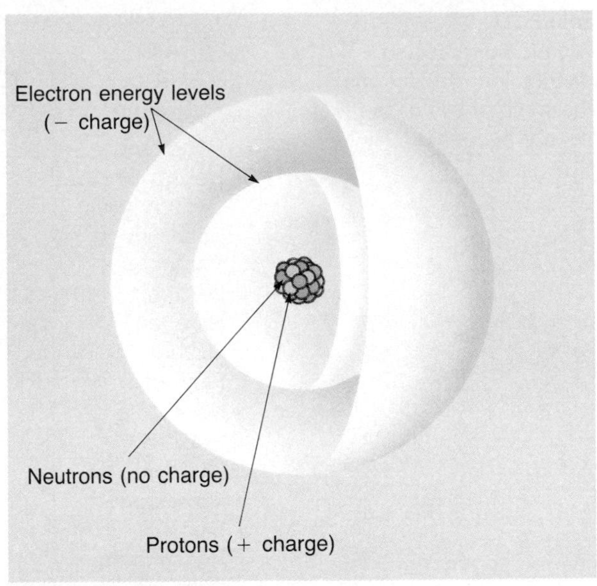

FIGURE 2.2
Simplified model of an atom. Atoms consist of a central nucleus composed of protons and neutrons which is encircled by electrons.

The simplest element, hydrogen, is composed of atoms that have only one proton in the nucleus and one electron surrounding the nucleus. Each successively heavier atom has one more proton and one more electron, in addition to a certain number of neutrons (Table 2.1). Studies of electron configurations have shown that each electron is added in a systematic fashion to a particular energy level or shell. In general, electrons enter higher energy levels only after lower energy levels have been filled to capacity. The first principal shell holds a maximum of two electrons, while each of the higher shells holds eight or more electrons. However, any shell that is an outermost shell (other than the first shell, which is filled with two electrons) will contain a maximum of only eight electrons. As we shall see, the outermost electrons are generally involved in chemical bonding.

Bonding

Elements combine with each other to form a wide variety of more complex substances called com-

TABLE 2.1
Atomic number and distribution of electrons in the main shells of the first twenty elements.

Element	Symbol	Atomic Number	Number of Electrons in Each Shell 1	2	3	4
Hydrogen	H	1	1			
Helium	He	2	2			
Lithium	Li	3	2	1		
Beryllium	Be	4	2	2		
Boron	B	5	2	3		
Carbon	C	6	2	4		
Nitrogen	N	7	2	5		
Oxygen	O	8	2	6		
Fluorine	F	9	2	7		
Neon	Ne	10	2	8		
Sodium	Na	11	2	8	1	
Magnesium	Mg	12	2	8	2	
Aluminum	Al	13	2	8	3	
Silicon	Si	14	2	8	4	
Phosphorus	P	15	2	8	5	
Sulfur	S	16	2	8	6	
Chlorine	Cl	17	2	8	7	
Argon	Ar	18	2	8	8	
Potassium	K	19	2	8	8	1
Calcium	Ca	20	2	8	8	2

pounds. A compound is a substance composed of two or more elements bonded together in definite proportions. When the elements separate, the bonds are broken and the compound is destroyed. Through experimentation it has been learned that the forces bonding the atoms together are electrical in nature. Further, it is known that chemical bonding results in a change in the electronic structures of the bonded atoms. Hence, the electron configurations of the atoms involved are important in determining the strength and nature of the chemical bonds that are produced.

As we noted earlier, the outermost electrons are generally involved in chemical bonding. Further, the atoms of most elements have less than the maximum number of electrons in their outermost shells. Only the noble gases such as neon and argon have a complete outer shell, which accounts for their chemical stability and the fact they do not react with other elements. However, every atom seeks a full outer shell to become chemically stable like the noble gases. The octet rule, literally meaning "a set of eight," refers to the concept of a completely filled outermost energy level. Simply, the **octet rule** states that atoms combine in order that each may have the electron arrangement of a noble gas, with the outer energy level containing eight electrons.

In order to satisfy the octet rule, an atom can either gain, lose, or share electrons with one or more atoms. The result of this process is the formation of an electrical "glue" that bonds the atoms. The electrons involved in the bonding process are commonly called **valence electrons**. The number of valence electrons that an element has determines the number of bonds it will form. For example, the element silicon has four valence electrons and forms four bonds in the process of completing its outer shell. On the other hand, oxygen forms two bonds, and hydrogen forms only one.

Ionic Bonds Perhaps the easiest type of bond to visualize is an **ionic bond.** In ionic bonding, one or more valence electrons are transferred from one atom to another. One atom becomes stable by giving up its valence electrons and the other uses them to complete its outer shell. An example of ionic bonding using sodium (Na) and chlorine (Cl) to produce sodium chloride (common table salt) is shown in Figure 2.3. Notice that sodium loses its single outer electron to chlorine. As a result, sodium acquires the electron configuration of the noble gas neon, which has two electrons in the first shell and eight in its outermost shell. By adding an electron, a chlorine atom fills its outermost shell to acquire the arrangement of argon. However, these atoms are no longer electrically neutral because neither contains an equal number of protons and electrons. Atoms such as these, which have an electrical charge because of unequal numbers of electrons and protons, are called **ions**. Sodium becomes a positively charged

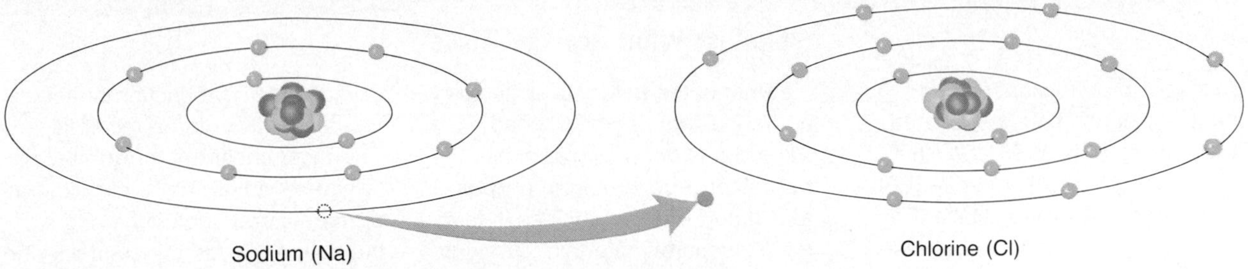

FIGURE 2.3
Chemical bonding of sodium and chlorine to produce sodium chloride. Through the transfer of one electron from sodium to chlorine, sodium becomes a positive ion and chlorine a negative ion.

ion and chlorine becomes a negatively charged ion. Studies have shown that particles with like charges repel and those with unlike charges attract. Thus, an ionic bond is one in which oppositely charged ions attract one another to produce a neutral chemical compound. Figure 2.4 illustrates the arrangement of sodium and chloride ions in ordinary table salt. Notice that salt consists of alternating sodium and chloride ions, positioned in such a manner that each positive ion is attracted to and surrounded on all sides by negative ions, and vice versa. This arrangement maximizes the attraction between ions with unlike electrical charges while minimizing the repulsion between ions with like charges. Further, ionic compounds consist of an orderly arrangement of oppositely charged ions assembled in a definite ratio that provides overall electrical neutrality.

This is an appropriate place in our discussion to point out that the properties of a chemical compound are dramatically different from the properties of the elements composing it. For example, chlorine is a green, poisonous gas that is so toxic it was used as a weapon during World War I. Sodium is a soft, silvery metal that reacts vigorously with water and, if held in your hand, could burn it severely. Together, however these atoms produce the compound sodium chloride (table salt) which is a clear crystalline solid that is essential for human life. This example also illustrates an important difference between a rock and a mineral. Most *minerals* are *chemical compounds* with unique properties that are very different from the elements which make them up. A *rock,* on the other hand, is a *mixture* of minerals, with each mineral retaining its own identity.

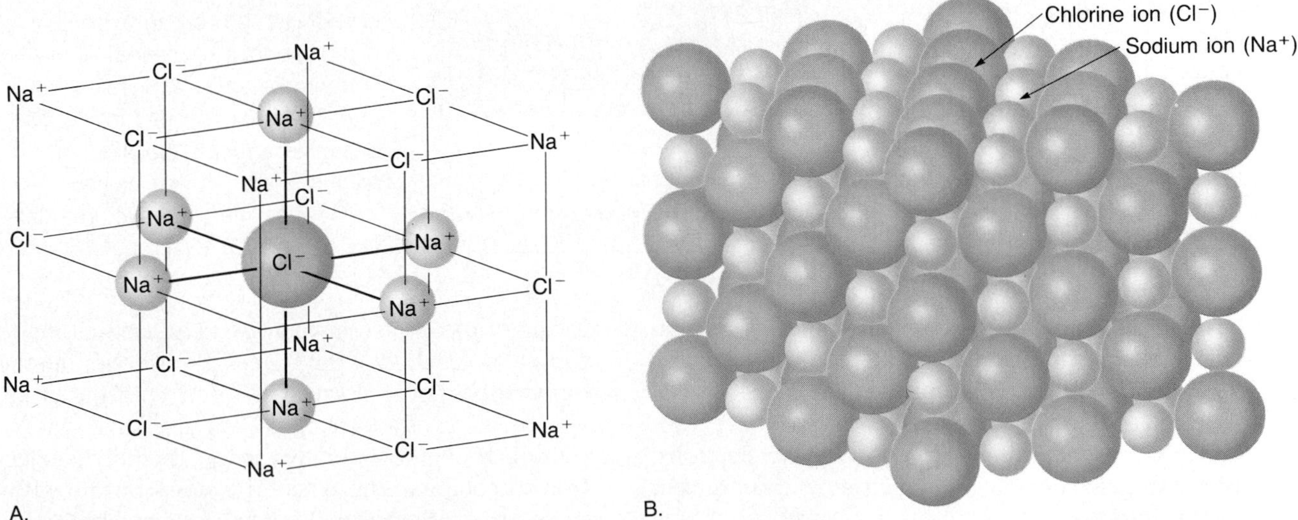

FIGURE 2.4
Schematic diagrams illustrating the arrangement of sodium and chloride ions in table salt. **A.** Structure has been opened up to show arrangement of ions. **B.** Actual ions are closely packed.

BOX 2.1

Asbestos: What are the Risks?

You have probably heard that there are health risks associated with asbestos. In 1986, the Environmental Protection Agency (EPA) instituted the Asbestos Hazard Emergency Response Act, a mandate that requires inspection of the nation's public and private schools for asbestos. This action brought asbestos to the attention of the public and instilled fears in parents that their children could contract asbestos-related cancers because of high levels of airborne fibers in schools. Since that time, billions of dollars have been spent on testing and asbestos removal. In addition, there have been significant, uncalculated costs related to not being able to use facilities during the rehabilitation process. Further, the EPA proposed a ban on all asbestos products that was intended to stop the use of these materials by 1996. Such actions fueled support for misconceptions such as the "one-fiber theory," which maintains that a single fiber of inhaled asbestos can cause a malignancy. Such is not the case. A number of researchers familiar with the "asbestos problem" disagree with the actions taken by the EPA. Although they admit that some forms of asbestos are a serious health risk in an occupational setting, they argue that most of the asbestos used in the United States is relatively harmless.

What is the nature of asbestos, and should its use be banned? Asbestos is not a single mineral but rather a general term applied to a group of silicate minerals that readily separate into thin, strong fibers (Figure 2.A). Because these fibers are flexible, heat resistant, chemically inert, and nonconducting, they have many uses. Asbestos has been widely used to strengthen concrete, make fireproof fabrics, and insulate around boilers and pipes. It is also employed as a component in floor tiles, and it constitutes the major ingredient in brake linings. In addition, wall coatings rich in asbestos fibers were used extensively during the 1950s and 1960s as substitutes for plaster.

It is estimated that over 30 million metric tons of asbestos have been used in the United States since the turn of the century. Most of the world's production of asbestos comes from three minerals. The most important, *chrysotile,* or "white asbestos," is a fibrous form of the mineral serpentine and is the only asbestos mineral mined in North America. *Crocidolite,* "blue asbestos," and *amosite,* "brown asbestos," are currently being mined in South Africa and make up about 5 percent of the world production.

FIGURE 2.A
Chrysotile asbestos. This sample is a fibrous form of the mineral serpentine.

The dangers of prolonged exposure to air that is laden with asbestos dust in an unregulated work place are well established. When very thin, rodlike fibers are inhaled into the lungs, they are neither broken down nor easily expelled but can remain for life. Three lung diseases may result: (1)asbestosis, a scarring of the tissue that decreases the lung's ability to absorb oxygen; (2) mesothelioma, a rare cancer that develops in the lining of the lungs; and (3) lung cancer. Evidence that incriminates one form of asbestos comes from health studies conducted at asbestos mines in South Africa and western Australia. Miners and millers who worked extensively with crocidolite showed an extremely high incidence of mesothelioma. In some individuals, exposure of less than one year resulted in the disease. As a result, the Australian mining operation was closed.

Despite this bleak picture, Mal-

Covalent Bonds Not all atoms combine by forming ions. For example, the gaseous elements oxygen (O_2), hydrogen (H_2), and chlorine (Cl_2) exist as stable molecules consisting of two atoms bonded together without the complete transfer of electrons. This is necessary because even if one of the atoms in each pair did accept one or more electrons to form a stable octet, the other atom would move farther away from such a stable condition. Instead, a stable octet is obtained when some of the outer electrons of both atoms are shared. Figure 2.5 illustrates the sharing of a pair of electrons between two chlorine atoms to form a molecule of chlorine gas. By overlapping the outer shells, one electron in each chlorine atom, which has seven electrons in its outer shell, has acquired, through cooperative action, the needed electron to complete the octet. The bond produced by the sharing of electrons to acquire the stable noble gas arrangement is called a **covalent bond**. The most common mineral group, the silicates, contains the element silicon that readily forms covalent bonds with oxygen.

colm Ross of the U.S. Geological Survey and others who have studied the issue conclude that the most widely used form of asbestos, chrysotile, is relatively harmless, unless it is breathed in high concentrations for long periods of time. They cite studies of miners and millers of white asbestos in Canada and northern Italy where mortality rates from mesothelioma and lung cancer differ very little from rates for the general public. One study that supports the nontoxicity of chrysotile was conducted on miner's wives in the area of Thetford Mines, Quebec. For many years when there were no dust controls on the mining and milling operations, these women were exposed to extremely high levels of airborne asbestos. Nevertheless, the subjects of the study exhibited below-normal levels of the diseases thought to be associated with asbestos exposure.

The various types of asbestos fibers differ in their chemical composition, shape, and durability. The thin, rodlike crocidolite fibers, which can easily penetrate the lining of the lungs, are certainly the most pathogenic. Chrysotile, which accounts for over 90 percent of the world's production, consists of fibers that are curly, occur in bundles, and can be intercepted in air passageways. Further, if inhaled, chrysotile fibers are expelled more rapidly from human lungs than crocidolite fibers. These differences are thought to explain the fact that the mortality rates for chrysotile workers differ very little from the rates for the general population.

Does asbestos present a risk to the health of the nation's students? Available data indicate that the levels of airborne asbestos in schools are approximately 0.01 of the permissible exposure levels for the U.S. work place. In addition, the indoor concentrations of those fibers that are the most biologically active, are comparable to outdoor levels. With few exceptions, the type of asbestos fibers found in schools is chrysotile, which in low concentrations has been shown to be relatively harmless. A comparison of the risk from asbestos exposure in schools to other risks in society is shown in Table 2.A. Playing high school football poses over 100 times the risk as exposure to airborne asbestos. These data clearly demonstrate that the asbestos panic was and is unwarranted.

TABLE 2.A
Estimates of risk from asbestos exposure in schools in comparison to other risks.

Cause	Annual Death Rate (per million)
Asbestos exposure in schools	0.005–0.093
Whooping cough vaccination (1970–80)	1–6
Aircraft accidents (1979)	6
High school football (1970–80)	10
Drowning (ages 5–14)	27
Motor vehicle accident, Pedestrian (ages 5–14)	32
Home accidents (ages 1–14)	60
Long-term smoking	1200

SOURCE: Data from Weill and Hughes.

A common analogy may help you visualize a covalent bond. Imagine two people at opposite ends of a dimly lit room, both reading under a lamp. By moving the lamps to the center of the room, they are able to combine their light sources so each can see better. Just as it is impossible to determine the source of the overlapping light, shared electrons are indistinguishable from each other.

It should be pointed out that many chemical bonds are actually a blend, consisting to some degree of electron sharing, as in covalent bonding, and to some degree of electron transfer, as in ionic bonding. Furthermore, both ionic and covalent bonds may occur within the same compound. This occurs, for example, in many silicate minerals, where silicon and oxygen atoms are covalently bonded to form the basic building block common to all silicates. These structures in turn are ionically bonded to metallic ions, producing various electrically neutral chemical compounds.

There also exists a chemical bond in which highly mobile valence electrons serve as the "electrical

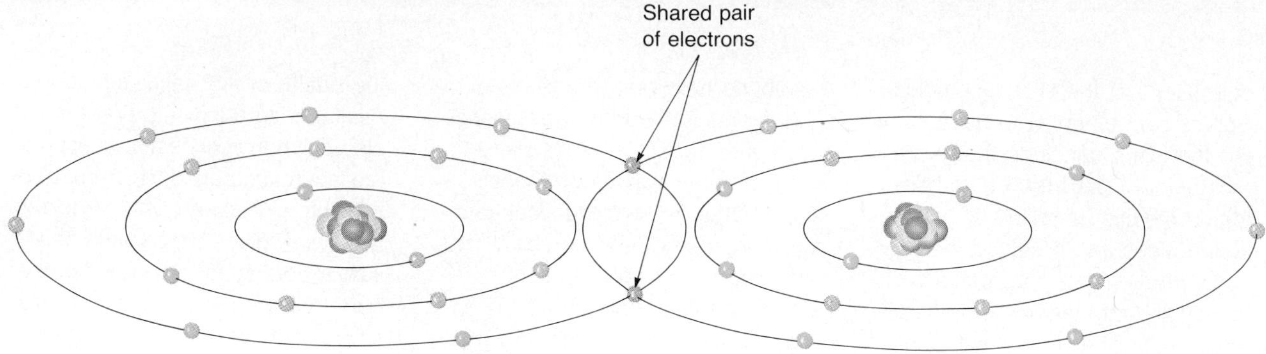

FIGURE 2.5
Illustration of the sharing of a pair of electrons between two chlorine nuclei to form a chlorine molecule.

glue." This type of electron sharing, found in metals such as copper, gold, aluminum, and silver, is called **metallic bonding**. Metallic bonding accounts for the high electrical conductivity of metals, the ease with which metals are reshaped, and numerous other special properties of metals.

Atomic Mass

Subatomic particles, such as protons, are so incredibly small that a special unit was devised to express their mass. A proton or a neutron has a mass just slightly more than one **atomic mass unit**, whereas an electron is only about one two-thousandths of an atomic mass unit. Thus, although electrons play an active role in chemical reactions, they do not contribute significantly to the mass of an atom. Because of this, the **mass number** of an atom is obtained simply by totaling the number of neutrons and the number of protons in the nucleus. Atoms of the same element commonly have varying numbers of neutrons, and therefore, different mass numbers. Such atoms are called **isotopes** of that element. For example, carbon has two well-known isotopes, one having a mass number of 12 (carbon-12), the other a mass number of 14 (carbon-14). Recall that all atoms of the same element must have the same number of protons (atomic number) and that carbon always has six. Hence, carbon-12 must have six neutrons to give it a mass number of 12, whereas carbon-14 must have eight neutrons to give it a mass number of 14. The term commonly used to express the average of the atomic masses of isotopes for a given element is **atomic weight.*** The atomic mass of carbon is much closer to 12 than 14, because carbon-12 is the more common isotope. Note that in a chemical sense all isotopes of the same element are nearly identical. To distinguish among them would be like trying to differentiate individual members from a group of similar objects, all having the same shape, size, and color, with only some being slightly heavier.

Although the vast majority of atoms are stable, many elements do have isotopes that are unstable. Unstable isotopes such as carbon-14 go through a process of natural disintegration called **radioactivity**, which occurs when the forces that bind the nucleus are not strong enough. The rate at which the unstable nuclei break apart (decay) is measurable and makes such isotopes useful "clocks" in dating the events of earth history. A discussion of radioactivity and its applications in dating events of the geologic past can be found in Chapter 8.

THE STRUCTURE OF MINERALS

A mineral is composed of an ordered array of atoms chemically bonded together to form a particular crystalline structure. This orderly stacking of atoms is reflected in the regularly shaped objects we call crystals (see chapter-opening photo).

What determines the particular crystalline structure a mineral will exhibit? For those compounds formed by ions, the internal atomic arrangement is determined partly by the charges on the ions, but more importantly by the size of the ions involved. In order to form stable ionic compounds, each positively charged ion is surrounded by the largest number of negative ions that will fit, while maintaining overall electrical neutrality, and vice versa. Figure 2.6 shows some ideal geometries for various-sized ions. We have already examined the geometric arrange-

*The term *weight* as used here is a misnomer that has been sanctioned by long usage. The correct term is atomic *mass*.

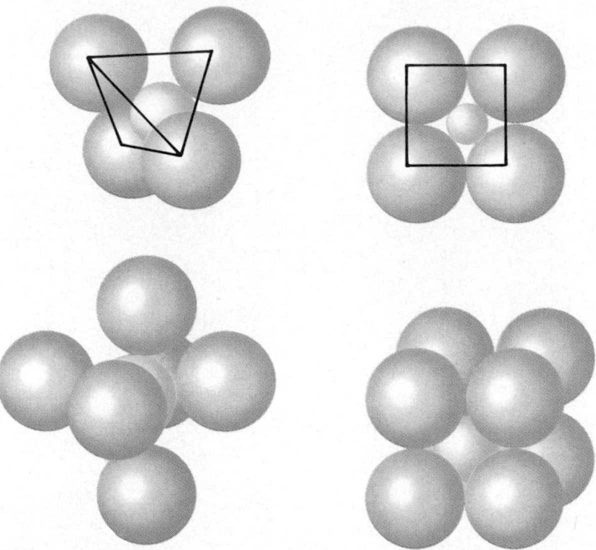

FIGURE 2.6
Ideal geometrical packing for various-sized positive and negative ions.

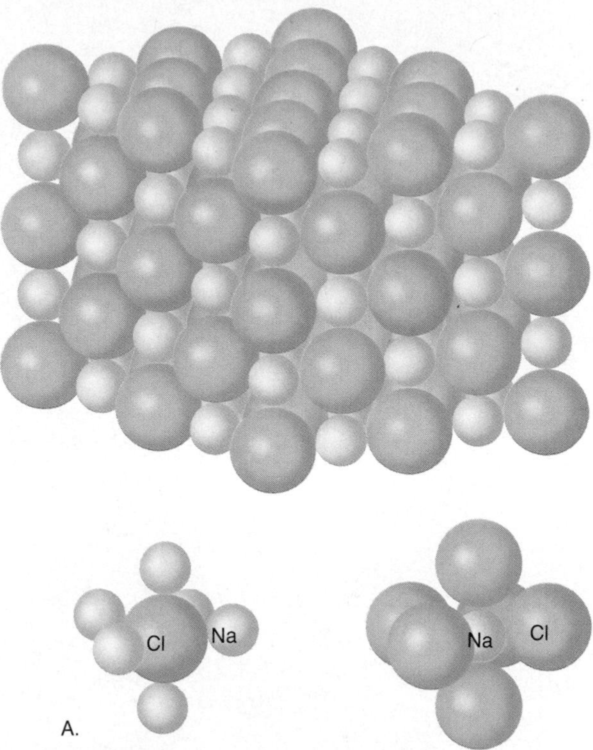

FIGURE 2.7
The structure of sodium chloride. **A.** The coordination of sodium and chloride ions in the mineral halite. **B.** The orderly arrangement at the atomic level produces regularly shaped crystals.

ment of sodium and chloride ions in the mineral halite. In Figure 2.7 we see that on a large scale the orderly packing of sodium and chloride ions produces cubic halite crystals. Like halite, all samples of a particular mineral contain the same elements, joined together in the same orderly arrangement.

Although it is true that every mineral has a particular internal structure, some elements are able to join together in more than one way. Thus, two minerals with totally different properties may have exactly the same chemical composition. Minerals of this type are said to be **polymorphs** (many forms). Graphite and diamond are particularly good examples of polymorphism because they consist exclusively of carbon yet are drastically different. Graphite is the soft gray material of which pencil lead is made, while diamond is the hardest known mineral. The differences between these minerals can be attributed to the conditions under which they formed. Diamonds are believed to form at depths approaching 200 kilometers, where extreme pressures produce the compact structure shown in Figure 2.8A. Graphite, on the other hand, consists of sheets of carbon atoms that are widely spaced and weakly held together (Figure 2.8B). Since these carbon sheets will easily slide past one another, graphite makes an excellent lubricant.

PHYSICAL PROPERTIES OF MINERALS

Minerals are solids formed by inorganic processes. Each mineral has an orderly arrangement of atoms (crystalline structure) and a definite chemical composition which give it a unique set of physical properties. Since the internal structure and chemical

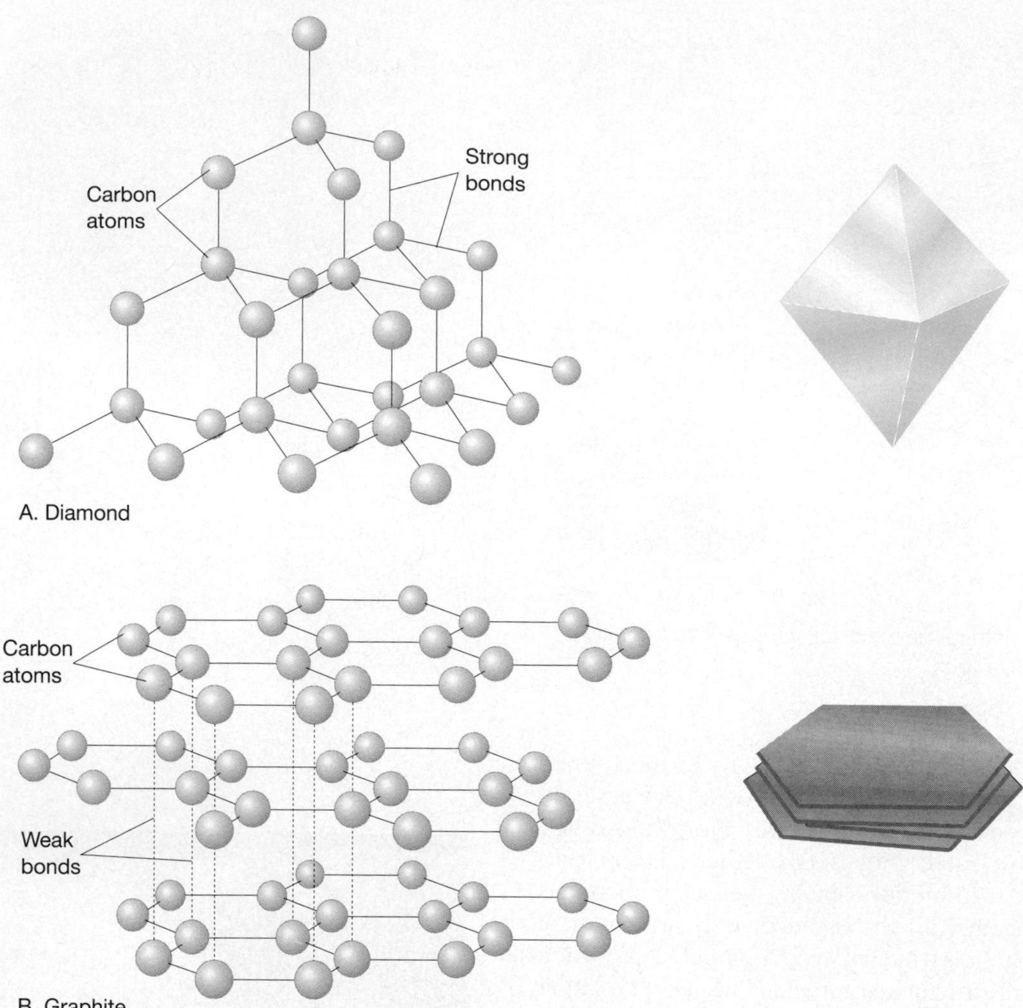

FIGURE 2.8
Comparing the structures of diamond and graphite. Diamond and graphite are naturally occurring forms of carbon that have the same chemical composition. Nevertheless, their internal structure and physical properties reflect the fact that each formed in a very different environment. **A.** All carbon atoms in diamond are covalently bonded into a compact, three-dimensional framework, which accounts for the extreme hardness of the mineral. **B.** In graphite the carbon atoms are bonded into sheets that are joined in a layered fashion by very weak electrical forces. These weak bonds allow the sheets of carbon to readily slide past each other, making graphite soft and slippery, and thus useful as a dry lubricant.

composition of a mineral are difficult to determine without the aid of sophisticated tests and apparatus, the more easily recognized physical properties are frequently used in identification. A discussion of some diagnostic physical properties follows.

Crystal Form

Most people think of a crystal as a rare commodity, when in fact most inorganic solid objects are composed of crystals. The reason for this misconception is that most crystals do not exhibit their crystal form. The **crystal form** is the external expression of a mineral that reflects the orderly internal arrangement of atoms. Figure 2.9A illustrates the characteristic form of the iron-bearing mineral pyrite. Generally, whenever a mineral is permitted to form without space restrictions, it will develop individual crystals with well-formed crystal faces. Some crystals such as those of the mineral quartz have a very distinctive crystal form that can be helpful in identification (Figure 2.9B). However, most of the time crystal growth is interrupted because of competition for space, resulting in an intergrown mass of crystals, none of which exhibits its crystal form.

Luster

Luster is the appearance or quality of light reflected from the surface of a mineral. Minerals that have the appearance of metals, regardless of color, are said to have a *metallic luster*. Minerals with a *nonmetallic luster* are described by various adjectives, including vitreous (glassy), pearly, silky, resinous, and earthy (dull). Some minerals appear partially metallic in luster and are said to be submetallic.

Color

Although **color** is an obvious feature of a mineral, it is often an unreliable diagnostic property. Slight impurities in the common mineral quartz, for example, give it a variety of colors, including pink, purple (amethyst), white, and even black (see Figure 2.23). When a mineral, such as quartz, exhibits a variety of colors, it is said to possess *exotic coloration*. Exotic coloration is usually caused by the inclusion of impurities, such as foreign ions, in the crystalline structure. Other minerals, for example, sulfur, which is yellow, and malachite, which is bright green, are said to have *inherent coloration*.

Streak

Streak is the color of a mineral in its powdered form and is obtained by rubbing the mineral across a piece of unglazed porcelain termed a *streak plate*. Although the color of a mineral may vary from sample to sample, the streak usually does not, and is therefore the more reliable property. Streak can also be an aid in distinguishing minerals with metallic lusters from those having nonmetallic lusters. Metallic minerals generally have a dense, dark streak, whereas minerals with nonmetallic lusters do not.

Hardness

One of the most useful diagnostic properties is **hardness**, a measure of the resistance of a mineral to abrasion or scratching. This property is determined by rubbing a mineral of unknown hardness against one of known hardness, or vice versa. A numerical value can be obtained by using the **Mohs scale** of hardness,

A.

B.

FIGURE 2.9
Crystal form is the external expression of a mineral's orderly internal structure. **A.** Pyrite, commonly known as "fool's gold," often forms cubic crystals that may contain parallel lines called striations. **B.** Quartz sample that exhibits well-developed hexagonal crystals with pyramidal-shaped ends.

which consists of ten minerals arranged in order from 1 (softest) to 10 (hardest) as follows:

Hardness	*Mineral*
1	Talc
2	Gypsum
3	Calcite
4	Fluorite
5	Apatite
6	Orthoclase
7	Quartz
8	Topaz
9	Corundum
10	Diamond

Any mineral of unknown hardness can be compared to these or to other objects of known hardness. For example, a fingernail has a hardness of 2.5, a copper penny 3, and a piece of glass 5.5. The mineral gypsum, which has a hardness of 2, can be easily scratched with your fingernail. On the other hand, the mineral calcite, which has a hardness of 3, will scratch your fingernail but will not scratch glass. Quartz, the hardest of the common minerals, will scratch a glass plate.

Cleavage

Cleavage is the tendency of a mineral to break along planes of weak bonding that exist between atoms in the crystalline structure. Minerals that possess cleavage are identified by the smooth surfaces which are produced when the mineral is broken. The simplest type of cleavage is exhibited by the micas (Figure 2.10). Because the micas have excellent cleavage in one direction, they break to form thin, flat sheets. Some minerals have several cleavage planes which produce smooth surfaces when broken, while others exhibit poor cleavage, and still others have no cleavage at all. When minerals break evenly in more than one direction, cleavage is described by the *number of different sets* of *planes* exhibited and the *angles at which the planes meet* (Figure 2.11).

Cleavage should not be confused with crystal form. When a mineral exhibits cleavage, it will break into pieces that have the same shape as the original sample. By contrast, the quartz crystals shown in the chapter-opening photo do not have cleavage and, if broken, would shatter into shapes that do not resemble each other or the original crystals.

Fracture

Minerals that do not exhibit cleavage when broken, such as quartz, are said to **fracture**. Those that break into smooth curved surfaces resembling broken glass have a *conchoidal fracture* (Figure 2.12). Others break into splinters or fibers, but most minerals fracture irregularly.

FIGURE 2.10
Sheet-type cleavage common to the micas.

Specific Gravity

Specific gravity is a number representing the ratio of the weight of a mineral to the weight of an equal volume of water. For example, if a mineral weighs three times as much as an equal volume of water, its specific gravity is 3. With a little practice, you can estimate the specific gravity of minerals by hefting

FIGURE 2.11
Smooth surfaces produced when a mineral with cleavage is broken. The sample on the left (fluorite) exhibits four planes of cleavage (eight sides), whereas the other two samples exhibit three planes of cleavage (six sides). Also notice that the mineral in the center (halite) has cleavage planes that meet at 90 degree angles, whereas the mineral on the right (calcite) has cleavage planes that meet at 75 degree angles.

FIGURE 2.12
Conchoidal fracture. The smooth curved surfaces result when minerals break in a glasslike manner. (Photo by E. J. Tarbuck)

FIGURE 2.13
Galena is lead sulfide and like other metallic ores has a relatively high specific gravity.

them in your hand. For example, if a mineral feels as heavy as the common rocks you have handled, its specific gravity will probably be somewhere between 2.5 and 3. Some metallic minerals have a specific gravity two or three times that of common rock-forming minerals. Galena, which is an ore of lead, has a specific gravity of roughly 7.5, while the specific gravity of 24 carat gold is approximately 20 (Figure 2.13).

Other Properties

In addition to the properties already discussed, some minerals can be recognized by other distinctive properties. For example, halite has a salty taste, thin sheets of mica will bend and elastically snap back, and gold is malleable and can be easily shaped. Talc and graphite both have distinctive feels; talc's is soapy and graphite's is greasy. A few minerals, such as magnetite, have a high iron content and can be picked up with a magnet, while some varieties (lodestone) are natural magnets and will pick up light, metal objects such as pins and paper clips. Some minerals exhibit special optical properties. For example, when a transparent piece of calcite is placed over printed material the letters appear twice. This optical property is known as *double refraction*. In addition, the streak of many sulfur-bearing minerals smells like rotten eggs.

One very simple chemical test involves placing a small drop of dilute hydrochloric acid on a fresh surface of a mineral. Certain minerals, including some carbonates, will effervesce (fizz) with hydrochloric acid. This test is useful in identifying the mineral calcite, which is a common carbonate mineral frequently found in sedimentary and metamorphic rocks.

In summary, a number of special physical and chemical properties are useful in identifying certain minerals. These include taste, smell, elasticity, malleability, feel, magnetism, double refraction, and chemical reaction to hydrochloric acid.

MINERAL GROUPS

Nearly four thousand minerals are presently known to exist and about 40 to 50 new ones are being identified each year.* Fortunately for students beginning to study minerals, no more than a few dozen are abundant. Collectively, these few make up most of the rocks of the earth's crust and as such, are classified as the *rock-forming minerals.* It is also interesting to note that only eight elements compose the bulk of these minerals and represent only 98 percent (by weight) of the continental crust (Table 2.2). The two most abundant elements are silicon and oxygen,

*Appendix C describes some of the common minerals found in the earth's crust.

TABLE 2.2
Relative abudance of the most common elements in the earth's crust.

Element	Approximate Percentage by Weight
Oxygen (O)	46.6
Silicon (Si)	27.7
Aluminum (Al)	8.1
Iron (Fe)	5.0
Calcium (Ca)	3.6
Sodium (Na)	2.8
Potassium (K)	2.6
Magnesium (Mg)	2.1
All others	1.5
Total	100

SOURCE: Data from Brian Mason.

which combine to form the framework of the most common mineral group, the **silicates.** Every silicate mineral contains oxygen and silicon. Moreover, except for a few minerals such as quartz, every silicate mineral contains one or more additional elements that are needed to produce electrical neutrality. Perhaps the next most common mineral group is the carbonates, of which calcite is the most prominent member. Other common rock-forming minerals include gypsum and halite (table salt).

In addition to the rock-forming minerals, a number of minerals are prized for their economic value. Included in this group are the ores of metals such as hematite (iron), sphalerite (zinc), and galena (lead); the native elements, including gold, silver, and carbon (diamonds); and a host of others, such as fluorite, corundum, and uraninite. Note that the rock-forming minerals themselves are not without economic value. For instance, quartz is used in the production of glass, calcite is the main constituent in portland cement, and plaster is composed of the mineral gypsum.

SILICATE STRUCTURES

All silicate minerals have the same fundamental building block, the **silicon-oxygen tetrahedron.** This structure consists of four oxygen atoms surrounding a much smaller silicon atom positioned in the space between them (Figure 2.14). The silicon-oxygen tetrahedron is not, however, a compound; rather it is a reactive ion with a charge of -4. This excess negative charge results because each of the four oxygen atoms contributes a charge of -2, whereas the one silicon atom has a charge of $+4$. In nature, one of the

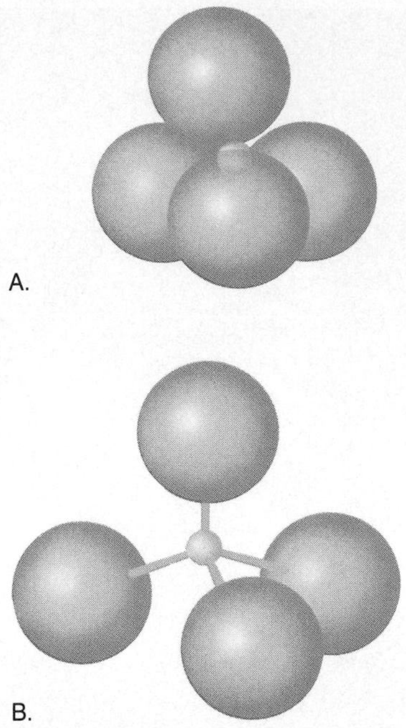

FIGURE 2.14
Two representations of the silicon-oxygen tetrahedron. **A.** The four large spheres represent oxygen atoms and the blue sphere represents a silicon atom. The spheres are drawn in proportion to the radii of the atoms. **B.** An expanded view of the tetrahedron using rods to depict the bonds that connect the atoms.

simplest ways in which these tetrahedra are neutralized is through the addition of positively charged ions. In this way a chemically stable structure consisting of individual tetrahedra linked together by positively charged ions is produced.

In addition to positive ions acting as the "glue" to bind the tetrahedra, the tetrahedra themselves may be linked in a variety of configurations. For example. the tetrahedra may be joined to form single chains, double chains, or sheet structures as shown in Figure 2.15. The joining of tetrahedra in each of these configurations results from the sharing of oxygen atoms between silicon atoms in adjacent tetrahedra. In order to better understand how this sharing takes place, select one of the silicon atoms (small spheres) near the middle of the single chain structure shown in Figure 2.15. Notice that this silicon atom is completely surrounded by four larger oxygen atoms. Also notice that two of the four oxygen atoms are joined to two silicon atoms, while the other two are not shared in this manner. It is the linkage across the shared oxygen atoms that joins the tetrahedra into a chain structure. Now examine a silicon atom near the middle of the sheet structure and count the number of shared and unshared oxygen atoms surrounding it. The increase in the degree of sharing accounts for

FIGURE 2.15
Three types of silicate structures. **A.** Single chains. **B.** Double chains. **C.** Sheet structures.

the sheet structure. Although they are not shown, other silicate structures exist. The most common silicate structure has all of the oxygen atoms shared to produce a complex three-dimensional framework.

By now we can see that the ratio of oxygen atoms to silicon atoms differs in each of the silicate structures. In the isolated tetrahedron there are 4 oxygen atoms for every silicon atom, in the single chain the oxygen to silicon ratio is 3 to 1, and in the three-dimensional framework this ratio is 2 to 1. Consequently, as more of the oxygen atoms are shared, the percentage of silicon in the structure increases. The silicate minerals are therefore described as having a high or low silicon content based on their ratio of oxygen to silicon. This difference in silicon content is important as we shall see later when we consider the formation of igneous rocks.

Most silicate structures, like those shown in Figure 2.15, are not neutral chemical compounds. Thus, like the individual tetrahedra, they all are neutralized by the inclusion of charged metallic ions that bond them together into a variety of crystalline configurations. The ions that most often link silicate structures are those of the elements iron (Fe), magnesium (Mg), potassium (K), sodium (Na), aluminum (Al), and calcium (Ca). Notice in Figure 2.16 that each of these positive ions has a particular atomic size and a particular charge. Generally, ions of approximately the same size are able to freely substitute for one another. For instance, the ions of iron (Fe^{2+}) and magnesium (Mg^{2+}) are nearly the same size and substitute for each other without altering the mineral structure. This also holds true for calcium and sodium, which can occupy the same site in a crystalline structure, and aluminum (Al), which substitutes for silicon in the silicon-oxygen tetrahedron.

Because of the ability of mineral structures to readily accommodate different ions at a given bonding site, individual specimens of a particular mineral may contain varying amounts of certain elements. A mineral of this type is often expressed by a chemical formula that uses parentheses to set apart the variable component. A good example is the mineral olivine $(Fe,Mg)_2SiO_4$. As we can see from the formula, it is the iron (Fe^{2+}) and magnesium (Mg^{2+}) ions in olivine that freely substitute for each other. At one

FIGURE 2.16
Relative sizes and electrical charges of ions commonly found in rock-forming minerals. Ionic radii are expressed in angstroms (one angstrom equals 10^{-8} cm).

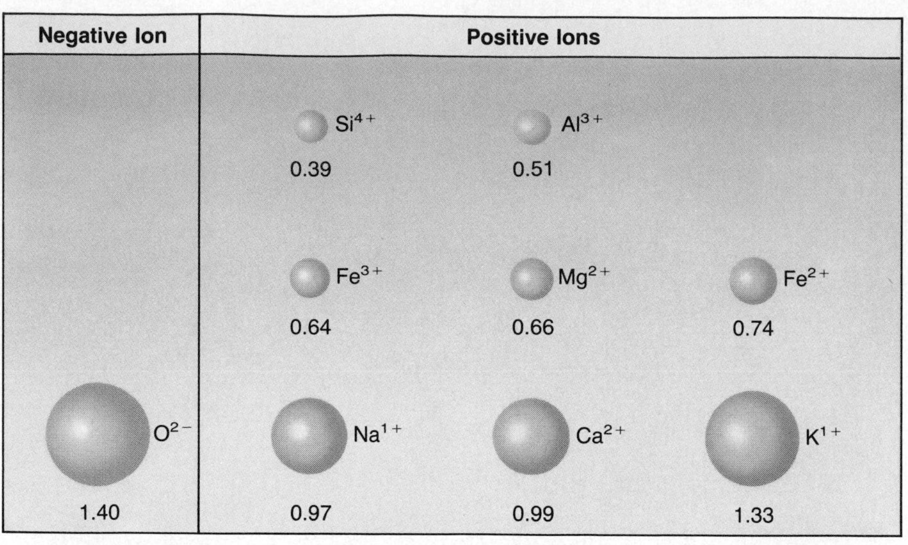

BOX 2.2

Carrot, Karat, or Carat

Like many other words in the English language, the words *carrot, karat,* and *carat* all have the same sound but are different in meaning (Figure 2.B). Such words are called *homonyms.* We all know that a *carrot* is a crunchy, orange vegetable that is supposed to be good for your eyesight. But what about the words *karat* and *carat?*

The term *karat* is used to indicate the purity of gold. Pure or fine gold is 24 karats. Gold that is less than 24 karats is actually an alloy (mixture) of gold and another metal, usually copper or silver. For example, 14-karat gold contains 14 parts of gold (by weight) mixed with 10 parts of other metals. Obviously, gold with a higher karat number is more expensive. Because 24-karat gold is much softer and more malleable than a gold-silver alloy, the alloy forms are more durable.

The third of our homonyms, *carat,* is a unit of weight used for precious gems such as diamonds, emeralds, and rubies. Throughout history, the size of a carat has varied somewhat. However, early in the twentieth century, a carat weight was established at 200 milligrams (or 0.2 gram). To put this in everyday terms, a 1-ounce diamond would be roughly 142 carats. The terms *karat* and *carat* are both derived from the Greek word *keration,* meaning "carob bean." The ancient Greeks used carob beans as a standard of weight.

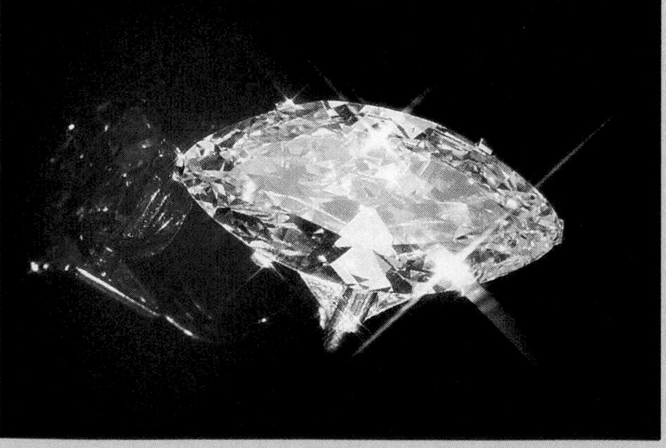

FIGURE 2.B
The purity of gold, such as this 82 ounce gold nugget, is measured in units called karats. Pure gold is 24 karats. By comparison, the weight of gemstones is measured in carat weight. Each carat equals 0.2 gram. This diamond ring is 29.3 carats. (Courtesy of Smithsonian Institution)

extreme, olivine may contain iron without any magnesium (Fe_2SiO_4) and at the other, iron is totally lacking (Mg_2SiO_4). Between these end members any ratio of iron to magnesium is possible. Thus olivine, as well as many other silicate minerals, is actually a family of minerals that has a range of composition between the two end members.

In certain substitutions the ions that interchange do not have the same electrical charge. For instance, when calcium (Ca^{2+}) substitutes for sodium (Na^{1+}), the structure gains a positive charge. In nature, one way in which this substitution is accomplished, while still maintaining overall electrical neutrality, is that a simultaneous substitution of aluminum (Al^{3+}) for silicon (Si^{4+}) takes place. This particular double substitution occurs in plagioclase feldspar, which is a member of the most abundant family of minerals found in the crust. The end members of this particular feldspar series are anorthite, $CaAl_2Si_2O_8$, and albite, $NaAlSi_3O_8$.

SILICATE MINERALS

The main groups of silicate minerals and common examples of each are given in Figure 2.17. The feldspars are by far the most plentiful group, comprising over 50 percent of the earth's crust. Quartz, the second most abundant mineral in the continental crust, is the only common mineral made completely of silicon and oxygen.

Notice in Figure 2.17 that each mineral group has a particular silicate structure. A relationship exists between the internal structure of a mineral and the

Mineral		Idealized Formula	Cleavage	Silicate Structure
Olivine		$(Mg,Fe)_2SiO_4$	None	Single tetrahedron 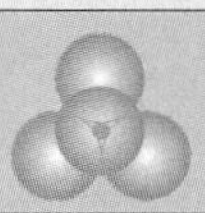
Pyroxene group (Augite)		$(Mg,Fe)SiO_3$	Two planes at right angles	Single chains 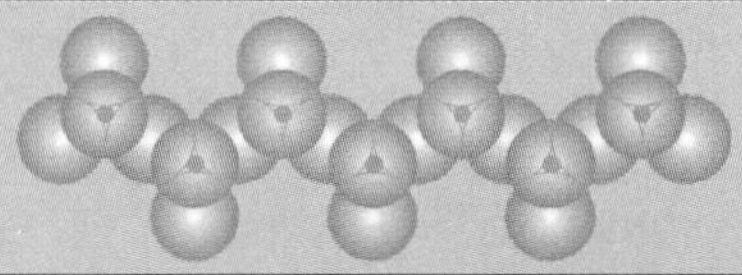
Amphibole group (Hornblende)		$(Ca_2Mg_5)Si_8O_{22}(OH)_2$	Two planes at 60° and 120°	Double chains 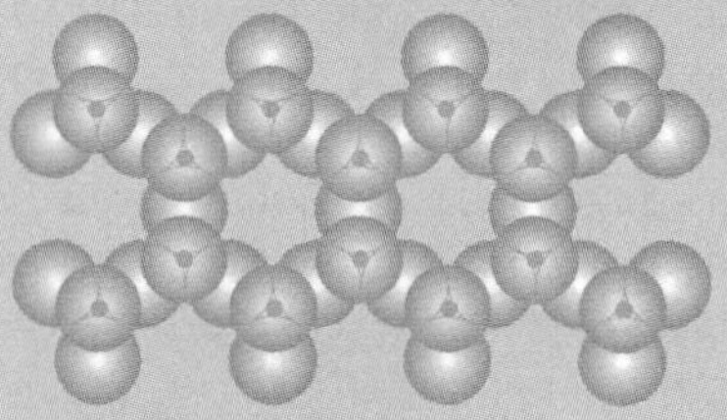
Micas	Muscovite	$KAl_2(AlSi_3O_{10})(OH)_2$	One plane	Sheets 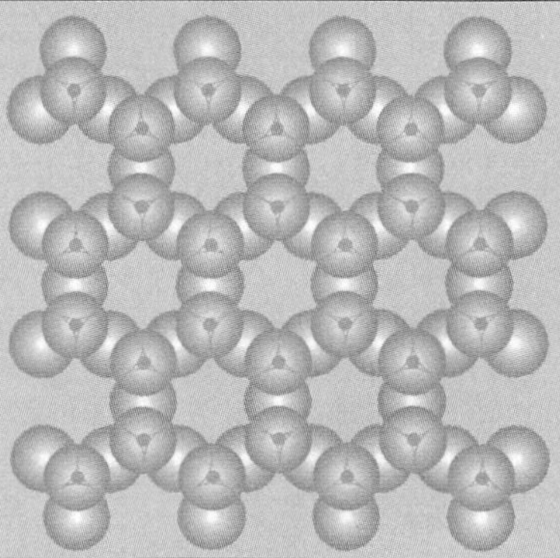
	Biotite	$K(Mg,Fe)_3AlSi_3O_{10}(OH)_2$		
Feld-spars	Orthoclase	$KAlSi_3O_8$	Two planes at 90°	Three-dimensional networks — Very complex structure
	Plagioclase	$(Ca,Na)AlSi_3O_8$		
Quartz		SiO_2	None	

FIGURE 2.17
Common silicate minerals. Note that the complexity of the silicate structure increases down the chart.

cleavage it exhibits. Because the silicon-oxygen bonds are strong, silicate minerals tend to cleave between the silicon-oxygen structures rather than across them. For example, the micas have a sheet structure and tend to cleave into flat plates (see Figure 2.10). Quartz, which has equally strong silicon-oxygen bonds in all directions, has no cleavage.

Most silicate minerals form when molten rock cools. This cooling can occur at or near the earth's surface or at great depths where temperatures and pressures are very high. The environment during crystallization and the chemical composition of the molten rock to a large degree determine the minerals that are produced. For example, the silicate mineral olivine crystallizes at high temperatures and possesses a chemical structure that is stable at high temperatures. Quartz, on the other hand, crystallizes at much lower temperatures. In addition, some silicate minerals are stable at the earth's surface and represent the weathered products of pre-existing silicate minerals. Still other silicate minerals are formed under the extreme pressures associated with metamorphism. Each silicate mineral therefore, has a structure and a chemical composition that indicate the conditions under which it formed. Thus, by carefully examining the mineral constituents of rocks, geologists can often determine the circumstances under which rocks formed.

The various silicate minerals can be divided on the basis of chemical makeup. The *ferromagnesian silicates* are those of minerals containing ions of iron and/or magnesium in their structure. Those minerals that do not contain these ions are simply called *nonferromagnesians.* Usually, ferromagnesian minerals are dark in color and have a specific gravity between 3.2 and 3.6. By comparison, nonferromagnesian silicates are generally light in color and have an average specific gravity of 2.7. These observed differences are mainly attributable to the presence or absence of iron.

Ferromagnesian Silicates

Olivine is a high-temperature silicate mineral that is black to olive green in color and has a glassy luster and a conchoidal fracture. Rather than developing large crystals, olivine commonly forms small, rounded crystals that give rocks consisting largely of olivine a granular appearance. Olivine is composed of individual tetrahedra which are bonded together by a mixture of iron and magnesium ions positioned so as to link the oxygen atoms together. Since the three-dimensional network generated in this fashion does not have its weak bonds aligned, olivine does not possess cleavage.

The members of the complex *pyroxene* group are thought to be important components of the mantle. The most common member, *augite,* is a black, opaque mineral with two directions of cleavage that meet at nearly a 90 degree angle. Its crystalline structure consists of single chains of tetrahedra bonded together by ions of iron and magnesium. Since the silicon-oxygen bonds are stronger than the bonds joining the silicate structures, augite cleaves parallel to the silicate chains. Augite is one of the dominant minerals in basalt, a common igneous rock of the oceanic crust and volcanic areas on the continents.

Hornblende is the most common member of a chemically complex group of minerals called *amphiboles* (Figure 2.18). Hornblende is usually dark green to black in color and except for its cleavage angles, which are about 60 degrees and 120 degrees, it is very similar in appearance to augite (Figure 2.19). The double chains of tetrahedra in the hornblende structure account for its particular cleavage. In a rock, hornblende often forms elongated crystals. This helps distinguish it from pyroxene, which forms rather blocky crystals. Hornblende is found predominantly in continental rocks, where it often makes up the dark portion of an otherwise light-colored rock.

Biotite is the dark iron-rich member of the mica family. Like other micas, biotite possesses a sheet structure which gives it excellent cleavage in one direction. Biotite also has a shiny black appearance that helps distinguish it from the other dark ferromagnesian minerals. Like hornblende, biotite is a common

FIGURE 2.18
Hornblende amphibole. Hornblende is a common dark silicate material having two cleavage directions that intersect at roughly 60° and 120°.

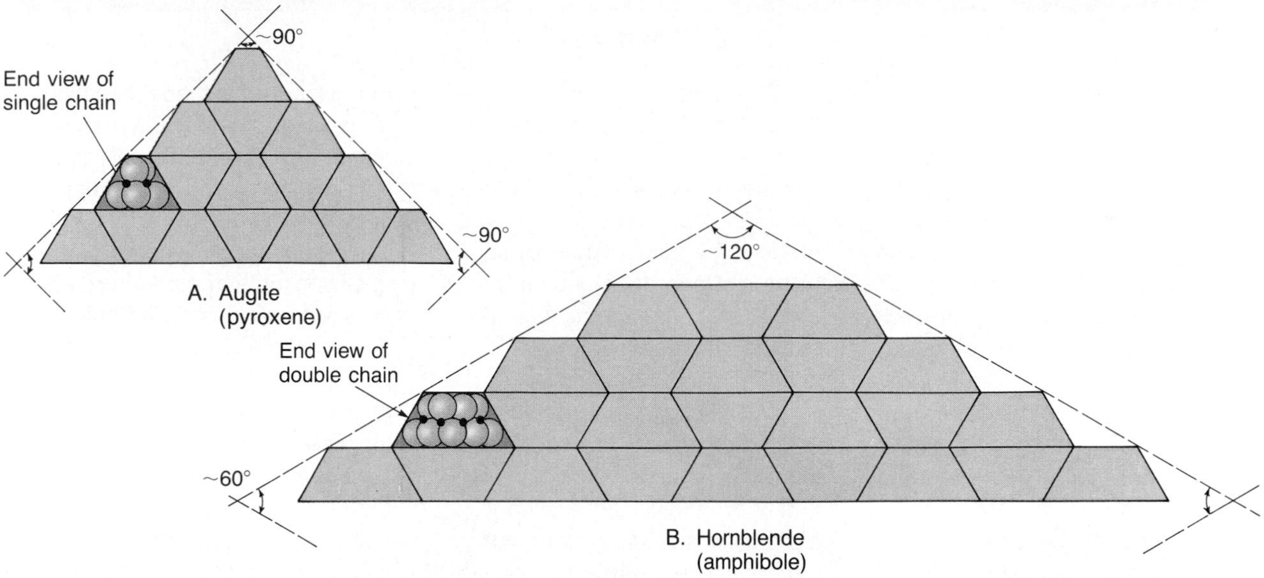

FIGURE 2.19
Cleavage angles for augite and hornblende.

constituent of continental rocks, including the igneous rock granite.

Garnet is similar to olivine in that its structure is composed of individual tetrahedra linked by metallic ions. Also like olivine, garnet has a glassy luster, lacks cleavage, and possesses conchoidal fracture. Although the colors of garnet are varied, this mineral is most often brown to deep red. Garnet readily forms equidimensional crystals that are most commonly found in metamorphic rocks (Figure 2.20). When garnets are transparent, they may be used as gemstones.

FIGURE 2.20
Garnet crystal in a common metamorphic rock. (Photo by E. J. Tarbuck)

Nonferromagnesian Silicates

Muscovite is a common member of the mica family. It is light in color and has a pearly luster. Like other micas, muscovite has excellent cleavage in one direction. In thin sheets muscovite is clear, a property which accounts for its use as window "glass" during the Middle Ages. Since muscovite is very shiny, it can often be identified by the sparkle it gives a rock. If you have ever looked closely at beach sand, you may have seen the glimmering brilliance of the mica flakes scattered among the other sand grains.

Feldspar, the most common mineral group, can form under a very wide range of temperatures and pressures, a fact that partially accounts for its abundance. All of the feldspars have similar physical properties. They have two planes of cleavage meeting at or near 90 degree angles, are relatively hard (6 on Mohs scale), and have a luster which ranges from glassy to pearly. As one component in a rock, feldspar crystals can be identified by their rectangular shape and rather smooth shiny faces (Figure 2.21).

The structure of the feldspar minerals is a three-dimensional framework formed when oxygen atoms are shared by adjacent silicon atoms. In addition, one-fourth to one-half of the silicon atoms in the feldspar structure are replaced by aluminum atoms. The difference in charge between aluminum (+3) and silicon (+4) is made up by the inclusion of one or more of the following ions into the crystal lattice: potassium (+1), sodium (+1), and calcium (+2). Because of the large size of the potassium ion as compared to the size of the sodium and calcium

BOX 2.3

Gemstones

Precious stones have been prized by humankind since antiquity. This tradition survives today. Gems are not just prized keepsakes of the rich but are possessed by people of modest means as well. Misinformation abounds about the nature of gems and the minerals of which they are composed. Part of the misinformation stems from the ancient practice of grouping precious stones by color rather than mineral makeup. For example, the more common red spinels were often passed off to royalty as rubies, which are more valuable gems. Even today, when modern techniques of mineral identification are commonplace, yellow quartz is frequently sold as topaz.

Compounding the confusion is the fact that many gems have names that are different from their mineral names. For example, diamond is composed of the mineral of the same name, whereas sapphire is a form of corundum, an aluminum oxide–rich mineral. Although pure aluminum oxide is colorless, a minute amount of a foreign element can produce a vividly colored gemstone. Hence, depending on the impurity, sapphires of nearly every color are known to exist. Pure aluminum oxide with trace amounts of titanium and iron produce the most prized blue sapphires. When the mineral corundum contains sufficient quantities of chromium, it exhibits a brilliant red color, and the gem is called *ruby.* Large, gem-quality rubies are much rarer than diamonds and thus can command a very high price.

To summarize, when a gem-quality sample of corundum exhibits a red hue, it is called *ruby,* but if it exhibits any other color, the gem is called *sapphire.* If the specimen is not suitable as a gem, it simply goes by the mineral name *corundum.* Although common corundum is not a gemstone, it does have value as an abrasive material. Whereas two gems, rubies and sapphires, are composed of the mineral corundum, quartz is the parent mineral of more than a dozen gems. Table 2.B lists some

TABLE 2.B
Important gemstones.

Gem	Mineral Source	Prized Hues
Precious		
Diamond	Diamond	Colorless, yellows
Emerald	Beryl	Greens
Opal	Opal	Brilliant hues
Ruby	Corundum	Reds
Sapphire	Corundum	Blues
Semiprecious		
Alexandrite	Chrysoberyl	Variable
Amethyst	Quartz	Purples
Cat's-eye	Chrysoberyl	Yellows
Chalcedony	Quartz (agate)	Banded
Citrine	Quartz	Yellows
Garnet	Garnet	Reds, greens
Jade	Jadeite or nephrite	Greens
Moonstone	Feldspar	Transparent blues
Peridot	Olivine	Olive greens
Smoky quartz	Quartz	Browns
Spinel	Spinel	Reds
Topaz	Topaz	Purples, reds
Tourmaline	Tourmaline	Reds, blue-greens
Turquois	Turquois	Blues
Zircon	Zircon	Reds

ions, two different feldspar structures exist. *Orthoclase feldspar* is a common member of a group of feldspar minerals that contains potassium ions in its structure. The other group, called *plagioclase feldspar,* contains both sodium and calcium ions that freely substitute for one another depending on the environment during crystallization.

Orthoclase feldspar is usually light cream to salmon pink in color. The plagioclase feldspars, on the other hand, range in color from white to medium gray. However, color should not be used to distinguish these groups. The only sure way to physically distinguish the feldspars is to look for a multitude of fine parallel lines, called *striations.* Striations are

well-known gemstones and their mineral names.

What constitutes a gemstone? In essence, certain mineral specimens when cut and polished possess beauty of such quality that they can command a price that makes the process of producing the gem profitable. Gemstones can be divided into two categories: precious and semiprecious. A *precious* gem has beauty, durability, size and rarity, whereas a *semiprecious* gem generally has only one or two of these qualities. The gems that have traditionally enjoyed the highest esteem are diamonds, rubies, sapphires, emeralds, and some varieties of opal (Table 2.B). All other gemstones are classified as semiprecious. It should be noted, however, that large, high-quality specimens of the so-called semiprecious stones can often command a very high price.

Obviously, beauty is the most important quality that a gem can possess. Today, we prefer translucent stones with evenly tinted colors. The most favored hues appear to be red, blue, green, purple, rose, and yellow. The most prized stones are pigeon-blood rubies, blue sapphires, grass-green emeralds, and canary-yellow diamonds. Colorless gems are generally less desirable except in the case of diamonds that display "flashes of color" known as *brilliance.* Note that gemstones in the "rough" are dull and would be passed over by most laypersons as "just another rock." Gems must be cut and polished by experienced craftsmen before their true beauty can be displayed (Figure 2.C).

The durability of a gem depends upon its hardness; that is, its resistance to abrasion by objects normally encountered in everyday living. For good durability, gems should be as hard as or harder than quartz as defined by the Mohs scale of hardness, which consists of ten minerals arranged in order from 1 (softest) to 10 (hardest). One notable exception is opal, which is comparatively soft (hardness 5 to 6.5) and brittle. Opal's esteem comes from its "fire," which is a display of a variety of brilliant colors including greens, blues, and reds.

It seems to be human nature to treasure that which is rare. In the case of gemstones, large, high-quality specimens are much rarer than smaller stones. Thus, large rubies, diamonds, and emeralds, which are rare in addition to being beautiful and durable, command the very highest prices.

FIGURE 2.C
Australian sapphires depicting variation in cuts and colors. (Photo by Fred Ward, Black Star)

found on some cleavage planes of plagioclase feldspar, but are not present on orthoclase feldspar (Figure 2.22).

Quartz is the only common silicate mineral consisting entirely of silicon and oxygen. As such, the term *silica* is applied to quartz, which has the chemical formula SiO_2. Since the structure of quartz contains a ratio of two oxygen ions (O^{2-}) for every one silicon ion (Si^{4+}), no other positive ions are needed to attain neutrality. In quartz, a three-dimensional framework is developed through the complete sharing of oxygen by adjacent silicon atoms. Thus, all of the bonds in quartz are of the strong silicon-oxygen type. Consequently, quartz is hard, resistant to weath-

FIGURE 2.21
Sample of the mineral orthoclase feldspar.

FIGURE 2.22
These parallel lines, called striations, are a distinguishing characteristic of the plagioclase feldspars. (Photo by E. J. Tarbuck)

ering, and does not have cleavage. When broken, quartz generally exhibits conchoidal fracture. In a pure form, quartz is clear and if allowed to solidify without interference, will form hexagonal crystals which develop pyramid-shaped ends (see Figure 2.9B). However, like most other clear minerals, quartz is often colored by the inclusion of various ions (impurities) and forms without developing good crystal faces (Figure 2.23). The most common varieties of quartz are milky (white), smoky (gray), rose (pink), amethyst (purple), and rock crystal (clear).

Clay is a term used to describe a variety of complex minerals which, like the micas, have a sheet structure. The clay minerals are generally very fine grained and can only be studied microscopically. Most clay minerals originate as products of the chemical weathering of the other silicate minerals. Thus, clay minerals make up a large percentage of the surface material we call soil. Because of the importance of soil in agriculture, and because of its role as a supporting material for buildings, clay minerals are extremely important to humans. One of the most common clay minerals is *kaolinite,* which is used in the manufacture of fine chinaware and in the production of high-gloss paper such as that used in this textbook. Further, some clay minerals absorb large

FIGURE 2.23
Quartz. Some minerals, such as quartz, occur in a variety of colors. These samples include crystal quartz (colorless), amethyst (purple quartz), citrine (yellow quartz), and smoky quartz (gray to black).

amounts of water, which allows them to swell to several times their normal size. These clays have been used commercially in a variety of ingenious ways, including as an additive to thicken milkshakes in fast-food restaurants.

NONSILICATE MINERALS

Although many are important from an economic standpoint, other mineral groups can be considered to be scarce when compared to the silicates. Table 2.3 lists examples of oxides, sulfides, sulfates, halides, and native elements of economic value. A discussion of a few more of the more common nonsilicate, rock-forming minerals follows.

The carbonate minerals are muc turally than the silicates. This miner posed of the complex carbonate ic one or more kinds of positive ions common carbonate minerals are *cal* *dolomite,* CaMg$(CO_3)_2$. Because these minerals are similar both physically and chemically, they are difficult to distinguish from one another. Both have a vitreous luster, a hardness between 3 and 4, and nearly perfect rhombic cleavage. They can, however, be distinguished by using dilute hydrochloric acid. Calcite reacts vigorously with this acid, whereas dolomite reacts much more slowly. Calcite and dolomite are usually found together as the primary constituents in the sedimentary rocks limestone and dolostone. When calcite is the dominant mineral, the

TABLE 2.3
Common nonsilicate mineral groups.

Group	Member	Formula	Economic Use
Oxides	Hematite	Fe_2O_3	Ore of iron
	Magnetite	Fe_3O_4	Ore of iron
	Corundum	Al_2O_3	Gemstone, abrasive
	Ice	H_2O	Solid form of water
	Chromite	$FeCr_2O_4$	Ore of chromium
	Ilmenite	$FeTiO_3$	Ore of titanium
Sulfides	Galena	PbS	Ore of lead
	Sphalerite	ZnS	Ore of zinc
	Pyrite	FeS_2	Sulfuric acid production
	Chalcopyrite	$CuFeS_2$	Ore of copper
	Bornite	Cu_5FeS_4	Ore of copper
	Cinnabar	HgS	Ore of mercury
Sulfates	Gypsum	$CaSO_4 \cdot 2H_2O$	Plaster
	Anhydrite	$CaSO_4$	Plaster
	Barite	$BaSO_4$	Drilling mud
Native elements	Gold	Au	Trade, jewelry
	Copper	Cu	Electrical conductor
	Diamond	C	Gemstone, abrasive
	Sulfur	S	Sulfa drugs, chemicals
	Graphite	C	Pencil lead, dry lubricant
	Silver	Ag	Jewelry, photography
	Platinum	Pt	Catlyst
Halides	Halite	NaCl	Common salt
	Fluorite	CaF_2	Used in steel making
	Sylvite	KCl	Fertilizer
Carbonates	Calcite	$CaCO_3$	Portland cement
	Dolomite	$CaMg(CO_3)_2$	Portland cement
	Malachite	$Cu_2(OH)_2CO_3$	Gemstone
	Azurite	$Cu_3(OH)_2(CO_3)_2$	Gemstone
Hydroxides	Limonite	$FeO(OH) \cdot nH_2O$	Ore of iron, pigments
	Bauxite*	$Al(OH)_3 \cdot nH_2O$	Ore of aluminum
Phosphates	Apatite	$Ca_5(F,Cl,OH)(PO_4)_3$	Fertilizer
	Turquoise	$CuAl_6(PO_4)_4(OH)_8$	Gemstone

*Bauxite is a mixture of hydrous aluminum oxides of indefinite composition.

FIGURE 2.24
Thick beds of halite (salt) are being tapped at an underground mine near Grand Saline, Texas. (Courtesy of Morton International, Morton Salt Division)

rock is called limestone, whereas dolostone results from a predominance of dolomite. Limestone has numerous economic uses, including as road aggregate, as building stone, and as the main ingredient in portland cement.

Two other nonsilicate minerals frequently found in sedimentary rocks are *halite* and *gypsum*. Both minerals are commonly found in thick layers, which are the last vestiges of ancient seas that have long since evaporated (Figure 2.24). Like limestone, both are important nonmetallic resources. Halite is the mineral name for common table salt (NaCl). Gypsum ($CaSO_4 \cdot 2H_2O$) is the mineral from which plaster and other similar building materials are composed.*

*For more on the economic significance of these and other minerals, see Chapter 21.

REVIEW QUESTIONS

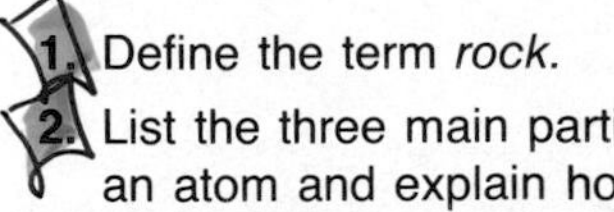

1. Define the term *rock*.
2. List the three main particles of an atom and explain how they differ from one another.
3. If the number of electrons in a neutral atom is 35 and its mass number is 80, calculate the following:
   (a) The number of protons.
   (b) The atomic number.
   (c) The number of neutrons.
4. What is the octet rule? What is the significance of valence electrons?
5. Briefly distinguish between ionic and covalent bonding.
6. What occurs in an atom to produce an ion?

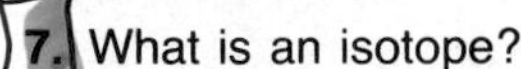

7. What is an isotope?
8. Although all minerals have an orderly internal arrangement of atoms (crystalline structure), most mineral samples do not demonstrate their crystal form. Why?
9. Why might it be difficult to identify a mineral by its color?
10. If you found a glossy-appearing mineral while rock hunting and had hopes that it was a diamond, what simple test might help you make a determination?
11. Explain the use of corundum as given in Table 2.3 in terms of the Mohs hardness scale.
12. Gold has a specific gravity of almost 20. If a 25-liter pail of water weighs 25 kilograms, how much would a 25-liter pail of gold weigh?
13. Explain the difference between the terms *silicon* and *silicate.*
14. What do ferromagnesian minerals have in common? List examples of ferromagnesian minerals.
15. What do muscovite and biotite have in common? How do they differ?
16. Should color be used to distinguish between orthoclase and plagioclase feldspar? What is the best means of distinguishing between the two types of feldspar?
17. Each of the following statements describes a silicate mineral or mineral group. In each case, provide the appropriate name.
   (a) The most common member of the amphibole group.
   (b) The most common nonferromagnesian member of the mica family.
   (c) The only silicate mineral made entirely of silicon and oxygen.
   (d) A high-temperature silicate with a name that is based on its color.
   (e) Characterized by striations.
   (f) Originates as a product of chemical weathering.
18. What simple test can be used to distinguish calcite from dolomite?

KEY TERMS

atom (p. 31)
atomic mass unit (p. 36)
atomic number (p. 31)
atomic weight (p. 36)
cleavage (p. 40)
color (p. 39)
compound (p. 31)
covalent bond (p. 34)
crystal form (p. 38)
electron (p. 31)
element (p. 31)
energy-level shell (p. 31)
fracture (p. 40)
hardness (p. 39)
ion (p. 32)
ionic bond (p. 32)
isotope (p. 36)
luster (p. 39)
mass number (p. 36)
metallic bond (p. 36)
mineral (p. 30)
Mohs scale (p. 39)
neutron (p. 31)
nucleus (p. 31)
octet rule (p. 32)
polymorph (p. 37)
proton (p. 31)
radioactivity (p. 36)
rock (p. 30)
silicate mineral (p. 42)
silicon-oxygen tetrahedron (p. 42)
specific gravity (p. 40)
streak (p. 39)
valence electron (p. 32)

3
Igneous Rocks

Opposite: Spectacular spires called Pingora, Wind River Range, Wyoming.
(Photo by Russell Lamb Photography).
(Top photo used by permission of Dennis Tasa)

In our discussion of the rock cycle, it was pointed out that igneous rocks form when **magma** cools and solidifies. This molten rock, which originates at depths as great as 200 kilometers within the earth, consists primarily of the elements found in silicate minerals, along with some gases, particularly water vapor, which are confined within the magma by the surrounding rocks. Because the magma body is lighter than the surrounding rocks, it works its way toward the surface, and on occasion breaks through, producing a volcanic eruption (Figure 3.1). The spectacular explosions that sometimes accompany an eruption are produced by the gases (volatiles) escaping as the confining pressure lessens near the surface. Sometimes blockage of the vent coupled with surface water seepage into the magma chamber can produce catastrophic explosions. Along with ejected rock fragments, a volcanic eruption can generate extensive lava flows. **Lava** is similar to magma, except that most of the gaseous component has escaped. The rocks which result when lava solidifies are classified as **extrusive**, or **volcanic**. The magma not able to reach the surface eventually crystallizes at depth. Igneous rocks produced in this manner are

FIGURE 3.1
Parícutin Volcano in eruption at night, 1943. (Photo by K. Segerstrom, U.S. Geological Survey)

termed **intrusive**, or **plutonic**, and would never be observed if not for the processes of erosion stripping away the overlying rocks.

CRYSTALLIZATION OF MAGMA

Magma is a hot fluid that may contain suspended crystals and a gaseous component. The liquid portion of a magma body is composed of ions that move about freely. However, as magma cools, the random movements of the ions slow and the ions begin to arrange themselves into orderly patterns. This process is called **crystallization**. Before we examine crystallization in more detail, let us first examine how a simple crystalline solid melts. In a crystalline solid, the ions are arranged in a closely packed regular pattern. However, they are not without some motion. They exhibit a sort of restricted vibration about fixed points. As the temperature rises, the ions vibrate more and more rapidly and consequently collide with ever-increasing vigor with their neighbors. Continued heating causes the ions to occupy additional space. This results in expansion of the solid and greater distance between ions. When the melting point is reached, the ions are far enough apart and are vibrating rapidly enough to overcome the force of the chemical bonds which had joined them. At this stage, the ions are able to slide past one another, destroying their orderly crystalline structure. Thus, what was once a solid has become a liquid composed of unordered ions moving randomly about.

In the process of crystallization, cooling reverses the events of melting. As the temperature of the liquid drops, the ions pack closer together and begin to lose their freedom of movement. When cooling is sufficient, the force of the chemical bonds will again confine the atoms to an orderly crystalline arrangement. Usually, all of the molten material does not solidify at the same time. Rather, as it cools, numerous embryo crystals develop. In a systematic fashion, ions are added to these centers of crystal growth. When the crystals grow large enough that their edges meet, their growth ceases and crystallization continues elsewhere. Eventually, all of the liquid is transformed into a solid mass of interlocking crystals (Figure 3.2).

The rate of cooling strongly influences the crystallization process, in particular the size of the crystals. When a magma cools slowly, relatively few centers of crystal growth develop. Slow cooling also allows ions to migrate over relatively great distances. Consequently, slow cooling results in the formation of rather large crystals. On the other hand, when cooling occurs rapidly, the ions quickly lose their motion

A.

B.

FIGURE 3.2
A. Close-up of interlocking crystals in a coarse-grained igneous rock. (Photo by E. J. Tarbuck)
B. Photomicrograph of interlocking crystals in a coarse-grained igneous rock.

and readily combine. This results in the development of large numbers of nuclei which all compete for the available ions. The result is a solid mass formed of small intergrown crystals.

When the molten material is quenched instantly, there is not sufficient time for the ions to arrange themselves into a crystalline network. Therefore, the solids produced in this manner consist of randomly distributed ions. Rocks that consist of unordered atoms are referred to as **glass** and are similar to ordinary manmade glass.

The crystallization of a magma, although more complex, occurs in a manner similar to that just described. Rather than being composed of only one or two elements, most magma consists of the eight elements that are the primary constituents of the silicate minerals (see Table 2.2). These are silicon, oxygen, aluminum, sodium, potassium, calcium, iron, and magnesium. In addition, trace amounts of many other elements, as well as volatiles, particularly water and carbon dioxide, are also found in magma. A *volatile* is a material that is commonly a gas at temperatures and pressures existing at the earth's surface.

When magma cools, it is generally the silicon and oxygen atoms that link together first, to form silicon-oxygen tetrahedra. As cooling continues, the tetrahedra join with each other and with other ions to form crystal nuclei of the various silicate minerals. Each crystal nucleus grows as ion after ion is added to the crystalline network in an unchanging pattern. However, the minerals which compose a magma do not all form at the same time or under the same conditions. As we shall see, certain minerals crystallize at much higher temperatures than others. Consequently, magmas often consist of solid crystals surrounded by molten material.

In addition to the rate of cooling, the mineral composition of a magma and the amount of volatile material influence the crystallization process. Since magmas differ in each of these aspects, the physical appearance and mineral composition of igneous rocks vary widely. Nevertheless, it is possible to classify igneous rocks based on their mode of origin and mineral constituents. The environment during crystallization can be inferred from the size and arrangement of the mineral grains, a property called texture. Consequently, igneous rocks are most often classified by their texture and mineral composition. We will consider both of these rock characteristics in the following sections.

IGNEOUS TEXTURES

The term *texture,* when applied to an igneous rock, is used to describe the overall appearance of the rock based on the size and arrangement of its interlocking crystals (Figure 3.3). Texture is an important characteristic since it reveals a great deal about the environment in which the rock formed. This fact allows geologists to make inferences about a rock's origin while working in the field where sophisticated equipment is not available.

The most important factor affecting the texture of a rock is the rate at which the magma cooled. From our discussion of crystallization, we learned that rapid cooling produces small crystals, whereas very slow cooling results in the formation of much larger crystals. As we might expect, the rate of cooling is quite slow in magma chambers lying deep within the crust, while a thin layer of lava extruded upon the earth's surface may chill in a matter of hours, and small molten blobs ejected into the air during a violent eruption can solidify almost instantly.

Igneous rocks that form at the earth's surface or as small masses within the upper crust possess a very fine-grained texture termed **aphanitic**. By definition, the grains of aphanitic rocks are too small for individual minerals to be distinguished with the unaided eye (Figure 3.3A). Although mineral identification is not possible, fine-grained rocks are commonly characterized as being light, intermediate, or dark in color. Using this system of grouping, light-colored aphanitic rocks are those composed primarily of light-colored nonferromagnesian silicate materials, and so forth.

Commonly seen in many aphanitic rocks are the voids left by gas bubbles trapped as the magma solidifies (Figure 3.4). These spherical or elongated openings are called **vesicles** and are limited to the upper portion of lava flows (Figure 3.5). It is in the upper zone of a lava flow that cooling occurs rapidly enough to "freeze" the lava, thereby preserving the openings produced by the expanding gas bubbles.

When large masses of magma solidify far below the surface, they form igneous rocks that exhibit a coarse-grained texture described as **phaneritic**. These coarse-grained rocks consist of a mass of intergrown crystals, which are roughly equal in size and large enough so that the individual minerals can be identified with the unaided eye (Figure 3.3B). Because phaneritic rocks form deep within the crust, their exposure at the surface results only after erosion removes the overlying rocks that once surrounded the magma chamber.

A large mass of magma located at depth may require tens of thousands, even millions, of years to solidify. Since all minerals within a magma do not crystallize at the same rate or at the same time during cooling, it is possible for some to become quite large before others even start to form. If magma containing some large crystals should change environments, by erupting at the surface, for example, the molten portion of the lava would cool quickly. The resulting rock, which has large crystals embedded in a matrix of smaller crystals, is said to have a **porphyritic texture** (Figure 3.3C). The large crystals in such a rock are referred to as **phenocrysts**, while the matrix of smaller crystals is called **groundmass**. A rock which has such a texture is called a **porphyry**.

During some volcanic eruptions, molten rock is ejected into the atmosphere where it is quenched quickly. Rapid cooling of this type may generate rock with a **glassy texture**. As was indicated earlier, glass results when the ions have not been permitted the time to unite into an orderly crystalline structure.

FIGURE 3.3
Igneous rock textures. **A.** Aphanitic. **B.** Phaneritic. **C.** Porphyritic. **D.** Glassy.

Obsidian, a common type of natural glass, is similar in appearance to a dark chunk of manufactured glass (Figure 3.3D).

Although the rate of cooling is the major factor that determines the texture of an igneous rock, other factors are also important. In particular, the composition of the magma influences the resulting texture. For example, basaltic magma, which is very fluid, will usually generate crystalline rocks when cooled quickly in a thin lava flow. Under the same conditions, granitic magma, which is quite viscous (resists flow), is much more likely to produce a rock with a glassy texture. Consequently, most of the lava flows which are composed of volcanic glass are granitic in composition. However, the surface of basaltic lava may be quenched rapidly enough to form a thin, glassy skin. Moreover, Hawaiian volcanoes sometimes generate lava fountains which spray basaltic lava tens of meters into the air. Such activity may produce strands of volcanic glass called *Pele's hair,* after the Hawaiian goddess of volcanoes.

Some igneous rocks are formed from the consolidation of individual rock fragments that are ejected during a violent eruption. The ejected particles may

FIGURE 3.4
Scoria is a volcanic rock that exhibits a vesicular texture. Vesicles form when gas bubbles are trapped near the top of a lava flow.

FIGURE 3.6
Pyroclastic texture. This volcanic rock found in the vicinity of Mount Rainier consists of angular rock fragments that have been fused together by hot glass shards. (Photo by E. J. Tarbuck)

be very fine ash, molten blobs, or large angular blocks which are torn from the walls of the vent during the eruption. Igneous rocks composed of these rock fragments are said to have a **pyroclastic texture** (Figure 3.6).

A common type of pyroclastic rock is composed of glass shards (thin strands) which remained hot enough during their flight to fuse together upon impact. Other pyroclastic rocks are composed of fragments that solidified before impact and became cemented together at some later time. Because pyroclastic rocks are made of individual rock fragments rather than interlocking crystals, their overall textures are often more similar to sedimentary rocks than to other igneous rocks.

MINERAL COMPOSITION

The mineral makeup of an igneous rock is ultimately determined by the chemical composition of the magma from which it crystallizes. Such a large variety of igneous rocks exists that it is logical to assume that an equally large variety of magmas must also exist.

FIGURE 3.5
Upper layer of a lava flow with characteristic voids called vesicles. (Photo by E. J. Tarbuck)

However, geologists have found that various eruptive stages of the same volcano often extrude lavas or pyroclastic material exhibiting somewhat different compositions, particularly if an extensive period of time separated the eruptions (Figure 3.7). Data of this type led them to look into the possibility that a single magma might produce rocks of varying mineral content.

Bowen's Reaction Series

A pioneering investigation into the crystallization of magma was carried out by N.L. Bowen in the first quarter of this century. Bowen discovered that as magma cools in the laboratory, those minerals with higher melting points crystallize before minerals with lower melting points. As shown in Figure 3.8, the first mineral to crystallize from a typical magma is the ferromagnesian mineral olivine. As the magma cools further, the minerals pyroxene and calcium-rich plagioclase begin to form. During the crystallization process, the composition of the melt (liquid portion of a magma, excluding any solid material) continually changes. For example, at the stage when about 50 percent of the magma has solidified, the melt will be depleted in iron, magnesium, and calcium because these elements are found in the earliest-formed minerals. At the same time, it will be enriched in aluminum, sodium, and potassium. Further, the silica component of the melt becomes enriched toward the latter stages of crystallization.

Bowen also demonstrated that as a magma cooled, the solid components (minerals) would react with the remaining melt and produce the next mineral in the sequence shown in Figure 3.8. For this reason, this arrangement of minerals became known as **Bowen's reaction series.** On the upper left branch of this reaction series, olivine, the first mineral to form, will react with the remaining melt to form pyroxene. As the magma body cools further, the pyroxene crystals will in turn react with the melt to generate amphibole. This reaction will continue until the last mineral in the series, biotite, is formed. The left branch of Bowen's reaction series is called a *discontinuous reaction series* because each mineral has a different crystalline structure. Olivine is composed of single tetrahedra whereas the other minerals in this sequence are composed of single chains, double chains, and sheet structures, respectively. Ordinarily, these reactions do not run to completion, so that various amounts of each of these minerals may exist at any given time.

The right branch of the reaction series, called the *continuous reaction series,* demonstrates that calcium-rich feldspar crystals react with the sodium ions contained in the melt to become progressively more sodium-rich. Often times the rate of cooling occurs rapidly enough to prohibit the reaction between early-formed, calcium-rich plagioclase crystals and the melt. In these instances, the feldspar crystals will have calcium-rich interiors surrounded by zones that are progressively richer in sodium.

During the last stage of crystallization, after most magma has solidified, the minerals muscovite and potassium feldspar are generated. Finally, the re-

FIGURE 3.7
Ash and pumice ejected during a large eruption of Mt. Mazama (Crater Lake). Notice the gradation from light-colored, silica-rich ash near the base to dark-colored rocks at the top. It is likely that prior to this eruption the magma began to segregate as the less-dense, silica-rich magma migrated toward the top of the magma chamber. The zonation seen in the rocks resulted because a sustained eruption tapped deeper and deeper levels of the magma chamber. Thus, this rock sequence is an inverted representation of the compositional zonation in the magma body; that is, the magma from the top of the chamber erupted first and is found at the base of these ash deposits and vice versa. (Photo by E. J. Tarbuck).

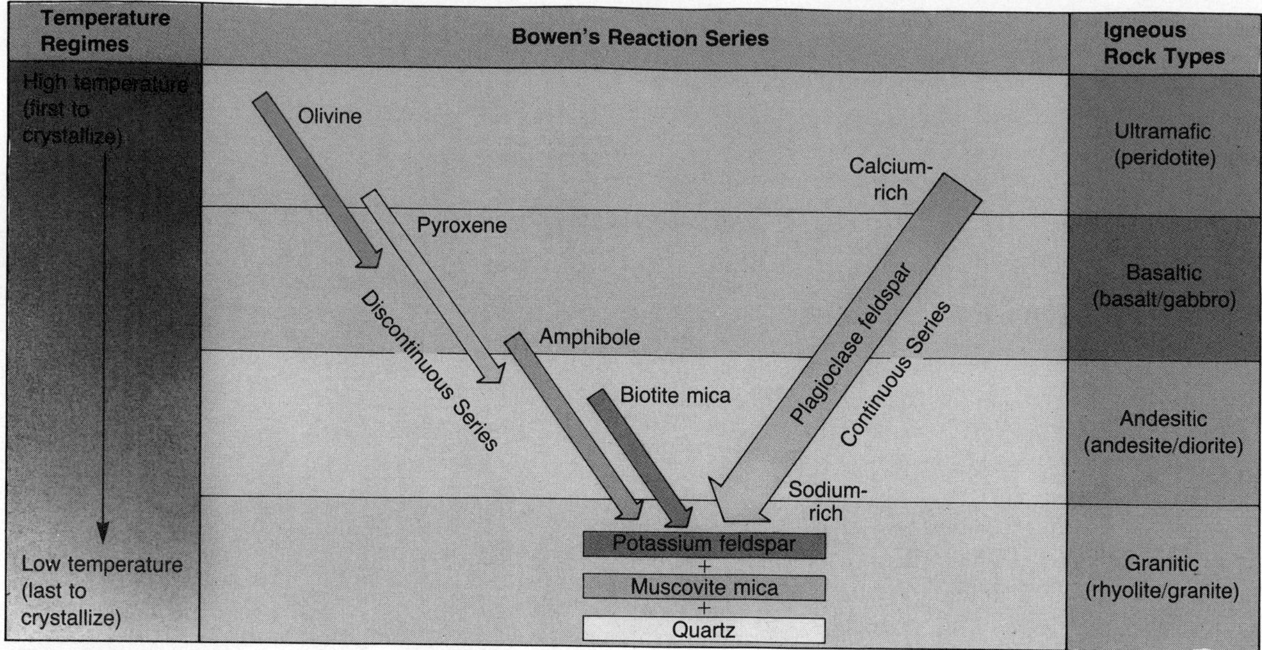

FIGURE 3.8
Bowen's reaction series shows the sequence in which minerals crystallize from a magma. Compare this figure to the mineral composition of the rock groups in Table 3.1. Note that each rock group consists of minerals that crystallize at the same time.

maining melt (if any) will have a high silica content that eventually crystallizes as quartz. Although these latter-formed minerals crystallize in the order shown, they do not react with the melt to produce other minerals.

Bowen's reaction series illustrates the sequence in which minerals crystallize from a magma. One of the consequences of this crystallization scheme is that minerals that form in the same general temperature regime are found together in the same igneous rock. For example, in Figure 3.8 notice that the minerals potassium feldspar, muscovite, and quartz are found in the same region of the diagram and are the major constituents of the igneous rock granite.

Magmatic Differentiation

Bowen demonstrated that minerals crystallize from magma in a systematic fashion. But how does Bowen's reaction series account for the great diversity of igneous rocks? It has been shown that at one or more stages in the crystallization process, a separation of the solid and liquid components of a magma can occur. This happens, for example, if the earlier-formed minerals are denser (heavier) than the liquid portion and settle to the bottom of the magma chamber as shown in Figure 3.9A. This process, called **crystal settling**, is thought to occur frequently with the dark silicates, such as olivine and pyroxene. When the remaining melt solidifies, either in place or in a new location if it migrates out of the magma chamber, it will form a rock with a chemical composition much different from the parent magma (Figure 3.9B). The process of developing more than one rock type from a common magma is called **magmatic differentiation**.

A classic example of magmatic differentiation is found in the Palisades Sill, which is a 300-meter-thick tabular mass of dark igneous rock exposed along the west shore of the Hudson River. Because of its great thickness and subsequent slow rate of crystallization, olivine, the first mineral to crystallize, sank toward the bottom and makes up about 25 percent of the lower portion of the sill. By contrast, near the top of this igneous body, olivine represents only one percent of the rock mass.*

At any stage in the evolution of a magma, it can be separated into two or more chemically distinct components. Further, each of these components may undergo further segregation by a variety of processes. Consequently, magmatic differentiation can

*Recent studies indicate that this igneous body was produced by multiple injections of magma and represents more than just a simple case of crystal settling.

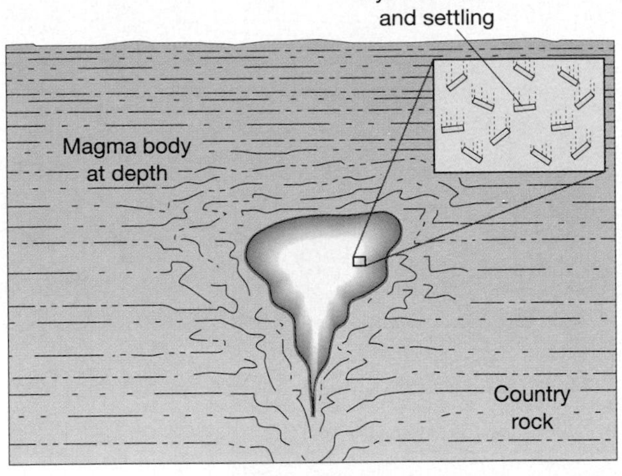

A.

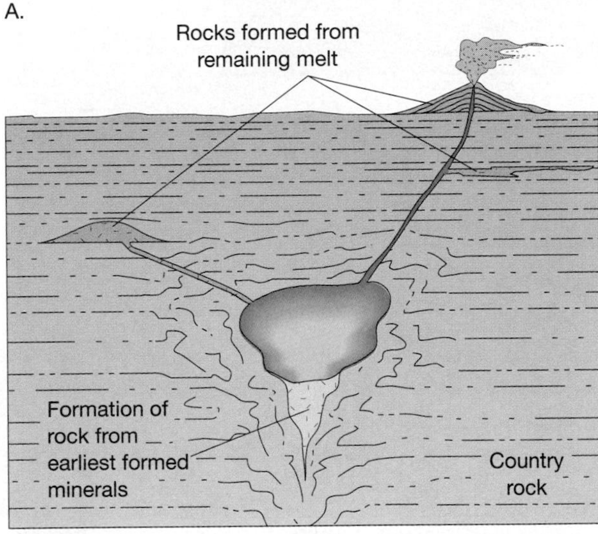

B.

FIGURE 3.9
Separation of minerals by crystal settling.
A. Illustration of how the earliest-formed minerals can be separated from a magma by settling. **B.** The remaining melt could migrate to a number of different locations and, upon further crystallization, generate rocks having a composition much different from that of the parent magma.

produce a variety of igneous rocks having a wide range of compositions (see Figure 3.7).

Assimilation and Magma Mixing

Bowen successfully demonstrated that through magmatic differentiation a single magma can generate several different igneous rocks. However, more recent work indicates that this process alone cannot account for the relative quantities of the rock types known to exist. Apparently other mechanisms also generate magmas of varied chemical compositions.

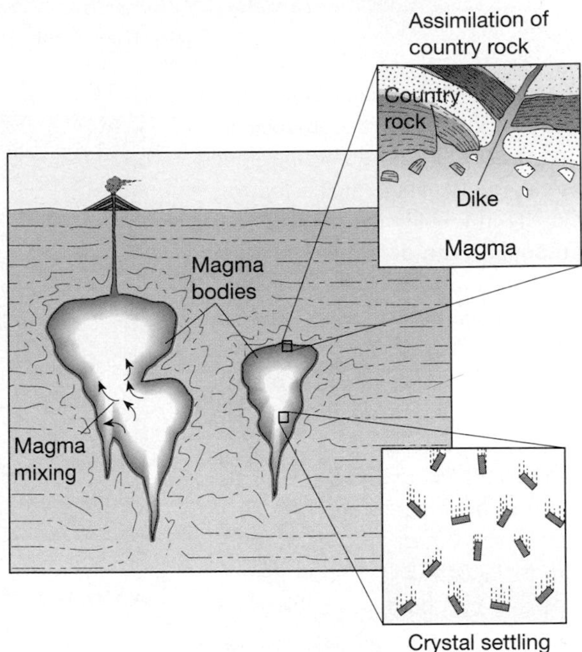

FIGURE 3.10
This illustration shows three ways that the composition of a magma body may be altered: Magma mixing; assimilation of country rock; and fractional crystallization and crystal settling.

Once a magma body forms, its composition can change through the incorporation of foreign material. For example, as magma migrates upward it may incorporate some of the surrounding country rock, a process called **assimilation** (Figure 3.10). One type of assimilation operates in a near-surface environment where rocks are brittle. As the magma pushes upward, stress causes numerous cracks in the overlying rock. The force of the injected magma is strong enough to dislodge blocks of the surrounding rock and incorporate them into the magma body. In other environments the magma may simply melt and assimilate some of the country rock.

Another means by which the composition of a magma body is altered is **magma mixing.** Simply, this process is believed to occur during ascent, when one magma body overtakes another (Figure 3.10). Once contact occurs, convective flow and other mechanisms mix the two magmas, generating a mass with a different composition. Both magma mixing and the assimilation of country rock are similar in that the magma body is contaminated through the incorporation of foreign material.

NAMING IGNEOUS ROCKS

As was stated previously, igneous rocks are most often classified, or grouped, on the basis of their tex-

BOX 3.1

Thin Sections and Rock Identification

Igneous rocks are classified on the basis of their mineral composition and texture. When analyzing specimens, geologists examine them closely to identify the minerals present and to determine the size and arrangement of the interlocking crystals. When out in the field, geologists use megascopic techniques to study rocks. The *megascopic* characteristics of rocks are those features that can be determined with the unaided eye or by using a low-magnification (×10) hand lens. When practical to do so, geologists collect hand samples that can be taken back to the laboratory where *microscopic,* or high-magnification, methods can be employed. Microscopic examination is important to identify trace minerals, as well as those textural features that are too small to be visible with the unaided eye.

Because most rocks are not transparent, microscopic work requires the preparation of a very thin slice of rock known as a *thin section* (Figure 3.A). First, a saw containing diamonds embedded in its blade is used to cut a narrow slab from the sample. Next, one side of the slab is polished using grinding powder and then cemented to a microscope slide. Once the mounted sample is firmly in place, the other side of it is ground to a thickness of about 0.03 millimeter. When a slice of rock is that thin, it is usually transparent. Nevertheless, some metallic minerals, such as pyrite and magnetite, remain opaque.

Once produced, thin sections are examined under a specially designed microscope called a *polarizing microscope.* Such an instrument has a light source beneath the stage so that light can be transmitted upward through the thin section. Because minerals have crystalline structures that influence polarized light in a measurable way, this procedure allows for the identification of even the smallest components of a rock. Part C of Figure 3.A is a photomicrograph (photo taken through a microscope) of a thin section of granite shown under polarized light. The mineral constituents are identified by their unique optical properties. In addition to assisting in the study of igneous rocks, microscopic techniques are used with great success in analyzing sedimentary and metamorphic rocks as well.

FIGURE 3.A
Thin sections are very useful in identifying the mineral constituents in rocks. **A.** A slice of rock is cut from a hand sample using a diamond saw. **B.** This slice is cemented to a microscope slide and ground until it is transparent to light (about 0.03 millimeter thick). This very thin slice of rock is called a *thin section.* **C.** A thin section of granite viewed under polarized light.

ture and mineral composition. The various igneous textures result from different cooling histories, whereas the mineral composition of an igneous rock is the consequence of the chemical makeup of the parent magma and the environment of crystallization. As we might expect from the results of Bowen's work, minerals that crystallize under similar conditions are most often found together composing the same igneous rock. Hence, the classification of igneous rocks closely corresponds to Bowen's reaction series (Figure 3.8).

The first minerals to crystallize—calcium feldspar, pyroxene, and olivine—are high in iron, magnesium, or calcium, and low in silicon. Since basalt is a common rock with this mineral makeup, the term *basaltic* is often used to denote any rock having a similar mineral composition. Moreover, because basaltic rocks contain a high percentage of ferromagnesian minerals, geologists may also refer to them as *mafic* rocks (from *ma*gnesium and *Fe,* the symbol for iron). Due to their iron content, mafic rocks are typically darker and slightly denser than other igneous rocks commonly found at the earth's surface.

Among the last minerals to crystallize are potassium feldspar and quartz, the primary components of the abundant rock granite. Igneous rocks in which these two minerals predominate are said to have a *granitic* composition. Geologists also refer to granitic rocks as being *felsic,* a term derived from *fel*dspar and *si*lica (quartz). Intermediate igneous rocks contain minerals found near the middle of Bowen's reaction series. Amphibole and the intermediate plagioclase feldspars are the main constituents of this compositional group. We will refer to rocks that have a mineral makeup between that of granite and basalt as being *andesitic,* after the rock andesite.

Although the rocks in each of these basic categories consist mainly of minerals located in a specific region of Bowen's reaction series, other constituents are usually present in lesser amounts. For example, granitic rocks are composed primarily of quartz and potassium feldspar (K feldspar), but may also contain muscovite, biotite, amphibole, and sodium feldspar (Na feldspar). See Table 3.1.

Thus far this discussion has focused on only three mineral compositions, yet it is important to note that gradations among these types also exist (Figure 3.11). For example, an abundant intrusive igneous rock called *granodiorite* has a mineral composition between that of granitic rocks and those with an andesitic composition. Another important igneous rock, *peridotite,* contains mostly olivine and pyrox-

TABLE 3.1
Classification of igneous rocks.

	Granitic (Felsic)	Andesitic (Intermediate)	Basaltic (Mafic)	Ultramafic
Phaneritic (coarse-grained)	Granite	Diorite	Gabbro	Peridotite
Aphanitic (fine-grained)	Rhyolite	Andesite	Basalt	
Mineral Composition	Quartz Potassium feldspar Sodium feldspar	Amphibole Intermediate plagioclase feldspar Biotite	Calcium feldspar Pyroxene	Olivine Pyroxene
Minor Mineral Constituents	Muscovite Biotite Amphibole	Pyroxene	Olivine Amphibole	Calcium feldspar
Rock color Based on % dark (mafic) minerals	Light-colored Less than 15% dark minerals	Medium-colored 15–40% dark minerals	Dark-gray to black More than 40% dark minerals	Dark-green to black Nearly 100% dark minerals

← **Increasing silica content**

Increasing iron and magnesium content →

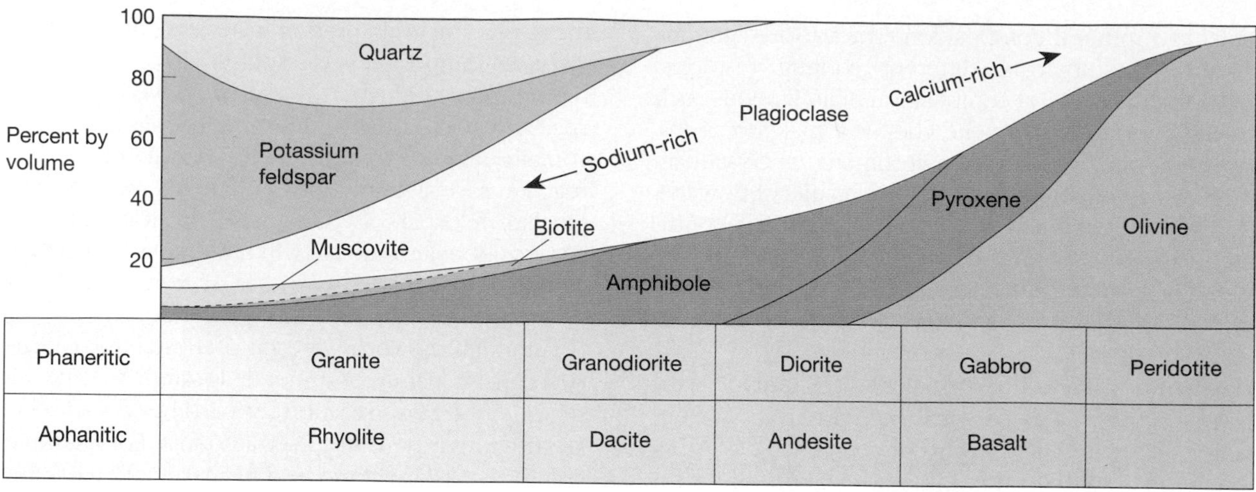

FIGURE 3.11
Mineralogy of the common igneous rocks. (After Dietrich)

ene and thus falls near the very beginning of Bowen's reaction series.

Since peridotite is composed almost entirely of ferromagnesian minerals, its chemical composition is often referred to as *ultramafic*. Although ultramafic rocks are rarely observed on the earth's surface, peridotite is believed to be a major constituent of the upper mantle.

An important aspect of the chemical composition of igneous rocks is silica (SiO_2) content. Recall that most of the minerals in igneous rocks contain some silica. Typically, the silica content of crustal rocks ranges from a low of 50 percent in basaltic rocks to a high of over 70 percent in granitic rocks. The percentage of silica in igneous rocks actually varies in a systematic manner which parallels the abundance of the other elements. For example, rocks low in silica contain large amounts of calcium, iron, and magnesium. Consequently, the chemical makeup of an igneous rock can be inferred directly from its silica content. Further, the amount of silica present in magma strongly influences its behavior. Granitic magma, which has a high silica content, is viscous and exists as a fluid at temperatures as low as 800°C. On the other hand, basaltic magmas are low in silica and generally fluid. Further, basaltic magmas are largely crystalline below 950°C.

Recall the igneous rocks are classified on the basis of their mineral composition and texture. Moreover, two rocks may have the same mineral constituents but have different textures and hence different names. For example, the coarse-grained intrusive rock granite has a fine-grained volcanic equivalent called rhyolite (Figure 3.12). Although these rocks

A.

B.

FIGURE 3.12
A. Granite, one of the most common coarse-grained igneous rocks. **B.** Rhyolite, the fine-grained equivalent of granite is less abundant. (Photos by E. J. Tarbuck)

are mineralogically the same, they have different textures and do not look at all alike.

Granitic Rocks

Granite is perhaps the best known of all igneous rocks (Figure 3.12A). This is partly because of its natural beauty, which is enhanced when it is polished, and partly because of its abundance. Slabs of polished granite are commonly used for tombstones and monuments and as building stones.

Granite is a phaneritic rock composed of about 25 to 35 percent quartz and over 50 percent potassium feldspar and sodium-rich feldspar. The quartz crystals, which are roughly spherical in shape, are most often clear to light gray in color. In contrast to quartz, the feldspar crystals in granite are not as glassy, but rectangular in shape and generally salmon pink to white in color. Other common constituents of granite are muscovite and the dark silicates, particularly biotite and amphibole. Although the dark components of granite make up less than 20 percent of most samples, dark minerals appear to be more prominent than their percentage would indicate. In some granites, K feldspar is dominant and dark pink in color, so that the rock appears almost reddish. This variety is popular as a building stone. However, most often the feldspar grains are white, so that when viewed at a distance granite appears light gray in color. Granite may also have a porphyritic texture, in which feldspar crystals a centimeter or more in length are scattered among the coarse-grained groundmass of quartz and amphibole.

FIGURE 3.13
El Capitan, a large igneous monolith located in Yosemite National Park, California. This location is one of many in the United States where a large mass of granite is exposed at the surface. (Photo by E. J. Tarbuck)

Granite is often produced by the processes which generate mountains. Because granite is a by-product of mountain building and is very resistant to weathering and erosion, it frequently forms the core of eroded mountains. For example, Pikes Peak in the Rockies, Mount Rushmore in the Black Hills, the White Mountains of New Hampshire, Stone Mountain in Georgia, and Yosemite National Park in the Sierra Nevada are all areas where large quantities of granite are exposed at the surface (Figure 3.13). As we can see from these examples, granite is a very abundant rock. However, it has become common practice among geologists to apply the term *granite* to any coarse-grained intrusive rock composed predominantly of light silicate materials. We will follow this practice for the sake of simplicity. The student should keep in mind that this use of the term *granite* covers rock having a range of mineral compositions.

Rhyolite is the volcanic equivalent of granite. Like granite, rhyolite is composed primarily of the light-colored silicates (Figure 3.12B). This fact accounts for its color, which is usually buff to pink or occasionally very light gray. Rhyolite is usually aphanitic and frequently contains glassy fragments and voids indicating rapid cooling in a surface environment. In those instances when rhyolite contains phenocrysts, they are usually small and composed of either quartz or potassium feldspar. In contrast to granite, rhyolite is rather uncommon. Yellowstone Park is one well-known exception. Here rhyolitic lava flows and ash deposits of similar composition are widespread.

Obsidian is a dark-colored, glassy rock which usually forms when silica-rich lava is quenched quickly (Figure 3.14). In contrast to the orderly arrangement of ions that is characteristic of minerals, the ions in glass are unordered. Consequently, glassy rocks like obsidian are not composed of minerals in the same sense as most other rocks.

Although usually black or reddish-brown in color, obsidian has a high silica content (Figure 3.15A).

FIGURE 3.14
Large pieces of fragmented obsidian found at the toe of an obsidian flow at Glass Mountain, California, near the south end of Mono Craters. (Photo by E. J. Tarbuck)

A.

B.

FIGURE 3.15
Igneous rocks that exhibit a glassy texture. **A.** Obsidian, a glassy volcanic rock. **B.** Pumice, a glassy rock containing numerous tiny voids.

Thus, its composition is more akin to the light igneous rocks such as graphite than to the dark rocks of basaltic composition. By itself, silica is clear like window glass; the dark color results from the presence of metallic ions. If you examine a thin edge of a piece of obsidian, it will be nearly transparent. Because of its excellent conchoidal fracture, obsidian was a prized material from which Native Americans made arrowheads and cutting tools.

Pumice is a volcanic rock which, like obsidian, has a glassy texture. Usually found with obsidian, pumice forms when large amounts of gas escape through lava to generate a gray, frothy mass (Figure 3.15B). In some samples, the voids are quite noticeable, while in others, the pumice resembles fine shards of intertwined glass. Because of the large percentage of voids, many samples of pumice will float when placed in water. Oftentimes flow lines are visible in pumice, indicating some movement before solidification was complete. Moreover, pumice and obsidian often form in the same rock mass, where they exist in alternating layers.

Andesitic Rocks

Andesite is a medium gray, fine-grained rock of volcanic origin. Its name comes from the Andes Mountains where numerous volcanoes are composed of this rock type. In addition to the volcanoes of the

A.

B.

FIGURE 3.16
Andesite porphyry. **A.** Hand sample of andesite porphyry, a common volcanic rock. **B.** Photomicrograph of a thin-section of andesite porphyry to illustrate texture. Notice that the few large crystals (phenocrysts) are surrounded by much smaller crystals (groundmass).

Andes, many of the volcanic structures encircling the Pacific Ocean are of andesitic composition. Andesite quite commonly exhibits a porphyritic texture (Figure 3.16). In these cases, the phenocrysts are often light, rectangular crystals of plagioclase feldspar or black, elongated hornblende crystals.

Diorite is a coarse-grained intrusive rock that looks somewhat similar to gray granite. However, it can be distinguished from granite by the absence of visible quartz crystals. The mineral makeup of diorite is primarily sodium-rich plagioclase and amphibole, with lesser amounts of biotite. Because the white feldspar grains and dark amphibole crystals are roughly equal in abundance, diorite has a "salt and pepper" appearance.

Basaltic Rocks

Basalt is a very dark green to black, fine-grained volcanic rock composed primarily of pyroxene and calcium-rich feldspar, with lesser amounts of olivine and amphibole present. When porphyritic, basalt commonly contains small, light-colored calcium feldspar phenocrysts or glassy-appearing olivine phenocrysts embedded in a dark groundmass.

Basalt is the most common extrusive igneous rock. Many volcanic islands, such as the Hawaiian Islands and Iceland, are composed mainly of basalt (Figure 3.17). Further, the upper layers of the oceanic crust consist of basalt. In the United States, large portions of central Oregon and Washington were the sites of extensive basaltic outpourings (see Figure 4.26). At some locations these once-fluid basaltic flows have accumulated to thicknesses approaching 2 kilometers.

Gabbro is the intrusive equivalent of basalt. Like basalt, it is very dark green to black in color and composed primarily of pyroxene and calcium-rich plagioclase. Although gabbro is not a common constituent of the continental crust, it undoubtedly makes up a significant percentage of the oceanic crust. Here large portions of the magma found in underground reservoirs that once fed basalt flows eventually solidified at depth to form gabbro.

Pyroclastic Rocks

Pyroclastic rocks are those which form from fragments ejected during a volcanic eruption. One of the most common pyroclastic rocks, called *tuff,* is composed of tiny ash-sized fragments which were later cemented together (Figure 3.18). In situations where the ash particles remained hot enough to fuse, the rock is generally called *welded tuff.* Although welded tuffs consist mostly of glass shards, they frequently contain pieces of obsidian or other rock fragments. Further, a microscope is often required to distinguish tuffs from other fine-grained igneous rocks. Deposits of partially welded tuffs are easily quarried and used as a durable building material. Several villages in Cappadocia in central Turkey, which date back as far as the fourth century, have been carved into vertical cliffs composed of this material.

Pyroclastic rocks composed of particles larger than ash are called *volcanic breccia.* The particles in volcanic breccia can consist of streamlined fragments

FIGURE 3.17
Lava of basaltic composition flowing from a lava tube, Hawaii Volcanoes National Park. (Photo by J. D. Griggs, U.S. Geological Survey)

FIGURE 3.18
Tuff. Perhaps the most common pyroclastic rock, tuff is composed mainly of ash-sized particles, and like this sample, may contain larger fragments of other rocks.

that solidified in air, blocks broken from the walls of the vent, crystals, and glass fragments (see Figure 3.6). Unlike the other igneous rock names, the terms *tuff* and *volcanic breccia* do not denote mineral composition.

OCCURRENCE OF IGNEOUS ROCKS

Although volcanic eruptions can be among the most violent and spectacular events in nature and therefore worthy of detailed study, most magma is believed to be emplaced at depth. Thus, an understanding of intrusive igneous activity is as important to geologists as the study of volcanic events. The structures that result from the emplacement of igneous material at depth are called **plutons.** Since all plutons form out of our view beneath the earth's surface, they can be studied only after uplifting and erosion have exposed them. The challenge lies in reconstructing the events that generated these structures millions or even hundreds of millions of years ago.

For the sake of clarity, we have separated our discussions of volcanism and plutonic activity. Volcanism will be treated in the following chapter; here we will concentrate on plutonic activity. Keep in mind, however, that these diverse processes occur simultaneously and involve basically the same earth materials.

Nature of Plutons

Plutons are known to occur in a great variety of sizes and shapes. Some of the most common types are illustrated in Figure 3.19. Notice that some of these structures have a tabular or sheet-like shape, while others are quite massive. Also, observe that some of these bodies cut across existing structures, such as the layering of sedimentary beds, whereas others form when magma is injected between sedimentary layers. Because of these differences, intrusive igneous bodies are generally classified according to their shape as either **tabular** (sheet-like) or **massive** and by their orientation with respect to the country (host) rock. Plutons are said to be **discordant** if they cut across existing structures and **concordant** if they

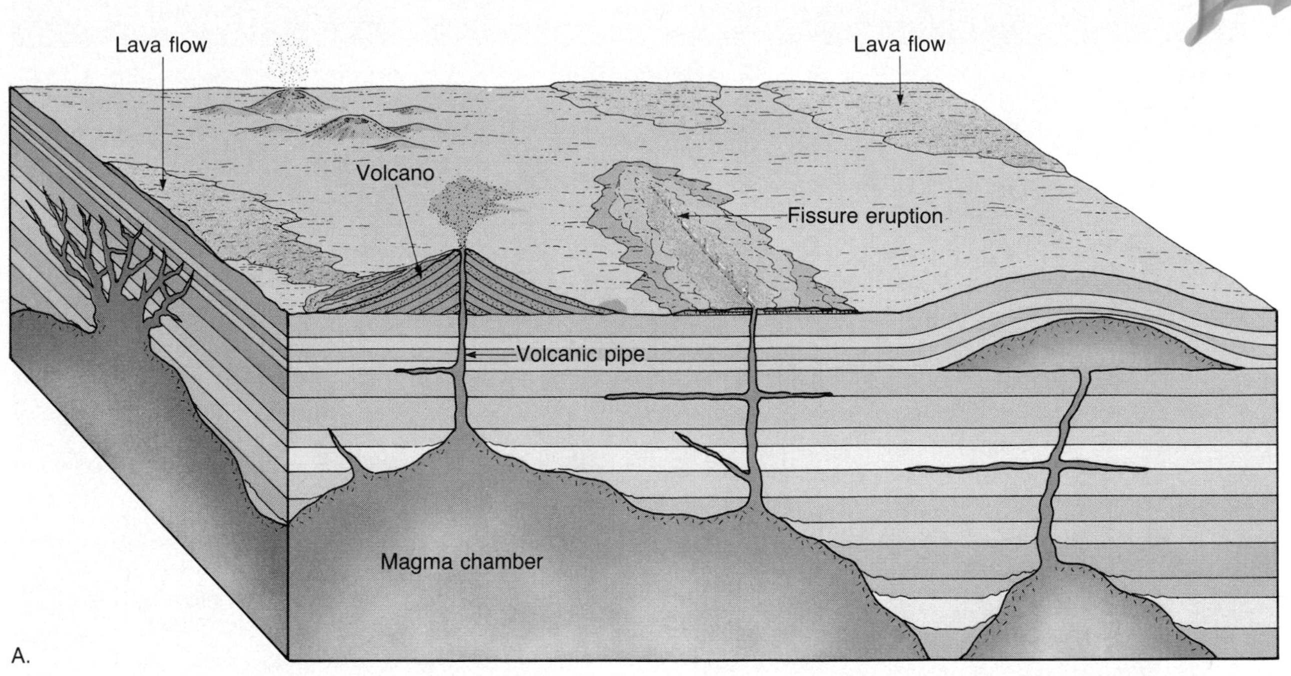

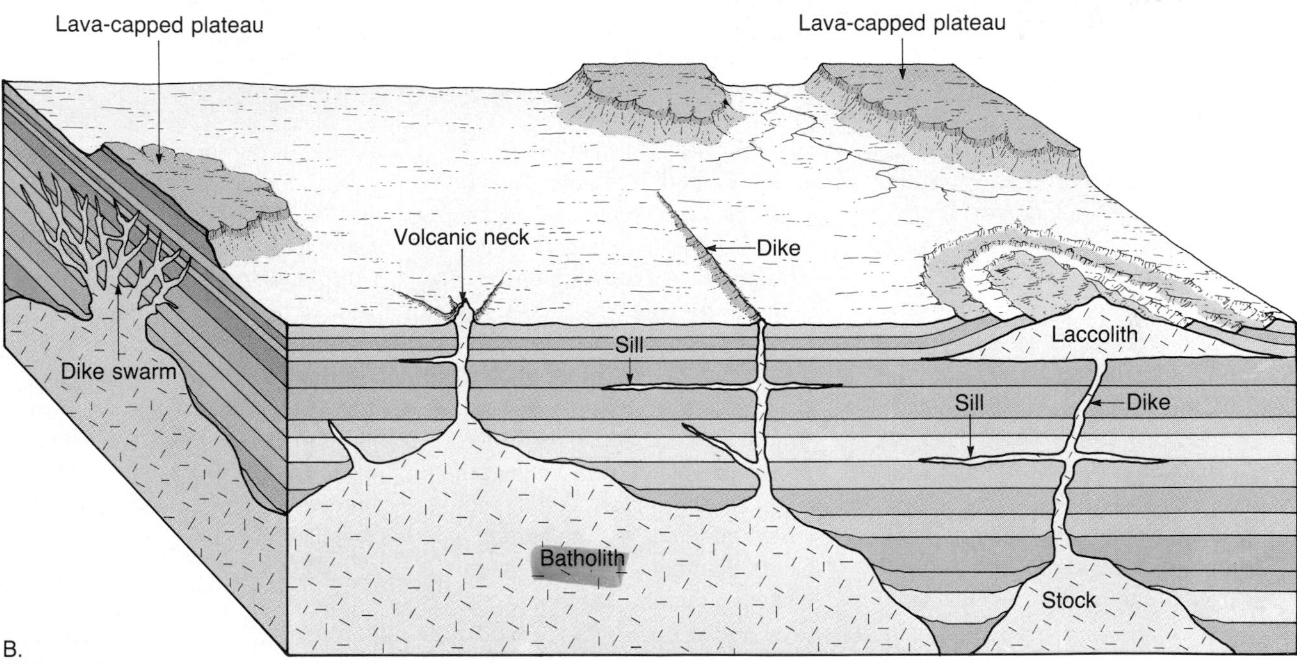

FIGURE 3.19
Illustrations showing basic igneous structures. **A.** This cross-sectional view shows the relationship between volcanism and plutonic activity. **B.** This view illustrates the basic intrusive igneous structures, some of which have been exposed by erosion long after their formation.

form parallel to features such as sedimentary layers. Further, as we can see in Figure 3.19, plutons are closely associated with volcanic activity. The largest intrusive bodies are thought to be the remnants of magma chambers which fed ancient volcanoes.

Dikes **Dikes** are discordant sheet-like bodies that are produced when magma is injected into fractures. The force exerted by the emplaced magma can be great enough to further separate the walls of the fracture. Once crystallized, these tabular structures have

BOX 3.2

Pegmatites

Pegmatite is a name given to an igneous rock composed of abnormally large crystals (Figure 3.B). How large is *large?* Crystals in most pegmatite samples are more than 1 centimeter in diameter. In some specimens, crystals that are 1 meter or more across are common. Gigantic hexagonal crystals of muscovite measuring a few meters across have been found in Ontario, Canada. In the Black Hills, crystals as large as telephone poles of the lithium-bearing mineral spodumene have been mined. The largest of these was more than 12 meters (40 feet) long. Further, feldspar masses the size of houses have been quarried from a pegmatite located in North Carolina.

FIGURE 3.B
Pegmatite excavated at Etta Mine, South Dakota. The elongated features in the center of the photo are molds of large spodumene crystals that were removed during the mining activity. (Photo by James G. Kirchner)

Most pegmatites have the composition of granite and contain unusually large crystals of quartz, feldspar, and muscovite. In addition to being an important source of excellent mineral specimens, large granitic pegmatites have been mined for their mineral constituents. Feldspar, for example, is used in the production of ceramics, and muscovite is used for electrical insulation and glitter. Although granitic pegmatites are most common, pegmatites with chemical makeups similar to those of other igneous rocks are also known. Further, pegmatites may contain significant amounts of some of the least abundant elements. Thus, in addition to the common silicates, pegmatites with minerals containing the elements lithium, cesium, uranium, and the rare earths are known. In addition, semiprecious gems such as beryl, topaz, and tourmaline are occasionally found.

Most pegmatite bodies are located within large igneous masses or as dikes or veins that cut into the country rock that surrounds a pluton. Pegmatites form in the late stages of magma crystallization. Because water and other volatile substances do not crystallize along with the bulk of the magma body, these fluids make up an unusually high percentage of the melt during the final phase of solidification. Crystallization in a fluid-rich environment where ion migration is enhanced is believed to result in the formation of crystals of abnormally large size. It is in this environment that the unusually large crystals of most pegmatites is thought to form.

---

thicknesses ranging from less than a centimeter to more than a kilometer. The largest have lengths of hundreds of kilometers. Most dikes, however, are a few meters thick and extend laterally for no more than a few kilometers. Dikes are often oriented vertically and represent pathways followed by molten rock which fed ancient lava flows. Some dikes end at depth; still others terminate at plutons.

Dikes may weather more slowly than the surrounding rock. When exposed, these dikes have the appearance of a wall as shown in Figure 3.20. Dikes are often found radiating, like spokes on a wheel, from an eroded volcanic neck (see Figure 4.24). In these situations the active ascent of magma is thought to have generated fractures in the volcanic cone.

Sills **Sills** are tabular plutons formed when magma is injected along sedimentary bedding surfaces (Figure 3.21). Horizontal sills are the most common, although all orientations, even vertical, are known to exist. Because of their relatively uniform thickness and large areal extent, sills are believed to form from very fluid magma. As we may expect, sills most often are composed of basaltic magma, which is typically quite fluid.

The emplacement of a sill requires that the overly-

FIGURE 3.20
The vertical structure in the foreground is a dike, which is more resistant to weathering than the surrounding rock. This dike is located west of Granby, Colorado, near Arapaho National Forest. (Photo by P. Jay Fleisher)

FIGURE 3.21
Salt River Canyon, Arizona. The dark, essentially horizontal band is a sill of basaltic composition that intruded into horizontal layers of sedimentary rock. (Photo by E. J. Tarbuck)

ing sedimentary rock be lifted to a height equal to the thickness of the sill. Although this seems to be a formidable task, it may require less energy than forcing the magma up the remaining distance to the surface. Consequently, sills form only at rather shallow depths, where the pressure exerted by the weight of overlying strata is relatively low. Although sills are intruded between existing layers, they can be locally discordant. Large sills frequently cut across sedimentary layers and resume their concordant nature at a higher level.

One of the largest and best-known sills in the United States is the Palisades Sill, which is exposed along the west shore of the Hudson River in southeastern New York and northeastern New Jersey. This sill is about 300 meters thick and, due to its resistant nature, has formed an imposing cliff that can be seen easily from the opposite side of the Hudson.

In many respects, sills closely resemble buried lava flows. Both are tabular and often exhibit *columnar jointing* (Figure 3.22). Further, because sills generally form in near-surface environments and may be

FIGURE 3.22
Devil's Post Pile National Monument, California, exhibits columnar joints and the columns that result. These five- to seven-sided columns are the consequence of contraction that occurs as a relatively thin layer of rock cools after it has solidified. The small photo insert shows a view looking down on these columns. (Courtesy of the National Park Service, U.S. Department of the Interior)

only a few meters thick, the emplaced magma is often chilled quickly enough to generate a fine-grained texture. When attempts are made to reconstruct the geologic history of a region, it becomes important to differentiate between sills and buried lava flows. Fortunately, under close examination these two structures can be readily distinguished. The upper portion of a buried lava flow usually contains voids produced by entrapped gas bubbles and only the rocks beneath a lava flow show evidence of metamorphic alteration. Sills, on the other hand, form when magma has been forcefully intruded between sedimentary layers. Thus, inclusions of the overlying wall rock occur only in sills, since lava flows are extruded before the overlying strata are deposited. Further, "baked" zones above and below are trademarks of a sill.

Laccoliths **Laccoliths** are similar to sills because they form when magma is intruded between sedimentary layers in a near-surface environment. However, the magma that generates laccoliths is more viscous. This nonfluid magma collects as a lens-shaped mass that arches the overlying strata upward (Figure 3.23). Consequently, a laccolith can occasionally be detected because of the dome-shaped structure it creates at the surface.

Most large laccoliths are probably not much wider than a few kilometers. The Henry Mountains in southeastern Utah are composed of several large laccoliths believed to have been fed by a much larger magma body emplaced nearby (Figure 3.23).

Batholiths By far the largest intrusive igneous bodies are **batholiths.** The largest batholiths are linear structures several hundred kilometers long and nearly one hundred kilometers wide as shown in Figure 3.24. The Idaho batholith, for example, encompasses an area of more than 40,000 square kilometers. Indirect evidence gathered from gravitational studies indicates that batholiths are also very thick, possibly extending tens of kilometers into the crust. Based on the amount exposed by erosion, some batholiths are at least several kilometers thick. By definition, a plutonic body must have a surface exposure of over 100 square kilometers (40 square miles) to be considered a batholith. Smaller plutons of this type are termed **stocks.** Many stocks appear to be portions of batholiths that are not yet fully exposed.

Batholiths are usually composed of rock types having chemical compositions near the granitic end of the spectrum, although diorite is commonly found. Smaller batholiths can be rather simple structures composed almost entirely of one rock type. However, studies of large batholiths have shown that they consist of several distinct parts that were intruded over a period of millions of years. The plu-

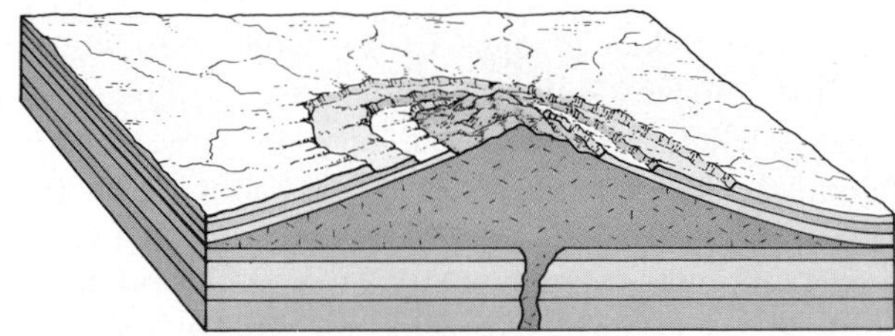

FIGURE 3.23
Laccolith exposed in the Henry Mountains of southern Utah. This dark, dome-shaped structure is one of several laccoliths exposed in the Henry Mountains. Notice the upturned sedimentary beds that flank this intrusion of mafic igneous rocks. (Photo by E. J. Tarbuck)

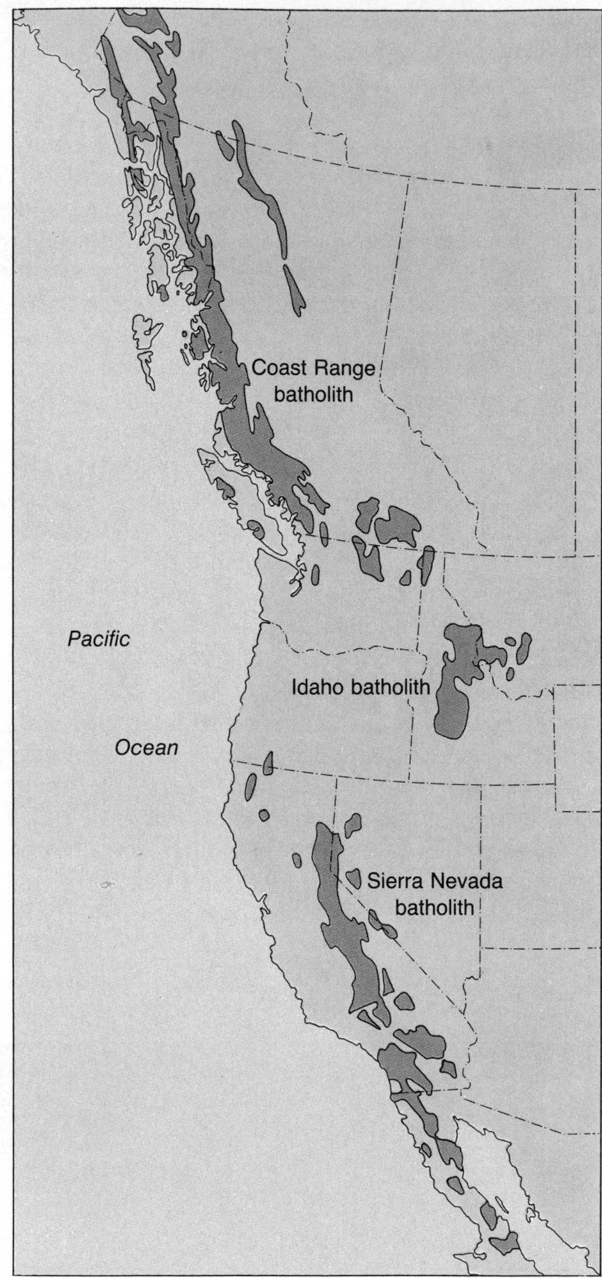

FIGURE 3.24
Location of granitic batholiths that occur along the western margin of North America. These gigantic, elongated bodies were emplaced during the last 150 million years of earth history.

tonic activity that created the Sierra Nevada batholith, for example, occurred nearly continuously over a 130-million-year span that ended about 80 million years ago during the Cretaceous period (Figure 3.25).

Batholiths frequently compose the core of mountain systems. Here uplifting and erosion have removed the surrounding rock, thereby exposing the resistant igneous body. Some of the highest peaks in the Sierra Nevada, such as Mount Whitney, are carved from such a granitic mass. Large expanses of granitic rock are also exposed in the stable interiors of the continents, such as the Canadian Shield of North America. These relatively flat outcrops are believed to be the remains of ancient mountains that have long since been leveled by erosion. Thus, the rocks composing the batholiths of youthful mountain ranges were generated near the top of a magma chamber, whereas in shield areas, the roots of former mountains and the lower portions of batholiths are exposed. We will consider the role of igneous activity as it relates to mountain building further in Chapter 20.

Emplacement of Batholiths

One ongoing and interesting debate in geology concerns the emplacement of granitic batholiths. One group of geologists supports the idea that batholiths formed from magma that migrated upward from great depths. This idea, however, presents a space problem. What happened to the rock originally in the location now occupied by these igneous masses? Further, the problem of explaining how magma is able to force its way through several kilometers of solid rock also plagued those supporting the magmatic origin of batholiths. The group opposing the magmatic origin hypothesis suggested that the granite in batholiths originated when hot ion-rich fluids and gases migrated through sedimentary rock and chemically altered the rock's composition. This essentially metamorphic process of converting country rock into granite without passing through a molten stage is called *granitization.* Although granitization undoubtedly generates small quantities of granite, the strongest evidence points to a magmatic origin for the largest intrusive bodies.

This controversy was resolved, for many at least, when careful studies were made of structures called *salt domes.* These structures are of economic importance as they are found in close association with major oil-producing areas in the Gulf Coast states and the Persian Gulf. Salt domes are produced in regions where extensive salt deposits were subsequently buried by thousands of meters of sediment. The salt, which is less dense than the overlying sediments, migrates very slowly upward. This is possible because salt behaves like a mobile fluid when it is subjected to differential stress over a long period of time. Since salt beds are not perfectly uniform, the zone of upward movement is thought to originate at a high spot along the layer. As the salt moves slowly

FIGURE 3.25
These massive igneous structures located near Yosemite National Park represent just a small portion of the Sierra Nevada batholith. (Photo by E. J. Tarbuck)

upward the stress exerted on the overlying sediments causes them to mobilize and be pushed aside (Figure 3.26A). Occasionally the salt breaches the surface, where it begins to flow outward not unlike a very thick lava flow.

It is now generally accepted that batholiths are emplaced in a manner similar to the formation of salt domes (Figure 3.26B). Because magma is less dense than the overlying rock, its buoyancy propels it upward. Also, like a salt dome, the mobile magma forcibly makes room for itself by pushing aside the country rocks. As the magma moves upward, some of the country rock which was shouldered aside will fill in the space left by the magma body as it passes. An analogous situation occurs when a can of oil-base paint is left in storage. The oil in the paint is less dense than the pigments used for coloration; thus, oil collects into drops that slowly migrate upward while the heavier pigments settle to the bottom. In the case of ascending granitic magma, gradual cooling results in a loss of mobility. Thus, much of the magma crystallizes at depth to form granite batholiths rather than extruding at the surface as a volcanic eruption.

The upper portions of batholiths often contain unmelted remnants of the country rock that are called **xenoliths** (Figure 3.27). These inclusions indicate that yet another process may operate during the emplacement of a batholith, at least in a near-surface environment where rocks are brittle. As the magma buoys upward, stress is believed to cause numerous cracks in the overlying rock. The force of the injected magma is strong enough to dislodge blocks of the surrounding rock and incorporate them into the magma body, a process called *stoping*. However, this process of assimilating country rock is only minor compared to the earlier-mentioned activity of mobilizing and displacing country rock.

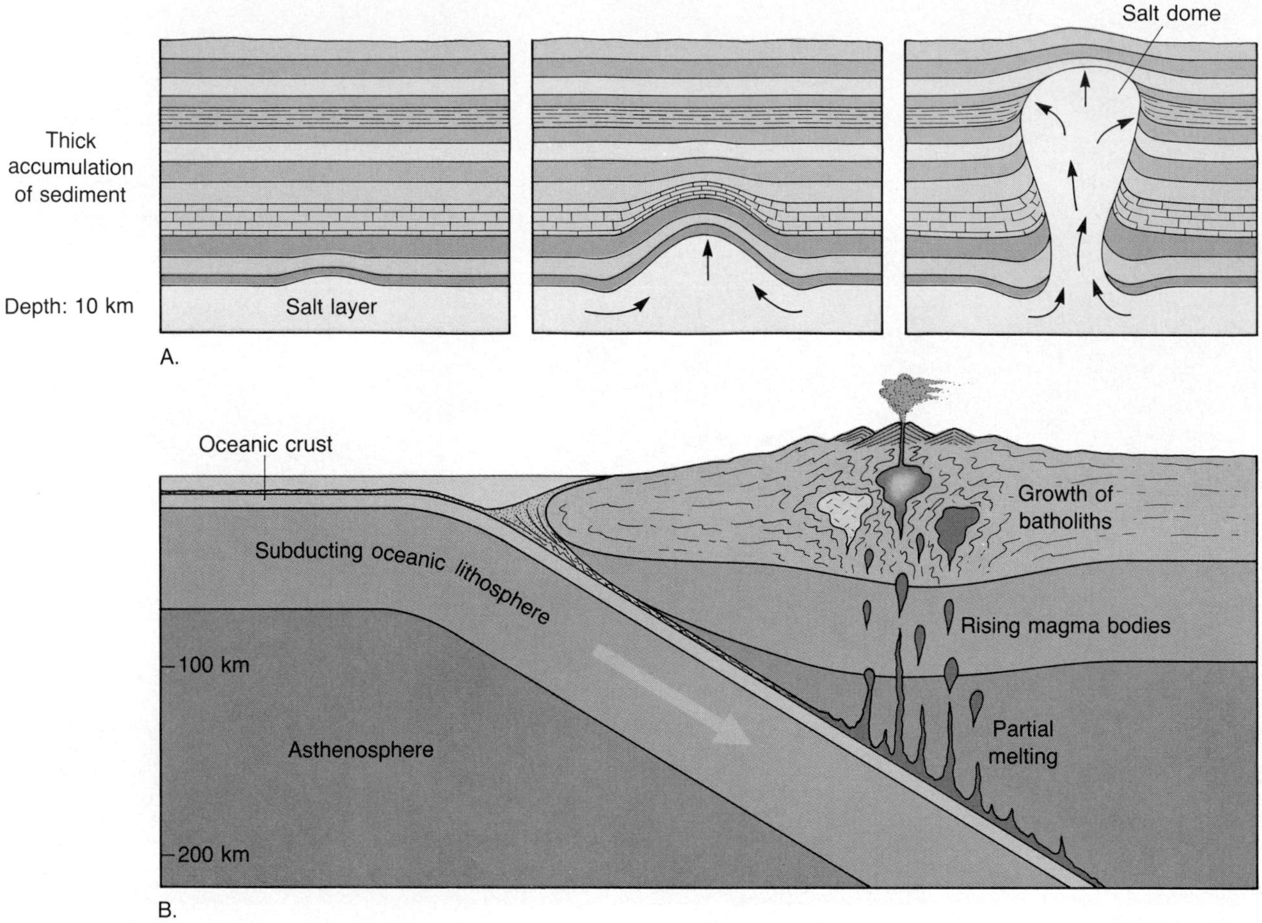

Salt dome
Thick accumulation of sediment
Depth: 10 km
Salt layer
A.
Oceanic crust
Growth of batholiths
Subducting oceanic lithosphere
Rising magma bodies
100 km
Partial melting
Asthenosphere
200 km
B.

FIGURE 3.26
Many geologists believe that the emplacement of a salt dome (Part A) is analogous to the processes involved in the emplacement of large magma bodies (Part B).

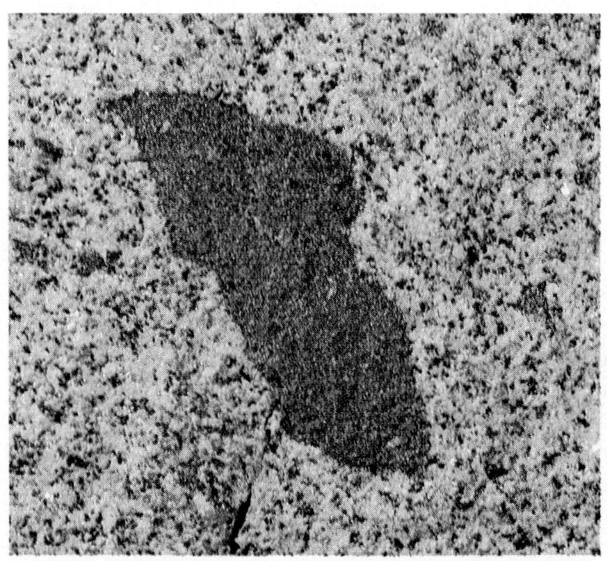

FIGURE 3.27
The dark xenolith is an unmelted remnant of country rock that was incorporated into the igneous rocks that compose this batholith. (Photo by E. J. Tarbuck)

REVIEW QUESTIONS

1. How does lava differ from magma?
2. How does the rate of cooling influence the crystallization process?
3. In addition to the rate of cooling, what other factors influence the crystallization process?
4. The classification of igneous rocks is based largely upon two criteria. Name these criteria.
5. The statements that follow relate to terms describing igneous rock textures. For each statement, identify the appropriate term.
   (a) Openings produced by escaping gases.
   (b) Obsidian exhibits this texture.
   (c) A matrix of fine crystals surrounding phenocrysts.
   (d) Crystals are too small to be seen with the unaided eye.
   (e) A texture characterized by two distinctly different crystal sizes.
   (f) Coarse grained, with crystals of roughly equal size.
6. What does a porphyritic texture indicate about igneous rock?
7. What is fractional crystallization? How might fractional crystallization lead to the formation of several different igneous rocks from a single magma?
8. Relate the classification of igneous rocks to Bowen's reaction series.
9. How are granite and rhyolite different? In what way are they similar?
10. Why are the crystals in pegmatites so large? (See Box 3.2.)
11. Compare and contrast each of the following pairs of rocks:
    (a) Granite and diorite.
    (b) Basalt and gabbro.
    (c) Andesite and rhyolite.
12. How do tuff and volcanic breccia differ from other igneous rocks such as granite and basalt?
13. What name is given to a tabular, discordant pluton?
14. Name a well-known tabular, concordant pluton.
15. What is the largest of all plutons? Is it tabular or massive? Concordant or discordant?
16. Relate the mechanism of batholith emplacement to the formation of salt domes.

KEY TERMS

aphanitic texture (p. 58)
assimilation (p. 63)
batholith (p. 75)
Bowen's reaction series (p. 61)
concordant (p. 70)
crystal settling (p. 62)
crystallization (p. 57)
dike (p. 71)
discordant (p. 70)
extrusive (p. 56)
glass (p. 57)
glassy texture (p. 58)
groundmass (p. 58)
intrusive (p. 57)
laccolith (p. 75)
lava (p. 56)
magma (p. 56)
magma mixing (p. 63)
magmatic differentiation (p. 62)
massive (p. 70)
phaneritic texture (p. 58)
phenocryst (p. 58)
pluton (p. 70)
plutonic (p. 57)
porphyritic texture (p. 58)
porphyry (p. 58)
pyroclastic texture (p. 60)
sill (p. 72)
stock (p. 75)
tabular (p. 70)
vesicle (p. 58)
volcanic (p. 56)
xenolith (p. 77)

4

Volcanic Activity

Opposite: Lava flow enters the sea, Hawaii Volcanoes National Park
(Photo by G. Brad Lewis, Douglas Peebles Photography)
(Top photo by T. J. Takahashi, U.S. Geological Survey)

At 8:32 A.M. on Sunday, May 18, 1980, one of the largest volcanic eruptions to occur in North America in recent times transformed a picturesque volcano into a decapitated remnant (Figure 4.1). On this date in southwestern Washington state, Mount St. Helens erupted with a force hundreds of times greater than that of the atomic bombs dropped on Japan during World War II. The blast blew out the entire north flank of the volcano, leaving a gaping hole. A once prominent volcano that had grown to more than 2900 meters (9500 feet) had, in one brief moment, been lowered by about 410 meters.

The early morning blast totally devastated a wide swath of timber-rich land on the north side of the mountain (Figure 4.2). Trees within a 400-square-kilometer area lay intertwined and flattened, stripped of their branches and appearing from the air like toothpicks strewn about. The immense force caused trees as far away as 25 kilometers to topple. The gases and ash unleashed from the volcano had temperatures that probably exceeded 800°C! Thirty-six persons were killed and 23 others were listed as missing. Some died from the intense heat and the suffocating cloud of ash and gases. Others perished as they were hurled from the mountain by the force of the blast. Still others were trapped by debris-laden mudflows.

The blast and accompanying mudflows carried ash, trees, and water-saturated rock debris 29 kilometers down the Toutle River. The river quickly became a mud-filled torrent and reached depths of 60 meters in some places. Further, a debris dam was deposited at the outlet of Spirit Lake, causing its level to rise by more than 30 meters. For several days, the threat of pent-up waters breaching the dam posed another potential hazard.

The eruption of May 18th ejected nearly one cubic kilometer of ash and rock debris (Figure 4.3). Following the devastating explosion, Mount St. Helens continued to emit great quantities of hot gases and ash. Only minutes after the eruption began, a dark plume rose from the volcano. The force of the blast

FIGURE 4.1
Before and after photographs show the transformation of Mount St. Helens caused by the May 18, 1980 eruption. ("Before" photo by Stephen Trimble, "after" photo by Jim Hughes, courtesy of USDA Forest Service)

was so strong that some of the ash was propelled high into the stratosphere, more than 18,000 meters above the ground. During the next hours and days, this very fine grained material was carried great distances by the strong upper air winds. Measurable deposits were reported from as far away as Oklahoma and Minnesota. Meanwhile, the ash fallout in the immediate vicinity accumulated to depths exceeding 2 meters, and the air over Yakima, Washington, 130 kilometers to the east, was so filled with ash that residents experienced midnight-like darkness at noon. Crop damage from the volcanic fallout was reported as far away as central Montana.

The events leading to the May 18th eruption began about two months earlier, on the 20th of March, as a series of minor earth tremors centered beneath the awakening mountain (Figure 4.4A). The first volcanic activity took place on March 27th, when a small amount of ash and steam rose from the summit. Over the next several weeks, sporadic eruptions of varied intensity occurred.

Prior to the main eruption, the primary concern had been the potential hazard of mudflows. These moving lobes of saturated debris were created when ice and snow were melted by heat from the magma within the volcano. The only sign of a potentially hazardous eruption was a bulge on the volcano's north flank (Figure 4.4B). Careful monitoring of this dome-shaped structure indicated a very slow but steady growth rate of a few meters per day. Geologists monitoring the activity suggested that if the growth rate of the bulge changed appreciably, an eruption might quickly follow. Unfortunately, no such variation was detected prior to the explosion. In fact, the seismic activity decreased during the two days preceding the huge blast.

"Vancouver, Vancouver, this is it!" was the only warning to precede the unleashing of tremendous quantities of pent-up gases. The trigger was an earth tremor with a rating of 5.1 on the Richter scale. The vibrations sent the north slope of the cone plummeting into the Toutle River, effectively removing the overburden which had trapped the magma below (Figure 4.4C). With the pressure reduced, the water-rich magma is thought to have ruptured like an overheated steam boiler. Since the eruption originated in the vicinity of the bulge, which was several hundred meters below the summit, the main impact of the eruption was directed laterally rather than vertically. Had the full force of the eruption been upward, far less destruction would have occurred.

Mount St. Helens is only one of the 15 large volca-

FIGURE 4.2
Blowdown of Douglas fir trees along Smith Creek caused by the lateral blast of Mount St. Helens on May 18, 1980. Note two U.S. Geological Survey scientists, lower right, for scale. (Photo by Lyn Topinka, David A. Johnston Cascades Volcano Observatory)

noes and innumerable smaller ones extending from British Columbia to northern California. Eight of the largest cones have been active in the past few hundred years, while the last eruptive phase of Mount St. Helens came to an end in 1857. Of the remaining seven "active" volcanoes, Mount Baker, Mount Shasta, Lassen Peak, Mount Hood, and Mount Rainier are believed most likely to erupt again. It is hoped that the eruptions of Mount St. Helens will provide geologists with enough data to more effectively evaluate the potential hazards of future volcanic eruptions.

THE NATURE OF VOLCANIC ACTIVITY

Volcanic activity is generally perceived as a process that produces a picturesque, cone-shaped structure which periodically erupts in a violent manner. However, although some eruptions may be cataclysmic, many are relatively quiescent. The primary factors which determine the nature of volcanic eruptions include the magma's composition, its temperature, and the amount of dissolved gases it contains. These factors affect the magma's mobility, or **viscosity**. The more viscous the material, the greater its resistance

FIGURE 4.3
May 18, 1980, eruption of Mount St. Helens. Taken a few seconds after the eruption began, the ash cloud is seen growing vertically (darker material) as it moves with even greater force laterally (lighter material). (Photo courtesy of Keith Ronnholm)

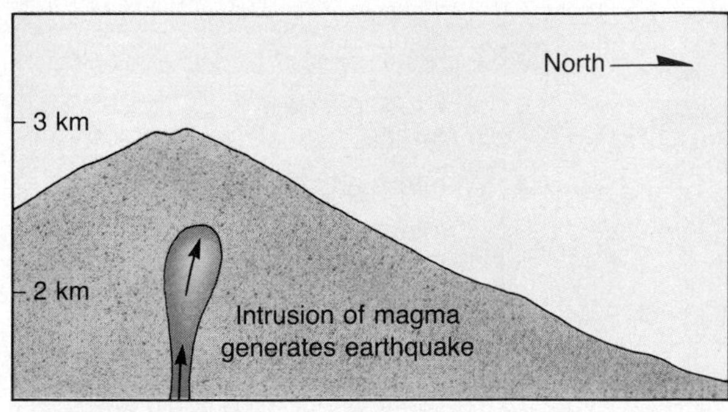

A. March 20, 1980

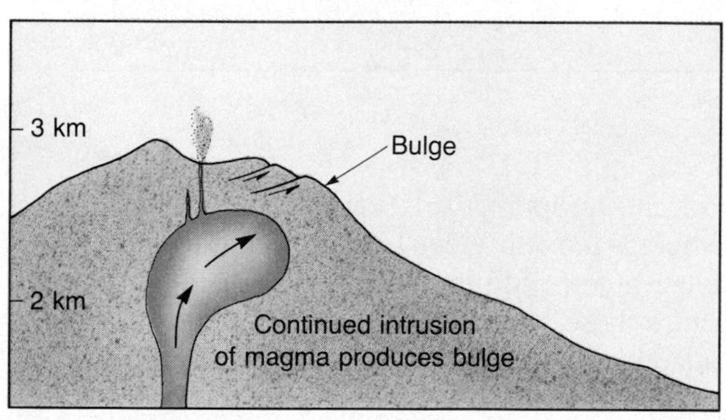

B. April 23, 1980

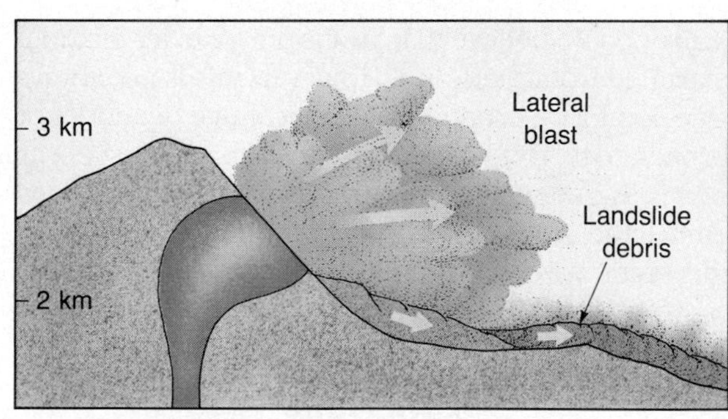

C. May 18, 1980

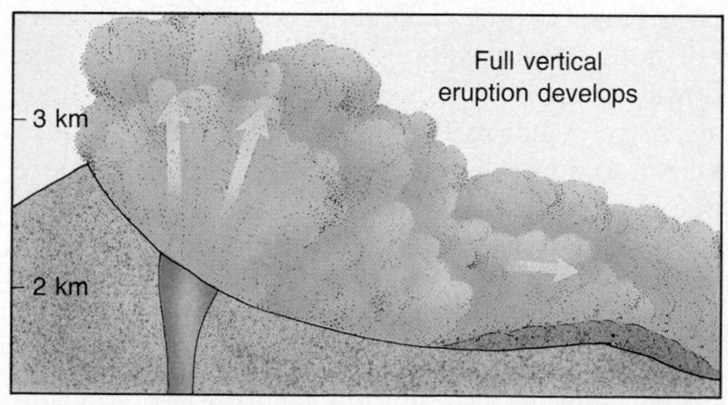

D. May 18, 1980

FIGURE 4.4
Idealized diagrams showing the events in the May 18, 1980, eruption of Mount St. Helens. **A.** First, a sizable earthquake recorded on Mount St. Helens indicates that renewed volcanic activity is possible. **B.** Alarming growth of a bulge on the north flank suggests increasing magma pressure below. **C.** Triggered by an earthquake, a giant landslide relieved the confining pressure on the magma body and initiated an explosive lateral blast. **D.** Within seconds, a large vertical eruption sent a column of volcanic ash to an altitude of about 19 kilometers. This phase of the eruption continued for over nine hours.

to flow. For example, molasses is more viscous than water. The effect of temperature on viscosity is easily seen. Just as heating molasses makes it more fluid, the mobility of lava is also influenced by temperature changes. As a lava flow cools and begins to congeal, its mobility decreases and eventually the flowing halts.

The chemical composition of magmas was discussed in Chapter 3 with the classification of igneous rocks. One major difference among various igneous rocks and therefore among the magmas from which they originate is their silica (SiO_2) content (Table 4.1). Magmas that produce basaltic rocks contain about 50 percent silica, whereas rocks of granitic composition (granite and its extrusive equivalent, rhyolite) contain over 70 percent silica. The intermediate rock types, andesite and diorite, contain around 60 percent silica. It is important to note that a magma's viscosity is directly related to its silica content. In general, the higher the percentage of silica in magma, the greater its viscosity. It has been shown that the flow of magma is impeded because the silicon-oxygen tetrahedra link into elongated structures even before crystallization begins. Due to their high silica content, granitic lavas are very viscous and tend to form comparatively short, thick flows. By contrast, basaltic lavas, which are lower in silica, tend to be quite fluid and have been known to travel distances of 150 kilometers or more before congealing.

The gas content of a magma also effects its mobility. Dissolved gases tend to increase the fluidity of magma. Of far greater consequence is the fact that escaping gases provide enough force to propel molten rock from a volcanic vent. As magma moves into a near-surface environment, such as within a volcano, the confining pressure in the uppermost portion of the magma body is greatly reduced. This reduction in confining pressure allows the gases, which had been dissolved when they were at greater depths, to be released suddenly. At temperatures of 1000°C and low, near-surface pressures, these gases will expand to occupy hundreds of times their origi-

TABLE 4.1
Variations in properties among magmas of differing compositions.

Property	Basaltic	Andesitic	Granitic
Silica content	Least (~50%)	Intermediate (~60%)	Most (~70%)
Viscosity	Least	Intermediate	Highest
Tendency to form lavas	Highest	Intermediate	Least
Tendency to form pyroclastics	Least	Intermediate	Highest
Density	Highest	Intermediate	Lowest
Melting temperature	Highest	Intermediate	Lowest
Typical minerals	Ca feldspar Pyroxene Olivine	Na feldspar Amphibole Pyroxene Mica	K feldspar Quartz Mica Amphibole

SOURCE: Modified from Peter Francis.

nal volume. Very fluid basaltic magmas allow the expanding gases to migrate upward and escape from the vent with relative ease. As they escape, the gases will often carry incandescent lava hundreds of meters into the air, producing lava fountains (Figure 4.5). Although spectacular, such fountains are not generally associated with major explosive events of the type which cause great loss of life and property. Rather, eruptions of fluid basaltic lavas, such as those that occur in Hawaii, are relatively quiescent.

At the other extreme, highly viscous magmas impede the upward migration of gases. As a consequence, gases collect as bubbles and pockets that increase in pressure until they explosively eject the semimolten rock from the volcano.

Once magma in the upper portion of the vent is ejected, the reduced pressure on the molten rock directly below is believed to cause it to blow out. Thus, rather than a single "bang," volcanic eruptions are really a series of explosions. This process might logically continue until the magma chamber is emptied, much like a geyser empties itself of water (see Chapter 11). However, this generally does not happen. The soluble gases in a viscous magma migrate quite slowly. Hence, only within the uppermost portion of the magma body, where the confining pressure is low, does the gas pressure build to explosive levels. Thus, an explosive event is commonly followed by the quiet emission of gas-free lavas. However, once this eruptive phase ceases, the process of gas buildup begins anew. This time lag may partially explain the sporadic eruptive patterns of volcanoes that eject viscous lavas.

To summarize, we have seen that the quantity of dissolved gases, as well as the ease with which the gases can escape, largely determine the nature of a volcanic eruption. We can now understand why the volcanic eruptions on Hawaii are relatively quiet, whereas the volcanoes bordering the Pacific are explosive and pose the greatest threat to people, because these latter volcanoes generally contain great quantities of gas and emit viscous lavas.

MATERIALS EXTRUDED DURING AN ERUPTION

Many people believe that lava is the primary material extruded from a volcano. However, this is not always true. Explosive eruptions that eject huge quantities of broken rock, lava bombs, and fine ash and dust occur nearly as frequently. Moreover, all volcanic eruptions emit large amounts of gas into the atmosphere. In this section we will examine each of these materials associated with a volcanic eruption.

Lava Flows

Due to their low silica content, basaltic lavas are usually very fluid and flow in thin, broad sheets or stream-like ribbons. On the island of Hawaii such lavas have been clocked at speeds of 30 kilometers per hour on steep slopes. These velocities are rare, however, and flow rates of 10 to 300 meters per hour are more common. Further, basaltic lavas have been known to travel distances of 150 kilometers or more before congealing. In contrast, the movement of silica-rich lava is on occasion too slow to be perceptible.

When fluid basaltic lavas of the Hawaiian type congeal, they commonly form a relatively smooth skin that wrinkles as the still-molten subsurface lava continues to advance (Figure 4.6A). These are known as **pahoehoe flows** (pronounced *pah-hoy-hoy*) and re-

FIGURE 4.5
Lava fountain along the east rift zone of Hawaii's Kilauea Volcano, 1983. (Photo by J. D. Griggs, U.S. Geological Survey)

semble the twisting braids in ropes. Another common type of basaltic lava has a surface of rough, jagged blocks with dangerously sharp edges and spiny projections (Figure 4.6B). The name **aa** (pronounced *ah ah*) is given to these flows. Active aa flows are relatively cool and thick and, depending upon the slope, advance at rates of from 5 to 50 meters per hour. Further, escaping gases fragment the cool surface and produce numerous voids and sharp spines in the congealing lava. As the molten interior advances, the outer crust is broken further, giving the flow the appearance of an advancing mass of lava rubble.

The lava which flowed from the famous Mexican volcano Parícutin and buried the city of San Juan Parangaricutiro was of the aa type (see Figure 4.17). At times one of the flows from Parícutin moved only one meter per day, but continued to advance day in and day out for more than three months.

Hardened lava flows commonly contain tunnels that once were horizontal conduits carrying lava from the vent to the flow's leading edge. These openings are found in the interior of a flow where temperatures remained high long after the surface had congealed. Under these conditions, the still-molten lava within the conduits continued its forward motion, leaving behind the voids called *lava tubes* (Figure 4.7). Lava tubes can play an important role in allowing fluid lavas to advance great distances from their source. The rock that surrounds the tunnels acts as excellent insulation. Therefore, the lava flowing through the tunnels cools very slowly and can travel far before congealing.

When lava flows enter the ocean, or when lava outpourings actually originate within an ocean basin, the flows' outer zones quickly congeal. The lava within the flows is usually able to move forward by breaking through the hardened surface. This process occurs over and over, generating a lava flow composed of elongated structures resembling large bed pillows stacked one upon the other. The identifica-

A.

B.

FIGURE 4.6
A. Typical pahoehoe (ropy) lava flow, Kilauea, Hawaii. **B.** Typical slow-moving aa flow. (Photos by J. D. Griggs, U.S. Geological Survey)

A.

B.

FIGURE 4.7
Lava streams that flow in confined channels often develop a solid crust and become flows within lava tubes. **A.** Thurston Lava Tube, Hawaii Volcanoes National Park. **B.** View of an active lava tube as seen through the collapsed roof. (Photo A by T. J. Takahashi and Photo B by Jeffrey B. Judd, both of the U.S. Geological Survey)

tion of such flows, called **pillow lavas**, aids in the interpretation of earth history, because when they are identified, they indicate that deposition occurred in an underwater environment.

Gases

Magmas contain variable amounts of dissolved gases held in the molten rock by confining pressure, just as carbon dioxide is held in soft drinks until you remove the cap. As with soft drinks, as soon as the pressure is reduced, the gases begin to escape. Because obtaining samples from an erupting volcano is often difficult and dangerous, geologists usually only estimate the amount of gas originally contained within the magma.

The gaseous portion of most magmas is believed to compose from 1 to 5 percent of the total weight, and most of this is in the form of water vapor. Although the percentage may be small, the actual quantity of emitted gas can exceed thousands of tons per day. The composition of the gases is also of interest to scientists, since much evidence points to these as the source of the earth's atmosphere and the oceans. The analysis of samples taken during Hawaiian eruptions indicated that the gases emitted there consist of about 70 percent water vapor, 15 percent carbon dioxide, 5 percent each of nitrogen and sulfur compounds, and lesser amounts of chlorine, hydrogen, and argon. Sulfur compounds are easily recognized by their pungent odor and because they readily form sulfuric acid, which when inadvertently inhaled produces a burning sensation.

In addition to propelling magma from a volcano, gases are thought to create the narrow conduit that connects the magma chamber to the surface. First, the intense heat from the magma body cracks the rock above. Then, hot streams of high-pressure gases expand the cracks and develop a passageway to the surface. Once completed, the hot gases armed with rock fragments erode the walls of the passageway, producing a larger conduit. Because these erosive forces will be concentrated on any protrusion along the pathway, the volcanic pipes that are produced have a circular shape. As the conduit enlarges, magma moves upward to produce surface activity. Following an eruptive phase the volcanic pipe often becomes choked with debris that was not thrown clear of the vent. Before the next eruption, a new surge of explosive gases may then clear the conduit.

Pyroclastic Materials

When basaltic lava is extruded, the dissolved gases escape quite freely and continually. As stated earlier, these gases carry incandescent blobs of lava to great heights, thereby producing spectacular lava fountains (see Figure 4.5). Some ejected material may land near the vent and produce a cone structure, while smaller particles will be carried great distances by the wind. The gases in highly viscous magmas, on the other hand, are less able to escape and may build up an internal pressure capable of producing a violent eruption. Upon release, these superheated gases expand a thousandfold as they blow pulverized rock and lava from the vent. The particles produced by these processes are called **pyroclastic materials.** These ejected lava fragments range in size from very fine dust and sand-sized volcanic ash, to large volcanic bombs and blocks.

The fine *ash* and *dust* particles are produced when the extruded lava contains so many gas bubbles that it resembles the froth flowing from a newly opened bottle of champagne. As the hot gases expand explosively, the lava is disseminated into very fine fragments. When the hot ash falls, the glassy shards often fuse to form welded tuff. Sheets of this material, as well as ash deposits that consolidate later, cover vast portions of the western United States. Sometimes the frothlike lava is ejected in larger pieces called *pumice* (Figure 4.8). This material has so many voids that it is often light enough to float in water.

Pyroclastics the size of walnuts, called *lapilli* ("little stones"), and pea-sized particles called *cinders* are also very common. Cinders contain numerous voids and form when ejected lava blobs are pulverized by the escaping gases. Particles larger than lapilli are called *blocks* when they are made of hardened lava and *bombs* when they are ejected as incandescent lava. Since volcanic bombs are semimolten upon ejection, they often take on a streamlined shape as shown in Figure 4.9. Due to their size, bombs and blocks usually fall on the slopes of a cone; however, bombs may occasionally be propelled far from the volcano by the force of escaping gases.

FIGURE 4.9
Volcanic bombs. Ejected lava fragments take on a streamlined shape as they sail through the air. This bomb is approximately 10 centimeters in length.

FIGURE 4.8
Pumice deposits near Crater Lake, Oregon. (Photo by E. J. Tarbuck)

Fine volcanic debris can be scattered great distances from its source. Dust in particular may be blasted high into the atmosphere, where it can remain for extended periods. While present, dust produces brilliant sunsets and has on occasion slightly lowered the earth's average temperature. The possible effects of volcanic eruptions on climate are discussed more thoroughly in Box 4.1 and Box 4.3.

VOLCANOES AND VOLCANIC ERUPTIONS

Successive eruptions from a central vent result in the formation of a mountainous accumulation of material known as a **volcano.** Located at the summit of many volcanoes is a steep-walled depression, a **crater,** which is connected to a magma chamber via a pipeline conduit, or **vent.** As was indicated earlier, the conduit as well as the crater is produced by the erosive force of gases and effervescent magma. Some volcanoes have unusually large summit depressions

BOX 4.1

The Year Without a Summer

The graph in Figure 4.A allows us to compare the volume of volcanic debris extruded during some well-known eruptions, beginning with Mount Vesuvius in 79 A.D. The eruption of the volcano named *Tambora* is clearly the largest of the modern era. Between April 7 and 12, 1815, this nearly 4000-meter Indonesian structure violently ejected an estimated 30,000 cubic kilometers of volcanic debris. That is 30 times more ash than was emitted during the eruption of Mount St. Helens in May, 1980.

Although the Tambora eruption occurred in an isolated part of the world, circulation patterns high in the atmosphere spread its influence far and wide. The impacts of the volcanic dust and gases on climate are believed to have been widespread in the Northern Hemisphere.* According to one researcher who studied the matter, "The extreme cold that prevailed during the spring and summer of 1816 in some regions of the world represents one of the most unusual climatic episodes that has occurred since the advent of instrumental weather observations."* The effects were especially severe in New England, where 1816 has come to be known as the "year without a summer."

From May through September, 1816, an unprecedented series of cold spells affected the northeastern United States and adjacent portions of Canada. The result was a late spring, a cold summer, and an early fall. There were heavy snows in June and frosts in July and August. Across the Northeast, crops were reported killed by the cold. Temperatures in New England averaged up to 3.5°C below normal in June and between 1°C and 2°C below normal in August. Although temperatures this cold had occurred before, there has never been such a protracted span of cold since records have been kept. The reduced averages may seem modest, but they took place in a region where even a small drop in minimum temperatures can mean severe frost.

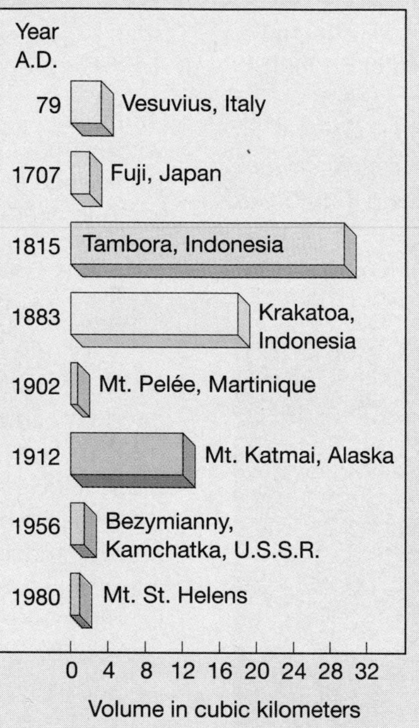

FIGURE 4.A
Approximate volume of volcanic debris emitted during some well-known eruptions. The 1815 eruption of Tambora, the largest-known eruption in historic time, ejected over 30 times more ash than did Mount St. Helens in 1980.

New England and adjacent Canada were not the only areas to experience a "year without a summer." Patrick Hughes writes, "Although the New England farmer considered it a local tragedy, the abnormal weather was widespread throughout the Northern Hemisphere. In England it was almost as cold as in the United States, and 1816 was a famine year there, as it was in France and Germany."†

The unusual meteorological events of 1816 which followed the massive 1815 eruption of Tambora are regarded by many as a spectacular example of the influence of explosive volcanism on climate. Although the effects were relatively short-lived, this geologic event had a significant atmospheric and human impact.

Box 4.3 "Volcanic Activity and Climate" is related to this topic and appears near the end of the chapter.

**American Weather Stories* (Washington, D.C.: National Oceanic and Atmospheric Administration, 1976), p. 43.

†*American Weather Stories,* p. 43.

that exceed one kilometer in diameter and are known as **calderas.*** (Figure 4.10).

When fluid lava leaves a conduit, it is often stored in the crater or caldera, until it overflows. On the other hand, lava that is very viscous forms a plug in the pipe which rises slowly or is blown out, often enlarging the crater. However, lava does not always issue from a central crater. Sometimes it is easier for the magma or escaping gases to push through fissures located on the volcano's flanks. Continued activity from a flank eruption may build a so-called **parasitic cone.** Mount Etna, for example, has more than 200 secondary vents. Some of these secondary vents emit only gases and are appropriately called **fumaroles.**

*The formation of calderas is discussed later in this chapter.

FIGURE 4.10
Summit Caldera of Mauna Loa, a shield volcano in Hawaii. The smaller depressions are known as pit craters. (Photo by D. W. Peterson, U.S. Geological Survey)

The eruptive history of each volcano is unique; consequently, all volcanoes are somewhat different in form and size (Figure 4.11). Nevertheless, volcanologists have recognized that volcanoes exhibiting somewhat similar eruptive styles can be grouped. Based on their "typical" eruptive patterns and characteristic form, three groups of volcanoes are generally recognized: shield volcanoes, cinder cones, and composite cones (stratovolcanoes).

Shield Volcanoes

When fluid lava is extruded, the volcano takes the shape of a broad, slightly domed structure called a **shield volcano** (Figure 4.12). Shield volcanoes are built primarily of basaltic lava flows and contain only a small percentage of pyroclastic material. Typically they have a slope of a few degrees at their flanks and generally do not exceed 15 degrees near their summits, as exemplified by the volcanoes of the Hawaiian Islands. Mauna Loa, probably the largest volcano on earth, is one of the five shield volcanoes that together make up the island of Hawaii (Figure 4.13). Its base rests on the ocean floor 5000 meters (16,400 feet) below sea level, and its summit reaches a height of 4170 meters (13,677 feet) above the water. Nearly one million years and numerous eruptive cycles were required to build this truly gigantic pile of volcanic rock. Many other volcanic structures, including Midway Island and the Galápagos Islands, have been built in a similar manner from the ocean's depths.

The 1959-60 eruption of the volcano Kilauea illustrates the "typical" cycle of a Hawaiian-type eruption.

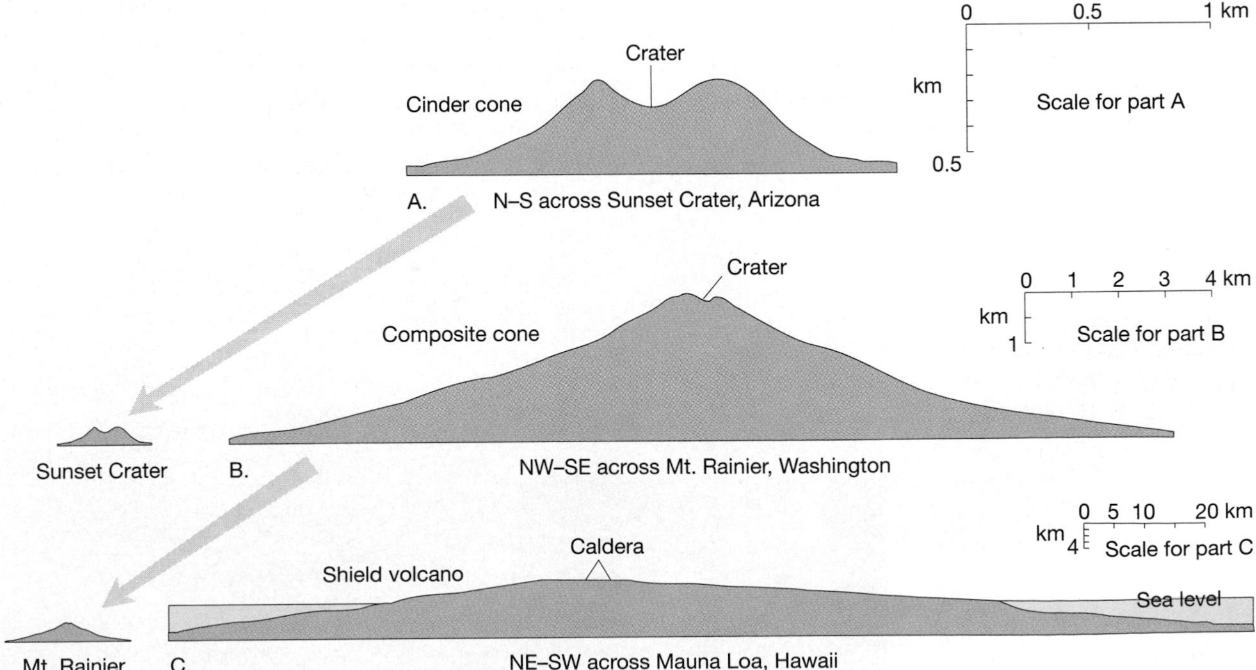

FIGURE 4.11
Profiles of volcanic landforms. **A.** Profile of Sunset Crater, Arizona, a typical steep-sided cinder cone. **B.** Sunset Crater compared with Mt. Rainier, Washington, a large composite cone of the Cascade Range. **C.** Mt. Rainier compared with Mauna Loa, Hawaii, the largest shield volcano in the Hawaiian Chain.

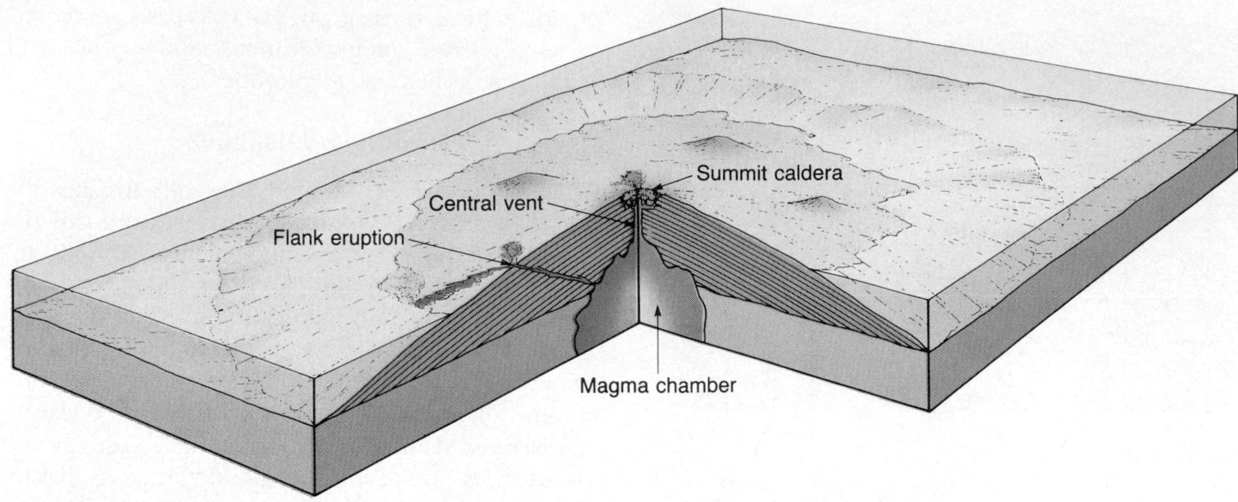

FIGURE 4.12
Shield volcanoes are built primarily of fluid basaltic lava flows and contain only a small percentage of pyroclastic materials. These broad, slightly domed structures, exemplified by the Hawaiian Islands, are the largest volcanoes on earth.

Kilauea is located on the island of Hawaii on the southeast flank of the larger volcano Mauna Loa. The summit caldera of Kilauea is an oval-shaped depression about 100 meters deep and roughly 5 kilometers long by 3 kilometers wide. Located within this crater is an even deeper depression called Halemaumau, or "firepit," which from time to time contains a "boiling" lava lake reaching depths upwards of 400 meters (1300 feet). In addition to the summit caldera, numerous smaller craters dot the landscape.

Kilauea has erupted over 50 times in recorded history. A testimonial to the quiescent nature of these eruptions is the location of the Hawaiian Volcano Observatory on the very rim of the summit caldera. Two years prior to the 1959-60 eruption, volcanologists equipped with tiltmeters discovered swelling in the summit area. Earth tremors produced by magma migrating toward the surface indicated that the material originated between 60 and 100 kilometers below the surface. This molten rock slowly worked its way upward and accumulated in smaller reservoirs lo-

FIGURE 4.13
Mauna Loa is one of five shield volcanoes that together make up the island of Hawaii. (Photo by J. D. Griggs, U.S. Geological Survey)

cated about 3 to 5 kilometers below the summit.

The first phase of the 1959-60 eruption began in November of 1959 in Kilauea Iki, a crater located just east of the summit caldera. The initial event consisted of lava fountaining from a one-kilometer-long fissure along the south wall of the crater. Within a day the eruption was restricted to a single vent, occasionally spraying lava to heights of more than 300 meters (see Figure 4.5). Most of the ejected lava fell back into the crater where it formed a lava lake. Around the vent, falling cinders built a small cone that frequently collapsed as the crater filled with lava. This phase of the eruption lasted about a month, during which a lava lake filled the crater to a depth of 125 meters. Immediately after this eruptive phase ceased, some of this lava drained back into the vent. The draining of this incandescent material resembled water spiraling into a drain.

The earth tremors persisted long after the activity at Kilauea Iki ceased. This seismic activity indicated that the magma was still flowing underground, but had been diverted toward the volcano's flank. Then abruptly on January 13, 1960, ground cracks were observed as a one-kilometer-long fissure opened near the village of Kapoho, located 45 kilometers east of Kilauea Iki. Here lava fountains played continuously day and night for several weeks. Within the first few days, molten lava collected into flows and advanced to the sea, 5 kilometers away. As the hot lava invaded the sea, boiling clouds of steam, occasionally darkened by explosive blasts of black ash, rose thousands of meters into the air (Figure 4.14). During the next few days the lava flows spread laterally, engulfing more of the countryside. Earthen dams were constructed in vain to protect the village of Kapoho. During this phase several developments were destroyed, while the island was enlarged by one-half kilometer in one location.

The activity at Kapoho was accompanied by gradual subsidence of the summit area and collapse of the floor of Halemaumau. About three months after the initial event, the fountain near Kapoho died, marking the end of this eruptive phase. However, geologists continued to monitor ground subsidence and to sample molten rock from beneath the crust of the lava lake in Kilauea Iki. From this type of information a better understanding of the formation of the Hawaiian Islands and volcanism in general has emerged.

There is now general agreement that the early stages of a shield volcano's formation consist of frequent eruptions of thin flows of very fluid basalts. As the structure enlarges, flank eruptions occur along with the summit eruptions. Collapse of the summit area frequently follows each eruptive phase. In the latter stages of growth, the activity is more sporadic and pyroclastic ejections are more prevalent. Further, the lavas increase in viscosity, resulting in thicker, shorter flows. These activities tend to steepen the slope of the summit area. This explains why Mauna Kea, an older and inactive volcano to the north, has a steeper summit than Mauna Loa.

FIGURE 4.14
Lava flow meets the sea, Hawaii Volcanoes National Park. (Photo by T. J. Takahashi, U.S. Geological Survey)

Cinder Cones

As the name suggests, **cinder cones** are built from ejected lava fragments (Figure 4.15). Because unconsolidated pyroclastic material maintains a high angle of repose (between 30 and 40 degrees), volcanoes of this type have very steep slopes. Cinder cones are rather small, usually less than 300 meters (1000 feet) high, and often form as parasitic cones on larger volcanoes. In addition, they frequently occur in groups,

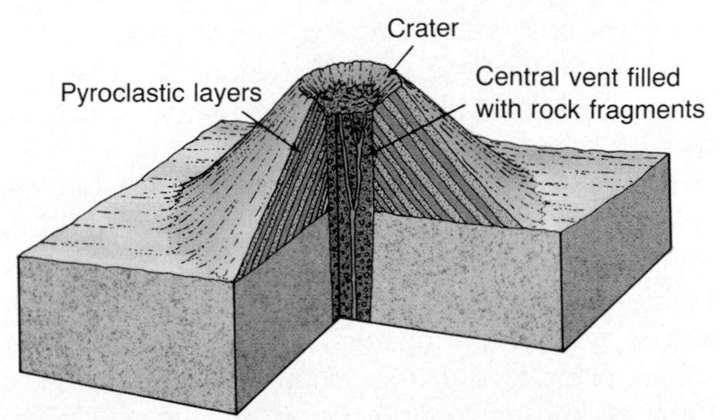

FIGURE 4.15
Cinder cones are built from ejected lava fragments and are usually less than 300 meters (1000 feet) high.

FIGURE 4.16
SP Crater, a cinder cone north of Flagstaff, Arizona. Note the road across the lava flow for scale. (Photo by Michael Collier)

where the cones apparently represent the last phase of activity in a region of older basaltic flows (Figure 4.16). This may result because the contributing magma has cooled and become more viscous.

One of the very few volcanoes whose formation has been observed by geologists from beginning to end is a cinder cone called Parícutin. This volcano's history serves to illustrate the formation and structure of a larger-than-usual cinder cone.

In 1943, about 200 miles west of Mexico City, the volcano Parícutin was born (see Figure 3.1). The eruption site was a cornfield owned by Dionisio Pulido, who with his wife Paula witnessed the event as they were preparing the field for planting. For two weeks prior to the first eruption, numerous earth tremors caused apprehension in the village of Parícutin about 3.5 kilometers away. Then around 4:00 P.M. February 20th, smoke with a sulfurous odor arose from a small hole that had been in the cornfield for as long as Señor Pulido could remember. During the night, hot, glowing rock fragments thrown into the air from the hole produced a spectacular fireworks display. By the next day the cone had grown to a height of 40 meters and by the fifth day it was over 100 meters high. At this time explosive eruptions were throwing hot fragments 1000 meters above the crater rim. The larger fragments fell near the crater, some remaining incandescent as they rolled down the slope. These fragments built an aesthetically pleasing cone, while finer ash fell over a much larger area, burning and eventually covering the village of Parícutin. Within two years the cone had grown to 400 meters high and would rise only a few tens of meters more.

The first lava flow came from a fissure that had opened just north of the cone, but after a few months of activity, flows began to emerge from the base of the cone itself. In June of 1944, a clinkery flow 10

FIGURE 4.17
The village of San Juan Parangaricutiro engulfed by lava from Parícutin, shown in the background. Only the church towers remain. (Photo by Tad Nichols)

meters thick moved over much of the village of San Juan Parangaricutiro, leaving the church steeple exposed (Figure 4.17). After nine years the activity ceased almost as quickly as it began. Now Parícutin is just another one of the numerous cinder cones dotting the landscape in this region of Mexico. Like the others, it will probably not erupt again.

Composite Cones

The earth's most picturesque volcanoes are termed **composite cones**, or **stratovolcanoes** (see Figure 4.1A). Most active composite cones are located in a narrow zone, appropriately named the *Ring of Fire,* which encircles the Pacific Ocean. Found in this region are Fujiyama in Japan, Mount Mayon in the Philippines, and the picturesque volcanoes of the Cascade Range in the northwestern United States, including Mount St. Helens, Mount Rainier, and Mount Hood.

A composite cone is a large, nearly symmetrical structure built of interbedded lavas and pyroclastic deposits emitted mainly from a central vent. Just as shield volcanoes owe their shape to the fluid nature of the extruded lavas, so too do composite cones reflect the nature of the erupted material. Composite cones are produced when relatively viscous lavas of andesitic composition are extruded. For long periods a composite cone may extrude viscous lava. Then suddenly, the eruptive style will change and the volcano will violently eject pyroclastic material. Most of the ejected pyroclastic material falls near the summit, building a steep-sided mound of cinders. In time this debris will be covered by lava. Occasionally both activities occur simultaneously. The resulting structure consists of alternating layers of lava and pyroclasts (Figure 4.18). Two of the most perfect cones, Mount Mayon in the Philippines and Fujiyama in Japan, exhibit the classic form of the stratovolcano with its steep summit area and rather gently sloping flanks.

Although composite cones are the most picturesque, they also represent the most violent type of volcanic activity. Their eruption can be unexpected and devastating as was the A.D. 79 eruption of the Italian volcano we now call Vesuvius. Prior to this eruption, Vesuvius was dormant for centuries. Although minor earthquakes probably warned of the events to follow, Vesuvius was covered with a heavy coat of vegetation and hardly looked threatening. On August 24th, however, the tranquility ended, and in the next three days the city of Pompeii (near Naples) and more than 2000 of its 20,000 residents were buried. They remained so for nearly seventeen centuries, until the city was rediscovered and excavated.

Although the destruction of Pompeii was truly catastrophic, eruptions of a more devastating nature occur when hot gases infused with incandescent ash are ejected, producing a fiery cloud called a **nuée ardente**. Also referred to as *glowing avalanches,* these turbulent steam clouds and companion ash

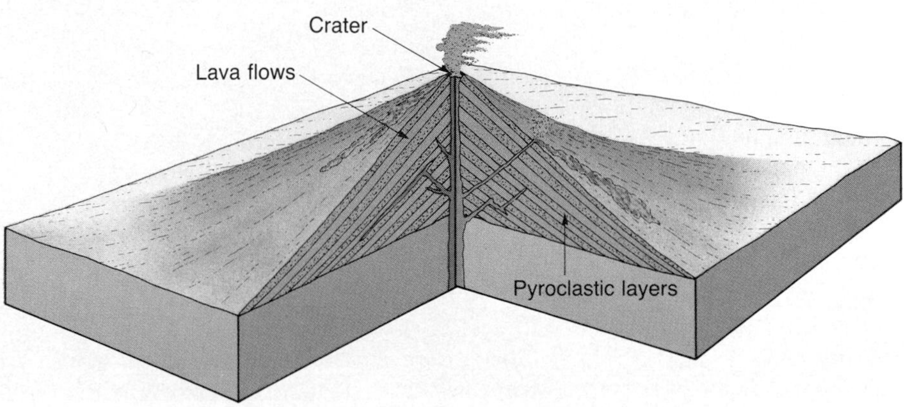

FIGURE 4.18
Composite cones are built of alternating layers of lava flows and pyroclastics. Generally, summit areas are relatively steep and the flanks have gentler slopes.

FIGURE 4.19
Nuée ardente races down the slope of Mount St. Helens on August 7, 1980, at speeds in excess of 100 kilometers (60 miles) per hour. (Photo by Peter W. Lipman, U.S. Geological Survey)

flows travel down steep volcanic slopes at speeds that can approach 200 kilometers (125 miles) per hour (Figure 4.19). The ground-hugging portions of glowing avalanches are rich in particulate matter that is suspended by hot, buoyant gases emitted from the volcanic debris within the moving flow. Thus, these flows, which can include larger rock fragments in addition to ash, travel downslope in a nearly frictionless environment cushioned by expanding volcanic gases. This fact explains why some nuée ardente deposits extend more than 100 kilometers from their source.

In 1902, a nuée ardente from Mount Pelée, a small volcano on the Caribbean island of Martinique, destroyed the port town of St. Pierre. The destruction was instantaneous and so devastating that almost all of St. Pierre's 28,000 inhabitants were killed. Only a prisoner protected in a dungeon, a shoemaker, and a few people on ships in the harbor were spared (Figure 4.20). Satis N. Coleman, in *Volcanoes, New and Old,* relates a vivid account of this event, which lasted less than five minutes:

> I saw St. Pierre destroyed. The city was blotted out by one great flash of fire. Nearly 40,000 people were killed at once. Of eighteen vessels lying in the roads, only one, the British steamship *Roddam* escaped and she, I hear, lost more than half of those on board. It

FIGURE 4.20
St. Pierre as it appeared shortly after the eruption of Mount Pelée, 1902. (Reproduced from the collection of the Library of Congress)

was a dying crew that took her out. Our boat, the *Roraima,* arrived at St. Pierre early Thursday morning. For hours before entering the roadstead we could see flames and smoke rising from Mt. Pelée. . . . The spectacle was magnificent. As we approached St. Pierre we could distinguish the rolling and leaping of red flames that belched from the mountain in huge volumes and gushed into the sky. Enormous clouds of back smoke hung over the volcano. There was a constant muffled roar. It was like the biggest oil refinery in the world burning up on the mountain top. There was a tremendous explosion about 7:45, soon after we got in. The mountain was blown to pieces. There was no warning. The side of the volcano was ripped out and there was hurled straight toward us a solid wall of flame. It sounded like a thousand cannons.

The wave of fire was on us and over us like a flash of lightning. It was like a hurricane of fire. I saw it strike the cable steamship *Grappler* broadside on, and capsize her. From end to end she burst into flames and then sank. The fire rolled in mass straight down upon St. Pierre and the shipping. The town vanished before our eyes.

The air grew stifling hot and we were in the thick of it. Wherever the mass of fire struck the sea, the water boiled and sent up vast columns of steam. . . . The blast of fire from the volcano lasted only a few minutes. It shriveled and set fire to everything it touched. Thousands of casks of rum were stored in St. Pierre, and these were exploded by the terrific heat. The burning rum ran in streams down every street and out into the sea. This blazing rum set fire to the Roraima several times. . . . Before the volcano burst, the landings of St. Pierre were covered with people. After the explosion, not one living soul was seen on land.*

When we compare the destruction of St. Pierre with that of Pompeii, several differences are noticed. Pompeii was totally buried by an event lasting three days, whereas St. Pierre was destroyed in a brief in-

*New York: John Day, 1946, pp. 80–81.

A.

B.

FIGURE 4.21
Following the May 18, 1980, eruption of Mount St. Helens, a large lava dome began to develop. **A.** Note the "steam cloud" produced by water released from the growing lava body. (Photo by Lyn Topinka, David A. Johnston Cascades Volcano Observatory) **B.** Close-up view of the lava dome. (Photo by Craig A. Chesner)

stant and its remains were mantled only by a thin layer of volcanic debris. Also, the structures of Pompeii remained intact except for the roofs that collapsed under the weight of the ash. In St. Pierre masonry walls nearly one meter thick were knocked over like dominoes; large trees were uprooted and cannons were torn from their mounts.

Nuée ardentes are usually associated with highly viscous magmas near the granitic end of the compositional spectrum (see Table 4.1). Bulbous masses called **lava domes** are another feature associated with volcanoes that extrude highly viscous material. These structures are produced when thick, viscous lava is slowly "squeezed" out of the vent (Figure 4.21). Lava domes often act as plugs, which deflect subsequent gaseous eruptions. Occasionally lava domes will form on a volcano's flanks. As these structures enlarge they pose a potential threat to inhabitants in the surrounding area. If the upper part of the dome slips downslope in response to the pull of gravity, the loss of confining pressure may trigger an explosive eruption of the magma below. These laterally directed eruptions, such as the May 18, 1980, eruption of Mount St. Helens, can be devastating.

In addition to their violent eruptions, volcanoes are potential hazards in other ways. In particular, large composite cones often generate a type of mudflow known as a *lahar.* These destructive events occur when deposits of volcanic ash and debris become saturated with water and flow down steep volcanic slopes, generally following stream valleys. Some lahars are produced when rainfall saturates volcanic deposits, whereas others are triggered as large volumes of ice and snow melt during an eruption.

When Mount St. Helens erupted in May 1980, several lahars formed. The flows and accompanying floods raced down the valleys of the north and south forks of the Toutle River at speeds in excess of 30 kilometers per hour. Water levels reached four meters above flood stage, and nearly all the homes and bridges along the river were destroyed or severely damaged (see Figure 9.13). Fortunately, the affected area was not densely populated. This was not the case in November, 1985, when Nevado del Ruiz, a volcano in the Andes, erupted and generated a lahar that killed nearly 20,000 people (see the section entitled "Mudflow" in Chapter 9).

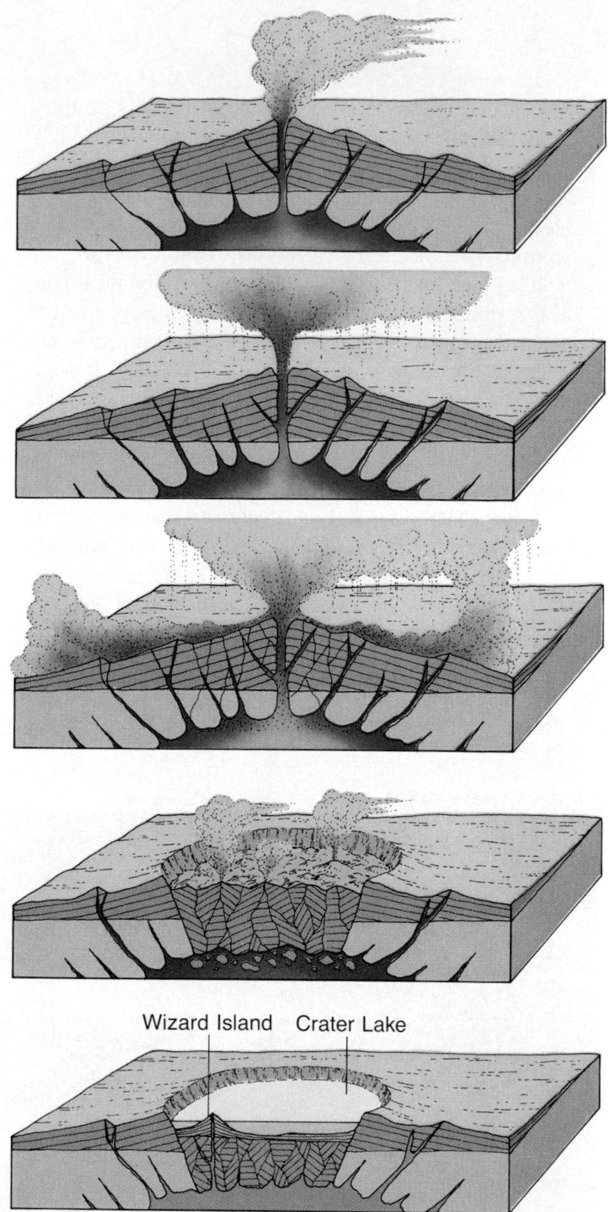

FIGURE 4.22
Sequence of events that formed Crater Lake, Oregon. About 7000 years ago, the summit of former Mount Mazama collapsed following a violent eruption which partly emptied the magma chamber. Subsequent eruptions produced the cinder cone called Wizard Island. Rainfall and groundwater contributed to form the lake. (After H. Williams, *The Ancient Volcanoes of Oregon,* p. 47. Courtesy of the University of Oregon)

Calderas

Earlier it was pointed out that some volcanoes have nearly circular depressions that exceed one kilometer in diameter known as *calderas.* Most calderas are thought to form when the summit of a volcanic structure collapses into the partially emptied magma chamber below (Figure 4.22). Crater Lake in Oregon, which is 8-10 kilometers wide and 1175 meters deep, is located in such a depression (Figure 4.23).

FIGURE 4.23
Crater Lake occupies a caldera about 10 kilometers (6 miles) in diameter. (Courtesy of the National Park Service)

BOX 4.2

The Lost Continent of Atlantis

One of the most popular legends of all time concerns the disappearance of the so-called continent of Atlantis. According to the accounts provided by Plato, an island empire named *Atlantis* disappeared beneath the sea in a single day and night. This event, which reportedly took place between 1500 B.C. and 1400 B.C. and caused the collapse of the Minoan civilization, is thought by many to have been the result of a cataclysmic volcanic eruption.

Research efforts in the eastern Mediterranean have provided evidence apparently linking Atlantis to the volcanic island of Santorin. Although once a majestic volcano, Santorin now consists of five islands located roughly midway between the island of Crete and Greece. Evidence collected from core samples taken around the remnants of Santorin revealed that a violent eruption occurred about 1500 B.C. This eruption generated great volumes of ash and pumice that reached a maximum depth of 60 meters. Many nearby Minoan cities were buried and their ruins preserved beneath the ash.* Even as far away as Crete, enough ash fell to kill crops and possibly livestock grazing on the ash-laden plants. Following the ejection of this large quantity of material, Santorin collapsed, producing a caldera 14 kilometers across. The eruption and collapse of Santorin undoubtedly generated large sea waves that caused widespread destruction to the coastal villages of Crete as well as those on the nearby islands to the north.

The connection between the eruption of Santorin and the disappearance of Atlantis is further supported by the fact that the Minoan civilization also disappeared between 1500 B.C. and 1400 B.C. About 1400 B.C., some of the Minoan traditions began to appear in the culture of Greece.

Most scholars agree that the eruption of Santorin contributed to the collapse of the Minoan civilization. Was this eruption the main cause of the dispersal of this great civilization or only one of many contributing factors? Was Santorin the island continent of Atlantis described by Plato? Whatever the answers to these questions, clearly volcanism can dramatically change the course of human events.

*One of these buried cities, Akrotiri, on the island of Thera, has been excavated.

The creation of Crater Lake began about 7000 years ago when the volcano, later to be named Mount Mazama, put forth a violent ash eruption much like that of Vesuvius. However, this ancient eruption was on a much larger scale, extruding an estimated 40-50 cubic kilometers of volcanic material. With the loss of support, 1500 meters of this once prominent 3600-meter cone collapsed. After the collapse, rainwater filled the caldera. Later activity built a small cinder cone called Wizard Island, which today provides a mute reminder of past activity.

Several very large calderas are known to exist, with the largest being more than 20 kilometers in diameter. One example, the Valles Caldera located west of Los Alamos, New Mexico, is about 16 kilometers across and 1 kilometer deep. Such features are usually associated with large granitic intrusive bodies located near the surface. Following a violent eruption that may emit tens of cubic kilometers of pyroclastic debris, the roof of the partially emptied magma chamber apparently collapses. Other examples of large calderas believed to have formed in a similar manner are California's Long Valley Caldera and Yellowstone Caldera in Wyoming.

Not all calderas are produced following explosive eruptions. For example, the summits of Hawaii's active shield volcanoes, Mauna Loa and Kilauea, have large calderas that formed as a result of relatively quiet activity. These calderas, which measure 3 to 5 kilometers across and nearly 200 meters deep, formed by collapse as magma was slowly drained from the summit magma chambers during flank eruptions.

FIGURE 4.24
Shiprock, New Mexico, a volcanic neck. This structure, which stands over 420 meters (1380 feet) high, consists of igneous rock which crystallized in the vent of a volcano that has long since been eroded away. The tabular structure in the foreground is a dike which undoubtedly served to feed lava flows along the flanks of the once active volcano. (Photo by Michael Collier)

Volcanic Necks and Pipes

Volcanoes, like all land areas, are continually being lowered by the forces of weathering and erosion. Cinder cones are easily eroded, because they are composed of unconsolidated materials. However, all volcanic structures will eventually be worn away. As erosion progresses, the rock occupying the vent is often more resistant, and may remain standing above the surrounding terrain long after most of the cone has vanished. Shiprock, New Mexico, is thought to be such a feature, called a **volcanic neck** (Figure 4.24). This structure, higher than many skyscrapers, is but one of many that protrude conspicuously from the desert landscape of the southwestern United States.

The tubelike conduits feeding most volcanoes serve as a vent for a magma body emplaced relatively close to the earth's surface. However, some conduits called **pipes** are thought to extend to depths of 200 kilometers or more and tap the upper mantle. Consequently, the ultramafic rocks associated with these structures are believed to be samples of the asthenosphere that have undergone very little alteration during their ascent. Geologists consider pipes, therefore, to be "windows" into the earth, since they allow us to view rocks found at great depths.

Occasionally, a volcanic pipe will reach the surface, but the eruptive phase will cease before any lava is extruded. The upper portions of these pipes often contain a jumble of lava fragments and fragments that were torn from the walls of the vent by the violently escaping gases. The best known of these structures are the diamond-bearing pipes of South Africa. Here the rocks filling the pipes are thought to have originated at depths of about 200 kilometers, where pressure is great enough to generate diamonds and other high-pressure minerals. In these instances the diamonds crystallized at depth and were carried upward by the still-molten portion of the magma.

FLOOD BASALTS AND PYROCLASTIC FLOW DEPOSITS

Although volcanic eruptions from a central vent are the most familiar, larger amounts of volcanic material are probably extruded from cracks or fractures in the crust called *fissures* (Figure 4.25). Rather than building an isolated cone, lava is extruded from several vents along these narrow cracks, resulting in the distribution of volcanic materials over a wide area. An extensive region in the northwestern United States known as the Columbia Plateau was formed in this manner (Figure 4.26). Here, numerous **fissure eruptions** extruded very fluid basaltic lava. Successive flows as thick as 50 meters buried the old landscape and built a lava plain, which in some places is 2 to 3 kilometers thick (Figure 4.27). The lava's fluidity is evident because some flows remained molten long enough to travel 150 kilometers from their source. The term **flood basalts** appropriately describes these flows. For a more complete discussion on the origin

FIGURE 4.25
Lava fountain along a fissure in the east rift zone of Kilauea, 1983. (Photo by J. D. Griggs, U.S. Geological Survey)

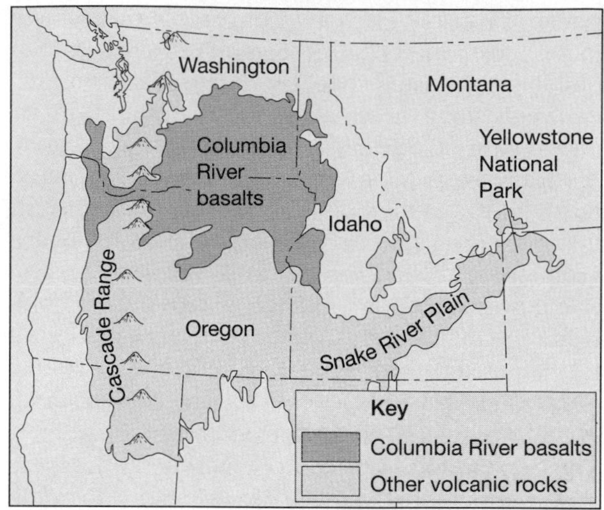

FIGURE 4.26
Volcanic areas in the northwestern United States. The Columbia River basalts cover an area of nearly 200,000 square kilometers (80,000 square miles). Activity here began about 17 million years ago as lava began to pour out of large fissures, eventually producing a basalt plateau with an average thickness of more than 1 kilometer. (After U.S. Geological Survey)

of flood basalt provinces see Box 18.2, "Hot Spots and Flood Basalts."

Although a few large continental areas are covered with basalt flows, the greatest activity of this type occurs on the ocean floor hidden from view. Along oceanic ridges, where sea-floor spreading is active, fissure eruptions generate new sea floor. Iceland, which is located astride the Mid-Atlantic Ridge, has experienced numerous fissure eruptions of the type believed to occur at spreading centers. The largest Icelandic eruptions in historic times occurred in 1783. A rift 25 kilometers long generated over 20 separate vents which initially extruded sulfurous gases and ash deposits that built several small cinder cones. This activity was followed by huge outpourings of very fluid basalt. The total volume of lava extruded by this so-called Laki eruption was in excess of 10 cubic kilometers. The fluid lava first collected in the nearby valleys and after repeated eruptive phases led to the creation of a flat plateaulike topography.

When silica-rich magma is extruded, **pyroclastic flows** consisting largely of ash and pumice fragments are the rule. When these pyroclastic materials are ejected, they move away from the vent at high speeds and may blanket extensive areas before coming to rest. Once deposited, the pyroclastic materials may closely resemble lava flows.

Extensive pyroclastic flow deposits are found in many parts of the world and are most often associated with large calderas. The first phase in the origin of these flows is thought to occur when a highly viscous magma body is emplaced near the surface, upwarping the overlying rocks. Next, fracturing of the roof allows the gas-rich magma to reach the surface where it produces short-lived, explosive eruptions. Finally, the loss of magma from beneath the surface allows the roof to collapse.

Perhaps the best-known region of pyroclastic flows is the Yellowstone Plateau in northwestern Wyoming. Here a large magma body, rich in silica, exists a few kilometers below the surface. Several times over the past two million years, fracturing of the rocks overlying the magma chamber has resulted in huge eruptions accompanied by the formation of calderas. In the northeastern portion of Yellowstone National Park, numerous fossil forests have been discovered, one resting upon another. During periods of inactivity, a forest developed upon the newly formed volcanic surface, only to be covered by ash

FIGURE 4.27
Basalt flows of the Columbia Plateau. (Photo by E. J. Tarbuck)

from the next eruptive phase. Very large quantities of pyroclastic materials have been extruded in the Yellowstone area; fortunately, no eruption of this type has occurred in modern times.

VOLCANISM AND PLATE TECTONICS

The origin of magma has been a controversial topic in geology almost from the very beginning of the science. How do magmas of different compositions arise? Why do volcanoes located in the deep-ocean basins primarily extrude basaltic lava, whereas those adjacent to oceanic trenches extrude mainly andesitic lava? Why are basaltic lavas common at the earth's surface, whereas most granitic magma is emplaced at depth? Why does an area of igneous activity commonly called the "Ring of Fire" surround the Pacific Ocean? New insights gained from the theory of plate tectonics are providing some answers to these questions. We will first examine the origin of magma and then look at the global distribution of volcanic activity as viewed from the model provided by plate tectonics.

Origin of Magma

Based on available scientific evidence, the earth's crust and mantle are composed primarily of solid rock. Further, although the outer core is a fluid, this iron-rich material is very dense and remains deep within the earth. If this is true, what is the source of magma that produces the earth's volcanic activity?

Since the molten outer core is not a source of magma, geologists conclude that magma must originate from essentially solid rock located in the crust and mantle. The most obvious way to generate magma from solid rock is to raise its temperature. In a near-surface environment, silica-rich rocks of granitic composition begin to melt at temperatures around 750°C, whereas basaltic rocks must reach temperatures above 1000°C before melting commences.

What is the source of heat that melts rock? One source is the heat liberated during the decay of radioactive elements that are found in the mantle and crust. Workers in underground mines have long recognized that temperatures get higher as they descend to greater depths. Although the rate of temperature increase varies from place to place, it averages between 20°C and 30°C per kilometer in the upper crust. This gradual increase in temperature with depth is known as the **geothermal gradient**.

The geothermal gradient is thought to contribute to magma production in two important ways. First, at deep-ocean trenches, slabs of cool oceanic lithosphere descend into the hot mantle. Here heat supplied by the surrounding rocks is thought to be sufficient to melt the subducting oceanic crust and produce basaltic magma. Second, a hot, mantle-derived magma body as just described could migrate to the base of the crust and intrude silica-rich rocks. Because granitic rocks have melting temperatures well below those required to melt basalt, heat derived from the hotter basaltic magma could melt the already warm crustal rocks. The volcanic activity that produced the vast ash flows in Yellowstone National Park is believed to have resulted from such activity. Here basaltic magma from the mantle transported heat to the crust, where the melting of silica-rich rocks generated explosive lavas.

If temperature were the only factor that determined whether or not rock melts, the earth would be a molten ball covered with a thin, solid outer shell. This, of course, is not the case. The reason is that pressure also increases with depth. Since rock must expand when heated, extra heat is required to melt buried rocks in order to overcome the effect of confining pressure. In general, an increase in the confining pressure causes an increase in the rock's melting temperature, and a reduction in confining pressure causes the rock's melting temperature to decline. Consequently, a drop in confining pressure can lower the melting temperature of rock sufficiently to trigger melting. This occurs whenever rock ascends, thereby moving into zones of lower pressures.

Another important factor affecting the melting temperature of rock is its water content. In general, the more water present, the lower the melting temperature. The effect of water on lowering the melting point is magnified by increased pressure. Consequently, "wet" rock under pressure has a much lower melting temperature that "dry" rock of the same composition. For example, notice in Figure 4.28 that at a depth of 10 kilometers, wet granite has a melting temperature of about 657°C, whereas dry granite begins to melt at temperatures approaching 1000°C. Therefore, in addition to a rock's composition, its temperature, confining pressure, and water content determine whether the rock exists as a solid or liquid.

One important difference exists between the melting of a substance that consists of a single compound, such as ice, and the melting of igneous rocks, which are mixtures of several different minerals. Whereas ice melts at a definite temperature, most igneous rocks melt over a temperature range of a few hundred degrees. As a rock is heated, the first liquid to form will contain a higher percentage of the low-melting-temperature minerals than the original

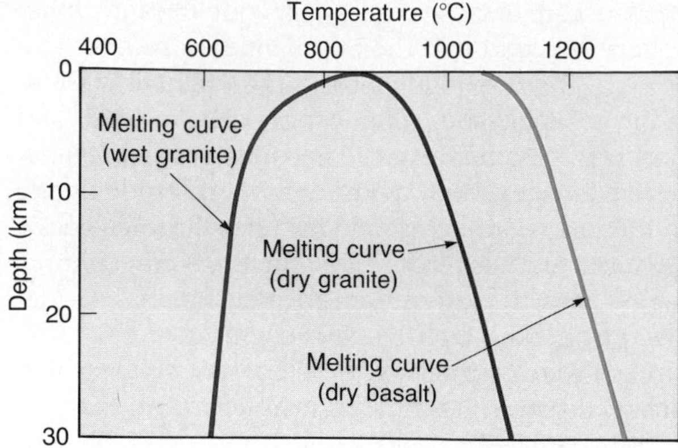

FIGURE 4.28
Idealized melting temperature curves. These curves portray the minimum temperatures required to melt rock within the earth's crust. Notice that dry granite and dry basalt melt at higher temperatures with increasing depth. By contrast, the melting temperature of wet granite actually decreases as the confining pressure increases.

rock. Should melting continue, the composition of the melt will steadily approach the overall composition of the rock from which it is derived. Most often, however, melting is not complete. This process, known as **partial melting** produces most, if not all, magma.

An important consequence of partial melting is that it generates a melt with a chemical composition that is different from the original rock. In particular, partial melting generates a melt that: (1) is enriched in the elements found in the low-melting-temperature silicate minerals; and (2) is higher in silica than the original material. Recall that ultramafic rocks contain mostly high-melting-temperature minerals that are comparatively low in silica, whereas granitic rocks are composed primarily of low-melting-point silicates that are enriched in silica. Consequently, magmas generated by partial melting are nearer the granitic end of the compositional spectrum than the materials from which they formed. This important relationship can best be illustrated by referring back to the diagram of Bowen's reaction series in Figure 3.8. Notice that this sequence is arranged in order from high-melting temperature minerals (ultramafic composition) on top to low-melting-temperature components (granitic composition) on the bottom. According to the relationship just described, partial melting of ultramafic rocks like peridotite will produce basaltic magmas. Further, partial melting of basalt will yield an andesitic magma, and so forth down the sequence. As we shall see, this concept will help us to understand the global distribution of various types of volcanic activity.

Most basaltic magmas are believed to originate from the partial melting of the rock peridotite, the major constituent of the upper mantle. Laboratory studies confirm that partial melting of this dry, silica-poor rock produces magma having a basaltic composition. Since mantle rocks exist in environments that are characterized by high temperatures and pressures, melting often results from a reduction in confining pressure. This can occur, for example, where mantle rock ascends as part of a slow-moving convection cell.

Due to the fact that basaltic magmas form many kilometers below the surface, we might expect that most of this material would cool and crystallize before reaching the surface. However, as dry basaltic magma flows upward, the confining pressure steadily diminishes and further reduces the melting temperature. Basaltic magmas appear to ascend rapidly enough so that as they enter cooler environments the heat loss is offset by a drop in the melting point. Consequently, large outpourings of basaltic magmas are common on the earth's surface.

Conversely, granitic magmas are thought to be generated by partial melting of water-rich rocks that were subjected to increased temperature. As a wet granitic melt rises, the confining pressure decreases, which in turn reduces the effect of water on lowering the melting temperature. Further, granitic melts are high in silica and thus more viscous than basaltic melts. Thus, in contrast to basaltic magmas that produce vast outpourings of lava, most granitic magmas lose their mobility before reaching the surface and therefore tend to produce large intrusive features such as batholiths. On those occasions when silica-rich magmas reach the surface, explosive pyroclastic flows, such as those that produced the Yellowstone Plateau, are the rule.

Even though most magma is thought to be generated by partial melting, once formed the composition of a magma body can change dramatically with time. For example, as a magma body migrates upward, it may incorporate some of the surrounding country rock. As the country rock is assimilated, the composition of the magma is altered. Further, recall from our discussion of Bowen's reaction series that during solidification, magma often undergoes magmatic differentiation. This produces a magma quite unlike the parent material. These processes account, at least in part, for the fact that a single volcano can extrude lavas with a wide range of chemical compositions.

Distribution of Igneous Activity

For many years geologists have realized that the global distribution of igneous activity is not random, but rather exhibits a very definite pattern. In particular, volcanoes that extrude mainly andesitic to granitic lavas are confined largely to continental margins or volcanic island chains located adjacent to deep-ocean trenches. By contrast, most volcanoes located within the ocean basins, such as those in Hawaii and Iceland, extrude lavas that are primarily of basaltic composition. Moreover, basaltic rocks are common in both oceanic and continental settings, whereas granitic rocks are rarely observed in oceanic regions. This pattern puzzled geologists until the development of the plate tectonics theory which greatly clarified the picture.

Most of the more than 600 active volcanoes that have been identified are located in the vicinity of convergent plate margins (Figure 4.29). Further, extensive volcanic activity occurs out of view along spreading centers of the oceanic ridge system. In this section we will examine three zones of volcanic activity and relate them to global tectonic activity. These active areas are found along the oceanic ridges, adjacent to ocean trenches, and within the plates themselves (Figure 4.30).

Spreading Center Volcanism As stated earlier, perhaps the greatest volume of volcanic rock is produced along the oceanic ridge system where seafloor spreading is active (Figure 4.30). As the rigid lithosphere pulls apart, the pressure on the underlying rocks is lessened. This reduced pressure, in turn, lowers the melting temperature of the mantle rocks. Partial melting of these rocks (primarily peridotite) generates large quantities of basaltic magma that move upward to fill the newly formed cracks.

Some of the molten basalt reaches the ocean floor, where it produces extensive lava flows or occasionally grows into a volcanic pile. Sometimes this activity produces a volcanic cone that rises above sea level as the island of Surtsey did in 1963 (Figure 4.31). Numerous submerged volcanic cones also dot the flanks of the ridge system and the adjacent deep-ocean floor. Many of these formed along the ridge crests and were moved away as new oceanic crust was created by the seemingly unending process of sea-floor spreading.

Subduction Zone Volcanism Recall that oceanic trenches are sites where slabs of water-rich oceanic crust are bent and descend into the mantle (Figure 4.30). Considerable evidence suggests that magma is generated within the descending slab, as well as in

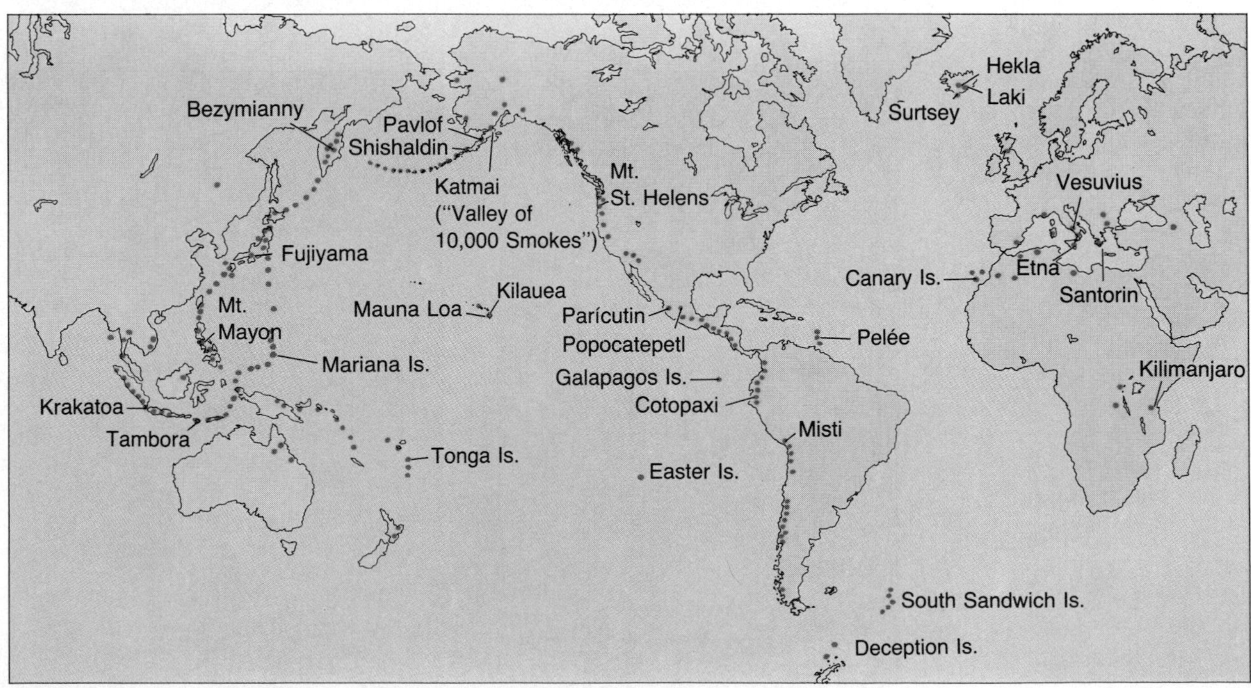

FIGURE 4.29
Locations of some of the more recently formed volcanoes.

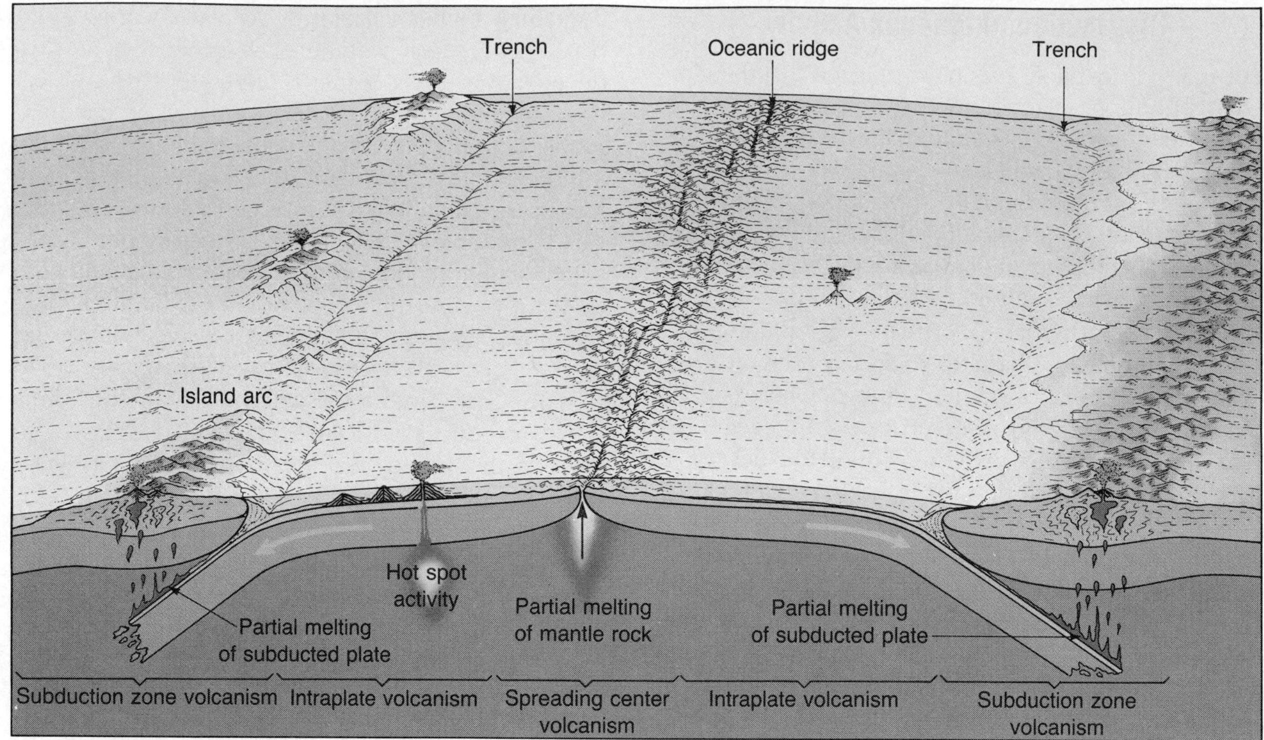

FIGURE 4.30
Three zones of volcanism. Two of these zones are plate boundaries, and the third includes areas within the plates themselves.

FIGURE 4.31
Surtsey emerged from the ocean just south of Iceland in 1963. (Courtesy of Sólarfilma)

the wedge-shaped area of the mantle overlying it. By the time the subducting plate reaches a depth of 100 kilometers, it has been heated enough to drive off water and other volatile components. The upward migration of water into the wedge of peridotite in the mantle above lowers the melting temperature of the rock sufficiently to promote partial melting. Ultimately, the descending slab is heated sufficiently that it too begins to melt. The processes just described are thought to generate basaltic as well as andesitic magmas.

After a sufficient quantity of magma has accumulated, the melt slowly migrates upward because it is less dense than the surrounding rock. When subduction volcanism occurs in the ocean, a chain of volcanoes called an *island arc* is produced (Figure 4.30). Examples are numerous in the Pacific and include the Aleutian Islands, the Tonga Islands, and the Mariana Islands.

When subduction occurs beneath continental crust, the magma generated is often altered before it solidifies. In particular, magmatic differentiation and the assimilation of crustal fragments into the ascending magma body can lead to a melt exhibiting an andesitic to granitic composition. The volcanoes of the Andes Mountains, from which andesite obtains its name, are examples of this mechanism at work.

Many subduction volcanoes border the Pacific Basin. Because of this pattern, the region has come to be called the *Ring of Fire.* Here volcanism is associated with subduction and partial melting of the Pacific sea floor. As oceanic plates sink, they carry sediments and oceanic crust containing abundant water downward. Because water reduces the melting point of rock, it aids the melting process. Further, the presence of water contributes to the high gas content and explosive nature of volcanoes that make up the Ring of Fire. The volcanoes of the Cascade Range in the

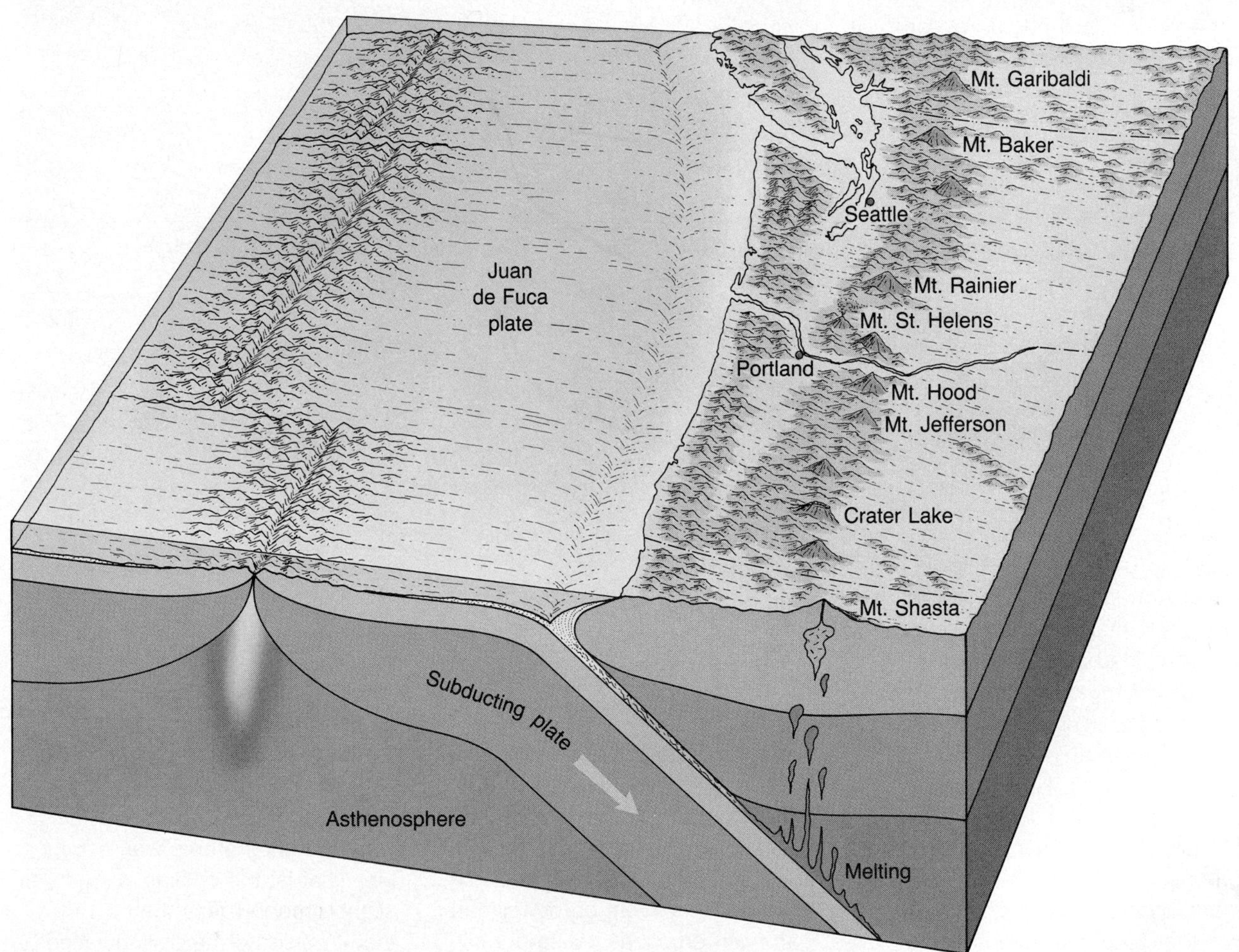

FIGURE 4.32
Locations of several of the larger composite cones that comprise the Cascade Range.

BOX 4.3

Volcanic Activity and Climate

The idea that explosive volcanic eruptions may cause changes in the earth's climate was first proposed many years ago and is still regarded as a plausible explanation for some aspects of climatic variability. Explosive eruptions emit huge quantities of gases and fine-grained debris into the atmosphere (Figure 4.B). The greatest eruptions are sufficiently powerful to inject material high into the atmosphere, where it spreads around the globe and remains for many months or even years. The basic premise is that this suspended volcanic material will filter out a portion of the incoming solar radiation, which, in turn, will lower temperatures in the troposphere. Two hundred years ago, Benjamin Franklin used this same idea to argue that material from the eruption of a large Icelandic volcano could have reflected sunlight back to space and therefore could have been responsible for unusually cold weather during the winter of 1783–1784.

Perhaps the most notable cool period linked to a volcanic event is the "year without a summer" that followed the 1815 eruption of Tambora in Indonesia (see Box 4.1). Similar, although apparently less dramatic, effects were associated with other explosive volcanoes, including Krakatoa in 1883 and Mount Agung in 1963. In recent years, two major volcanic events have provided considerable data and insight regarding the impact of volcanoes on global temperatures. The eruptions of Washington State's Mount St. Helens in 1980 and the Mexican volcano El Chichón in 1982 gave scientists an opportunity to study the atmospheric effects of volcanic eruptions with the aid of more sophisticated technology than was available in the past. Satellite images and remote sensing instruments allowed scientists to monitor closely the effects of the clouds of gases and ash that these volcanoes emitted.

FIGURE 4.B
When Mount St. Helens erupted on May 18, 1980, huge quantities of volcanic ash were blown into the atmosphere. (Photo by Austin Post, U.S. Geological Survey)

When Mount St. Helens erupted, there was almost immediate speculation about the possible effects of this event on our climate. Can such eruptions cause our climate to change? There is no doubt that the large quantity of volcanic ash emitted by the explosive eruption had significant local and regional effects for a short period. Still, studies indicated that any longer-term lowering of hemispheric temperatures was negligible. That is, the cooling was so slight, probably less than 0.1°C, that it could not be distinguished from other natural temperature fluctuations.

Two years of monitoring and studies following the El Chichón eruption indicated that its cooling effect on global mean temperatures was on the order of 0.3°C to 0.5°C. Because the eruption of El Chichón was less explosive than the Mount St. Helens blast, why did it have a greater impact on global temperatures? The reason seems to be that the material emitted by Mount St. Helens was largely fine ash that settled out in a relatively short time. El Chichón, on the other hand, emitted far greater quantities of sulfur-rich gases (an estimated 40 times more) than did Mount St. Helens. These gases combined with water vapor in the stratosphere to produce a dense cloud of tiny sulfuric acid droplets.* Such clouds take several years to settle out completely and are capable of decreasing the troposphere's mean temperature because the droplets both absorb solar radiation and scatter it back to space. It now appears that volcanic clouds that remain in the stratosphere for a year or more are composed largely of sulfuric acid droplets and not of dust as was once thought. Thus, the volume of fine debris emitted during an explosive event is not the best criterion for predicting the global atmospheric effects of an eruption.

It may be true that the impact on global temperatures of an eruption like that of El Chichón is relatively minor, but many scientists agree that the cooling produced could alter the general pattern of atmospheric circulation for a limited period. Such a change, in turn, could have an effect on the weather in some regions. Predicting or even identifying specific regional effects still presents a considerable challenge to scientists, however.

*The stratosphere is the atmospheric layer that extends from about 12 to 50 kilometers above the earth's surface. Beneath the stratosphere is the lowermost layer of the atmosphere known as the troposphere.

northwestern United States, including Mount St. Helens, Mount Rainier, and Mount Shasta, are all of this type (Figures 4.32 and 4.33).

Intraplate Volcanism The processes that actually trigger volcanic activity within a rigid plate are difficult to establish. Activity such as in the Yellowstone region and other nearby areas produced rhyolitic pumice and ash flows, while extensive basaltic flows cover vast portions of our Northwest. Yet these rocks of greatly varying compositions actually overlie one another in several locations.

Since basaltic extrusions occur on the continents as well as within the ocean basins, the partial melting of mantle rocks is the most probable source for this activity. Recall that earth tremors indicate that the island of Hawaii does in fact tap the mantle. One proposal suggests that the source of some intraplate basaltic magma comes from rising plumes of hot mantle material. These hot plumes which may extend to the core-mantle boundary, produce volcanic regions a few hundred kilometers across called **hot spots**. More than 100 hot spots have been identified

FIGURE 4.33
Mount Shasta, California, one of the largest composite cones in the Cascade Range. (Photo by John S. Shelton)

and most appear to have persisted for a few tens of millions of years. A hot spot is believed to be located beneath the island of Hawaii and one may have formerly existed beneath the Columbia Plateau.

Generally lavas and ash of granitic composition are extruded from vents located landward of the continental margins. This suggests that remelting of the continental crust may be one of the mechanisms responsible for the formation of these silica-rich magmas. But what mechanism causes large quantities of continental material to be melted? One proposal suggests that a thick segment of continental crust occasionally becomes situated over a rising plume of hot mantle material. Rather than producing vast outpourings of basaltic lava as occurs at oceanic sites such as Hawaii, the magma from the rising plume is emplaced at depth. Here the incorporation and melting of the surrounding country rock, coupled with magmatic differentiation, result in the formation of a secondary, silica-rich magma which slowly migrates upward. Continued hot spot activity supplies heat to the rising mass, thereby aiding its ascent. The activity in the Yellowstone region may have resulted from just this type of activity.

Although the plate tectonics theory has answered many of the questions which have plagued volcanologists for decades, many new questions have arisen; for example, Why does sea-floor spreading occur in some areas and not others? How do hot spots originate? These are just two of many unanswered questions that are the subject of continuing scientific research.

REVIEW QUESTIONS

1. What triggered the May 18, 1980, eruption of Mount St. Helens?
2. List three factors that determine the nature of a volcanic eruption. What role does each play?
3. Why is a volcano fed by highly viscous magma likely to be a greater threat than a volcano supplied with very fluid magma?
4. Contrast pahoehoe and aa lava.
5. List the main gases released during a volcanic eruption.
6. How do volcanic bombs differ from blocks of pyroclastic debris?
7. Compare a volcanic crater to a caldera.
8. Compare and contrast the main types of volcanoes as to size, shape, and eruptive style.
9. Name a prominent volcano for each of the three types.
10. Briefly compare the eruptions of Kilauea and Parícutin.
11. Contrast the destruction of Pompeii with the destruction of St. Pierre.
12. Describe the formation of Crater Lake. Compare it to the caldera formed during the eruption of Kilauea.
13. What is Shiprock, New Mexico, and how did it form?
14. How do the eruptions that created the Columbia Plateau differ from eruptions that create volcanic peaks?
15. Where are fissure eruptions most common?
16. Extensive pyroclastic flow deposits are most often associated with which volcanic structures?
17. What is partial melting? What three factors influence the melting temperatures of rocks?
18. In what two ways are rocks melted in nature?
19. Since basaltic magma forms at great depths, why does it often reach the surface rather than crystallize below the surface?
20. Spreading center volcanism is associated with which rock type? What causes rocks to melt in regions of spreading center volcanism?
21. What is the Ring of Fire?
22. Are volcanoes in the Ring of Fire generally described as quiescent or violent? Name a volcano that would support your answer.
23. Describe the situation that generates magma in subduction zone volcanism.
24. The Hawaiian Islands and Yellowstone are thought to be associated with which of the three zones of volcanism?
25. Briefly describe the mechanism by which explosive volcanic eruptions are thought to influence the earth's climate.

KEY TERMS

aa lava (p. 87)
caldera (p. 90)
cinder cone (p. 93)
composite cone (p. 95)
crater (p. 89)
fissure eruption (p. 101)
flood basalt (p. 101)
fumarole (p. 90)
geothermal gradient (p. 103)
hot spot (p. 109)
lava dome (p. 98)
nuée ardente (p. 95)
pahoehoe lava (p. 86)
parasitic cone (p. 90)
partial melting (p. 104)
pillow lava (p. 88)
pipe (p. 100)
pyroclastic flow (p. 102)
pyroclastic material (p. 89)
shield volcano (p. 91)
stratovolcano (p. 95)
vent (p. 89)
viscosity (p. 84)
volcanic neck (p. 100)
volcano (p. 89)

5
Weathering and Soil

Opposite: Bryce Canyon National Park, Utah. Weathering accentuates differences in rocks to produce some of our most spectacular scenery. (Photo by Tom Till) (Top photo by E. J. Tarbuck)

To the casual observer the face of the earth may appear to be without change, unaffected by time. For that matter, less than 200 years ago most people believed that mountains, lakes, and deserts were permanent features of an earth that was thought to be no more than a few thousand years old. Today, however, we know that mountains eventually succumb to weathering and erosion and are washed into the sea, lakes fill with sediment and vegetation or are drained by streams, and deserts come and go as climates change.

The earth is indeed a dynamic body. Volcanic and tectonic activities are elevating parts of the earth's surface, while opposing processes are continually removing materials from higher elevations and moving them to lower elevations. The latter processes include:

1. **Weathering**—the disintegration and decomposition of rock at or near the earth's surface.
2. **Mass wasting**—the transfer of rock material downslope under the influence of gravity.
3. **Erosion**—the incorporation and transportation of material by mobile agents such as water, wind, or ice.

The primary focus of this chapter is on rock weathering and the products generated by this activity. However, weathering cannot be easily separated from the other two processes because as weathering breaks rocks apart, it encourages the movement of rock debris by erosion and mass wasting (Figure 5.1). On the other hand, the transport of material by erosion and mass wasting furthers the disintegration and decomposition of rock.

FIGURE 5.1
Delicate Arch, Arches National Park, Utah. When weathering accentuates differences in rocks, spectacular landforms are sometimes created. As the rock gradually disintegrates and decomposes, mass wasting and erosion remove the products of weathering. (Photo by Stephen Trimble)

WEATHERING

All materials are susceptible to weathering. Consider, for example, the fabricated product concrete, which closely resembles a sedimentary rock called conglomerate. A newly poured concrete sidewalk has a smooth, fresh, unweathered look. However, not many years later the same sidewalk will appear chipped, cracked, and rough, with pebbles exposed at the surface. If a tree is nearby, its roots may heave and buckle the concrete as well. The same natural processes which eventually destroy a concrete sidewalk also act to disintegrate rock.

Weathering occurs when rock is mechanically fragmented (disintegrated) and chemically altered (decomposed). Mechanical weathering is accomplished by physical forces which break rock into smaller and smaller pieces without changing the rock's mineral composition. Chemical weathering, on the other hand, involves a chemical transformation of the rock into one or more new compounds. A simple example of these two concepts can be made using a piece of paper. Disintegration of the paper is accomplished by tearing it into smaller and smaller pieces, whereas decomposition occurs when the paper is set afire and burned.

Why does rock weather? Simply, weathering is the response of earth materials to a changing environment. For instance, after millions of years of uplift and erosion, the rocks overlying a large intrusive igneous body may be removed, exposing it at the surface. This mass of crystalline rock, which formed in a high-temperature, high-pressure environment perhaps several kilometers below ground, is now subjected to a very different and comparatively hostile surface environment. In response, this rock mass will gradually change. This transformation of rock is what we call weathering.

In the following sections we will discuss the various modes of mechanical and chemical weathering. Although we will consider these two processes separately, keep in mind that they usually work simultaneously in nature.

Mechanical Weathering

When a rock undergoes **mechanical weathering** it is broken into smaller and smaller pieces, each retaining the characteristics of the original material. The end result is many small pieces from a single large one. Figure 5.2 shows that breaking a rock into smaller pieces increases the surface area available for chemical attack. An analogous situation occurs when sugar is added to a liquid. In this situation, a cube of sugar will dissolve much slower than an

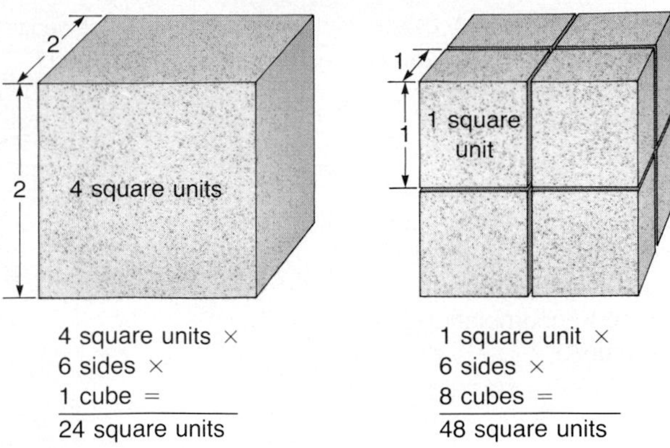

FIGURE 5.2
Mechanical weathering increases the surface area available for chemical attack.

equal volume of granules because of the vast difference in surface area. Hence, by breaking rocks into smaller pieces, mechanical weathering increases the amount of surface area available for chemical weathering.

In nature four important physical processes lead to the fragmentation of rock: frost wedging, expansion resulting from unloading, thermal expansion, and organic activity. In addition, although the work of erosional agents such as wind, glacial ice, and running water is usually considered separately from mechanical weathering, it is nevertheless important. As these mobile agents move rock debris, they relentlessly disintegrate the materials they carry.

Frost Wedging Alternate freezing and thawing is one of the most important processes of mechanical weathering. Water has the unique property of expanding about 9 percent as it freezes. This increase in volume occurs because as water solidifies, the water molecules arrange themselves into a very open crystalline structure. As a result, when water freezes it expands and exerts a tremendous outward force. This can be verified by filling a container with water and freezing it. When the water turns to ice, it shatters the container.

In nature, water works its way into cracks or voids in rock and, upon freezing, expands and breaks the rock into angular fragments. This process is appropriately called **frost wedging** (Figure 5.3). Frost wedging is most pronounced in mountainous regions in the middle latitudes where a daily freeze-thaw cycle often exists. Here, sections of rock are wedged loose and may tumble into large piles called **talus slopes** that often form at the base of steep rock outcrops (Figure 5.4).

Frost wedging also causes great destruction to the highways in the northern United States, particularly

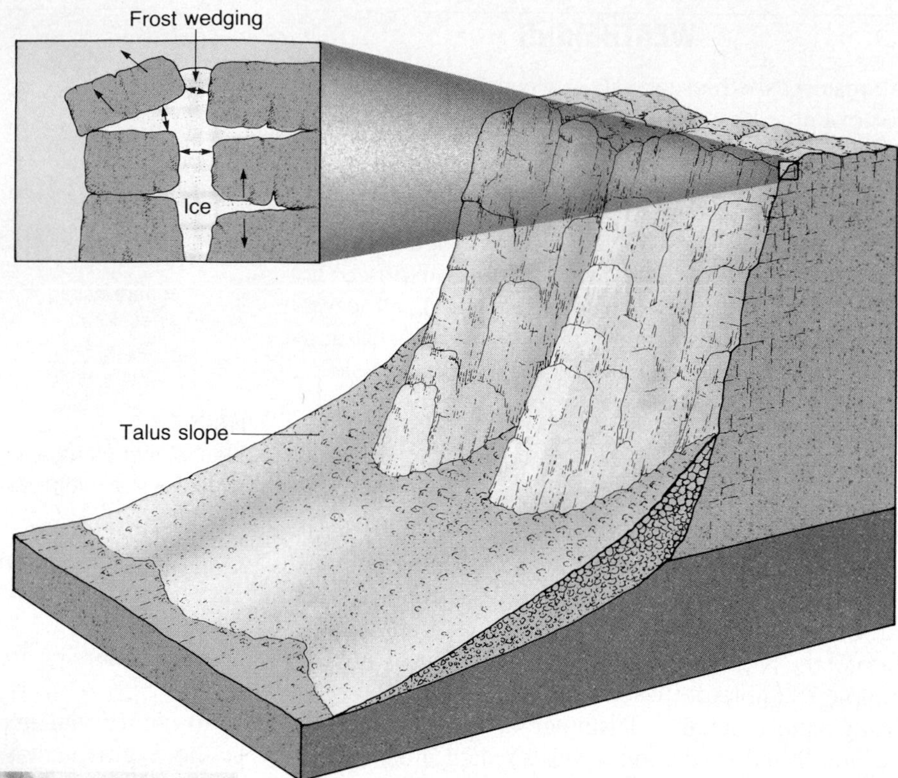

FIGURE 5.3
Frost wedging. As water freezes it expands, exerting a force great enough to break rock. When frost wedging occurs in a setting such as this, the broken rock fragments fall to the base of the cliff and create a cone-shaped accumulation known as talus.

FIGURE 5.4
After frost wedging loosens rock fragments on a steep rock exposure, the angular pieces fall to the base of the cliff and produce a talus slope. (Photo by E. J. Tarbuck)

in the early spring when the freeze-thaw cycle is well established. Roadways acquire numerous potholes and are occasionally heaved and buckled by this destructive force.

Unloading When large igneous bodies, particularly those composed of granite, are exposed by erosion, concentric slabs begin to break loose. The process generating these onionlike layers is called **sheeting** and is thought to occur, at least in part, because of the great reduction in pressure when the overlying rock is stripped away. Accompanying the unloading, the outer layers expand more than the rock below, and thus separate from the rock body (Figure 5.5). The fractures separating the individual slabs are usually more closely spaced near the earth's surface and therefore layers that result are generally less than a meter thick. Further, the fractures typically develop parallel to the surface topography and give the exhumed igneous body a domed shape. Continued weathering eventually causes the slabs produced by sheeting to separate and break off from these large structures, which are known as **exfoliation domes**. Excellent examples of exfoliation domes are Stone Mountain, Georgia, and Half Dome and Liberty Cap in Yosemite National Park (Figure 5.6).

Mine shafts provide us with a view of how rocks behave once the confining pressure is removed. Large rock slabs have been known to explode off the walls of newly cut mine shafts because of the re-

FIGURE 5.5
Sheeting is caused by the expansion of crystalline rock as erosion removes the overlying material. Fractures that roughly parallel the surface topography are common in large, intrusive granitic masses. (Photo by E. J. Tarbuck)

duced pressure. Evidence of this type, plus the fact that fracturing occurs parallel to the floor of a rock quarry when large blocks are removed, strongly supports the process of unloading as the cause of sheeting.

Although many fractures are created by expansion, others are produced by contraction during the crystallization of magma, and still others by tectonic forces during mountain building. Fractures produced by these activities generally form a definite pattern and are called **joints** (Figure 5.7). Joints are important rock structures which allow water to penetrate to depth and start the process of weathering long before the rock is exposed.

FIGURE 5.6
Summit of Half Dome, an exfoliation dome in Yosemite National Park. (Photo by Stephen Trimble)

Thermal Expansion The daily cycle of temperature change is thought to weaken rocks, particularly in hot deserts where daily variations may exceed 30°C. Heating a rock causes expansion and cooling causes contraction. Repeated swelling and shrinking of minerals with different expansion rates should logically exert some stress on the rock's outer shell.

Although this process was once thought to be of major importance in the disintegration of rock, laboratory experiments have not substantiated this. In one test, unweathered rocks were heated to temperatures much higher than those normally experienced on the earth's surface and then cooled. This procedure was repeated many times to simulate hundreds of years of weathering, but the rocks showed little apparent change.

Nevertheless, in desert areas pebbles do exhibit unmistakable evidence of shattering from what appears to be temperature changes (Figure 5.8). A proposed solution to this dilemma suggests that rocks must first be weakened by chemical weathering before they can be broken down by thermal activity. Further, this process may be aided by the rapid cooling of a desert rainstorm. Additional data are needed before a definite answer can be given as to the impact of temperature variation on rock disintegration.

Organic Activity Weathering is also accomplished by the activities of organisms, including plants, burrowing animals, and human beings. Plant roots in search of minerals and water grow into fractures, and as the roots grow, they wedge the rock apart (Figure 5.9). Burrowing animals further break down rock by moving fresh material to the surface, where physical and chemical processes can more effectively attack it. Further, decayed organisms produce acids which contribute to chemical weathering. Where rock has been blasted in search of minerals or for road construction, the impact of humans is quite noticeable, but on a worldwide scale, humans probably rank behind burrowing animals in earth-moving accomplishments.

Although usually considered separately from mechanical weathering, the activities of the erosional agents—wind, water, and glaciers—are nonetheless important. For as these mobile agents move rock debris, they relentlessly disintegrate the earth materials they carry.

FIGURE 5.8
These stones were once rounded stream gravels; however, long exposure in a hot desert climate disintegrated them. (Photo by C. B. Hunt, U.S. Geological Survey)

FIGURE 5.7
Aerial view of nearly parallel joints near Moab, Utah. (Photo by Michael Collier)

Chemical Weathering

Chemical weathering involves the complex processes that alter the internal structures of minerals by removing and/or adding elements. During this transformation, the original rock decomposes into substances that are stable in the surface environment. Consequently, the products of chemical weathering will remain essentially unchanged as long as they remain in an environment similar to the one in which they formed.

Water is by far the most important agent of chemical weathering. Although pure water is nonreactive, a small amount of dissolved material is generally all that is needed to activate it. The major processes by which water decomposes rock are solution, oxidation, and hydrolysis.

Solution Perhaps the easiest type of decomposition to envision is the process of **solution**. Just as sugar dissolves in water, so too do certain minerals. One of the most water-soluble minerals is halite (common salt), which as you may recall, is composed of sodium and chloride ions. The reason halite readily dissolves in water has to do with the fact that, although this compound maintains overall electrical neutrality, the individual ions retain their respective charges. Moreover, the surrounding water molecules are polar; that is, the oxygen end of the molecule has a small residual negative charge whereas the end with hydrogen has a small positive charge. As the water molecules collide with a halite crystal, their negative ends contact and disrupt the oppositely charged sodium ions (Figure 5.10). The attractive

FIGURE 5.9
Root wedging widens fractures in rock and aids the process of mechanical weathering. (Photo by E. J. Tarbuck)

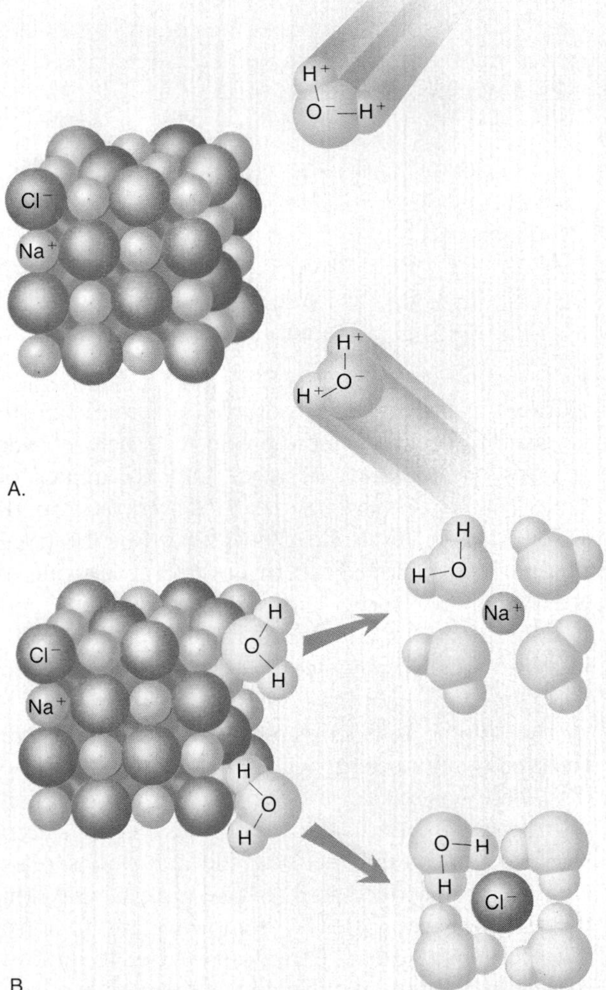

FIGURE 5.10
Illustration of halite dissolving in water. **A.** Sodium and chloride ions are attacked by the polar water molecules. **B.** Once removed, these ions are surrounded and held by a number of water molecules as shown.

force of the water pulls the sodium ions from the crystalline lattice. The chloride ions are similarly removed layer after layer by the positive end of the water molecules.

Although most minerals are, for all practical purposes, insoluble in pure water, the presence of even a small amount of acid dramatically increases the corrosive force of water. (An acidic solution contains the reactive hydrogen ion, H^+.) In nature, acids are produced by a number of processes. For instance, carbonic acid is generated when carbon dioxide is dissolved in droplets of rain. Various organic acids are also released as organisms decay, and sulfuric acid is produced by the weathering of pyrite and other sulfide minerals. Regardless of the source of the acid, this highly reactive substance readily decomposes most rocks and produces certain products that are water soluble. For example, the mineral calcite, $CaCO_3$, which composes the common building stones marble and limestone, is easily attacked by even a weakly acidic solution:

$$\underset{\text{calcium carbonate (insoluble)}}{CaCO_3} + \underset{\text{aqueous acid}}{2[H^+(H_2O)]} \longrightarrow \underset{\text{calcium ion (soluble)}}{Ca^{2+}} + \underset{\text{carbon dioxide}}{CO_2\uparrow} + \underset{\text{water}}{3H_2O}$$

During this process, the insoluble calcium carbonate is transformed into soluble products. In nature, over periods of thousands of years, large quantities of limestone are dissolved and carried away by groundwater. This activity is clearly evidenced by the large number of subsurface caverns found in every one of the contiguous forty-eight states. Monuments and buildings made of limestone or marble are also subjected to the corrosive work of acids, particularly in industrial areas that have smoggy, polluted air.

The soluble ions from reactions of this type are retained in our underground water supply. It is these dissolved ions that are responsible for the so-called "hard water" found in many locales. Simply, hard water is undesirable because the active ions react with soap to produce an insoluble material that renders soap nearly useless in removing dirt. To solve this problem a water softener can be used to remove these ions, generally by replacing them with others that do not chemically react with soap.

Oxidation The process of rusting occurs when oxygen combines with iron to form iron oxide as follows:

$$\underset{\text{iron}}{4Fe} + \underset{\text{oxygen}}{3O_2} \longrightarrow \underset{\text{iron oxide (hematite)}}{2Fe_2O_3}$$

This type of chemical reaction, called **oxidation**,* occurs when electrons are lost from one element during the reaction. In this case, we say that iron was oxidized because it lost electrons to oxygen. Although the oxidation of iron progresses very slowly in a dry environment, the addition of water greatly speeds the reaction.

Oxidation is important in decomposing such ferromagnesian minerals as olivine, pyroxene, and hornblende. Oxygen readily combines with the iron in these minerals to form the reddish-brown iron oxide called *hematite* (Fe_2O_3) or in other cases a yellowish-colored rust called *limonite* [FeO(OH)]. These products are responsible for the rusty color on the surfaces of dark igneous rocks, such as basalt, as they begin to weather. However, oxidation can occur only after iron is freed from the silicate structure by another process called hydrolysis.

Another important oxidation reaction occurs when sulfide minerals such as pyrite decompose. Sulfide minerals are major constituents in many metallic ores, and pyrite is frequently associated with coal deposits as well. In a moist environment, chemical weathering of pyrite (FeS_2) yields sulfuric acid (H_2SO_4) and iron oxide [FeO(OH)]. In many mining locales this weathering process creates a serious environmental hazard; particularly in humid areas where abundant rainfall infiltrates spoil banks (waste material left after coal or other minerals are removed). This so-called *mine acid* eventually makes its way to streams, killing fish and making the water unfit for human consumption.

Hydrolysis The most common mineral group, the silicates, is decomposed primarily by the processes of **hydrolysis**, which in the broadest sense is the reaction of any substance with water. Ideally, the hydrolysis of a mineral could take place in pure water since some of the water molecules dissociate to form the very reactive hydrogen (H^+) and hydroxyl (OH^-) ions. It is the hydrogen ion which attacks and replaces other positive ions found in the crystal lattice. With the introduction of hydrogen ions into the crystalline structure, the original orderly arrangement of atoms is destroyed and the mineral decomposes.

*The reader should note that *oxidation* is a term referring to any chemical reaction in which a compound or radical loses electrons. The element oxygen is not necessarily present.

In nature, water usually contains other substances that contribute additional hydrogen ions, thereby greatly accelerating hydrolysis. The most common of these substances is carbon dioxide, CO_2, which dissolves in water to form carbonic acid, H_2CO_3. Rain dissolves some carbon dioxide in the atmosphere, and additional amounts, released by decaying organic matter, are acquired as the water percolates through the soil. In water, carbonic acid ionizes to form hydrogen ions (H^+) and bicarbonate ions (HCO_3^-). To illustrate how a rock undergoes hydrolysis in the presence of carbonic acid, we will examine the chemical weathering of granite, the most abundant continental rock. Recall that granite consists mainly of quartz and potassium feldspar. The weathering of the potassium feldspar component of granite is as follows:

$$\underset{\text{potassium feldspar}}{2KAlSi_3O_8} + \underset{\text{carbonic acid}}{2(H^+ + HCO_3^-)} + \underset{\text{water}}{H_2O} \longrightarrow$$

$$\underset{\substack{\text{kaolinite}\\\text{(residual clay)}}}{Al_2Si_2O_5(OH)_4} + \underbrace{\underset{\text{potassium ion}}{2K^+} + \underset{\text{bicarbonate ion}}{2HCO_3^-} + \underset{\text{silica}}{4SiO_2}}_{\text{in solution}}$$

In this reaction, the hydrogen ions (H^+) attack and replace potassium ions (K^+) in the feldspar structure, thereby disrupting the crystalline network. Once removed, the potassium is available as a nutrient for plants or becomes the soluble salt potassium bicarbonate ($KHCO_3$), which may be incorporated into other minerals or carried to the ocean.

The most abundant product of the chemical breakdown of potassium feldspar is the clay mineral kaolinite. Clay minerals are the end products of weathering and are very stable under surface conditions. Consequently, clay minerals make up a high percentage of the inorganic material in soils. Moreover, the most abundant sedimentary rock, shale, usually contains a high proportion of clay minerals. In addition to the formation of clay minerals during this reaction, some silica is removed from the feldspar structure and carried away by groundwater. This dissolved silica will eventually precipitate, producing nodules of chert or flint, or fill in the pore spaces between grains of sediment, or be carried to the ocean, where microscopic animals will remove it to build hard silica shells.

To summarize, the weathering of potassium feldspar generates a residual clay mineral, a soluble salt (potassium bicarbonate), and some silica which enters into solution.

TABLE 5.1
Products of weathering.

Mineral	Residual Products	Material in Solution
Quartz	Quartz grains	Silica
Feldspars	Clay minerals	Silica K^+, Na^+, Ca^{2+}
Amphibole (hornblende)	Clay minerals Limonite Hematite	Silica Ca^{2+}, Mg^{2+}
Olivine	Limonite Hematite	Silica Mg^{2+}

Quartz, the other main component of granite, is very resistant to chemical weathering; hence it remains substantially unaltered when attacked by weakly acidic solutions. As a result, when granite weathers, the feldspar crystals dull and slowly turn to clay, releasing the once-interlocked quartz grains, which still retain their fresh, glassy appearance. Although some of the quartz remains in the soil, much is transported to the sea, where it becomes the main constituent of sandy beaches and in time is often converted to the sedimentary rock sandstone.

Table 5.1 lists the weathered products of some of the most common silicate minerals. Remember that silicate minerals make up most of the earth's crust and that these minerals are essentially composed of only eight elements. When chemically weathered, the silicate minerals yield sodium, calcium, potassium, and magnesium ions that form soluble products which may be removed by groundwater. The element iron combines with oxygen, producing relatively insoluble iron oxides, most notably hematite and limonite, which give soil a reddish-brown or yellowish color. Under most conditions the three remaining elements, aluminum, silicon, and oxygen, join with water to produce residual clay minerals. However, even the highly insoluble clay minerals are very slowly removed by subsurface water.

Alterations Caused by Chemical Weathering As noted earlier the most significant result of chemical weathering is the decomposition of unstable minerals and the generation or retention of those materials which are stable at the earth's surface. This accounts for the predominance of certain minerals in the surface material we call soil.

In addition to altering the internal structure of minerals, chemical weathering causes physical changes as well. For instance, when angular rock fragments are attacked by water flowing through

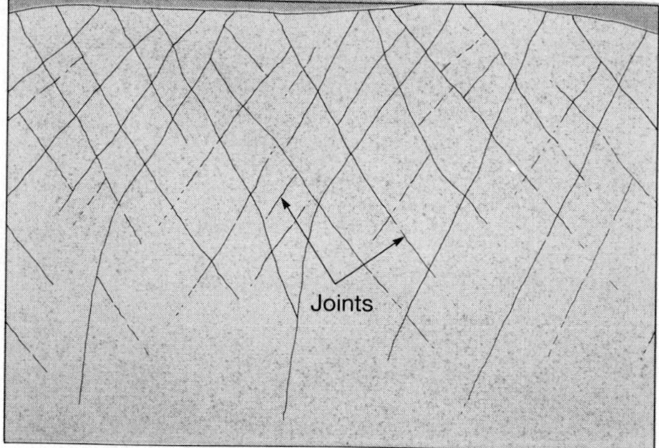

A.

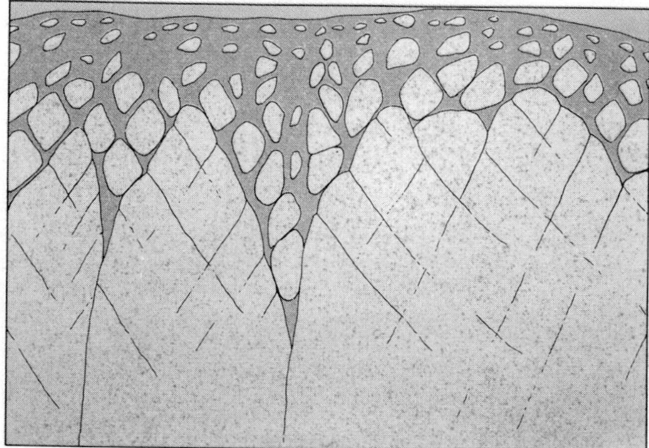

B.

C.

FIGURE 5.11
Spheroidal weathering of extensively jointed rock. Water moving through the joints begins to enlarge them. Since the rocks are attacked more on the corners and edges, they take on a spherical shape.

joints, the fragments tend to take on a spherical shape. The gradual rounding of the corners and edges of angular blocks is illustrated in Figure 5.11. The corners are attacked most readily because of the greater surface area for their volume as compared to the edges and faces. This process, called **spheroidal weathering**, gives the weathered rock a more rounded or spherical shape (Figure 5.12).

Commonly during the formation of spheroidal boulders, successive shells separate from the rock's main body (Figure 5.13). Eventually the outer shells spall off, allowing the chemical weathering activity to penetrate deeper into the boulder. This spherical scaling results because, as the minerals in the rock weather to clay, they increase in size through the addition of water to their structure. This increased bulk exerts an outward force that causes concentric layers of rock to break loose and fall off. Hence, chemical weathering does produce forces great enough to cause mechanical weathering. This type of spheroidal weathering in which shells spall off should not be confused with the phenomenon of sheeting discussed earlier. In sheeting, the fracturing occurs as a result of unloading and the rock layers which separate from the main body are largely unaltered at the time of separation.

Rates of Weathering

Several factors influence the type and rate of rock weathering. Most important of these are rock structure, climate, and topography.

Rock Structure Rock structure encompasses all of the chemical characteristics of rocks, including mineral composition and solubility, as well as any physical features that may be present, such as fractures, bedding planes, and voids. The variations in weathering rates, attributable to the mineral constituents, can be demonstrated by comparing old headstones carved from different rock types. Headstones made of granite, which is composed of silicate minerals, are relatively resistant to chemical weathering as we can see by examining the inscriptions on the headstones shown in Figure 5.14. This is not true of the marble headstone, which shows signs of extensive chemical alteration over a relatively short period. Marble is composed of calcite (calcium carbonate), which readily dissolves even in a weakly acidic solution.

The most abundant mineral group, the silicates, weathers in the order shown in Figure 5.15. This arrangement of minerals is identical to that of Bowen's reaction series. The order in which the silicate minerals weather is essentially the same as their order of crystallization. The minerals that crystallize first form under much higher temperatures than those that crystallize last. Consequently, the early-formed minerals are not as stable at the earth's surface, where

FIGURE 5.12
Spheroidal weathering of a massive granite outcrop in Joshua Tree National Monument. (Photo by E. J. Tarbuck)

the temperature and pressure are drastically different from the environment in which they formed. By examining Figure 5.15, we see that olivine crystallizes first and is therefore the least resistant to chemical weathering, while quartz, which crystallizes last, is the most resistant.

Climate Climatic factors, particularly temperature and moisture, are of primary significance to the rate of rock weathering. These climatic elements largely determine the weathering rate and strongly influence the kind and amount of vegetation present. Regions with lush vegetation generally have a thick mantle of soil rich in decayed organic matter from which chemically active fluids such as carbonic and humic acids are derived.

The optimum environment for chemical weathering is a combination of warm temperatures and abundant moisture. In polar regions chemical weathering is ineffective because frigid temperatures keep

FIGURE 5.13
Successive shells are loosened as the weathering process continues to penetrate ever deeper into the rock. (Photo by Martin Schmidt, Jr.)

FIGURE 5.14
An examination of headstones reveals the rate of chemical weathering on diverse rock types. The granite headstone (left) was erected a few years before the marble headstone (right). The inscription date of 1892 on the marble monument is nearly illegible. (Photos by E. J. Tarbuck)

the available moisture locked up as ice, whereas in arid regions there is insufficient moisture to foster rapid chemical weathering. A classic example of how climate affects the rate of weathering was provided when Cleopatra's Needle, a granite obelisk, was moved from Egypt to New York City. After withstanding approximately 3500 years of exposure in the dry climate of Egypt, the hieroglyphics were almost completely removed from the windward side in less than 75 years in the wet and chemical-laden air of New York City (Figure 5.16).

Topography Topography greatly influences the amount of rock exposed to the forces of weathering. In addition, the topographic setting may indirectly determine the amount of precipitation as well as influence the kind and amount of vegetation present.

Angular topography with prominent rock outcrops is most common in arid regions whereas more subdued topography mantled with soil and vegetation is most characteristic in humid areas. These differences are often attributed to the predominance of chemical weathering in humid regions and mechanical weathering in arid regions. However, it is probably more correct to say that chemical weathering is most important in both environments. Although mechanical weathering may be relatively more important in arid regions than in humid places, chemical weathering is still very important in dry lands.

The sum of these factors determines the type and rate of rock weathering for a given place. There is generally enough variation, even within a relatively small area, for the rocks to exhibit some differential weathering. **Differential weathering** simply relates to the fact that rocks exposed at the earth's surface usually do not weather at the same rate. Because of variations in such factors as mineral makeup, degree of jointing, and exposure to the elements, significant differences occur. Consequently differential weathering and subsequent erosion create many unusual and often spectacular rock formations and landforms (see Box 5.1). Included are features such as the natural bridges found in Arches National Park (see Figure 5.1) and the sculptured rock pinnacles found in Bryce Canyon National Park (see chapter-opening photo).

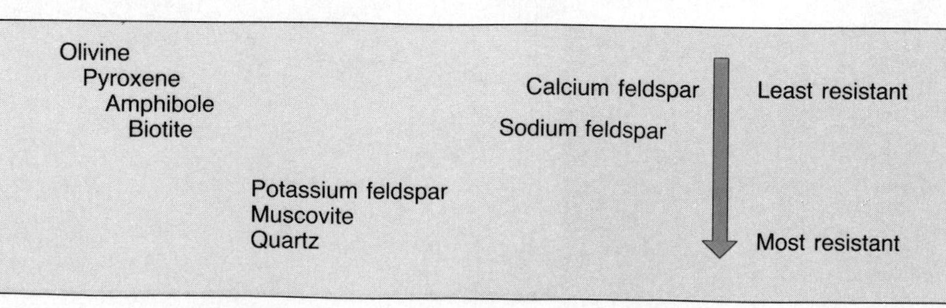

FIGURE 5.15
The stability of common silicate minerals in relation to chemical weathering. Those minerals which crystallize at high temperatures decompose most rapidly and vice versa.

A.

B.

FIGURE 5.16
Chemical weathering of Cleopatra's Needle, a granite obelisk. **A.** Before it was removed from Egypt. (Courtesy of the Metropolitan Museum of Art) **B.** After a span of 75 years in New York City's Central Park. After surviving intact for about 35 centuries in Egypt, the windward side has been almost completely defaced in less than a century. (Courtesy of New York City Parks)

SOIL

Soil has accurately been called "the bridge between life and the inanimate world." All life owes its existence to a dozen or so elements that must ultimately come from the earth's crust. Once weathering and other processes create soil, plants carry out the intermediary role of assimilating the necessary elements and making them available to animals and people.

With few exceptions, the earth's land surface is covered by **regolith**, the layer of rock and mineral fragments produced by weathering. Some would call this material soil, but soil is more than an accumulation of weathered debris. **Soil** is a combination of mineral and organic matter, water, and air—that portion of the regolith that supports the growth of plants. Although the proportions vary, the major components do not (Figure 5.17). About one-half of the total volume of a good quality surface soil is a mixture of disintegrated and decomposed rock (mineral matter) and **humus**, the decayed remains of animal and plant life (organic matter). The remaining half consists of pore spaces, where air and water circulate.

Although the mineral portion of the soil is usually much greater than the organic portion, humus is a very significant component. In addition to being an important source of plant nutrients, humus enhances the soil's ability to retain water. Since plants require air and water to live and grow, the portion of the soil consisting of pore spaces that allow for the circulation of these fluids is as vital as the solid soil constituents. Soil water is not "pure" water; instead, it is a complex solution containing many soluble nutrients.

BOX 5.1

Australia's Ayers Rock: An Example of Differential Weathering

When travelers contemplating a trip to Australia consult brochures and other tourist literature, they are bound to see a photograph or read a description of Ayers Rock. As Figure 5.A illustrates, this well-known attraction is a massive feature that rises steeply from the surrounding plain. Located southwest of Alice Springs in the dry center of the continent, the roughly circular monolith is more than 350 meters (1200 feet) high, and its base is more than 9.5 kilometers (6 miles) in circumference. Its summit is flattened, its sides furrowed. The rock type is sandstone, and the hues of red and orange change with the light of day. In addition to being a striking geological attraction, Ayers Rock is of interest because it is a sacred place for the aboriginal tribes of the region.

Ayers Rock is a spectacular example of a feature known as an inselberg. *Inselberg* is a German word meaning "island mountain" and seems appropriate because these masses clearly resemble rocky islands standing above the surface of a broad sea. Similar features are scattered throughout many other arid and semiarid regions of the world. Ayers Rock is a special type of inselberg that consists of a very resistant rock mass exhibiting a rounded or domed form. Such masses are termed *bornhardts* for the nineteenth-century German explorer Wilhelm Bornhardt who described similar features in parts of Africa.

Bornhardts form in regions where massive or resistant rock such as granite or sandstone is surrounded by rock that is more susceptible to weathering. The greater susceptibility of the adjacent rock is often the result of its being more highly jointed. Joints allow water and therefore weathering processes to penetrate to greater depths. When the adjacent, deeply weathered rock is stripped away by erosion, the far less weathered rock mass remains standing high. After a bornhardt forms, it tends to shed water. By contrast, the surrounding debris-covered plains absorb water and weather more rapidly. Therefore, once formed, a bornhardt helps to perpetuate its existence by reinforcing the processes that created it. In fact, masses such as Ayers Rock can remain a part of the landscape for tens of millions of years.

Bornhardts are more common in the lower latitudes because the weathering that is responsible for their formation proceeds more rapidly in warmer climates. In regions that are now arid or semiarid, bornhardts may reflect times when the climate was wetter than it is today.

FIGURE 5.A
Ayers Rock rises conspicuously above the dry plains of central Australia. It is a type of inselberg known as a *bornhardt.* As erosion gradually lowers the surface, the less weathered massive rock remains standing high above the more jointed and more easily weathered rock that surrounds it. (Photo courtesy of Australian Overseas Information Service)

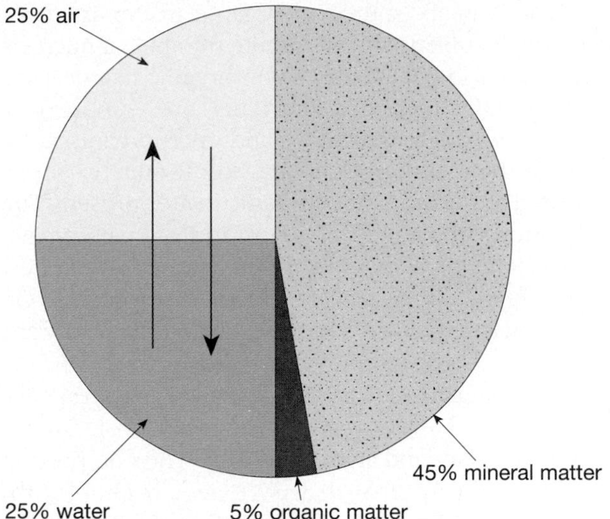

FIGURE 5.17
Composition (by volume) of a soil in good condition for plant growth. Although the percentages vary, each soil is composed of mineral and organic matter, water, and air.

Soil water not only provides the necessary moisture for the chemical reactions that sustain life, it also supplies plants with nutrients in a form they can use. The pore spaces not filled with water contain air. This air is the source of necessary oxygen and carbon dioxide for most microorganisms and plants that live in the soil.

Controls of Soil Formation

Soil is the product of the complex interplay of several factors, including parent material, time, climate, plants and animals, and slope. Although all of these factors are interdependent, it will be helpful to examine their roles separately.

Parent Material The **parent material** from which a soil has evolved may be either the underlying bedrock or a layer of unconsolidated deposits. Soils formed on bedrock are termed **residual soils**, whereas those developed on unconsolidated deposits are called **transported soils** (Figure 5.18).

The nature of the parent material influences soils in two ways. First, the type of parent material to some degree will affect the rate of weathering, and thus the

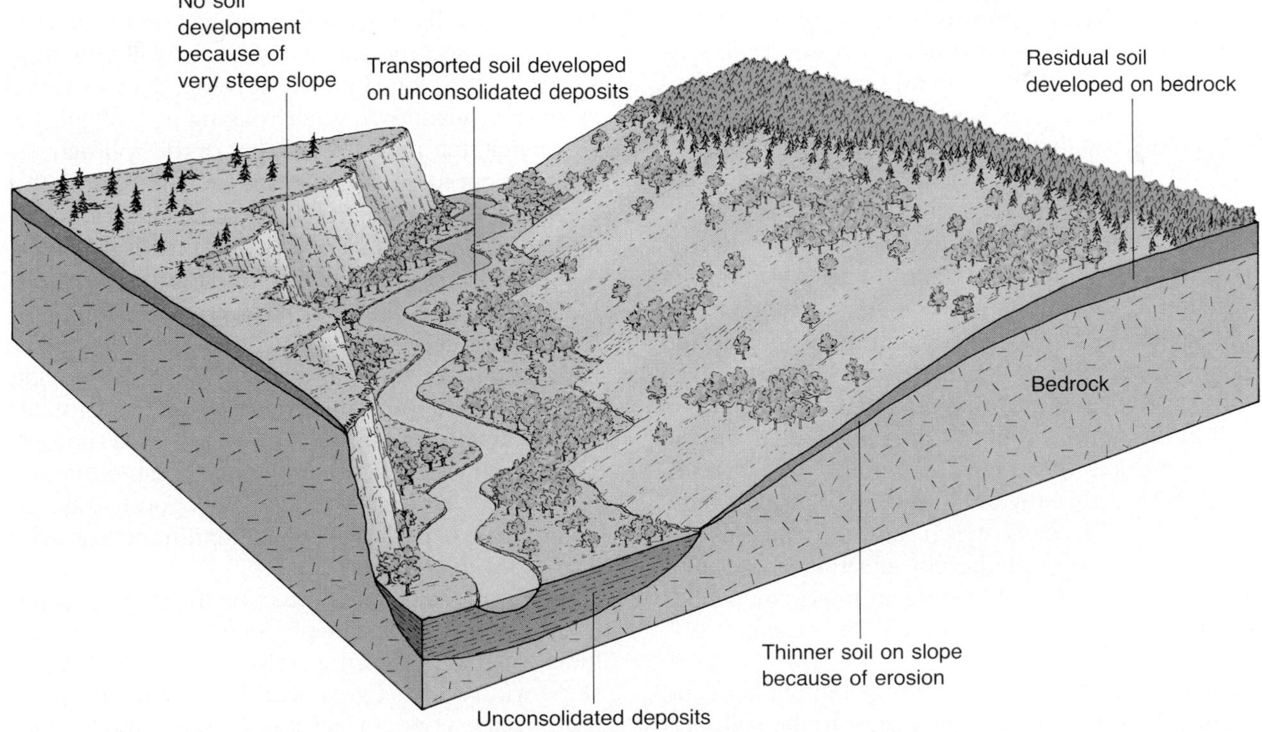

FIGURE 5.18
The parent material for residual soils is the underlying bedrock, whereas transported soils form on unconsolidated deposits. Also note that as slopes become steeper, soils become thinner.

rate of soil formation. Also, since unconsolidated deposits are already partly weathered, soil development on such material will likely progress more rapidly than when bedrock is the parent material. Second, the chemical makeup of the parent material will affect the soil's fertility. For instance, if it lacks the elements necessary for plant growth, its usefulness is obviously diminished.

At one time the parent material was believed to be the primary factor causing differences among soils. Today soil scientists realize that other factors, especially climate, are more important. In fact, it has been found that similar soils are often produced from different parent materials and that dissimilar soils have developed from the same parent material. Such discoveries reinforce the importance of the other soil-forming factors.

Time Time is an important component of every geological process and soil formation is no exception. The nature of soil is strongly influenced by the length of time processes have been operating. If weathering has been going on for a comparatively short time, the character of the parent material determines to a large extent the characteristics of the soil. As weathering processes continue, the influence of parent material on soil is overshadowed by the other soil-forming factors, especially climate. The amount of time required for various soils to evolve cannot be listed because the soil-forming processes act at varying rates under different circumstances. However, as a rule the longer a soil has been forming, the thicker it becomes and the less it resembles the parent material.

Climate Climate is considered to be the most important control of soil formation, since it determines whether chemical or mechanical weathering will predominate and also greatly influences the rate and depth of weathering. For instance, a hot, wet climate may produce a thick layer of chemically weathered soil in the same amount of time that a cold, dry climate produces a thin mantle of mechanically weathered debris. Also, the amount of precipitation influences the degree to which various materials are leached from the soil, thereby affecting soil fertility. Finally, climatic conditions are an important control on the type of plant and animal life present.

Plants and Animals The chief function of plants and animals is to furnish organic matter to the soil. Certain bog soils are composed almost entirely of organic matter, whereas desert soils may contain as little as a small fraction of one percent. Although the quantity of organic matter varies substantially among soils, it is the rare soil that completely lacks it.

The primary source of organic matter is plants, although animals and an infinite number of microorganisms also contribute. When organic matter is decomposed, important nutrients are supplied to plants, as well as to animals and microorganisms living in the soil. Consequently, soil fertility is in part related to the amount of organic matter present. Furthermore, the decay of plant and animal remains causes the formation of various organic acids. These complex acids hasten the weathering process. Organic matter also has a high water-holding ability and thus aids water retention in a soil.

Microorganisms, including fungi, bacteria, and single-celled protozoa, play an active role in the decay of plant and animal remains. The end product is humus, a material that no longer resembles the plants and animals from which it formed. In addition, certain microorganisms aid soil fertility because they have the ability to *fix* (change) atmospheric nitrogen into soil nitrogen.

Earthworms and other burrowing animals act to mix the mineral and organic portions of a soil. Earthworms, for example, feed on organic matter and thoroughly mix soils in which they live, often moving and enriching many tons per acre each year. Burrows and holes also aid the passage of water and air through the soil.

Slope Slope has a significant impact on the amount of erosion and the water content of soil. On steep slopes soils are often poorly developed. In such situations the quantity of water soaking in is slight, and as a result, the moisture content of the soil may not be sufficient for vigorous plant growth. Further, because of accelerated erosion on steep slopes, the soils are thin, or in some cases nonexistent (Figure 5.18). On the other hand, poorly drained and waterlogged soils found in bottomlands have a much different character. Such soils are usually thick and dark. The dark color results from the large quantity of organic matter that accumulates because saturated conditions retard the decay of vegetation. The optimum terrain for soil development is a flat-to-undulating upland surface. Here we find good drainage, minimum erosion, and sufficient infiltration of water into the soil.

Slope orientation, the direction the slope is facing, is another aspect worthy of mention. In the mid-latitudes of the northern hemisphere, a south-facing slope will receive a great deal more sunlight than a north-facing slope. In fact, a steep north-facing slope may receive no direct sunlight at all. The difference in the amount of solar radiation received will cause differences in soil temperature and moisture, which in turn may influence the nature of the vegetation and the character of the soil.

Although this section dealt separately with each of the soil-forming factors, remember that all work together to form soil. No single factor is responsible for a soil being as it is, but rather it is the combined influence of parent material, time, climate, plants and animals, and slope that determines a soil's character.

The Soil Profile

Since soil-forming processes operate from the surface downward, variations in composition, texture, structure, and color gradually evolve at varying depths. These vertical differences, which usually become more pronounced as time passes, divide the soil into zones or layers known as **horizons.** If you were to dig a trench in soil, you would see that its walls are layered. Such a vertical section through all of the soil horizons constitutes the **soil profile** (Figure 5.19).

Four basic horizons are identified and from top to bottom are designated as *O, A, B,* and *C,* respectively. The three upper layers may be further divided.

Unlike the layers beneath it which consist mainly of mineral matter, the *O* horizon consists largely of organic material. The upper portion of this horizon is primarily plant litter such as loose leaves and other organic debris that are still recognizable. By contrast, the lower portion of the *O* horizon is made up of partly decomposed organic matter (humus) in which plant structures can no longer be identified.

Underlying the organic-rich *O* horizon is the *A* horizon. This zone is largely mineral matter, yet biological activity is high and humus is generally present—up to 30 percent in some instances. Together the *O* and *A* horizons make up what is commonly called the *topsoil.* As water percolates downward from the surface through the *A* horizon, finer particles are carried away with it. This washing out of fine soil components is termed **eluviation.** As a consequence of eluviation, the texture of the *A* horizon gradually becomes coarser, because a portion of the fine particles is removed. Water percolating downward also dissolves soluble inorganic soil components and carries them to deeper zones. This deple-

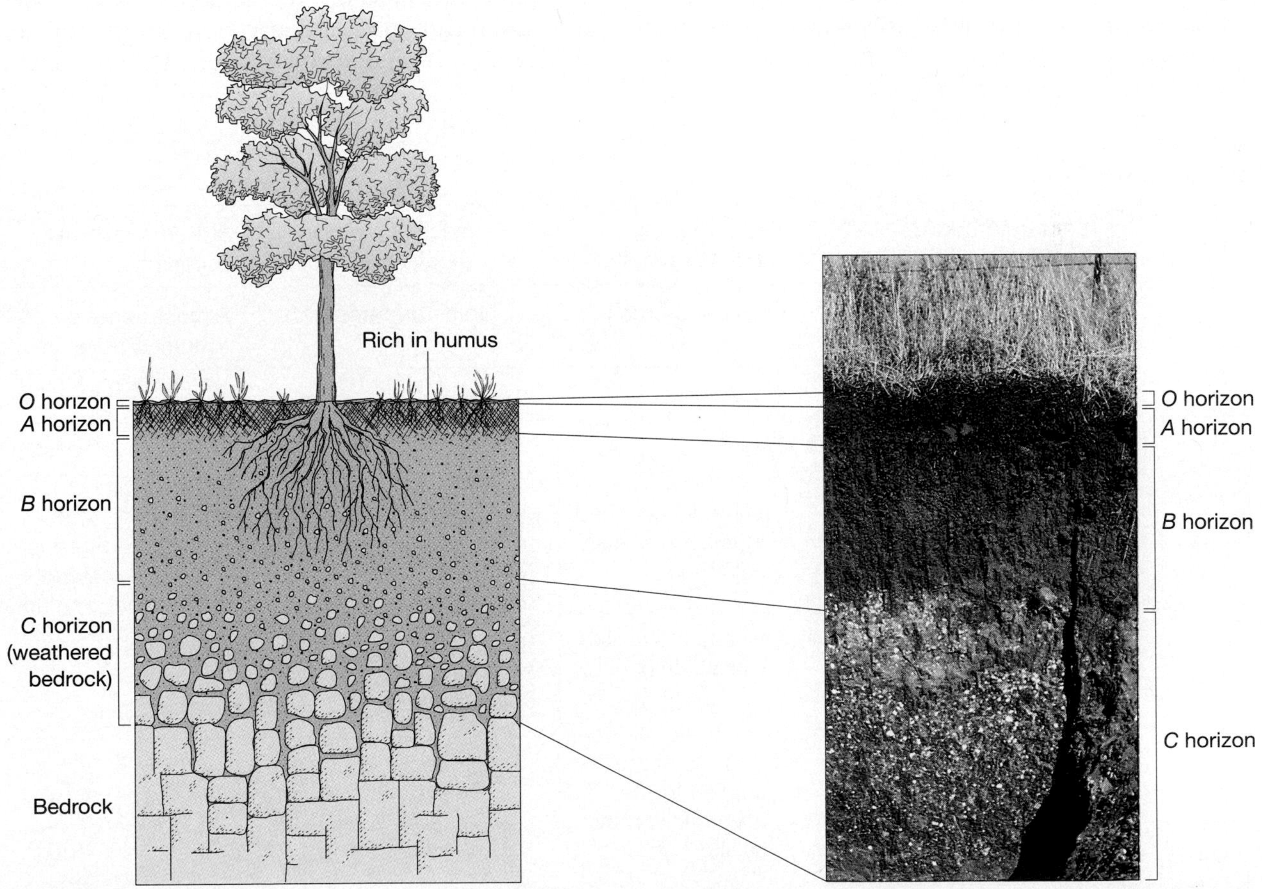

FIGURE 5.19
Soil profile. Mature soils are characterized by a series of horizontal layers called horizons, which comprise the soil profile.

tion of soluble materials from the upper soil is termed **leaching.**

Immediately below the *A* horizon is the *B* horizon, or *subsoil.* Much of the material removed from the *A* horizon by eluviation is deposited in the *B* horizon, which is often referred to as the *zone of accumulation.* The accumulation of fine clay particles derived from the *A* horizon enhances water retention in the subsoil. However, in extreme cases, clay accumulation can form a very compact and impermeable layer called *hardpan.* Since the *B* horizon has an intermediate position in the soil profile, it may be considered, at least in part, a transitional zone. For example, living organisms and organic matter are more abundant in the *B* than in the *C* horizon, but considerably less so than in the *A* horizon. The *O, A,* and *B* horizons together constitute the **solum,** or "true soil." It is in the solum that the soil-forming processes are active and that living roots and other plant and animal life are largely confined.

Below the solum is the *C* horizon, a layer characterized by partially altered rock debris and little if any organic matter (Figure 5.19). Thus, while the material from which the soil formed may be so dramatically altered in the solum that its original character is not recognizable, it is easily identifiable in the *C* horizon.

The boundaries between soil horizons may be very sharp, or the horizons may blend gradually from one to another. Furthermore, some soils lack horizons altogether. Such soils are called **immature,** because soil building has been going on for only a short time. Immature soils are also characteristic of steep slopes where erosion continually strips away the soil, preventing full development.

Soil Types

In the following discussion, we will briefly examine some common soil types. As you read, notice that the characteristics of each soil type are primarily manifestations of the prevailing climatic conditions. A summary of the characteristics of the soils discussed in this section is provided in Table 5.2.

The term **pedalfer** gives a clue to the basic characteristic of this soil type. The word is derived from the Greek **ped***on,* meaning "soil," and the chemical symbols **Al** (aluminum) and **Fe** (iron). Pedalfers are characterized by an accumulation of iron oxides and aluminum-rich clays in the *B* horizon. In mid-latitude areas where the annual rainfall exceeds 63 centimeters (25 inches) most of the soluble materials, such as calcium carbonate, are leached from the soil and carried away by underground water. The less soluble

TABLE 5.2
Summary of soil types.

Climate	Temperate humid (>63 cm rainfall)	Temperate dry (<63 cm rainfall)	Tropical (heavy rainfall)	Extreme arctic or desert
Vegetation	Forest	Grass and brush	Grass and trees	Almost none, so no humus develops
Typical Area	Eastern U.S.	Western U.S.		
Soil Type	Pedalfer	Pedocal	Laterite	
Topsoil	Sandy, light colored; acid	Commonly enriched in calcite; whitish color	Zones not developed Enriched in iron (and aluminum); brick red color	No real soil forms because there is no organic material. Chemical weathering is very slow
Subsoil	Enriched in aluminum, iron and clay; brown color	Enriched in calcite; whitish color	All other elements removed by leaching	
Remarks	Extreme development in conifer forests, because abundant humus makes groundwater very acid. Produces light gray soil because of removal of iron	*Caliche* is name applied to the accumulation of calcite	Apparently bacteria destroy humus, so no acid is available to remove iron	

iron oxides and clays are carried from the *A* horizon and deposited in the *B* horizon, giving it a brown to red-brown color. These soils are best developed under forest vegetation where large quantities of decomposing organic matter provide the acid conditions necessary for leaching. In the United States pedalfers are found east of a line extending from northwestern Minnesota to south-central Texas.

Pedocal is derived from the Greek **ped***on,* meaning "soil," and the first three letters of **calc***ite* (calcium carbonate). As the name implies, pedocals are characterized by an accumulation of calcium carbonate. This soil type is found in the drier western United States in association with grassland and brush vegetation. Since chemical weathering is less intense in drier areas, pedocals generally contain a smaller percentage of clay minerals than pedalfers.

In the arid and semiarid western states a calcite-enriched layer called **caliche** may be present in the soils. In these areas little of the rain that falls penetrates to great depths. Rather it is held by the soil particles near the surface until it evaporates. As a result, the soluble materials, chiefly calcium carbonate, are removed from the uppermost layer and redeposited below, forming the caliche layer.

In the hot, wet climates of the tropics, soils called **laterites** may develop. Since chemical weathering is intense under such climatic conditions, these soils are usually deeper than soils developing over a similar period in the mid-latitudes. Not only does leaching remove the soluble materials such as calcite, but the great quantities of percolating water also remove much of the silica, with the result that oxides of iron and aluminum become concentrated in the soil. The iron gives the soil a distinctive red color (Figure 5.20). When dried, laterites are very hard. In fact, some people use this soil for making bricks. Since bacterial activity is very high in the tropics, laterites contain practically no humus. This fact, coupled with the highly leached and bricklike nature of these soils, makes laterites poor for growing crops. The infertility of these soils has been borne out repeatedly in tropical countries where cultivation has been expanded into such areas.

In cold or dry climates soils are generally very thin and poorly developed. The reasons for this are fairly obvious. Chemical weathering progresses very slowly in such climates, and the scanty plant life yields very little organic matter.

Soil Erosion

Soils represent just a tiny fraction of all earth materials, yet they are a vital resource. Because soils are necessary for the growth of rooted plants, they are the very foundation of the human life-support system. Just as human ingenuity can increase the agricultural productivity of soils through such practices as fertilization and irrigation, soils can be damaged or destroyed by the careless activities of people. Despite their basic role in providing food, fiber, and other basic materials, soils have been one of our most abused resources. Perhaps this neglect and indifference has occurred because a substantial amount of soil seems to remain even in places where soil erosion is serious. Nevertheless, although the loss of fertile topsoil may not be obvious to the untrained eye, it is a growing problem as human activities expand and disturb more and more of the earth's surface.

FIGURE 5.20
Characteristic red color of a well-developed lateritic soil.

Soil erosion is a natural process; it is part of the constant recycling of earth materials that we call the rock cycle. Once soil forms, erosional forces, especially water and wind, move soil components from one place to another. Every time it rains, raindrops strike the land with a surprising amount of force (Figure 5.21). Each drop acts like a tiny bomb, blasting movable soil particles out of their positions in the soil mass. Then water flowing across the surface carries away the dislodged soil particles. Because the soil is moved by thin sheets of water, this process is

FIGURE 5.21
When it is raining, millions of water drops are falling at velocities approaching 10 meters per second (35 kilometers per hour). When water drops strike an exposed surface, soil particles may splash as high as one meter into the air and land more than a meter away from the point of raindrop impact. Soil dislodged by splash erosion is more easily moved by sheet erosion. (Photo courtesy of U.S. Department of Agriculture)

termed *sheet erosion*. After flowing as a thin, unconfined sheet for a relatively short distance, threads of current typically develop and tiny channels called *rills* begin to form. Still deeper cuts in the soil, known as *gullies,* are created as rills enlarge (Figure 5.22). When cultivation cannot eliminate the channels, we know the rills have grown large enough to be called gullies. Although most dislodged soil particles move only a short distance during each rainfall, substantial quantities eventually leave the fields and make their way downslope to a stream. Once in the stream channel, these soil particles, which can now be called *sediment,* are transported downstream and eventually deposited.

We know that soil erosion is the ultimate fate of practically all soils. In the past, soil erosion occurred at slower rates than it does today because more of the land surface was covered and protected by trees, shrubs, grasses, and other plants. However, human activities such as farming, logging, and construction, which remove or disrupt the natural vegetation, have greatly accelerated the rate of soil erosion. Without the stabilizing effect of plants, the soil is more easily swept away by the wind or carried downslope by sheet wash.

Natural rates of soil erosion vary greatly from one place to another and depend on soil characteristics as well as such factors as climate and slope. Over a broad area, erosion caused by surface runoff may be estimated by determining the sediment loads of the streams that drain the region. When studies of this kind were made on a global scale they indicated that, prior to the appearance of humans, sediment transport by rivers to the ocean amounted to just over 9 billion metric tons per year. By contrast, the amount of material currently transported to the sea by rivers is about 24 billion metric tons per year, or more than two and one-half times the earlier rate.

FIGURE 5.22
Gully erosion in poorly protected soil. (Photo by James E. Patterson)

BOX 5.2

Dust Bowl: Soil Erosion in the Great Plains

During a span of dry years in the 1930s, large dust storms plagued the Great Plains. Because of the size and severity of these storms, the region came to be called the "Dust Bowl," and the time period, the "dirty thirties." The heart of the Dust Bowl consisted of nearly 100 million acres in the panhandles of Texas and Oklahoma as well as adjacent parts of Colorado, New Mexico, and Kansas (Figure 5.B). To a lesser extent, dust storms were also a problem over much of the Great Plains, from North Dakota to west central Texas.

At times dust storms were so severe that they were called "black blizzards" and "black rollers" because visibility was reduced to only a few feet. Examples of storms that lasted for hours and stripped huge volumes of topsoil from the land are numerous. In the spring of 1934, a wind storm that lasted for a day and a half created a dust cloud that extended for 2000 kilometers (1200 miles). As the sediment moved east, "muddy rains" were experienced in New York, and "black snows" in Vermont. Less than a year later, another storm carried dust more than 3 kilometers (2 miles) into the atmosphere and transported it 3000 kilometers from its source in Colorado to create twilight conditions in the middle of the day in parts of New England and New York.

What caused the Dust Bowl? Clearly, the fact that portions of the Great Plains experience some of North America's strongest winds is important. However, it was the expansion of agriculture that set the stage for the disastrous period of soil erosion. Mechanization allowed the rapid transformation of the grass-covered prairies of this semiarid region into farms. Between the 1870s and 1930, the area of cultivation in the region expanded nearly tenfold, from about 10 million acres to more than 100 million acres.

As long as precipitation was adequate, the soil remained in place. However, when a prolonged drought struck in the 1930s, the unprotected fields were vulnerable to the wind. The results were severe soil loss, crop failures, and economic hardship.

Beginning in 1939, a return to rainier conditions brought relief. Moreover, farming practices that were designed to reduce soil loss by wind had also been instituted. Although dust storms are less numerous and not as severe as in the "dirty thirties," soil erosion by strong winds still occurs periodically whenever the combination of drought and unprotected soil exists.

FIGURE 5.B
An abandoned farmstead shows the disastrous effects of wind erosion and deposition during the "Dust Bowl" period. This photo of a previously prosperous farm was taken in Oklahoma in 1937. (Photo courtesy of Soil Conservation Service, U.S. Department of Agriculture)

It is more difficult to measure the loss of soil due to wind erosion. However, the removal of soil by wind is generally much less significant than erosion by flowing water except during periods of prolonged drought. When dry conditions prevail, strong winds can remove large quantities of soil from unprotected fields. Such was the case in the 1930s in the portions of the Great Plains that came to be called the Dust Bowl (see Box 5.2).

In many regions the rate of soil erosion is significantly greater than the rate of soil formation. This means that a potentially renewable resource has become nonrenewable in these places. At present, it is estimated that topsoil is eroding faster than it forms on more than one-third of the world's croplands. The result is lower productivity, poorer crop quality, and reduced agricultural income.

Another problem related to excessive soil erosion involves the deposition of sediment. Each year in the United States hundreds of millions of tons of eroded soil is deposited in lakes, reservoirs, and streams. The detrimental impact of this process can be significant. For example, as more and more sediment is deposited in a reservoir, the capacity of the reservoir is diminished, limiting its usefulness for flood control, water supply, and/or hydroelectric power generation. In addition, sedimentation in streams and other waterways can restrict navigation and lead to costly dredging operations.

In some cases soil particles are contaminated with pesticides used in farming. When these chemicals are introduced into a lake or reservoir, the quality of the water supply is threatened and aquatic organisms may be endangered. In addition to pesticides, nutrients found naturally in soils as well as those added by agricultural fertilizers make their way into streams and lakes, where they stimulate the growth of plants. Over a period of time, excessive nutrients accelerate the process by which plant growth leads to the depletion of oxygen and an early death of the lake.

The availability of good soils is critical if the world's rapidly growing population is to be fed. On every continent unnecessary soil loss is occurring because appropriate conservation measures are not being used. Although it is a recognized fact that soil erosion can never be completely eliminated, soil conservation programs can substantially reduce the loss of this basic resource. Windbreaks, terracing, and contour farming techniques are some of the effective measures as are special tillage practices and crop rotation.

REVIEW QUESTIONS

1. Differentiate between the products of mechanical weathering and chemical weathering.
2. In what type of environment is frost wedging most effective?
3. Describe the processes of sheeting and spheroidal weathering. How are they different and how are they similar?
4. How does mechanical weathering add to the effectiveness of chemical weathering?
5. Granite and basalt are exposed at the surface in a hot, wet region.
   (a) Which type of weathering will predominate?
   (b) Which of these rocks will weather most rapidly? Why?
6. Heat speeds up a chemical reaction. Why then does chemical weathering proceed slowly in a hot desert?
7. How is carbonic acid (H_2CO_3) formed in nature? What results when this acid reacts with potassium feldspar?
8. What is the difference between soil and regolith?
9. What factors might cause different soils to develop from the same parent material, or similar soils to form from different parent materials?
10. Which of the controls of soil formation is most important? Explain.
11. How can slope affect the development of soil? What is meant by the term *slope orientation?*
12. List the characteristics associated with each of the horizons in a well-developed soil profile. Which of the horizons constitute the solum? Under what circumstances do soils lack horizons?
13. Distinguish between pedalfers and pedocals.
14. Soils formed in the humid tropics and the Arctic both contain little organic matter. Do both lack humus for the same reasons?
15. List three detrimental effects of soil erosion, other than the loss of topsoil from croplands.

KEY TERMS

caliche (p. 131)
chemical weathering (p. 119)
differential weathering (p. 124)
eluviation (p. 129)
erosion (p. 114)
exfoliation dome (p. 116)
frost wedging (p. 115)
horizon (p. 129)
humus (p. 125)
hydrolysis (p. 120)
immature soil (p. 130)
joints (p. 117)
laterite (p. 131)
leaching (p. 130)
mass wasting (p. 114)
mechanical weathering (p. 115)
oxidation (p. 120)
parent material (p. 127)
pedalfer (p. 130)
pedocal (p. 131)
regolith (p. 125)
residual soil (p. 127)
sheeting (p. 116)
soil (p. 125)
soil profile (p. 129)
solum (p. 130)
solution (p. 119)
spheroidal weathering (p. 122)
talus slope (p. 115)
transported soil (p. 127)
weathering (p. 114)

6

Sedimentary Rocks

TYPES OF SEDIMENTARY ROCKS
Detrital Sedimentary Rocks ■ Chemical Sedimentary Rocks

TURNING SEDIMENT INTO SEDIMENTARY ROCK

CLASSIFICATION OF SEDIMENTARY ROCKS

SEDIMENTARY ENVIRONMENTS

SEDIMENTARY STRUCTURES

FOSSILS

Opposite: This series of weathered and eroded sedimentary layers in Capitol Reef National Park, Utah, is called The Castle. (Photo by Tom Till)
(Top photo by Michael Collier)

In Chapter 1, the rock cycle (see Figure 1.20) provided a brief glimpse of the origin of sedimentary rocks. Recall that weathering begins the process. Next, erosional agents such as running water, wind, waves, and ice remove the products of weathering and carry them to a new location where they are deposited. Usually the particles are broken down further during the transport phase. Following deposition, this material, which is now called **sediment**, is lithified. Commonly compaction and cementation transform the sediment into solid sedimentary rock.

The products of mechanical and chemical weathering constitute the raw materials for sedimentary rocks. The word *sedimentary* indicates the nature of these rocks, for it is derived from the Latin *sedimentum,* which means "settling," a reference to solid material settling out of a fluid. Most but not all sediment is deposited in this fashion. Weathered debris is constantly being swept from bedrock, carried away, and eventually deposited in lakes, river valleys, seas, and countless other places. The particles in a desert sand dune, the mud on the floor of a swamp, the gravels in a stream bed, and even household dust are examples of this never-ending process. Since the weathering of bedrock and the transport and deposition of the weathering products are continuous, sediment is found almost everywhere. As piles of sediment accumulate, the materials near the bottom are compacted. Over long periods, these sediments are cemented together by mineral matter deposited in the spaces between particles to form solid rock.

Geologists estimate that sedimentary rocks account for only about 5 percent (by volume) of the earth's outer 16 kilometers (10 miles). However, the importance of this group of rocks is far greater than this percentage would imply. If we were to sample the rocks exposed at the earth's surface, we would find that the great majority are sedimentary. Indeed, about 75 percent of all rock outcrops on the continents are sedimentary (Figure 6.1). Therefore, we may think of sedimentary rocks as comprising a relatively thin and somewhat discontinuous layer in the uppermost portion of the crust. This fact is readily understood when we consider that sediment accumulates at the surface of the earth.

It is from sedimentary rocks that geologists reconstruct many of the details of earth history. Because sediments are deposited in a variety of different settings at the earth's surface, the rock layers that they eventually form hold many clues to past surface environments. They may also exhibit characteristics that allow geologists to decipher information about the method and length of sediment transport. In addition, this group of rocks frequently contains fossils, which are very important tools in the study of the geologic past.

Finally, it should be mentioned that many sedimentary rocks are very important economically. Coal, for example, is classified as a sedimentary rock, whereas our other major energy sources, petroleum and natural gas, are found in association with sedimentary rocks. Still others represent major sources of iron, aluminum, manganese, and fertilizer as well as numerous materials essential to the construction industry.

TYPES OF SEDIMENTARY ROCKS

Materials accumulating as sediment have two principal sources. First, sediments may be accumulations of materials that originate and are transported as solid particles derived from both mechanical and chemical weathering. Deposits of this type are termed *detrital* and the sedimentary rocks that they form are called **detrital sedimentary rocks.** The second major source of sediment is soluble material produced largely by chemical weathering. When these dissolved substances are precipitated by either inorganic or organic processes, the material is known as chemical sediment and the rocks formed from it are called **chemical sedimentary rocks.**

Detrital Sedimentary Rocks

Though a wide variety of minerals and rock fragments may be found in detrital rocks, clay minerals and quartz are the chief constituents of most sedimentary rocks in this category. Recall that clay minerals are the most abundant product of the chemical weathering of silicate minerals, especially the feldspars. Clays are fine-grained minerals with sheetlike crystalline structures similar to the micas. The other common mineral, quartz, is abundant because it is extremely durable and very resistant to chemical weathering. Thus, when igneous rocks such as granite are attacked by weathering processes, individual quartz grains are freed.

Other common minerals in detrital rocks are the feldspars and micas. Since chemical weathering rapidly transforms these minerals into new substances, their presence in sedimentary rocks indicates that erosion and deposition were fast enough to preserve some of the primary minerals from the source rock before they could be decomposed.

Particle size is the primary basis for distinguishing among various detrital sedimentary rocks. Table 6.1 presents the size categories for particles making up detrital rocks. Note that in this context the term *clay* refers only to a particular size and not to the minerals of the same name. Although most clay minerals are of clay size, not all clay-sized sediment consists of clay minerals.

TABLE 6.1
Particle size classification for detrital rocks.

Size Range (millimeters)	Particle Name	Common Sediment Name	Detrital Rock
>256 64–256 4–64 2–4	Boulder Cobble Pebble Granule	Gravel	Conglomerate or breccia
1/16-2	Sand	Sand	Sandstone
1/256–1/16 <1/256	Silt Clay	Mud	Shale or mudstone

Particle size is not only a convenient method of dividing detrital rocks: the sizes of the component grains also provide useful information about environments of deposition. Currents of water or air sort the particles by size; the stronger the current, the larger the particle size carried. Gravels for example, are moved by swiftly flowing rivers as well as by landslides and glaciers. Less energy is required to transport sand, thus it is common to such features as windblown dunes, as well as some river deposits and beaches. Because clays settle very slowly, accumulations of these tiny grains are generally associated with the quiet waters of a lake, lagoon, swamp, or certain marine environments.

FIGURE 6.1
At sunset this exposure of sedimentary rock in Arizona's Monument Valley is especially colorful. Harder, more resistant sandstone stands above weaker crumbling shale. Sedimentary rocks are exposed at the earth's surface far more than igneous and metamorphic rocks. Because they contain fossils and other clues about the geologic past, sedimentary rocks are important in the study of earth history. (Photo by James E. Patterson)

Shale *Shale* is a sedimentary rock consisting of silt- and clay-sized particles (Figure 6.2). These fine-grained detrital rocks account for well over half of all sedimentary rocks. The particles in these rocks are so small that they cannot be readily identified without great magnification and for this reason make shale more difficult to study and analyze than most other sedimentary rocks. Much of what can be learned is based on particle size. The tiny grains in shale indicate that deposition occurs as the result of gradual settling from relatively quiet, non-turbulent currents. Such environments include lakes, river floodplains, lagoons, and portions of the deep-ocean basins. Even in these "quiet" environments, there is usually enough turbulence to keep clay-sized particles suspended almost indefinitely. Consequently, much of the clay is deposited only after the individual particles coalesce to form larger aggregates. Sometimes the chemical composition of the rock provides additional information to aid interpretation. One example is black shale, which contains abundant organic matter. When such a rock is found, it strongly implies that deposition occurred in an oxygen-poor environment such as a swamp, where organic materials do not readily oxidize and decay.

As silt and clay accumulate, they tend to form thin layers which are commonly referred to as *laminae.* Initially the particles in the laminae are oriented randomly. This disordered arrangement leaves a high percentage of open space (called *pore space*) that is filled with water. However, this situation usually changes with time as additional layers of sediment pile up and compact the sediment below. During this phase the clay and silt particles take on a more nearly parallel alignment and become tightly packed. This rearrangement of grains reduces the size of the pore spaces and forces out much of the water. Once the grains are pressed closely together, the tiny spaces between particles do not readily permit solutions containing cementing material to circulate. Therefore shales are often described as being weak because they are not well lithified. The inability of water to penetrate its microscopic pore spaces explains why shale often forms barriers to the subsurface movement of water and petroleum. Indeed, layers containing groundwater are commonly underlain by shale beds that do not allow further downward movement. Petroleum reservoirs, on the other hand, are often capped by shale beds which effectively prevent oil and gas from escaping to the surface.*

FIGURE 6.2
Shale is a fine-grained detrital rock that is by far the most abundant of all sedimentary rocks. Dark shales containing plant remains are relatively common.

It is common to apply the term *shale* to all fine-grained sedimentary rocks, especially in a nontechnical context. However, it should be noted that there is a more restricted use of the term. In this narrower usage, shale must exhibit the ability to split into thin layers along well-developed, closely spaced planes. This property is termed **fissility.** If the rock breaks into chunks or blocks, the name *mudstone* is applied. Another fine-grained sedimentary rock which, like mudstone, is often grouped with shale but lacks fissility is *siltstone.* As its name implies, siltstone is composed largely of silt-sized particles and contains less clay than shale and mudstone.

Although shale is far more common than other sedimentary rocks, it does not usually attract as much notice as other less abundant members of this group. The reason is that shale does not form prominent outcrops as sandstone and limestone often do. Rather, shale crumbles easily and usually forms a cover of soil that hides the unweathered rock below. This is illustrated nicely in the Grand Canyon where the gentler slopes of weathered shale contrast sharply with the bold cliffs produced by more durable layers (Figure 6.3).

Although shale beds may not form striking cliffs and prominent outcrops, some deposits have economic value. Certain shales serve as a basic raw ma-

*The relationship between impermeable beds and the occurrence and movement of groundwater is examined in Chapter 11. Shale beds as cap rocks in oil traps are discussed in Chapter 21.

FIGURE 6.3
Sedimentary rock layers exposed in the walls of the Grand Canyon, Arizona. Beds of resistant sandstone and limestone produce bold cliffs. By contrast, weaker, poorly cemented shale crumbles easily and produces a gentler slope of weathered debris in which some vegetation is growing. (Photo by Tom Till)

terial for pottery, brick, tile, and china. Moreover, when mixed with limestone, shale is used to make portland cement. In the future, one type of shale, called oil shale, may become a valuable energy resource. This possibility will be explored in greater detail in Chapter 21.

Sandstone *Sandstone* is the name given rocks in which sand-sized grains predominate (Figure 6.4). After shale, sandstone is the most abundant sedimentary rock, accounting for approximately 20 percent of the entire group.

Sandstones form in a variety of environments and often contain significant clues about their origin. Sorting is one conspicuous result of the movement of sediment grains by currents of turbulent air or water. **Sorting** refers to the degree of similarity in particle size in a sedimentary rock. For example, if all the grains in a sample of sandstone are about the same size, the sand is considered well sorted. Conversely, if the rock contains many particles that are much larger or much smaller than the average, the sand is said to be poorly sorted. By studying the degree of sorting, the kind of depositing current may be determined. Deposits of windblown sand are usually better sorted than deposits sorted by wave activity, and particles washed by waves are commonly better sorted than materials deposited by streams. Accumulations of sediments that exhibit poor sorting usually result when particles are transported for only a relatively short time and then rapidly deposited. For example, when a turbulent stream reaches the

FIGURE 6.4
This quartz sandstone is cemented with iron oxide, which gives the rock its rusty color. After shale, sandstone is the most abundant sedimentary rock.

gentler slopes at the base of a steep mountain, its velocity is quickly reduced and poorly sorted sands and gravels are deposited.

The shapes of sand grains can also help decipher the history of a sandstone. When streams, winds, or waves move sand and other sedimentary particles, the grains lose their sharp edges and corners and become more rounded as they collide with other particles during transport. Consequently, the degree of rounding indicates the distance or time involved in the transportation of sediment by currents of air or water. Highly rounded grains indicate that a great deal of abrasion and hence a great deal of transport has occurred. Very angular grains, on the other hand, imply that the materials were transported only a short distance before they were deposited.

In addition to affecting the degree of rounding and the amount of sorting that particles undergo, the length of transport by turbulent air and water currents also influences the mineral composition of a sedimentary deposit. In turn, the mineral composition of sediment can provide clues to the length of transport. Substantial weathering and long transport lead to the gradual destruction of weaker and less stable minerals including the feldspars and ferromagnesians. Because quartz is very durable, it is usually the mineral that survives the long trip in a turbulent environment.

The preceding discussion has shown that the origin and history of a sandstone can often be deduced by examining the sorting, roundness, and mineral composition of its constituent grains. Knowing this information allows us to infer that a well-sorted, quartz-rich sandstone consisting of highly rounded grains must be the result of a great deal of transport. Such a rock, in fact, may represent several cycles of weathering, transport, and deposition. We may also conclude that a sandstone containing a significant amount of feldspar and angular grains of ferromagnesian minerals underwent little chemical weathering and transport and was probably deposited close to the source area of the particles.

Due to its durability, quartz is the predominant mineral in most sandstones. When this is the case, the rock may simply be called *quartz sandstone.* When a sandstone contains appreciable quantities of feldspar, the rock is called *arkose.* In addition to feldspar, arkose usually contains quartz and mica, which implies the gradual erosion of a landscape underlain by granitic rocks. A third variety of sandstone is known as *graywacke.* Along with quartz and feldspar, this dark-colored rock contains abundant rock fragments and clay. The poor sorting and angular grains characteristic of graywacke suggest that the particles were transported only a relatively short distance from their source area and then rapidly deposited. Before the sediment could be reworked and sorted further, it was buried by additional layers of material. Graywacke is frequently associated with submarine deposits made by dense, sediment-choked torrents called turbidity currents.*

Conglomerate *Conglomerate* consists largely of gravels (Figure 6.5). As Table 6.1 indicates, these particles may range in size from large boulders to particles as small as garden peas. The particles are commonly large enough to be identified as distinctive rock types; thus, they can be valuable in identifying the source areas of sediments. More often than not conglomerates are poorly sorted because the openings between the large gravel particles contain sand or mud.

Gravels accumulate in a variety of environments and usually indicate the existence of steep slopes or very turbulent currents. The coarse particles in a conglomerate may reflect the action of energetic mountain streams or result from strong wave activity along a rapidly eroding coast. Some glacial and landslide deposits also contain much gravel. If the large particles are angular rather than rounded, the rock is called *breccia* (Figure 6.6). Since large particles abrade and become rounded very rapidly during transport, the pebbles and cobbles in a breccia indi-

*For more on this, see the discussion of *graded beds* in the section on "Sedimentary Structures" later in this chapter.

FIGURE 6.5
Conglomerate is composed primarily of rounded gravel-sized particles.

cate that they did not travel far from their source area before they were deposited. Thus, as with many other sedimentary rocks, conglomerates and breccias contain clues to their history. Their particle sizes reveal the strength of the currents that transported them, whereas the degree of rounding indicates how far the particles traveled. The fragments within a sample identify the source rocks that supplied them.

FIGURE 6.6
When the gravel-sized particles in a detrital rock are angular, the rock is called *breccia.* (Photo by E. J. Tarbuck)

Chemical Sedimentary Rocks

In contrast to detrital rocks, which form from the solid products of weathering, chemical sediments derive from material that is carried in solution to lakes and seas. This material does not remain dissolved in the water indefinitely, however. Rather, some of it precipitates to form chemical sediments. This precipitation of material may occur directly as the result of inorganic processes or indirectly as the result of the life processes of water-dwelling organisms. Sediment formed in this second way is said to have a **biochemical** origin (Figure 6.7).

An example of a deposit resulting from inorganic chemical processes is the salt left behind as a body of seawater evaporates. In contrast, many water-dwelling animals and plants extract dissolved mineral matter to form shells and other hard parts. After the organisms die, their skeletons collect on the floor of a lake or ocean.

Limestone Representing about 10 percent of the total volume of all sedimentary rocks, *limestone* is the most abundant chemical sedimentary rock. It is composed chiefly of the mineral calcite ($CaCO_3$) and forms by either inorganic means or as the result of biochemical processes. The mineral composition of all limestone is similar, yet many different types exist. This is true because limestones are produced under a variety of conditions. Those forms having a marine biochemical origin are by far the most common.

FIGURE 6.7
This rock, called coquina, consists of shell fragments; therefore, it has a biochemical origin.

Corals are one important example of organisms that are capable of creating large quantities of marine limestone. These relatively simple invertebrate animals secrete a calcareous (calcite-rich) external skeleton. Although they are small, corals are capable of creating massive structures called *reefs*. Reefs consist of coral colonies made up of great numbers of individuals that live side by side on a calcite structure secreted by the animals. In addition, calcium carbonate-secreting algae live in association with the corals and help cement the entire structure into a solid mass. A wide variety of other organisms are also found living in and near the reef. Certainly the best known modern reef is Australia's Great Barrier Reef, but many lesser reefs also exist. They develop in the shallow, warm waters of the tropics and subtropics equatorward of about 30° latitude. Striking examples exist in the Bahamas and Florida Keys. Of course, it is not only modern corals that build reefs. Corals have been responsible for producing appreciable quantities of limestone in the geologic past as well. In the United States, reefs of Silurian age are prominent features in Wisconsin, Illinois, and Indiana. In West Texas and adjacent southeastern New Mexico, a massive reef complex formed during the Permian Period is strikingly exposed in Guadalupe Mountains National Park (Figure 6.8).

Although most limestone is the product of biological processes, this origin is not always evident because shells and skeletons may undergo considerable change before being converted to rock. However, one easily identified biochemical limestone is *coquina,* a coarse rock composed of poorly cemented shells and shell fragments (see Figure 6.7).

FIGURE 6.8
El Capitan Peak, a massive limestone cliff in Guadalupe Mountains National Park, Texas. The rocks of the Guadalupe Mountains are an exposed portion of a large reef that formed during the Permian period. (Photo by E. J. Tarbuck)

FIGURE 6.9
The White Chalk Cliffs of Dover. This prominent chalk deposit underlies large portions of southern England as well as parts of northern France. (Photo by Kevin Schafer)

Another less obvious, but nevertheless familiar example is *chalk,* a soft, porous rock made up almost entirely of the hard parts of microscopic marine organisms no larger than the head of a pin. Among the most famous chalk deposits are those exposed in the well-known White Cliffs of Dover in England (Figure 6.9).

Limestones having an inorganic origin form when chemical changes or high water temperatures increase the concentration of calcium carbonate to the point that it precipitates. *Travertine,* the type of limestone commonly seen in caves, is an example, as is *oolitic limestone.* When travertine is deposited in caves, groundwater is the source of the material. As water droplets reach the air in a cavern, some of the carbon dioxide dissolved in the water escapes, causing calcium carbonate to precipitate. The formation of oolitic limestone, a rock composed of small spherical grains called *ooids,* is somewhat more complex. Ooids form in shallow marine waters as tiny "seed" particles that commonly consist of small shell fragments that are moved back and forth by currents. As the grains are rolled about in the warm water that is supersaturated with calcium carbonate, they become coated with layer upon layer of the precipitate.

Dolomite Closely related to limestone is *dolomite,* a rock composed of the calcium-magnesium carbonate mineral of the same name. Dolomite is one of the rare cases where the same term is used to designate both a mineral and a rock. To avoid confusion, some geologists refer to the rock as *dolostone.* Although dolomite can form by direct precipitation from seawater, it is thought that most originates when magnesium in seawater replaces some of the calcium in limestone. The latter hypothesis is reinforced by the fact that there are practically no young dolomite rocks. Rather, most dolomites are ancient rocks in which there was ample time for magnesium to replace calcium.

Chert *Chert* is a name used for a number of very compact and hard rocks made of microcrystalline silica (SiO_2). One well-known form is *flint,* whose dark color results from the organic matter it contains. *Jasper,* a red variety, gets its bright color from the iron oxide it contains. The banded form is usually referred to as *agate* (Figure 6.10).

Chert deposits are commonly found in one of two situations: as irregularly shaped nodules in limestone and as layers of rock. The silica composing most chert nodules is believed to have been deposited directly from water. Thus, these nodules have an inorganic origin. However, it is unlikely that a very high percentage of chert layers was precipitated directly from seawater, because seawater is seldom saturated with silica. Hence, beds of chert are thought to have originated largely as biochemical sediment. Although most water-dwelling organisms that produce hard parts secrete shells made of calcium carbonate, some, such as diatoms and radiolarians, produce glasslike silica skeletons. These tiny organisms are able to extract silica even though seawater contains only tiny quantities of the dissolved material. It

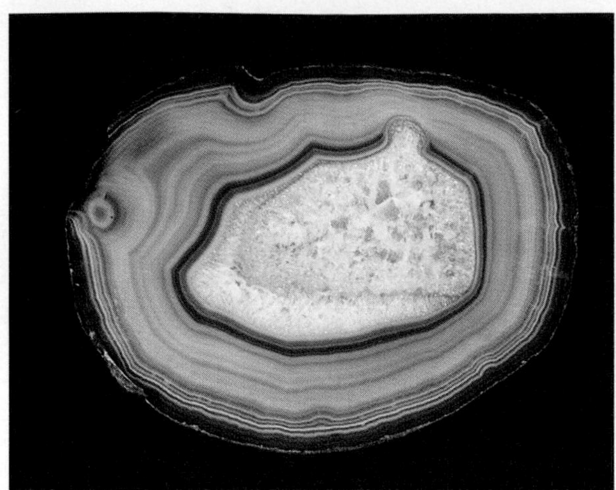

FIGURE 6.10
Agate is the banded form of chert. (Photo by E. J. Tarbuck)

is from these remains that most chert beds are believed to originate. Some bedded cherts occur in association with lava flows and layers of volcanic ash. For these occurrences it is probable that the silica was derived from the decomposition of the volcanic ash and not from biochemical sources. Note that when a hand specimen of chert is being examined, there are few reliable criteria by which the mode of origin (inorganic versus biochemical) can be determined. However, such a determination can often be made by microscopic examination.

Evaporites Very often evaporation is the mechanism triggering deposition of chemical precipitates. Minerals commonly precipitated in this fashion include halite (sodium chloride), the chief component of *rock salt,* and gypsum (hydrous calcium sulfate), the main ingredient of *rock gypsum.* Both materials have significant commercial importance. Halite is familiar to everyone as the common salt used in cooking and seasoning foods. Of course, it has many other uses and has been considered important enough that people have sought, traded, and fought over it for much of human history. Gypsum is the basic ingredient of plaster of Paris. This material is used most extensively in the construction industry for wallboard and plaster for interior use.

In the geologic past, many areas that are now dry land were basins drowned by shallow arms of a sea that had only narrow connections to the open ocean. Under these conditions, water continually moved into the bay to replace water lost by evaporation. Eventually the waters of the bay became saturated and salt deposition began. Such deposits are called **evaporites.** When a body of seawater evaporates, the minerals that precipitate do so in a sequence which is determined by their solubility. That is to say, less-soluble minerals precipitate first and more-soluble minerals precipitate later when salinities are higher. Gypsum is precipitated when about two-thirds to three-quarters of the seawater has evaporated, and halite settles out when nine-tenths of the water has been removed. During the last stages of this process, potassium and magnesium salts precipitate. One of these last-formed salts, the mineral *sylvite,* is mined as a significant source of potassium for fertilizer.

On a smaller scale, evaporite deposits may be seen in such places as Death Valley, California. Here, following rains or periods of snowmelt in the mountains, streams flow from the surrounding mountains into an enclosed basin. As the water evaporates, **salt flats** form from dissolved materials left behind as a white crust on the ground (Figure 6.11).

Coal is a brown to black combustible rock that is commonly grouped with the biochemical sedimentary rocks. Unlike most other rocks in this category, which are calcite- or silica-rich, coal is made from organic matter. It is the end product of the burial and subsequent alteration of large quantities of vegetation.

Under certain circumstances, great volumes of plants accumulate rather than decompose. This commonly happens in an oxygen-poor swamp. Before the plants become coal they turn into peat, a brown spongy mass of organic matter in which roots, stems, and other plant structures are still recognizable. With time and the pressure that accompanies burial, the peat undergoes chemical and physical changes that transform it into coal. The particular type of coal depends to a large extent on how deeply the deposit was buried and whether or not it was subjected to the strong structural deformation associated with the forces that build mountains. Coal is a major energy resource which will be discussed further in Chapter 21.

TURNING SEDIMENT INTO SEDIMENTARY ROCK

Lithification refers to the processes by which unconsolidated sediments are transformed into solid sedimentary rocks. One of the most common processes affecting sediments is **compaction.** As sediments accumulate through time, the weight of overlying material compresses the deeper sediments. As the grains are pressed closer and closer, there is a considerable reduction in pore space. For example, when clays are buried beneath several thousand meters of material, the volume of the clay may be re-

FIGURE 6.11
These salt flats in Death Valley, California, are examples of evaporite deposits and are common in basins located in the arid Southwest. A temporary lake is created when water drains from the adadjacent mountains into the basin following periods of rain or snowmelt. When the water evaporates, salts are left behind. (Photo by Michael Collier)

duced by as much as 40 percent. Since sands and other coarse sediments are only slightly compressible, compaction is most significant as a lithification process in fine-grained sedimentary rocks such as shale.

Cementation is the most important process by which sediments are converted to sedimentary rocks. The cementing materials are carried in solution by water percolating through the open spaces between particles. Through time, the cement precipitates onto the sediment grains, fills the open spaces, and joins the particles. Calcite, silica, and iron oxide are the most common cements. It is often a relatively simple matter to identify the cementing material. Calcite cement will effervesce with dilute hydrochloric acid. Silica is the hardest cement and thus produces the hardest sedimentary rocks. When a sedimentary rock has an orange or dark red color, this usually means that iron oxide is present.

Although most sedimentary rocks are lithified by compaction and cementation, some are held together because they consist of interlocking crystals. Certain chemical sedimentary rocks, such as the evaporites, initially form as solid masses of intergrown crystals. That is, they do not begin as accumulations of separate particles that later become solid. Other crystalline sedimentary rocks do not begin that way but are transformed into masses of interlocking crystals sometime after the sediment is deposited. For example, with time and burial, loose sediment consisting of delicate calcareous skeletal debris may be recrystallized into a relatively dense crystalline limestone. Because crystals grow until they fill all the available space, pore spaces are frequently lacking in crystalline sedimentary rocks. Unless the rocks later develop joints and fractures, they will be relatively impermeable to fluids like water and oil.

CLASSIFICATION OF SEDIMENTARY ROCKS

The classification scheme in Table 6.2 divides sedimentary rocks into two major groups: detrital and chemical. Further, we can see that the main criterion for subdividing the detrital rocks is particle size, whereas the primary basis for distinguishing among different rocks in the chemical group is their mineral composition.

As is the case with many (perhaps most) classifications of natural phenomena, the categories presented in Table 6.2 are more rigid than the actual state of nature. In reality, many of the sedimentary rocks classified into the chemical group also contain at least small quantities of detrital sediment. Many limestones, for example, contain varying amounts of mud or sand, giving them a "sandy" or "shaly" quality. On the other hand, since practically all detrital rocks are cemented with material that was originally dissolved in water, they too are far from being "pure."

As was the case with the igneous rocks examined in Chapter 3, texture is a part of sedimentary rock classification. There are two major textures used in the classification of sedimentary rocks: clastic and nonclastic. The term **clastic** is taken from a Greek word meaning "broken." Rocks that display a clastic

BOX 6.1

Harvesting Salt from the Sea

Each year about 30 percent of the world's supply of salt is extracted from seawater. In this process, salt water is held in shallow ponds while the sun evaporates the water. The nearly pure salt deposits that eventually form are essentially artificial evaporite deposits.

Although seawater is certainly plentiful, there are only a relatively few places around the world where the solar evaporation process is undertaken. The criteria that a site must meet are a combination of physical and economic factors. In addition to an abundant source of salt water, other necessary conditions include long, uninterrupted periods of sunshine; little rainfall; ample air movement; flat, impervious soil; and proximity to a major market or access to inexpensive transportation. In the United States, these conditions exist at the Great Salt Lake in Utah and near the cities of San Diego and San Francisco in California.

At the southern end of San Francisco Bay, it takes nearly 38,000 liters (10,000 gallons) of water to produce 900 kilograms (1 ton) of salt. Each year about 1 million tons of salt are harvested in the Bay Area. The process involves a large amount of land and 560 kilometers (350 miles) of earthen dikes to confine the evaporating water in ponds of the right size and depth. The cycle, from bay water to harvested salt, requires about five years.

Bay water is admitted to the first of the concentrating ponds in early summer when the discharge of the rivers entering the bay is reduced and the water in the bay is at its highest salinity. Once in the ponds, the brine is moved from pond to pond as it reaches certain pre-established levels of salinity. Evaporation continues until the solution is concentrated enough to crystallize virtually pure sodium chloride (NaCl).

During the first stages in the evaporation process, small fish thrive and are netted commercially for use as bait by sport fishermen. Toward the middle of the cycle, the brine becomes too concentrated for fish to survive. Instead, tiny brine shrimp find this environment to their liking. They are harvested for sale as fish food for aquariums. As the brine becomes more concentrated in the final concentrating pond, most plant and animal life begins to die. Gypsum (hydrous calcium sulfate) precipitates during this stage, which is now just about fully saturated with respect to sodium chloride.

In early spring, crystallization ponds are rolled extremely flat, with only a slight incline to permit drainage. In April or May, the saturated brine is pumped into these carefully cleaned 20- to 60-acre crystallizers. Shortly after these ponds are filled to a depth of about 0.5 meter, tiny crystals of pure salt begin to grow. By fall, a 12- to 15-centimeter-thick layer of salt crystals has collected on the bottom. The remaining fluid, called *bittern,* is rich in magnesium, bromine, and potassium. The bittern is drained, and the salt is ready for harvest, five years after the seawater first entered the process.

The salt harvest takes place between September and December, before the rainy season starts. It is accomplished by a special harvester that picks up the salt, leaving the compacted-earth bottom virtually undisturbed (Figure 6.A). The salt is loaded by the harvester into miniature railroad cars and is transported by small locomotives to a processing site.

FIGURE 6.A
Harvesting salt produced by the evaporation of seawater. (Photo courtesy of the Leslie Salt Company)

TABLE 6.2
Classification of sedimentary rocks.

DETRITAL ROCKS			
Texture	**Sediment Name and Particle Size**	**Comments**	**Rock Name**
Clastic	Gravel (>2 mm)	Rounded rock fragments	Conglomerate
		Angular rock fragments	Breccia
	Sand (1/16–2 mm)	Quartz predominates	Quartz sandstone
		Quartz with considerable feldspar	Arkose
		Dark color; quartz with considerable feldspar, clay, and rock fragments	Graywacke
	Mud (<1/16 mm)	Splits into thin layers	Shale
		Breaks into clumps or blocks	Mudstone
CHEMICAL ROCKS			
Group	**Texture**	**Composition**	**Rock Name**
Inorganic	Clastic or nonclastic	Calcite, $CaCO_3$	Limestone
	Nonclastic	Dolomite, $CaMg(CO_3)_2$	Dolomite (dolostone)
	Nonclastic	Microcrystalline quartz, SiO_2	Chert
	Nonclastic	Halite, NaCl	Rock salt
	Nonclastic	Gypsum, $CaSO_4 \cdot 2H_2O$	Rock gypsum
Biochemical	Clastic or nonclastic	Calcite, $CaCO_3$	Limestone
	Nonclastic	Microcrystalline quartz, SiO_2	Chert
	Nonclastic	Altered plant remains	Coal

texture consist of discrete fragments and particles that are cemented and compacted together. Although cement is present in the spaces between particles, these openings are rarely filled completely. An examination of Table 6.2 reveals that *all* detrital rocks have a clastic texture. The table also shows that some chemical sedimentary rocks may also exhibit this texture. For example, coquina, the limestone composed of shells and shell fragments, is obviously as clastic as a conglomerate or sandstone. The same applies for some varieties of oolitic limestone.

Some chemical sedimentary rocks have a **nonclastic** texture in which the minerals form a pattern of interlocking crystals. The crystals may be microscopically small or large enough to be visible without magnification. Common examples of rocks with nonclastic textures are those deposited when seawater evaporates (Figure 6.12). The materials composing many other nonclastic rocks may actually have originated as detrital deposits. In these instances, the particles probably consisted of shell fragments and other hard parts rich in calcium carbonate or silica. The clastic nature of the grains was subsequently obliterated or obscured because the particles recrystallized when they were consolidated into limestone or chert.

Nonclastic rocks consist of intergrown crystals, and some may resemble igneous rocks, which are also crystalline. The two rock types are usually easily distinguished since the minerals that compose nonclastic sedimentary rocks are quite unlike those found in most igneous rocks. For example, rock salt, rock gypsum, and some forms of limestone consist of intergrown crystals, but the minerals that compose these rocks (halite, gypsum, and calcite) are seldom associated with igneous rocks.

BOX 6.2

Sea-Floor Sediments and Climatic Change

Although direct measurements of atmospheric conditions are a relatively recent development, it is nevertheless possible to study the earth's climatic history by indirect means. Sea-floor sediments represent one source of data that allows us to better understand atmospheric changes through time.

The earth's pattern of climates not only has a strong influence on human activities but also is a major factor affecting the geologic development of landscapes. Whether it is the selection of crops that are grown in an agricultural region or the type and amount of weathering influencing a rocky outcrop, climate plays a significant role. The earth's climates have not always exhibited the same geographic distribution as exists today. To the contrary, patterns gradually shift with the passage of time. Today, scientists recognize that climate is variable on virtually all scales of time and space. In order to understand the behavior of the atmosphere and to predict future trends in climate, it is important that past fluctuations be documented and comprehended as fully as possible.

A great deal of high technology and precision instrumentation is currently available to study the present composition and dynamics of the atmosphere. But these tools are only recent inventions and therefore provide data for only a short time span. Moreover, instrumental records of any kind go back only a couple of centuries at best, and the farther back we go, the more incomplete and unreliable the data become. To overcome this lack of direct measurements, scientists must decipher and reconstruct past climates by using indirect evidence; that is, they must examine and analyze phenomena that respond to and reflect changing atmospheric conditions. One such reservoir of data is the sediments on the ocean floor.

Although sea-floor sediments are of many types, most contain the remains of organisms that once lived near the sea surface (the ocean-atmosphere interface). When such near-surface organisms die, their shells slowly settle to the floor of the ocean, where they become part of the sedimentary record. One reason why sea-floor sediments are useful recorders of worldwide climatic change is that the numbers and types of organisms living near the sea surface change as the climate changes. This principle is explained by Richard Foster Flint as follows:

> We would expect that in any area of the ocean/atmosphere interface the average annual temperature of the surface water of the ocean would approximate that of the contiguous atmosphere. The temperature equilibrium established between surface seawater and the air above it should mean that . . . changes in climate should be reflected in changes in organisms living near the surface of the deep sea. . . . When we recall that the sea-floor sediments in vast areas of the ocean consist mainly of shells of pelagic foraminifers, and that these

FIGURE 6.12
Like other evaporites, this sample of rock salt is said to have a nonclastic texture because it is composed of intergrown crystals.

SEDIMENTARY ENVIRONMENTS

As stated at the beginning of the chapter, sedimentary rocks are important in the interpretation of earth history. By understanding the conditions under which sedimentary rocks form, geologists can often deduce the history of a rock, including information about the origin of its component particles, the method and length of its transport, and the nature of the place where the grains eventually came to rest; that is, the environment of deposition.

animals are sensitive to variations in water temperature, the connection between such sediments and climatic change becomes obvious.*

Thus, in seeking to understand climatic change, as well as other environmental transformations, scientists have become increasingly interested in the huge reservoir of data in sea-floor sediments. Ever since the late 1960s, the National Science Foundation has been involved in a major international ocean-drilling program. The Deep Sea Drilling Project and its successor, the Ocean Drilling Program, have used specially designed research vessels capable of drilling into the ocean floor and collecting cores of deep-sea sediments (Figure 6.B). The sediment cores have proven to be excellent sources of useful data that have greatly expanded our understanding of past climates. One notable example of the importance of sea-floor sediments to our understanding of climate relates to unraveling the fluctuating atmospheric conditions of the most recent ice age. The record of temperature changes contained in cores of sediment from the ocean floor have proven critical to our present understanding of this recent span of earth history.†

*Richard Foster Flint, *Glacial and Quaternary Geology* (New York: John Wiley & Sons, Inc., 1971), p. 718.

†For more on this topic, see "Causes of Glaciation" in Chapter 12.

FIGURE 6.B
The *JOIDES Resolution,* the drilling ship of the Ocean Drilling Program. The sea-floor sediments recovered by this and other research vessels provide scientists with data that allows them to reconstruct past climates. (Photo courtesy of the Ocean Drilling Program)

Sediments are deposited at the earth's surface. Thus, they hold many clues about the physical, chemical, and biological conditions that existed in the areas where the materials accumulated. By applying a thorough knowledge of present-day conditions, geologists attempt to reconstruct the ancient environments and geographical relationships of an area at the time a particular set of sedimentary layers were deposited. Such analyses often lead to the creation of maps, which depict the distribution of land and sea, mountains and plains, deserts and glaciers, and other environments of deposition.

Sedimentary environments are commonly placed into one of two broad categories: *terrestrial* (continental) or *marine*. Since the shore zone exhibits characteristics of both, it can be considered transitional between land and sea. Figure 6.13 divides the two broad categories of terrestrial and marine into several major sedimentary environments. Chapters 10 through 14, as well as portions of Chapter 19, will describe these environments in detail. Each is an area where sediment accumulates and where organisms live and die. Each produces a characteristic sedimentary rock or assemblage that reflects prevailing conditions.

When a series of sedimentary layers is studied, we can see the successive changes in environmental conditions that occurred at a particular place with the passage of time. Changes in past environments may also be seen when a single unit of sedimentary rock is traced laterally. This is true because at any one time many different depositional environments can exist over a broad area. For example, when sand is accumulating in a beach environment, finer muds are often being deposited in quieter offshore waters. Still farther out, perhaps in a zone where biological activity is high and land-derived sediments are scarce, the deposits consist largely of the calcareous remains of small organisms. In this example different sediments are accumulating adjacent to one another

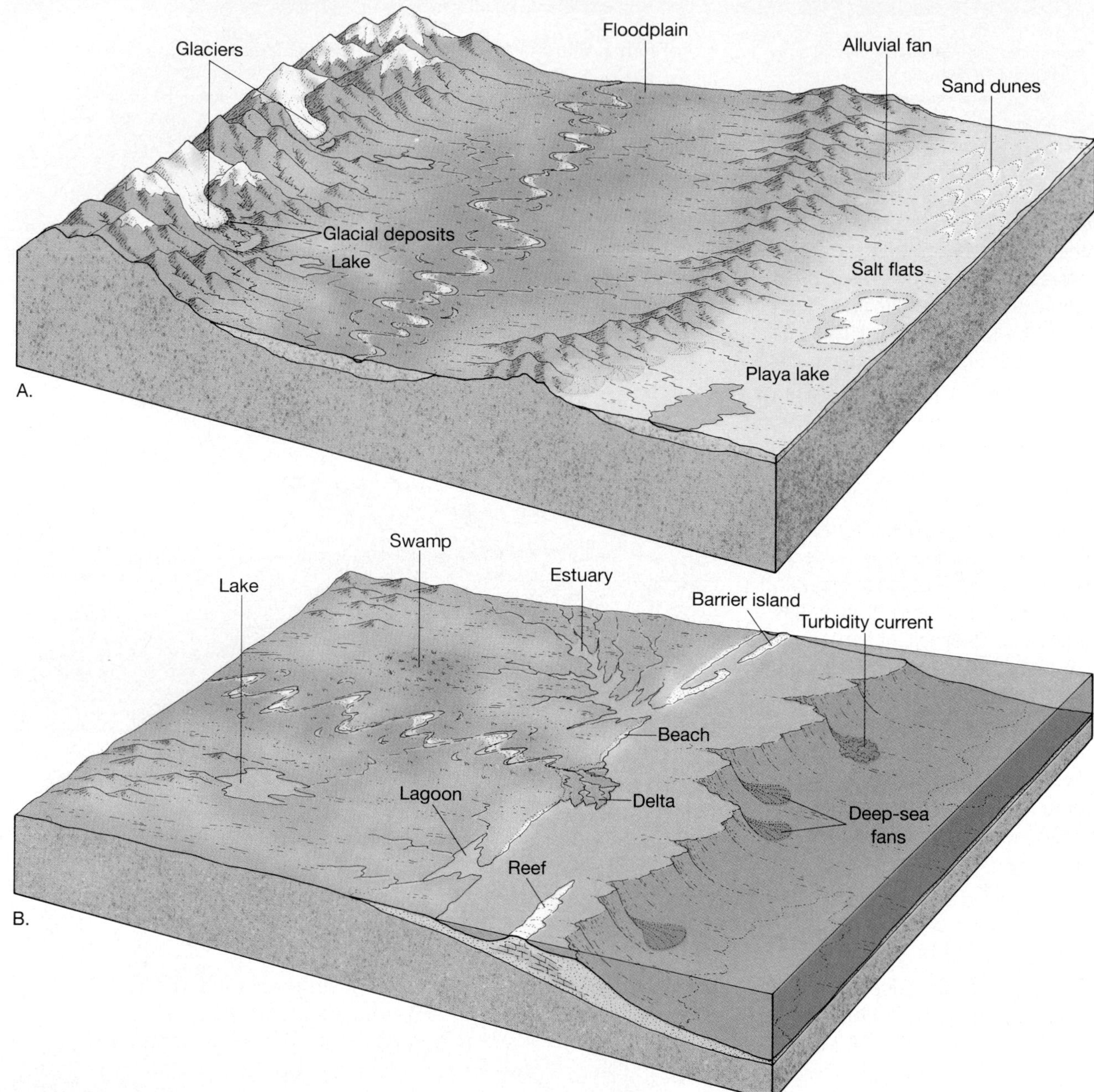

FIGURE 6.13
Sedimentary environments are those places where sediment accumulates. Each is characterized by certain physical, chemical, and biological conditions. Because each sediment contains clues about the environment in which it was deposited, sedimentary rocks are important in the interpretation of earth history. A large number of important sedimentary environments are represented in these two idealized diagrams. Those labeled in part **A** are all considered terrestrial or continental environments. Shoreline (transitional) and marine environments are depicted in part **B.**

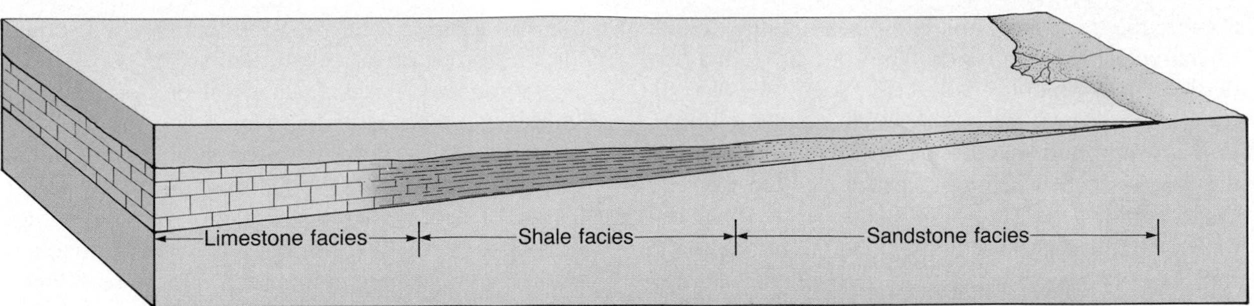

FIGURE 6.14
When a single sedimentary layer is traced laterally, we may find that it is made up of several different rock types. This can occur because many sedimentary environments can exist at the same time over a broad area. The term *facies* is used to describe such sets of sedimentary rocks. Each facies grades laterally into another that formed at the same time but in a different environment.

FIGURE 6.15
This outcrop of sedimentary strata illustrates the characteristic layering of this group of rocks. (Photo by Stephen Trimble)

at the same time. Each unit possesses a distinctive set of characteristics that reflects the conditions in a particular environment. To describe such sets of sediments, the term **facies** is used. When a sedimentary unit is examined in cross-section from one end to the other, each facies grades laterally into another that formed at the same time but which exhibits different characteristics (Figure 6.14). Commonly the merging of adjacent facies tends to be a gradual transition rather than a sharp boundary, but abrupt changes do sometimes occur.

SEDIMENTARY STRUCTURES

In addition to variations in grain size, mineral composition, and texture, sediments exhibit a variety of structures. Some, such as graded beds, are created when sediments are accumulating and are a reflection of the transporting medium. Others, such as mud cracks, form after the materials have been deposited and result from processes occurring in the environment. When present, sedimentary structures provide additional information that can be useful in the interpretation of earth history.

Sedimentary rocks form as layer upon layer of sediment accumulates in various depositional environments. These layers, called **strata**, or **beds**, are probably the single most common and characteristic feature of sedimentary rocks (Figure 6.15). The thickness of beds ranges from microscopically thin to tens of meters thick. Separating the strata are **bedding planes**, flat surfaces along which rocks tend to separate or break. Changes in the grain size or in the composition of the sediment being deposited can create bedding planes. Pauses in deposition can also lead to layering because chances are slight that newly deposited material will be exactly the same as previously deposited sediment. Generally each bedding plane marks the end of one episode of sedimentation and the beginning of another.

Since sediments usually accumulate as particles settle from a fluid, most strata were originally deposited as horizontal layers. There are circumstances, however, when sediments do not accumulate in horizontal beds. Sometimes when a bed of sedimentary

FIGURE 6.16
The cross-bedding in this sandstone indicates it was once a sand dune. (Photo by Stephen Trimble)

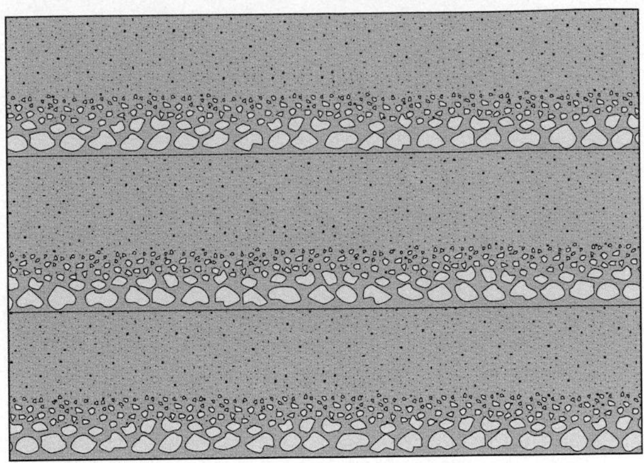

FIGURE 6.17
Graded beds. Each layer grades from coarse at its base to fine at the top.

rock is examined, we see layers inclined to the horizontal. Such layering is termed **cross-bedding** and is most characteristic of sand dunes, river deltas, and certain stream channel deposits (Figure 6.16).

Graded beds represent another special type of bedding. In this case the particles within a single sedimentary layer gradually change from coarse at the bottom to fine at the top (Figure 6.17). Graded beds are most characteristic of rapid deposition from water containing sediment of varying sizes. When a current experiences a rapid energy loss, the largest particles settle first, followed by successively smaller grains. The deposition of a graded bed is most often associated with a turbidity current, a mass of sediment-choked water that is denser than clear water and moves downslope along the bottom of a lake or the ocean.*

As geologists examine sedimentary rocks, much can be deduced. A conglomerate, for example, may indicate a high-energy environment, such as a rushing stream, where only coarse materials can settle out (Figure 6.18). If the rock is arkose, it may signify a dry climate where little chemical alteration of feldspar is possible. Carbonaceous shale is a sign of a low-energy, organic-rich environment, such as a swamp or lagoon.

Other features found in some sedimentary rocks also give clues to past environments. Ripple marks are such a feature. **Ripple marks** are small waves of sand that develop on the surface of a sediment layer by the action of moving water or air (Figure 6.19A). The ridges form at right angles to the direction of motion. If the ripple marks were formed by air or water moving in essentially one direction, their form will be asymmetrical. These *current ripple marks* will have steeper sides in the downcurrent direction and more gradual slopes on the upcurrent side. Ripple marks produced by a stream flowing across a sandy channel or by wind blowing over a sand dune are two common examples of current ripples. When present in solid rock, they may be used to determine the direction of movement of ancient wind or water currents. Other ripple marks have a symmetrical form. These features, called *oscillation ripple marks,*

*More on these currents and graded beds may be found in the section on "Submarine Canyons and Turbidity Currents" in Chapter 19.

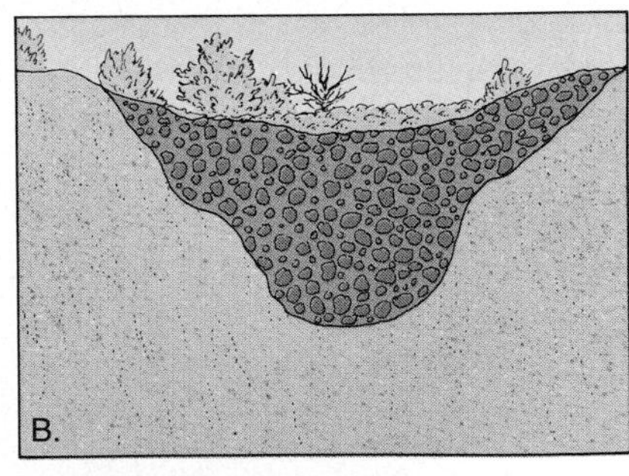

FIGURE 6.18
Cross section of an ancient stream channel filled with conglomerate. In a high-energy environment, such as a rushing stream, only the coarse materials can settle out. (Photo by W. R. Hansen, U.S. Geological Survey)

BOX 6.3

Types of Fossilization

Fossils are of many types. The remains of relatively recent organisms may not have been altered at all. Such objects as teeth, bones, and shells are common examples. Far less common are entire animals, flesh included, that have been preserved because of rather unusual circumstances. Remains of prehistoric elephants called *mammoths* that were frozen in the Arctic tundra of Siberia and Alaska are examples, as are the mummified remains of sloths preserved in a dry cave in Nevada.

Given enough time, the remains of an organism are likely to be modified. Often fossils become *petrified* (literally, "turned into stone"), meaning that the small internal cavities and pores of the original structure are filled with precipitated mineral matter (Figure 6.C). In other instances, *replacement* may occur. Here, the cell walls and other solid material are removed and replaced with mineral matter. Sometimes the microscopic details of the replaced structure are faithfully retained.

Molds and casts constitute another common class of fossils. When a shell or other structure is buried in sediment and then dissolved by underground water, a *mold* is created. The mold faithfully reflects only the shape and surface markings of the organism; it does not reveal any information concerning its internal structure. If these hollow spaces are subsequently filled with mineral matter, *casts* are created (Figure 6.D).

A type of fossilization called *carbonization* is particularly effective in preserving leaves and delicate animal forms. It occurs when fine sediment encases the remains of an organism. As time passes, pressure squeezes out the liquid and gaseous components and leaves behind a thin residue of carbon (Figure 6.E). Black shales deposited as organic-rich mud in oxygen-poor environments often contain abundant carbonized remains. If the film of carbon is lost from a fossil preserved in fine-grained sediment, a replica of the surface, called an *impression,* may still show considerable detail (Figure 6.F).

Delicate organisms, such as insects, are difficult to preserve and consequently are quite rare in the fossil record. Not only must they be protected from decay; they also must not be subjected to any pressure that would crush them. One way in which some insects have been preserved is in *amber,* the hardened resin of ancient trees. The fly in Figure 6.G was preserved after being trapped in a drop of sticky resin. Resin sealed off the insect from the atmosphere and protected the remains from damage by water and air. As the resin hardened, a protective, pressure-resistant case was formed.

In addition to the fossils already mentioned, there are numerous other types, many of them only traces of prehistoric life. Examples of such indirect evidence include:

1. *Tracks*—footprints made by animals in soft sediment that was later lithified (Figure 6.H).
2. *Burrows*—tubes in sediment, wood, or rock made by an animal. These holes may later become filled with mineral matter and preserved. Some of the oldest-known fossils are believed to be worm burrows.
3. *Coprolites*—fossil dung and stomach contents that can provide useful information pertaining to food habits of organisms.
4. *Gastroliths*—highly polished stomach stones that were used in the grinding of food by some extinct reptiles.

FIGURE 6.C–6H
There are many types of fossilization. Six examples are shown here. **6.C** Petrified wood in Petrified Forest National Park, Arizona. **6.D** Natural casts of shelled invertebrates. **6.E** A fossil bee preserved as a thin carbon film. **6.F** Impressions are common fossils and often show considerable detail. **6.G** Insect in amber.
6.H Dinosaur footprint in fine-grained limestone near Tuba City, Arizona. (Photos 6.C, 6.D, 6.F, and 6.H, by E. J. Tarbuck; Photo 6.E courtesy of the National Park Service; Photo 6.G courtesy of Ward's Natural Science Establishment, Inc., Rochester, N.Y.)

6C.

6D.

6E.

6F.

6G.

6H.

result from the back and forth movement of surface waves in a shallow nearshore environment.

Mud cracks (Figure 6.19B) indicate that the sediment in which they were formed was alternately wet and dry. When exposed to air, wet mud dries out and shrinks, producing cracks. Mud cracks are associated with such environments as shallow lakes and desert basins.

FOSSILS

Fossils, the remains or traces of prehistoric life, are perhaps the most important inclusions in sedimentary rocks. Fossils are important tools used to interpret the geologic past. Knowing the nature of the life forms that existed at a particular time helps researchers understand past environmental conditions. Furthermore, fossils are important time indicators and play a key role in correlating rocks of similar ages but from different places.*

Only a tiny fraction of the organisms that lived during the geologic past have been preserved as fossils. Normally, the remains of an animal or plant are totally destroyed. Under what circumstances are they preserved? Two special conditions seem to be necessary: rapid burial and the possession of hard parts.

Usually when an organism perishes, its remains are quickly eaten by scavengers or decomposed by bacteria. Occasionally, however, the remains are buried by sediment. When this occurs the remains are removed from the environment where destructive forces operate most effectively. Rapid burial therefore is an important condition favoring preservation.

In addition, organisms have a much better chance of being preserved as part of the fossil record if they have hard parts. Although traces and imprints of soft-bodied animals such as jellyfish, worms, and insects exist, they are far less common. Flesh usually decays so rapidly that the chance of preservation is exceedingly remote. Shells, bones, and teeth as well as similar hard parts predominate in the record of past life.

Because preservation is contingent on special conditions, the record of life in the geologic past is biased. The fossil remains of those organisms with hard parts that lived in areas of sedimentation are quite abundant. However, we get only an occasional glimpse of the vast array of other life forms that did not meet the special conditions favoring preservation.

*The section entitled "Correlation" in Chapter 8 contains a more detailed discussion of the role of fossils in the interpretation of earth history.

FIGURE 6.19
A. Asymmetrical ripple marks are produced by currents of water or wind. In this photograph the ripple marks indicate that the current moved from right to left. (Photo by Stephen Trimble) **B.** Mud cracks preserved in limestone. Mud cracks form when wet mud or clay dries out and shrinks. (Photo by E. J. Tarbuck)

A.

B.

REVIEW QUESTIONS

1. How does the volume of sedimentary rocks in the earth's crust compare with the volume of igneous rocks in the crust? Are sedimentary rocks evenly distributed throughout the crust?

2. What minerals are most common in detrital sedimentary rocks? Why are these minerals so abundant?

3. What is the primary basis for distinguishing among various detrital sedimentary rocks?

4. The term *clay* can be used in two different ways. Describe the two meanings.

5. Why does shale usually crumble quite easily?

6. How are the degree of sorting and the amount of rounding related to the transportation of sand grains?

7. Distinguish between conglomerate and breccia.

8. Distinguish between the two categories of chemical sedimentary rocks.

9. What are evaporite deposits? Name a rock that is an evaporite.

10. When a body of seawater evaporates, minerals precipitate in a certain order. What determines this order?

11. Each of the following statements describes one or more characteristics of a particular sedimentary rock. For each statement, name the sedimentary rock that is being described.
    (a) An evaporite used to make plaster.
    (b) A fine-grained detrital rock that exhibits *fissility*.
    (c) Dark-colored sandstone containing angular rock particles as well as clay, quartz, and feldspar.
    (d) The most abundant chemical sedimentary rock.
    (e) A dark-colored, hard rock made of microcrystalline quartz.
    (f) A variety of limestone composed of small spherical grains.

12. How is coal different from other biochemical sedimentary rocks?

13. Compaction is an important lithification process with which sediment size?

14. List three common cements for sedimentary rocks. How might each be identified?

15. What is the primary basis for distinguishing among different chemical sedimentary rocks?

16. Distinguish between clastic and nonclastic textures. What type of texture is common to all detrital sedimentary rocks?

17. Some nonclastic sedimentary rocks closely resemble igneous rocks. How might the two be distinguished easily?

18. Why are sea-floor sediments useful in studying climates of the past? (See Box 6.2).

19. What is probably the single most characteristic feature of sedimentary rocks?

20. Distinguish between cross-bedding and graded bedding.

21. Examine the photograph of ripple marks in Figure 6.19A. The caption indicates that the current which produced these features moved from right to left. How is this fact determined?

22. List two conditions that favor the preservation of organisms as fossils.

23. What type of fossilization is indicated by each of the following statements? Which one is an example of indirect evidence? (See Box 6.3.)
    (a) A leaf preserved as a thin carbon film.
    (b) Small internal cavities and pores of a log are filled with mineral matter.
    (c) Fossil dung.
    (d) This is created when a mold is filled with mineral matter.

KEY TERMS

beds (strata) (p. 154)
bedding plane (p. 154)
biochemical (p. 143)
cementation (p. 147)
chemical sedimentary rock (p. 138)
clastic (p. 147)
compaction (p. 146)
cross-bedding (p. 155)
detrital sedimentary rock (p. 138)
evaporite deposit (p. 146)
facies (p. 154)
fissility (p. 140)
fossil (p. 158)
graded bed (p. 155)
lithification (p. 146)
mud crack (p. 158)
nonclastic (p. 149)
ripple mark (p. 155)
salt flat (p. 146)
sediment (p. 138)
sorting (p. 141)
strata (beds) (p. 154)

7

Metamorphic Rocks

Deformed and metamorphased rocks in the Panamint Mountains east of Salina Valley, California. (Photo by Michael Collier) (Top photo by Michael Collier)

Extensive areas of metamorphic rocks are exposed on every continent in the relatively flat regions known as shields. Moreover, metamorphic rocks are an important component of many mountain belts, where they make up a large portion of a mountain's crystalline core (see chapter opening photos). Even the stable continental interiors, which are generally covered by sedimentary rocks, are underlain by metamorphic basement rocks (Figure 7.1). In all of these settings the metamorphic rocks are usually highly deformed and intruded by igneous masses. Indeed, significant parts of the earth's continental crust are composed of metamorphic and associated igneous rocks.

Metamorphic rocks provide important clues about the geologic processes that operate within the earth's crust. Unlike some igneous and sedimentary processes which take place in surface or near-surface environments, metamorphism always occurs deep within the earth beyond our direct observation. Notwithstanding this significant obstacle, geologists have developed techniques that have allowed them to learn a great deal about the conditions under which metamorphic rocks form.

METAMORPHISM

As we learned in the section on the rock cycle in Chapter 1, metamorphism involves the transformation of pre-existing rocks by heat, pressure, and chemically active fluids. Metamorphic rocks can form from igneous, sedimentary, or even from other metamorphic rocks. The term for this process is very appropriate because it literally means to "change form." The agents of change include heat, pressure, and chemically active fluids, while the changes that occur are textural as well as mineralogical.

In some instances metamorphic rocks are only slightly changed. For example, under low-grade met-

FIGURE 7.1
In the stable interior of a continent, metamorphic rocks are generally covered by layers of sedimentary rocks. The Precambrian Vishnu Schist shown here exposed in the inner gorge of the Grand Canyon is an example of metamorphic basement rocks. (Photo by E. J. Tarbuck)

amorphism the common sedimentary rock shale becomes the metamorphic rock called slate. Although slate is more compact than shale, hand samples of these rocks are sometimes difficult to distinguish. In other cases the transformation is so complete that the identity of the original rock cannot be determined. In high-grade metamorphism, such features as bedding planes, fossils, and vesicles that may have existed in the parent rock are completely destroyed. Further, when rocks at depth are subjected to directional pressure, they exhibit ductile flow and bend into a series of intricate folds (Figure 7.2). Such deformation often exhibits remarkable uniformity no matter what the scale of observation. For example, in Figure 7.3 the orientation and shape of the folds are the same from mountain range down to thin section. In the most extreme metamorphic environments, the temperatures approach those at which rocks melt. However, during metamorphism the deformed material must remain solid, for once melting occurs, we have entered the realm of igneous activity.

The process of metamorphism takes place when rock is subjected to conditions unlike those in which it formed; the rock becomes unstable and gradually changes until a state of equilibrium with the new environment is reached. The changes occur at the temperatures and pressures existing in the region extending from a few kilometers below the earth's surface to the crust-mantle boundary.

Metamorphism most often occurs in one of three settings. First, during mountain building great quantities of rock are subjected to the intense stresses and temperatures associated with large-scale deformation. The end result may be extensive areas of metamorphic rocks that are said to have undergone **regional metamorphism**. The greatest volume of metamorphic rock is produced in this fashion. Second, when rock is in contact with, or near, a mass of magma, **contact metamorphism** takes place. In this circumstance the changes are caused primarily by the high temperatures of the molten material, which in effect "bake" the surrounding rock. The third and least common type of metamorphism occurs along *fault zones*. Here rock is broken and distorted as crustal blocks on opposite sides of a fault grind past one another.

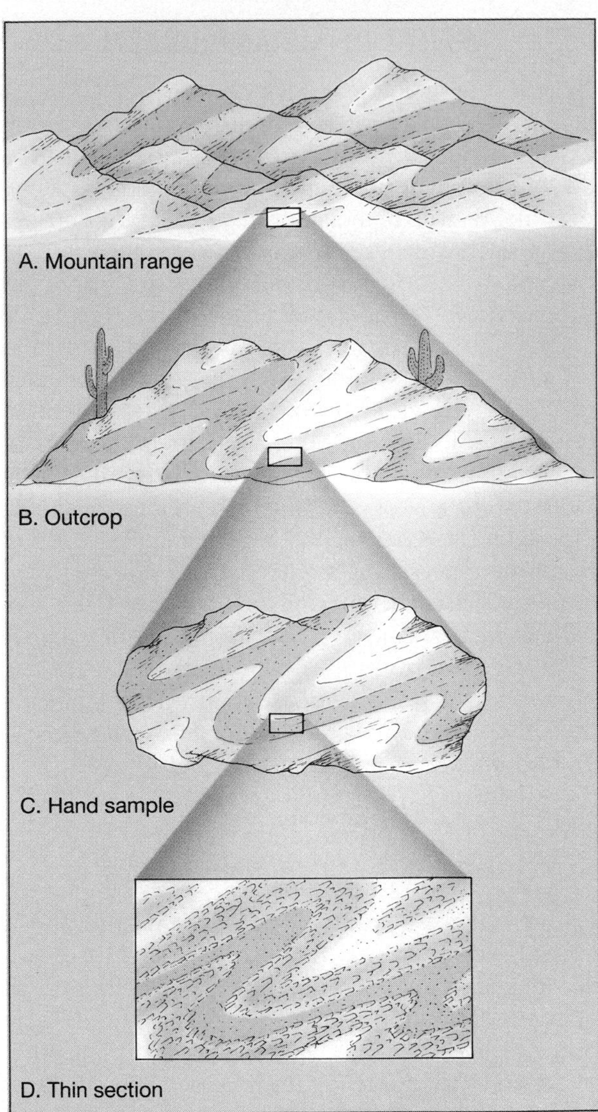

FIGURE 7.3
Folds produced during metamorphism are often remarkably uniform in orientation and shape on all scales of observation, from mountain range to thin section. (After Myron G. Best)

FIGURE 7.2
Intricate folds in metamorphic rocks exposed in a road cut at Cedar Grove, Kings Canyon National Park, California. (Photo courtesy of Kings Canyon National Park)

AGENTS OF METAMORPHISM

As stated earlier, the agents of metamorphism include *heat, pressure,* and *chemically active fluids.* During metamorphism, rocks are often subjected to all three metamorphic agents simultaneously. However, the degree of metamorphism and the contribution of each agent vary greatly from one environment to another. In low-grade metamorphism, rocks are subjected to temperatures and pressures only slightly greater than those associated with the lithification of sediments. High-grade metamorphism, on the other hand, involves extreme conditions closer to those at which rocks melt. In addition, the mineralogy of the parent rock determines, to a large extent, the degree to which various metamorphic agents will cause change. For example, when an intruding igneous mass enters a rock unit, hot, ion-rich fluids circulate through the host rock. If the host rock is quartz sandstone, very little alteration may take place. On the other hand, if the host rock is limestone, the impact of these fluids can be dramatic and the effects of metamorphism may extend for several kilometers from the igneous mass.

Heat as a Metamorphic Agent

Perhaps the most important agent of metamorphism is *heat*, because it provides the energy to drive chemical reactions that result in the recrystallization of minerals. Rocks formed near the earth's surface may be subjected to intense heat when they are intruded by molten material rising from below. The effects of contact metamorphism are most apparent when it occurs at or near the surface where the temperature contrast between the molten, intrusive rock and the country rock is most pronounced. Here the adjacent country rock is "baked" by the emplaced magma. In this high-temperature and high-pressure environment, the boundary that forms between the intrusive igneous body and the altered rocks is usually quite distinct.

Rocks that originate in a surface environment may also be subjected to extreme temperatures if they are subsequently buried deep within the earth. Recall that temperatures increase with depth at a rate known as the geothermal gradient. In the upper crust, this increase in temperature averages between 20°C to 30°C per kilometer. As we discussed earlier, earth materials are continually being transported to great depths at convergent plate boundaries. When buried to a depth of only a few kilometers, certain minerals, such as clay, become unstable and begin to recrystallize into minerals, such as muscovite, that are stable in this environment. Other minerals, particularly those found in crystalline igneous rocks, are stable at relatively high temperatures and pressures and therefore require burial to 20 kilometers or more before metamorphism will occur.

Pressure as a Metamorphic Agent

Pressure, like temperature, also increases with depth. Buried rocks are subjected to the force, or **stress**, exerted by the load above. This confining pressure is analogous to air pressure where the force is applied equally in all directions.

In addition to the confining pressure exerted by the load of material above, rocks are also subjected to stresses during the process of mountain building. Most often these directional forces are compressional and act to shorten a rock body. In some environments, however, the stresses are tensional and tend to elongate, or pull apart, the rock mass (Figure 7.4). Directional stress can also cause a rock to **shear**. Shearing is similar to the slippages that occur between individual cards when a deck is held between your hands and the top of the deck is moved relative to the bottom. In near-surface environments shearing results when relatively brittle rock is broken into thin slabs that are forced to slide past one another. This deformation grinds and pulverizes the original mineral grains into small fragments. By contrast, rock located at great depths is warmer and behaves plastically during deformation. This accounts for its ability to flow and bend into intricate folds when subjected to shearing (see Figure 7.2).

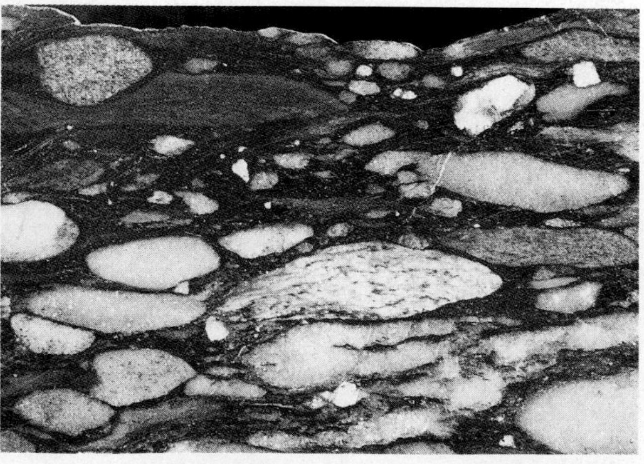

FIGURE 7.4
Metaconglomerate. These once nearly spherical pebbles have been heated and stretched into elongated structures. (Photo by E. J. Tarbuck)

Chemical Activity and Metamorphism

Chemically active fluids, most commonly water containing ions in solution, also enhance the metamorphic process. Some water is contained in the pore spaces of virtually every rock. In addition, many minerals are hydrated, and thus contain water within their crystalline structures. When deep burial occurs, water is forced out of the mineral structures and is then available to aid in chemical reactions. Further, as heat is applied the dehydration of minerals releases water. Water that surrounds the crystals acts as a catalyst by aiding ion migration. In some instances the minerals recrystallize to form more stable configurations. In other cases, ion exchange among minerals results in the formation of completely new minerals. Complete alteration of rock by hot, mineral-rich water has been observed in the near-surface environment of Yellowstone National Park. Further, along oceanic ridges seawater circulates through the still-hot basaltic rocks, transforming iron- and magnesium-rich minerals into metamorphic minerals such as serpentine and talc.

TEXTURAL AND MINERALOGICAL CHANGES

The degree of metamorphism is reflected in the texture and mineralogy of metamorphic rocks. When rocks are subjected to low-grade metamorphism, they become more compact and thus more dense. A common example is the metamorphic rock slate, which forms when shale is subjected to temperatures and pressures only slightly greater than those associated with sedimentary processes. In this case, directed pressure causes the microscopic clay minerals in shale to align into the more compact arrangement found in slate. Under more extreme conditions, pressure causes certain minerals to recrystallize. As

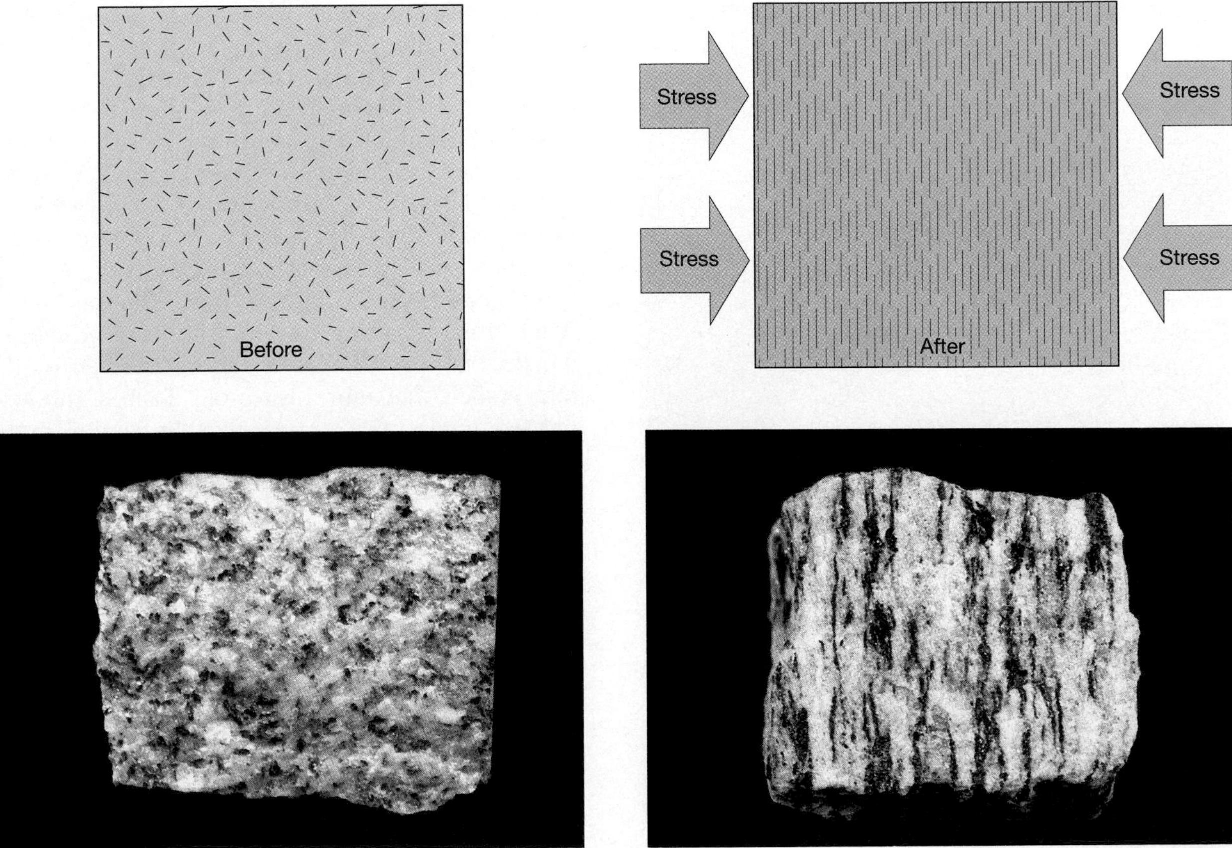

FIGURE 7.5
Under the pressures of metamorphism, some mineral grains become reoriented and aligned at right angles to the stress. The resulting planar orientation of mineral grains gives the rock a foliated texture. If the coarse-grained igneous rock (granite) on the left underwent intense metamorphism, it could end up closely resembling the metamorphic rock on the right (gneiss). (Photos by E. J. Tarbuck)

described earlier, water is believed to play a very important part in the recrystallization process by aiding the migration of ions. In general, recrystallization encourages the growth of larger crystals. Consequently, many metamorphic rocks consist of visible crystals, much like phaneritic igneous rocks. The crystals of some minerals, such as micas, which have a sheet structure, and hornblende, which has an elongated structure, recrystallize with a preferred orientation. The new orientations will be essentially perpendicular to the direction of the compressional force. The resulting mineral alignment usually gives the rock a layered or banded appearance termed **foliation** (Figure 7.5). Simply, foliation results whenever the minerals and structural features of a rock are brought into parallel alignment.

Various types of foliation exist, depending to a large extent upon the degree of metamorphism and the mineralogy of the parent rock. For example, during the transformation of shale to slate, clay minerals, which are stable at the surface, recrystallize into minute mica flakes, which are stable at much higher temperatures and pressures. Further, during recrystallization these fine-grained mica crystals become aligned so that their flat surfaces are nearly parallel. Consequently, slate can be split easily along these layers of mica grains into rather flat slabs. This property is called **rock cleavage** or **slaty cleavage,** to differentiate it from the type of cleavage exhibited by minerals (Figure 7.6). Since the mica flakes composing slate are minute, slate is not usually visibly foliated. However, because it exhibits excellent rock cleavage, which is evidence that its minerals are aligned, slate is considered to be foliated.

FIGURE 7.6
Rock cleavage in schist, northeastern California. (Photo by E. J. Tarbuck)

Under more extreme temperature-pressure regimes, the very fine mica grains found in slate will grow many times larger. These mica crystals, which are about a centimeter in diameter, will give the rock a platy or scaly appearance. This type of foliation is called **schistosity**, and a rock having this texture is called *schist.* Many types of schist exist and are named according to their mineral constituents. By far the most abundant are the mica schists, which usually contain muscovite and/or biotite plus some other minerals (Figure 7.7).

During high-grade metamorphism, ion migrations can be extreme enough to cause minerals to segregate. An example of a metamorphic rock that exhibits mineral segregation is shown in Figure 7.5. Notice that the dark and light silicate minerals have separated, giving the rock a banded appearance. Metamorphic rocks with this texture are called *gneiss* (pronounced "nice") and are quite common. Gneiss often forms from the metamorphism of granite or diorite, but can form from gabbro or even by the high-grade metamorphism of shale. Although banded, gneiss will not usually split parallel to the crystals as easily as slate.

Not all metamorphic rocks have a foliated texture. Such rocks are said to be **nonfoliated.** Metamorphic rocks composed of only one mineral which forms equidimensional crystals are as a rule not visibly foliated. For example, when a fine-grained limestone is metamorphosed, the small calcite crystals combine to form relatively large interlocking crystals. The resulting rock has a texture similar to that of a coarse-grained igneous rock. This metamorphic equivalent of limestone is *marble.* Although most marbles are nonfoliated, microscopic investigation of marble may reveal some flattening and parallelism of the grains. Some limestones contain thin layers of clay minerals which may become distorted during metamorphism. These "impurities" will often appear as curved bands of dark material flowing through the marble.

In the metamorphism of shale to slate we saw that clay minerals recrystallized to form mica crystals. In

FIGURE 7.7
Mica schist, a common metamorphic rock composed of shiny mica flakes. (Photo by E. J. Tarbuck)

most instances, including this example, the chemical composition of the rock does not change during recrystallization, except for the loss of water and carbon dioxide. Rather, the existing minerals and available ions in the water will recombine to form minerals which are stable in the new environment. A common example is the formation of the metamorphic mineral *wollastonite.* Wollastonite is generated when limestone ($CaCO_3$), which contains abundant sandy material in the form of quartz (SiO_2), is subjected to high temperatures during contact metamorphism. The calcite and quartz crystals chemically react to form wollastonite ($CaSiO_3$), while carbon dioxide is liberated.

In some environments, however, new materials are actually introduced during the metamorphic process. For example, country rock adjacent to a large magma body would be altered by **hydrothermal** (hot water) **solutions** released during the latter stages of crystallization. Many metallic ore deposits were formed by the deposition of minerals from hydrothermal solutions. With the development of the theory of plate tectonics, it became clear that some metal-rich hydrothermal deposits had their origins along ancient oceanic ridges. As seawater percolated through newly formed oceanic crust, it dissolved metallic sulfides from the mafic rocks. The hot, metal-rich fluids rose along fractures and gushed from the sea floor as particle-filled clouds called *black smokers.* Upon mixing with the cold seawater, the sulfides precipitated to form massive metallic deposits like the copper ores mined on the island of Cyprus. Some of the earth's richest copper deposits have formed in this manner.

In summary, the metamorphic process causes many changes in rocks, including increased density, growth of larger crystals, reorientation of the mineral grains into a layered or banded appearance known as foliation, and the transformation of low-temperature minerals into high-temperature minerals. Further, the introduction of ions generates new minerals, some of which are economically important.

COMMON METAMORPHIC ROCKS*

Foliated Rocks

Slate is a very fine-grained foliated rock composed of minute mica flakes. The most noteworthy characteristic of slate is its excellent rock cleavage. This property has made slate a most useful rock for roof and floor tile, blackboards, and billiard tables. Slate is most often generated by the low-grade metamorphism of shale, although less frequently it forms from the metamorphism of volcanic ash. Slate can be almost any color depending on its mineral constituents. Black (carbonaceous) slate contains organic material, red slate gets its color from iron oxide, and green slate is usually composed of chlorite, a mica-like mineral formed by the metamorphism of iron-rich silicates. Because slate forms during low-grade metamorphism, evidence of the original bedding planes is often preserved. However, the orientation of slate's rock cleavage generally trends across the original sedimentary layering (Figure 7.8). Thus, unlike shale, which splits along bedding planes, slate splits across them.

*Please refer to Table 7.1 throughout this discussion.

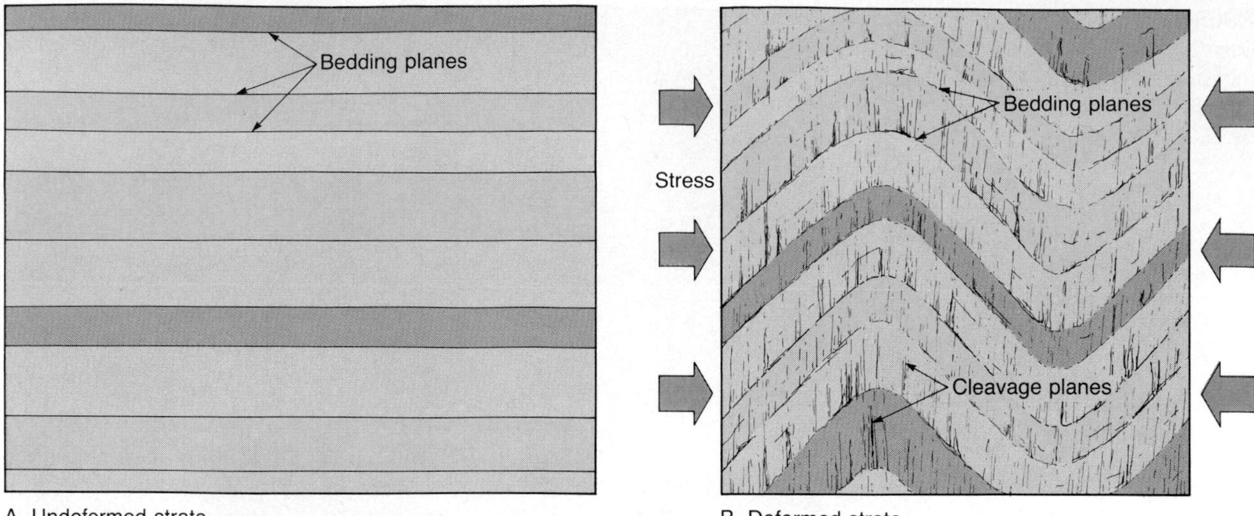

FIGURE 7.8
Illustration showing the relationship between slate cleavage and original bedding planes. **A.** Undeformed strata. **B.** Deformed strata.

Phyllite represents a gradation in metamorphism between slate and schist. Its constituent platy minerals are larger than those in slate, but not yet large enough to be clearly identifiable. Although phyllite appears similar to slate, it can be easily distinguished from slate by its glossy sheen (Figure 7.9). Phyllite usually exhibits rock cleavage and is composed mainly of very fine crystals of either muscovite or chlorite.

Schists are strongly foliated rocks, formed by regional metamorphism, that can be readily split into thin flakes or slabs. By definition, schists contain more than 50 percent platy and elongated minerals that commonly include muscovite, biotite, and amphibole. Like slate, the parent material from which many schists originate is shale, but in the case of schist, the metamorphism is more intense. If the parent rock contained abundant silica, schist will often contain thin layers of quartz and possibly feldspar as well.

The term schist describes the texture of a rock regardless of composition unless specifically included in the rock name. For example, schists composed primarily of muscovite and biotite with lesser amounts of quartz and feldspar are called *mica schists.* Depending upon the degree of metamor-

TABLE 7.1
Common Metamorphic Rocks

Metamorphic Rock	Texture	Parent Rock	Comments
Slate	Foliated	Shale	Very fine grained
Phyllite	Foliated	Shale	Fine- to medium-grained
Schist	Foliated	Shale, granitic and volcanic rocks	Coarse-grained micaceous minerals
Gneiss	Foliated	Shale, granitic and volcanic rocks	Coarse-grained (non-micaceous)
Marble	Nonfoliated	Limestone, dolostone	Composed of interlocking calcite grains
Quartzite	Nonfoliated	Quartz sandstone	Composed of interlocking quartz grains
Hornfels	Nonfoliated	Any fine-grained material	Fine-grained
Migmatite	Weakly foliated	Mixture of granitic and mafic rocks	Composed of contorted layers
Mylonite	Weakly foliated	Any material	Hard, fine-grained rock
Metaconglomerate	Weakly foliated	Quartz-rich conglomerate	Strongly stretched pebbles
Amphibolite	Weakly foliated	Mafic volcanic rocks	Coarse-grained

FIGURE 7.9
Phyllite (left) can be distinguished from slate by its glossy sheen. (Photo by E. J. Tarbuck)

phism and composition of the parent rock, mica schists often contain accessory minerals unique to metamorphic rocks. Some common accessory minerals include garnet, staurolite, and sillimanite, in which case the rock is called garnet-mica schist and so forth (Figure 7.10). Some schists contain graphite which is recovered for use as pencil lead, graphite fibers, and lubricant. In addition, schists may be composed largely of the minerals chlorite or talc, in which case they are called *chlorite schist* and *talc schist,* respectively. Both chlorite and talc schists can form when rocks with a basaltic composition undergo metamorphism.

Gneiss is the term applied to banded metamorphic rocks that contain mostly elongated and granular, as opposed to platy, minerals (see Figure 7.5). The most common minerals found in gneisses are quartz, potassium feldspar, and sodium feldspar. In addition, lesser amounts of muscovite, biotite, and hornblende are common. The segregation of light and dark silicates is developed in gneisses, giving them a characteristic banded texture. Thus, most gneisses consist of alternating bands of white or reddish feldspar-rich zones and layers of dark ferromagnesian minerals. These banded gneisses are often deformed while in a plastic state into rather intricate folds (Figure 7.11). Some gneisses will split readily along the layers of platy minerals, but most break in an irregular fashion, like other crystalline rocks.

FIGURE 7.10
Garnet-mica schist.

Those gneisses that have a composition similar to that of granite are probably derived from granite or its aphanitic equivalent. However, they may also form from the high-grade metamorphism of shale. In this instance, gneiss represents the last rock in the sequence of shale, slate, phyllite, schist, and gneiss. Like schists, gneisses may also include large crystals of accessory minerals such as garnet and staurolite. Gneisses can also be made up primarily of dark minerals such as those that compose basalt. For example, an amphibole-rich rock that exhibits a gneissic texture is called *amphibolite gneiss.*

Nonfoliated Rocks

Marble is a coarse, crystalline rock whose parent rock was limestone or dolomite (Figure 7.12). When a hand sample is examined, marble closely resembles crystalline limestone. Pure marble is white and composed essentially of the mineral calcite. Because of its color and relative softness (hardness of 3), marble is a popular building stone. White marble is particularly prized as a stone from which to carve monu-

FIGURE 7.11
Deformed Precambrian rocks exposed near Colorado National Monument. (Photo by Stephen Trimble)

ments and statues, such as the famous statue of David by Michelangelo.

Often the limestone from which marble forms contains impurities that color the marble. Thus, marble can be pink, gray, green, or even black. Also, when impure limestone is metamorphosed, it may yield a variety of accessory minerals including chlorite, mica, garnet, and commonly wollastonite. When marble forms from limestone interbedded with shales, it will appear banded. Occasionally, marble will split along these bands and reveal the mica minerals that recrystallized from clay minerals. Under extreme deformation, the bands in marble may become highly contorted and give the rock a rather artistic design.

Quartzite is a very hard metamorphic rock most often formed from quartz sandstone (Figure 7.13). Under moderate-to-high-grade metamorphism, the quartz grains in sandstone fuse. The recrystallization is so complete that when broken, quartzite will split across the original quartz grains, rather than between them. In some instances, such sedimentary structures as crossbedding are preserved and give the rock a banded appearance.

Quartzite is typically white, but iron oxide may produce reddish or pinkish stains. Dark mineral grains may impart a gray color. Being mainly quartz, with its equidimensional crystals, quartzite seldom develops a strong metamorphic foliation.

FIGURE 7.12
Marble, a crystalline rock formed by the metamorphism of limestone. (Photo by E. J. Tarbuck)

Quartz sand grains

Photomicrograph (× 26.6)
Original sample width is 1.23 mm

FIGURE 7.13
Quartzite is a nonfoliated metamorphic rock formed from quartz sandstone. The photomicrograph shows the interlocking quartz grains typical of quartzite.

OCCURRENCES OF METAMORPHIC ROCKS

Recall that metamorphic rocks commonly form in one of three environments: along fault zones, in contact with igneous bodies, or during dynamic episodes associated with mountain building.

Metamorphism along Fault Zones

When faulting occurs near the surface where rock behaves like a brittle solid, movement along the fault zone crushes, pulverizes, and generally mills the rock into fine components. The result is a loosely coherent rock called *fault breccia* that is composed of broken and distorted rock fragments (Figure 7.14). Most of the intense deformation associated with fault zones occurs at great depths. In this environment the rocks deform by ductile flow, which generates elongated grains that often give the rock a foliated or lineated appearance. Rocks formed in this manner are termed *mylonites.* Because mylonites may closely resemble hard, fine-grained rocks formed by other processes, their identification requires the use of microscopic techniques.

The quantity of metamorphic rock generated solely by faulting is relatively insignificant when compared to the amount generated by the other two processes. Nevertheless, in some areas these granulated rocks are quite abundant. For example, movements along California's San Andreas fault have created a zone of fault breccia and related rock types nearly 1000 kilometers long and up to 3 kilometers wide.

FIGURE 7.14
Fault breccia consisting of large angular fragments. This outcrop, located in Titus Canyon, Death Valley, California, was produced along a fault zone. The largest dark fragments are about 2 to 3 meters in diameter. (Photo by E. J. Tarbuck)

Contact Metamorphism

Contact metamorphism occurs when molten rock comes into contact with cooler rock. Contact metamorphism is clearly distinguishable only when it occurs at the surface or in a near-surface environment where the temperature contrast between the magma and host rock is great. Undoubtedly, contact metamorphism is also an active process at great depth. However, its effect is blurred due to the general alteration of regional metamorphism.

During contact metamorphism, a zone of alteration called an **aureole** (or halo) forms around the emplaced magma. Small intrusive bodies such as dikes and sills have aureoles only a few centimeters thick. On the other hand, large igneous bodies such as batholiths and laccoliths may form zones of metamorphic rocks a few kilometers or more in thickness (Figure 7.15). These large aureoles often consist of definite zones of metamorphism. Near the magma body, high-temperature minerals such as garnet may form, while farther away such low-grade minerals as chlorite are produced. In addition to the size of the intrusive magma body, the mineral composition of the country rock and the availability of water greatly affect the size of the aureole produced. In chemically active rock such as limestone, the zone of alteration can extend to distances of 10 kilometers or more from the igneous body. Here the occurrence of minerals such as garnet and wollastonite marks the extent of metamorphism.

Most contact metamorphic rocks are fine-grained, dense, tough rocks of various chemical compositions. For example, during contact metamorphism, clay minerals are baked, as if placed in a kiln, and can generate a very hard, fine-grained rock not unlike porcelain. Since directional pressure is not a major factor in the formation of these rocks, they generally are not foliated. The name applied to a wide variety of rather hard, nonfoliated metamorphic rocks is *hornfels*.

When large igneous masses are involved in contact metamorphism, hydrothermal solutions which originate within the magma can migrate great distances. As these solutions percolate through the host rock, chemical reactions with this rock greatly enhance the metamorphic process. Further, ores of numerous metals are thought to result from the emplacement of metallic ions whose source is hydrothermal solutions. These deposits include ores of copper, zinc, lead, iron, and gold.

Regional Metamorphism

By far the greatest quantity of metamorphic rock is produced during regional metamorphism. As stated earlier, regional metamorphism takes place at considerable depths over an extensive area and is associ-

FIGURE 7.15
The dark layer, called a roof pendant, consists of metamorphosed country rock adjacent to the upper part of an igneous pluton. The term *roof pendant* implies that the rock was once the roof of a magma chamber. This roof pendant is in the Sierra Nevada of eastern California, near Split Mountain. (Photo by John S. Shelton)

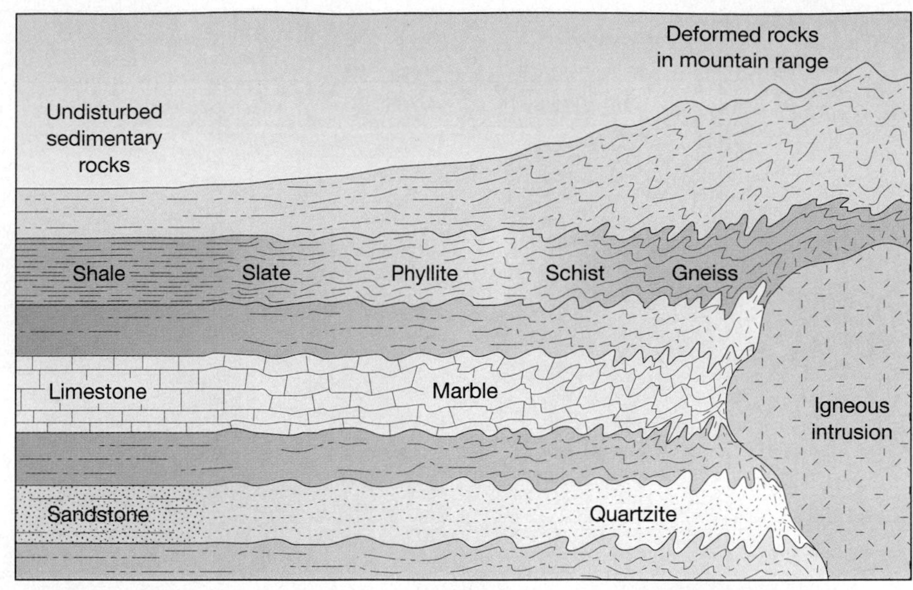

FIGURE 7.16
Idealized cross section illustrating progressive regional metamorphism. Going from the left side of the diagram toward the right, we progress from low-grade metamorphism, typified by the transformation of shale to slate, to high-grade metamorphism.

ated with the process of mountain building. During mountain building, a large segment of the earth's crust is intensely squeezed into a highly deformed mass. As the rocks are folded and faulted, the crust is shortened and thickened. This general thickening of the crust results in mountain terrains that stand high above sea level. Although material is obviously elevated to great heights during mountain building, an equally large quantity of rock is forced downward where it is exposed to high temperatures and pressures. Here in the "roots" of mountains, the most intense metamorphic activity occurs. Some of the deformed rock is thought to be heated enough to melt and thereby generate magma. As we shall see in the next section magma is also generated by other mountain-building processes. This magma, being less dense than the surrounding rock, will migrate upward. Magmas emplaced in a near-surface environment will cause contact metamorphism within the zone of regional metamorphism. Consequently, the cores of many mountain ranges consist of intrusive igneous bodies surrounded by high-grade metamorphic rocks. As these deformed rock masses are uplifted, erosion removes the overlying material to expose the igneous and metamorphic rocks composing the central core of the mountain range.

Since metamorphic rocks that form during regional metamorphism are deformed by directional stress, they are usually foliated. Further, in regional metamorphism, there usually exists a gradation in the intensity of metamorphism. As we progress from areas of low-grade metamorphism to areas of high-grade metamorphism, changes in mineralogy and rock texture can be observed.

A somewhat oversimplified example of progressive metamorphism can be made using the sedimentary rock shale, which under low-grade metamorphism yields the metamorphic rock slate (Figure 7.16). In high-temperature, high-pressure environments, slate will turn into mica schist. Under more extreme conditions, the micas in schist will recrystallize into minerals such as feldspar and hornblende, and eventually generate a gneiss. We can see certain aspects of this transition as we approach the Appalachian Mountains from the west. Beds of shale, which once extended over large areas of the eastern United States, can still be found as nearly flat-lying strata in Ohio. However, in the broadly folded Appalachians of central Pennsylvania, these beds are inclined and composed of low-grade slate. As we progress farther eastward to the intensely deformed crystalline Appalachians, we find large outcrops of schist and gneiss, some of which are perhaps remnants of once flat-lying shale beds. The most intense zones of metamorphism are found in regions such as Vermont and New Hampshire, often in close association with igneous intrusions.

In addition to the textural changes already considered, changes in mineralogy will be encountered as we progress from areas of low-grade metamorphism to areas of high-grade metamorphism. The typical transition in mineralogy that would result from the regional metamorphism of shale is shown in Figure 7.17. The first new mineral to be produced in the formation of slate is chlorite, which as we move toward the region of high-grade metamorphism would be replaced by ever greater amounts of muscovite and biotite. Mica schists are formed under more extreme conditions and may contain garnet and staurolite crystals as well. At temperatures and pressures

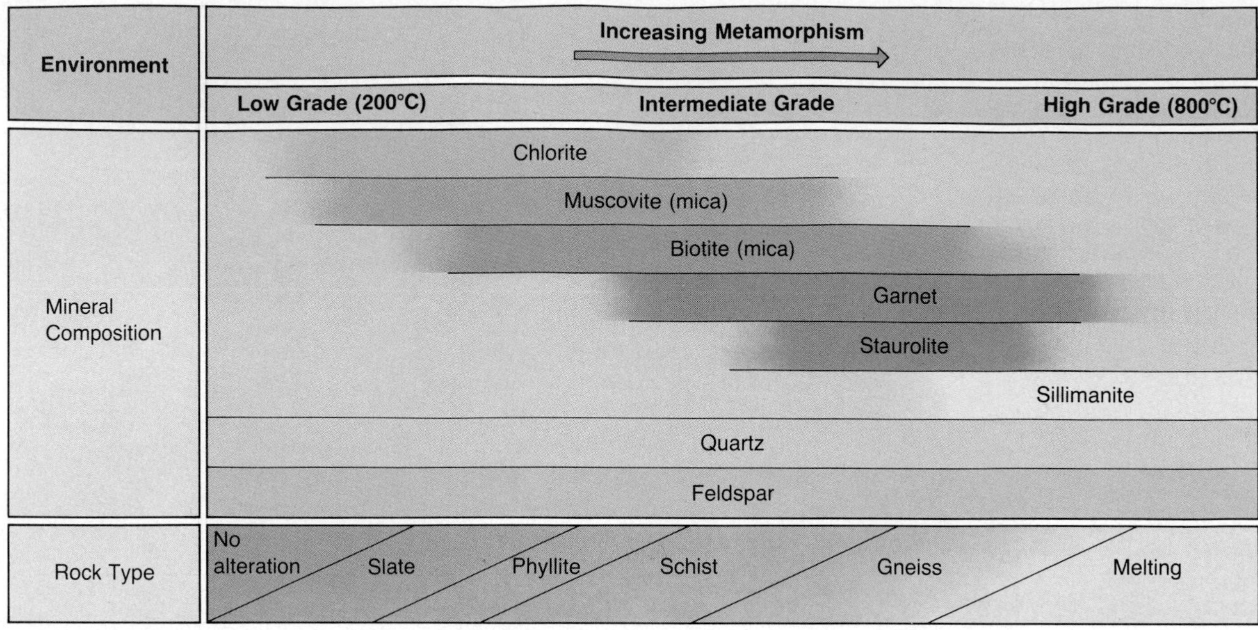

FIGURE 7.17
The typical transition in mineralogy that results from progressive metamorphism of shale.

approaching the melting point of rock, sillimanite forms. Sillimanite is a high-temperature metamorphic mineral used to make refractory porcelains such as those used in the casing of spark plugs.

Through the study of metamorphic rocks and laboratory experiments, researchers have learned that certain minerals are good indicators of the metamorphic environment in which they formed. Using these **index minerals**, geologists distinguish among different zones of regional metamorphism. For example, the mineral chlorite is produced when temperatures are relatively low, about 200°C (Figure 7.17). The mineral sillimanite on the other hand forms in very extreme environments where temperatures exceed 600°C. By mapping the occurrences of index minerals, geologists in effect map zones of varying metamorphic intensities. Figure 7.18 outlines the zones of metamorphic intensities that crop out in New England and the Maritime Provinces of Canada.

In the most extreme environments even the highest-grade metamorphic rocks are subjected to change. For example, in a near-surface environment where temperatures exceed 750°C, a schist or gneiss having a chemical composition similar to granite will begin to melt. However, recall from our discussion of igneous rocks that not all minerals melt at the same temperature. The light-colored silicates, usually quartz and potassium feldspar, will melt first, whereas the mafic silicates, such as amphibole and biotite, will remain solid. If this partially melted rock cools, the light bands will be made of crystalline igneous rock while the dark bands will consist of unmelted metamorphic material. Rocks of this type fall into a transitional zone somewhere between "true" igneous rocks and "true" metamorphic rocks and are called **migmatites** (Figure 7.19).

METAMORPHISM AND PLATE TECTONICS

Most of the present knowledge of metamorphism appears to conform well to the dynamics of the earth as proposed by the plate tectonics theory. In this model, mountain building and associated metamorphism occur along convergent zones where slabs of lithosphere are moving toward one another (Figure 7.20). It is at these locations that compressional forces squeeze and generally deform the converging plates and the sediments that have accumulated along the margins of continents. The plate tectonics model also accounts for the igneous activity associated with mountain building. At convergent zones, material is being thrust to depths where temperatures and pressures are high (Figure 7.21). The eventual melting of some subducted material creates magma that migrates upward to crystallize in the cores of mountain masses.

A close examination of Figure 7.20 shows that more than one type of metamorphic environment exists along convergent boundaries. Near ocean

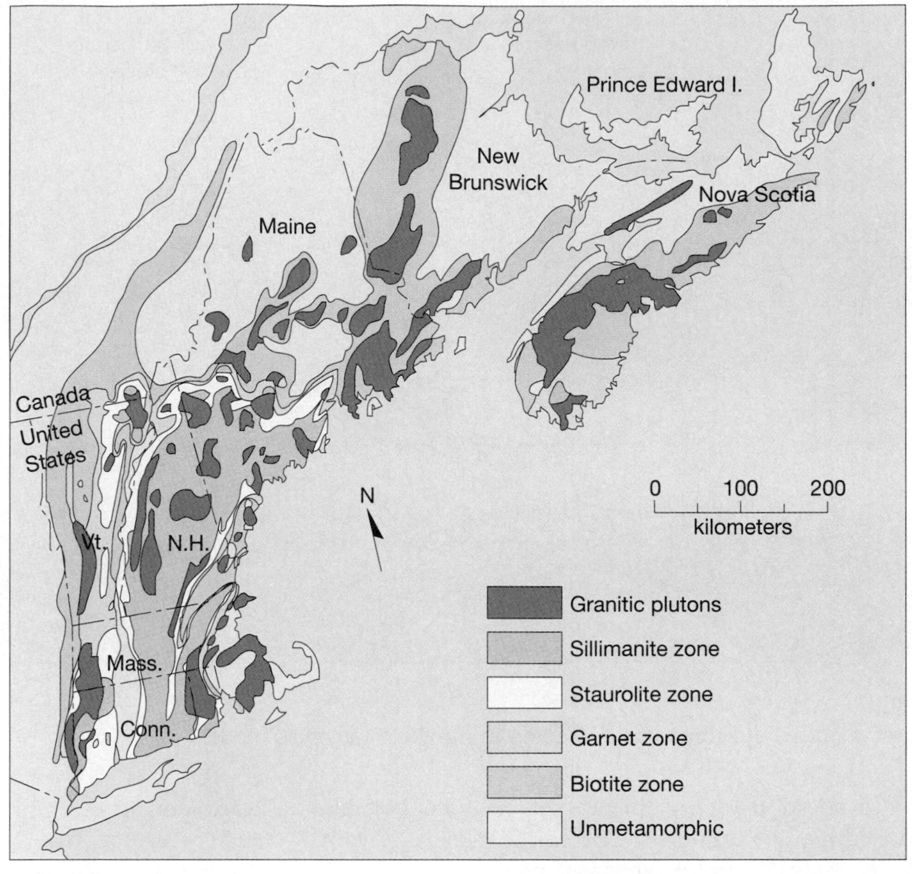

FIGURE 7.18
Zones of metamorphic intensities in New England and the Maritime Provinces. (After Donald W. Hyndman, *Petrology of Igneous and Metamorphic Rocks,* New York: McGraw-Hill, 1972)

trenches, slabs of cold lithosphere are being subducted to great depths. As the lithosphere descends, rocks are subjected to steadily increasing pressures (Figure 7.21). However, temperatures in the slab remain cooler than the surrounding mantle because rock is a poor heat conductor. As a result, the descending slab heats slowly. Rock formed in this high-pressure, low-temperature environment is called *blueschist,* after the presence of the blue-colored amphibole glaucophane. The Coast Range of California was once a subduction zone of this type. Here highly deformed rocks that were once deeply buried have been elevated as a result of a change in the plate boundary.

In near-surface zones landward of trench areas, the metamorphic environment consists of high temperatures and low to moderate pressures (Figure 7.21). Here the emplacement of molten rock from below alters the existing bedrock. The Sierra Nevada, which consists of igneous intrusions and associated metamorphic rocks, exemplifies this type of environment (see Figure 7.15).

Apparently, most of the deformed material found adjacent to oceanic trenches consists of two distinctive linear belts of metamorphic rocks. Nearest to the trench we find a high-pressure, low-temperature metamorphic regime similar to that of the Coast Range of California. Farther inland, in the region of plutonic emplacement, metamorphism is dominated by high temperatures and low to moderate pressures; that is, environments similar to those that generated the Sierra Nevada Batholith.

FIGURE 7.19
Migmatite. The lightest colored layers are igneous rock composed of quartz and feldspar, while the darker layers have a metamorphic origin. (Photo by Hal Roepke)

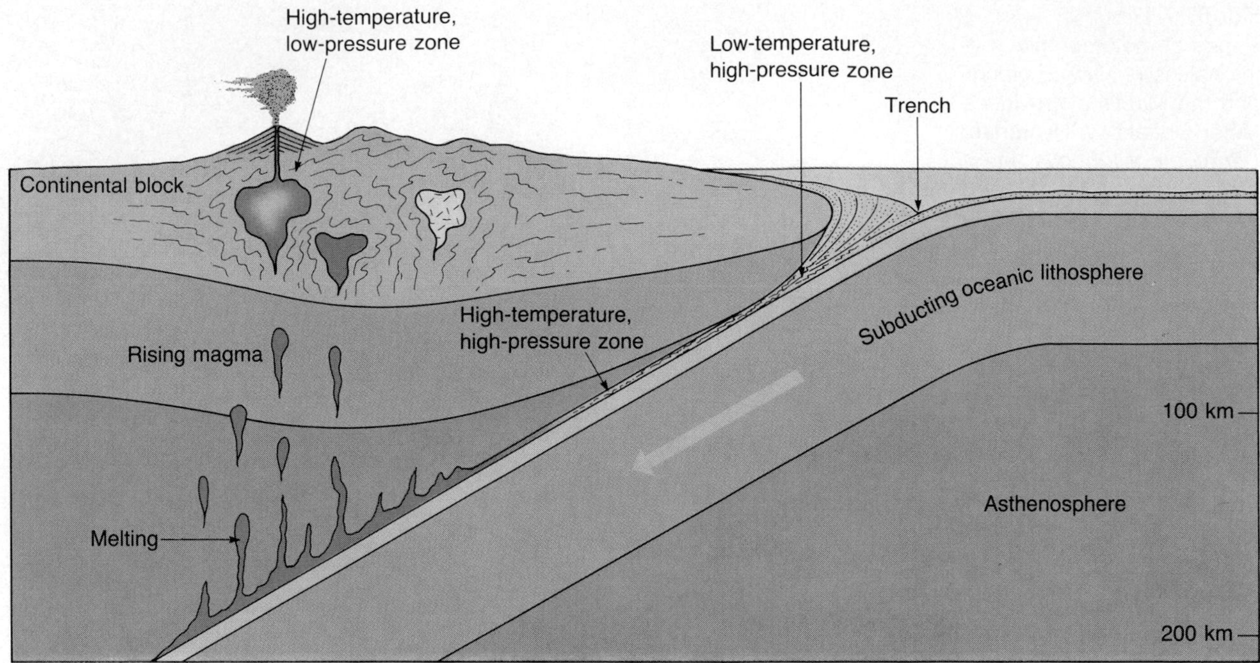

FIGURE 7.20
Metamorphic environments according to the plate tectonics model.

In addition to the linear belts of metamorphic rocks that are found in the axes of most mountain belts, even larger expanses of metamorphic rocks exist within the stable continental interiors (Figure 7.22). These relatively flat expanses of metamorphic rocks and igneous plutons are called **shields**. One such structure, the Canadian Shield, has very little topographic expression and forms the bedrock over much of central Canada, extending from Hudson Bay to northern Minnesota. Radiometric dating of the

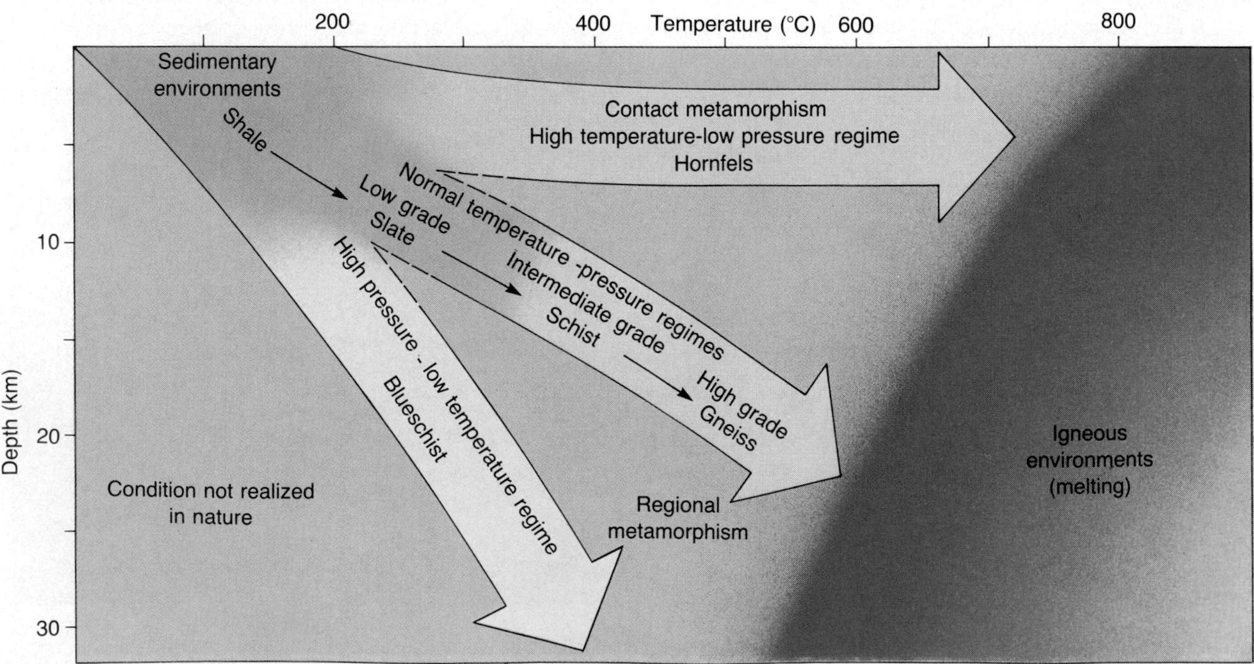

FIGURE 7.21
Illustration of various temperature-pressure regimes that produce different types of metamorphic rocks.

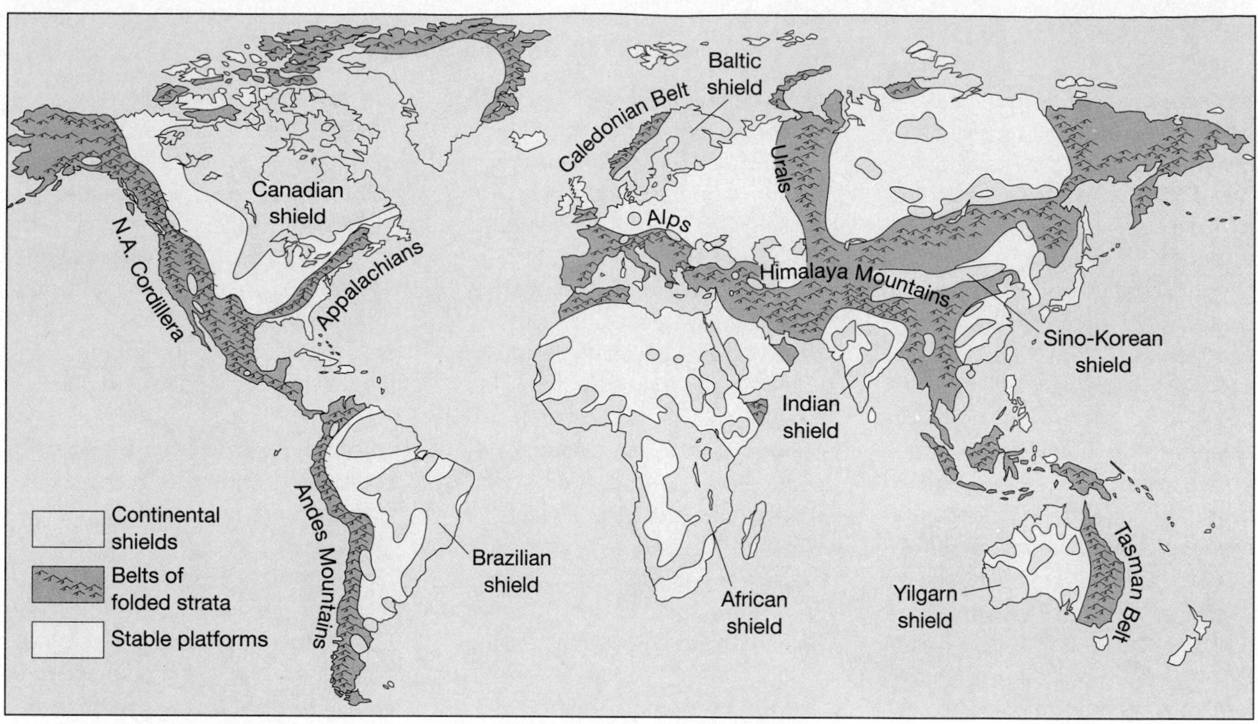

FIGURE 7.22
The continental shields of the world are composed largely of metamorphic rocks. In addition, the deformed portions of many mountain belts are also metamorphic. The light tan area shown in this map is the stable continental interior, which generally consists of undeformed sedimentary beds that overlie metamorphic and igneous basement rocks.

Canadian Shield rocks indicates that they are among the oldest rocks on earth. Because shields are old and since their rock structure is similar to that found in the central core of existing mountains, they are believed to be remnants of much earlier periods of mountain building. If this concept is correct, it indicates that the earth has been a dynamic planet throughout a great expanse of its history. Additional study of these vast areas of metamorphism, in the context of the plate tectonics model, should give geologists new insights into the problem of discerning just how the continents came to exist. We will consider that topic further in Chapter 20.

BOX 7.1

Impact Metamorphism and Tektites

In the last few decades, it has become increasingly clear that comets and asteroids have collided with Earth far more frequently than had previously been thought. The evidence for these ancient catastrophic events is in the form of giant impact structures known as *astroblems* (Figure 7.A) More than 100 astroblems have been identified to date. Many were once considered to be the result of some poorly understood volcanic process. Although most astroblems are so old that they no longer resemble an impact crater, evidence of an intense shock remains. One notable exception is a very fresh looking crater located near Winslow, Arizona, known as Meteor Crater.

One phenomenon associated with astroblems is impact or shock metamorphism. When hypervelocity projectiles, such as comets or asteroids, impact the earth's surface, pressures that reach millions of atmospheres and temperatures in excess of 2000°C are generated. The result is pulverized, shattered, and melted rock. The products of high-velocity impacts, called *impactites,* include mixtures of fragmented rock and melted or practically melted material fused together, as well as glass-rich ejecta that resemble volcanic bombs. In some cases, particles of coesite, a very dense form of quartz and minute diamonds, are found at impact sites. These high-pressure minerals are indicators of shock metamorphism. The best-known occurrences of impactites in North America are found at Meteor Crater and in association with astroblems located at Kentland Crater in northern Indiana and at Sinking Spring in south-central Ohio.

In places where impact craters are relatively fresh, shock-melted ejecta and rock fragments can be found ringing the impact site. Although most of this material is deposited close to its source, some ejecta can travel great distances. One example is *tektites,* globular or teardrop-shaped beads of silica-rich glass, some of which show the effects of aerodynamic shaping during flight. Most tektites are small, typically no more than a few centimeters across, and range in color from jet-black to dark green or yellowish. In Australia, millions of tektites are strewn over an area of roughly 5 million square kilometers. Several of these tektite groupings or strewnfields have been identified worldwide, one stretching nearly halfway around the globe.

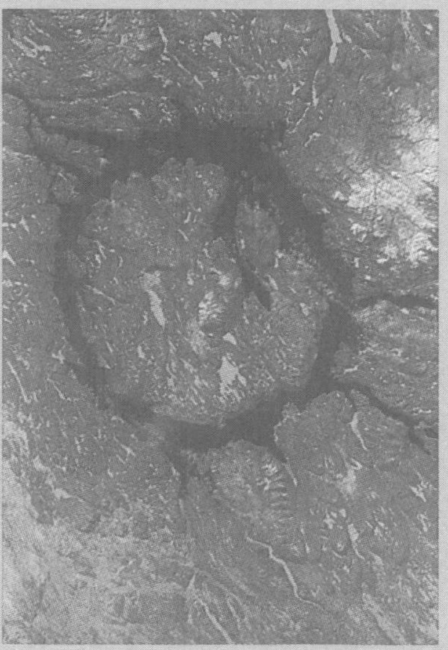

FIGURE 7.A
Manicouagan, Quebec, is a 200-million-year old eroded impact crater. The lake, which outlines the crater, is 70 kilometers (42 miles) wide. Fractures related to this event extend outward for an additional 30 kilometers. (Courtesy of U.S. Geological Survey)

No tektite falls have ever been observed and recorded, and thus their origin is not known with certainty. Since tektites are much higher in silica than volcanic glass (obsidian), a volcanic origin is unlikely. Most researchers agree that tektites are the result of impacts of large projectiles. One hypothesis suggests that tektites have an extraterrestrial origin. Supporters of this idea propose that asteroids impacted the moon with such force that ejecta "splashed" outward at sufficient velocity to escape the moon's gravitational field. Others disagree and argue that tektites have a terrestrial origin. One objection to the latter view is that some strewnfields, such as those found in Australia, lack an identifiable impact crater. It is possible, however, that the meteorite that produced the tektites in Australia impacted the continental shelf. If this occurred, the remnant crater may lie out of sight below sea level. Evidence in support of a terrestrial origin includes tektite falls in the Ivory Coast of Africa. These tektites appear to be the same age as the Bosumtwi Crater located in the same region.

REVIEW QUESTIONS

1. What is metamorphism? What are the agents of change?

2. What is foliation? Distinguish between *slaty cleavage* and *schistosity*.

3. List some changes that might occur to a rock in response to metamorphic processes.

4. Slate and phyllite resemble each other. How might you distinguish one from the other?

5. Each of the following statements describes one or more characteristics of a particular metamorphic rock. For each statement, name the metamorphic rock that is being described.

(a) Calcite-rich and nonfoliated.

(b) Foliated and composed mainly of granular minerals.

(c) Represents a grade of metamorphism between slate and schist.

(d) Very fine-grained and foliated; excellent rock cleavage.

(e) Foliated and composed of more than 50 percent platy minerals.

(f) Often composed of alternating bands of light and dark silicate minerals.

(g) Hard, nonfoliated rock resulting from contact metamorphism.

6. Distinguish between contact metamorphism and regional metamorphism. Which creates the greatest quantity of metamorphic rock?

7. What feature would easily distinguish schist and gneiss from quartzite and marble?

8. Briefly describe the textural and mineralogical differences among slate, mica schist, and gneiss. Which one of these rocks represents the highest degree of metamorphism?

9. Are migmatites associated with high-grade or low-grade metamorphism?

10. With what type of plate boundary is regional metamorphism associated?

KEY TERMS

aureole (p. 172)
contact metamorphism (p. 163)
foliation (p. 166)
hydrothermal solution (p. 167)
index mineral (p. 174)
migmatite (p. 174)
nonfoliated (p. 166)
regional metamorphism (p. 163)
rock cleavage (p. 166)
schistosity (p. 166)
shear (p. 164)
shield (p. 176)
slaty cleavage (p. 166)
stress (p. 164)

8

Geologic Time

RELATIVE DATING

CORRELATION

RADIOACTIVITY AND RADIOMETRIC DATING

THE GEOLOGIC TIME SCALE

DIFFICULTIES IN DATING THE GEOLOGIC TIME SCALE

Opposite: The inner gorge of the Grand Canyon as seen from Toroweep Overlook. The rocks exposed in the Grand Canyon represent vast spans of earth history. (Photo by Jack W. Dykinga)
(Top photo by Michael Collier)

In 1869 John Wesley Powell, who was later to head the U.S. Geological Survey, led a pioneering expedition down the Colorado River and through the Grand Canyon (Figure 8.1). Writing about the strata that were exposed by the downcutting of the river, Powel said that, ". . . the canyons of this region would be a Book of Revelations in the rock-leaved Bible of geology." He was undoubtedly impressed with the millions of years of earth history exposed along the walls of the Grand Canyon (see chapter-opening photo). Powell realized that the evidence for an ancient earth is concealed in its rocks. Like the pages in a long and complicated history book, rocks record the geological events and changing life forms of the past. The book, however, is not complete. Many pages, especially in the early chapters, are missing. Others are tattered, torn, or smudged. Yet enough of the book remains to allow much of the story to be deciphered. Interpreting earth history is a prime goal of the science of geology. Like a modern-day sleuth, the geologist must interpret the clues found preserved in the rocks. By studying rocks, especially sedimentary rock, and the features they contain, geologists can unravel the complexities of the past.

Geological events by themselves, however, have little meaning until they are put into a time perspective. Studying history, whether it be the Civil War or the Age of Dinosaurs, requires a calendar. Among the major contributions that geology has made to the knowledge of humankind is the geologic time scale and the concept that earth history is exceedingly long. Over many years geologists put together a time scale of earth history in which geologic events can be placed in their proper sequence. Geologists, recognizing that earth history has spanned an immense amount of time, worked at finding out just how old the earth is.

RELATIVE DATING

During the late nineteenth and early twentieth centuries a number of attempts were made to determine the age of the earth. Although some of the methods appeared promising at the time, none of these early efforts proved to be reliable (see Box 8.1). What these scientists were seeking was an **absolute date**. Such dates pinpoint the time in history when something took place. Today radiometric dating allows us to accurately determine absolute dates for rock units that represent important events in the earth's distant past.* However, prior to the discovery of radioactiv-

*Radiometric dating is the subject of a later section in this chapter.

A.

B.

FIGURE 8.1
A. Start of the expedition from Green River station. A drawing from Powell's 1875 book. **B.** Major John Wesley Powell, pioneering geologist and the second director of the U.S. Geological Survey. (Courtesy of the U.S. Geological Survey)

A.

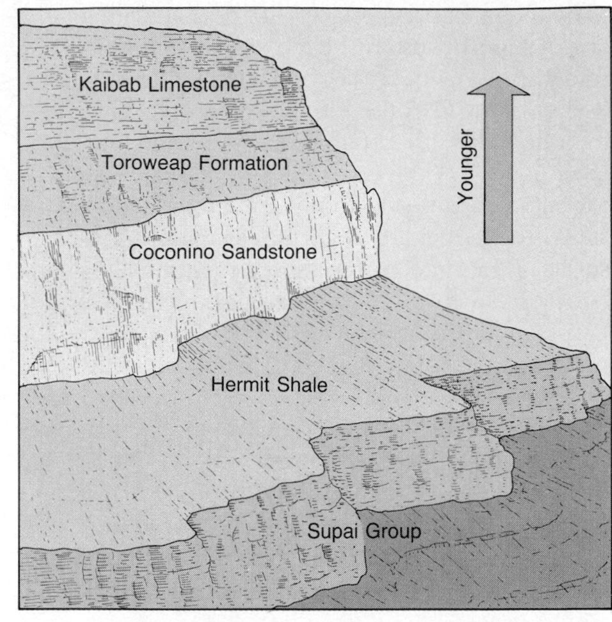

B.

FIGURE 8.2
Applying the law of superposition to these layers exposed in the upper portion of the Grand Canyon, the Supai Group is oldest and the Kaibab Limestone is youngest. (Photo by E. J. Tarbuck)

ity, geologists had no reliable method of absolute dating and had to rely solely on relative dating. **Relative dating** means that rocks are placed in their proper sequence or order. Relative dating will not tell us how long ago something took place, only that it followed one event and preceded another. The relative dating techniques which were developed are still widely used. Absolute dating methods did not replace these techniques; they simply supplemented them. To establish a relative time scale, a few simple principles or rules had to be discovered and applied. Although they may seem obvious to us today, their discovery was an important scientific achievement.

Nicolaus Steno, a physician in Florence, Italy, is credited with being the first to recognize a sequence of historical events in an outcrop of sedimentary rock layers. Working in the mountains of western Italy, Steno applied a very simple rule that has come

FIGURE 8.3
Most layers of sediment are deposited in a nearly horizontal position. Thus, when we see rock layers that are inclined, we can assume they must have been moved into that position by crustal disturbances after their deposition. (Photo by E. J. Tarbuck)

to be the most basic principle of relative dating—the **law of superposition**. The law simply states that in an undeformed sequence of sedimentary rocks, each bed is older than the one above and younger than the one below. Although it may seem obvious that a layer could not be deposited with nothing beneath it for support, it was not until 1669 that Steno clearly stated the principle. This rule also applies to other surface-deposited materials such as lava flows and beds of ash from volcanic eruptions. Applying the law of superposition to the beds exposed in the upper portion of the Grand Canyon (Figure 8.2), we can easily place the layers in their proper order. Among those that are pictured, the sedimentary rocks in the Supai Group are the oldest, followed in order by the Hermit Shale, Coconino Sandstone, Toroweap Formation, and Kaibab Limestone.

Steno is also credited with recognizing the importance of another basic principle, called the **principle of original horizontality**. Simply stated, it means that layers of sediment are generally deposited in a nearly horizontal position. Thus, if we observe rock layers that are folded or inclined at a steep angle they must have been moved into that position by crustal disturbances sometime after their deposition (Figure 8.3).

When igneous intrusions or faults cut through other rocks, they are assumed to be younger than the rocks they cut. For example, when two dikes intersect, the older one must have been opened up in order to allow the younger one to cut through it. The younger dike would be continuous, whereas the older dike would be interrupted at the point of their intersection. A fault is a fracture in the earth along which movement occurs. When rocks are cut and offset by a fault, we know that they must be older than the fault. Figure 8.4 illustrates this **principle of cross-cutting relationships**. For example, by apply-

BOX 8.1

Early Methods of Dating the Earth

Today, thanks to reliable radiometric dating methods, we know that the age of the earth is about 4.6 billion years. However, this great age is a relatively recent discovery. In the late eighteenth century James Hutton recognized the immensity of earth history and the importance of time as a component in all geological processes. In the nineteenth century Sir Charles Lyell and others effectively demonstrated that the earth had experienced many episodes of mountain building and erosion, which must have required great spans of geologic time. Although these pioneering scientists believed that the earth was very old, they had no way of knowing its true age. Was it tens of millions, hundreds of millions, or even billions of years old? During the latter half of the nineteenth and the early twentieth centuries, solutions to this problem were sought and several methods were subsequently devised.

One method that was attempted several times involved the rate at which sediment is deposited. Some geologists reasoned that if they could determine the rate that sediment accumulates, and could further ascertain the total thickness of sedimentary rock that had been deposited during earth history, they could estimate the length of geologic time. All that was necessary was to divide the rate of sediment accumulation into the total thickness of sedimentary rock. Unfortunately, this method was riddled with difficulties, some of which are as follows:

1. Different sediments accumulate at different rates under varying conditions. Thus, determining an overall rate of sediment accumulation is difficult. Further, if such a rate is determined, it does not necessarily mean that the same rate can be applied to the past.
2. Since no single locality has a complete geologic column, estimates of the total thickness of sedimentary rocks had to be compiled by adding together the maximum known thickness of rocks of each age. These estimates had to be revised each time a thicker section was discovered.
3. Sediment compacts when it is lithified; thus, a correction for compaction had to be made.

Needless to say, estimates of the earth's age varied considerably as different scientists attempted this method. The figure representing the maximum thickness of sedimentary rock ranged from 9600 meters (32,000 feet) to over 100,500 meters (330,000 feet). The amount of time for 0.3 meter (1 foot) of sediment to accumulate varied from 100 years to over 8600 years. The age of the earth as calculated by this method therefore ranged from 3 million to 1.5 billion years!

ing the cross-cutting principle we can determine that fault A occurred after the sandstone layer was deposited but before the conglomerate was laid down. It is possible to say this because the sandstone is offset by the fault and the conglomerate is not. We can also state that dike B and its associated sill are older than dike A because dike A cuts the sill. In the same manner we know that the batholith was emplaced after movement occurred along fault B but before dike B was formed. This is true because the batholith cuts across fault B while dike B cuts across the batholith.

Sometimes inclusions can aid the relative dating process. **Inclusions** are pieces of one rock unit that are contained within another. The basic principle is logical and straightforward. The rock mass adjacent to the one containing the inclusions must have been there first in order to provide the rock fragments. Therefore, the rock mass containing inclusions is the younger of the two. Figure 8.5 provides an example. Here the inclusions of granite in the adjacent sedimentary layer indicate that the sedimentary layer was deposited on top of an eroded mass of granite rather than being intruded from below by a younger granite.

Layers of rock are said to be **conformable** when they are found to have been deposited essentially without interruption. Although particular sites may exhibit conformable beds representing significant spans of geologic time, there is no place on earth that contains a full set of conformable strata. Throughout earth history, the deposition of sediment has been interrupted over and over again. All such breaks in the rock record are termed unconformities. An **unconformity** represents a long period of time during which deposition ceased, erosion removed previously formed rocks, and then deposition resumed. In each case uplift and erosion are followed by subsidence and renewed sedimentation. Unconformities

Another method for dating the earth involved the salinity of the oceans, which were assumed to originally have been fresh water. Scientists felt that if they could accurately estimate the quantity of salt being carried to the ocean each year by rivers and the total amount of salt currently in the oceans, they could determine the length of geologic time by dividing the latter figure by the former. Near the turn of the twentieth century, John Joly calculated the age of the earth at about 90 million years using this method. Joly, however, had no accurate notion of the amount of salt lost from the oceans because of deposition and winds blowing salt inland. It is also probable that the rate of salt accumulation has not always been constant. Thus, Joly's estimate for the age of the earth was not accurate. However, both of the methods for dating the earth that have just been described indicated that the earth was considerably older than the 6000 years given it by Archbishop Ussher.*

Perhaps the most influential estimates of the age of the earth were compiled by the well-known and highly respected physicist Lord Kelvin in the latter part of the nineteenth century. Since Kelvin's estimates required few assumptions and were based on precise measurements, they were widely accepted for a time. One of Kelvin's methods was founded on the widely held assumption that the earth had originally been molten and had cooled to its present condition. Although his data and calculations were limited, Kelvin still made it quite obvious that the earth could not be more than 100 million years old, and likely much less. The second of Kelvin's estimates was based on the assumption that the source of the sun's tremendous output of energy was of a conventional nature (nuclear fusion and radioactivity had not yet been discovered). His calculations indicated that the sun could have illuminated the earth for only a few tens of millions of years. Furthermore, he said that in the past it had been much hotter and in the future it would become much cooler. He believed that the earth was inhabitable for organisms for a period of only 20–40 million years. Kelvin's apparently irrefutable estimates had a rather profound impact:

> Evolutionists found it virtually impossible to accept these figures, but all they had were educated guesses in the face of Kelvin's potent mathematics. Darwin and others compromised their original theories in their later years in an effort to reconcile evolution and uniformitarianism with the physicists' estimates. Eventually, however, they were vindicated.†

*See the section on catastrophism in Chapter 1.

†Leigh W. Mintz, *Historical Geology: The Science of a Dynamic Earth,* 2nd ed. (Columbus, Ohio: Merrill/Macmillan 1977), pp. 84–85.

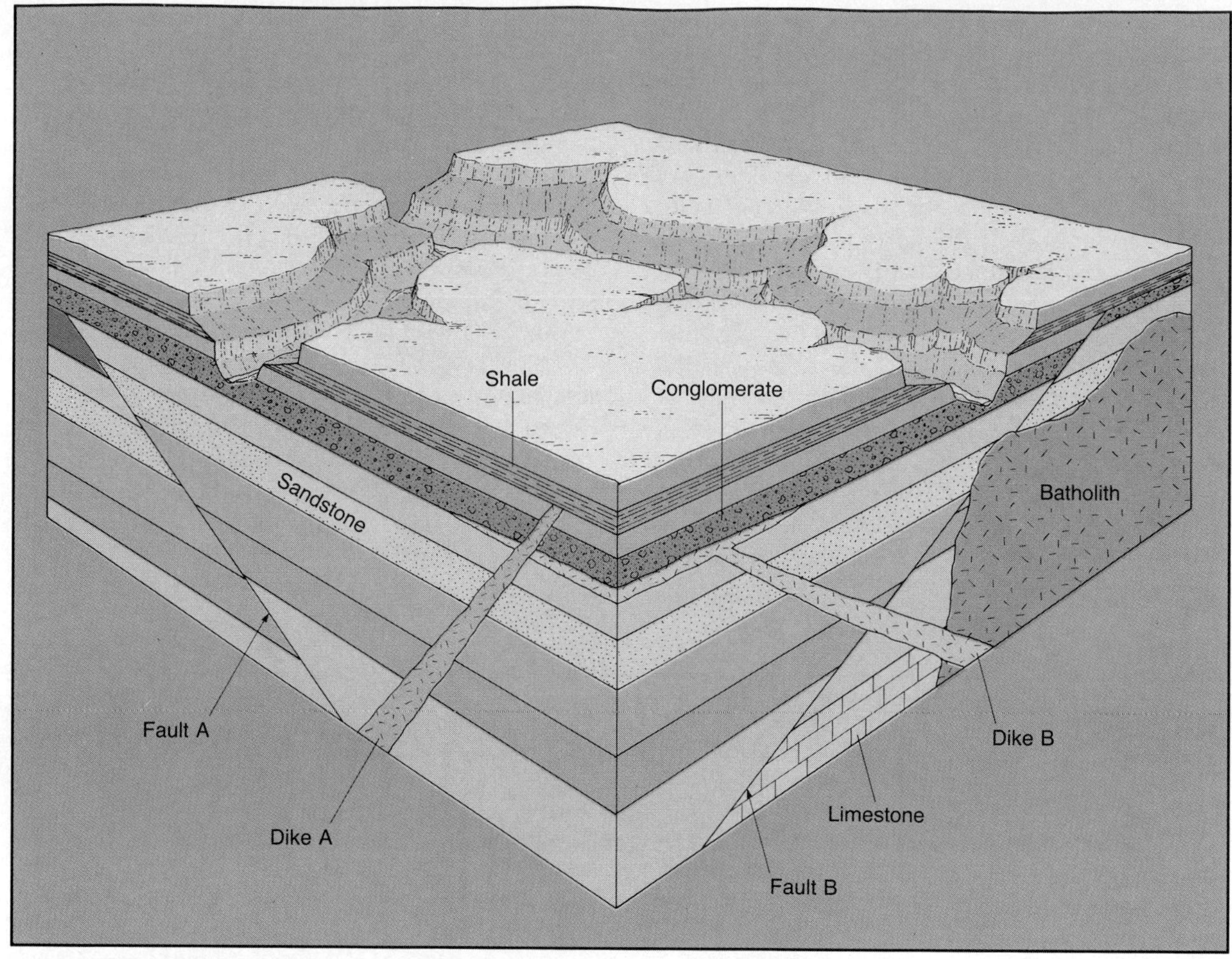

FIGURE 8.4
Cross-cutting relationships represent one principle used in relative dating. An intrusive rock body is younger than the rocks it intrudes. A fault is younger than the rock layers it cuts. (The Tasa Collection: Geologic Time. Published by Merrill Publishing Co., Columbus, OH. Copyright © 1986, by Tasa Graphic Arts, Inc. All rights reserved)

are important features because they represent significant geologic events in earth history. Moreover, their recognition helps us identify what intervals of time are not represented by strata.

Perhaps the most easily recognized type of unconformity consists of tilted or folded sedimentary rocks that are overlain by younger, more flat-lying strata. These are called **angular unconformities** and indicate that during the pause in deposition, a period of deformation (folding or tilting) as well as erosion occurred (Figure 8.6). The angular unconformity pictured in Figure 8.7 is well known to many geologists because it was originally studied and described by James Hutton and John Playfair more than 200 years ago.* It was clear to these men that the relationship between the nearly vertical lower layers and the gently inclined beds overlaying them was evidence for a significant episode of geologic change. Hutton and Playfair also appreciated the immense time span implied by such relationships. When Playfair later wrote of their visit to this site he stated that, "The mind seemed to grow giddy by looking so far into the abyss of time. . . ."

When contrasted with angular unconformities, **disconformities** are more common, but usually far less conspicuous because the strata on either side are essentially parallel. Many disconformities are difficult to identify because the rocks above and below are similar and there is little evidence of erosion. Such a break often resembles an ordinary bedding plane. Other disconformities are easier to identify because the ancient erosion surface is cut deeply into the older rocks below.

*These two pioneering geologists are discussed in the section on the birth of modern geology in Chapter 1.

Nonconformities are the third basic type of unconformity. Here the break separates older metamorphic or intrusive igneous rocks from younger sedimentary strata (see Figure 8.5). Just as angular unconformities and disconformities imply crustal movements, so too do nonconformities. Intrusive igneous masses and metamorphic rocks originate far below the surface. Thus, for a nonconformity to develop, there must be a period of uplift and the erosion of overlying rocks. Once exposed at the surface, the igneous or metamorphic rocks are subjected to weathering and erosion prior to subsidence and the renewal of sedimentation.

By applying the principles of relative dating to the hypothetical geologic cross section shown in Figure 8.8, the rocks and the events in earth history they represent can be placed in their proper sequence. The following statements summarize the logic used to interpret the cross section:

1. Applying the law of superposition, beds *A, B, C,* and *E* were deposited, in that order. Since bed *D* is a sill (a concordant igneous intrusion), it is younger than the rocks that were intruded. Further evidence that the sill is younger than beds *C* and *E* are the inclusions in the sill of fragments from these beds. If the igneous mass contains pieces of surrounding rock, the surrounding rock must have been there first.
2. Following the intrusion of the sill *(D)* the intrusion of the dike *(F)* occurred. Since the dike cuts through beds *A* through *E,* it must be younger than all of them.
3. Next, the rocks were tilted and then eroded. We know the tilting happened first because the upturned ends of the strata have been eroded. The tilting and erosion, followed by further deposition, produced an angular unconformity.

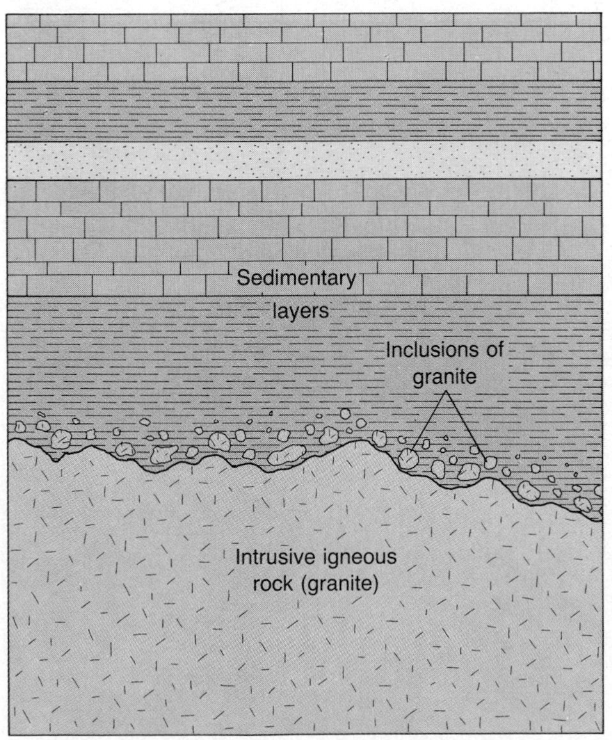

FIGURE 8.5
Since pieces of granite are contained within the overlying sedimentary bed, we know the granite must be older. When older intrusive igneous rocks are overlain by younger sedimentary layers, a type of unconformity termed a nonconformity is said to exist.

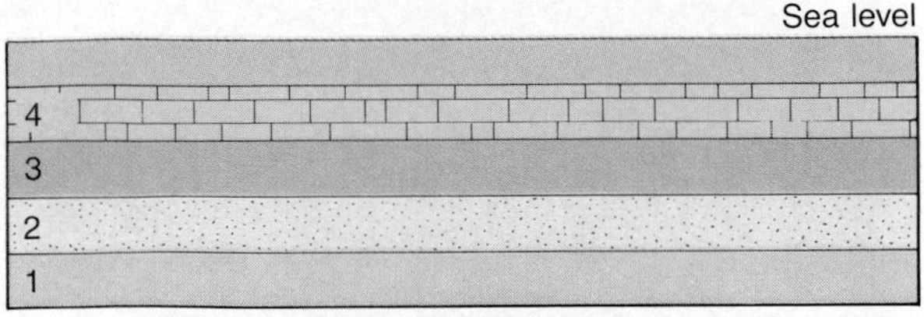

A. Deposition

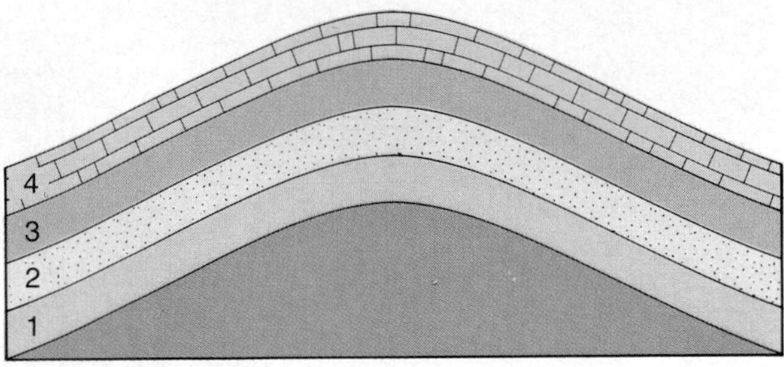

B. Folding and uplifting

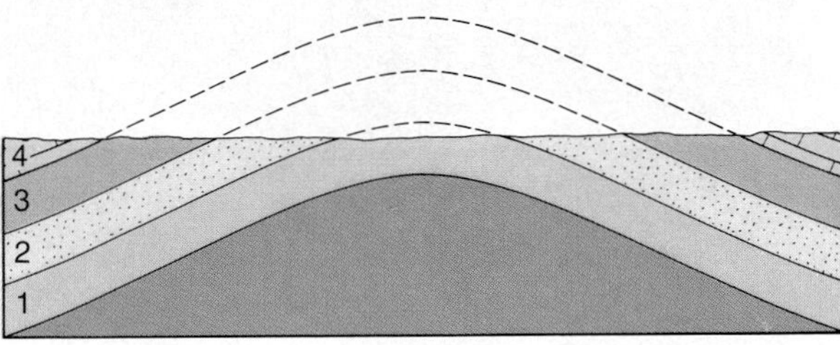

C. Erosion

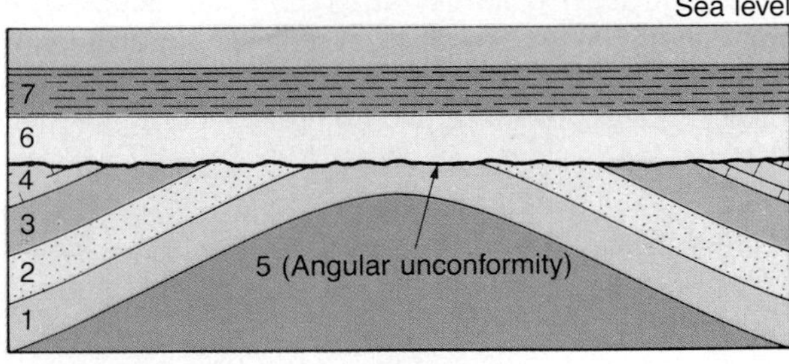

D. Subsidence and renewed deposition

FIGURE 8.6
Formation of an angular unconformity. An angular unconformity represents an extended period during which deformation and erosion occurred.

A.

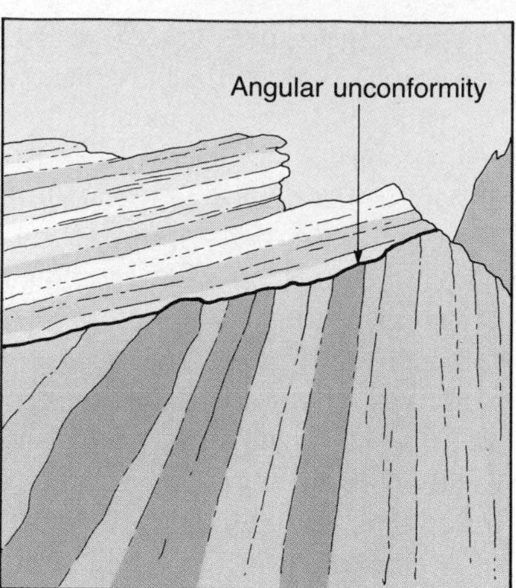

B.

FIGURE 8.7
This angular unconformity at Siccar Point, Scotland, was first described more than 200 years ago by James Hutton and John Playfair. To produce this feature, the lower layers were deposited, then uplifted, tilted, and eroded. Later, another period of sedimentation produced the flatter-lying beds on top. (Photo by Edward A. Hay)

4. Beds *G, H, I, J,* and *K* were deposited in that order, again using the law of superposition. Although the lava flow (bed *H*) is not a sedimentary rock layer, it is a surface-deposited layer, and thus superposition may be applied.

5. Finally, the irregular surface and the stream valley indicate that another gap in the rock record is being produced by erosion.

FIGURE 8.8
Geologic cross section of a hypothetical region.

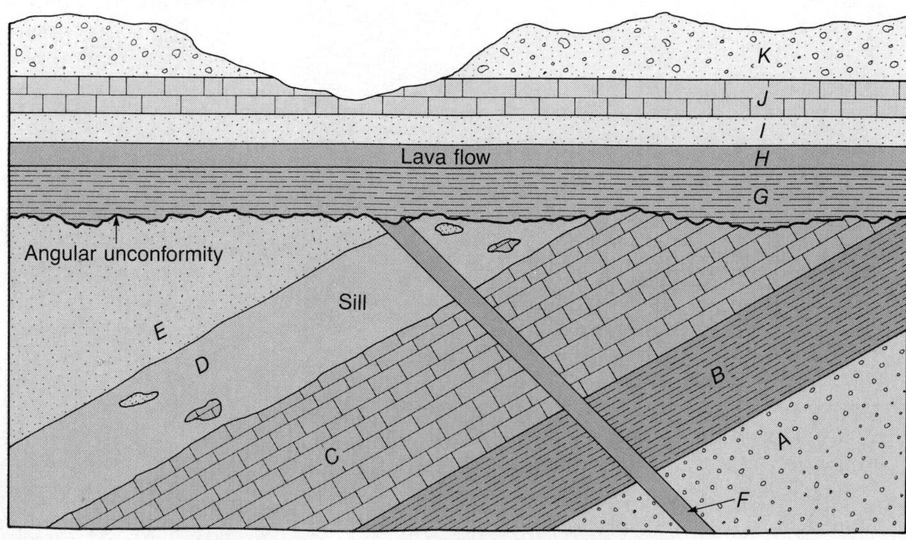

In the foregoing example our goal was to establish a relative time scale for the rocks and events in the area of the cross section. Remember, we do not have any idea how many years of earth history are represented, nor do we know how the ages of the strata in this area compare to ages in any other area.

CORRELATION

In order to develop a geologic time scale that is applicable to the whole earth, rocks of similar age in different regions must be matched up. Such a task is referred to as **correlation.** Within a limited area there are several methods of correlating the rocks of one locality with those of another. A bed or series of beds may be traced simply by walking along the outcrop. However, this may not be possible when the bed is not continuously exposed. Correlation over short distances is often achieved by noting the place of a bed in a sequence of strata, or a bed may be identified in another location if it is composed of distinctive minerals (Figure 8.9). By correlating the rocks from one place to another, a more comprehensive view of the geologic history of a region is possible. Figure 8.10, for example, shows the correlation of strata at three sites on the Colorado Plateau. No single locale exhibits the entire sequence, but correlation reveals a more complete picture of the sedimentary rock record.

Many geologic studies involve relatively small areas. Although they are important in their own right,

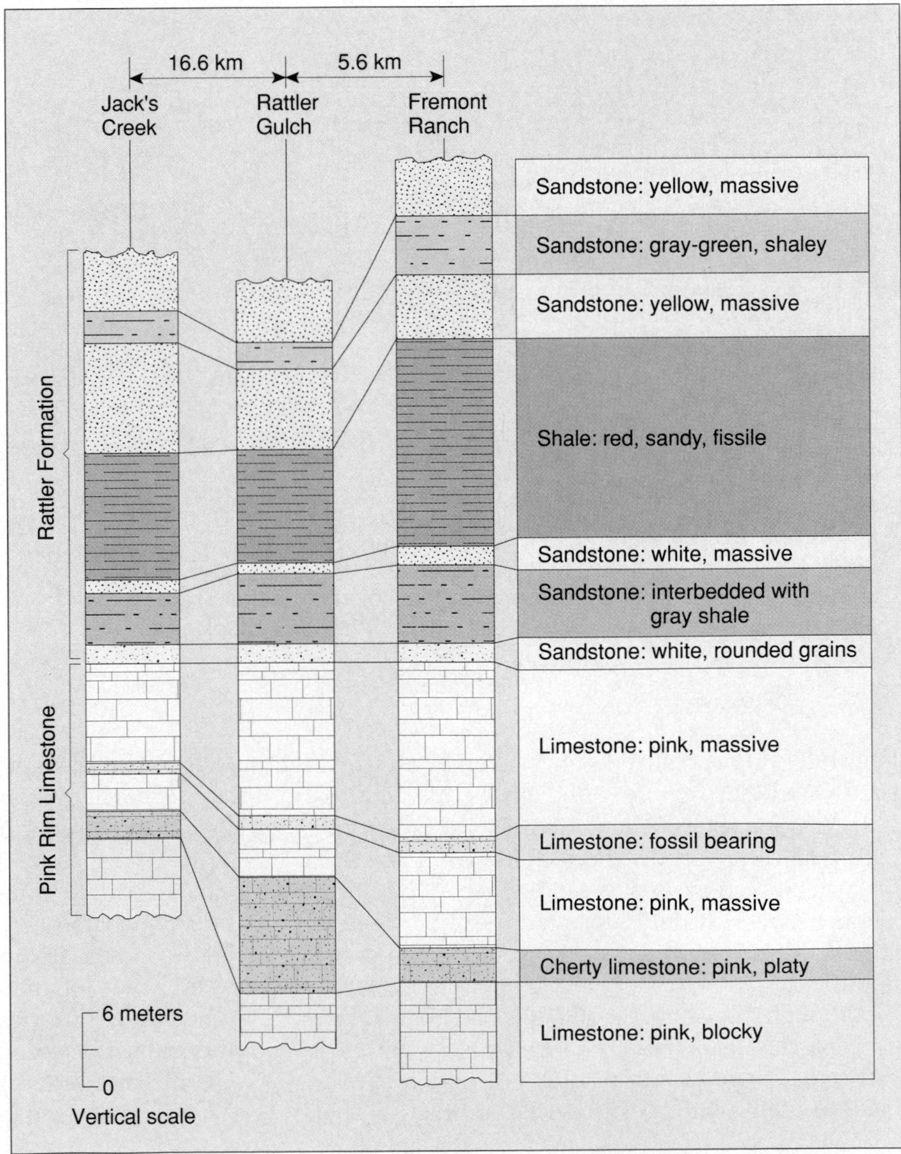

FIGURE 8.9
Correlation of strata within a small area. (After U.S. Geological Survey)

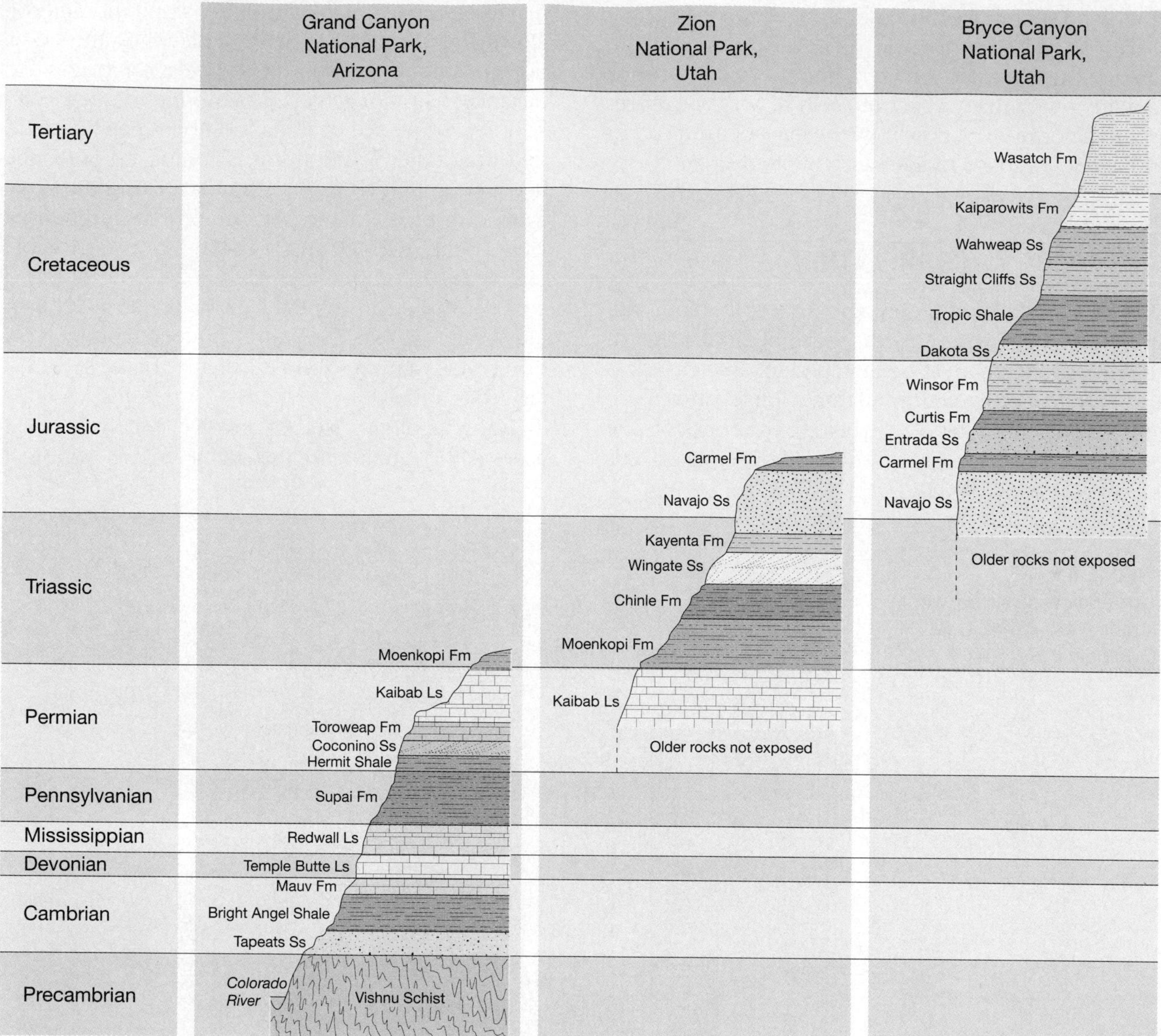

FIGURE 8.10
Correlation of strata at three locations on the Colorado Plateau reveals the total extent of sedimentary rocks in the region. (After U.S. Geological Survey)

their full value is realized only when they are correlated with other regions. Although the methods just described may be sufficient to trace a rock formation over relatively short distances, they are not adequate for matching up rocks at great distance. When correlation between widely separated areas or between continents is the objective, the geologist must rely upon fossils.

Although the existence of fossils had been known for centuries, it was not until the late 1700s and early 1800s that their significance as geologic tools was made evident. During this period an English engineer and canal builder, William Smith, discovered that each rock formation in the canals contained fossils unlike those in the beds either above or below. Further, he noted that sedimentary strata in widely separated areas could be identified by their distinctive fossil content. Based upon Smith's classic observations and the findings of many geologists who followed, one of the most important and basic principles in historical geology was formulated: fossil organisms succeed one another in a definite and determinable order, and therefore any time period can be recognized by its fossil content. This has come to be known as the **principle of faunal succession.*** In other words, when fossils are arranged according to their age, they do not present a random or haphazard picture. To the contrary, fossils show progressive changes from simple to complex and reveal the advancement of life through time. For example, an Age of Trilobites is recognized quite early in the fossil record. Then, in succession, paleontologists recognize an Age of Fishes, an Age of Coal Swamps, an Age of Reptiles, and an Age of Mammals. These "ages" pertain to groups that were especially plentiful and characteristic during particular time periods. Within each of the "ages" there are many subdivisions based, for example, on certain species of trilobites, and certain types of fish, reptiles, and so on. This same succession of dominant organisms, never out of order, is found on every major landmass.

Since fossils were found to be time indicators, they became the most useful means of correlating rocks of similar age in different regions. Geologists pay particular attention to certain fossils called **index fossils.** Since these fossils are widespread geographically and are limited to a short span of geologic time, their presence provides an important method of matching rocks of the same age. Rock formations, however, do not always contain a specific index fossil. In such situations, groups of fossils are used to establish the age of the bed. Figure 8.11 illustrates how an assemblage of fossils may be used to date rocks more precisely than could be accomplished by the use of any one of the fossils.

In addition to being important and often essential tools for correlation, fossils are important environmental indicators. Although much can be deduced about past environments by studying the nature and characteristics of sedimentary rocks, a close examination of the fossils present can usually provide a great deal more information. For example, when the remains of certain clam shells are found in limestone, the geologist quite reasonably assumes that the region was once covered by a shallow sea. Also, by using what we know of living organisms, we can conclude that fossil animals with thick shells capable of withstanding pounding and surging waves inhabited shorelines. On the other hand, animals with thin, delicate shells probably indicate deep, calm offshore waters. Hence by taking a closer look at the types of fossils, the approximate shoreline may be identified. Further, fossils can be used to indicate the former temperature of the water. Certain kinds of present-day corals must live in warm and shallow tropical seas like those around Florida and the Bahamas. When similar types of coral are found in ancient limestones, they indicate the marine environment that must have existed when they were alive. These examples illustrate how fossils can help unravel the complex story of earth history.

*Although the term *fauna* refers only to animals, the principle is equally applicable to fossil plants. For this reason, the term *biotic succession* is used by some instead of the more traditional term *faunal succession*.

RADIOACTIVITY AND RADIOMETRIC DATING

In addition to establishing relative dates by using the principles described in the preceding sections, it is also possible to obtain reliable absolute dates for events in the geologic past. For example, we know that the earth is about 4.6 billion years old and that the dinosaurs became extinct about 66 million years ago. Dates that are expressed in millions and billions of years truly stretch our imaginations because our personal calendars involve time measured in hours, weeks, and years. Nevertheless, the vast expanse of geologic time is a reality and it is radiometric dating

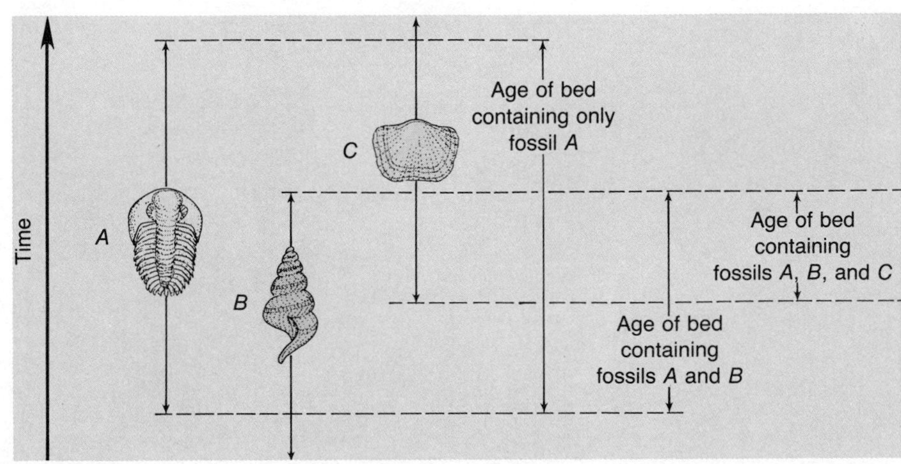

FIGURE 8.11
Overlapping ranges of fossils help date rocks more exactly than using a single fossil.

which allows us to accurately measure it. In this section we will learn about radioactivity and its application in radiometric dating.

Most atoms are stable and do not change. However, some are unstable, constantly releasing heat as their nuclei break apart or decay. This is the heat that helps maintain the high temperatures in the earth's interior and is the source of the heat which Kelvin was measuring when he thought he was measuring the "cooling" earth.

In Chapter 2 we learned that an atom is composed of electrons, protons, and neutrons. *Electrons* have a negative charge and *protons* have a positive charge. Since a *neutron* is actually a proton and electron combined, it has no charge. Protons and neutrons are found in the center, or *nucleus,* of the atom, and electrons spin around the nucleus in definite paths, or orbits. Practically all (99.9 percent) of the mass of an atom is found in the nucleus, indicating that electrons have practically no mass at all. By adding together the number of protons and neutrons in the nucleus, the mass *number* of the atom is determined. The *atomic number* (the atom's identifying number) is equal to the number of protons. Each one of the more than 100 known elements has a different number of protons in the nucleus, and thus a different atomic number. Atoms of the same element may have different numbers of neutrons in the nucleus. Such atoms, called *isotopes,* have different mass numbers but the same atomic number.

The forces which bind protons and neutrons together in the nucleus are strong; however, the nature of these forces is still not fully understood. Some isotopes have unstable nuclei; that is, the forces which bind the protons and neutrons together are not sufficiently strong. As a result, the nuclei spontaneously break apart, or decay, a process called **radioactivity**. What happens when unstable nuclei break

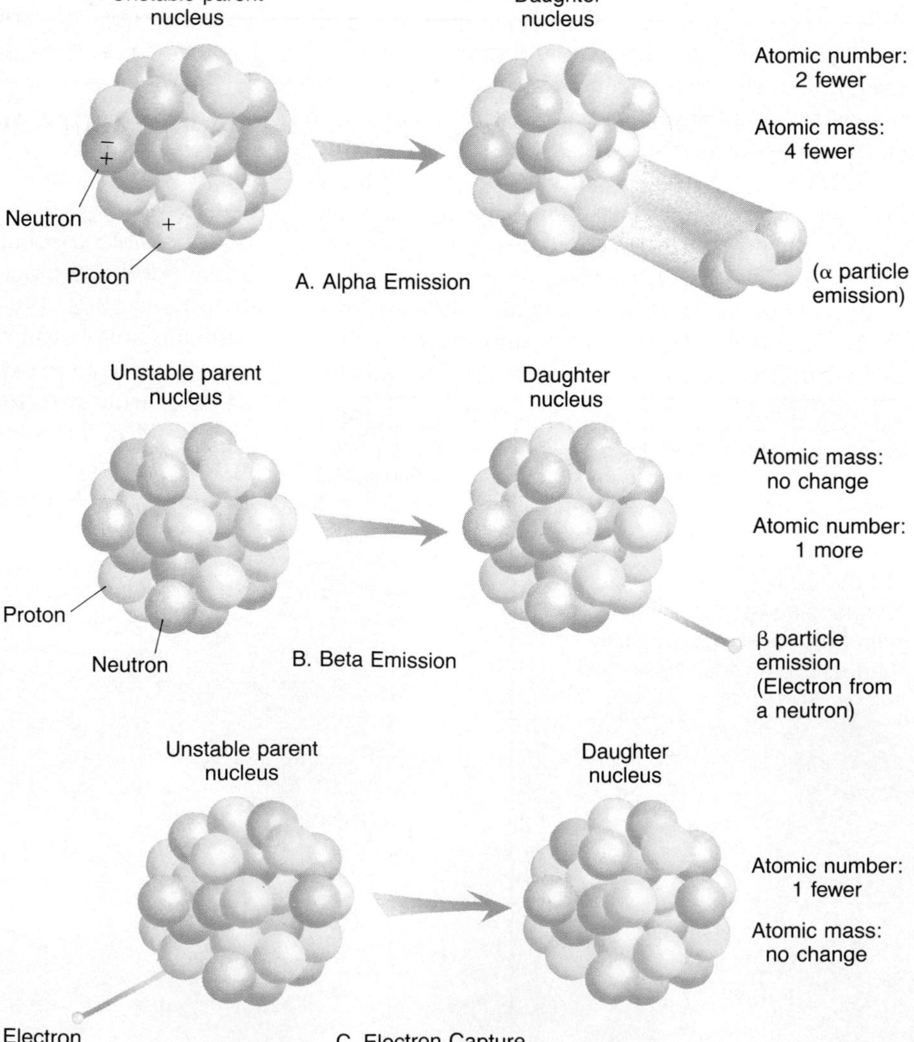

FIGURE 8.12
Common types of radioactive decay. Notice that in each case the number of protons (atomic number) in the nucleus changes, thus producing a different element.

apart? Three common types of radioactive decay are illustrated in Figure 8.12 and are summarized as follows:

1. *Alpha particles* (α particles) may be emitted from the nucleus. An alpha particle is composed of 2 protons and 2 neutrons. Thus, the emission of an alpha particle means that the mass number of the isotope is reduced by 4 and the atomic number is lowered by 2.
2. When a *beta particle* (β particle), or electron, is given off from a nucleus, the mass number remains unchanged, because electrons have practically no mass. However, since the electron must have come from a neutron (remember, a neutron is a combination of a proton and an electron), the nucleus contains one more proton than before. Therefore, the atomic number increases by 1.
3. Sometimes an electron is captured by the nucleus. The electron combines with a proton and forms a neutron. As in the last example, the mass number remains unchanged. However since the nucleus now contains one less proton, the atomic number drops by 1.

The radioactive isotope is referred to as the **parent**, and the isotopes resulting from the decay of the parent are termed the **daughter products.** Figure 8.13 provides an example of radioactive decay. Here it may be seen that when the radioactive parent, uranium-238 (atomic number 92, mass number 238) decays, it emits 8 alpha particles and 6 beta particles before becoming the stable daughter product lead-206 (atomic number 82, mass number 206).

Certainly among the most important results of the discovery of radioactivity is that it provided a reliable means of calculating the ages of rocks and minerals which contain particular radioactive isotopes, a procedure referred to as **radiometric dating.** Why is radiometric dating reliable? The answer lies in the fact that the rates of decay for many isotopes have been precisely measured and do not vary under the physical conditions that exist in the outer layers of the earth. Therefore, each radioactive isotope used for dating has been decaying at a fixed rate since the formation of the rocks in which it occurs and the products of decay have been accumulating at a corresponding rate. For example, when uranium is incorporated into a mineral that crystallizes from magma, there is no lead (the stable daughter product) from previous decay. As the uranium in this newly formed mineral disintegrates, atoms of the daughter product are trapped and measurable amounts of lead eventually accumulate.

The time required for one-half of the nuclei in a sample to decay, called **half-life**, is a common way to express the rate of radioactive disintegration. Figure 8.14 illustrates what occurs when a radioactive parent decays directly into the stable daughter product. When the quantities of parent and daughter are equal (ratio 1:1), we know that one half-life has transpired. When one-quarter of the original parent atoms remain and three-quarters have decayed to the daughter product, the parent/daughter ratio is 1:3 and we know that two half-lives have passed. After three half-lives, the ratio of parent atoms to daughter atoms is 1:7 (1 parent atom for every 7 daughter atoms). If the half-life of a radioactive isotope is known and the parent/daughter ratio can be determined, the age of the sample can be calculated. For example, assume that the half-life of a hypothetical unstable isotope is 1 million years and the parent/daughter ratio in a sample is 1:15. Such a ratio indicates that four half-lives have passed and that the sample must be 4 million years old.

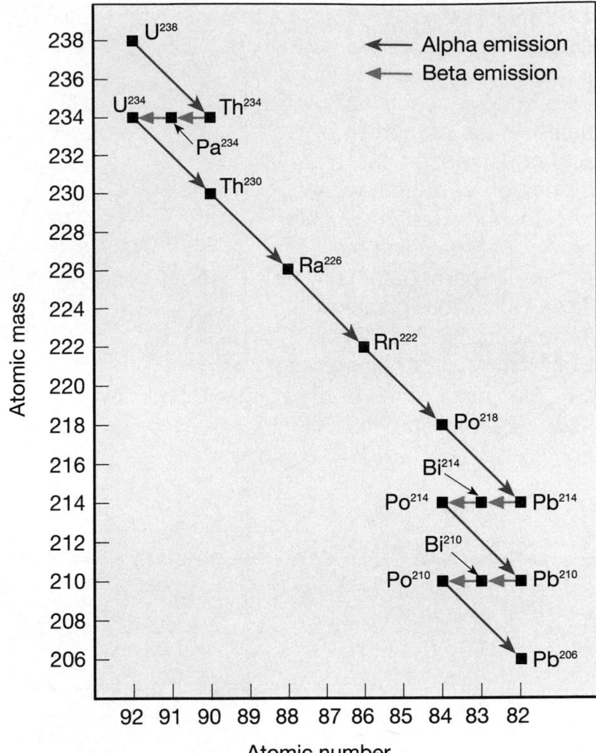

FIGURE 8.13
The most common isotope of uranium (U-238) is an example of a radioactive decay series. Before the stable end product (Pb-206) is reached, many different isotopes are produced as intermediate steps.

Notice that the percentage of radioactive atoms that decay during one half-life is always the same. However, the actual number of atoms that decay with the passing of each half-life continually decreases. Thus, as the percentage of radioactive parent atoms

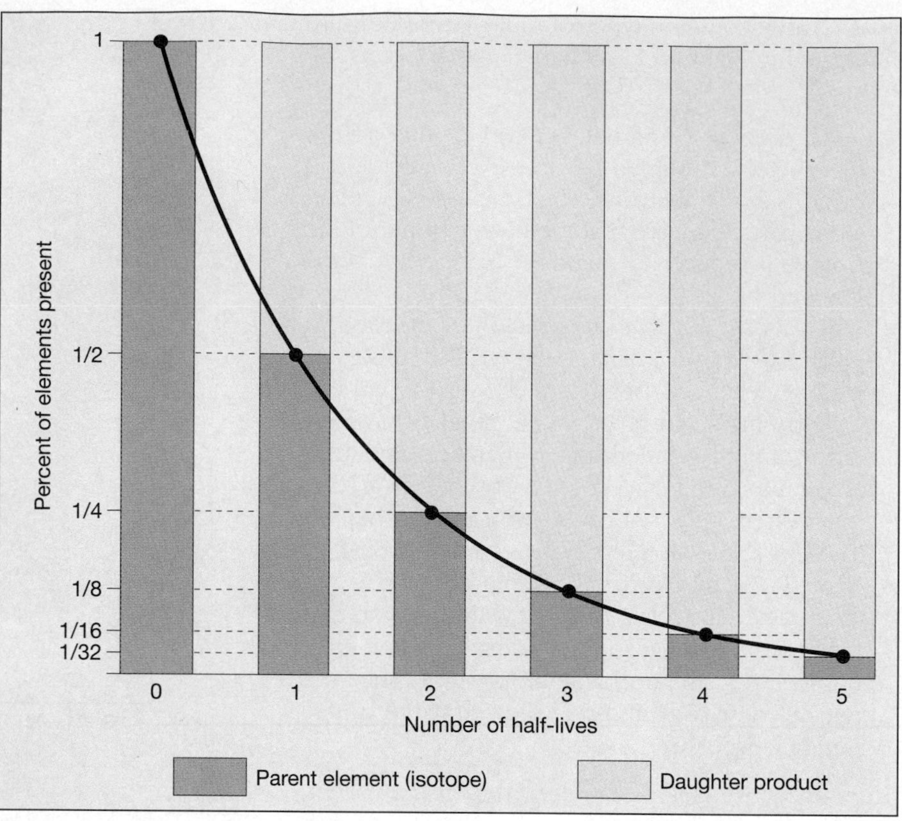

FIGURE 8.14
Decay of a radioactive isotope. The parent/daughter ratio changes continually with time. As the proportion of parent decreases, the proportion of daughter rises, with the increase in daughter atoms just matching the decline in parent atoms. (The Tasa Collection: Geologic Time. Published by Merrill Publishing Co., Columbus, OH. Copyright © 1986, by Tasa Graphic Arts, Inc. All rights reserved)

declines, the proportion of stable daughter atoms rises, with the increase in daughter atoms just matching the drop in parent atoms. This fact is the key to radiometric dating.

Of the many radioactive isotopes that exist in nature, five have proven important in providing radiometric ages for ancient rocks. Table 8.1 summarizes these most commonly used isotopes. Others are either rare or have half-lives that are too short or much too long to be useful. Rubidium-87 and the two isotopes of uranium are used only for dating rocks that are millions of years old, but potassium-40 (K^{40}) is more versatile and is therefore used more often to obtain radiometric dates. With a half-life of 1.3 billion years, potassium-40 can be used to date very ancient rocks. Moreover, sophisticated analytical techniques make possible the detection of tiny amounts of its stable daughter product in some rocks that are younger than 100,000 years. Another important reason for its frequent use is that potassium is an abundant constituent of many common minerals, particularly micas and feldspars.

Although potassium has three natural isotopes, K^{39}, K^{40}, and K^{41}, only K^{40} is radioactive. When K^{40} decays, it does so in two ways. About 11 percent changes to argon-40 (Ar^{40}) by means of electron capture (see Figure 8.12C). The remaining 89 percent of K^{40} decays to calcium-40 (Ca^{40}) by beta emission (see Figure 8.12B). The decay of K^{40} to Ca^{40}, however, is not useful for radiometric dating, because the Ca^{40} produced by radioactive disintegration cannot be distinguished from calcium that may have been present when the rock formed.

The potassium-argon clock begins when potassium-bearing minerals crystallize from a magma or form within a metamorphic rock. At this point the new minerals will contain K^{40} but will be free of Ar^{40}, because this element is an inert gas that does not chemically combine with other elements. As time passes, the K^{40} steadily decays, producing Ar^{40}, which remains trapped within the mineral's crystal lattice. Since there was no Ar^{40} present when the mineral formed, all of the daughter atoms trapped in the mineral must have come from the decay of K^{40}. To determine a sample's age, the K^{40}/Ar^{40} ratio must be measured precisely and the known half-life for K^{40} applied.

It is important to realize that an accurate radiometric date can be obtained only if the mineral remained a closed system during the entire period since its formation; that is, a correct date is not possible unless there was neither the addition nor loss of parent or daughter isotopes. This is not always the case. In fact, an important limitation of the potassium-argon method arises from the fact that argon is a gas and may leak from the minerals in which it

TABLE 8.1
Isotopes most frequently used for radiometric dating.

Radioactive Parent	Stable Daughter Product	Currently Accepted Half-life Values
Uranium-238	Lead-206	4.5 billion years
Uranium-235	Lead-207	713 million years
Thorium-232	Lead-208	14.1 billion years
Rubidium-87	Strontium-87	47.0 billion years
Potassium-40	Argon-40	1.3 billion years

forms. Indeed, losses can be significant if the rock is subjected to relatively high temperatures. Of course, a reduction in the amount of Ar^{40} leads to an underestimation of the rock's age. Sometimes temperatures are high enough for a sufficiently long period that all argon escapes. When this happens, the potassium-argon clock is reset and dating the sample will give only the time of thermal resetting, not the true age of the rock. For other radiometric clocks, a loss of daughter atoms can occur if the rock has been subjected to weathering or leaching. To avoid such a problem, one simple safeguard is to use only fresh, unweathered material and not samples that may have been chemically altered.

If parent/daughter ratios are not always reliable, how can meaningful radiometric dates be obtained? One common precaution against unknown errors is the use of cross checks. Often this simply involves subjecting a sample to two different radiometric methods. If the two dates agree, the likelihood is high that the date is reliable. If, on the other hand, there is an appreciable difference between the two dates, other cross checks must be employed to determine which, if either, is correct.

To date very recent events, carbon-14 (also called **radiocarbon**), the radioactive isotope of carbon, is used. Because it has a half-life of only 5730 years, it can be used for dating events from the historic past, as well as those from recent geologic history. Until the late 1970s radiocarbon was useful in dating events only as far back as 40,000–50,000 years. However, the development of more sophisticated analytical techniques has increased the usefulness of this "clock." In some instances carbon-14 can be used to date events as far back as 75,000 years. This is a significant accomplishment because it means that geologists can now date many ice-age phenomena that previously could not be dated accurately.

Carbon-14 is continuously produced in the upper atmosphere as a consequence of cosmic ray bombardment, in which cosmic rays (high-energy nuclear particles) shatter the nuclei of gases to release neutrons (Figure 8.15). Some of the neutrons are absorbed by nitrogen (atomic number 7, mass num-

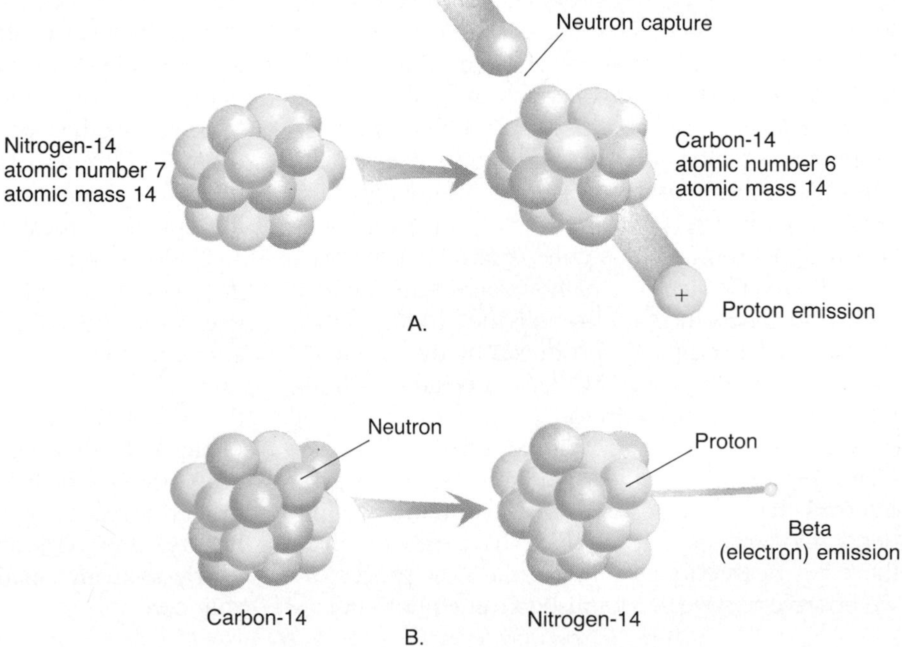

FIGURE 8.15
A. Production and **B.** decay of carbon-14. These sketches represent the nuclei of the respective atoms.

ber 14), causing its nucleus to emit a proton. Thus, the atomic number drops by 1 (to 6), and a different element, carbon-14, is created (Figure 8.15A). This isotope of carbon is quickly incorporated into carbon dioxide, circulates in the atmosphere, and is absorbed by living matter. As a result, all organisms contain a small amount of carbon-14.

While an organism is alive, the decaying radiocarbon is continually replaced. As a result, the ratio of carbon-14 to carbon-12 (the most common isotope of carbon) remains constant. However, when the plant or animal dies, the amount of carbon-14 gradually decreases as it decays to nitrogen-14 by beta emission (Figure 8.15B). Therefore, by comparing the proportions of carbon-14 and carbon-12 in a sample, radiocarbon dates can be determined. Although carbon-14 is only useful in dating the last small fraction of geologic time, it has become a very valuable tool for anthropologists, archeologists, and historians, as well as for geologists who study very recent earth history. In fact, the development of radiocarbon dating was considered so important that the chemist who discovered this application, Willard F. Libby, received a Nobel Prize.

Bear in mind that although the basic principle of radiometric dating is relatively simple, the actual procedure is quite complex, for the chemical analysis which determines the quantities of parent and daughter that are present must be painstakingly precise. In addition, some radioactive materials do not decay directly into the stable daughter product, a fact which may further complicate the analysis. In the case of uranium-238, there are thirteen intermediate unstable daughter products formed before the fourteenth, and last, daughter product, the stable isotope lead-206, is produced (see Figure 8.13).

Radiometric dating methods have produced literally thousands of dates for events in earth history. Rocks from several localities have been dated at more than 3 billion years, and geologists realize that still older rocks exist. For example, a granite from South Africa which has been dated at 3.2 billion years contains inclusions of quartzite. Quartzite is a metamorphic rock which originally was the sedimentary rock sandstone. Since sandstone is the product of the lithification of sediments produced by the weathering of pre-existing rocks, we have a positive indication that older rocks existed.

Radiometric dating has vindicated the ideas of Hutton, Darwin, and others who over 150 years ago inferred that geologic time must be immense. Indeed, modern dating methods have proven that there has been enough time for the slow processes we observe to have accomplished tremendous tasks.

THE GEOLOGIC TIME SCALE

The whole of geologic history has been divided into units of varying magnitude which together comprise the time scale of earth history (Figure 8.16). The major units of the time scale were delineated during the nineteenth century, principally by workers in western Europe and Great Britain. Since absolute dating was unavailable at that time, the entire time scale was created using methods of relative dating. It has only been in this century that absolute dates have been added.

The geologic time scale subdivides the 4.6-billion-year history of the earth into many different units and provides a meaningful time frame within which the events of the geologic past are arranged. As shown in Figure 8.16, **eons** represent the greatest expanses of time. The eon that began about 570 million years ago is known as the **Phanerozoic**, a term derived from the Greek words meaning *visible life*. It is an appropriate description because the rocks and deposits of the Phanerozoic eon contain an abundance of fossils that document major evolutionary trends. Another glance at the time scale reveals that the Phanerozoic eon is divided into units called **eras**. The eras recognized for this span are the **Paleozoic** ("ancient life"), the **Mesozoic** ("middle life"), and the **Cenozoic** ("recent life"). As the names imply, the eras are bounded by profound worldwide changes in life forms. Each era is subdivided into time units known as **periods**. The Paleozoic has seven, the Mesozoic three, and the Cenozoic two. Each period is characterized by a somewhat less profound change in life forms as compared with the eras. The major divisions, with brief explanations of each, are shown in Table 8.2. Finally, each of the twelve periods is further divided into still smaller units called **epochs**. Seven epochs have been named for the periods of the Cenozoic era. The epochs of other periods, however, are not commonly referred to by specific names. Instead, the terms *early, middle,* and *late* are generally applied to the epochs of these earlier periods.

Notice that the detail of the geologic time scale does not begin until about 570 million years ago, the date for the beginning of the first period of the Paleozoic era, the Cambrian period. The more than 4 billion years prior to the Cambrian is divided into three eons, the **Hadean**, the **Archean** and the **Proterozoic**. It is also common for this vast expanse of time to simply be referred to as the **Precambrian**. Although it represents more than 85 percent of earth history, the Precambrian is not divided into nearly as many smaller time units as the Phanerozoic eon.

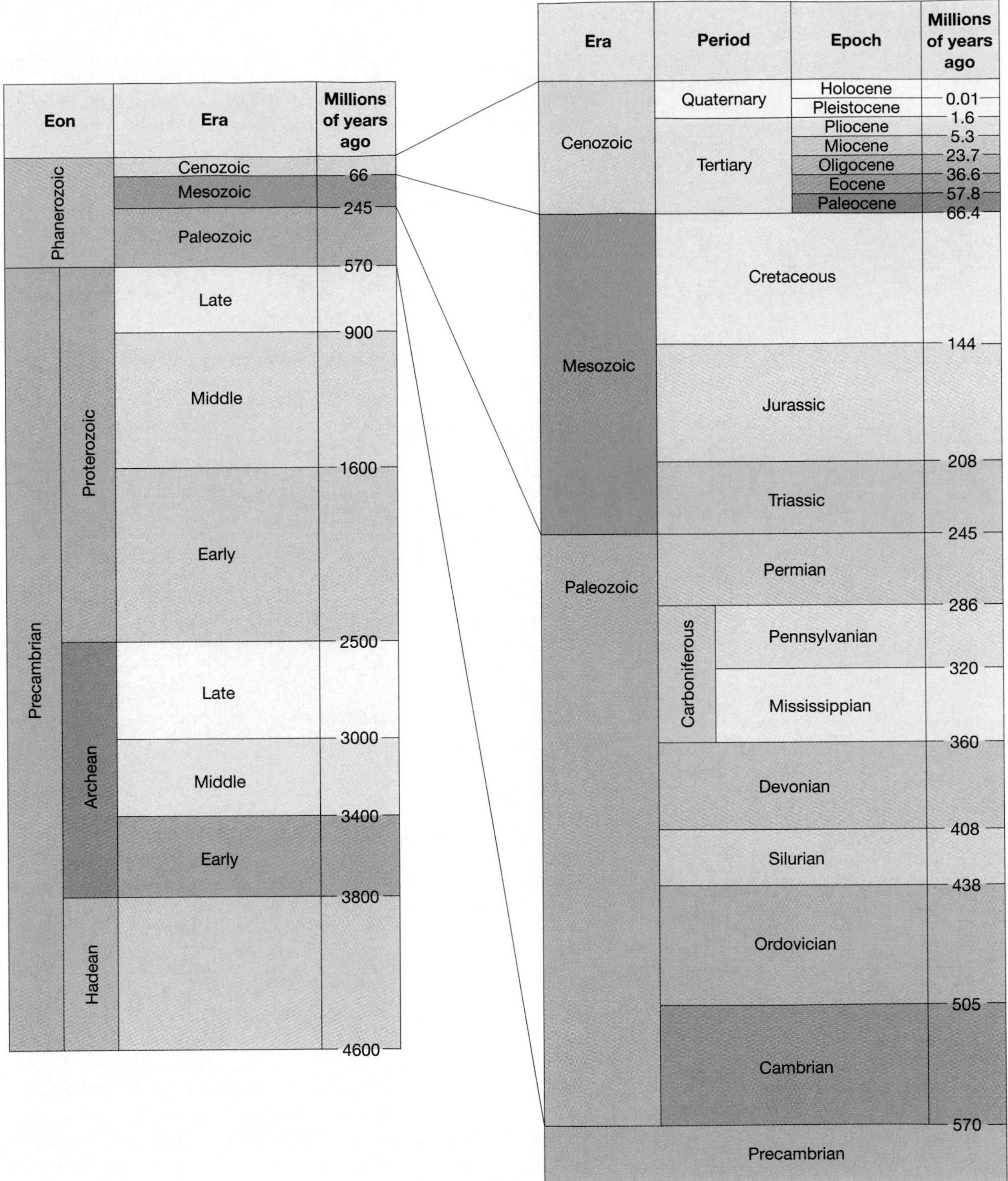

FIGURE 8.16
The geologic time scale. The absolute dates were added long after the time scale had been established using relative dating techniques. (Data from Geological Society of America)

TABLE 8.2
Major divisions of geologic time.

Era	Period	Origin of name
CENOZOIC ERA (Age of Recent Life)	Quaternary period	The several geologic eras were originally named Primary, Secondary, Tertiary, and Quaternary. The first two names are no longer used; Tertiary and Quaternary have been retained but used as period designations.
	Tertiary period	
MESOZOIC ERA (Age of Middle Life)	Cretaceous period	Derived from Latin word for chalk (creta) and first applied to extensive deposits that form white cliffs along the English Channel (see Figure 6.9)
	Jurassic period	Named for the Jura Mountains, located between France and Switzerland, where rocks of this age were first studied.
	Triassic period	Taken from word "trias" in recognition of the threefold character of these rocks in Europe.
PALEOZOIC ERA (Age of Ancient Life)	Permian period	Named after the province of Perm, U.S.S.R., where these rocks were first studied.
	Pennsylvanian period*	Named for the state of Pennsylvania where these rocks have produced much coal.
	Mississippian period*	Named for the Mississippi River valley where these rocks are well exposed.
	Devonian period	Named after Devonshire County, England, where these rocks were first studied.
	Silurian period	Named after Celtic tribes, the Silures and the Ordovices, that lived in Wales during the Roman Conquest.
	Ordovician period	
	Cambrian period	Taken from Roman name for Wales (Cambria), where rocks containing the earliest evidence of complex forms of life were first studied
PRECAMBRIAN		The time between the birth of the planet and the appearance of complex forms of life. More than 85 percent of the earth's estimated 4.6 billion years falls into this span.

SOURCE: U.S. Geological Survey.

*Outside of North America, the Mississippian and Pennsylvania periods are combined into the Carboniferous period.

Why is the huge expanse of Precambrian time not divided into numerous eras, periods, and epochs? The reason is that Precambrian history is not known in great enough detail. The quantity of information geologists have deciphered about the earth's past is somewhat analogous to the detail of human history. The farther back we go, the less that is known. Certainly more data and information exist about the past ten years than for the first decade of the twentieth century; the events of the nineteenth century have been documented much better than the events of the first century A.D.; and so on. So it is with earth history. The more recent past has the freshest, least disturbed, and most observable record. The farther back in time the geologist goes, the more fragmented the record and clues become. There are other reasons to explain our lack of a detailed time scale for this vast segment of earth history:

1. The first abundant fossil evidence does not appear in the geologic record until the beginning of the Cambrian period. Prior to the Cambrian, simple life forms such as algae, bacteria, fungi, and worms predominated. All of these organisms lack hard parts, an important prerequisite for fossilization. For this reason, there is only a meager Pre-

cambrian fossil record. Many exposures of Precambrian rocks have been studied in some detail, but correlation is often difficult when fossils are lacking.

2. Because Precambrian rocks are very old, most have been subjected to a great many changes. Much of the Precambrian rock record is composed of highly distorted metamorphic rocks. This makes the interpretation of past environments difficult, because many of the clues present in the original sedimentary rocks have been destroyed.

With the development of radiometric dating methods, a partial solution to the troublesome task of dating and correlating Precambrian rocks now exists. Untangling the complex Precambrian record, however, is still in its early stages.

DIFFICULTIES IN DATING THE GEOLOGIC TIME SCALE

Although reasonably accurate dates have been worked out for the periods of the geologic time scale (Figure 8.16), the task is not without its difficulties. The primary difficulty in assigning absolute dates to units of time is the fact that not all rocks can be dated by radiometric methods. Recall that for a radiometric date to be useful, all the minerals in the rock must have formed at about the same time. For this reason, radioactive isotopes can be used to determine when minerals in an igneous rock crystallized and when pressure and heat created new minerals in a metamorphic rock. However, samples of sedimentary rock can only rarely be dated directly by radiometric means. Although a detrital sedimentary rock may include particles that contain radioactive isotopes, the rock's age cannot be accurately determined because the grains composing the rock are not the same age as the rock in which they occur. Rather, the sediments have been weathered from rocks of diverse ages. Radiometric dates obtained from metamorphic rocks may also be difficult to interpret, since the age of a particular mineral in a metamorphic rock does not necessarily represent the time when the rock initially formed. Instead, the date may indicate any one of a number of subsequent metamorphic phases.

If samples of sedimentary rocks rarely yield reliable radiometric ages, how can absolute dates be assigned to sedimentary layers? Usually the geologist must relate the strata to igneous masses, as in Figure 8.17. In this example, the ages of the volcanic ash bed

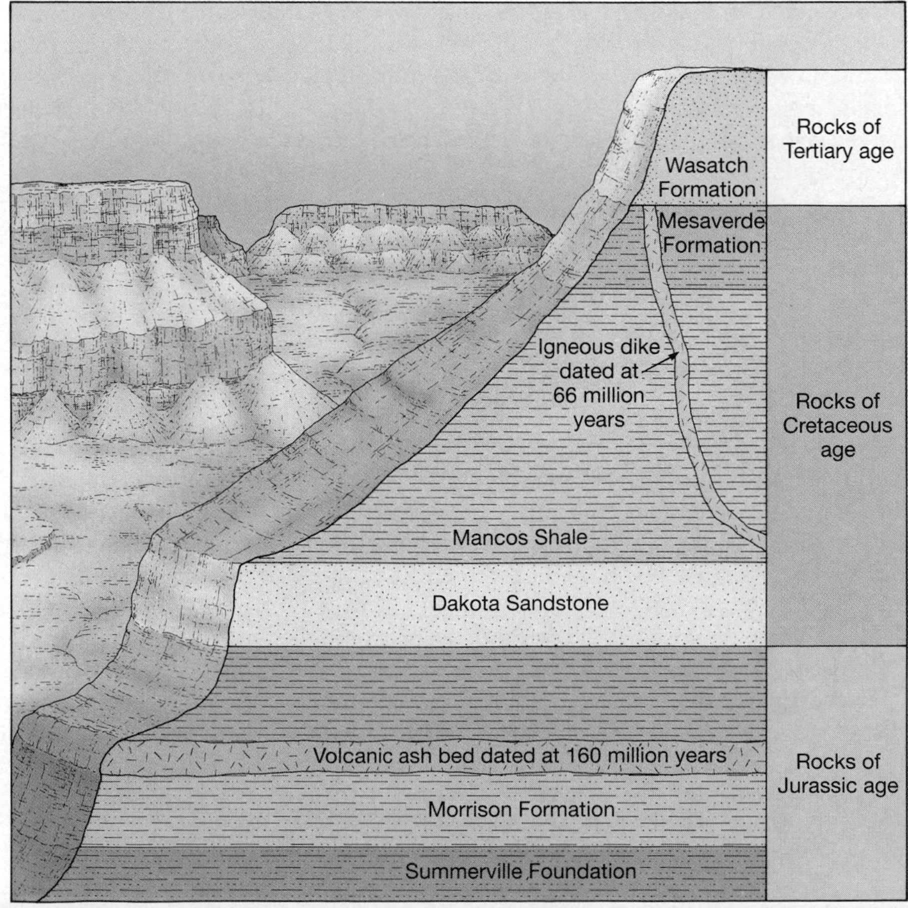

FIGURE 8.17
Absolute dates for sedimentary layers are usually determined by examining their relationship to igneous rocks. (After U.S. Geological Survey)

BOX 8.2

The Cretaceous-Tertiary Extinction

The boundaries between divisions on the geologic time scale represent times of significant geological and/or biological change. About 65 million years ago, more than half of all plant and animal species died out. The event marks the end of the Mesozoic era and the beginning of the Cenozoic era. It represents the end of the era in which dinosaurs and other reptiles dominated the landscape and the beginning of the era when mammals became very important (Figure 8.A). Since the Cretaceous (abbreviated K to avoid confusion with other "C" periods) is the last period of the Mesozoic, and the Tertiary (abbreviated T) is the first period of the Cenozoic, the time of this mass extinction is called the Cretaceous-Tertiary or KT boundary.

The dinosaurs met their demise at the KT boundary, along with large numbers of other animal and plant groups, both terrestrial and marine. Equally important, of course, is the fact that many species survived the disaster. Human beings are descended from these survivors. Perhaps this fact explains why an event that occurred 65 million years ago has captured the interest of so many people. The extinction of the great reptiles is generally attributed to this group's inability to adapt to some radical change in environmental conditions. What event could have triggered the sudden extinction of the dinosaurs—the most successful group of land animals ever to have lived?

One modern view proposes that about 65 million years ago, a large asteroid or comet about 10 kilometers in diameter collided with the earth. The impact of such a body would have produced a cloud of dust thousands of times greater than that released during the 1980 eruption of Mount St. Helens. For many months the dust-laden atmosphere would have greatly restricted the amount of sunlight that penetrated to the earth's surface. Without sunlight for photosynthesis, delicate food chains would collapse. It is further hypothesized that large dinosaurs would be affected more adversely by this chain of events than would smaller life forms. In addition, acid rains and global fires may have added to the environmental disaster. It is estimated that when the sunlight returned, more than half of the species on earth, including numerous marine organisms, had become extinct.

What evidence points to this catastrophic collision 65 million years ago? First, a thin layer of sediment nearly 1 centimeter thick has been discovered at the KT boundary. This particular sediment is found worldwide and contains a high level of the element iridium, which is rare in the earth's crust but is found in similar proportions in stony meteorites. Could this layer represent the scattered remains of an asteroid that was responsible for the environmental changes which led to the demise of many reptile groups? Second, this period of mass extinction appears to have affected all land animals larger than dogs. Supporters of this catastrophic-event scenario suggest that small, ratlike mammals could survive a breakdown of food chains lasting perhaps several months. Large animals, they contend, could not survive such an event.

Other scientists disagree with the foregoing hypothesis. They claim that what appears to be a mass extinction over a short period of time did, in fact, occur over a much broader time span. Based upon an examination of the fossil record at the Cretaceous-Tertiary boundary, these geologists conclude that the decline of the dinosaurs and numerous other organisms was gradual.

Those who disagree with the impact hypothesis have suggested

within the Morrison Formation and the dike cutting the Mancos Shale and Mesaverde Formation are known. The sedimentary beds below the ash are obviously older than the ash, and all the layers above the ash are younger. The dike is younger than the Mancos Shale and the Mesaverde Formation but older than the Wasatch Formation because the dike does not intrude the Tertiary rocks. From this kind of evidence, geologists estimate that part of the Morrison Formation was deposited about 160 million years ago as indicated by the ash bed. Further, they conclude that the Tertiary period began after the intrusion of the dike, 66 million years ago. This is one example of literally thousands that illustrate how dated materials bracket the various episodes in earth history within specific time periods and show the necessity of combining laboratory dating methods with geological information.

FIGURE 8.A
A composite Mesozoic landscape showing large carnivorous and herbivorous dinosaurs. (Courtesy of the Peabody Museum of Natural History, Yale University)

that volcanism could account for the mass extinction. This competing hypothesis is based on the fact that enormous outpourings of basaltic lavas occurred in India approximately 65 million years ago. Today, these lava flows make up a region known as the Deccan Traps. Although much more extensive, the Deccan Traps resemble the flood basalts of the Columbia Plateau in the Pacific Northwest of the United States. Advocates of the volcanism hypothesis argue that the consequences of an asteroid impact would be quite similar to those associated with extensive volcanism. The first effect would be darkness resulting from large quantities of dust and ash in the atmosphere and causing the food chain to collapse. Furthermore, volcanic eruptions are capable of emitting large quantities of sulfur that could form toxic sulfuric acid rains. To support this scenario, they point to the 1783 eruption at Laki, Iceland. Although small by comparison to the eruptions that produced the Deccan Traps, this volcanic activity killed 75 percent of all livestock and ultimately 24 percent of the inhabitants of Iceland.

Whichever group is correct, the fact remains that one of the more successful groups ever to inhabit the earth died out at the close of the Mesozoic. The decline of the reptiles provided vacancies in habitats for the mammals. Although small and inconspicuous during the Mesozoic, the mammals rose to dominance during the Cenozoic.

REVIEW QUESTIONS

1. Distinguish between absolute and relative dating.
2. Describe two early methods for dating the earth. How old was the earth thought to be according to these estimates? List some weaknesses of each method. (See Box 8.1.)
3. What is the law of superposition? How are cross-cutting relationships used in relative dating?
4. Refer to Figure 8.4 and answer the following questions:
   (a) Is fault A older or younger than the sandstone layer?
   (b) Is dike A older or younger than the sandstone layer?
   (c) Was the conglomerate deposited before or after fault A?
   (d) Was the conglomerate deposited before or after fault B?

(e) Which fault is older, A or B?
(f) Is dike A older or younger than the batholith?

5. When you observe an outcrop of steeply inclined sedimentary layers, what principle allows you to assume that the beds were tilted after they were deposited?
6. A mass of granite is in contact with a layer of sandstone but does not cut across it. How might you determine whether the sandstone was deposited on top of the granite or the granite was intruded from below after the sandstone was deposited?
7. Distinguish among angular unconformity, disconformity, and nonconformity.
8. What is meant by the term *correlation?*
9. Describe William Smith's important contribution to the science of geology.
10. Why are fossils such useful tools in correlation?
11. What subdivisions make up the geologic time scale?
12. Explain the lack of a detailed time scale for the vast span known as the Precambrian.
13. Figure 8.18 is a block diagram of a hypothetical area in the American Southwest. Place the lettered features in the proper sequence, from oldest to youngest. Identify an angular unconformity and a nonconformity.
14. If a radioactive isotope of thorium (atomic number 90, mass number 232) emits 6 alpha particles and 4 beta particles during the course of radioactive decay, what are the atomic number and mass number of the stable daughter product?
15. Why is radiometric dating the most reliable method of dating the geologic past?
16. A hypothetical radioactive isotope has a half-life of 10,000 years. If the ratio of radioactive parent to stable daughter product is 1:3, how old is the rock containing the radioactive material?
17. Why is potassium-40 used more frequently in radiometric dating than other isotopes?
18. Why is the ratio between potassium-40 and calcium-40 not used for radiometric dating?
19. In order to provide a reliable radiometric date, a mineral must remain a closed system from the time of its formation until the present. Why is this true?
20. What precautions are taken to insure reliable radiometric dates?
21. Assume that the age of the earth is 5 billion years.
    (a) What fraction of geologic time is represented by recorded history (assume 5000 years for the length of recorded history)?
    (b) The first abundant fossil evidence does not appear until the beginning of the Cambrian period (570 million years ago). What percent of geologic time is represented by abundant fossil evidence?
22. Briefly describe the difficulties in assigning absolute dates to layers of sedimentary rock.

KEY TERMS

absolute date (p. 182)
angular unconformity (p. 186)
Archean eon (p. 196)
Cenozoic era (p. 196)
conformable (p. 185)
correlation (p. 189)
cross-cutting relationships, principle of (p. 184)
daughter product (p. 193)
disconformity (p. 186)
eon (p. 196)
epoch (p. 196)
era (p. 196)
faunal succession, principle of (p. 191)
Hadean eon (p. 196)
half-life (p. 193)
inclusions (p. 185)
index fossil (p. 191)
Mesozoic era (p. 196)
nonconformity (p. 187)
original horizontality, principle of (p. 184)
Paleozoic era (p. 196)
parent (p. 193)
period (p. 196)
Phanerozoic eon (p. 196)
Precambrian (p. 196)
Proterozoic eon (p. 196)
radioactivity (p. 192)
radiocarbon (p. 195)
radiometric dating (p. 193)
relative dating (p. 183)
superposition, law of (p. 184)
unconformity (p. 185)

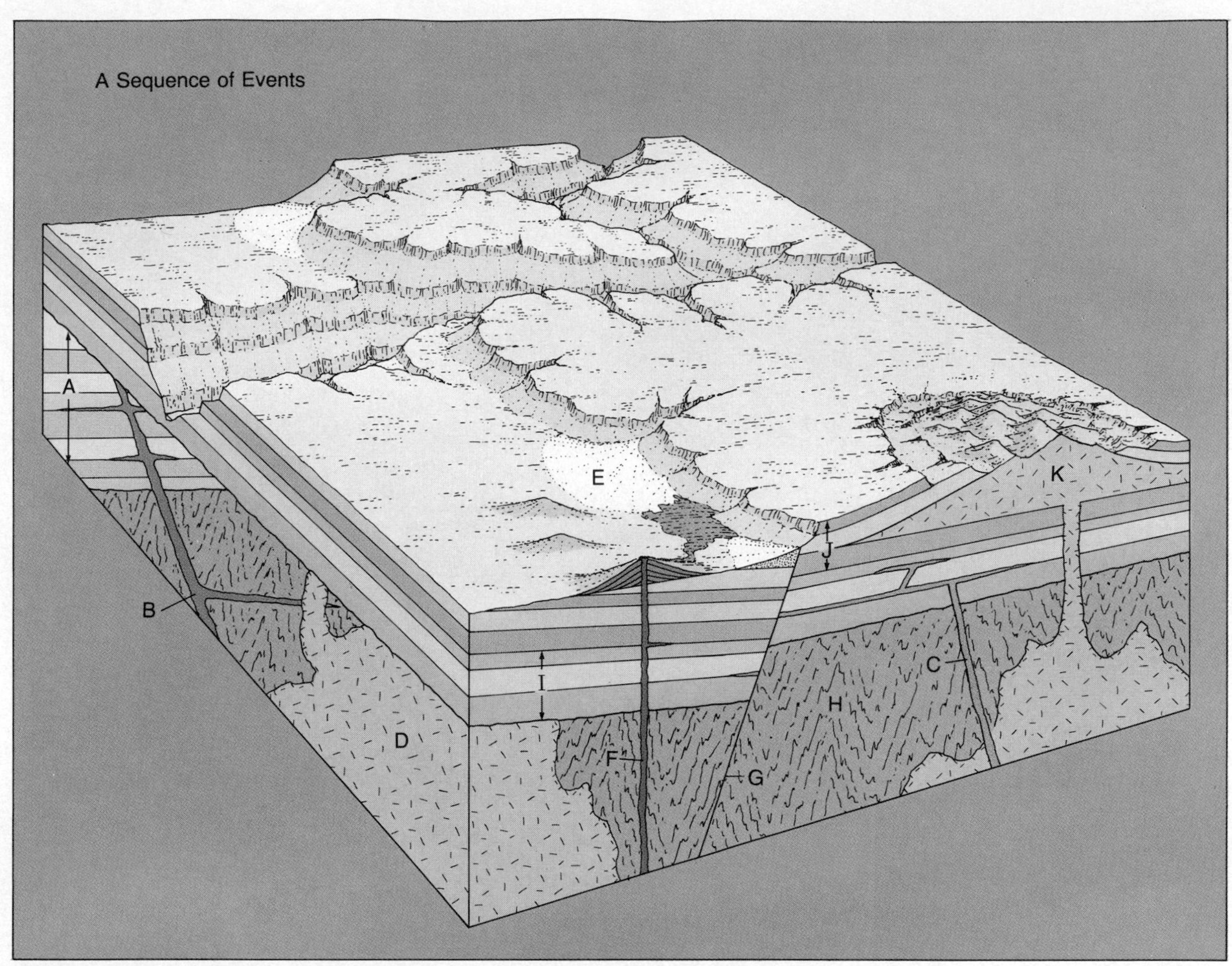

FIGURE 8.18
Use this block diagram in conjunction with Review Question 13. (The Tasa Collection: Geologic Time. Published by Merrill Publishing Co., Columbus, OH. Copyright © 1986, by Tasa Graphic Arts, Inc. All rights reserved)

9
Mass Wasting

Opposite: Rockslide near Lake Tahoe, California. (Photo by James E. Patterson) (Top photo by P. Jay Fleisher)

The earth's surface is never perfectly flat but instead consists of slopes. Some are steep and precipitous; others are moderate or gentle. Some are long and gradual; others are short and abrupt. Some slopes are mantled with soil and covered by vegetation; others consist of barren rock and rubble. Their form and variety are great. Taken together, slopes are the most common elements in our physical landscape. Although most slopes appear to be stable and unchanging, they are not static features because the force of gravity causes material to move downslope. At one extreme, the movement may be gradual and practically imperceptible. At the other extreme it may consist of a thundering rockfall or avalanche.

The news media periodically relate the terrifying and often grim details of landslides. On May 31, 1970, one such event occurred when a gigantic rock avalanche buried more than 20,000 people in Yungay and Ranrahirca, Peru (Figure 9.1). There was little warning of the impending disaster; it began and ended in just a matter of a few minutes. The avalanche started 14 kilometers from Yungay, near the summit of the 6700-meter (22,000-foot) Nevado Huascaran, the loftiest peak in the Peruvian Andes. Triggered by the ground motion from a strong offshore earthquake, a huge mass of rock and ice broke free from the precipitous north face of the mountain. After plunging nearly one kilometer, the material pulverized on impact and immediately began rushing down the mountainside, made fluid by trapped air and melted ice. The initial mass ripped loose additional millions of tons of debris as it roared downhill. The shock waves produced by the event created thunderlike noise and stripped nearby hillsides of vegetation. Although the material followed a previously eroded gorge, a portion of the debris jumped a 200—300-meter bedrock ridge that had protected Yungay from similar events in the past and buried the entire city. After inundating another town in its path, Ranrahirca, the mass of debris finally reached the bottom of the valley where its momentum carried it across the Rio Santa and tens of meters up the valley wall on the opposite side.

A.

B.

FIGURE 9.1
This Peruvian valley was devastated by the rock avalanche that was triggered by an off-shore earthquake in May, 1970. **A.** Before. **B.** After the rock avalanche. (Photos courtesy of Iris Lozier)

BOX 9.1

The Vaiont Dam Disaster

A massive rock avalanche in Peru is described at the beginning of this chapter. As with most occurrences of mass wasting, this tragic episode was triggered by a natural event—in this case, an earthquake. However, disasters also result from the mass movement of surface material triggered by the actions of humans. For example, in 1960 a large dam almost 265 meters tall was built across the Vaiont Canyon in the Italian Alps. Three years later, on the night of October 9, 1963, a violent disaster occurred and the dam was largely responsible.

The bedrock in Vaiont Canyon slanted steeply downward toward the lake impounded behind the dam and was composed of weak, highly fractured limestone strata that contained beds of clay and numerous solution cavities. As the reservoir filled, the lower portions of these rocks became saturated and the clays became swollen and more plastic. The rising water reduced the internal friction that had kept the rock in place. Measurements made shortly after the reservoir was filled hinted at the problem, because they indicated that a portion of the mountain was slowly creeping downhill at the rate of 1 centimeter per week. In September, 1963, the rate increased to 1 centimeter per day, then 10–20 centimeters per day, and eventually to as much as 80 centimeters on the day of the disaster. Finally, the mountainside let loose. In just an instant, 240 million cubic meters of rock and rubble slid down the face of Mount Toc and filled nearly 2 kilometers of the gorge to heights of 150 meters above the reservoir level (Figure 9.A). The filling of the reservoir pushed the water completely over the dam in a wave more than 90 meters high. More than 1.5 kilometers downstream, the wall of water was still 70 meters high and everything in its path was completely destroyed. The entire event lasted less than seven minutes, yet it claimed an estimated 2600 lives. This is known as the worst dam disaster in history, but when it was over, the Vaiont Dam was still standing intact. Although the catastrophe was triggered by human interference with the Vaiont River, the slide would have eventually occurred on its own; however, the effects would not have been nearly as tragic.

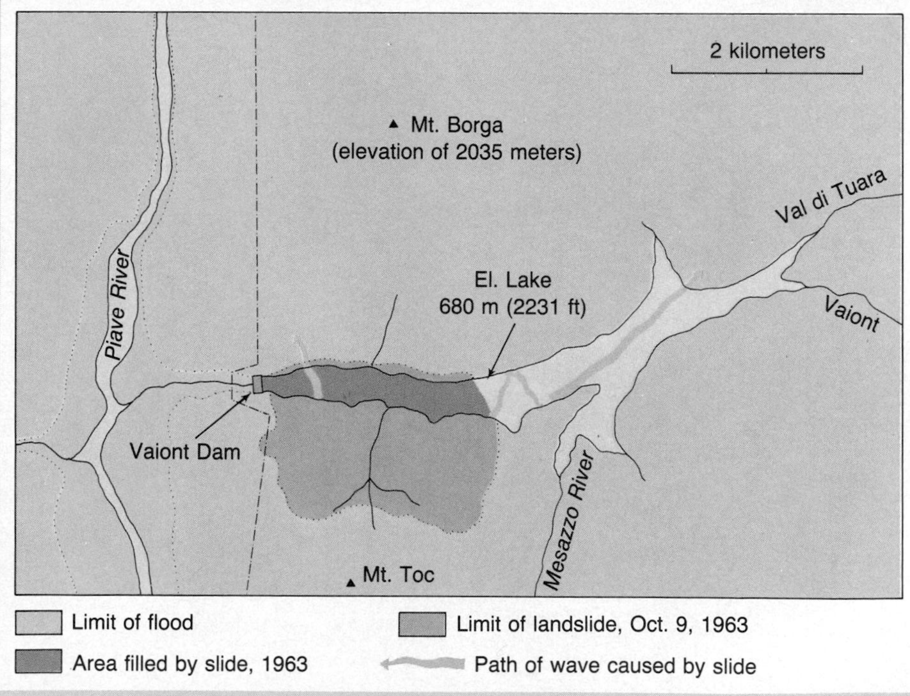

FIGURE 9.A
Sketch map of the Vaiont River area showing the limits of the landslide, the portion of the reservoir that was filled with debris, and the extent of flooding downstream. [After G. A. Kiersh, "Vaiont Reservoir Disaster," *Civil Engineering* 34 (1964): 32–39]

This was not the first such disaster in the region and will no doubt not be the last. Just eight years earlier, a less spectacular, but nevertheless devastating, rock avalanche took the lives of an estimated 3500 people on the heavily populated valley floor at the base of the mountain. Fortunately, mass movements such as the one just described are infrequent and only occasionally affect large numbers of people.

CONTROLS OF MASS WASTING

Landslides are spectacular examples of a basic geologic process called **mass wasting**. Mass wasting refers to the downslope movement of rock, regolith, and soil under the direct influence of gravity. It is distinct from the erosional processes that are examined in subsequent chapters because mass wasting does not require a transporting medium.

In the evolution of most landforms, mass wasting is the step that follows weathering. By itself weathering does not produce significant landforms. Rather landforms develop as the products of weathering are removed from the places where they originate. Once weathering weakens and breaks rock apart, mass wasting transfers the debris downslope, where a stream, acting as a conveyor belt, usually carries it away. Although there may be many intermediate stops along the way, the sediment is eventually transported to its ultimate destination, the sea. It is the combined effects of mass wasting and running water that produce stream valleys, which are the most common and conspicuous landforms at the earth's surface. If streams alone were responsible for creating the valleys in which they flow, the valleys would be very narrow features. However, the fact that most river valleys are much wider than they are deep is a strong indication of the significance of mass wasting processes in supplying material to streams. This is illustrated by the Grand Canyon (Figure 9.2). The walls of the canyon extend far from the Colorado River due to the transfer of weathered debris downslope to the river and its tributaries by mass wasting processes. In this manner, streams and mass wasting combine to modify and sculpture the earth's surface. Of course, glaciers, groundwater, waves, and wind are also important agents in shaping landforms and developing landscapes.

Although gravity is the controlling force of mass wasting, other factors play an important part in bringing about the downslope movement of material. Water is one of these factors. When the pores in sediment become filled with water, the cohesion among particles is destroyed, allowing them to slide past one another with relative ease. For example, when sand is slightly moist, it sticks together quite well.

FIGURE 9.2
The walls of the Grand Canyon extend far from the channel of the Colorado River. This results primarily from the transfer of weathered debris downslope to the river and its tributaries by mass wasting processes. (Photo by James E. Patterson)

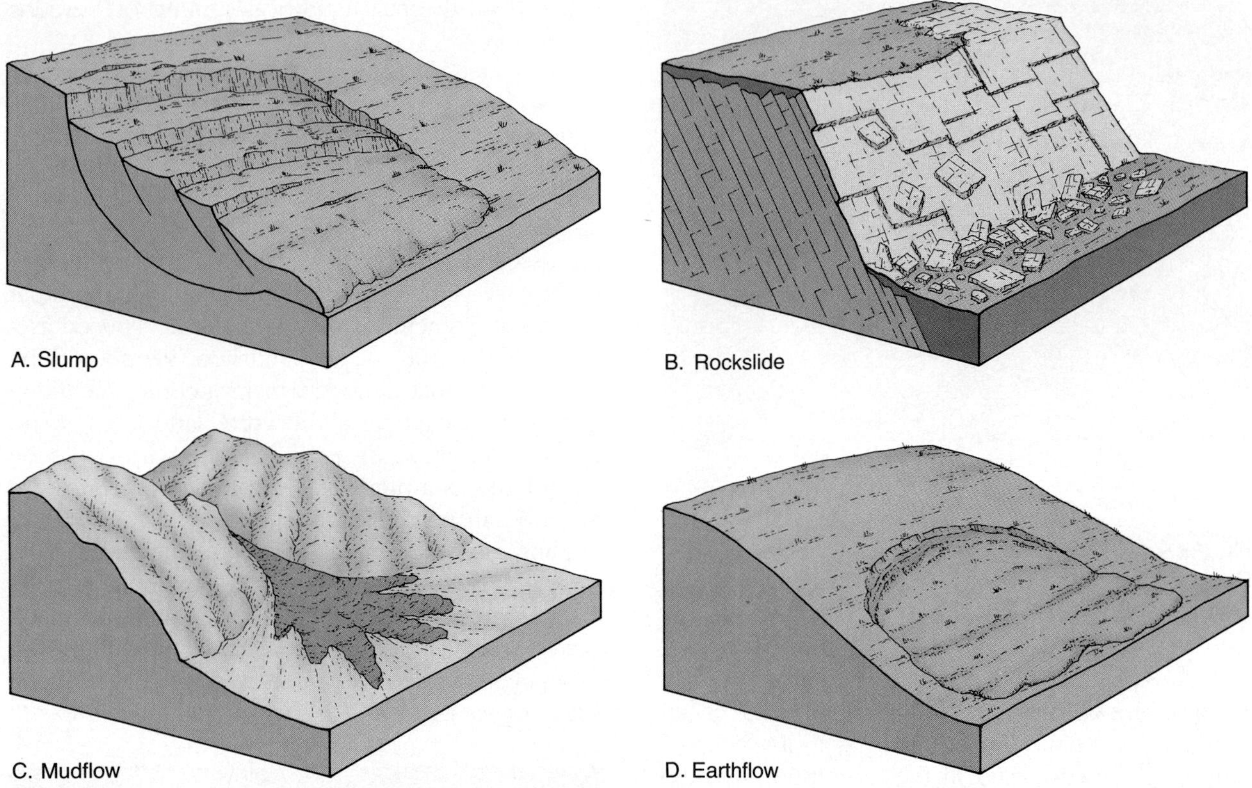

FIGURE 9.3
The four processes illustrated here are all considered to be relatively rapid forms of mass wasting. Because material in slumps **(A)** and rockslides **(B)** move along well-defined surfaces, they are said to move by sliding. By contrast, when material moves downslope as a viscous fluid, the movement is described as a flow. Mudflow **(C)** and earthflow **(D)** advance downslope in this manner.

However, if enough water is added to fill the openings between the grains, the sand will ooze out in all directions. Thus, saturation reduces the internal resistance of materials, which are then easily set in motion by the force of gravity. When clay is wetted, it becomes very slick—another example of the "lubricating" effect of water. Water also adds considerable weight to a mass of material. The added weight in itself may be enough to cause the material to slide or flow downslope.

Oversteepening of slopes is another cause of many mass movements. Loose, undisturbed particles assume a stable slope called the **angle of repose**, the steepest angle at which material remains stable. Depending upon the size and shape of the particles, the angle varies from 25 to 40 degrees. The larger, more angular particles maintain the steepest slopes. If the angle is increased, the rock debris will adjust by moving downslope. There are many situations in nature where this takes place. A stream undercutting a valley wall and waves pounding against the base of a cliff are but two familiar examples. Furthermore, through their activities, people often create oversteepened and unstable slopes that become prime sites for mass wasting.

CLASSIFICATION OF MASS WASTING PROCESSES

There is a broad array of processes that geologists call mass wasting. Four are illustrated in Figure 9.3. Generally, the different types are divided and described on the basis of the type of material involved, the kind of motion displayed, and the velocity of the movement.

Type of Material

The classification of mass wasting processes on the basis of material involved in the movement depends upon whether the descending mass began as uncon-

FIGURE 9.4
Rockfall blocking a highway. (Photo courtesy of U.S. Geological Survey)

solidated material or as bedrock. If soil and regolith dominate, we typically see terms such as "debris," "mud," or "earth" used in the description. On the other hand, when a mass of bedrock breaks loose and moves downslope, the term "rock" may be part of the description.

Type of Motion

In addition to characterizing the type of material involved in a mass wasting event, the way in which the material moves may also be important. Generally, the kind of motion is described as either a fall, a slide, or a flow.

When the movement involves the free-fall of detached individual pieces of any size, it is termed a **fall.** Fall is a common form of movement on slopes that are so steep that loose material cannot remain on the surface. The rock may fall directly to the base of the slope or move in a series of leaps and bounds over other rocks along the way. Many falls result when freeze and thaw cycles or the action of plant roots loosen rock to the point that gravity takes over. Although signs along bedrock cuts on highways warn of falling rock, few of us have actually witnessed such an event. However, as Figure 9.4 illustrates, they do indeed occur. In fact, this is the primary way in which talus slopes are built and maintained (Figure 9.5). Sometimes falls may trigger other forms of downslope movement. For example, recall that the Yungay disaster described at the beginning of the chapter was initiated by a mass of freefalling material that broke from the nearly vertical summit of Nevados Huascaran.

Many mass wasting processes are described as **slides.** Slides occur whenever material remains fairly coherent and moves along a well-defined surface. Sometimes the surface is a joint, a fault, or a bedding plane that is approximately parallel to the slope. However, in the case of the movement called slump, the descending mass moves along a curved surface of rupture. A note of clarification is appropriate at this point. Sometimes the word "slide" is used as a synonym for the word "landslide." It should be pointed out that although many people, including geologists, use the term, the word "landslide" has no specific definition in geology. Rather it should be considered as a popular nontechnical term used to describe all perceptible forms of mass wasting, including those in which sliding does not occur.

The third type of movement common to mass wasting processes is termed **flow.** Flow occurs when material moves downslope as a viscous fluid. Most flows are saturated with water and typically move as lobes or tongues.

FIGURE 9.5
Talus is a slope built of angular rock fragments. Mechanical weathering, especially frost wedging, loosens the pieces of bedrock, which then fall to the base of the cliff. With time, a series of steep, cone-shaped accumulations builds up at the base of the mountain. (Photo by E. J. Tarbuck)

FIGURE 9.6
This 4-kilometer long tongue of rubble was deposited atop Alaska's Sherman Glacier by a rock avalanche. The event was triggered by a tremendous earthquake in March, 1964. (Photo by Austin Post, U.S. Geological Survey)

Rate of Movement

The event described at the beginning of this chapter clearly involved rapid movement. The rock and debris moved downslope at speeds well in excess of 200 kilometers (125 miles) per hour. This most rapid type of mass movement is termed a **rock avalanche.** Many researchers believe that rock avalanches, such as the one that produced the scene in Figure 9.6, must literally "float on air" as they move downslope. That is, high velocities result when air becomes trapped and compressed beneath the falling mass of debris, allowing it to move as a buoyant, flexible sheet across the surface.

Most mass movements, however, do not move with the speed of a rock avalanche. In fact, a great deal of mass wasting is imperceptibly slow. One process that we will examine later, termed creep, results in particle movements that are usually measured in millimeters or centimeters per year. Thus, as you can see, rates of movement can be spectacularly sudden or exceptionally gradual. Although various types of mass wasting are often classified as either rapid or slow, such a distinction is highly subjective because there is a wide range of rates between the two extremes. Even the velocity of a single process at a particular site can vary considerably from one time to another.

SLUMP

Slump refers to the downward sliding of a mass of rock or unconsolidated material moving as a unit along a curved surface (Figure 9.7). Usually the slumped material does not travel spectacularly fast nor very far. This is a common form of mass wasting, especially in thick accumulations of cohesive materials such as clay. The surface of rupture beneath the slump block is characteristically spoon-shaped and concave upward or outward. As the movement occurs, a crescent-shaped scarp (cliff) is created at the

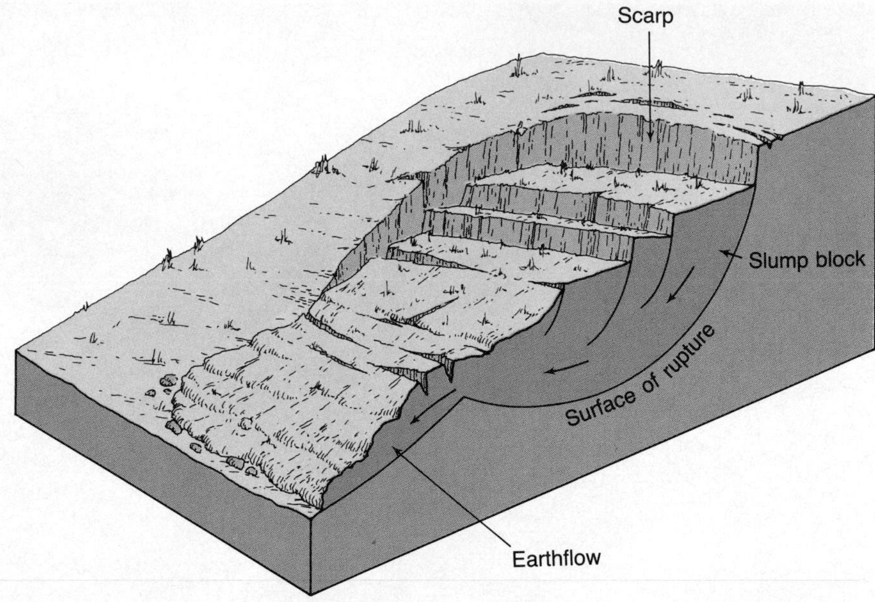

FIGURE 9.7
Slump occurs when material slips downslope en masse along a curved surface of rupture.

head and the block's upper surface is sometimes tilted backwards. Although slump may involve a single mass, it often consists of multiple blocks. Sometimes water is impounded between the base of the scarp and the top of the tilted block. As this water percolates downward along the surface of rupture, it may promote further instability and additional movement.

FIGURE 9.8
Slump at Point Fermin, California. Slump is often triggered when slopes become oversteepened by erosional processes such as wave action. (Photo by John S. Shelton)

FIGURE 9.9
On August 17, 1959, an earthquake triggered a massive rockslide in the canyon of Montana's Madison River. An estimated 27 million cubic meters of debris slid down the canyon wall, forming a dam that created Earthquake Lake. (Photo by John Montagne)

Slump commonly occurs because a slope has been oversteepened. The material on the upper portion of a slope is held in place by the material at the bottom of the slope. As this anchoring material at the base is removed, the material above is made unstable and reacts to the pull of gravity. One relatively common example is a valley wall that becomes oversteepened by a meandering river. Figure 9.8 provides another example in which a cliffed seashore has been undercut by wave action at its base. Slumping may also occur when a slope is overloaded, causing internal stress on the material below. This type of slump often occurs where weak, clay-rich material underlies layers of stronger, more resistant rock such as sandstone. The seepage of water through the upper layers reduces the strength of the clay below and slope failure results.

ROCKSLIDE

Rockslides occur when blocks of bedrock break loose and slide down a slope (see Figure 9.3B). If the material involved is largely unconsolidated, the term **debris slide** is used instead. Such events are among the fastest and most destructive mass movements. Usually rockslides take place in a geologic setting where the rock strata are inclined, or joints and fractures exist parallel to the slope. When such a rock unit is undercut at the base of the slope, it loses support and the rock eventually gives way. Sometimes the rockslide is triggered when rain or melting snow lubricates the underlying surface to the point that friction is no longer sufficient to hold the rock unit in place. As a result, rockslides tend to be most likely during the spring, when heavy rains and melting snow are most prevalent. Earthquakes are another mechanism which may trigger rockslides and other mass movements. The 1811 earthquake at New Madrid, Missouri, for example, caused slides in an area of more than 13,000 square kilometers (5000 square miles) along the Mississippi River valley. A more recent example occurred on August 17, 1959, when a severe earthquake west of Yellowstone National Park triggered a massive slide in the canyon of the Madison River in southwestern Montana (Figure 9.9). In a matter of moments an estimated 27 million cubic meters of rock, soil, and trees slid into the canyon.

The debris dammed the river and buried a campground and highway. More than 20 unsuspecting campers perished.

The Gros Ventre River flows west from the Wind River Range in northwestern Wyoming, through the Grand Teton National Park, and eventually empties into the Snake River. On June 23, 1925, a classic rockslide took place in its valley, just east of the small town of Kelly. In the span of just a few minutes a great mass of sandstone, shale, and soil crashed down the south side of the valley, carrying with it a dense pine forest. The volume of debris, estimated at 38 million cubic meters (50 million cubic yards), created a 70-meter high dam on the Gros Ventre River (Figure 9.10). Because the river was completely blocked a lake was created. It filled so quickly that a house that had been 18 meters (60 feet) above the river was floated off its foundation 18 hours after the slide. In 1927 the lake overflowed the dam, partially draining the lake and resulting in a devastating flood downstream.

Why did the Gros Ventre rockslide take place? Figure 9.11 is a diagrammatic cross-sectional view of the geology of the valley. Notice the following points: (1) the sedimentary strata in this area dip (tilt) 15–21 degrees; (2) underlying the bed of sandstone is a relatively thin layer of clay; and (3) at the bottom of the valley the river had cut through much of the sandstone layer. During the spring of 1925 water from heavy rains and melting snow seeped through the sandstone, saturating the clay below. Since much of the sandstone layer had been cut through by the Gros Ventre River, the layer had virtually no support at the bottom of the slope. Eventually the sandstone could no longer hold its position on the wetted clay, and gravity pulled the mass down the side of the

FIGURE 9.10
Although the Gros Ventre rockslide occurred in 1925, the scar left on the side of Sheep Mountain is still a prominent feature. (Photo by Stephen Trimble)

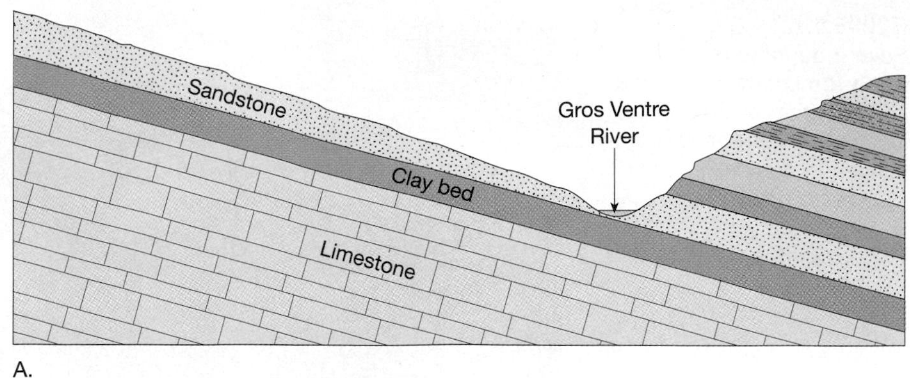

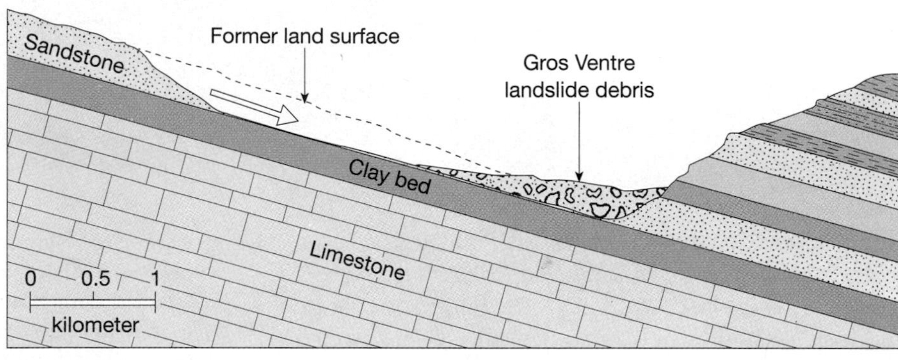

FIGURE 9.11
Cross-sectional view of the Gros Ventre rockslide. The slide occurred when the tilted and undercut sandstone bed could no longer maintain its position atop the saturated bed of clay. [After W. C. Alden, "Landslide and Flood at Gros Ventre, Wyoming." *Transactions* (AIME) 76 (1928): 348]

valley. The circumstances at this location were such that the event was inevitable.

MUDFLOW

Mudflow is a relatively rapid type of mass wasting that involves a flowage of debris containing a large amount of water (Figure 9.3C). Mudflows are most characteristic of semiarid mountainous regions and, because of their high water content and the predominance of fine particles, tend to follow canyons and gullies. Although rains in semiarid regions are infrequent, they are typically heavy when they occur. When a cloudburst or rapidly melting mountain snows create a sudden flood, large quantities of soil and regolith are washed into nearby stream channels because there is usually little or no vegetation to anchor the surface material. The end product is a flowing tongue of well-mixed mud, soil, rock, and water. Its consistency may range from that of wet concrete to a soupy mixture not much thicker than muddy water. The rate of flow therefore depends not only on the slope but on the water content as well. When dense, mudflows are capable of carrying or pushing large boulders, trees, and even houses with relative ease.

Mudflows pose a serious hazard to development in dry mountainous areas such as southern California. Here construction of homes on canyon hillsides and the removal of native vegetation by brush fires and other means have increased the frequency of these destructive events (Figure 9.12). Moreover, when a mudflow reaches the end of a steep, narrow canyon, it spreads out, covering the area beyond the mouth of the canyon with a mixture of wet debris. This material contributes to the buildup of fanlike deposits at canyon mouths.* The fans are relatively easy to build on, often have nice views, and are close to the mountains; thus, like the nearby canyons, many have become preferred sites for development. Since mudflows occur only sporadically, the public is often unaware of the potential hazard of such sites. This ignorance has led to many serious problems.

Mudflows are also common on the slopes of some volcanoes, in which case they are termed **lahars**. The word originated in Indonesia, a volcanic region that has experienced many of these often destructive events. Lahars result when highly unstable layers of ash and debris become saturated with water and flow down steep volcanic slopes, generally following ex-

*These structures are called *alluvial fans* and will be discussed in greater detail in Chapters 10 and 13.

FIGURE 9.12
Severe damage resulted when a mudflow buried the lower portion of this house located near the mouth of a canyon in southern California. (Photo by James E. Patterson)

isting stream channels. Some are initiated when heavy rainfalls erode volcanic deposits. Others are triggered when large volumes of ice and snow are suddenly melted by heat flowing to the surface from within the volcano or by the hot gases and near-molten debris emitted during a violent eruption.

When Mount St. Helens erupted in May, 1980, several lahars were created. The flows and accompanying floods raced down the valleys of the north and south forks of the Toutle River at speeds that were often in excess of 30 kilometers per hour. Fortunately the affected area was not densely settled. Nevertheless, more than 200 homes were destroyed or severely damaged (Figure 9.13). Most bridges met a similar fate. According to the U.S. Geological Survey:

> Even after traveling many tens of miles from the volcano and mixing with cold waters, the mudflows maintained temperatures in the range of about 84° to 91°F; they undoubtedly had higher temperatures closer to the eruption source. . . . Locally the mudflows surged up the valley walls as much as 360 feet and over hills as high as 250 feet. From the evidence left by the "bathtub-ring" mudlines, the larger mudflows at their peak averaged from 33 to 66 feet deep.*

Eventually the lahars in the Toutle River drainage area carried more than 50 million cubic meters of

*Robert I. Tilling, *Eruptions of Mount St. Helens: Past, Present, and Future,* Washington, D.C., U.S. Government Printing Office, 1987.

FIGURE 9.13
A house damaged by a lahar along the Toutle River, west-northwest of Mount St. Helens. The end section of the house was torn free and lodged against trees. (Photo by D. R. Crandell, U.S. Geological Survey)

material to the lower Cowlitz and Columbia rivers. The deposits temporarily reduced the water-carrying capacity of the Cowlitz River by 85 percent and the depth of the Columbia River navigational channel was decreased from 12 meters to less than 4 meters.

In November, 1985, lahars were produced when Nevado del Ruiz, a 5300-meter (17,400-foot) volcano in the Andes Mountains of Colombia erupted. The eruption melted much of the snow and ice that capped the uppermost 600 meters of the peak, producing torrents of hot viscous mud, ash, and debris. The lahars moved outward from the volcano, following the valleys of three rain-swollen rivers that radiate from the peak. The flow that moved down the valley of the Lagunilla River was the most destructive, devastating the town of Armero, 48 kilometers from the mountain. Most of the more than 25,000 deaths caused by the event occurred in this once-thriving agricultural community. Death and property damage also occurred in 13 other villages within the 180-square-kilometer disaster area. Although a great deal of pyroclastic material was explosively ejected from Nevado del Ruiz, it was the lahars triggered by this eruption that made this such a devastating natural disaster. In fact, it was the worst volcanic disaster since 28,000 people died following the 1902 eruption of Mount Pelée on the Caribbean island of Martinique.*

*A discussion of the Mount Pelée eruption can be found in the section on composite cones in Chapter 4.

EARTHFLOW

Unlike mudflows, which are usually confined to channels in semiarid regions, **earthflows** most often form on hillsides in humid areas during times of heavy precipitation or snowmelt (see Figure 9.3D). When water saturates the soil and regolith on a hillside, the material may break away, leaving a scar on the slope and forming a tongue- or teardrop-shaped mass that flows downslope (Figure 9.14). The materials most commonly involved are rich in clay and silt and contain only small proportions of sand and coarser particles. Earthflows range in size from bodies a few meters long, a few meters wide, and less than one meter deep to masses more than 1 kilometer long, several hundred meters wide, and more than 10 meters deep. Because earthflows are quite viscous, they generally move at slower rates than the more fluid mudflows described in the preceding section. They are characterized by a slow and persistent movement and may remain active for periods ranging from days to years. Depending upon the steepness of the slope and the material's consistency, measured velocities range from less than 1 millimeter per day up to several meters per day. Over the time span that earthflows are active, movement is typically faster during wet periods than during drier times. In addition to occurring as isolated hillside phenomena, earthflows commonly take place in association with large slumps. In this situation, they may be seen as tonguelike flows at the base of the slump block (see Figure 9.7).

FIGURE 9.14
This small, tongue-shaped earthflow occurred on a newly formed slope along a recently constructed highway. It formed in clay-rich material following a period of heavy rain. Notice the small slump at the head of the earthflow. (Photo by E. J. Tarbuck)

BOX 9.2

Permafrost

Many of the mass wasting disasters described in Chapter 9 had sudden and disastrous impacts on people. When the activities of people cause ice contained in permanently frozen ground to melt, the impact is more gradual and less deadly. Nevertheless, because permafrost regions are sensitive and fragile landscapes, the scars resulting from poorly planned actions can remain for generations.

Permanently frozen ground, known as *permafrost,* occurs in regions where summertime temperatures do not get sufficiently high for long enough periods to melt more than a shallow surface layer. Deeper ground remains frozen throughout the year. Strictly speaking, permafrost is defined only on the basis of temperature; that is, it is ground with temperatures that have remained below 0°C (32°F) continuously for two years or more. The degree to which ice is present in the ground strongly affects the behavior of the surface material. Knowing how much ice is present and where it is located is very important when it comes to constructing roads, buildings, and other projects in areas underlain by permafrost.

Permafrost underlies about 20 percent of the earth's land area. In addition to occurring in Antarctica and in some high mountain areas, permafrost is extensive in the lands surrounding the Arctic Ocean. It covers more than 80 percent of Alaska and about 50 percent of Canada as well as a substantial portion of northern Siberia (Figure 9.B). Near the southern margins of the region, the permafrost consists of relatively thin, isolated masses. Farther north the area and thickness gradually increase to the point where the permafrost is essentially continuous and its thickness may approach or even exceed 500 meters. In the discontinuous zone, land-use planning is frequently more difficult than in the continuous zone farther north because the occurrences of permafrost are patchy and difficult to predict.

When changes occur in the surface environment, such as the clearing of the insulating vegetation mat or the building of roads and other structures, the delicate thermal balance is disturbed, and thawing of the permafrost can result. This thawing, in turn, produces unstable ground that may be susceptible to slides, slumps,

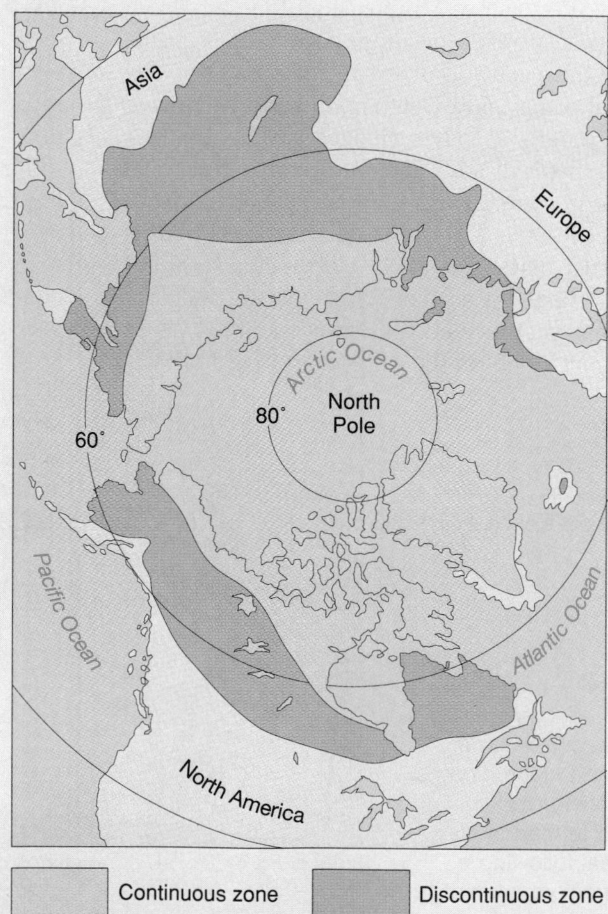

FIGURE 9.B
Distribution of permafrost in the Northern Hemisphere. More than 80 percent of Alaska and about 50 percent of Canada are underlain by permafrost. Two zones are recognized. In the continuous zone, the only ice-free areas are beneath deep lakes or rivers. In the higher-latitude portions of the discontinuous zone, there are only scattered islands of thawed ground. Moving southward, the percentage of unfrozen ground increases until all the ground is unfrozen. (Courtesy of U.S. Geological Survey)

subsidence, and severe frost heaving. The many environmental problems stemming from human activities in the Arctic during and following World War II demonstrated the need for a thorough understanding of the nature of permafrost.

As Figure 9.C illustrates, when a heated structure is built directly on permafrost that contains a high proportion of ice, thawing can cause the building to sink into the soggy material beneath the foundation. One solution is to place buildings and other structures on piles. Such piles allow subfreezing air to circulate between the floor of the building and the soil and therefore cause minimal disturbance to the frozen ground.

When oil was discovered on Alaska's North Slope, many people were concerned about the building of a pipeline linking the oil fields at Prudhoe Bay to the ice-free port of Valdez 1300 kilometers to the south. There was serious concern about the impact of such a massive project on the sensitive permafrost environment. Many were also worried about the effects of possible oil spills after the pipeline was completed.

Because the oil that was to move through the pipeline had to be hot (about 60°C) in order to flow properly, special engineering procedures had to be developed to minimize thawing. The procedures that were used included insulating the pipe to reduce heat flow, elevating portions of the pipeline above ground level in areas of ice-rich permafrost, and, in some instances, placing cooling devices in the ground. The Alaska pipeline is clearly one of the most complex and costly projects ever built in the Arctic tundra. Detailed studies and careful engineering helped minimize adverse effects resulting from the disturbance of frozen ground.

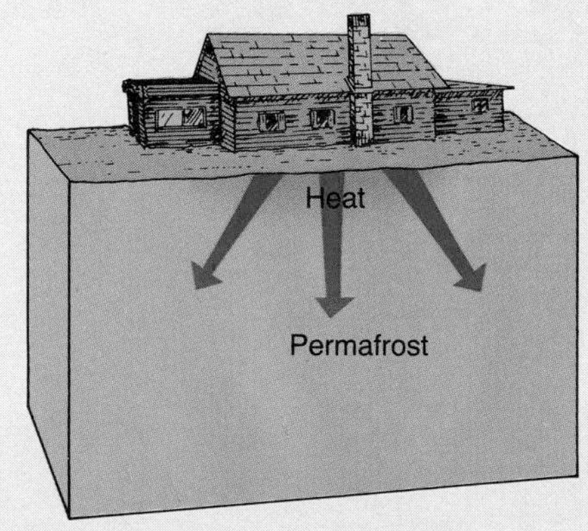

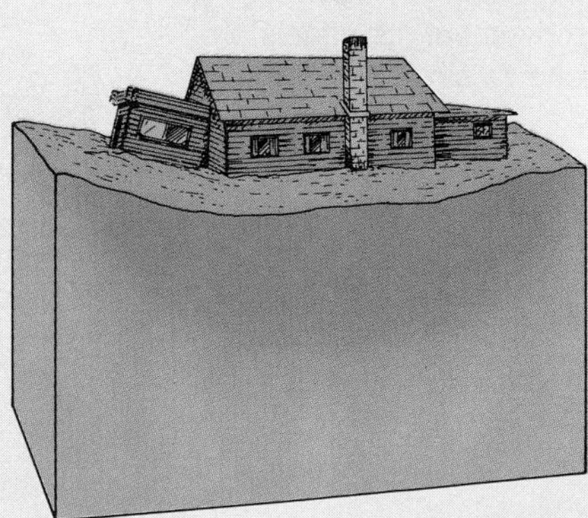

FIGURE 9.C
This building, located south of Fairbanks, Alaska, subsided because of thawing permafrost. Notice that the right side, which was heated, settled much more than the unheated porch on the left. (Photo courtesy of O. J. Ferrians, Jr., U.S. Geological Survey)

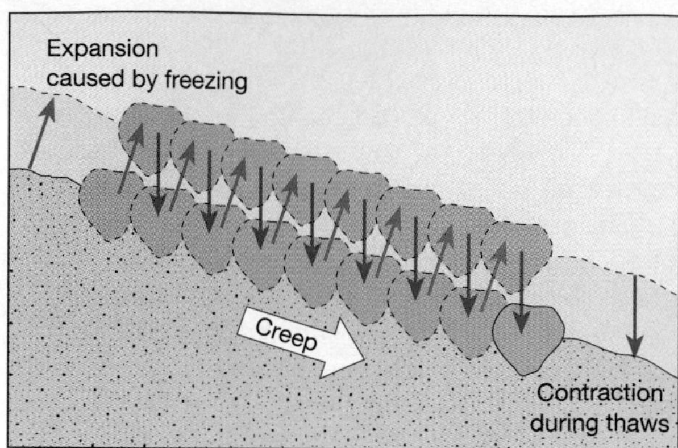

FIGURE 9.15
The repeated expansion and contraction of the surface material causes a net downslope migration of rock and soil particles—a process called creep.

CREEP

Movements such as rockslides, rock avalanches, and lahars are certainly the most spectacular and catastrophic forms of mass wasting. Since these events have been known to kill thousands, they deserve intensive study so that, through more effective prediction, timely warnings and better controls can help save lives. However, because of their large size and spectacular nature they give us a false impression of their importance as a mass wasting process. Indeed, sudden movements are responsible for moving less material than the slower and far more subtle action of creep. Whereas rapid types if mass wasting are characteristic of mountains and steep hillsides, creep can take place on gentle slopes and is thus much more widespread.

Creep is a type of mass wasting that involves the gradual downhill movement of soil and regolith. One of the primary causes of creep is the alternate expansion and contraction of surface material caused by freezing and thawing or wetting and drying. As shown in Figure 9.15, freezing or wetting lifts particles at right angles to the slope, and thawing or drying allows the particles to fall back to a slightly lower level. Each cycle therefore moves the material a short distance downhill. Creep may also be initiated if the ground becomes saturated with water. Following a heavy rain or snowmelt, a waterlogged soil may lose its internal cohesion, allowing gravity to pull the material downslope. Because creep is imperceptibly slow, the process cannot be observed in action. What can be observed, however, are the effects of creep. Creep causes fences and utility poles to tilt and retaining walls to be displaced (Figure 9.16).

Solifluction is a form of mass wasting that is common in regions underlain by permafrost. **Permafrost** refers to the permanently frozen ground that occurs in association with the earth's harsh tundra and ice cap climates (see Box 9.2). Solifluction may be regarded as a form of creep in which unconsolidated, water-saturated material gradually moves downslope. Solifluction occurs in a zone above the permafrost called the *active layer,* which thaws in summer and refreezes in winter. During the summer season, water is unable to percolate into the impervious permafrost layer below. As a result, the active layer becomes saturated and slowly flows. The process can occur on slopes as gentle as 2–3 degrees. Where there is a well-developed mat of vegetation, a solifluction sheet may move downward in a series of well-defined lobes or as a series of partially overriding folds (Figure 9.17).

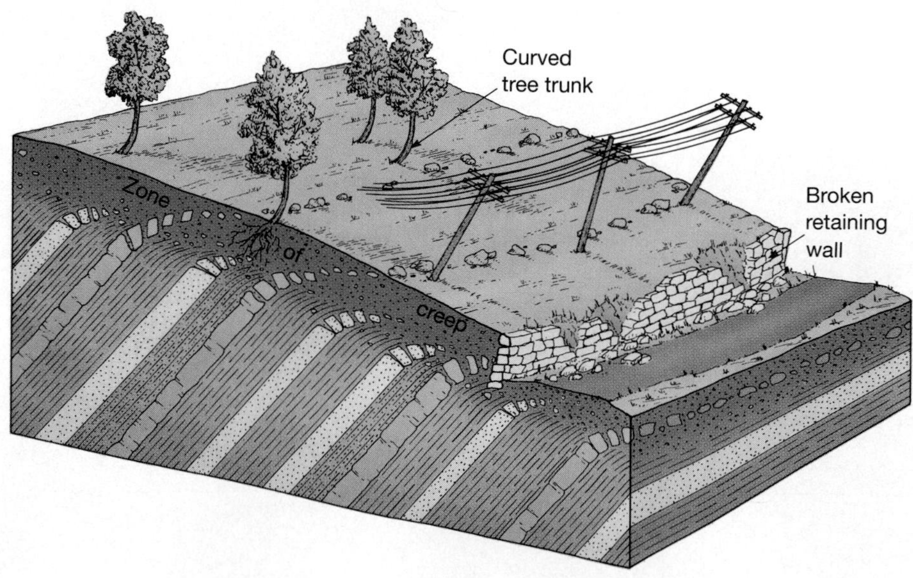

FIGURE 9.16
Although creep is an imperceptibly slow movement, its effects are often visible.

FIGURE 9.17
Solifluction lobes northeast of Fairbanks, Alaska. Solifluction occurs when the active layer thaws in summer. (Photo by James E. Patterson)

REVIEW QUESTIONS

1. Describe how mass wasting processes contribute to the development of stream valleys.
2. How did the building of a dam contribute to the Vaiont Canyon disaster? Was the disaster avoidable? (See Box 9.1.)
3. What is the controlling force of mass wasting?
4. How does water affect mass wasting processes?
5. Describe the significance of the angle of repose.
6. Distinguish among fall, slide, and flow.
7. Why can rock avalanches move at such great speeds?
8. Both slump and rockslide move by sliding. In what ways do these processes differ?
9. What factors led to the massive rockslide at Gros Ventre, Wyoming?
10. Compare and contrast mudflow and earthflow.
11. Describe the mass wasting that occurred at Mount St. Helens during its active period in 1980 and at Nevado del Ruiz in 1985.
12. Since creep is an imperceptibly slow process, what evidence might indicate that this phenomenon is affecting a slope? Describe the mechanism that creates this slow movement.
13. Why is solifluction only a summertime process?
14. What is permafrost? What portion of the earth's land surface is affected? (See Box 9.2).

KEY TERMS

angle of repose (p. 209)
creep (p. 220)
debris slide (p. 213)
earthflow (p. 217)
fall (p. 210)
flow (p. 210)
lahar (p. 215)
mass wasting (p. 208)
mudflow (p. 215)
permafrost (p. 220)
rock avalanche (p. 211)
rockslide (p. 213)
slide (p. 210)
slump (p. 211)
solifluction (p. 220)

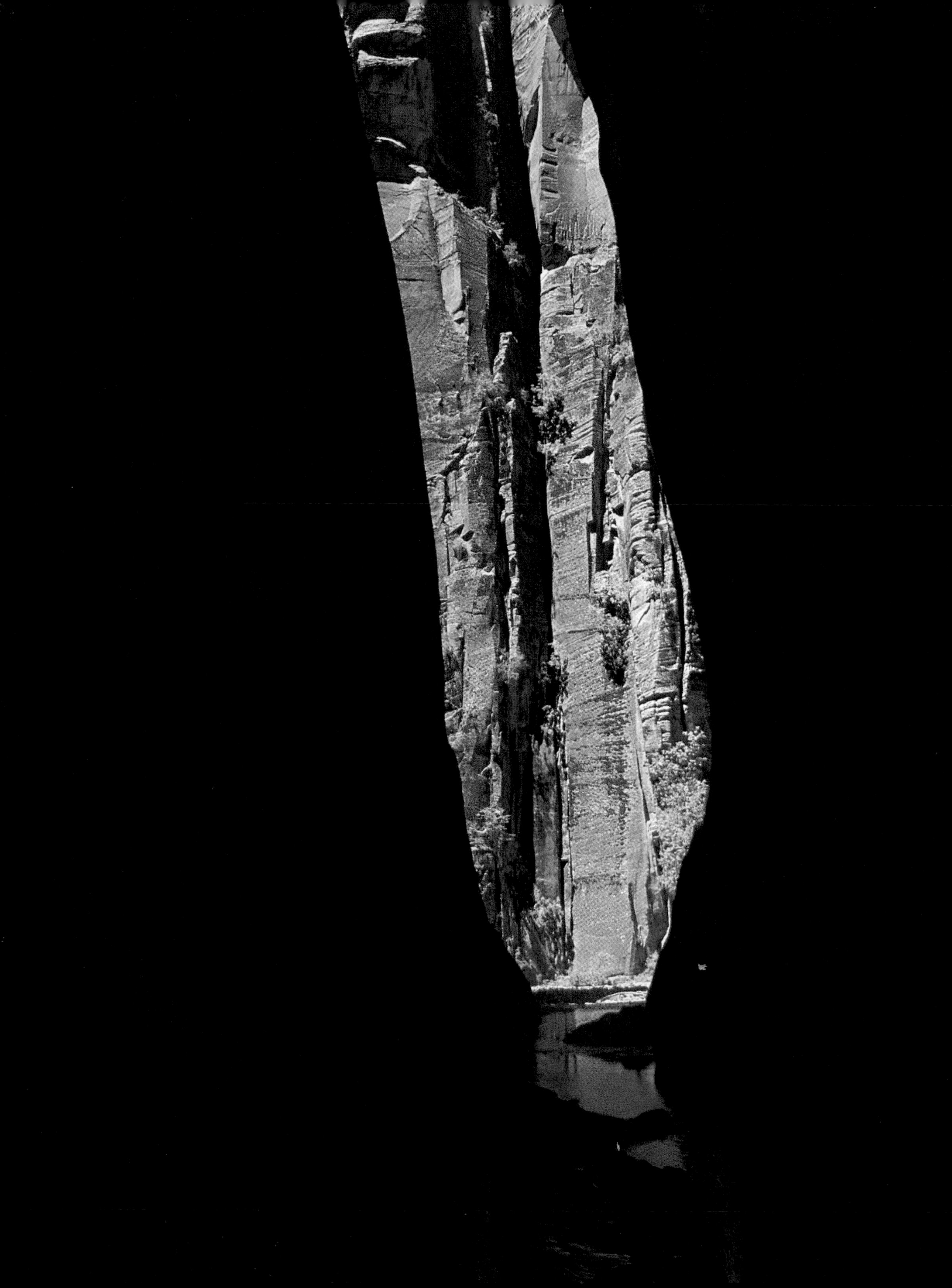

10

Running Water

Opposite: Virgin River Narrows at Zion National Park, Utah. (Photo by Stephen Trimble) (Top photo by P. Jay Fleisher)

THE HYDROLOGIC CYCLE

The amount of water on earth is immense: an estimated 1.36 billion cubic kilometers (326 million cubic miles). Of this total, the vast bulk—97.2 percent—is part of the world ocean. Icecaps and glaciers account for another 2.15 percent, leaving only 0.65 percent to be divided among lakes, streams, subsurface water, and the atmosphere (Figure 10.1). Although the percentage of the earth's total water found in each of the latter sources is but a small fraction of the total inventory, the absolute quantities are great.

An adequate supply of water is vital to life on earth. With increasing demands on this finite resource, science has given a great deal of attention to the exchanges of water among the oceans, the atmosphere, and the continents. This unending circulation of the earth's water supply has come to be called the **hydrologic cycle**, a gigantic system powered by energy from the sun in which the atmosphere provides the vital link between the oceans and continents. Water from the oceans, and to a much lesser extent from the continents, is constantly evaporating into the atmosphere. Wind transports the moisture-laden air, often great distances, until the complex processes of cloud formation are set in motion. Precipitation that falls into the ocean has ended its cycle and is ready to begin another. The water that falls on the continents, however, must still make its way back to the ocean.

What happens to precipitation once it has fallen on the land? A portion of the water soaks into the ground, some of it moving downward, then laterally,

FIGURE 10.1
The Delores River in Colorado represents one part of the hydrologic cycle. The amount of water in streams at any one time is just a tiny fraction of the earth's total water supply. Yet the absolute quantities of water that flow through this part of the hydrologic cycle annually are great. (Photo by Robert Winslow)

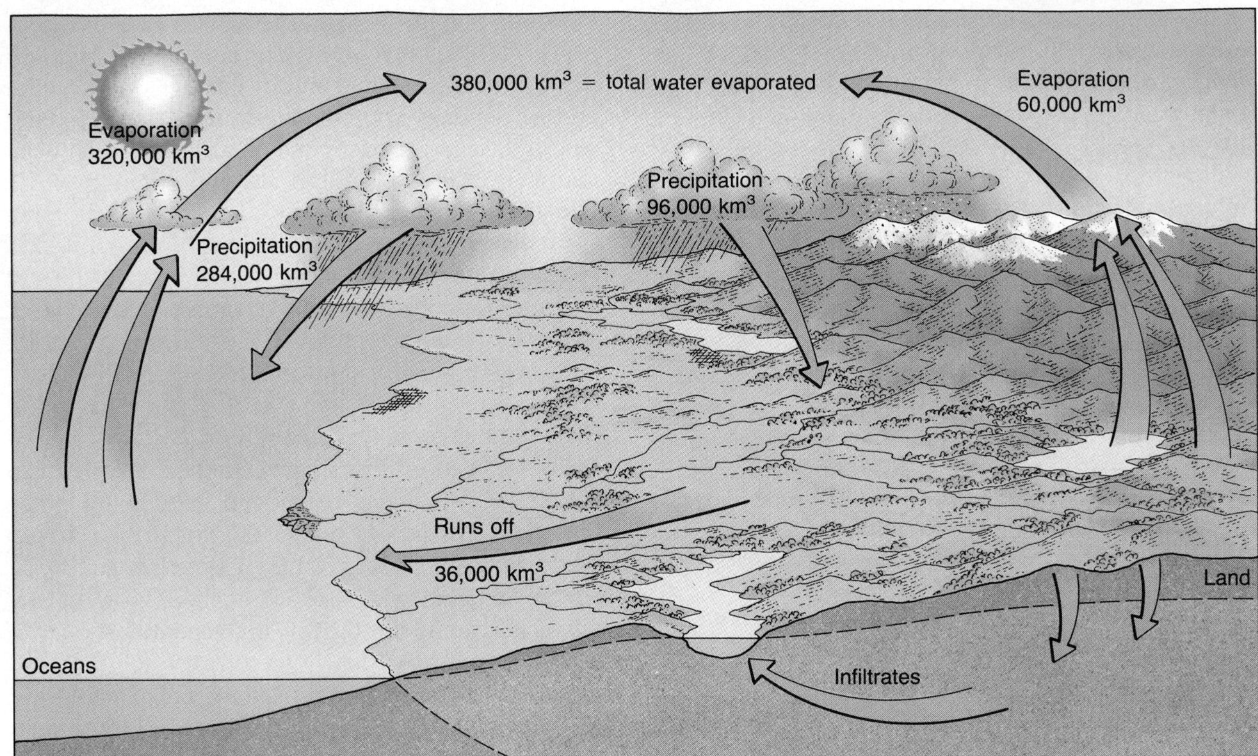

FIGURE 10.2
The earth's water balance. About 320,000 cubic kilometers of water are evaporated each year from the oceans, while evaporation from the land (including lakes and streams) contributes 60,000 cubic kilometers of water. Of this total of 380,000 cubic kilometers of water, about 284,000 cubic kilometers fall back to the ocean, and the remaining 96,000 cubic kilometers fall on the earth's land surface. Since 60,000 cubic kilometers of water evaporate from the land, 36,000 cubic kilometers of water remain to erode the land during the journey back to the oceans.

finally seeping into lakes, streams, or directly into the ocean. When the rate of rainfall is greater than the earth's ability to absorb it, the additional water flows over the surface into lakes and streams. Much of the water which soaks in (**infiltration**) or runs off (**runoff**) eventually finds its way back to the atmosphere because of evaporation from the soil, lakes, and streams. Also, some of the water that infiltrates the ground surface is absorbed by plants, which later release it into the atmosphere. This process is called **transpiration**. Each year a field of crops may transpire an amount of water equivalent to a layer 60 centimeters deep over the entire field, while a forest may pump twice this amount into the atmosphere. Because we cannot clearly distinguish between the amount of water that is evaporated and the amount that is transpired by plants, the term **evapotranspiration** is often used for the combined effect.

When precipitation falls at high elevations or high latitudes, the water may not immediately soak in, run off, or evaporate. Instead it becomes part of a snowfield or glacier. Glaciers store large quantities of water on land. If present-day glaciers were to melt and release their storage of water, sea level would rise by several tens of meters and submerge many heavily populated coastal areas. As we shall see in Chapter 12, over the past two million years, huge continental glaciers have formed and melted on several occasions, each time upsetting the hydrologic cycle.

A diagram of the earth's water balance, a quantitative view of the hydrologic cycle, is shown in Figure 10.2. While the amount of water vapor in the air at any one time is but a minute fraction of the earth's total water supply, the absolute quantities that are cycled through the atmosphere over a one-year period are immense—some 380,000 cubic kilometers.

Since the total amount of water vapor in the atmosphere remains about the same, average annual precipitation over the earth must be equal to the quantity of water evaporated. However, for all of the continents taken together, precipitation exceeds evaporation. Conversely, over the oceans, evaporation exceeds precipitation. Since the level of the world ocean is not dropping, runoff from land areas must balance the deficit of precipitation over the oceans.

The work accomplished by the 36,000 cubic kilometers of water that flows annually over the land to the sea is enormous. Arthur Bloom effectively described it as follows:

> The average continental height is 823 meters above sea level. . . . If we assume that the 36,000 cubic kilometers of annual runoff flow downhill an average of 823 meters, the potential mechanical power of the system can be calculated. Potentially, the runoff from all lands would continuously generate almost 9×10^9 kW. If all this power were used to erode the land, it would be comparable to having . . . one horse-drawn scraper or scoop at work on each 3-acre piece of land, day and night, year round. Of course, a large part of the potential energy of runoff is wasted as frictional heat by turbulent flow and splashing of water.*

Although only a small percentage of the energy of running water is used to erode the earth's surface, running water nevertheless represents the single most important agent sculpturing our planet's landscapes.

To summarize, the hydrologic cycle represents the continuous movement of water from the oceans to the atmosphere, from the atmosphere to the land, and from the land back to the sea. The wearing down of the earth's land surface is largely attributable to the last of these steps and is the primary focus of the remaining pages of this chapter.

RUNNING WATER

Of all the geologic processes, running water may have the greatest impact on people. We depend upon rivers for energy, travel, and irrigation; their fertile floodplains have fostered human progress since the dawn of civilization. As the dominant agent of landscape alteration, streams have shaped much of our physical environment.

Although people have always depended to a great extent on running water, its source eluded them for centuries. Not until the sixteenth century did they realize that streams were supplied by runoff and underground water, which ultimately had their sources as rain and snow.

**Geomorphology: A Systematic Analysis of Late Cenozoic Landforms* (Englewood Cliffs, N.J.: Prentice-Hall, 1978) p. 97.

Runoff initially flows in broad, thin sheets, appropriately termed **sheet flow**. The amount of water that runs off in this manner rather than sinking into the ground depends upon the **infiltration capacity** of the soil. Infiltration capacity is controlled by many factors including: (1) the intensity and duration of the rainfall, (2) the prior wetted condition of the soil, (3) the soil texture, (4) the slope of the land, and (5) the nature of the vegetative cover. When the soil becomes saturated, sheet flow commences as a layer only a few millimeters thick. After flowing as a thin, unconfined sheet for only a short distance, threads of current typically develop and tiny channels called **rills** begin to form and carry the water to a stream.

To some, the term *stream* implies relative size. That is to say, streams are thought of as being larger than creeks or brooks but smaller than rivers. In geology, however, this is not the case. Here the word **stream** is used to denote channelized flow of any size, from the smallest trickle to the mightiest river. It should be pointed out, however, that although the terms *river* and *stream* are used interchangeably, the term *river* is often preferred when describing a main stream into which several tributaries flow.

The remainder of this chapter will concentrate on that part of the hydrologic cycle in which the water moves in stream channels. Further, the discussion will deal primarily with the characteristics of streams in humid regions. Streams are also important in arid landscapes, as we shall see in Chapter 13.

Streamflow

Water may flow in one of two ways, either as **laminar flow** or **turbulent flow**. When the movement is laminar, the water particles flow in straight-line paths that are parallel to the channel. The water particles move steadily downstream without mixing. By contrast, when the flow is turbulent, the water moves in a confused and erratic fashion that is often characterized by swirling, whirlpool-like eddies (Figure 10.3).

The stream's velocity is a primary factor that determines whether the flow is laminar or turbulent. Laminar flow is possible only when water is moving very slowly through a smooth channel. If the velocity increases or the channel becomes rough, laminar flow changes to turbulent flow. The movement of water in streams is usually fast enough that most flow is turbulent. The multidirectional movement of turbulent flow is very effective both in eroding a stream's channel and in keeping sediment suspended within the water so that it can be transported downstream.

Flowing water makes its way to the sea under the influence of gravity. The time required for the journey depends upon the velocity of the stream, which

FIGURE 10.3
Most streamflow is turbulent, although it is usually not as rough as that experienced by these rafters on the Colorado River. (Photo by Michael Collier)

is measured in terms of the distance the water travels in a given unit of time. Some sluggish streams travel at less than 1 kilometer per hour, whereas a few rapid ones may reach speeds in excess of 30 kilometers per hour. Velocities are determined at gauging stations where measurements are taken at several locations across the channel and then averaged. This is done because the rate of water movement is not uniform within a stream channel. When the channel is straight, the highest velocities occur in the center just below the surface. It is here that friction is least. Minimum velocities occur along the sides and bottom (bed) of the channel where friction is always greatest. When a stream channel is crooked or curved, the fastest flow is not in the center. Rather, the zone of maximum velocity shifts toward the outside of each bend. As we shall see later, this shift plays an important part in eroding the bank on that side.

The ability of a stream to erode and transport material is directly related to its velocity. Even slight variations in velocity can lead to significant changes in the load of sediment transported by the water. Several factors determine the velocity of a stream and therefore control the amount of erosional work a stream may accomplish. These factors include: (1) the gradient, (2) the shape, size, and roughness of the channel, and (3) the discharge.

Certainly one of the most obvious factors controlling stream velocity is the **gradient**, or slope, of a stream channel. Gradient is typically expressed as the vertical drop of a stream over a fixed distance. Gradients may vary considerably from one stream to another as well as along the course of a given stream.

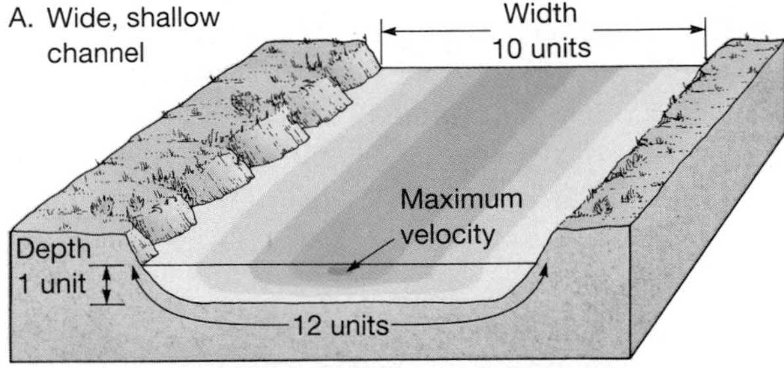

Cross-sectional area = 10 square units
Perimeter = 12 units

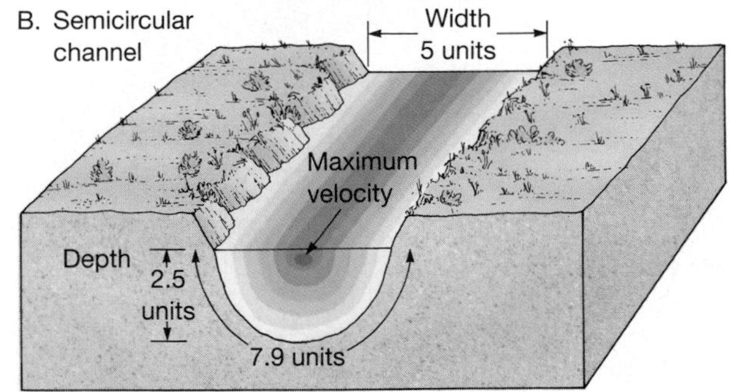

Cross-sectional area = 10 square units
Perimeter = 7.9 units

FIGURE 10.4
Influence of channel shape on velocity. Although the cross-sectional area of these channels is the same, the semicircular channel has less water in contact with the channel, and hence less frictional drag. As a result, the water will flow more rapidly in this channel, all other factors being equal.

Portions of the lower Mississippi River, for example, have gradients of 10 centimeters per kilometer and less. By way of contrast, some mountain stream channels decrease in elevation at a rate of more than 40 meters per kilometer, 400 times more abruptly than the lower Mississippi. The higher the gradient, the more energy available for streamflow. If two streams were identical in every respect except gradient, the stream with the higher gradient would obviously have the greater velocity.

The cross-sectional shape of a channel determines the amount of water in contact with the channel and hence affects the frictional drag. The most efficient channel is one with the least perimeter for its cross-sectional area. Figure 10.4 compares two types of channels. Although the cross-sectional area of both is identical, the semicircular shape has less water in contact with the channel and therefore less frictional drag. As a result, if all other factors are equal, the water will flow more rapidly in the semicircular channel.

The size and roughness of the channel also affect the amount of friction. An increase in the size of a channel reduces the ratio of perimeter to cross-sectional area and therefore increases the efficiency of flow. The effect of roughness is obvious. A smooth channel promotes a more uniform flow, whereas an irregular channel filled with boulders creates enough turbulence to significantly retard the stream's forward motion.

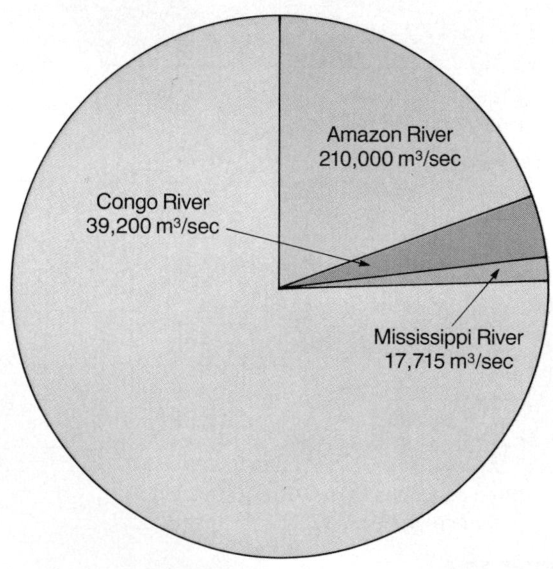

FIGURE 10.5
The Amazon's average flow is 12 times greater than the flow of the Mississippi, North America's largest river, and represents 15 percent of all the fresh water discharged into the oceans by all the world's rivers. (After U.S. Geological Survey)

The **discharge** of a stream is the amount of water flowing past a certain point in a given unit of time. This is usually measured in cubic meters per second or cubic feet per second. Discharge is determined by multiplying a stream's cross-sectional area by its velocity:

discharge (m^3/second) =
channel width (meters) × channel depth (meters) × velocity (meters/second)

The largest river in North America, the Mississippi, discharges an average of 17,715 cubic meters per second. Although this is a huge quantity of water, it is nevertheless dwarfed by the mighty Amazon, the world's largest river. Draining an area that is nearly three-quarters the size of the conterminous United States and that averages about 200 centimeters of rain per year, the Amazon discharges 12 times more water than the Mississippi (Figure 10.5). In fact, the

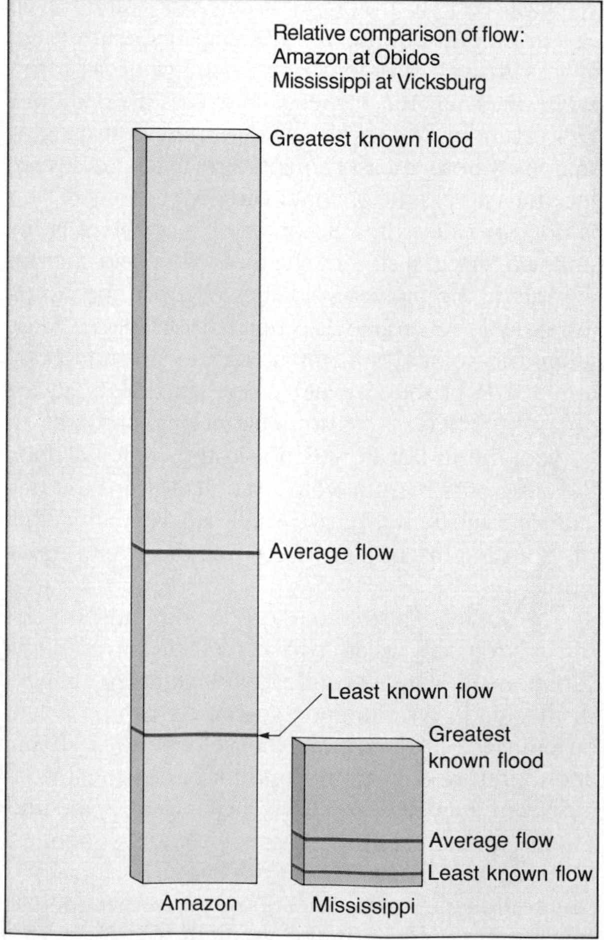

FIGURE 10.6
As the data for the Amazon and Mississippi rivers illustrate, the discharge of a river can be highly variable. (After U.S. Geological Survey)

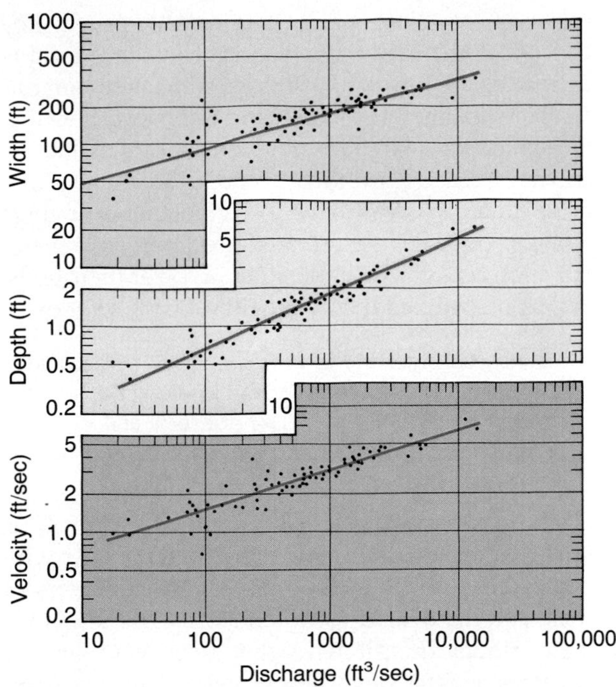

FIGURE 10.7
Relationship of width, depth, and velocity to discharge of the Powder River at Locate, Montana. As discharge increases, width, depth, and velocity all increase in an orderly fashion. (From L. B. Leopold and Thomas Maddock, Jr., U.S. Geological Survey Professional Paper 252, 1953)

flow of the Amazon accounts for about 15 percent of all the fresh water discharged into the ocean by all of the world's rivers. Just one day's discharge would supply the water needs of New York City for 9 years!

As Figure 10.6 illustrates, the discharges of the Amazon and Mississippi rivers are far from constant. This is true for most rivers because of such variables as rainfall and snowmelt. If discharge changes, then the factors noted earlier must also change. When discharge increases, the width or depth of the channel must increase or the water must flow faster, or some combination of these factors must change. Indeed, measurements show that when the amount of water in a stream increases, the width, depth, and velocity all increase in an orderly fashion (Figure 10.7). In order to handle the additional water, the stream will increase the size of its channel by widening and deepening it. As we saw earlier, when the size of the channel increases, proportionally less of the water is in contact with the bed and banks of the channel. This means that friction, which acts to retard the flow, is relatively decreased. The less friction, the more swiftly the water will flow.

Changes Downstream

One useful way of studying a stream is to examine its **longitudinal profile**. Such a profile is simply a cross-sectional view of a stream from its source area (called the **head** or **headwaters**) to its **mouth**, the point downstream where the river empties into another water body. By examining Figure 10.8, you can see that the most obvious feature of a typical longitudinal profile is a constantly decreasing gradient from the head to the mouth. Although many local irregularities may exist, the overall profile is a smooth concave-upward curve.

The longitudinal profile shows that the gradient decreases downstream. To see how other factors change in a downstream direction, observations and measurements must be made. When data are collected from successive gauging stations along a river,

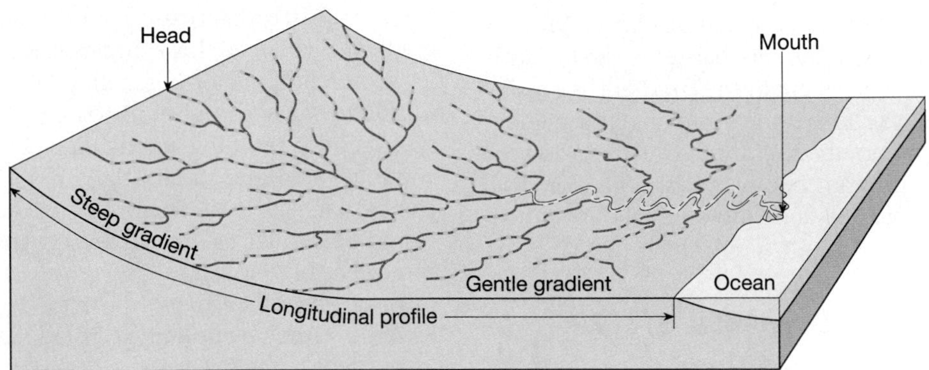

FIGURE 10.8
A longitudinal profile is a cross section along the length of a stream. Note the concave-upward curve of the profile, with a steeper gradient upstream and a gentler gradient downstream.

they show that discharge increases toward the mouth. This should come as no surprise since, as we move downstream, more and more tributaries contribute water to the main channel. In the case of the Amazon, for example, about 1000 tributaries join the main river along its 6500-kilometer course across South America. Furthermore, in most humid regions, additional water is continually being added from the groundwater supply. Since this is the case, the width, depth, and velocity must change in response to the increased volume of water carried by the stream. Indeed, the downstream changes in these variables have been shown to vary in a manner similar to what occurs when discharge increases at one place; that is, width, depth, and velocity all increase systematically.

The observed increase in velocity that occurs downstream contradicts our impressions about wild, rushing mountain streams and wide, placid rivers. The mental picture that we may have of "old man river just rollin along" is just not so. Although a mountain stream may have the appearance of a raging torrent, its average velocity is often less than for the river near its mouth. The difference is primarily attributable to the greater efficiency of the larger channel in a downstream direction.

In the headwaters region where the gradient is steepest, the water must flow in a relatively small and often boulder-strewn channel. The small channel and rough bed create great drag and inhibit movement by sending water in all directions with almost as much backward motion as forward motion. However, as one progresses downstream, the material on the bed of the stream becomes much smaller, offering less resistance to flow, and the width and depth of the channel increase to accommodate the greater discharge. These factors, especially the wider and deeper channel, permit the water to flow more freely and hence more rapidly.

In summary, we have seen that there is an inverse relationship between gradient and discharge. Where the gradient is high, the discharge is small, and where the discharge is great, the gradient is small. Stated another way, a stream can maintain a higher velocity near its mouth even though it has a lower gradient than upstream because of the greater discharge, larger channel, and smoother bed.

BASE LEVEL AND GRADED STREAMS

In 1875 John Wesley Powell, the pioneering geologist who first explored the Grand Canyon and later headed the U.S. Geological Survey, introduced the concept that there is a downward limit to stream erosion, which he called **base level**. Although the idea is relatively straightforward, it is nevertheless a key concept in the study of stream activity. Base level is defined as the lowest elevation to which a stream can erode its channel. Essentially this is the level at which the mouth of a stream enters the ocean, a lake, or another stream. Base level accounts for the fact that most stream profiles have low gradients near their mouths, because the streams are approaching the elevation below which they cannot lower their beds. Powell recognized that two types of base level exist:

> We may consider the level of the sea to be a grand base level, below which the dry lands cannot be eroded; but we may also have, for local and temporary purposes, other base levels of erosion. . . .*

Sea level, which Powell called "grand base level," is now referred to as **ultimate base level. Local** or **temporary base levels** include lakes, resistant layers of rock, and main streams which act as base levels for their tributaries. All have the capacity to limit a stream at a certain level. For example, when a stream enters a lake, its velocity quickly approaches zero and its ability to erode ceases. Thus the lake prevents the stream from eroding below its level at any point upstream from the lake (Figure 10.9). Although lakes and layers of resistant rock may not seem like temporary features, over the long span of geologic time they are indeed only passing phenomena. Even the largest lakes are eventually drained by the downcutting of their outlets and even the hardest rock layers will ultimately be cut through by the grinding action of sediment in a swiftly moving stream. Thus lakes or rock layers are only temporary hindrances to a stream's ability to downcut its channel.

Any change in base level will cause a corresponding readjustment of stream activities. When a dam is built along a stream course, the reservoir which forms behind it raises the base level of the stream (Figure 10.10). Upstream from the dam the stream gradient is reduced, lowering its velocity and, hence, its sediment-transporting ability. The stream, now unable to transport all of its load, will deposit material, thereby building up its channel. This process continues until the stream again has a gradient sufficient to carry its load. The profile of the new channel would be similar to the old, except that it would be somewhat higher.

If, on the other hand, the base level should be lowered, either by uplifting of the land or by a drop in sea level, the stream would again readjust. The

***Exploration of the Colorado River of the West* (Washington, D.C.: Smithsonian Institution, 1875), p. 203.

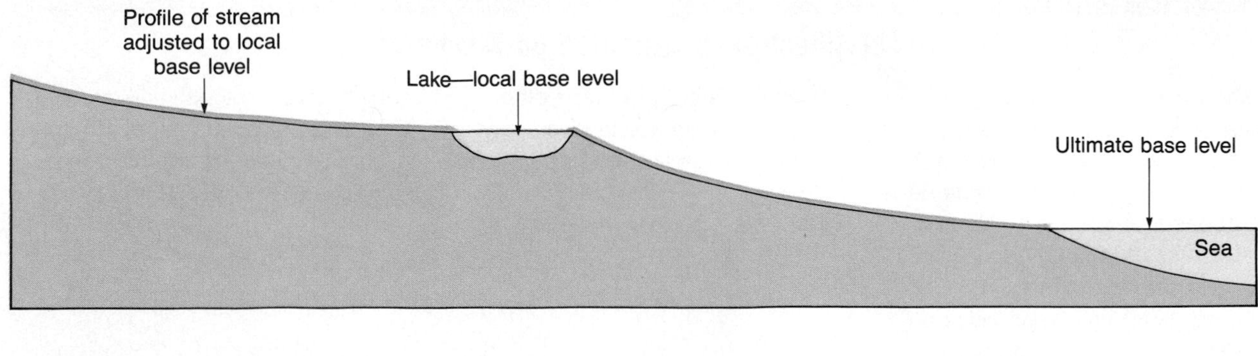

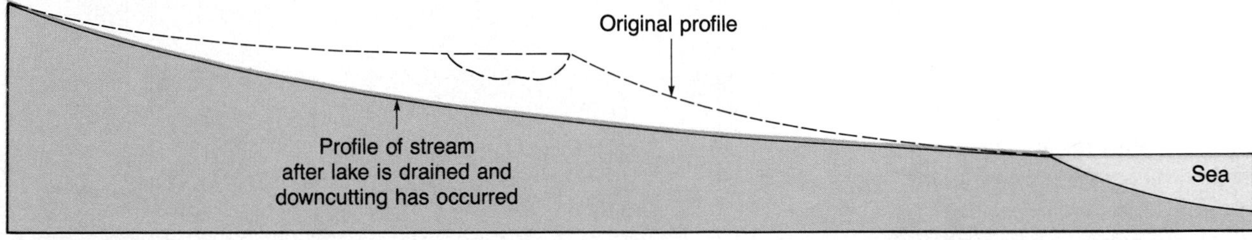

FIGURE 10.9
The lake acts as local base level. As long as the lake is present, the stream cannot erode below the level of the lake. If the lake is drained, the stream will downcut toward sea level (ultimate base level).

stream, now above base level, would have excess energy and downcut its channel to establish a balance with its new base level. Erosion would first progress near the mouth, then work upstream until the stream profile was adjusted along its full length.

The observation that streams adjust their profiles for changes in base level led to the concept of a graded stream. A **graded stream** has the correct slope and other channel characteristics necessary to maintain just the velocity required to transport the material supplied to it. On the average, a graded system is not eroding or depositing material but is simply transporting it. Once a stream has reached this state of equilibrium, it becomes a self-regulating system in which a change in one characteristic causes an adjustment in the others to counteract the effect. Refer-

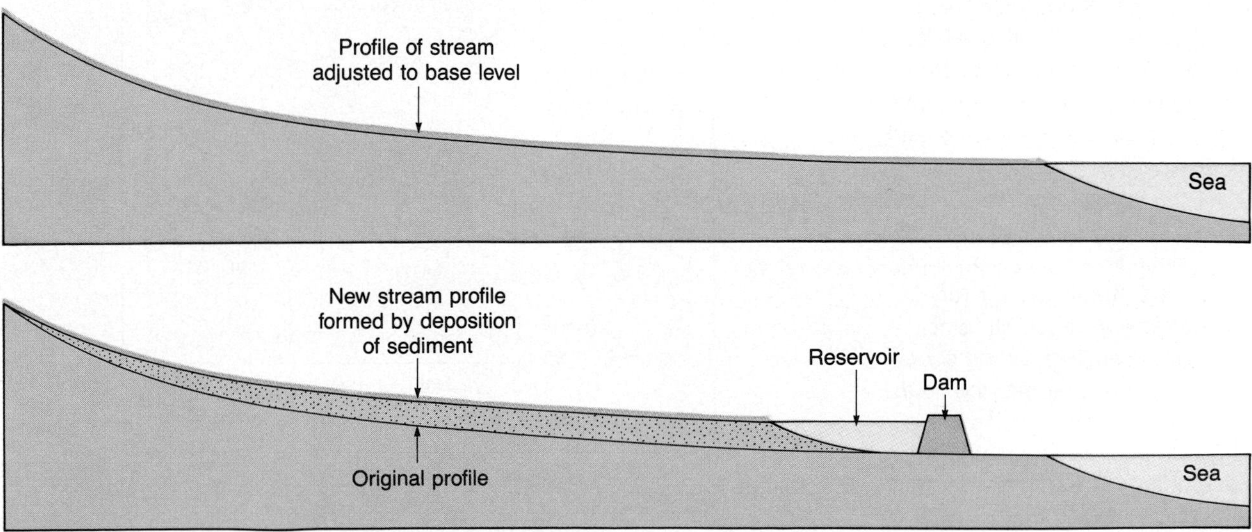

FIGURE 10.10
When a dam is built and a reservoir forms, the stream's base level is raised. This reduces the stream's velocity and leads to deposition and a reduction of the gradient upstream from the reservoir.

BOX 10.1

The Effect of Urbanization on Discharge

When rains occur, stream discharge increases. If the rains are sufficiently heavy, the ability of the channel to contain the discharge is exceeded, and water spills over the banks as a flood. Floods are natural events that should be expected. However, when cities are built, the magnitude and frequency of flooding increases.

The top portion of Figure 10.A is a hypothetical hydrograph that shows the time relationship between a rainstorm and the occurrence of flooding. Notice that the water level in the stream does not rise at the onset of precipitation because time is needed for water to move from the place where it fell to the stream. This time difference is called the **lag time.**

When an area changes from being predominantly rural to largely urban, streamflow is affected. The effect of urbanization on streamflow is illustrated by the bottom hydrograph in Figure 10.A. Notice that after urbanization the peak discharge during a flood is greater, and that the lag time between precipitation and flood peak is shorter than before urbanization. The explanation for this effect is relatively simple. The construction of streets, parking lots, and buildings covers over the ground that once soaked up water. Thus, less water infiltrates the ground, and the rate and amount of runoff increase. Further, since much less water soaks into the ground, the low-water (dry-season) flow in urban streams, which is maintained by the seepage of groundwater into the channel, is greatly reduced. As one might expect, the magnitude of these effects is a function of the percentage of land that is covered by impermeable surfaces.

Urbanization is just one example of human interference with streams. There are many other ways that land use inadvertently influences the flow of streams and the work they carry out. Moreover, there are also many ways by which people intentionally attempt to manipulate and control streams. Some of these are discussed at appropriate points in this chapter.

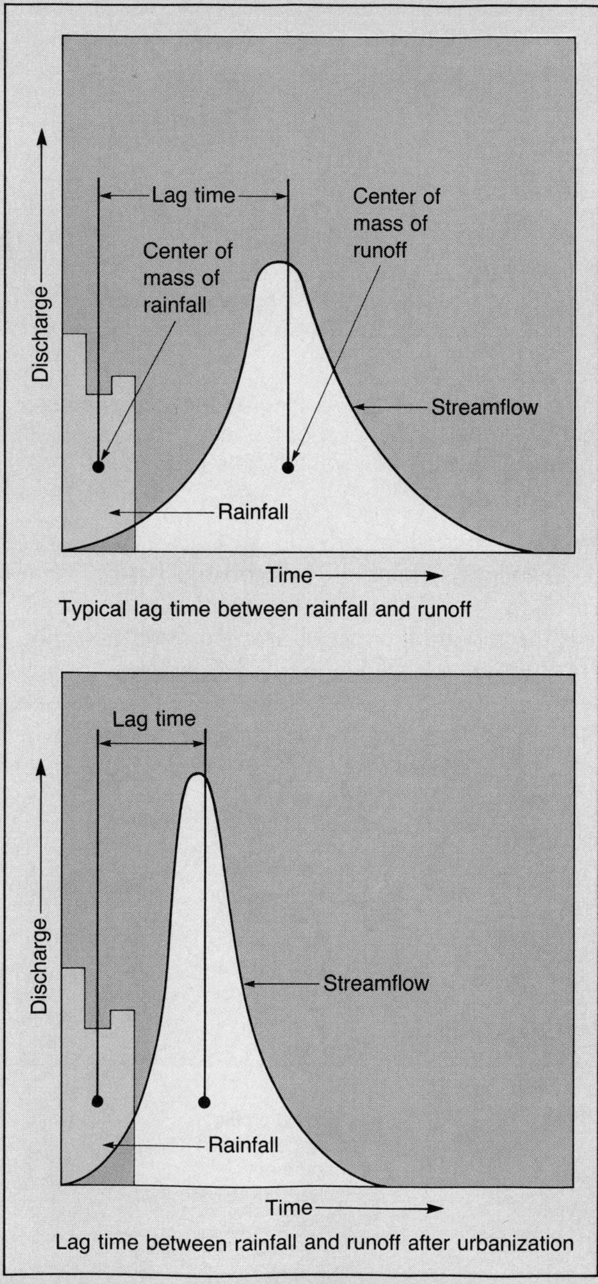

FIGURE 10.A
When an area changes from rural to urban, the lag time between rainfall and flood peak is shortened. The flood peak is also higher following urbanization. (After L. B. Leopold, U.S. Geological Survey Circular 559, 1968)

ring again to our example of a stream adjusting to a lowering of its base level, the stream would not be graded while it was cutting its new channel but would achieve this state after downcutting had ceased.

STREAM EROSION

Streams may erode their channels in several different ways: by lifting loosely consolidated particles, by abrasion, and by solution activity. The last of these is by far the least significant. Although some erosion results from the solution of soluble bedrock and channel debris, most of the dissolved material in a stream is contributed by groundwater.

As we learned earlier, when the flow of water is turbulent, the water whirls and eddies. When an eddy is sufficiently strong, it can dislodge particles from the channel and lift them into the moving water. In this manner, the force of running water swiftly erodes poorly consolidated materials on the bed and banks of the stream. The stronger the current, the more effectively the stream will lift particles. In some instances water is forced into cracks and bedding planes with sufficient strength to actually pry up pieces of rock from the bed of the channel.

FIGURE 10.11
This steep-sided gorge in southern Utah has been cut through solid rock by stream erosion. Canyons such as this are narrow due to rapid downcutting by the stream and very slow weathering of the canyon walls. (Photo by Stephen Trimble)

Observing a muddy stream will show that currents of water can lift and carry debris. However, it is not as obvious that a stream is capable of eroding solid rock. However, just as the particles on sandpaper can effectively wear down a piece of wood, so too the solid particles carried by a stream (especially sand and gravel) are capable of abrading a bedrock channel. Many steep-sided gorges cut through solid rock by the ceaseless bombardment of particles against the bed and banks of a channel serve as testimony to this erosional strength (Figure 10.11). In addition, the individual sediment grains are also abraded by the many impacts with the channel and with one another. Thus, by scraping, rubbing, and bumping, abrasion erodes a bedrock channel and simultaneously smoothes and rounds the abrading particles.

Common features on some river beds are rounded depressions known as **potholes** (Figure 10.12). They are created by the abrasive action of particles swirling in fast-moving eddies. The rotational motion of the sand and pebbles acts like a drill to bore the holes. As the particles wear down to nothing, they are replaced by new ones that continue to drill the stream bed. Eventually, smooth depressions several meters across and just as deep may result.

TRANSPORT OF SEDIMENT BY STREAMS

Streams are the most important erosional agent not only because they have the ability to downcut their channels, but also because they have the capacity to transport the enormous quantities of sediment produced by weathering. Although erosion by running water in the channel contributes significant amounts of material for transport, by far the greatest quantity of sediment carried by a stream is derived from the products of weathering. Weathering produces tremendous amounts of material that are delivered to the stream by sheet flow, mass wasting, and groundwater.

Streams transport their loads of sediment in three ways: (1) in solution (**dissolved load**), (2) in suspension (**suspended load**), and (3) along the bottom of the channel (**bed load**).

We have already seen that some of the material carried in solution may be acquired as a stream dissolves bedrock in its channel. However, the greatest portion of the dissolved load transported by most streams is supplied by groundwater. As water percolates through the ground, it first acquires soluble soil compounds. As the water seeps deeper through cracks and pores in the bedrock below, additional mineral matter may be dissolved. Eventually much of this mineral-rich water finds its way into streams.

The velocity of stream flow, which is very important to the transportation of solid particles, has essen-

FIGURE 10.12
Potholes in the bed of a small stream in Cataract Falls State Park, Indiana. The rotational motion of swirling pebbles acts like a drill to create potholes. (Photo by Tom Till)

tially no effect upon a stream's ability to carry its dissolved load. After material is in solution, it goes wherever the stream goes, regardless of velocity. Precipitation occurs only when the chemistry of the water changes.

The quantity of material carried in solution is highly variable and depends upon such factors as climate and the geologic setting. Usually the dissolved load is expressed as parts of dissolved material per million parts of water (parts per million, or ppm). Although some rivers may have a dissolved load of 1000 ppm or more, the average figure for the world's rivers is estimated at between 115 and 120 ppm. Almost 4 billion metric tons of dissolved mineral matter are supplied to the oceans each year by streams.

Most streams (but not all) carry the largest part of their load in suspension. Indeed, the visible cloud of sediment suspended in the water is the most obvious portion of a stream's load. Usually only fine sand-, silt-, and clay-sized particles can be carried this way, but during floodstage larger particles are carried as well. Also during floodstage, the total quantity of material carried in suspension increases dramatically, as can be verified by persons whose homes have been sites for the deposition of this material. During floodstage the Hwang Ho (Yellow River) of China is reported to carry an amount of sediment equal in weight to the water that carries it. Rivers like this are appropriately described as "too thick to drink but too thin to cultivate."

The type of material and amount of material carried in suspension are controlled by two factors: the velocity of the water and the settling velocity of each sediment grain. **Settling velocity** is defined as the speed at which a particle falls through a still fluid. The larger the particle, the more rapidly it settles toward the stream bed. In addition to size, the shape and specific gravity of particles also influence settling velocity. Flat grains sink through water more slowly than spherical grains and dense particles fall toward the bottom more rapidly than less dense particles. As long as the velocity of a stream exceeds the settling velocity of a sediment grain, that particle will be carried downstream with the flowing water.

A portion of a stream's load of solid material consists of sediment that is too large to be carried in suspension. These coarser particles move along the bottom of the stream and constitute the bed load (Figure 10.13). In terms of the erosional work accomplished by a downcutting stream, the grinding action of the bed load is of great importance.

The particles composing the bed load move along the bottom by rolling, sliding, and saltation. Sediment moving by **saltation** appears to jump or skip along the stream bed. This occurs as particles are propelled upward by collisions or sucked upward by the current and then carried downstream a short distance until gravity pulls them back to the bed of the stream. Particles that are too large or heavy to move by saltation either roll or slide along the bottom, depending upon their shapes.

Unlike the suspended and dissolved loads, which are constantly in motion, the bed load is in motion only intermittently, when the force of the water is sufficient to move the larger particles. Although the bed load may constitute up to 50 percent of the total load of a few streams, it usually does not exceed 10 percent of a stream's total load. For example, consider the distribution of the 750 million tons of material carried to the Gulf of Mexico by the Mississippi River each year. Of this total, it is estimated that approximately 500 million tons are carried in suspension, 200 million tons in solution, and the remaining 50 million tons as bed load. Estimates of a stream's bed load, however, should be viewed cautiously because this fraction of the load is very difficult to measure accurately. Not only is the bed load more inaccessible than the suspended and dissolved loads, but it moves primarily during periods of flooding when the bottom of a stream channel is most difficult to study.

A stream's ability to carry solid particles is typically described using two criteria. First, the maximum load of solid particles that a stream can transport is termed its **capacity**. The capacity of a stream is directly related to its discharge. The greater the

FIGURE 10.13
Although the bed load of many rivers consists of sand, the bed load of this stream is made up of boulders and is easily seen during periods of low water. During floods the seemingly immovable rocks in this channel are rolled along the bed of the stream. The maximum-size particle a stream can move is determined by the velocity of the water. (Photo by E. J. Tarbuck)

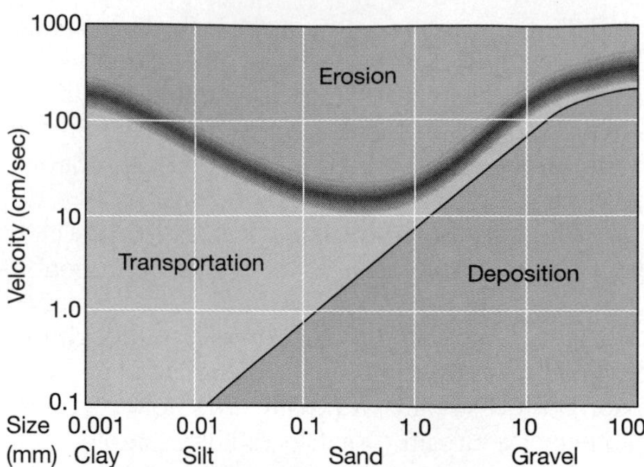

FIGURE 10.14
Relationship of stream velocity to erosion, transportation, and deposition of various particle sizes.

amount of water flowing in a stream, the greater the stream's capacity for hauling sediment. Second, the **competence** of a stream is a measure of the maximum size of particles it is capable of transporting. The stream's velocity determines its competence; the stronger the flow, the larger the particles it can carry in suspension or as bed load. It is a general rule that the competence of a stream increases as the square of its velocity. Thus, if the velocity of a stream doubles, the impact force of the water increases four times; if the velocity triples, the force increases nine times, and so forth. Hence, the large boulders that are often visible during a low-water stage and seem immovable can, in fact, be transported during floodstage because of the stream's increased velocity (Fig. 10.13).

Figure 10.14 illustrates the controlling influence of velocity on stream erosion, transportation, and deposition. The velocity necessary to lift a particle of a specific size is shown on the upper curve. Notice that the velocity necessary to lift clay-sized particles is greater than that needed to set sand in motion. This unexpected upturn in the curve is due primarily to strong cohesive forces which cause the tiny clay-sized particles to cling tightly together. However, once these fine particles are set in motion, they can remain suspended at very low velocities. Coarse particles, on the other hand, remain in motion in only a narrow range of velocities.

By now it should be relatively clear why the greatest erosion and transportation of sediment must occur during floods (Figure 10.15). The increase in discharge not only results in a greater capacity but in an increased velocity. With rising velocity the water becomes more turbulent, and larger and larger particles are set in motion. In the course of just a few days, or perhaps just a few hours, a stream in floodstage can erode and transport more sediment than it does during months of normal flow.

DEPOSITION OF SEDIMENT BY STREAMS

Whenever a stream's velocity subsides, its competence is reduced. The lower curve in Figure 10.14 shows that as the velocity of a river diminishes, particles of sediment are deposited according to size. As streamflow drops below the critical settling velocity of a certain particle size, sediment in that category begins to settle out. Thus stream transport provides a

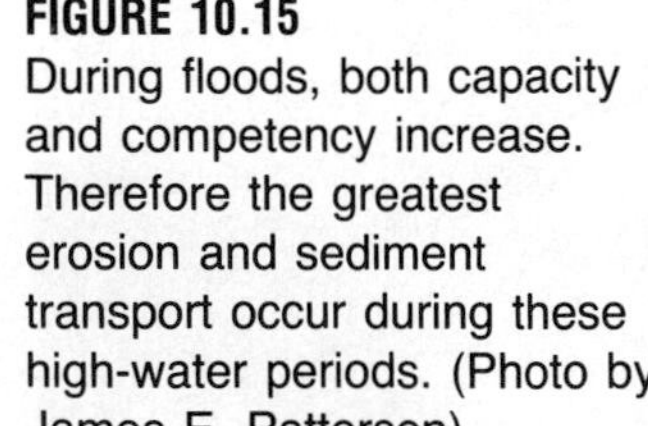
FIGURE 10.15
During floods, both capacity and competency increase. Therefore the greatest erosion and sediment transport occur during these high-water periods. (Photo by James E. Patterson)

mechanism by which solid particles of various sizes are separated. This process, called **sorting**, explains why particles of similar size are deposited together.

The well-sorted material typically deposited by a stream is called **alluvium**, a general term applicable to any stream-deposited sediment. Many different depositional features are composed of alluvium. Some of these features may be found within stream channels, some occur on the valley floor adjacent to the channel, and some exist at the mouth of the stream.

Channel Deposits

As a river transports sediment toward the sea, some material may be deposited within the channel. Channel deposits are most often composed of sand and gravel, the coarser components of a stream's load, and are commonly referred to as **bars**. Such features, however, are only temporary, for the material will be picked up again by the running water and be transported farther downstream. Eventually, most of the material will be carried to its ultimate destination, the ocean.

Sand and gravel bars can form in a number of situations. For example, they are common where streams flow in a series of bends, called meanders. As a stream flows around a bend, the velocity of the water on the outside increases, leading to erosion at that site. At the same time, the water on the inside of the meander slows, which causes some of the sediment load to settle out. Since these deposits occur on the inside "point" of the bend, they are called **point bars.** Actually these deposits would be better described as crescent-shaped accumulations of sand and gravel.

Sometimes a stream deposits materials on the floor of its channel. As these accumulations begin to choke the channel, they force the stream to split and follow several paths. What results is a complex network of converging and diverging channels that thread their way among the bars. Because such channels have an interwoven appearance, the stream is said to be **braided** (Figure 10.16). Braided patterns

FIGURE 10.16
Braided stream choked with sediment near the edge of a melting glacier. (Photo by Bradford Washburn)

most often form when the load supplied to a stream exceeds its competency or capacity. For example, if a steeper, more turbulently flowing tributary enters a main stream, its rocky bed load may be deposited at the junction. Excessive load may also be provided when debris from barren slopes is flushed into a channel during a heavy downpour, or at the end of a glacier where ice-eroded sediment is dumped into a meltwater stream flowing away from the glacier. Braided streams also form when there is an abrupt decrease in gradient or a decrease in the stream's discharge. The latter situation could result from a drop in rainfall in the area drained by the stream. It also commonly occurs when a stream leaves a humid area with many tributaries and enters a dry region with few tributaries. In this case the loss of water to evaporation and seepage into the channel results in a diminished discharge.

Floodplain Deposits

As its name implies, a **floodplain** is that part of a valley that is inundated during a flood. Most streams are bordered by floodplains. Although some are impressive features that are many kilometers across, others are very modest, having widths of just a few meters. If we were to sample the alluvium covering a floodplain, we would find that some of it consists of coarse

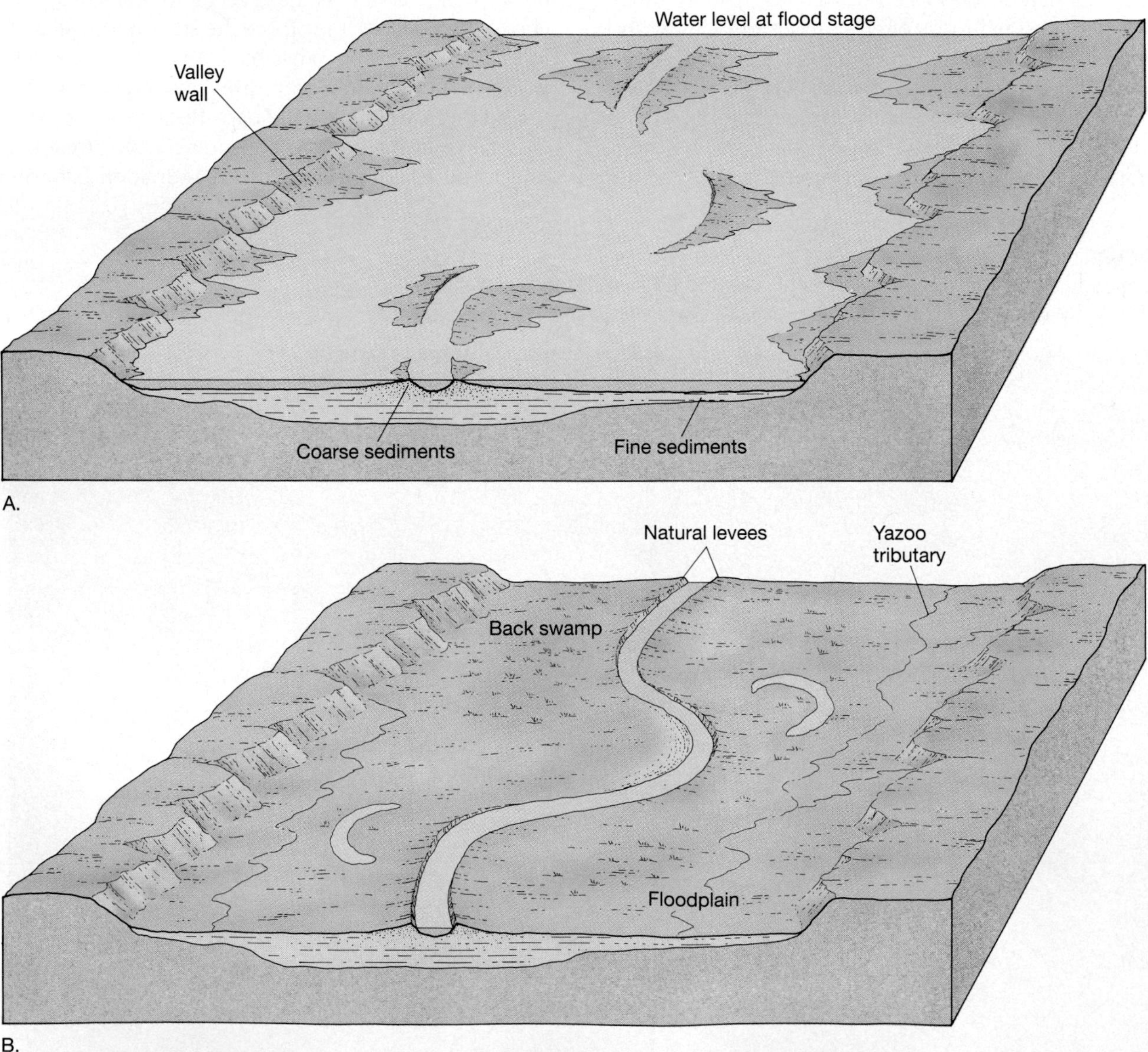

FIGURE 10.17
Formation of natural levees. After repeated flooding, streams may build very gently sloping levees.

sand and gravel that were originally deposited as point bars by meanders shifting laterally across the valley floor. Other sediments would be composed of fine sands, silts, and clays that were spread across the floodplain whenever water left the channel during floodstage.

Rivers that occupy valleys with broad, flat floors sometimes create a landform called a **natural levee** that parallels the stream channel. Natural levees are built by successive floods over a period of many years. When a stream overflows its banks onto the floodplain, the water moves over the surface as a broad sheet. Since such a flow pattern significantly reduces the water's velocity and turbulence, the coarser portion of the suspended load is deposited in strips bordering the channel. As the water spreads out over the floodplain, a lesser amount of finer sediment is laid down over the valley floor. This uneven distribution of material produces the gentle, almost imperceptible slope of the natural levee (Figure 10.17). The natural levees of the lower Mississippi River rise 6 meters above the lower portions of the valley floor. The area behind the levee is characteristically poorly drained for the obvious reason that water cannot flow up the levee and into the river. Marshes called **back swamps** often result. When a tributary stream enters a valley having substantial natural levees, it may not be able to make its way into the main channel. As a consequence the tributary may be forced to flow through the back swamp zone parallel to the main river for many kilometers before it eventually joins the main river. Such streams are called **yazoo tributaries**, after the Yazoo River, which parallels the lower Mississippi River for more than 300 kilometers.

Sometimes artificial levees are built along rivers as a means of flood control. These are usually easy to distinguish from natural levees because their slopes are much steeper. When a river is confined by levees during periods of high water, it deposits material in its channel as the discharge diminishes. This is sediment that otherwise would have been dropped on the floodplain. Thus, each time there is a high flow, deposits are left on the river bed and the bottom of the channel is built up. With the buildup of the bed, less water is required to overflow the original levee. As a result, the height of the levee must be raised to protect the floodplain. For this reason, many levees along the lower Mississippi River have had to be raised to cope with the increasing height of the water in the channel. As you can see, artificial levees are not a permanent solution to the problem of flooding. If protection is to be maintained, the structures must be heightened periodically, a process that cannot go on indefinitely.

Deltas and Alluvial Fans

Two of the most common landforms composed of alluvium are deltas and alluvial fans. They are sometimes similar in shape and are deposited for essentially the same reason: an abrupt loss of competence in a stream. Although similar in these respects, the two features are distinct. Alluvial fans are deposited on land; deltas are deposited in a body of water. In addition deltas are relatively flat, barely protruding above the level surface of the ocean or lake in which they formed. Conversely, alluvial fans can be quite steep.

Alluvial fans typically develop where a high-gradient stream leaves a narrow valley in mountainous terrain and comes out suddenly onto a broad, flat plain or valley floor. Alluvial fans form in response to the abrupt drop in gradient combined with the change from a narrow channel of a mountain stream to the unconfined flow on the slopes of the plain. The sudden drop in velocity causes the stream to dump its load of sediment quickly in a distinctive cone- or fan-shaped accumulation. As illustrated by Figure 10.18, the surface of the fan slopes outward in a broad arc from an apex at the mouth of the steep valley. Usually, coarse material is dropped near the apex of the fan, while finer material is carried toward the base of the deposit. As we learned in Chapter 9, steep canyons in dry regions are prime locations for mud flows. Therefore, it should be expected that many alluvial fans in arid areas have mudflow deposits interbedded with the alluvium.

In contrast to an alluvial fan, a **delta** forms when a stream enters an ocean or a lake. Figure 10.19A depicts the structure of a simple delta that might form in the relatively quiet waters of a lake. As the stream's forward motion is checked upon entering the lake, the dying current deposits its load of sediments. The finer silts and clays settle out some distance from the mouth in nearly horizontal layers called *bottomset beds*. Prior to the accumulation of the bottomset beds, *foreset beds* begin to form. These beds are composed of coarser particles which drop almost immediately upon entering the lake to form layers that slope downcurrent from the delta front. The foreset beds are usually covered by thin, horizontal *topset beds* that are deposited during floodstage. As the delta grows outward, the stream's gradient continually lessens. This circumstance eventually causes the channel to become choked with sediment from the slowing water. As a consequence, the river seeks a shorter route to base level, as illustrated in Figure 10.19B. This illustration also shows the main channel dividing into several smaller ones called **distributaries**. Most deltas are characterized by these shifting

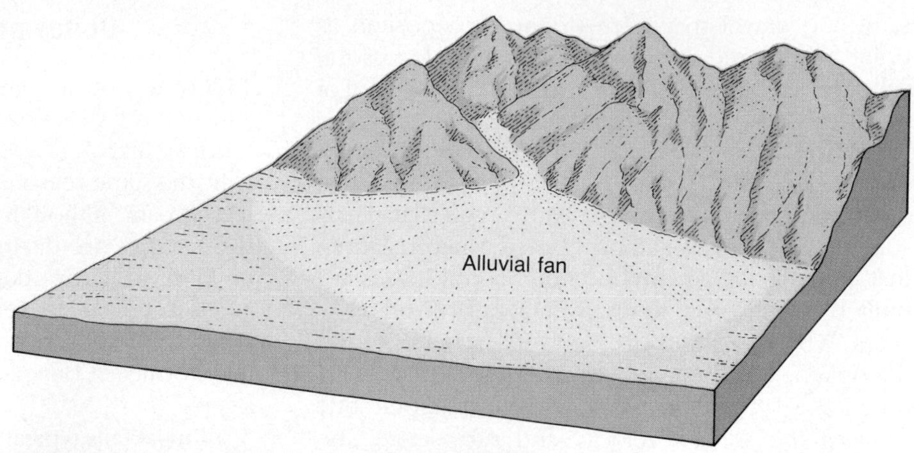

A.

B.

FIGURE 10.18
A. Alluvial fans develop where the gradient of a stream changes abruptly from steep to flat. **B.** Such a situation exists in Death Valley, California, where streams emerge from the mountains into a flat basin. As a result, Death Valley has many large alluvial fans. (Photo by Michael Collier)

channels that act in an opposite way to that of tributaries. Rather than carrying water into the main channel, distributaries carry water away from the main channel in varying paths to base level. After numerous shifts of the channel, the simple delta grows into the idealized triangular shape of the Greek letter delta (Δ), for which it was named. Note, however, that many deltas do not exhibit this idealized shape. Differences in the configurations of shorelines, and variations in the nature and strength of wave activity, result in many different shapes.

Although deltas that form in the ocean generally exhibit the same basic form as the simple lake-deposited feature just described, most large marine deltas are far more complex and have foreset beds that are inclined at a much lower angle than those depicted in Figure 10.19A. Indeed, many of the world's great rivers have created massive deltas, each with its own peculiarities and none as simple as the one illustrated in Figure 10.19.

Many large rivers have deltas that extend over thousands of square kilometers. The delta of the Mississippi River is one such feature. It resulted from the accumulation of huge quantities of sediment derived from the vast region drained by the river and its tributaries. Today New Orleans rests where there was ocean less than 5000 years ago. Figure 10.20 shows that portion of the Mississippi delta which has been

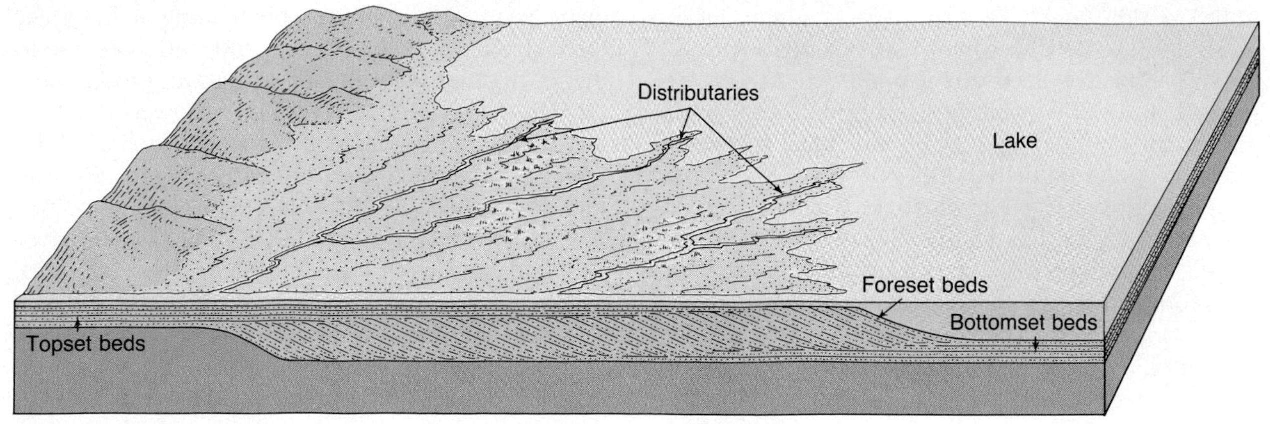

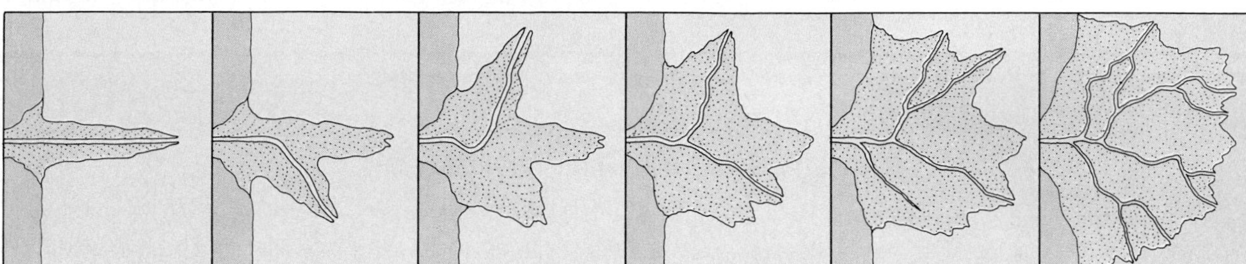

FIGURE 10.19
A. Structure of a simple delta that forms in the relatively quiet waters of a lake. **B.** Growth of a simple delta. As a stream extends its channel, the reduced gradient causes it to find a shorter route to its base level. (After Ward's Natural Science Establishment, Inc., Rochester, N.Y.)

FIGURE 10.20
During the past 5000–6000 years, the Mississippi River has built a series of seven coalescing subdeltas. The numbers indicate the order in which the subdeltas were deposited. The present bird-foot delta (number 7) represents the activity of the past 500 years. Without ongoing human efforts, the present course will shift and follow the path of the Atchafalaya River. The inset shows the point where the Mississippi may someday break through (arrow) and the shorter path it would take to the Gulf of Mexico. (After C. R. Kolb and J. R. Van Lopik)

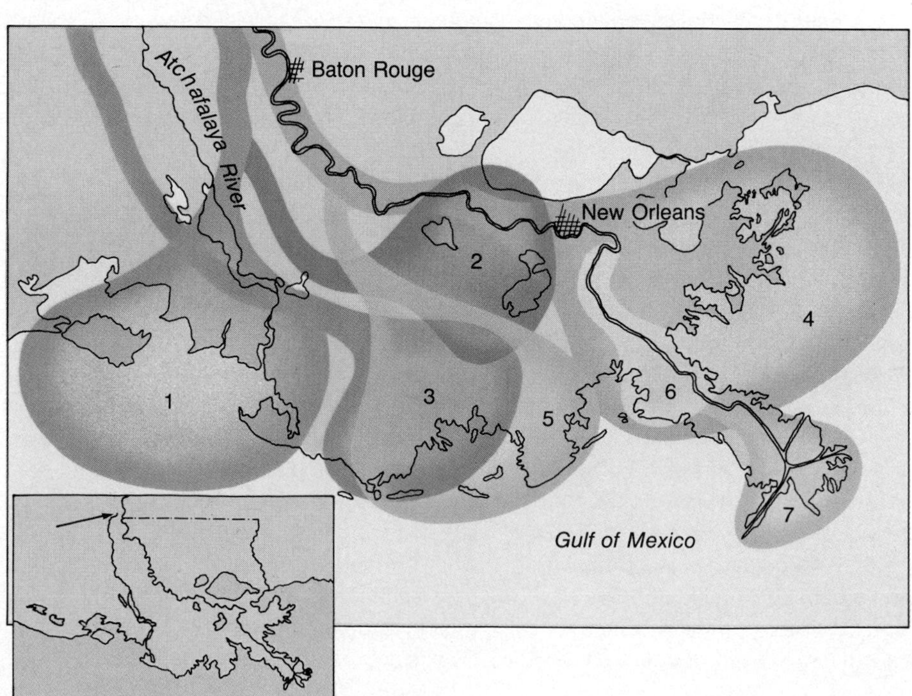

built over the past 5000–6000 years. As the figure illustrates, the delta is actually a series of seven coalescing subdeltas. Each was formed when the river left its existing channel to find a shorter, more direct path to the Gulf of Mexico. The individual subdeltas interfinger and partially cover one another to produce a very complex structure. It is also apparent from Figure 10.20 that after each portion was abandoned, coastal erosion processes acted to reduce and modify the features. The present subdelta, called a *bird-foot* delta because of the configuration of its distributaries, has been built by the Mississippi in the last 500 years.

At present this active bird-foot delta has reached about as far as natural forces will allow. In fact, for many years the river has been struggling to cut through a narrow neck of land and shift its course to that of the Atchafalaya River. If this were to happen, the Mississippi would abandon its lowermost 500-kilometer path in favor of a much shorter 225-kilometer route. From the early 1940s until the 1950s, an increasing portion of the Mississippi's discharge was diverted to this new path, indicating that the river was ready to shift and begin building a new subdelta. To prevent such an event, and keep the Mississippi following its present course, a damlike structure was erected at the site where the channel was trying to break through. Floods in the early 1970s weakened the control structure, and the river again threatened to shift until a massive auxiliary dam was completed

BOX 10.2

Louisiana's Vanishing Wetlands

Coastal wetlands form in sheltered environments that include swamps, tidal flats, coastal marshes, and bayous. They are rich in wildlife and provide nesting grounds and important stopovers for waterfowl and migratory birds, as well as spawning areas and valuable habitats for commercial and recreational fish.

The delta of the Mississippi River in Louisiana contains about 40 percent of all coastal wetlands in the lower 48 states. Louisiana's wetlands are sheltered from the wave action of hurricanes and winter storms by low-lying barrier islands. Both the wetlands and the protecting islands have formed as a result of the shifting of the Mississippi River during the past 5000–6000 years. The deltaic processes that control the movement of water and sediment have resulted in a system of complex drainage patterns, natural ridges and levees, and offshore barrier beaches—all of which restrict the advance and encroachment of salt water. The dependence of Louisiana's coastal wetlands on the Mississippi River and its distributaries as a direct source of sediment and fresh water leaves them vulnerable to changes in the river system. Moreover, the reliance on barrier islands for protection from storm waves leaves coastal wetlands vulnerable when these narrow offshore islands are eroded.*

Today the coastal wetlands of Louisiana are disappearing at an alarming rate. The U.S. Army Corps of Engineers estimates that about 90 square kilometers (35 square miles) of Louisiana's wetlands are lost annually. Fifty years from now the shoreline is predicted to be many kilometers from where it is today. As the shoreline retreats, southern Louisiana's wetlands will be covered by the Gulf of Mexico. If a predicted rise in sea level occurs, it will only add to the coastal land losses.†

Why are Louisiana's wetlands shrinking? First, the Mississippi delta and its wetlands are subject to ongoing natural changes. As sediment accumulates and builds the delta in one area, erosion and subsidence cause losses elsewhere (Figure 10.B). When the river shifts, the zones of delta growth and destruction also shift. Ever since people arrived on the scene, the rate at which the delta and its wetlands are destroyed has accelerated.

Before Europeans settled the delta, the Mississippi River regularly overflowed its banks. The huge quantities of sediment that were deposited renewed the soil and kept the delta from sinking below sea level. However, with settlement came flood-control efforts and the desire to maintain and improve navigation on the river. Artificial levees were constructed to contain the rising river during flood stage. Over time, the levees were extended all the way to the mouth of the Mississippi to keep the channel open for navigation. The effects have been straightforward. The levees prevent sediment and fresh water from being dispersed into the wetlands. Instead, the river is forced to carry its load to the deep waters at the mouth. Meanwhile, the processes of compaction, subsidence, and wave erosion continue. Because not enough sediment is added to offset these forces, the size of the delta and the extent of its wetlands gradually shrink.

in the mid-1980s. For the time being, at least, the inevitable has been avoided, and the Mississippi River will continue to flow past Baton Rouge and New Orleans on its way to the Gulf of Mexico.

Although deltas are deposited by many large rivers, not all rivers create these features. Even streams that transport large loads of sediment may lack deltas because powerful currents and waves quickly redistribute the material as soon as it is deposited. The Columbia River in the Pacific Northwest is one such situation. In other cases, rivers do not carry sufficient quantities of sediment to build up a delta. The St. Lawrence River, for example, has little opportunity to pick up much sediment between Lake Ontario and its mouth in the Gulf of St. Lawrence.

STREAM VALLEYS

Valleys are the most common landforms on the earth's surface. In fact they exist in such large numbers that they have never been counted except in limited areas used for study. Prior to the turn of the nineteenth century it was generally believed that valleys were created by catastrophic events that pulled the crust apart and created avenues for streams to follow. Today, however, we know that with a few exceptions, streams create the valleys through which they flow.

Among the first meaningful statements that related streams to the creation of their valleys was one made by the English geologist, John Playfair in 1802. In his

FIGURE 10.B
Satellite image of the Mississippi delta. For the past 600 years or so, the main flow of the river has been along its present course, extending southeast from New Orleans. During that span, the delta advanced into the Gulf of Mexico at a rate of about 10 kilometers (6 miles) per century. (Photo courtesy of NASA)

The problem has been aggravated by a decline in the amount of sediment transported by the Mississippi. The rate of sediment supply to the delta has decreased by approximately 50 percent over the past 100 years. A substantial portion of the reduction is a result of the trapping of sediment in large reservoirs created by dams built on tributaries to the Mississippi, especially the Missouri and Arkansas rivers. The diversion of part of the Mississippi's flow to the Atchafalaya distributary also reduces the amount of sediment available to build and maintain the delta.

Understanding and modifying the impact of people is a necessary basis for any plan to reduce the loss of wetlands in the Mississippi delta. Although some projects are underway to divert more sediment and fresh water to the wetlands, plans to cope effectively with the problem have yet to be implemented.

*Barrier islands and their dynamics are discussed in Chapter 14.
†For more on this topic, see the section "Sea Level Is Rising" in Chapter 14.

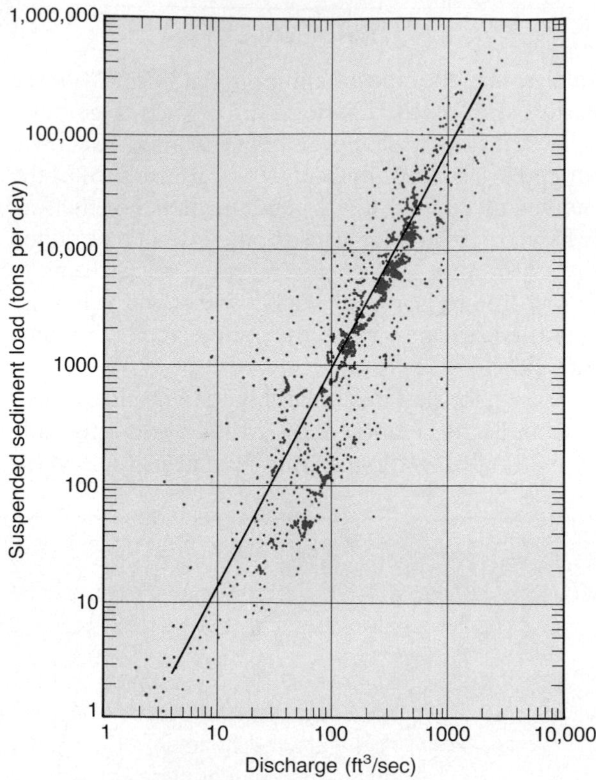

FIGURE 10.21
Relationship between suspended load and discharge on the Powder River at Arvada, Wyoming. (From L. B. Leopold and Thomas Maddock, Jr., U.S. Geological Survey Professional Paper 252, 1953)

well-known work, *Illustrations of the Huttonian Theory of the Earth,* Playfair stated the principle that has come to be called **Playfair's law**:

> Every river appears to consist of a main trunk, fed from a variety of branches, each running in a valley proportioned to its size, and all of them together forming a system of valleys, communicating with one another, and having such a nice adjustment of their declivities, that none of them join the principal valley, either on too high or too low a level; a circumstance that would be indefinitely improbable, if each of these valleys were not the work of the stream that flows in it.

Not only were Playfair's observations essentially correct, they were written in a style that is seldom achieved in scientific prose.

Stream valleys can be divided into two general types. Narrow, V-shaped valleys and wide valleys with flat floors exist as the ideal forms, with many gradations between. In some arid regions, where downcutting is rapid and weathering is slow, and in places where rock is particularly resistant, narrow valleys may not be V-shaped but rather many have nearly vertical walls (see Figure 10.11). However, most valleys, even those that are narrow at the base, are much broader at the top than the width of the channel at the bottom. This would not be the case if the only agent responsible for eroding valleys were the streams flowing through them.

The sides of most valleys are shaped primarily as the result of weathering, sheet flow, and mass wasting. Consider the following example of this process. Figure 10.21 shows the relationship between suspended load and discharge at a gauging station on the Powder River in Wyoming. Notice that as the discharge increases, the quantity of suspended sedi-

FIGURE 10.22
V-shaped valley of the Yellowstone River. The rapids and waterfalls indicate that the river is vigorously downcutting. (Photograph used by permission of Dennis Tasa)

ment increases. In fact, the increase is exponential; that is to say, if the discharge at the gauging station increases tenfold, the suspended load may increase by a factor of 100 or more. Measurements and calculations have shown that stream channel erosion during periods of increased discharge can account for only a portion of the additional sediment transported by a stream. Therefore, much of the increased load must be delivered to the stream by sheet flow and mass wasting.

A narrow, V-shaped valley indicates that the primary work of the stream has been downcutting toward base level. The most prominent features in such a valley are **rapids** and **waterfalls** (Figure 10.22). Both occur where the stream profile drops rapidly, a situation usually caused by variations in the erodibility of the bedrock into which the stream channel is cutting. A resistant bed produces a rapids by acting as a temporary base level upstream while allowing downcutting to continue downstream. Once erosion has eliminated the resistant rock, the stream profile smoothes out again. Waterfalls are places where the stream makes a vertical drop. One type of waterfall is exemplified by Niagara Falls (Figure 10.23). Here, the falls are supported by a resistant bed of dolomite that is underlain by a less resistant shale. As the water plunges over the lip of the falls it erodes the less resistant shale, undermining a section of dolomite, which eventually breaks off. In this manner the waterfall retains its vertical cliff while slowly but continually retreating upstream. Since its formation Niagara Falls has retreated approximately 11 kilometers (about 7 miles) upstream.

Once a stream has cut its channel closer to base level, it approaches a graded condition, and downward erosion becomes less dominant. At this point more of the stream's energy is directed from side to side. The reason for this change is not fully understood, but the reduced gradient probably is an important factor. Nevertheless it does occur, and the result is a widening of the valley as the river cuts away first at one bank and then the other (Figure 10.24). In this manner the flat valley floor, or floodplain, is produced. This is an appropriate name because the river is confined to its channel except during floodstage, when it overflows its banks and inundates the floodplain.

When a river erodes laterally and creates a floodplain as just described, it is called an *erosional floodplain*. Floodplains can be depositional in nature as

FIGURE 10.23
The smaller American Falls at Niagara Falls. The river plunges over the falls and erodes the shale beneath the more resistant Lockport Dolomite. As a section of dolomite is undercut, it loses support and breaks off. (Photo by James E. Patterson)

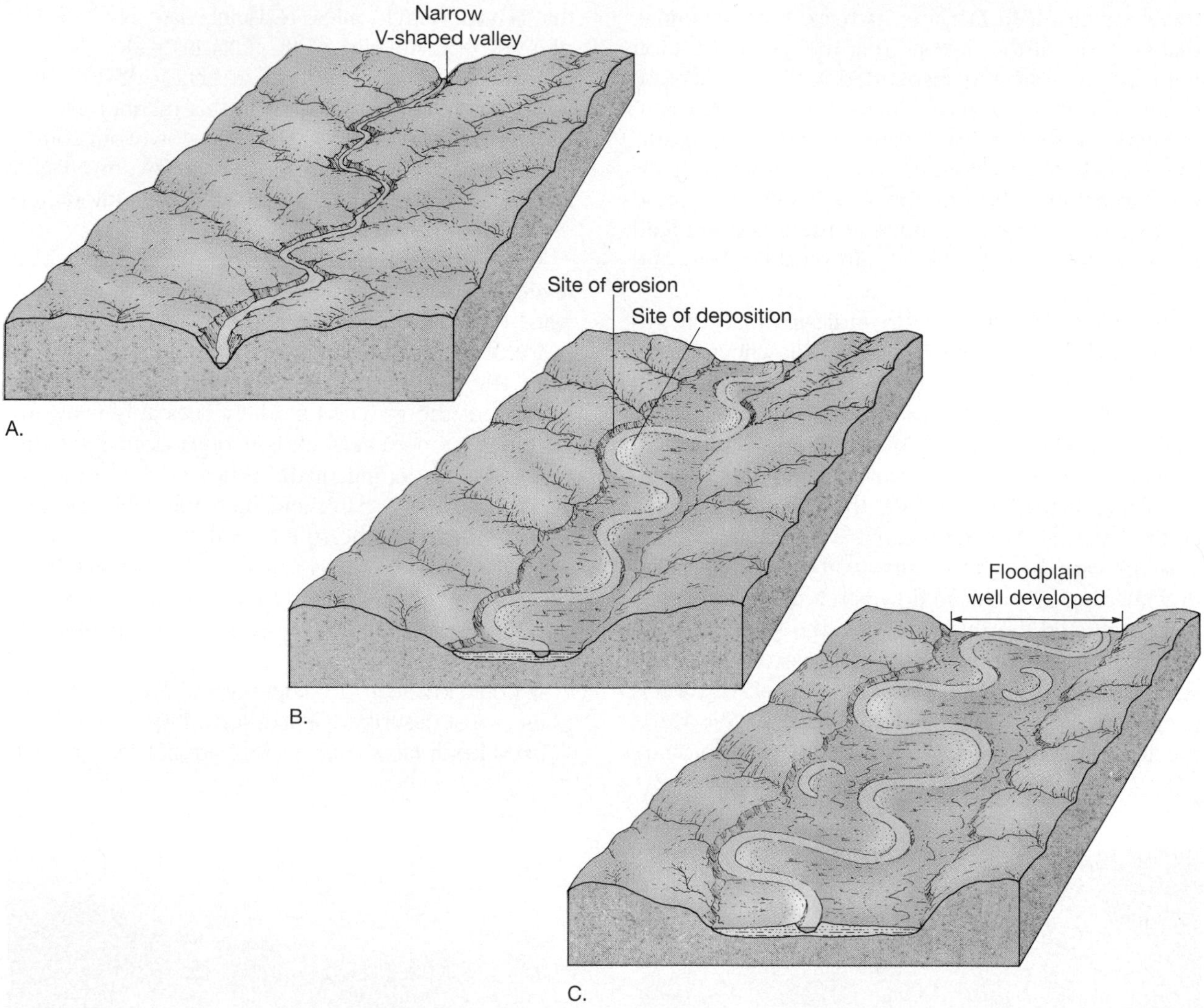

Narrow
V-shaped valley

A.

Site of erosion
Site of deposition

B.

Floodplain
well developed

C.

FIGURE 10.24
Stream eroding its floodplain.

A.

B.

FIGURE 10.25
Erosion of a cut bank along the Newaukum River, Washington. **A.** January, 1965. **B.** March, 1965. (Photos by P. A. Glancy, U.S. Geological Survey)

well. *Depositional floodplains* are produced by a major fluctuation in conditions, such as a change in base level. The floodplain in Yosemite Valley is one such feature, and was produced when a glacier gouged the former stream valley about 300 meters (1000 feet) deeper than it had been. After the glacial ice melted, the stream readjusted itself to its former base level by refilling the valley with alluvium.

Streams that flow upon floodplains, whether erosional or depositional, move in sweeping bends called **meanders**. The term is derived from a river in western Turkey, the Menderes, which has a very sinuous course. Once a bend in a channel begins to form, it grows larger. Erosion occurs on the outside of the meander where velocity and turbulence are greatest. Commonly the outside bank is undermined, especially during periods of high water. As the bank becomes oversteepened, it fails by slumping into the channel. Because the outside of a meander is a zone of active erosion, it is often referred to as the **cut bank** (Figure 10.25). Much of the debris detached by the stream at the cut bank moves downstream and is soon deposited as point bars in zones of decreased velocity on the insides of meanders. Thus, meanders migrate laterally, while keeping the same cross-sectional area, by eroding on the outside of the bends and depositing on the inside (Figure 10.26). Growth ceases when the meander reaches a critical size that is determined by the size of the stream. The larger the stream, the larger its meanders can be.

Due to the slope of the channel, erosion is more effective on the downstream side of a meander. Therefore, in addition to growing laterally, the bends also gradually migrate down the valley. Sometimes the downstream migration of a meander is slowed when it reaches a more resistant portion of the floodplain. This allows the next meander upstream to "catch up." Gradually the neck of land between the meanders is narrowed. When they get close enough, the river may erode through the narrow neck of land to the next loop (Figure 10.27). The new, shorter channel segment is called a **cutoff** and, because of its shape, the abandoned bend is called an **oxbow lake** (Figure 10.28). Over a period of time, the oxbow lake fills with sediment to create a **meander scar**.

The process of meander cutoff formation has the effect of shortening the river and was described humorously by Mark Twain in *Life on the Mississippi*.

> In the space of one hundred and seventy-six years the lower Mississippi has shortened itself two hundred and forty-two miles. This is an average of a trifle over one mile and a third per year. Therefore, any calm person, who is not blind or idiotic, can see that in the Old Oolitic Silurian Period, just a million years ago next November, the Lower Mississippi River was upwards of one million three hundred

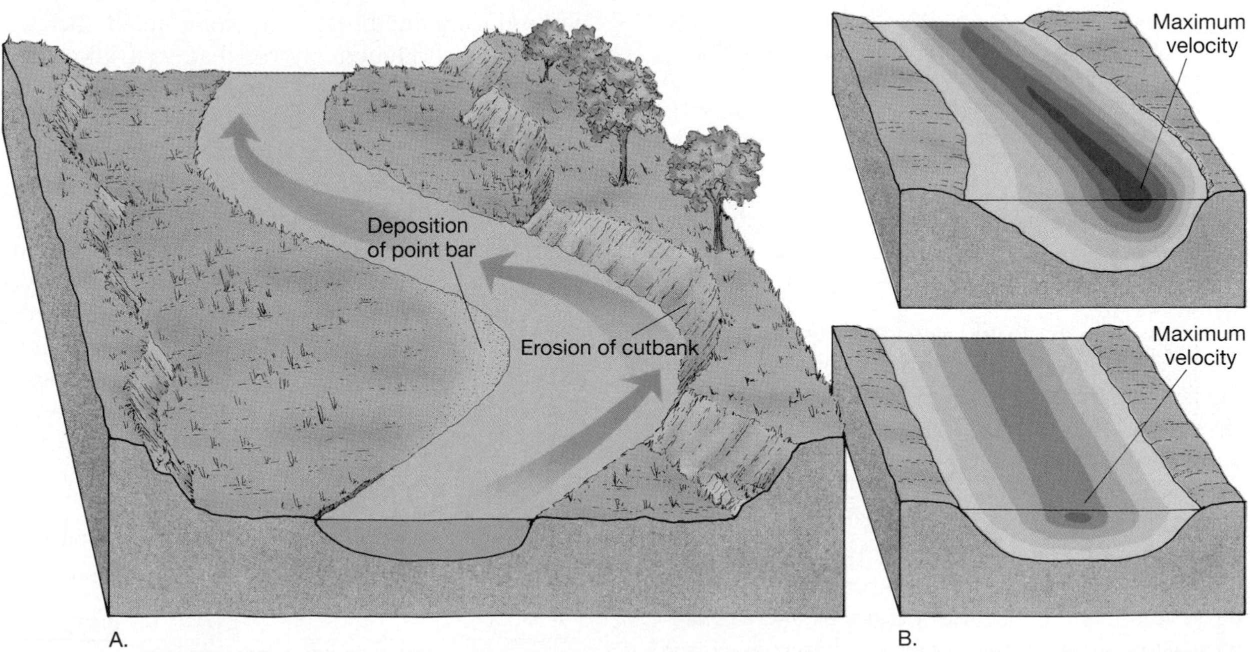

FIGURE 10.26
Lateral movement of meanders. By eroding its outer bank and depositing material on the inside of the bend, a stream is able to shift its channel.

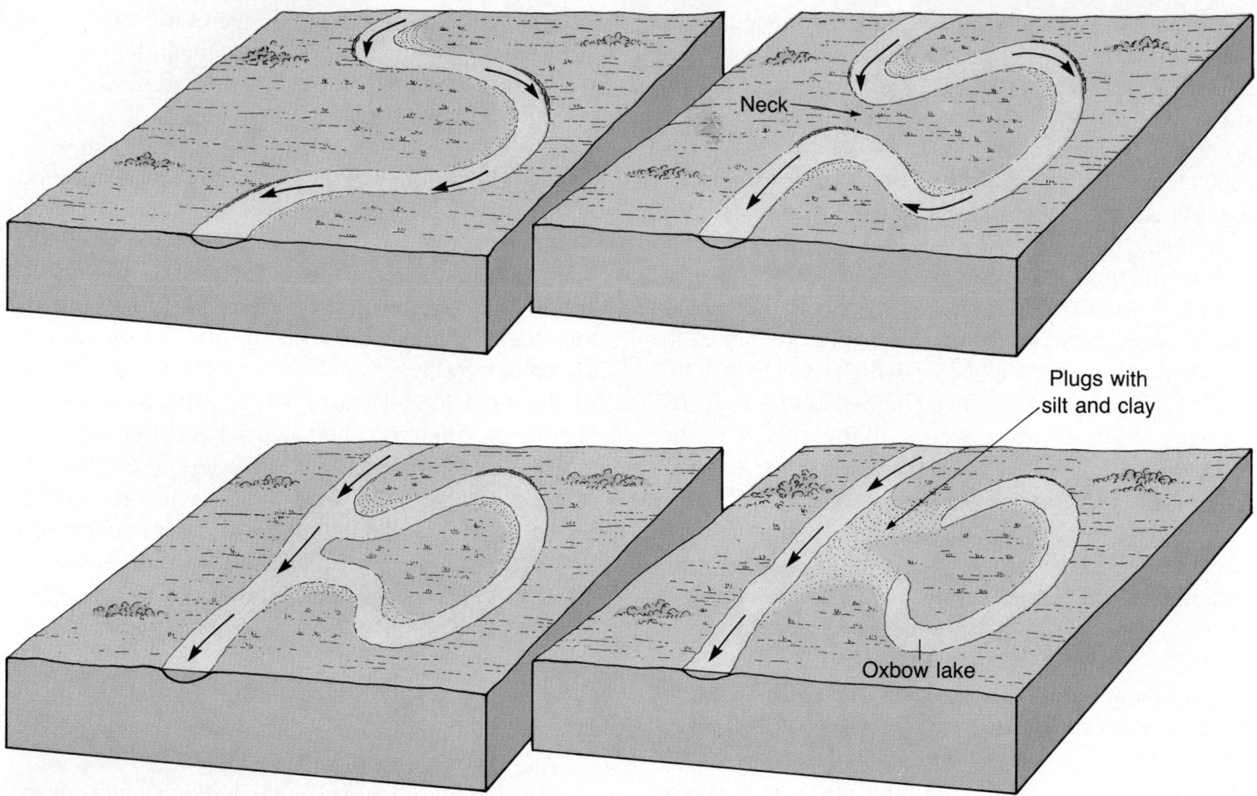

FIGURE 10.27
Formation of a cutoff and oxbow lake.

> thousand miles long, and stuck out over the Gulf of Mexico like a fishing rod. And by the same token any person can see that seven hundred and forty-two years from now the Lower Mississippi will be only a mile and a three quarters long, and Cairo and New Orleans will have joined their streets together, and be plodding comfortably along under a single mayor and a mutual board of aldermen. There is something fascinating about science. One gets such wholesale returns of conjecture out of such a trifling investment of fact.

Although the data used by Mark Twain may be reasonably accurate, he intentionally forgot to include the fact that the Mississippi created many new meanders, thus lengthening its course by a similar amount. In fact, with the growth of its delta, the Mississippi is actually getting longer, not shorter.

One method of flood control that is sometimes used is to straighten a channel by creating cutoffs artificially. The idea is that by shortening the stream, the gradient and hence the velocity are increased. By increasing velocity, the larger discharge associated with flooding can be dispersed more rapidly. Since the early 1930s the Army Corps of Engineers has created many artificial cutoffs on the Mississippi for the purpose of increasing the efficiency of the channel and reducing the threat of flooding. In all, the river has been shortened more than 240 kilometers (150 miles). The program has been somewhat successful in reducing the height of the river in flood. However, since the river's tendency toward meandering still exists, preventing the river from returning to its previous course has been difficult.

Artificial cutoffs increase a stream's velocity and may also accelerate erosion of the bed and banks of the channel. A case in point is the Blackwater River in Missouri, whose meandering course was shortened in 1910. Among the many effects of this project was a dramatic increase in the width of the channel caused by the increased velocity of the stream. One particular bridge over the river collapsed because of bank erosion in 1930. Over the next 17 years the same bridge was replaced on three more occasions, each time with a wider span.

DRAINAGE NETWORKS

A stream is just a small component of a larger system. Each system consists of a **drainage basin**, the land

A.

B.

FIGURE 10.28
A. Oxbow lakes occupy abandoned meanders. As they fill with sediment, oxbow lakes gradually become swampy meander scars. (Photo by Peter Kresan). **B.** Aerial view of oxbow lakes created by the meandering Mara River in Kenya. (Photo by Peter Arnold, Inc.)

area that contributes water to the stream. The drainage basin of one stream is separated from another by an imaginary line called a **divide** (Figure 10.29). Divides range in size from a ridge separating two small gullies to continental divides, which split continents into enormous drainage basins. For example, the continental divide that runs somewhat north-south through the Rocky Mountains separates the drainage which flows west to the Pacific Ocean from that which flows to the Atlantic via the Gulf of Mexico. Although divides separate the drainage of two streams, if they are tributaries of the same river, they are both a part of that larger drainage system.

Drainage Patterns

All drainage systems are made up of an interconnected network of streams which together form particular patterns. The nature of a drainage pattern can vary greatly from one type of terrain to another, primarily in response to the kinds of rock on which the streams developed or the structural pattern of faults and folds.

The most commonly encountered drainage pattern is the **dendritic** pattern (Figure 10.30A). This pattern is characterized by irregular branching of tributary streams that resembles the branching pat-

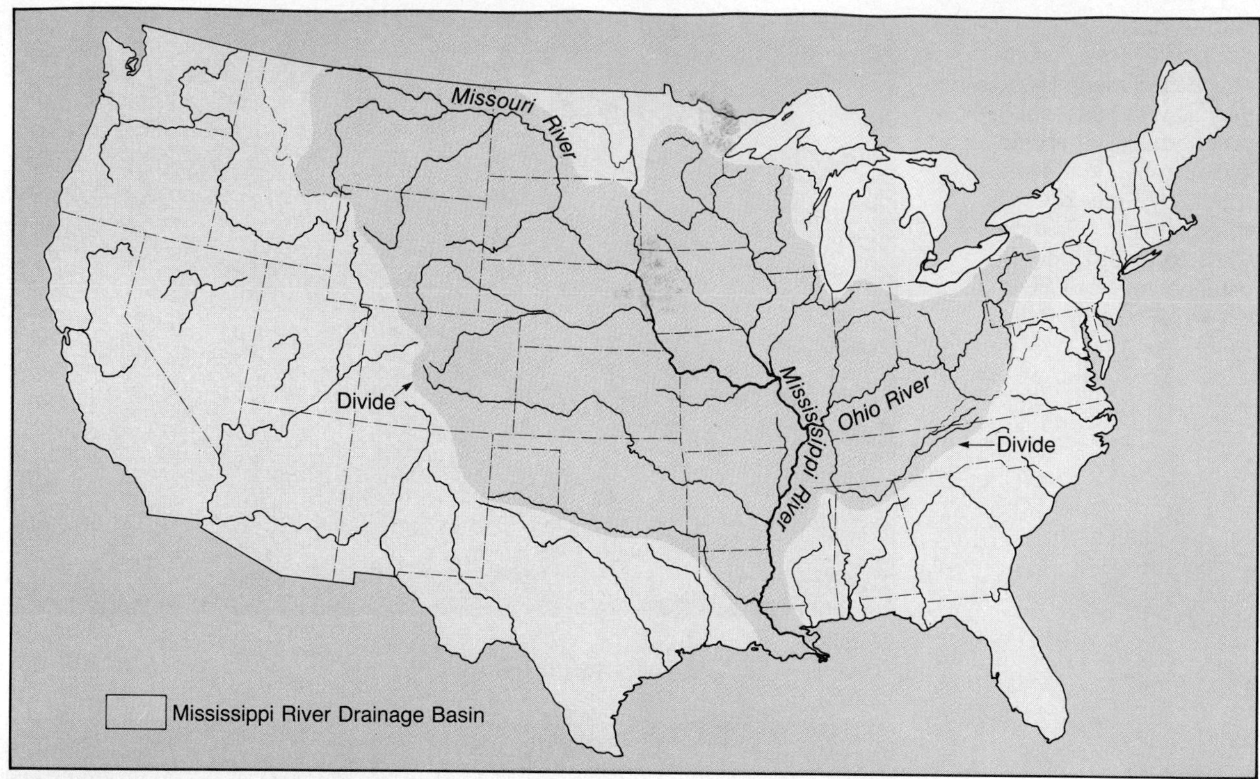

FIGURE 10.29
The drainage basin of the Mississippi River, North America's largest river, covers about 3 million square kilometers. Divides are the boundaries that separate drainage basins from each other. Drainage basins and divides exist for all streams.

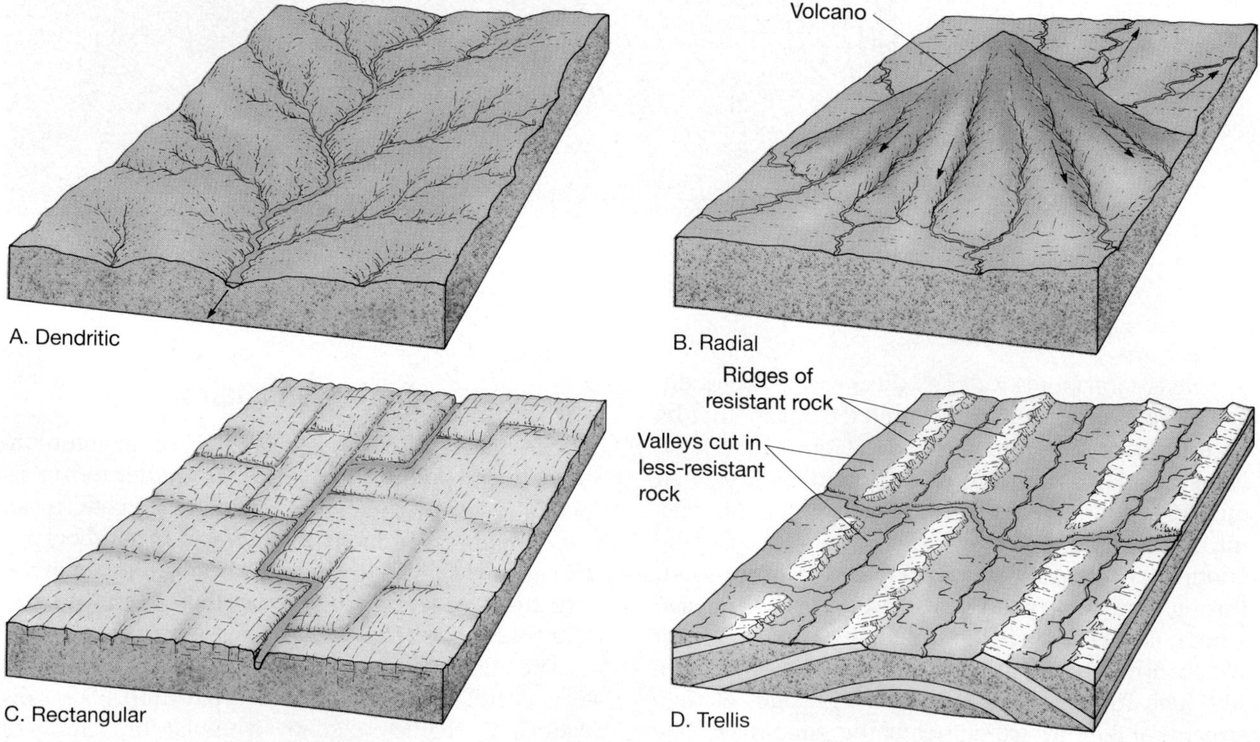

FIGURE 10.30
Drainage patterns. **A.** Dendritic. **B.** Radial. **C.** Rectangular. **D.** Trellis.

tern of a deciduous tree. In fact, the work *dendritic* means "treelike." The dendritic pattern forms where underlying bedrock is relatively uniform, such as flat-lying strata or massive igneous rocks. Since the underlying material is essentially uniform in its resistance to erosion, it does not control the pattern of streamflow. Rather, the pattern is determined chiefly by the direction of slope of the land.

When streams diverge from a central area like spokes from the hub of a wheel, the pattern is said to be **radial** (Figure 10.30B). This pattern typically develops on isolated volcanic cones and domal uplifts.

Figure 10.30C illustrates a **rectangular** pattern, with many right-angle bends. This pattern develops when the bedrock is crisscrossed by a series of joints and faults. Since these structures are eroded more easily than unbroken rock, their geometric pattern guides the directions of valleys.

Figure 10.30D illustrates a **trellis** drainage pattern, a rectangular pattern in which tributary streams are nearly parallel to one another and have the appearance of a garden trellis. This pattern forms in areas underlain by alternating bands of resistant and less resistant rock and is particularly well displayed in the folded Appalachians, where both weak and strong strata outcrop in nearly parallel belts.

FIGURE 10.31
Gullies extending headward into undissected terrain. (Photo by John S. Shelton)

Headward Erosion and Stream Piracy

We have seen that a stream can lengthen its course by building a delta at its mouth. A stream can also lengthen its course by **headward erosion**; that is, by extending the head of its valley upslope. As sheet flow converges, a valley collects water at its headward end. As the water becomes concentrated into a channel, its velocity, and hence its power to erode, increase. The result can be vigorous erosion at the head of the valley. Thus, through headward erosion, the valley extends itself into previously undissected terrain (Figure 10.31).

As we shall see, headward erosion by streams plays a major role in the dissection of upland areas. In addition, an understanding of this process helps explain changes that take place in drainage patterns. One cause for changes that occur in the pattern of streams is **stream piracy**, the diversion of the drainage of one stream because of the headward erosion of another stream. Piracy can occur, for example, if a stream on one side of a divide has a steeper gradient than a stream on the other side. Since the stream with the steeper gradient has more energy, it can extend its valley headward, eventually breaking down the divide and capturing part or all of the drainage of the slower stream. In Figure 10.32, the flow of stream *A* was captured when the more swiftly flowing stream *B* shifted the divide at its head until the divide intersected and diverted stream *A*.

Stream piracy also explains the existence of narrow, steep-sided gorges that have no active streams running through them. These abandoned water gaps (called *wind gaps*) form when the stream that cut the notch has its course changed by a pirate stream. In Figure 10.32, two water gaps that had been created by stream *A* became wind gaps as a result of stream piracy.

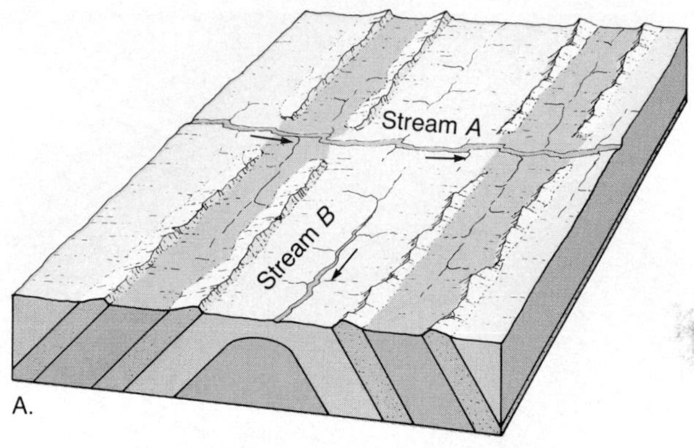

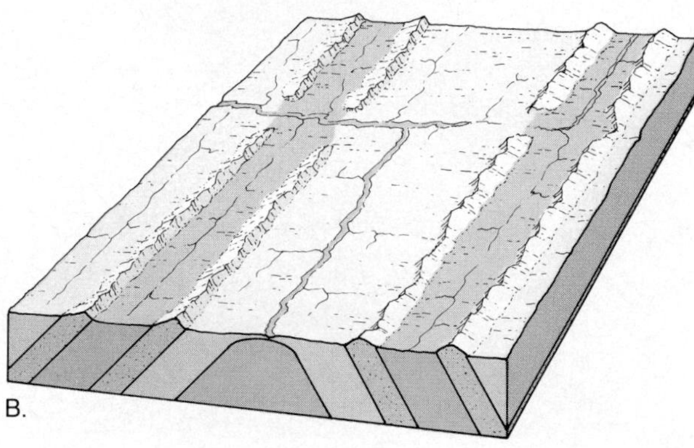

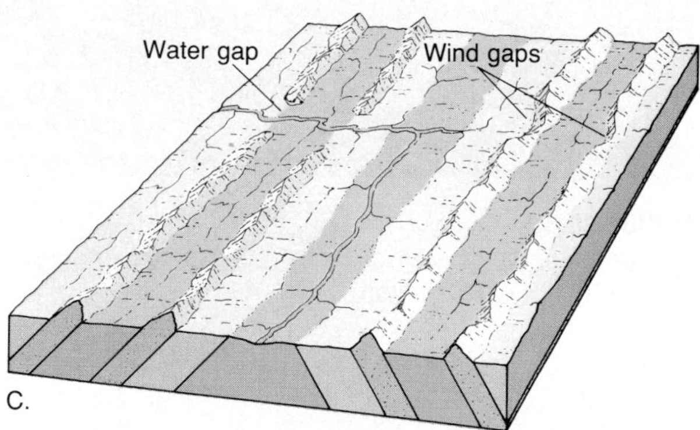

FIGURE 10.32
Stream piracy and the formation of wind gaps. Stream *B* erodes headward until it eventually captures and diverts stream *A.* Two water gaps through which stream *A* flowed are abandoned because of the piracy. As a result, these features are now wind gaps. In this Valley and Ridge-type setting, the softer rocks in the valleys are eroded more easily than the resistant ridges. Consequently, as the valleys are lowered, the ridges and wind gaps become elevated relative to the valleys.

STAGES OF VALLEY DEVELOPMENT

Contrary to the popular belief of their day, James Hutton and John Playfair proposed that streams were responsible for cutting the valleys in which they flowed. Later geologic work conducted in stream valleys substantiated Hutton's and Playfair's pronouncements and further revealed that the development of stream valleys progresses in a somewhat orderly fashion. The evolution of a valley has been arbitrarily divided into three sequential stages: youth, maturity, and old age.

As long as the stream is downcutting to establish a graded condition with its base level, its valley is considered youthful. Rapids, an occasional waterfall, and a narrow, V-shaped valley are all visible signs of the vigorous downcutting that is going on. Other features of a youthful valley include a steep gradient, little or no floodplain, and a rather straight course without meanders (Figure 10.33A).

When a stream reaches maturity, downward erosion diminishes and lateral erosion dominates. Thus the mature stream begins actively cutting its floodplain and meandering upon it (Figure 10.33B). During the mature stage cutoffs occur, producing oxbows, and a few streams may even produce natural levees (Figure 10.33C). In contrast to the gradient of a youthful stream, the gradient of a mature stream is much lower, and the profile is much smoother, since all rapids and waterfalls have been eliminated.

A stream enters old age after it has cut its floodplain several times wider than its meander belt, which is the width of the meander (Figure 10.33D). When this stage is reached the stream is rarely near the valley walls; hence, it ceases to significantly enlarge the floodplain. For that reason, the primary work of a river in an old-age valley is the reworking of unconsolidated floodplain deposits. Because this task is easier than cutting bedrock, a stream in an old-age valley shifts more rapidly than a stream in a mature valley. For example, some meanders in the lower Mississippi valley move 20 meters (66 feet) a year, and its large floodplain is dotted with oxbow lakes and old cutoffs. Natural levees are also common features of old-age valleys and, when present, are accompanied by back swamps and yazoo tributaries.

Thus far we have assumed that the base level of a stream remains constant as a river progresses from youth to old age. On many occasions, however, the land is uplifted. The effect of uplifting on a youthful stream is to increase its gradient and accelerate its

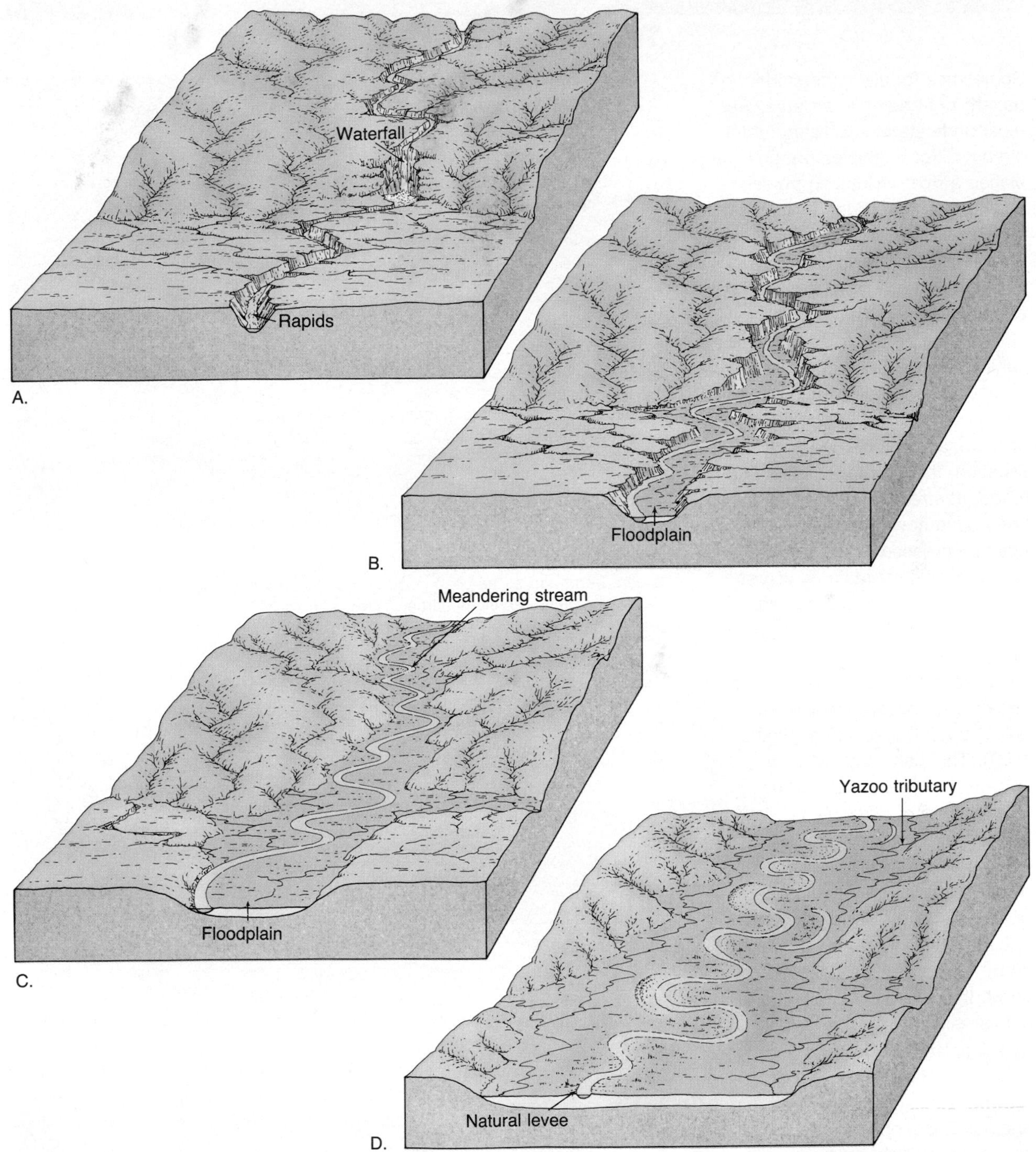

FIGURE 10.33
Stages of valley development. **A.** Youth. The youthful stage is characterized by downcutting and a V-shaped valley. **B.** and **C.** Maturity. Once a stream has sufficiently lowered its gradient, it begins to erode laterally, producing a wide valley. **D.** Old age. After the valley has been cut several times wider than the width of the meander belt, it has entered old age. (After Ward's Natural Science Establishment, Inc., Rochester, N.Y.)

BOX 10.3

Formation of a Water Gap

Sometimes to fully understand the pattern of streams in an area, we must understand the history of the streams. For example, in many places a river valley can be seen cutting through a ridge or mountain that lies across its path. The steep-walled notch followed by the river through the structure is called a *water gap* (Figure 10.C).

Why does a stream cut across such a structure and not flow around it? One possibility is that the stream existed before the ridge or mountain was formed. In this situation, the stream, called an *antecedent stream,* would have to keep pace downcutting while the uplift progressed. That is, the stream would maintain its course as folding or faulting raised an area of the crust across the path of the stream.

A second possibility is that the stream was *superposed,* or let down, upon the structure (Figure 10.D). This can occur when a ridge or mountain is buried beneath layers of relatively flat-lying sediments or sedimentary strata. Streams originating on this cover would establish their courses without regard to the structures below. Then, as the valley was deepened and the structure was encountered, the river would continue to cut its valley into it. The folded Appalachians provide some good examples. Here, a number of major rivers, such as the Potomac and the Susquehanna, are superposed streams that cut across the folded strata on their way to the Atlantic.

FIGURE 10.C
Harpers Ferry gap at the confluence of the Shenandoah and Potomac rivers near the West Virginia–Maryland border. Water gaps such as this one are common in parts of the Appalachians. (Photo by John S. Shelton)

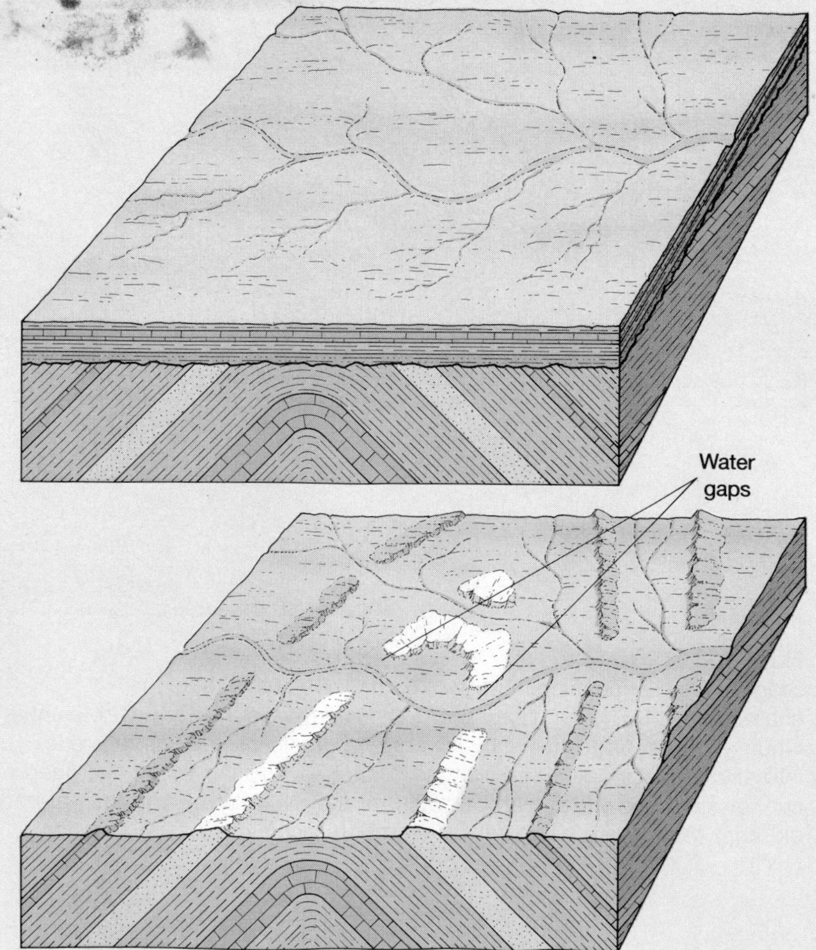

FIGURE 10.D
Development of a superposed stream. The river establishes its course on relatively uniform strata; it then encounters and cuts through the underlying structure.

A.

B.

FIGURE 10.34
Entrenched meanders. **A.** This high-altitude image shows entrenched meanders of the Delores River in western Colorado. (Courtesy of USDA-ASCS) **B.** A close-up view of entrenched meanders of the San Juan River in southern Utah. (Photo by John S. Shelton) In both places, meandering streams began downcutting because of the uplift of the Colorado Plateau.

rate of downcutting. However, uplifting of a mature stream would cause it to abandon lateral erosion and revert to downcutting. Rivers of this type are said to be **rejuvenated**, and the meanders are known as **entrenched meanders** (Figure 10.34). Mature streams may eventually readjust to uplift by cutting a new floodplain at a level below the old one. The remnants of the old floodplain are often present in the form of flat surfaces called **terraces**.

Two additional points concerning valley development should be made. First, the time required for a stream to reach any given stage depends on several factors, including the erosive ability of the stream, the nature of the material through which the stream must cut, and its height above base level. Consequently, a stream which starts out very near base level and has to cut through only unconsolidated sediments may reach maturity in a matter of a few hundred years. On the other hand, the Colorado River, where it is actively cutting the Grand Canyon, has retained its youthful nature for an estimated 5 to 6 million years. Second, individual portions of a stream reach each stage at different times. Often the lower reaches of a stream attain maturity or old age while the headwaters are still youthful in character.

CYCLE OF LANDSCAPE EVOLUTION

While streams are cutting their valleys they simultaneously sculpture the land. To describe this unending process we will need a starting point. For this

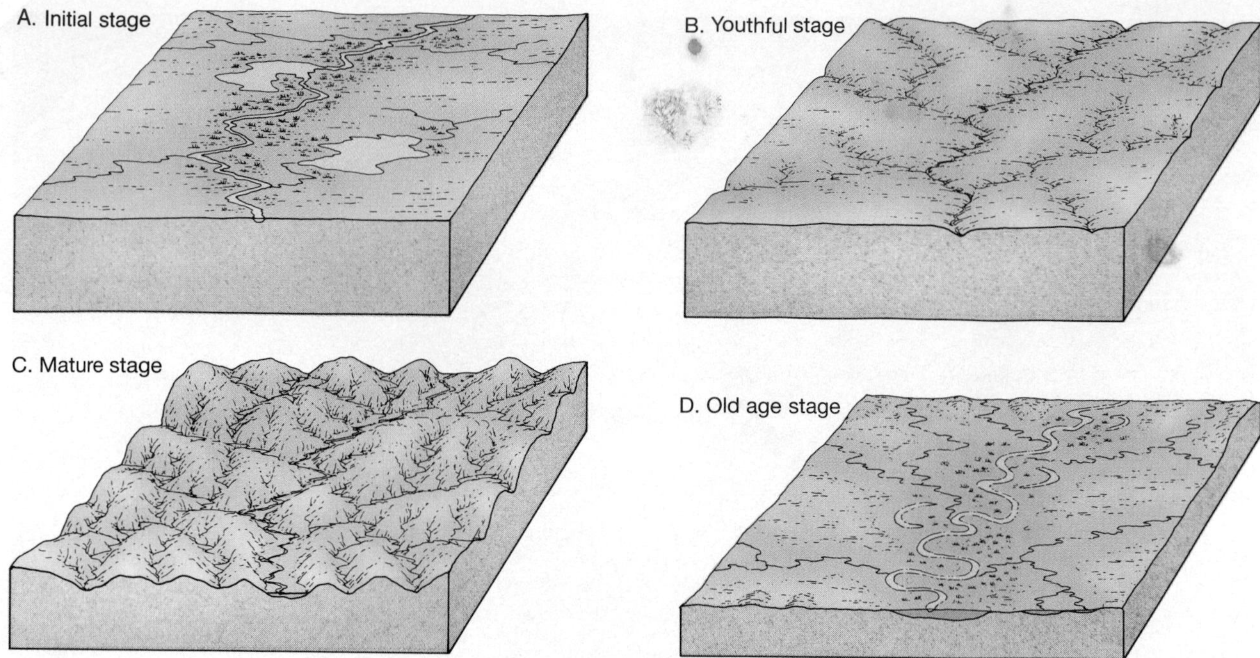

FIGURE 10.35
Idealized cycle of landscape evolution. **A.** Initial stage. The land is poorly drained and situated well above base level. **B.** Youthful stage. Streams have cut downward and drained the lakes. The landscape is still relatively flat between stream valleys. **C.** Mature stage. All of the area between the initial streams has been eroded by running water. Most of the landscape is in slope, and maximum relief exists. **D.** Old-age stage. Lateral erosion, mass wasting, and slope wash have lowered most of the hills to the level of the floodplains. The entire surface has become an undulating plain near base level. (Courtesy of Ward's Natural Science Establishment, Inc., Rochester, N.Y.)

reason only, we will assume the existence of a relatively flat upland area in a humid region. Until a well-established drainage system forms, lakes and ponds will occupy any depressions that exist (Figure 10.35A). As streams form and begin to downcut toward base level they will drain the lakes. During the youthful stage the landscape retains its relatively flat surface, interrupted only by narrow stream valleys (Figure 10.35B). As downcutting continues, relief increases, and the flat, youthful landscape is transformed into one consisting of the hills and valleys which characterize the mature stage (Figure 10.35C).

Eventually some of the streams will approach base level, and downcutting will give way to lateral erosion. As the cycle nears the old-age stage the effects of sheet flow and mass wasting, coupled with the lateral erosion by streams, will reduce the land to a **peneplain** ("near plain"), an undulating plain near base level (Figure 10.35D). Although no peneplains are known to exist today, there is evidence that they formed in the past and have since been uplifted. Once the peneplain has formed, uplifting starts the cycle over again. Most often, uplifting interrupts the cycle before it reaches old age.

REVIEW QUESTIONS

1. Describe the movement of water through the hydrologic cycle. Once precipitation has fallen on land, what paths are available to it?
2. Over the oceans the quantity of water lost to evaporation is not equaled by the amount gained from precipitation. Why then does the sea level not drop?
3. List several factors that influence infiltration capacity.
4. "Water in streams moves primarily in laminar flow." Briefly explain whether this statement is true or false.
5. A stream originates at 2000 meters above sea level and travels 250 kilometers to the ocean. What is its average gradient in meters per kilometer?
6. Suppose that the stream mentioned in Question 5 developed extensive meanders so that its course was lengthened to 500 kilometers. Calculate its new gradient. How does meandering affect gradient?
7. When the discharge of a stream increases, what happens to the stream's velocity?
8. What typically happens to channel width, channel depth, velocity, and discharge from the point where a stream begins to the point where it ends? Briefly explain why these changes take place.
9. When an area changes from being predominantly rural to largely urban, how is streamflow affected? (See Box 10.1.)
10. Define *base level*. Name the main river in your area. For what streams does it act as base level? What is the base level for the Mississippi River?
11. Why do most streams have low gradients near their mouths?
12. Describe three ways in which a stream may erode its channel. Which one of these is responsible for creating potholes?
13. If you were to collect a jar of water from a stream, what part of the load would settle to the bottom of the jar? What portion would remain in the water? What part of a stream's load would probably not be present in your sample?
14. What is settling velocity? What factors influence settling velocity?
15. Distinguish between capacity and competency.
16. The water velocity needed to lift a tiny clay particle is greater than that required to set much larger sand grains in motion. Explain.
17. Describe a situation that might cause a stream channel to become braided.
18. Briefly describe the formation of a natural levee. How is this feature related to back swamps and yazoo tributaries?
19. In what way is a delta similar to an alluvial fan? In what way are they different?
20. Why does a river flowing across a delta eventually change course?
21. How has the construction of artificial levees and dams on the Mississippi River and its tributaries contributed to a shrinking of the Mississippi's delta and its extensive wetlands? (See Box 10.2.)
22. For what purpose are artificial cutoffs made?
23. Each of the following statements refers to a particular drainage pattern. Identify the pattern.
    (a) Streams diverging from a central high area such as a dome.
    (b) Branching, "treelike" pattern.
    (c) A pattern that develops when bedrock is crisscrossed by joints and faults.
24. Describe how a water gap might form. (See Box 10.3.)
25. Why is it possible for a youthful valley to be older (in years) than a mature valley?
26. Do rivers flowing in mature and old-age valleys make good political boundaries? Explain.

KEY TERMS

alluvial fan (p. 239)
alluvium (p. 237)
back swamp (p. 239)
bar (p. 237)
base level (p. 230)
bed load (p. 233)
braided stream (p. 237)
capacity (p. 235)
competence (p. 236)
cut bank (p. 247)
cutoff (p. 247)
delta (p. 239)
dendritic pattern (p. 249)
discharge (p. 228)
dissolved load (p. 233)
distributary (p. 239)
divide (p. 249)
drainage basin (p. 248)
entrenched meander (p. 255)
evapotranspiration (p. 225)
floodplain (p. 238)
graded stream (p. 231)
gradient (p. 227)
head (headwaters) (p. 229)
headward erosion (p. 251)
hydrologic cycle (p. 224)
infiltration (p. 225)
infiltration capacity (p. 226)
lag time (p. 232)
laminar flow (p. 226)
local (temporary) base level (p. 230)
longitudinal profile (p. 229)
meander (p. 247)
meander scar (p. 247)
mouth (p. 229)
natural levee (p. 239)
oxbow lake (p. 247)
peneplain (p. 256)
Playfair's law (p. 244)
point bar (p. 237)
pothole (p. 233)
radial pattern (p. 251)
rapids (p. 245)
rectangular pattern (p. 251)
rejuvenated stream (p. 255)
rills (p. 226)
runoff (p. 225)
saltation (p. 235)
settling velocity (p. 235)
sheet flow (p. 226)
sorting (p. 237)
stream (p. 226)
stream piracy (p. 251)
suspended load (p. 233)
temporary (local) base level (p. 230)
terrace (p. 255)
transpiration (p. 225)
trellis pattern (p. 251)
turbulent flow (p. 226)
ultimate base level (p. 230)
waterfall (p. 245)
yazoo tributary (p. 239)

11
Groundwater

Opposite: Castle Geyser erupts in Yellowstone National Park, Wyoming. (Photo by Peter Kresan) (Top photo by E. J. Tarbuck)

Of all the world's water only about six-tenths of one percent is found underground. Nevertheless, the amount of water stored in the rocks and sediments beneath the earth's surface is vast. When the oceans are excluded and only sources of freshwater are considered, the significance of groundwater becomes more apparent. Table 11.1 contains estimates of the distribution of freshwater in the hydrosphere. Clearly the largest volume occurs as glacial ice. Second in rank is groundwater, with slightly more than 14 percent of the total. However, when ice is excluded and just liquid water is considered, more than 94 percent is groundwater.

In many parts of the world, wells and springs provide the water needs not only for great numbers of people but also for crops, livestock, and industry. In the United States, groundwater is the source of about 40 percent of the water used for all purposes exclusive of hydroelectric power generation and power-plant cooling. It provides drinking water for more than 50 percent of the population as well as 40 percent of the water for irrigation and 26 percent of industry's needs. In some areas, however, overuse of this basic resource has resulted in streamflow depletion, land subsidence, saltwater intrusion, and increased pumping costs. In addition, groundwater

FIGURE 11.1
The cathedral-like beauty of the Big Room at Carlsbad Caverns National Park, New Mexico, is the handiwork of groundwater. (Photo courtesy of the National Park Service)

TABLE 11.1
Freshwater of the hydrosphere.

Parts of the Hydrosphere	Volume of Freshwater (km^3)	Share of Total Volume of Freshwater (percent)	Rate of Water Exchange
Ice sheets and glaciers	24,000,000	84.945	8000 years
Groundwater	4,000,000	14.158	280 years
Lakes and reservoirs	155,000	0.549	7 years
Soil moisture	83,000	0.294	1 year
Water vapor in the atmosphere	14,000	0.049	9.9 days
River water	1,200	0.004	11.3 days
Total	28,253,200	100.000	

SOURCE: U.S. Geological Survey Water Supply Paper 2220, 1987.

contamination due to human activities is a real and growing threat in many places.

Geologically, groundwater is important as an erosional agent. The dissolving action of groundwater is responsible for producing the surface depressions known as sinkholes as well as creating subterranean caverns (Figure 11.1). Another significant role is as an equalizer of streamflow. Much of the water that flows in rivers is not transmitted directly to the channel after falling as rain. Rather, a large percentage soaks in and then moves slowly underground to stream channels. Groundwater is thus a form of storage that sustains streams during periods when rain does not fall. The information in Table 11.1 reinforces this point. Here we see that the rate of exchange for groundwater is 280 years. This figure represents the amount of time required to replace the water now stored underground. By contrast, the rate of water exchange for rivers is just slightly more than 11 days: if the groundwater supply to a river were cut off and no rain fell, the river would run dry in just over 11 days. Thus, when we see water flowing in a river during a dry period, it represents rain that fell at some earlier time and was stored underground.

DISTRIBUTION OF UNDERGROUND WATER

When rain falls, some of the water runs off, some evaporates, and the remainder soaks into the ground. This last path is the primary source of practically all subsurface water. The amount of water that takes each of these paths, however, varies greatly both in time and space. Several influential factors include steepness of slope, nature of surface material, intensity of rainfall, and type and amount of vegetation. Heavy rains falling upon steep slopes underlain by impervious materials will obviously result in a high percentage of runoff. On the other hand, if rain falls steadily and gently upon more gradual slopes composed of materials that are easily penetrated by the water, a much larger percentage of water soaks into the ground.

Some of the water that soaks in does not travel far, because molecular attraction holds it as a film on the surface of soil particles. A portion of this moisture evaporates back into the atmosphere, while much of the remainder serves as a source of water for use by plants between rains. Water that is not held in this **belt of soil moisture** penetrates downward until it reaches a zone where all of the open spaces in sediment and rock are completely filled with water (Figure 11.2). The water in this saturated zone is called **groundwater**. The upper limit of this zone is known as the **water table**. Extending upward from the water table is the **capillary fringe**. Here groundwater is lifted against gravity by surface tension in tiny threadlike passages between grains of soil or sediment. Capillary action is easily demonstrated by placing the corner of a paper towel in some water and watching the liquid move upward. The area above the water table that includes the capillary fringe and the belt of soil moisture is called the **zone of aeration**. Unlike the **zone of saturation** below, the spaces in the zone of aeration are unsaturated and filled mainly with air.

THE WATER TABLE

The water table, the upper limit of the zone of saturation, is a very significant feature of the groundwater system. The water table level is important in predicting the productivity of wells, explaining the changes in the flow of springs and streams, and accounting for fluctuations in the levels of lakes.

Although we cannot observe the water table directly, its position can be mapped and studied in de-

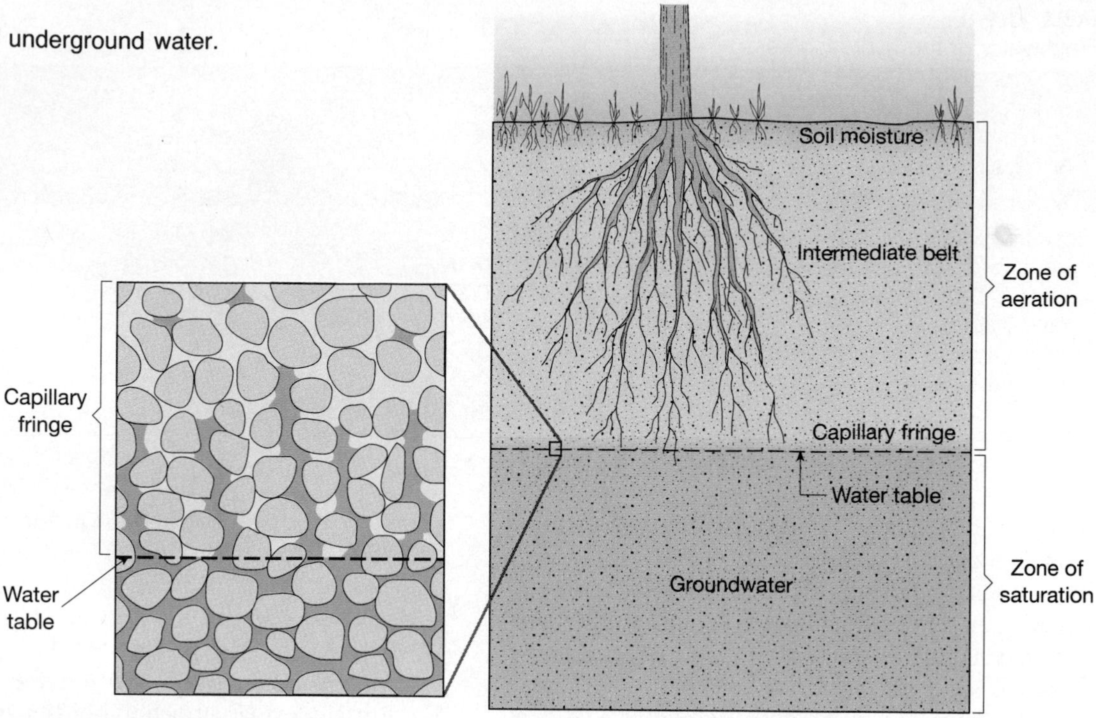

FIGURE 11.2
Distribution of underground water.

tail in areas where wells are numerous because the water level in wells coincides with the upper boundary of the saturated zone. Such maps reveal that the water table is rarely level as we might expect a table to be. Instead, its shape is usually a subdued replica of the surface topography, reaching its highest elevations beneath hills and then descending toward valleys (Figure 11.3). Where a swamp is encountered, the water table is right at the surface, while lakes and streams generally occupy areas where the water table is above the land surface.

A number of factors contribute to the irregular surface of the water table. For example, variations in rainfall and permeability from place to place can lead to uneven infiltration and thus to differences in the water table level. However, the most important cause is simply the fact that groundwater moves very slowly and at varying rates under different conditions. Because of this, water tends to "pile up" beneath high areas between stream valleys. If rainfall were to cease completely, these water table "hills" would slowly subside and gradually approach the level of the val-

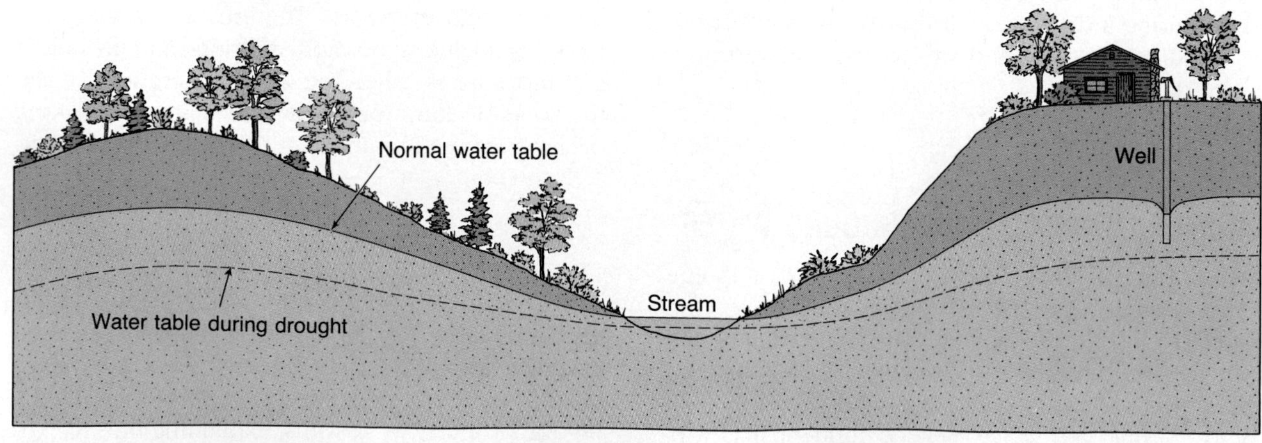

FIGURE 11.3
The shape of the water table is usually a subdued replica of the surface topography. During periods of drought, the water table falls, reducing streamflow and drying up some wells.

leys. However, new supplies of rainwater are usually added frequently enough to prevent this. Nevertheless, in times of extended drought, the water table may drop enough to dry up shallow wells (Figure 11.3).

The relationship between the water table and a stream in a humid region is illustrated in Figure 11.4A. Even during dry periods, the movement of groundwater into the channel maintains a flow in the stream. In situations such as this, streams are said to be **effluent.** By contrast, in arid regions, where the water table is far below the surface, groundwater does not contribute to streamflow. Therefore, the only permanent streams in such areas are those that originate in wet regions and then happen to traverse the desert. Under these conditions the zone of saturation below the valley floor is supplied by downward seepage from the stream channel which, in turn, produces an upward bulge in the water table. Streams that provide water to the water table in this manner are called **influent streams** (Figure 11.4B).

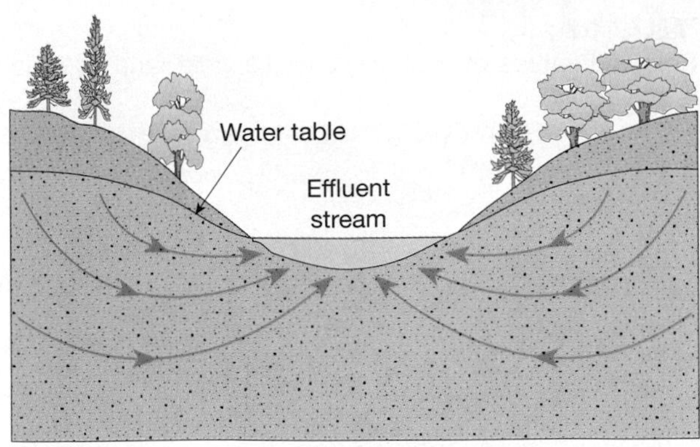

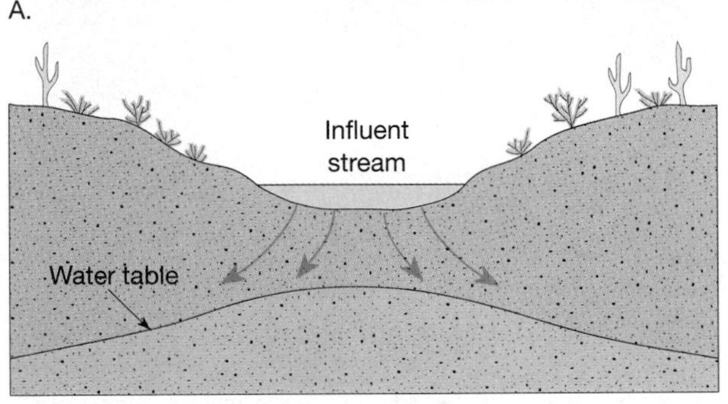

FIGURE 11.4
A. Effluent streams are characteristic of humid areas and are supplied by water from the zone of saturation. **B.** Influent streams are found in deserts. Seepage from such streams produces an upward bulge in the water table.

POROSITY AND PERMEABILITY

Depending upon the nature of the subsurface material, the flow of groundwater and the amount of water that can be stored are highly variable. Water soaks into the ground because bedrock, sediment, and soil contain voids or openings. These openings are similar to those of a sponge and are often called *pore spaces.* The quantity of groundwater that can be stored depends on the **porosity** of the material; that is, the percentage of the total volume of rock or sediment that consists of pore spaces. Although these openings often consist of spaces between particles of sediment or sedimentary rock, such features as vesicles (voids left by gases escaping from lava), joints and faults, and cavities formed by the solution of soluble rocks like limestone are also common.

As one might expect, variations in porosity can be great. Sediment is commonly quite porous, and open spaces may occupy from 10 to 50 percent of the sediment's total volume. The amount of pore space depends on the size and shape of the grains, as well as the packing, degree of sorting, and in the case of sedimentary rocks, the amount of cementing material. For example, clay may have a porosity as high as 50 percent, whereas in some gravels, voids make up only 20 percent of the material's volume. Where sediments of various sizes are mixed, the porosity is reduced because the finer particles tend to fill the openings between the larger grains. Most igneous and metamorphic rocks, as well as some sedimentary rocks, are composed of tightly interlocking crystals. The amount of open space between the grains may be negligible. Therefore, if these rocks are to have greater porosity, fractures must provide a significant proportion of the open space.

Porosity alone is not a satisfactory measure of a material's ability or capacity to yield groundwater. Rock or sediment may be very porous and still not allow water to move through it. The **permeability** of a material, its ability to transmit a fluid, is also very important. Groundwater moves by twisting and turning through small openings. The smaller the pore spaces, the slower the water moves. If the spaces between particles are very small, the films of water clinging to the grains will come in contact or overlap. As a result, the force of molecular attraction binding the water to the particles extends across the opening and the water is held firmly in place. This idea is illustrated nicely by examining the information about the water-yielding potential of different materials in Table 11.2. Here groundwater is divided into two categories: (1) that portion which will drain under the influence of gravity (called *specific yield*) and (2) that part which is retained as a film on particle and rock surfaces and in tiny openings (called *specific retention*). Specific yield represents how

TABLE 11.2
Selected values of porosity, specific yield, and specific retention.*

Material	Porosity	Specific Yield	Specific Retention
Soil	55	40	15
Clay	50	2	48
Sand	25	22	3
Gravel	20	19	1
Limestone	20	18	2
Sandstone (semiconsolidated)	11	6	5
Granite	0.1	0.09	0.01
Basalt (fresh)	11	8	3

*Values in percent by volume
SOURCE: U.S. Geological Survey Water Supply Paper 2220, 1987.

much water is actually available for use, whereas specific retention indicates how much water remains bound to the material. For example, although clay's ability to store water is high, its pore spaces are so small that water is unable to move. That is, its porosity is high but because its permeability is poor, clay has a very low specific yield. Impermeable layers composed of materials such as clay that hinder or prevent water movement are termed **aquicludes.** On the other hand, larger particles, such as sand or gravel, have larger pore spaces. Therefore, the water in the centers of the openings is not bound to the particles by molecular attraction and can move with relative ease. Permeable rock strata or sediment that transmit groundwater freely are called **aquifers.**

In summary, we have seen that porosity is not always a reliable guide to the amount of groundwater that can be produced and the property of permeability is a significant factor in determining the rate of groundwater movement and the quantity of water that might be pumped from a well.

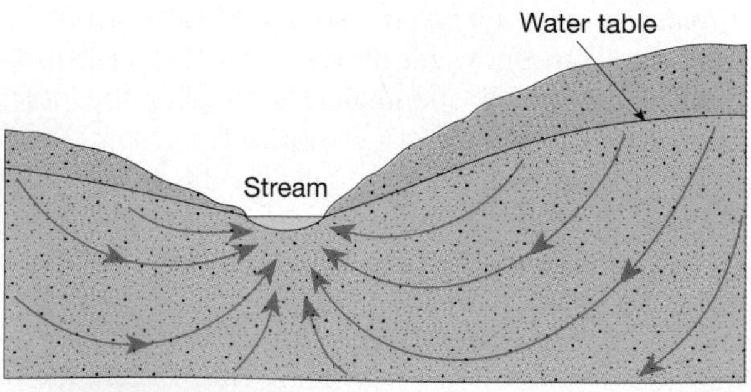

FIGURE 11.5
Arrows indicate groundwater movement through uniformly permeable material. The looping curves may be thought of as a compromise between the downward pull of gravity and the tendency of water to move toward areas of reduced pressure.

MOVEMENT OF GROUNDWATER

One relatively common misconception regarding groundwater is that it occurs in underground rivers that resemble surface streams. Although subsurface streams do exist, they are not common. Rather, as we learned in the preceding sections, most groundwater must migrate through the pore spaces in rock and sediment. Thus, contrary to any impressions of rapid flow that an underground river might evoke, the movement of most groundwater is exceedingly slow.

The energy responsible for groundwater movement is provided by the force of gravity. In response to gravity, water moves from areas where the water table is high to zones where the water table is lowest; that is, toward a stream channel, lake, or spring. Although some water takes the most direct path down the slope of the water table, much of the water follows long curving paths toward the zone of discharge. As illustrated in Figure 11.5, water percolates into the stream from all possible directions, with some of the paths turning upward, apparently against the force of gravity, and entering through the bottom of the channel. Such paths are followed because differences in the height of the water table create differences in the groundwater pressure at a particular height. Stated another way, the water at any given height is under greater pressure beneath a hill than beneath a stream channel and the water tends to migrate toward points of lower pressure. Thus, the looping curves followed by water in the saturated zone may be thought of as a compromise between the downward pull of gravity and the tendency of water to move toward areas of reduced pressure.

The modern concepts of groundwater movement were formulated in the middle of the nineteenth century. During this period Henry Darcy, a French engineer studying the water supply of the city of Dijon in east-central France, formulated a law that now bears his name and is basic to an understanding of groundwater movement. Darcy found that if permeability remains uniform, the velocity of groundwater will increase as the slope of the water table increases. The water table slope, known as the **hydraulic gradient**, is determined by dividing the vertical difference between the recharge and discharge points (a quantity known as the **head**) by the length of flow between these points. **Darcy's law** can be expressed by the following formula:

$$V = K\frac{h}{l}$$

FIGURE 11.6
Thousand Springs, Gooding County, Idaho. (Photo by James E. Patterson)

where V represents velocity, h the head, l the length of flow, and K a coefficient that accounts for a material's permeability.

Rates of groundwater movement have been determined directly in a number of ways. In one experiment dye is introduced into a well and the time is measured until the coloring agent appears in another well at a known distance from the first. Measurements of groundwater movement have also been made by applying radiometric dating techniques, specifically by using the radioactive isotope of carbon, carbon-14. Upon entering the ground the carbon dioxide dissolved in rainwater contains a characteristic amount of carbon-14. As the water gradually percolates through the ground, the radioactive carbon decays. The rate of movement is determined by measuring the distance between the well where the water was withdrawn and the recharge area, and dividing this by the radiometrically determined age.*

Experiments such as those just described have shown that the rate of groundwater movement is highly variable. Although a typical rate for many aquifers is thought to be about 15 meters per year (about 4 centimeters per day), velocities more than 15 times this figure have been measured in exceptionally permeable materials.

SPRINGS

Springs have aroused the curiosity and wonder of people for thousands of years. The fact that springs were, and to some people still are, rather mysterious phenomena is not difficult to understand, for here is water flowing freely from the ground in all kinds of weather in seemingly inexhaustible supply, but with no obvious source. As a result, some rather interesting (although incorrect) explanations for the source of springs were proposed. Some of these have managed to live on to the present. One such explanation is that springs draw their water from the ocean. However, just how the salt is removed and how the water is elevated to the great heights it reaches in mountain springs remain unanswered questions. Another explanation for springs, one supported by Aristotle, suggested that the water originated in cold subterranean caverns where water vapor condensed from the air that had penetrated the earth and produced the needed water supply.

Not until the middle of the 17th century did the French physicist Pierre Perrault invalidate the age-old assumption that precipitation could not adequately account for the amount of water emanating from springs and flowing in rivers. Over a period of years, Perrault computed the quantity of water that fell on the Seine River basin. He then calculated the mean annual runoff by measuring the river's discharge, and after allowing for the loss of water by evaporation, he showed that there was sufficient water remaining to feed the springs. Thanks to Perrault's pioneering efforts and the measurements by many afterward, we now know that the source of springs is water from the zone of saturation and that the ultimate source of this water is precipitation.

When the water table intersects the earth's surface, a natural flow of groundwater results, which we call a **spring** (Figure 11.6). Springs such as those pictured in Figure 11.6 form when an aquiclude blocks the downward movement of groundwater and forces it to move laterally. Where the permeable bed outcrops in a valley, a line of springs results. In this particular example the aquifer consists of jointed and vesicular basaltic lava flows that form the well-known

*A discussion of radiometric dating appears in Chapter 8.

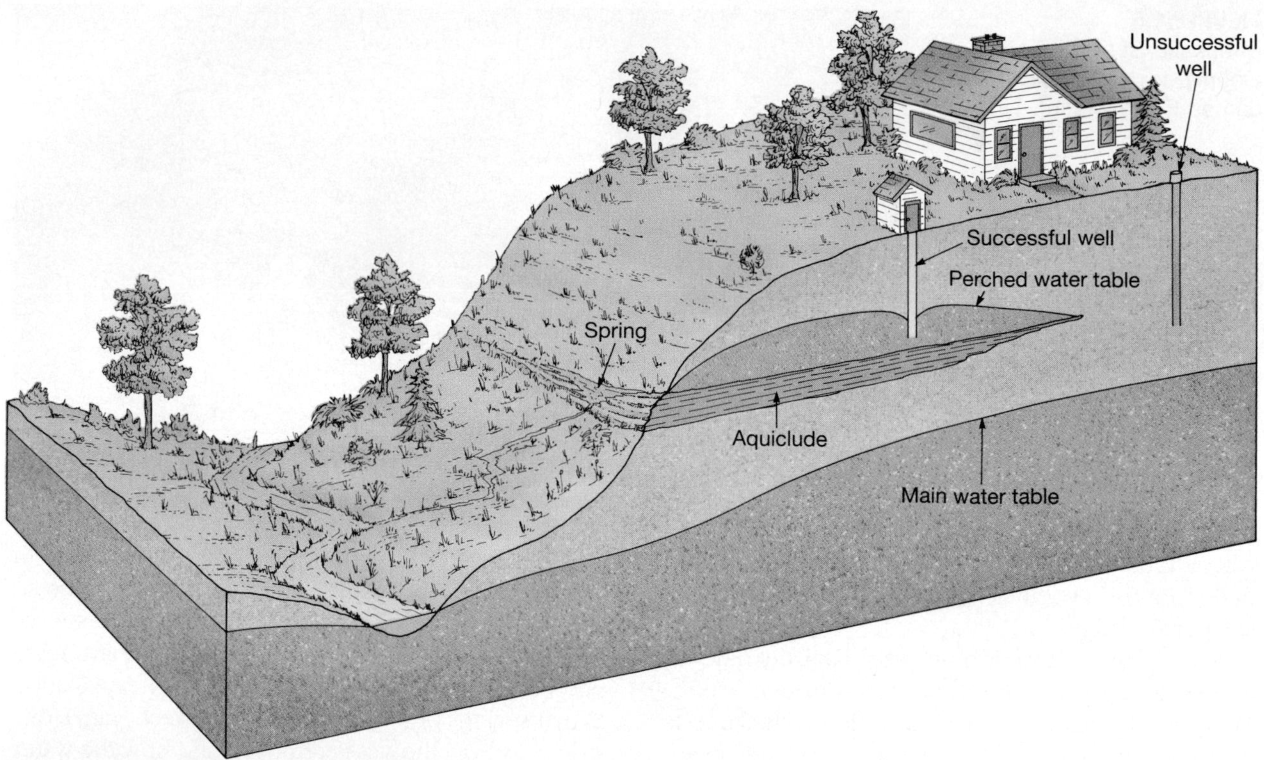

FIGURE 11.7
When an aquiclude is situated above the main water table, a localized zone of saturation may result. Where the perched water table intersects the side of the valley, a spring flows. The perched water table also caused the well on the left to be successful, whereas the well on the right will be unsuccessful unless it is drilled to a greater depth.

Thousand Springs along the Snake River in southern Idaho. Another situation leading to the formation of a spring is illustrated in Figure 11.7. Here an aquiclude is situated above the main water table. As water percolates downward, a portion of it is intercepted by the aquiclude, thereby creating a localized zone of saturation and a **perched water table**. Springs, however, are not confined to places where a perched water table creates a flow at the surface. Indeed, there are a wide variety of spring types because subsurface conditions vary greatly from place to place. Even in areas underlain by impermeable crystalline rocks, permeable zones may exist in the form of fractures or solution channels. If these openings fill with water and intersect the ground surface along a slope, a spring will result.

WELLS

The most common device used by humans for removing groundwater is the **well**, an opening bored into the zone of saturation. Wells serve as reservoirs into which groundwater moves and from which it can be pumped to the surface. Digging for water dates back many centuries and continues to be an important method of obtaining water today. By far the single greatest use of this water in the United States is for irrigation. More than 65 percent of the groundwater used each year is for this purpose. Industrial uses rank a distant second, followed by the amount used in cities and rural areas.

The water table level may fluctuate considerably during the course of a year, dropping during dry seasons and rising following periods of rain. Therefore, to insure a continuous supply of water, a well should penetrate below the water table. Whenever water is withdrawn from a well, the water table in the vicinity of the well is lowered. The extent of this effect, which is termed **drawdown**, decreases with increasing distance from the well. The result is a depression in the water table, roughly conical in shape, known as a **cone of depression** (Figure 11.8). Since the cone of depression increases the hydraulic gradient near the well, groundwater will flow more rapidly toward the opening. For most small domestic wells, the cone of

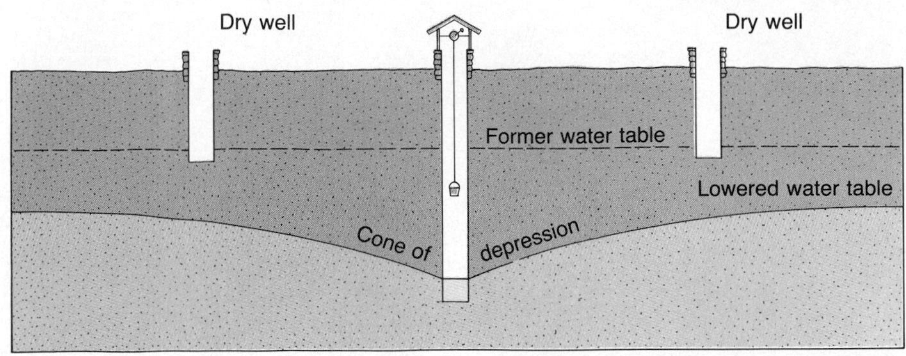

FIGURE 11.8
A cone of depression in the water table often forms around a pumping well. If heavy pumping lowers the water table, the shallow wells may be left dry.

depression is hardly appreciable. However, when wells are used for irrigation or for industrial purposes, the withdrawal of water can be great enough to create a very wide and steep cone of depression that may substantially lower the water table in an area and cause nearby shallow wells to become dry (Figure 11.8).

Digging a successful well is a familiar problem for people in areas where groundwater is the primary source of supply. One well may be successful at a depth of 10 meters (33 feet) whereas a neighbor may have to go twice as deep to find an adequate supply. Still others may be forced to go deeper or try a different site altogether. When subsurface materials are heterogenous, the amount of water a well is capable of providing may vary a great deal over short distances. For example, when two nearby wells are drilled to the same level and only one is successful, it may be caused by the presence of a perched water table beneath one of them. Such a case is shown in Figure 11.7. Massive igneous and metamorphic rocks provide a second example. These crystalline rocks are usually not very permeable except where they are cut by many intersecting joints and fractures. Therefore, when a well drilled into such rock does not intersect an adequate network of fractures, it is likely to be unproductive.

ARTESIAN WELLS

To many people the term *artesian* is applied to any well drilled to great depths. This use of the term is incorrect. Others believe that an artesian well must flow freely at the surface (Figure 11.9). Although this is a more correct notion than the first, it represents too narrow a definition. The term **artesian** may be applied correctly to any situation in which groundwater under pressure rises above the level of the aquifer. As we shall see, this does not always mean a free-flowing surface discharge.

FIGURE 11.9
Sometimes water flows freely at the surface when an artesian well is developed. However, for most artesian wells the water must be pumped to the surface. (Photo by James E. Patterson)

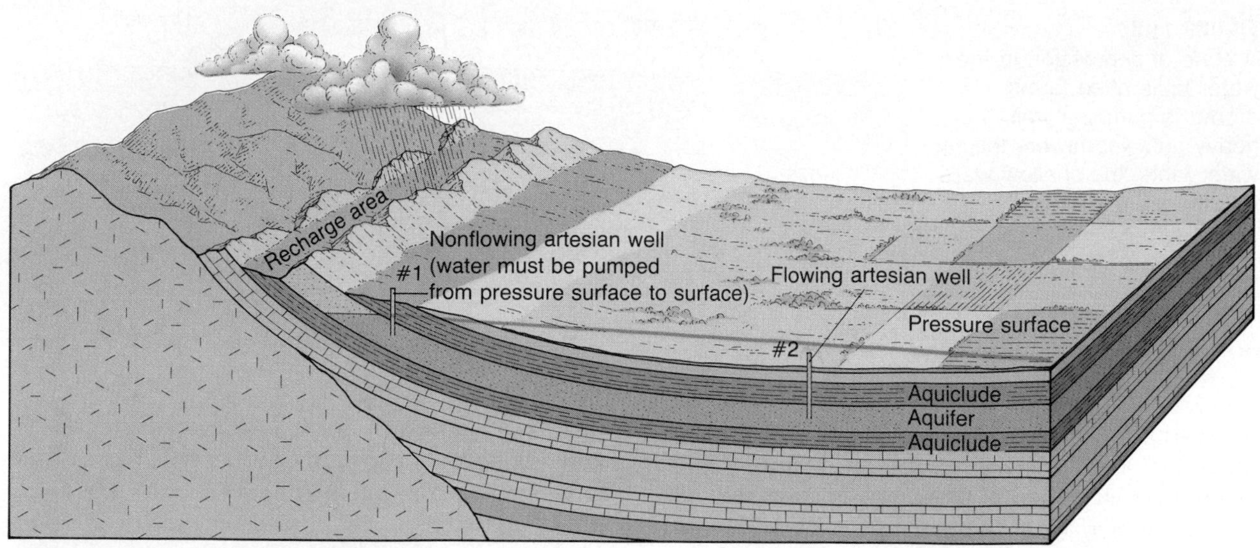

FIGURE 11.10
Artesian systems occur when an inclined aquifer is surrounded by impermeable beds.

For an artesian system to exist, two conditions must be met (Figure 11.10): (1) water must be confined to an aquifer that is inclined so that one end can receive water; and (2) aquicludes, both above and below the aquifer, must be present to prevent the water from escaping. When such a layer is tapped, the pressure created by the weight of the water above will cause the water to rise. If there were no friction, the water in the well would rise to the level of the water at the top of the aquifer. However, friction reduces the height of this pressure surface. The greater the distance from the recharge area (area where water enters the inclined aquifer), the greater the friction and the less the rise of water. In Figure 11.10, well 1 is a **nonflowing artesian well,** because at this location the pressure surface is below ground level. When the pressure surface is above the ground and a well is drilled into the aquifer, a **flowing artesian well** is created (well 2, Figure 11.10). It is important to realize that not all artesian systems are wells. Artesian springs also exist. Here groundwater reaches the surface by rising through a natural fracture rather than through an artificially produced hole.

Artesian systems act as conduits, often transmitting water great distances from remote areas of recharge to points of discharge. A well-known artesian system in South Dakota is a good example of this. In the western part of the state, the edges of a series of sedimentary layers have been bent up to the surface along the flanks of the Black Hills. One of these beds, the permeable Dakota Sandstone, is sandwiched between impermeable strata and gradually dips into the ground toward the east. When the aquifer was

FIGURE 11.11
A "gusherlike" flowing artesian well in South Dakota in the early part of the century. Thousands of additional wells now tap the same confined aquifer; thus, the pressure has dropped to the point that many wells stopped flowing altogether and have to be pumped. (Photo by N. H. Darton, U.S. Geological Survey)

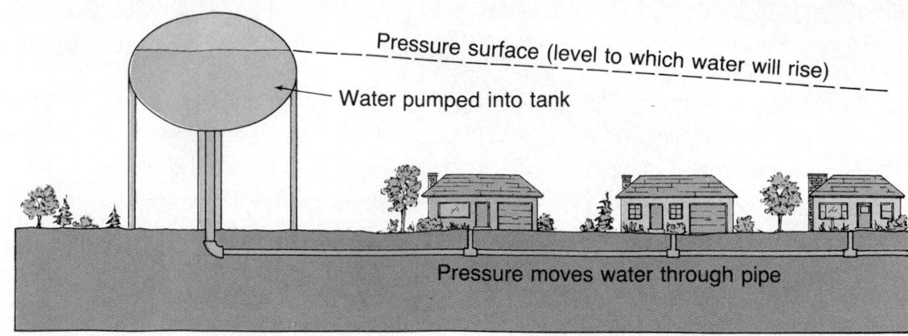

FIGURE 11.12
City water systems can be considered to be artificial artesian systems.

first tapped, water poured from the ground surface, creating fountains many meters high (Figure 11.11). In some places, the force of the water was sufficient to power waterwheels. Scenes such as the one pictured in Figure 11.11, however, can no longer occur, because thousands of additional wells now tap the same source. This depleted the reservoir and the water table in the recharge area was lowered. As a consequence, the pressure dropped to the point where many wells stopped flowing altogether and had to be pumped. On a different scale, city water systems can be considered examples of artificial artesian systems (Figure 11.12). The water tower, into which water is pumped, would represent the area of recharge, the pipes the confined aquifer, and the faucets in homes the flowing artesian wells.

PROBLEMS ASSOCIATED WITH GROUNDWATER WITHDRAWAL

As with many of our valuable natural resources, groundwater is being exploited at an increasing rate. In some areas, overuse threatens the groundwater supply. In addition, a number of costly problems related to the withdrawal of groundwater may accompany and compound the difficulties associated with tapping groundwater resources.

The tendency for many natural systems is to establish or attempt to establish a condition of equilibrium. The groundwater system is no exception. The water table level represents the balance between the rate of infiltration and the rate of discharge and withdrawal. Any imbalance will either raise or lower the water table. Long-term imbalances can lead to a significant drop in the water table if there is either a decrease in groundwater recharge, such as occurs during a prolonged drought, or an increase in the rate of groundwater discharge or withdrawal. As we would expect, water tables have gradually dropped in many areas where withdrawal has increased steadily.

The High Plains, a relatively dry region that extends from South Dakota to western Texas, provides an example. Here an extensive agricultural economy is largely dependent on irrigation. The widespread occurrence of permeable layers of sand and gravel has permitted the construction of high-yield wells almost anywhere in the region. As a result there are an estimated 168,000 wells being used to irrigate more than 65,000 square kilometers (16 million acres) of land. In the southern part of this region, which includes the Texas panhandle, the natural recharge of the aquifer is very slow and the problem of declining groundwater levels is acute. In fact, in years of average or below-average precipitation, recharge is negligible because all or nearly all of the meager rainfall is returned to the atmosphere by evaporation and transpiration. Thus, it is only during especially wet years that significant recharge of the aquifer occurs, and this averages only about 5 millimeters per year. Therefore, where intense irrigation has been practiced for an extended period, depletion of groundwater is severe. Declines in groundwater levels at rates as high as 1 meter per year have led to an overall drop in the water table of between 15 and 60 meters (50 and 200 feet) in some areas.

Subsidence

As we shall see later in this chapter, surface subsidence can result from natural processes related to groundwater. However, the ground may also sink when water is pumped from wells faster than natural recharge processes can replace it. This effect is particularly pronounced in areas underlain by thick layers of unconsolidated sediments. As the water is withdrawn, the water pressure drops and the weight of the overburden is transferred to the sediment. The greater pressure packs the sediment grains tightly together and the ground subsides.

Many areas may be used to illustrate land subsidence resulting from the excessive pumping of groundwater from relatively loose sediment. A classic example in the United States occurred in the San Joaquin Valley of California and is discussed in Box 11.1. Many other cases of land subsidence due to groundwater pumping exist in the United States, in-

BOX 11.1

Land Subsidence in the San Joaquin Valley

The San Joaquin Valley is a broad structural basin that contains a thick fill of sediments. Occupying 26,000 square kilometers (10,000 square miles), it constitutes the southern two-thirds of California's Central Valley, a flatland separating the Coast Ranges to the west and the Sierra Nevada to the east (Figure 11.A). The aquifer system consists of a heterogenous mixture of alluvial materials derived from the surrounding mountains. Sediment thicknesses average about 870 meters (2900 feet). The climate of the region is arid to semiarid, with average annual precipitation ranging from 12 to 35 centimeters (5 to 14 inches).

The San Joaquin Valley has a strong agricultural economy that requires large quantities of water for irrigation. For many years a substantial portion (up to 50 percent) of this need was met by groundwater. In addition, nearly every city in the region uses groundwater as the principal source for municipal and industrial supplies.

Although development of the valley's groundwater for irrigation began in the late nineteenth century, land subsidence did not begin until the mid-1920s after withdrawals were substantially increased. By the early 1970s, maximum subsidence exceeded 8.5 meters (29 feet) at one point in the region (Figure 11.B), and water levels had declined as much as 120 meters (400 feet). At that time, areas within the valley were subsiding at rates in excess of 0.3 meter (1 foot) per year. Then, because of the importation of surface water and a decrease in groundwater pumping, water levels in the aquifer recovered and subsidence ceased. However, during a drought in 1976–1977, heavy groundwater pumping led to a period of renewed subsidence. This time, water levels dropped at a much faster rate than during the previous period because of the reduced storage capacity caused by earlier compaction of sediments. In all, more than 13,400 square kilometers (5200 square miles) of irrigable land, one-half the entire valley, were affected by subsidence. According to the U.S. Geological Survey:

> Subsidence in the San Joaquin Valley probably represents one of the greatest single manmade alterations in the configuration of the Earth's surface. . . . It has caused serious and costly problems in construction and maintenance of water-transport structures, highways, and highway structures; also many millions of dollars have been spent on the repair or replacement of deep water wells. Subsidence, besides changing the gradient and course of valley creeks and streams, has caused unexpected flooding, costing farmers many hundreds of thousands of dollars in recurrent land leveling.*

Similar effects have been documented in the San Jose area of the Santa Clara Valley, California, where, between 1916 and 1966, subsidence in an area of intensive groundwater development approached 4 meters. Flooding of lands bordering the southern part of San Francisco Bay was one of the results. As was the case in the San Joaquin Valley, the subsidence stopped when imports of surface water were increased, allowing groundwater withdrawal rates to be decreased.

*R. L. Ireland, J. F. Poland, and F. S. Riley, *Land Subsidence in the San Joaquin Valley, California, as of 1980,* U.S. Geological Survey Professional Paper 437-I (Washington, D.C.: U.S. Government Printing Office, 1984), p. 11.

FIGURE 11.A
The shaded area shows California's San Joaquin Valley.

FIGURE 11.B
The marks on this utility pole indicate the level of the surrounding land in preceding years. Between 1925 and 1975 this part of the San Joaquin Valley subsided almost 9 meters because of the withdrawal of groundwater and the resulting compaction of sediments. (Photo courtesy of U.S. Geological Survey)

cluding the Houston and Panhandle areas of Texas as well as Las Vegas and New Orleans.

Outside the United States, one of the most spectacular examples of subsidence occurred in Mexico City, which is built upon a former lake bed. In the first half of this century thousands of wells were sunk into the water-saturated sediments beneath the city. As water was withdrawn, portions of the city subsided by as much as 6 to 7 meters. In some places buildings have sunk to such a point that access to them from the street is at what used to be the second floor level!

Saltwater Contamination

In many coastal areas the groundwater resource is being threatened by the encroachment of salt water. In order to understand this problem, we must examine the relationship between fresh groundwater and salt groundwater. Figure 11.13A is a diagrammatic cross section that illustrates this relationship in a coastal area underlain by permeable homogenous materials. Since fresh water is less dense than salt water, it floats on the salt water and forms a large, lens-shaped body that may extend to considerable depths below sea level. In such a situation, if the water table is 1 meter above sea level, the base of the freshwater body will extend to a depth of about 40 meters below sea level. Stated another way, the depth of the fresh water below sea level is about 40 times greater than the elevation of the water table above sea level. Thus, when excessive pumping lowers the water table by a certain amount, the bottom of the freshwater zone will rise by 40 times that amount. Therefore, if groundwater withdrawal continues to exceed recharge, there will come a time when the elevation of the salt water will be sufficiently high to be drawn into wells, thus contaminating the freshwater supply (Figure 11.13B). Deep wells and wells near the shore are usually the first to be affected.

In urbanized coastal areas, the problems created by excessive pumping are compounded by a decrease in the rate of natural recharge. As more and more of the surface is covered by streets, parking lots, and buildings, infiltration into the soil is diminished.

In an attempt to correct the problem of saltwater contamination of groundwater resources, a network of recharge wells may be used. These wells allow waste water to be pumped back into the groundwater system. A second method of correction is accomplished by building large basins. These basins collect surface drainage and allow it to seep into the ground. On Long Island, where the problem of saltwater contamination was recognized more than 40 years ago, both of these methods have been employed with considerable success.

Although contamination of freshwater aquifers by salt water is primarily a problem in coastal areas, it can threaten noncoastal locations as well. Many ancient sedimentary rocks of marine origin were deposited when the ocean covered places that are now far inland. In some instances, significant quantities of sea-water were trapped and still remain in the rock. These strata sometimes contain important quantities of fresh water and may be pumped for use by people. However, if fresh water is removed more rapidly than it is replenished, saline water may encroach and render the wells unusable. Such a situation threatened users of a deep (Cambrian age) sandstone aquifer in the Chicago area. To counteract this, water from Lake Michigan was allocated to the affected communities to offset the rate of withdrawal from the aquifer.

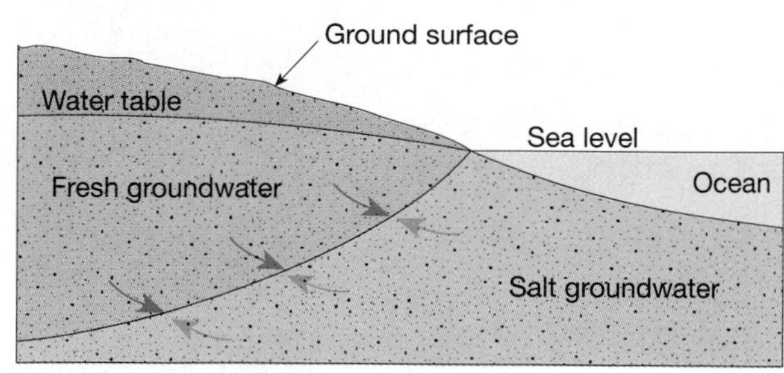

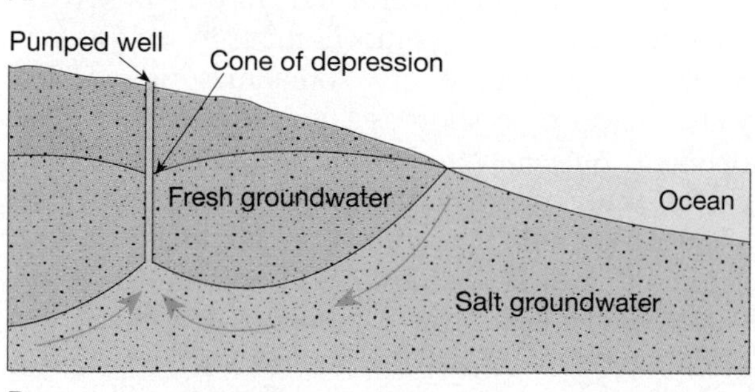

FIGURE 11.13
A. Since freshwater is less dense than salt water, it floats on the salt water and forms a lens-shaped body that may extend to considerable depths below sea level. **B.** When excessive pumping lowers the water table, the base of the freshwater zone will rise by 40 times that amount. The result may be saltwater contamination of wells.

GROUNDWATER CONTAMINATION

The pollution of groundwater is a serious matter, particularly in areas where aquifers supply a large part of the water supply. A very common type of

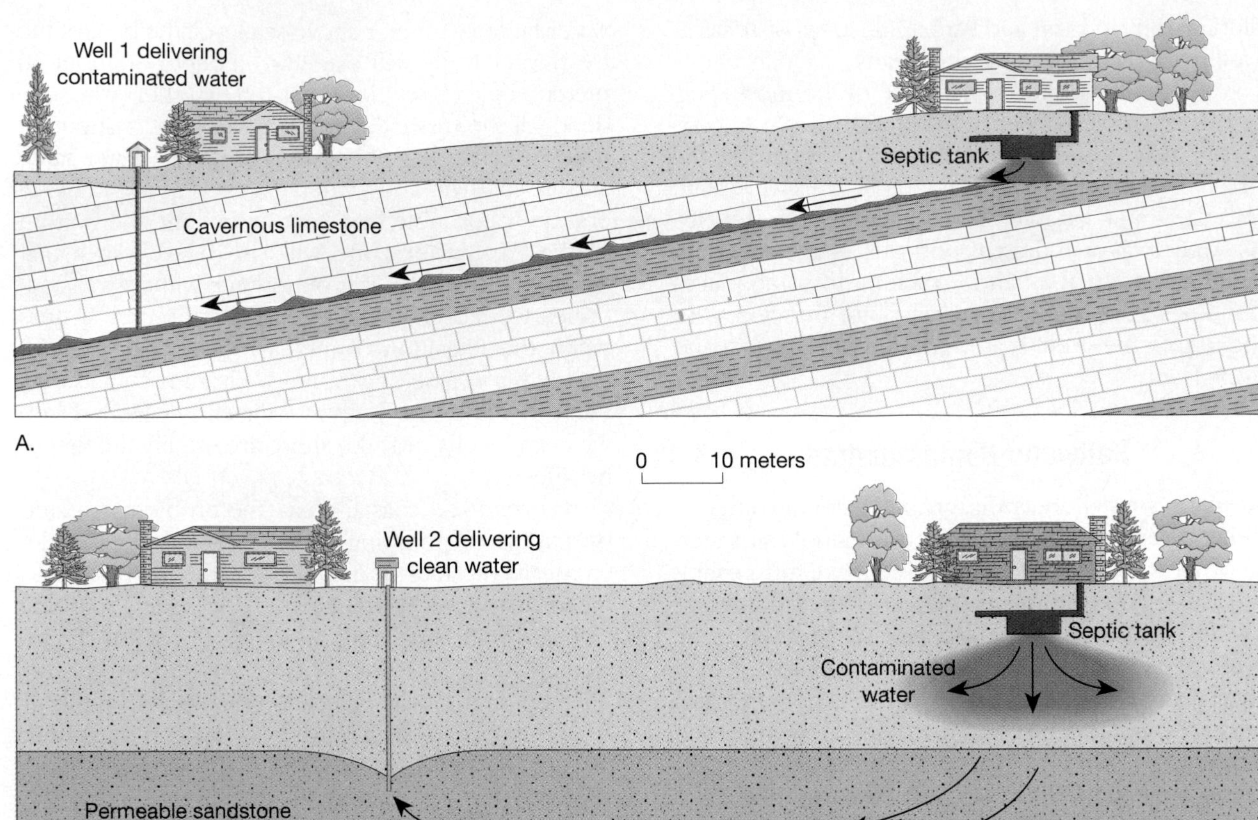

FIGURE 11.14
A. Although the contaminated water has traveled more than 100 meters before reaching well 1, the water moves too rapidly through the cavernous limestone to be purified. **B.** As the discharge from the septic tank percolates through the permeable sandstone, it is purified in a relatively short distance.

groundwater pollution is sewage. Its sources include an ever-increasing number of septic tanks, as well as inadequate or broken sewer systems and barnyard wastes.

If water contaminated with bacteria from sewage enters the groundwater system, it may become purified through natural processes. The harmful bacteria may be mechanically filtered out by the sediment through which the water percolates, destroyed by chemical oxidation, and/or assimilated by other organisms. In order for purification to occur, however, the aquifer must be of the correct composition. For example, extremely permeable aquifers such as highly fractured crystalline rock, coarse gravel, or cavernous limestone have such large openings that contaminated groundwater may travel long distances without being cleansed. In this case, the water flows too rapidly and is not in contact with the surrounding material long enough for purification to occur. This is the problem at well 1 in Figure 11.14. On the other hand, when water moves through sand or permeable sandstone, it can sometimes be purified within distances as short as a few tens of meters. The openings between sand grains are large enough to permit water movement, yet the movement of the water is slow enough to allow ample time for its purification (well 2, Figure 11.14).

Sometimes sinking a well can lead to groundwater pollution problems. If the well pumps a sufficient quantity of water, the cone of depression will locally increase the slope of the water table. In some instances, the original slope may even be reversed. This could lead to the contamination of wells that yielded unpolluted water before heavy pumping began (Figure 11.15). Also recall that the rate of groundwater movement increases as the slope of the water table steepens. This could produce problems because a faster rate of movement allows less time

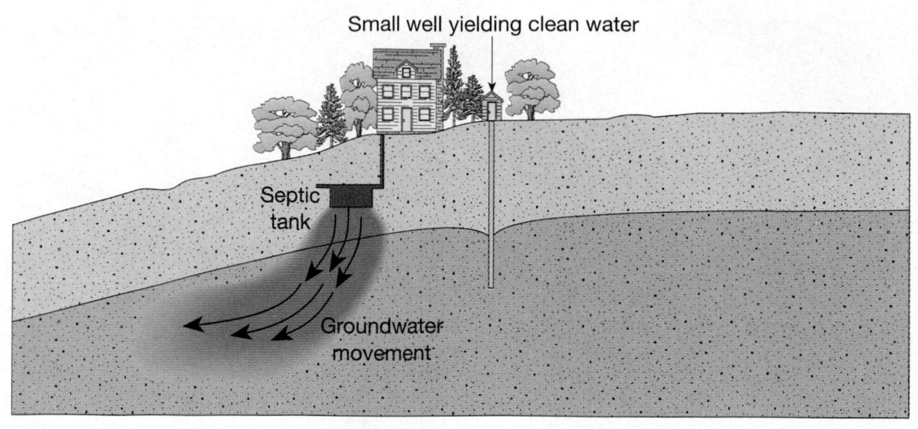

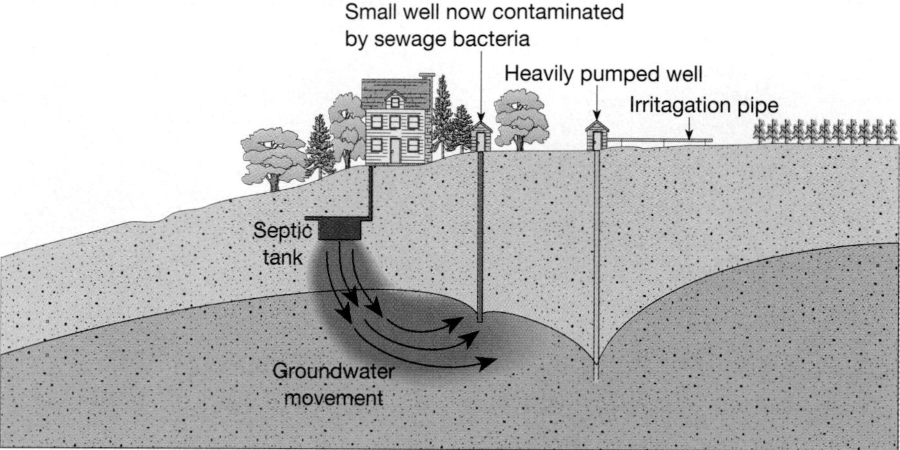

FIGURE 11.15
A. Originally the outflow from the septic tank moved away from the small well. **B.** The heavily pumped well changed the slope of the water table, causing contaminated groundwater to flow toward the small well.

for the water to be purified in the aquifer before it is pumped to the surface.

Other sources and types of contamination also threaten groundwater supplies. These include widely used substances such as highway salt, fertilizers that are spread across the land surface, and pesticides. In addition, a wide array of chemicals and industrial materials may leak from pipelines, storage tanks, landfills, and holding ponds. Some of these pollutants are classified as *hazardous,* meaning that they are either flammable, corrosive, explosive, or toxic. In land disposal, potential contaminants are heaped onto mounds or spread directly over the ground. As rainwater oozes through the refuse, it may dissolve a variety of organic and inorganic materials. If the leached material reaches the water table, it will mix with the groundwater and contaminate the supply. Similar problems may result from leakage of shallow excavations called holding ponds into which a variety of liquid wastes are disposed.

Since groundwater movement is usually slow, polluted water may go undetected for a considerable time. Most contamination is discovered only after drinking water has been affected. By this time, the volume of polluted water may be very large, and even if the source of contamination is removed immediately, the problem is not solved. Although the sources of groundwater contamination are numerous, the solutions are relatively few. Once the source of the problem has been identified and eliminated, the most common practice in dealing with contaminated aquifers is simply to abandon the water supply and allow the pollutants to be flushed away gradually. This is the least costly and easiest solution, but aquifers must remain unused for many years. To accelerate this process, polluted water is sometimes pumped out and treated. Following removal of the tainted water, the aquifer is allowed to recharge naturally or, in some cases, the treated water or other fresh water is pumped back in. This process, however, is costly and time consuming. It may also be somewhat risky because there is no way to be certain that all of the contamination has been removed. Clearly the most effective solution to groundwater contamination is prevention.

HOT SPRINGS AND GEYSERS

By definition, the water in **hot springs** is 6–9°C (10–15°F) warmer than the mean annual air temperature

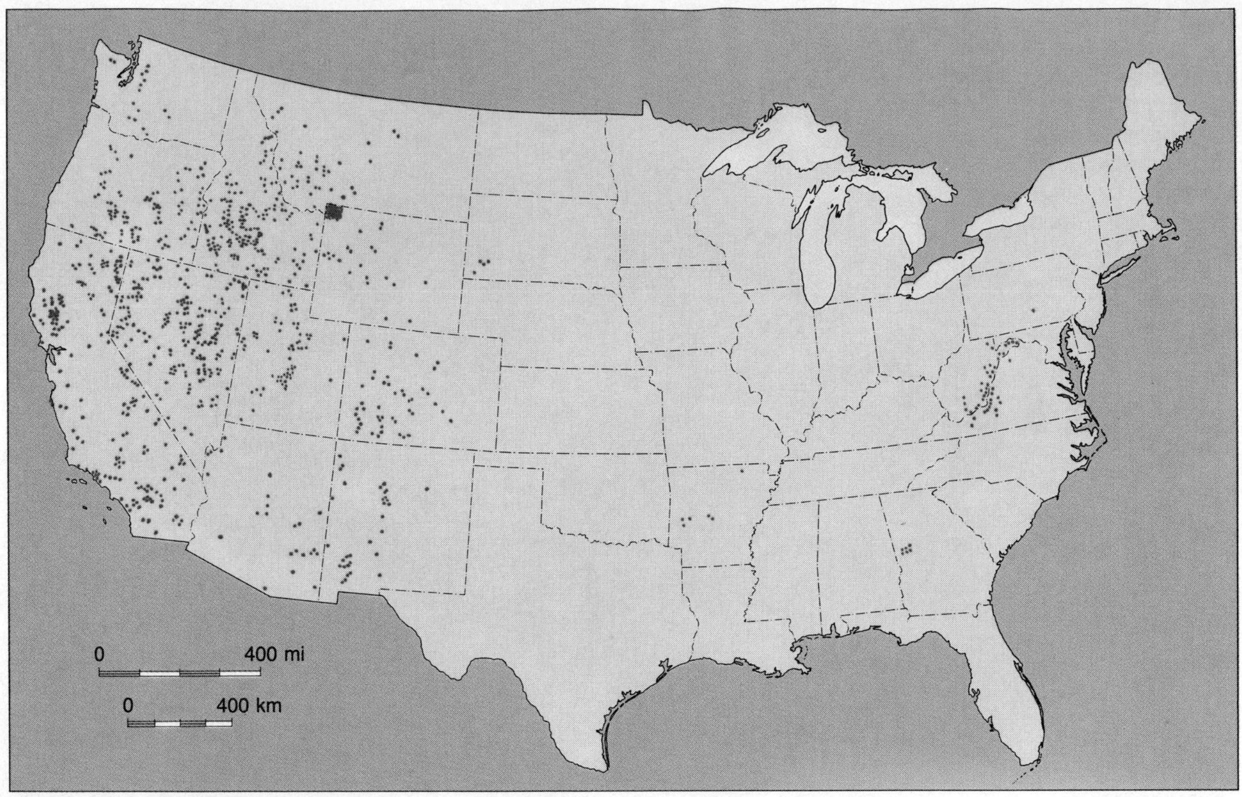

FIGURE 11.16
Distribution of hot springs and geysers in the United States. Note the concentration in the West, where igneous activity has been most recent. (After G. A. Waring, U.S. Geological Survey Professional Paper 492, 1965)

for the localities where they occur. In the United States alone, there are over 1000 such springs (Figure 11.16).

Mineral explorations over the world have shown that temperatures in deep mines and oil wells usually rise with an increase in depth below the surface. Temperatures in such situations increase an average of about 2°C per 100 meters (1°F per 100 feet). Therefore when groundwater circulates at great depths, it becomes heated, and if it rises to the surface, the water may emerge as a hot spring. The water of some hot springs in the United States, particularly in the East, is heated in this manner. However, the great majority (over 95 percent) of the hot springs (and geysers) in the United States are found in the West (Figure 11.16). The reason for such a distribution is that the source of heat for most hot springs is cooling igneous rock, and it is in the West that igneous activity has occurred more recently.

Geysers are intermittent hot springs or fountains in which columns of water are ejected with great force at various intervals, often rising 30–60 meters (100–200 feet). After the jet of water ceases, a column of steam rushes out, usually with a thundering roar. Perhaps the most famous geyser in the world is Old Faithful in Yellowstone National Park, which erupts about once each hour (Figure 11.17). Although Yellowstone is geologically outstanding in many ways, the great abundance, diversity, and spectacular nature of its geysers and other thermal features were undoubtedly the primary reasons for its being set aside as the first national park in the United States. Geysers are also found in other parts of the world, including New Zealand and Iceland, where the term *geyser,* meaning "spouter" or "gusher," originated.

Geysers occur where extensive underground chambers exist within hot igneous rocks. As relatively cool groundwater enters the chambers, it is heated by the surrounding rock. At the bottom of the chambers, the water is under great pressure because of the weight of the overlying water. Consequently this water must reach temperatures well above 100°C (212°F) before it will boil. For example, water at the bottom of a 300-meter (1000-foot) water-filled chamber must attain a temperature of nearly 230°C to boil. As the temperature rises, water expands and some flows out at the surface. This loss of water reduces the pressure on the remaining water in the chamber. The reduced pressure lowers the boiling point and a small portion of the water deep within the chamber quickly turns to steam and the geyser erupts (Figure 11.18). Following the eruption, cool groundwater again seeps into the chamber and the cycle begins anew.

When groundwater from hot springs and geysers flows out at the surface, material in solution is often precipitated, producing an accumulation of chemical sedimentary rock. The material deposited at any given place commonly reflects the chemical makeup of the rock through which the water circulated. When the water contains dissolved silica, a material called *siliceous sinter* or *geyserite* is deposited around the spring. When the water contains dissolved calcium carbonate, a form of limestone called *travertine* or *calcareous tufa* is deposited. The latter term is used if the material is spongy and porous. The deposits at Mammoth Hot Springs in Yellow-

FIGURE 11.17
Old Faithful, one of the world's most famous geysers, emits as much as 45,000 liters (almost 12,000 gallons) of hot water and steam about once each hour. (Photo by James E. Patterson)

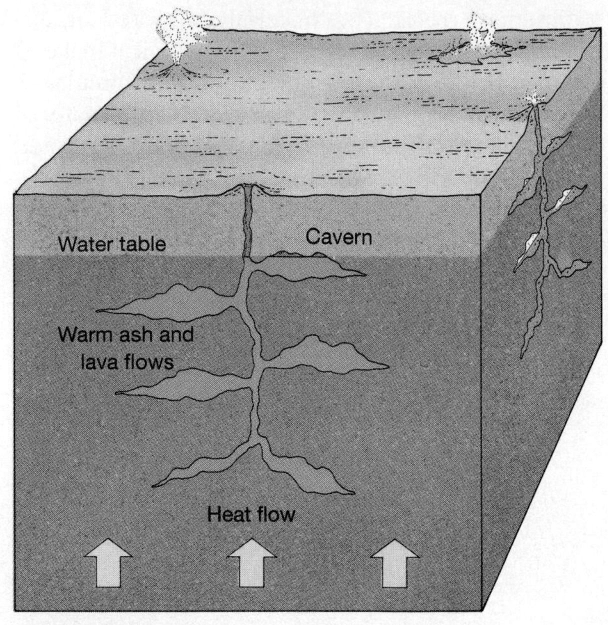

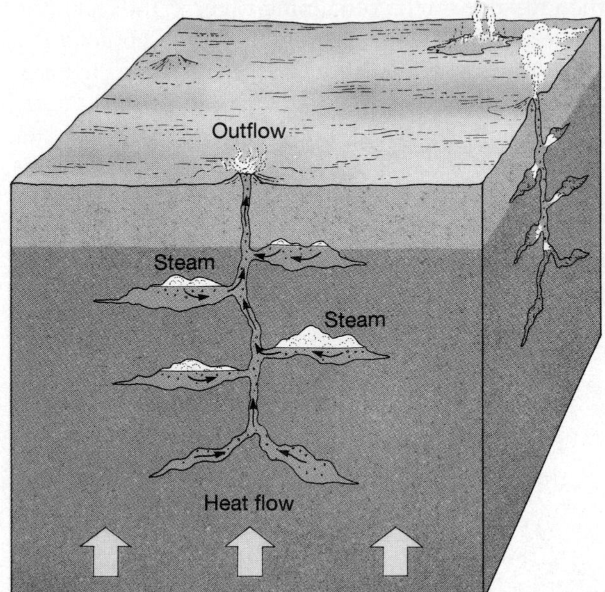

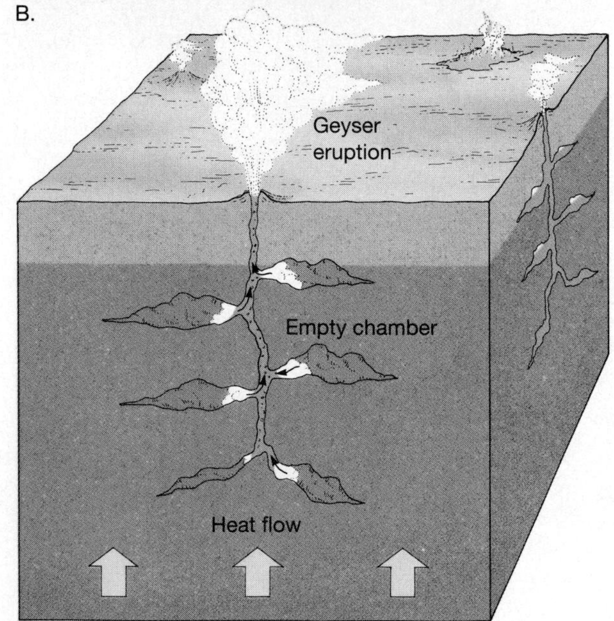

FIGURE 11.18
Idealized diagrams of a geyser. A geyser can form if the heat is not distributed by convection. **A.** In this figure, the water near the bottom is heated to near its boiling point. The boiling point is higher there than at the surface because the weight of the water above increases the pressure. **B.** The water higher in the geyser system is also heated; therefore, it expands and flows out at the top, reducing the pressure on the water at the bottom. **C.** At the reduced pressure on the bottom, boiling occurs. Some of the bottom water flashes into steam, and the expanding steam causes an eruption.

stone National Park are more spectacular than most (Figure 11.19). As the hot water flows upward through a series of channels and then out at the surface, the reduced pressure allows carbon dioxide to separate and escape from the water. The loss of carbon dioxide causes the water to become supersaturated with calcium carbonate, which then precipitates. In addition to containing dissolved silica and calcium carbonate, some hot springs contain sulfur, which gives water a poor taste and unpleasant odor. Undoubtedly Rotten Egg Spring, Nevada, is such a situation.

THE GEOLOGIC WORK OF GROUNDWATER

The primary erosional work carried out by groundwater is that of dissolving rock. Since soluble rocks, especially limestone, underlie millions of square kilometers of the earth's surface, it is here that groundwater carries on its rather unique and important role as an erosional agent. Although nearly insoluble in pure water, limestone is quite easily dissolved by water containing small quantities of carbonic acid. Most natural water contains this weak acid because rainwater readily dissolves carbon dioxide from the air and from decaying plants. Therefore, when groundwater comes in contact with limestone, the carbonic acid reacts with the calcite in the rocks to form calcium bicarbonate, a soluble material that is then carried away in solution.

Caverns

Among the most spectacular results of groundwater's erosional handiwork is the creation of limestone **caverns**. In the United States alone about 17,000 caves have been discovered. Although most are relatively small, some have spectacular dimensions. Carlsbad Caverns in southeastern New Mexico and Mammoth Cave in Kentucky are famous examples. The Mammoth Cave system is the most extensive in the world, with more than 500 kilometers of interconnected passages. The dimensions at Carlsbad Caverns are impressive in a different way. Here we find the largest and perhaps most spectacular single chamber. The Big Room at Carlsbad Caverns has an area equivalent to fourteen football fields and enough height to accommodate the U.S. Capitol Building (see Figure 11.1).

Most caverns are created at or just below the water table in the zone of saturation. Here the groundwater follows lines of weakness in the rock, such as joints and bedding planes. As time passes, the dissolving process slowly creates cavities and gradually enlarges them into caverns. The material that is dissolved by the groundwater is carried away and discharged into streams. In many caves, development

FIGURE 11.19
Mammoth Hot Springs at Yellowstone National Park. Although most of the deposits associated with geysers and hot springs in Yellowstone Park are silica-rich geyserite, the deposits at Mammoth Hot Springs consist of a form of limestone called travertine. (Photo by Stephen Trimble)

FIGURE 11.20
Speleothems are of many types, including stalactites, stalagmites, and columns. Virgin Cave, Lincoln National Forest, New Mexico. (Photo by John and Ed Burke)

has occurred at several levels, with the current cavern-forming activity occurring at the lowest elevation. This situation reflects the close relationship between the formation of major subterranean passages and the river valleys into which they drain. As streams cut their valleys deeper, the water table drops as the elevation of the river drops. Consequently, during periods when surface streams are rapidly downcutting, surrounding groundwater levels drop rapidly and cave passages are abandoned by the water while the passages are still relatively small in cross-sectional area. Conversely, when the entrenchment of streams is slow or negligible, there is time for large cave passages to form.

Certainly the features that arouse the greatest curiosity for most cavern visitors are the stone formations that often exhibit bizarre patterns and give some caverns a wonderland appearance. These features are created by the seemingly endless dripping of water over great spans of time. The calcite that is left behind produces the limestone we call travertine. These cave deposits, however, are also commonly called *dripstone,* an obvious reference to their mode of origin. Although the formation of caverns takes place in the zone of saturation, the deposition of dripstone is not possible until the caverns are above the water table in the zone of aeration. As soon as the chamber is filled with air, the stage is set for the decoration phase of cavern building to begin.

The various dripstone features found in caverns are collectively called **speleothems**, no two of which are exactly alike (Figure 11.20). Perhaps the most familiar speleothems are **stalactites**. These icicle-like pendants hang from the ceiling of the cavern and form where water seeps through cracks above. When the water reaches air in the cave, some of the dissolved carbon dioxide escapes from the drop and calcite begins to precipitate. Deposition occurs as a ring around the edge of the water drop. As drop after drop follows, each leaves an infinitesimal trace of calcite behind, and a hollow limestone tube is created. Water then moves through the tube, remains suspended momentarily at the end, contributes a tiny ring of calcite, and falls to the cavern floor. The stalactite just described is appropriately called a *soda straw* (Figure 11.21). Often the hollow tube of the soda straw becomes plugged or its supply of water increases. In either case, the water is forced to flow, and hence deposit, along the outside of the tube. As deposition continues, the stalactite takes on the more common conical shape.

Speleothems that form on the floor of a cavern and reach upward toward the ceiling are called **stalagmites.** The water supplying the calcite for stalagmite growth falls from the ceiling and splatters over the surface. As a result, stalagmites do not have a central tube and are usually more massive in appearance and rounded on their upper ends than stalactites.

Karst Topography

Many areas of the world have landscapes that to a large extent have been shaped by the dissolving power of groundwater. Such areas are said to exhibit **karst topography**. The term is derived from a plateau region located along the northeastern shore of the Adriatic Sea in the part of Yugoslavia called Slovenia where such topography is strikingly developed. The most common geologic setting for karst development is an area where limestone is present at the surface beneath a mantle of soil. In the United States, karst landscapes occur in areas of Kentucky, Tennessee, Alabama, southern Indiana, and central and northern Florida. Generally, arid and semiarid areas do not develop karst topography. When solution features exist in such regions, they are likely to be remnants of a time when more humid conditions prevailed.

Karst areas characteristically exhibit an irregular terrain punctuated with many depressions, called **sinkholes** or **sinks**. In the limestone areas of Florida, Kentucky, and southern Indiana, there are literally tens of thousands of these depressions varying in depth from just a meter or two to a maximum of more than 50 meters (Figure 11.22).

BOX 11.2

The Winter Park Sinkhole

The craterlike sinkhole in Figure 11.C began forming in Winter Park, Florida, on May 8, 1981, just one day before this photograph was taken. Newspaper accounts, such as the one that follows, were front page news and made this sinkhole one of the most publicized ever.

FIGURE 11.C
Aerial view of a large sinkhole that formed in Winter Park, Florida, in May, 1981. (Photo courtesy of George Remaine, *Orlando Sentinel Star*)

SINKHOLE NIBBLES AWAY AT FLORIDA CITY

WINTER PARK, Fla. (AP)—A giant sinkhole—already several hundred feet wide after swallowing a three-bedroom bungalow, half a swimming pool and six Porsches—nibbled away at a side street yesterday and threatened a main thoroughfare.

"It has slowed down, but it hasn't quit," said Winter Park Fire Capt. Gus LaGarde.

The crater, estimated at between 450 and 600 feet wide and 125 to 170 feet deep, grew by eight to 10 feet yesterday and was filling with water, authorities said.

It developed Friday night and opened rapidly Saturday, when it gulped the wood-frame cottage, cars and part of a foreign car lot and wrecked the city's $150,000 municipal swimming pool.

It was slowly eating its way west yesterday, leaving a group of businesses, their backs lost in Saturday's slide, hanging at the edge of the pit, LaGarde said.

The hole devoured most of a side street yesterday and was about 50 feet from one main thoroughfare and moving closer to several others, he said.

"We're still losing some of the perimeter," he said. "It doesn't appear to get any deeper . . . it continues to eat up the roadway, power poles, anything that gets in the way."

Residents and owners of nearby homes and businesses were warned on Saturday to leave until the sinkhole stopped growing. Some people rented trucks and began moving furniture and other property.*

Sinkhole formation is not uncommon in northern and central Florida. In fact, the Winter Park event was just one of three that occurred in the area over a two-week span. In each case the collapse at the surface was probably triggered by a lowering of the water table brought about by a severe drought. As the water table dropped, the roofs of the underground cavities lost support and fell into the voids below.

*Courtesy of Associated Press.

A.

B.

FIGURE 11.21
A. A "live" solitary soda straw stalactite. (Photo by Clifford Stroud, National Park Service). **B.** A soda straw "forest" in Carlsbad Caverns. (Photo courtesy of the National Park Service)

FIGURE 11.22
This high-altitude infrared image shows an area of karst topography in Florida. The numerous lakes occupy sinkholes. (Courtesy of USDA-ASCS)

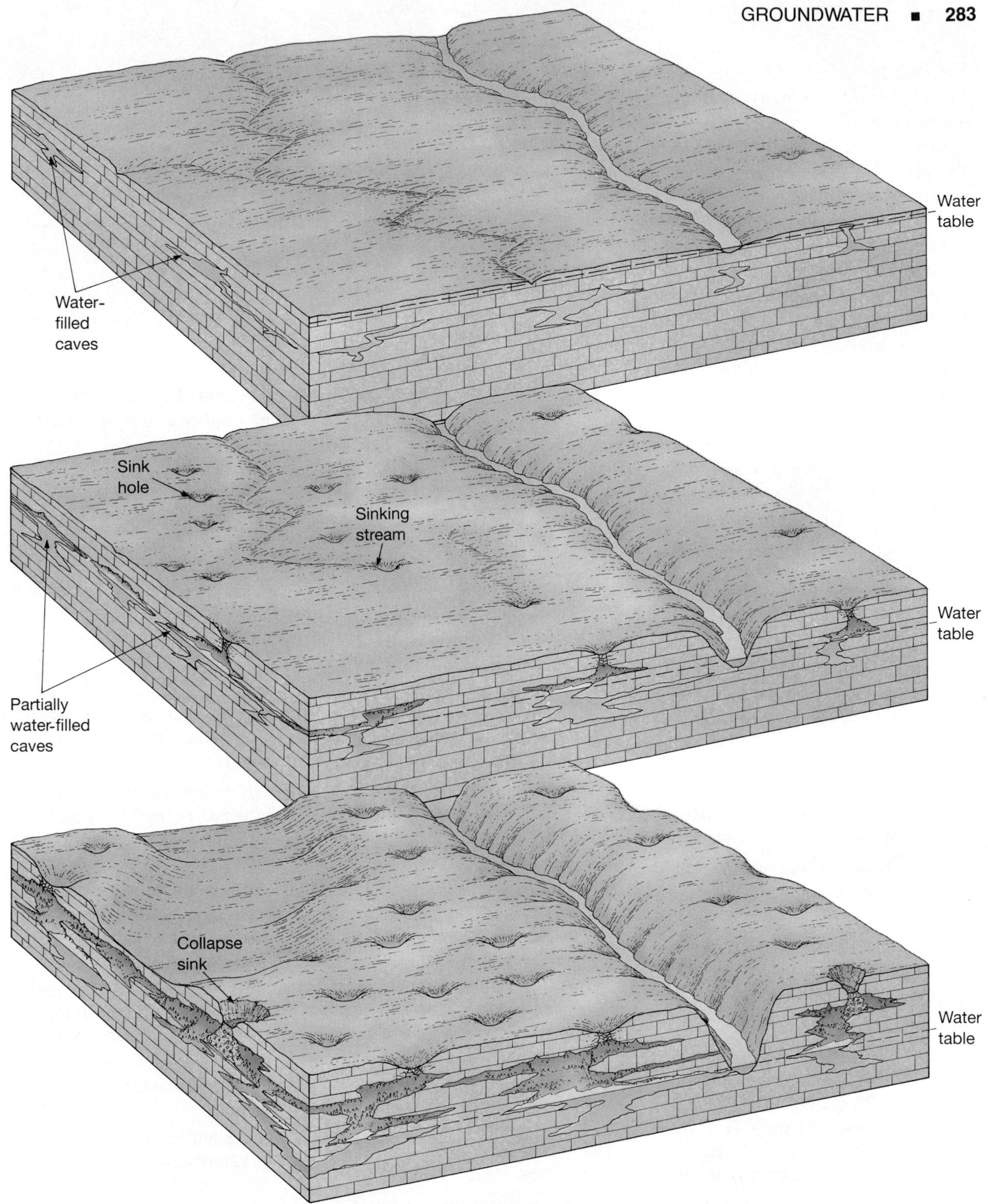

FIGURE 11.23
Development of a karst landscape. **A.** During early stages, groundwater percolates through limestone along joints and bedding planes. Solution activity creates and enlarges caverns at and below the water table. **B.** Sinkholes are well developed and surface streams are funneled below ground. **C.** Caverns grow larger and the number and size of sinkholes increase. Collapse of caverns and coalescence of sinkholes form larger, flat-floored depressions. Eventually solution activity may remove most of the limestone from the area, leaving only isolated remnants.

Sinkholes commonly form in one of two ways. Some develop gradually over many years without any physical disturbance to the rock. In these situations, the limestone immediately below the soil is dissolved by downward-seeping rainwater that is freshly charged with carbon dioxide. With time, the bedrock surface is lowered and the fractures into which the water seeps are enlarged. As the fractures grow in size, soil subsides into the widening voids, from which it is removed by groundwater flowing in the passages below. These depressions are usually not deep and are characterized by relatively gentle slopes. By contrast, sinkholes can also form suddenly and without warning when the roof of a cavern collapses under its own weight. Typically, the depressions created in this manner are steep-sided and deep. When they form in populous areas, they may represent a serious geologic hazard. Such a case is described in Box 11.2.

In addition to a surface pockmarked by sinkholes, karst regions characteristically show a striking lack of surface drainage. Following a rainfall, the runoff is funneled below ground by way of sinks where it then flows through caverns until it finally reaches the water table. When streams do exist at the surface, their paths are usually short. The names of such streams often give a clue to their fate. In the Mammoth Cave area of Kentucky, for example, there is Sinking Creek, Little Sinking Creek, and Sinking Branch. Other sinkholes become plugged with clay and debris to create small lakes or ponds. The development of a karst landscape is depicted in Figure 11.23.

REVIEW QUESTIONS

1. Compare and contrast the zones of aeration and saturation. Which of these zones contains groundwater?
2. Although we usually think of tables as being flat, the water table generally is not. Explain.
3. What is an effluent stream? How does an influent stream differ?
4. Distinguish between porosity and permeability.
5. What is the difference between an aquiclude and an aquifer?
6. Under what circumstances can a material have a high porosity but not be a good aquifer?
7. As illustrated in Figure 11.5, groundwater moves in looping curves. What factors cause the water to follow such paths?
8. Briefly describe the important contribution to our understanding of groundwater movement made by Henry Darcy.
9. When an aquiclude is situated above the main water table, a localized saturated zone may be created. What term is applied to such a situation?
10. Two neighbors each dig a well. Although both wells penetrate to the same depth, one neighbor is successful and the other is not. Describe a circumstance that might explain what happened.
11. What is meant by the term *artesian?*
12. In order for artesian wells to exist, two conditions must be present. List these conditions.
13. When the Dakota Sandstone was first tapped, water poured freely from many artesian wells. Today these wells must be pumped. Explain.
14. What problem is associated with the pumping of groundwater for irrigation in the southern part of the High Plains?
15. Briefly explain what happened in the San Joaquin Valley as the result of excessive groundwater withdrawal. (See Box 11.1.)
16. In a particular coastal area the water table is 4 meters above sea level. Approximately how far below sea level does the fresh water reach?
17. Why does the rate of natural groundwater recharge decrease as urban areas develop?
18. Which aquifer would be most effective in purifying polluted groundwater: coarse gravel, sand, or cavernous limestone?
19. What is meant when a groundwater pollutant is classified as hazardous?
20. What is the source of heat for most hot springs and geysers? How is this reflected in the distribution of these features?
21. Name two common speleothems and distinguish between them.
22. Speleothems form in the zone of saturation. True or False? Briefly explain your answer.
23. Areas whose landscapes largely reflect the erosional work of groundwater are said to exhibit what kind of topography?
24. Describe two ways in which sinkholes are created.

KEY TERMS

aquiclude (p. 266)
aquifer (p. 266)
artesian (p. 269)
belt of soil moisture (p. 263)
capillary fringe (p. 263)
cavern (p. 279)
cone of depression (p. 268)
Darcy's law (p. 266)
drawdown (p. 268)
effluent stream (p. 265)
flowing artesian well (p. 270)
geyser (p. 276)
groundwater (p. 263)
head (p. 266)
hot spring (p. 275)
hydraulic gradient (p. 266)
influent stream (p. 265)
karst topography (p. 280)
nonflowing artesian well (p. 270)
perched water table (p. 268)
permeability (p. 265)
porosity (p. 265)
sinkhole (sink) (p. 281)
speleothem (p. 280)
spring (p. 267)
stalactite (p. 280)
stalagmite (p. 280)
water table (p. 263)
well (p. 268)
zone of aeration (p. 263)
zone of saturation (p. 263)

12
Glaciers and Glaciation

Opposite: Active valley glaciers continue to sculpture this rugged mountain area. (Photo by Michael Collier) (Top photo by Michael Collier)

Many present-day landscapes were modified by the widespread glaciers of the most recent ice age and still strongly reflect the handiwork of ice (Figure 12.1). The basic character of such diverse places as the Alps, Cape Cod, and Yosemite Valley was fashioned by now vanished masses of glacial ice. Moreover, Long Island, the Great Lakes, and the fiords of Norway and Alaska all owe their existence to glaciers. Glaciers, of course, are not just a phenomenon of the geologic past. As we shall see, they are still sculpturing and depositing in many regions today.

TYPES OF GLACIERS

A **glacier** is a thick ice mass that originates on land from the accumulation, compaction, and recrystallization of snow. Because glaciers are agents of erosion, they must also flow. Indeed, like running water, groundwater, wind, and waves, glaciers are dynamic forces that are capable of accumulating, transporting, and depositing sediment. Although glaciers are found in many parts of the world today, most are located in remote areas.

Literally thousands of relatively small glaciers exist

FIGURE 12.1
Glacier National Park, Montana. St. Mary Lake occupies a glacially eroded valley (glacial trough) and exists because of glacial deposits that act as a dam. The sharp ridge on the left (an arête) and the peak in the background (a horn) were also sculptured by glacial ice. (Photo by Russel Lamb Photography)

FIGURE 12.2
Aerial view of South Cascade Glacier, a valley glacier about 3 kilometers long in the North Cascade Range, Washington. The snowline and cracks called crevasses are clearly visible. (Photo by Austin Post, U.S. Geological Survey)

in lofty mountain areas, where they usually follow valleys originally occupied by streams. Unlike the rivers that previously flowed in these valleys, the glaciers advance slowly, perhaps only a few centimeters per day. Because of their setting, these moving ice masses are termed **valley glaciers** or **alpine glaciers** (Figure 12.2). Each is a stream of ice, bounded by precipitous rock walls, that flows downvalley from an accumulation center near its head. Like rivers, valley glaciers can be long or short, wide or narrow, single or with branching tributaries. Generally the widths of alpine glaciers are small compared to their lengths. Some extend for just a fraction of a kilometer, whereas others go on for many tens of kilometers. The west branch of the Hubbard Glacier, for example, runs through 112 kilometers of mountainous terrain in Alaska and the Yukon Territory.

In contrast to valley glaciers, **ice sheets** exist on a much larger scale. These enormous masses flow out in all directions from one or more centers and completely obscure all but the highest areas of underlying terrain. Even sharp variations in the topography beneath the glacier usually appear as relatively subdued undulations on the surface of the ice. Such topographic differences, however, do affect the behavior of the ice sheets, especially near their margins, by guiding flow in certain directions and creating zones of faster and slower movement. Although many ice sheets have existed in the past, just two achieve this status at present (Figure 12.3). In the Northern Hemisphere, Greenland is covered by an imposing ice sheet that occupies 1.7 million square kilometers, or about 80 percent of this large island. Averaging nearly 1500 meters thick, in places the ice extends 3000 meters above the island's bedrock floor. In the south polar realm, the huge Antarctic Ice Sheet attains a maximum thickness of nearly 4300 meters and covers an area of more than 13.9 million square kilometers. Because of the proportions of these huge features, the term *ice sheet* is often preceded by the word *continental.* Indeed the combined areas of present-day continental ice sheets represent almost 10 percent of the earth's land area.

Along portions of the Antarctic coast, glacial ice

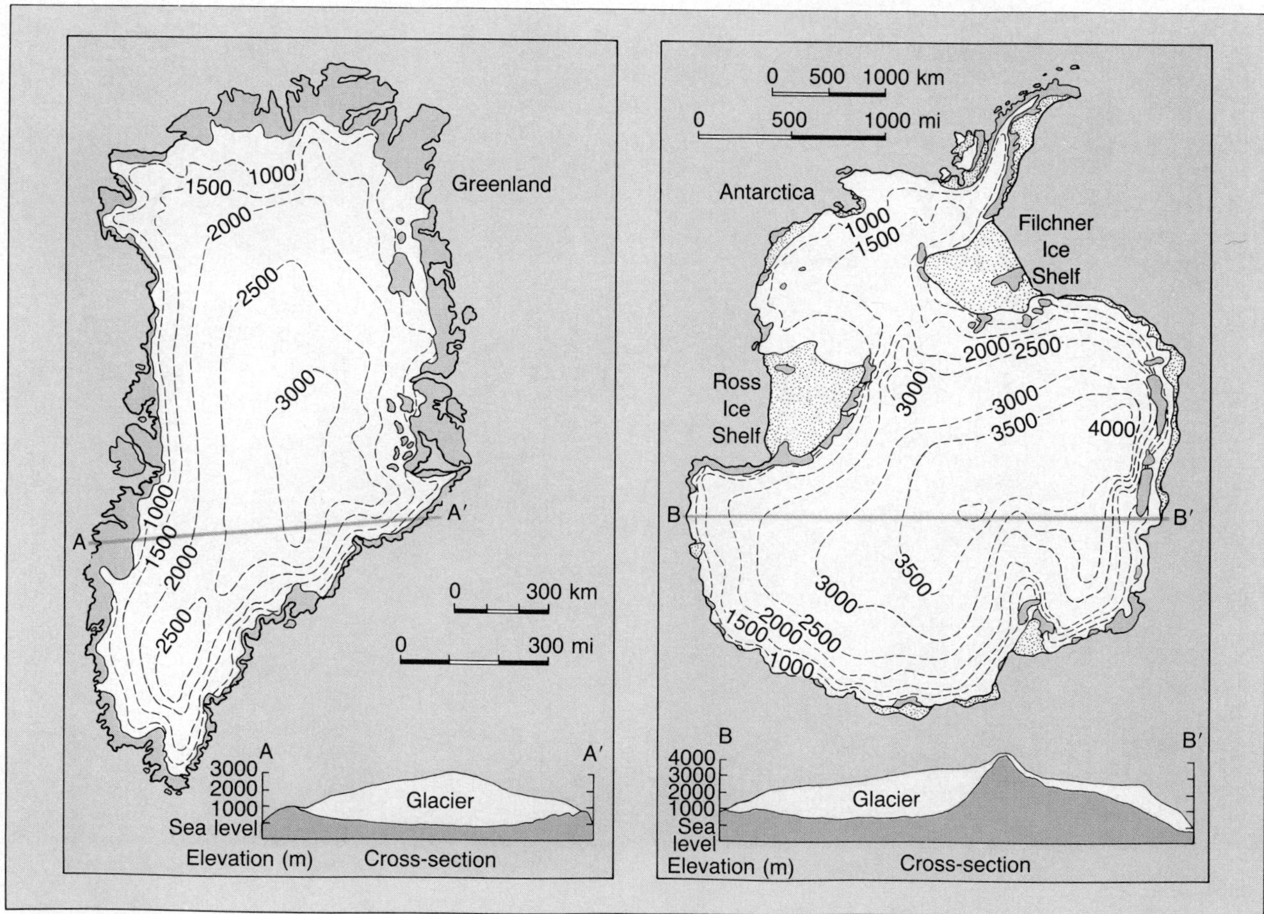

FIGURE 12.3
The only present-day continental ice sheets are those covering Greenland and Antarctica. Their combined areas represent almost 10 percent of the earth's land area. Greenland's ice sheet occupies 1.7 million square kilometers, or about 80 percent of the island. The area of the Antarctic Ice Sheet is almost 14 million square kilometers. Ice shelves occupy an additional 1.4 million square kilometers adjacent to the Antarctic Ice Sheet. The contours indicate the thickness of the ice in meters.

flows into bays, creating features called **ice shelves.** They are large, relatively flat masses of floating ice that extend seaward from the coast but remain attached to the land along one or more sides. The shelves are thickest on their landward sides and thin seaward. They are sustained by ice from the adjacent ice sheet as well as being nourished by snowfall and the freezing of seawater to their bases. Antarctica's ice shelves extend over nearly 1.4 million square kilometers. The Ross and Filchner ice shelves are the largest, with the Ross Ice Shelf alone covering an area nearly the size of Texas (Figure 12.3).

In addition to valley glaciers and ice sheets, other types of glaciers are also identified. Covering some uplands and plateaus are masses of glacial ice called **ice caps.** They resemble ice sheets but are much smaller than the continental-scale features. Ice caps occur in many places, including Iceland and several of the large islands in the Arctic Ocean. Often ice caps and ice sheets feed **outlet glaciers.** These tongues of ice flow down valleys extending outward from the margins of these larger ice masses. The tongues are essentially valley glaciers that are avenues for ice movement from an ice cap or ice sheet

BOX 12.1

Glaciers and the Hydrologic Cycle

Earlier we learned that the earth's water is in constant motion. Time and time again the same water is evaporated from the oceans into the atmosphere, precipitated upon the land, and carried by rivers and underground flow back to the sea. However, when precipitation falls at high elevations or high latitudes, the water may not immediately make its way toward the sea. Instead, it may become part of a glacier. Although the ice will eventually melt and continue its path to the sea, it can be stored as glacial ice for many tens, hundreds, or even thousands of years. For example, data collected from Greenland show that portions of its glacier are more than 25,000 years old.

How much water is stored as glacial ice? Estimates by the U.S. Geological Survey indicate that only slightly more than 2 percent of the world's water is accounted for by glaciers. But this small figure may be misleading when the actual amounts of water are considered. The total volume of all valley glaciers is about 210,000 cubic kilometers, comparable to the combined volume of the world's largest saline and freshwater lakes. Furthermore, 80 percent of the world's ice and nearly two-thirds of the earth's fresh water are represented by Antarctica's ice sheet, which covers an area almost one and one-half times that of the United States. If this ice melted, sea level would rise an estimated 60 to 70 meters, and the ocean would inundate many densely populated coastal areas (Figure 12.A). The hydrologic importance of the continent and its ice can be illustrated in another way. If Antarctica's ice sheet were melted at a uniform rate, it could feed (1) the Mississippi River for more than 50,000 years, (2) all the rivers in the United States for about 17,000 years, (3) the Amazon River for approximately 5000 years, or (4) all the rivers of the world for about 750 years.

As the foregoing discussion illustrates, the quantity of ice on earth today is truly immense. However, present glaciers occupy only slightly more than one-third the area they did in the very recent geologic past.

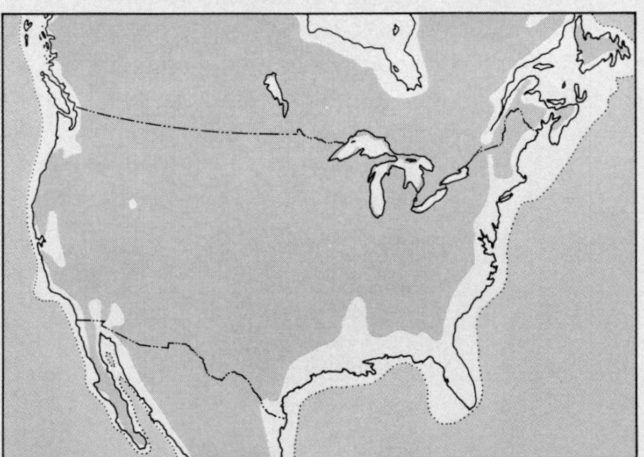

FIGURE 12.A
This map of a portion of North America shows the present-day coastline compared to the coastline that existed during the last ice-age maximum (18,000 years ago) and the coastline that would exist if present ice sheets in Greenland and Antarctica melted. (After R. H. Dott, Jr., and R. L. Battan, *Evolution of the Earth,* New York: McGraw-Hill, 1971. Reprinted by permission of the publisher)

through mountainous terrain to the sea. Where they encounter the ocean, some outlet glaciers spread out as floating ice shelves. Often large numbers of icebergs are produced.

Piedmont glaciers occupy broad lowlands at the bases of steep mountains and form when one or more alpine glaciers emerge from the confining walls of mountain valleys. Here the advancing ice spreads out to form a broad sheet. The size of individual piedmont glaciers varies greatly. Among the largest is the broad Malaspina Glacier along the coast of southern Alaska. It covers more than 5000 square kilometers of the flat coastal plain at the foot of the lofty St. Elias range.

FORMATION OF GLACIAL ICE

Snow is the raw material from which glacial ice originates; therefore, glaciers form in areas where more snow falls in winter than melts during the summer. Before a glacier is created, snow must be converted into glacial ice. When temperatures remain below freezing following a snowfall, the fluffy accumulation of delicate hexagonal crystals soon changes. As air infiltrates the spaces between the crystals, the extremities of the crystals evaporate and the water vapor condenses near the centers of the crystals. In this manner snowflakes become smaller, thicker, and more spherical, and the large pore spaces disappear. By this process air is forced out and what was once light, fluffy snow is recrystallized into a much denser mass of small grains having the consistency of coarse sand. This granular recrystallized snow is called **firn** and is commonly found making up old snowbanks near the end of winter. As more snow is added, the pressure on the lower layers increases, compacting the ice grains at depth. Once the thickness of ice and snow exceeds 50 meters, the weight is sufficient to fuse firn into a solid mass of interlocking ice crystals. Glacial ice has now been formed.

MOVEMENT OF A GLACIER

The movement of glacial ice is generally referred to as flow. The fact that glacial movement is described in this way would seem to constitute a paradox—ice is solid, yet it is capable of flow. The way in which ice flows is complex and is of two basic types. The first of these, **plastic flow**, involves movement within the ice. Ice behaves as a brittle solid until the pressure or load upon it is equivalent to the weight of about 50 meters of ice. Once that load is surpassed, ice behaves as a plastic material and flow begins. Such flow

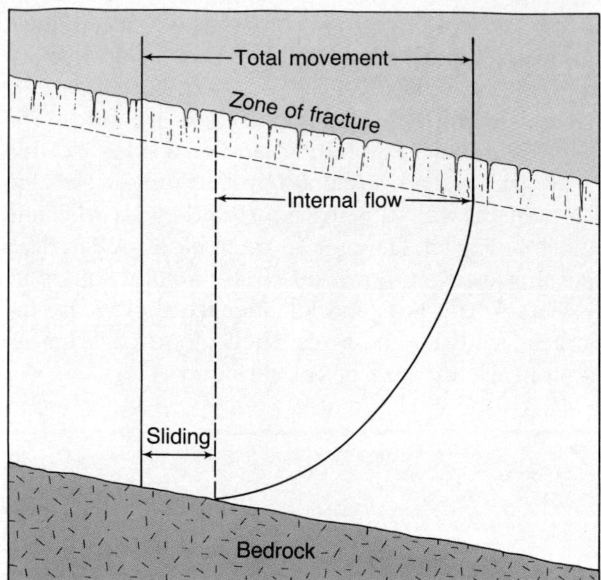

FIGURE 12.4
Vertical cross section through a glacier to illustrate the ice movement. Glacial movement is divided into two components. Below about 50 meters, ice behaves plastically and flows. In addition, the entire mass of ice may slide along the ground. The ice in the zone of fracture is carried along "piggyback" style. Notice that the rate of movement is slowest at the base of the glacier where frictional drag is greatest.

occurs because of the molecular structure of ice. Glacial ice consists of layers of molecules stacked one upon the other. The bonds between layers are weaker than those within each layer. Therefore, when a stress exceeds the strength of the bonds between the layers, the layers remain intact and slide over one another.

A second, and often equally important, mechanism of glacial movement consists of the entire ice mass slipping along the ground. With the exception of some glaciers located in polar regions where the ice is probably frozen to the solid bedrock floor, most glaciers are thought to move by this sliding process, called **basal slip**. In this process, meltwater is believed to act as a hydraulic jack and perhaps as a lubricant helping the ice over the rock. The source of the liquid water is related in part to the fact that the melting point of ice decreases as pressure increases. Therefore, deep within a glacier the ice may be at the melting point even though its temperature is below 0°C. In addition, other factors may contribute to the presence of meltwater deep within the glacier. Temperatures may be increased by plastic flow (an effect similar to frictional heating), by heat added from the earth below, and by the refreezing of meltwater that has seeped down from above. This last process relies

FIGURE 12.5
Crevasses form in the brittle ice of the zone of fracture. They do not continue down into the zone of flow. (Photo by P. Jay Fleisher)

on the property that as water changes state from liquid to solid, heat (termed latent heat of fusion) is released.

Figure 12.4 illustrates the effects of these two basic types of glacial motion. This vertical profile through a glacier also shows that not all the ice flows forward at the same rate. Just as in streams, frictional drag with the bedrock floor causes the lower portions of the glacier to move more slowly.

In contrast to the lower portion of the glacier, the upper 50 meters or so is not under sufficient pressure to exhibit plastic flow. Consequently, the ice in this uppermost zone is brittle and is appropriately referred to as the **zone of fracture**. Incapable of flow, the ice in the zone of fracture is carried along "piggyback" style by the ice below. When the glacier moves over irregular terrain, the zone of fracture is subjected to tension, resulting in cracks called **crevasses** (see Figures 12.2 and 12.5). These gaping cracks can make travel across glaciers dangerous and may extend to depths of 50 meters. Below this depth, plastic flow seals them off.

Rates of Glacial Movement

Unlike streamflow, the movement of glaciers is not readily apparent to the casual observer. If we could watch an alpine glacier move, we would see that, like

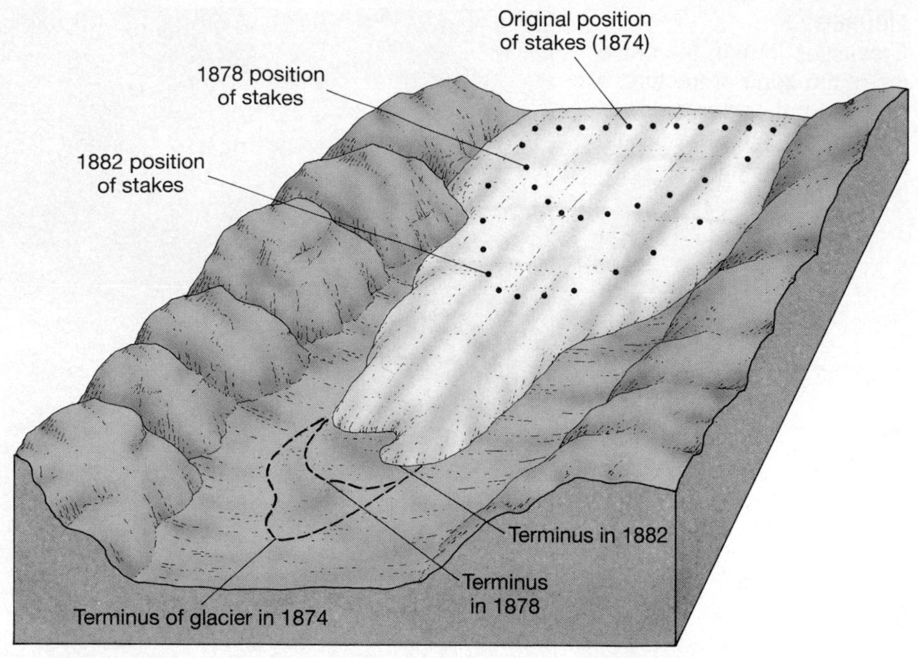

FIGURE 12.6
Ice movement and changes in the terminus at Rhone Glacier, Switzerland. In this classic study of a valley glacier, the movement of stakes clearly showed that ice along the sides of the glacier moves slowest. Also notice that even though the ice front was retreating, the ice within the glacier was advancing.

the water in a river, all of the ice in the valley does not move downvalley at an equal rate. Just as friction with the bedrock floor slows the movement of the ice at the bottom of the glacier, the drag created by the valley walls leads to the flow being greatest in the center of the glacier.

The first measurements of glacial movement were made in the late nineteenth century. In this experiment, stakes were carefully placed in a straight line across the top of a valley glacier. Periodically, the positions of the stakes were recorded, revealing the type of movement just described (Figure 12.6).

How rapidly does glacial ice move? Average velocities vary considerably from one glacier to another. Some move so slowly that trees and other vegetation may become well established in the debris that has accumulated on the glacier's surface, whereas others may move at rates of up to several meters per day. For example, Byrd Glacier, an outlet glacier in Antarctica that was the subject of a 10-year study using satellite images, moved at an average rate of 750 to 800 meters per year (about 2 meters per day). Other glaciers in the study advanced at one-fourth that rate. The advance of some glaciers is characterized by periods of extremely rapid movements called **surges**. Glaciers that exhibit such movement may flow along in an apparently normal manner, then speed up for a relatively short time before returning to the normal rate again. The flow rates during surges are as much as 100 times the normal rate. Evidence indicates that many glaciers may be of the surging type.

Budget of a Glacier

Glaciers are constantly gaining and losing ice. Snow accumulation and ice formation occur in a region termed the **zone of accumulation**, the outer limits of which are defined by the **snowline**. The elevation of the snowline varies greatly. In polar regions, it may be sea level, whereas in tropical areas, the snowline exists only high in mountain areas, often at altitudes exceeding 4500 meters. Above the snowline, the addition of snow thickens the glacier and promotes movement. Below the snowline, snow from the previous winter as well as some of the glacial ice melts.

In addition to melting, glaciers also waste as large pieces of ice break off the front of the glacier in a process called **calving**. Calving creates icebergs in places where the glacier has reached the sea or a lake. Because icebergs are just slightly less dense than seawater, they float very low in the water, with more than 80 percent of their mass submerged. Along the margins of Antarctica's ice shelves, calving is the primary means by which these masses lose ice. The relatively flat icebergs produced here can be several kilometers across and 600 meters thick. By comparison, thousands of irregularly shaped icebergs are produced by outlet glaciers flowing from the margins of the Greenland Ice Sheet. Many drift southward and find their way into the North Atlantic, where they can pose a hazard to navigation.

Whether the margin of a glacier is advancing, retreating, or remaining stationary depends upon the budget of the glacier. That is, it depends upon the balance or lack of balance between accumulation on the one hand and wastage (also termed **ablation**) on the other. If ice accumulation exceeds ablation, the glacial front advances until the two factors balance. At this point, the terminus of the glacier is stationary. Should a warming trend occur that causes ablation to exceed accumulation, the ice front will retreat. As the terminus of the glacier retreats, the extent of the zone of ablation diminishes. Therefore, in time a new balance will be reached between accumulation and ablation, and the ice front will again become stationary.

Whether the margins of a glacier are advancing, retreating, or stationary, the ice within the glacier continues to flow forward. In the case of a receding glacier, the ice simply does not flow forward rapidly enough to offset ablation. This point is illustrated rather well in Figure 12.6. As the line of stakes within Rhone Glacier continued to move downvalley, the terminus of the glacier slowly retreated upvalley.

GLACIAL EROSION

Glaciers are capable of great amounts of erosion. For anyone who has observed the terminus of an alpine glacier, the evidence of its erosive force is clear. The observer can witness firsthand the release of rock material of various sizes from the ice as it melts. All signs lead to the conclusion that the ice has scraped, scoured, and torn rock from the floor and walls of the valley and carried it downvalley. Indeed, as a medium of sediment transport, ice has no equal. Once rock debris is acquired by the glacier, the enormous competency of ice will not allow the debris to settle out like the load carried by a stream or by the wind. Consequently, glaciers can carry huge blocks that no other erosional agent could possibly budge (Figure 12.7). Although today's glaciers are of limited importance as erosional agents, many landscapes that were modified by the widespread glaciers of the most recent ice age still reflect to a high degree the work of ice.

FIGURE 12.8
Results of glacial abrasion. Scratches and grooves in limestone, St. Johns Bay, Newfoundland. (Photo by Peter Kresan)

FIGURE 12.7
A giant, glacially transported boulder in Yellowstone National Park. (Photo by E. J. Tarbuck)

Glaciers erode the land primarily in two ways. First, as a glacier flows over a fractured bedrock surface, it loosens and lifts blocks of rock and incorporates them into the ice. This process, known as **plucking**, occurs when meltwater penetrates the cracks and joints of bedrock beneath a glacier and freezes. As the water expands, it exerts tremendous leverage that pries the rock loose. In this manner sediment of all sizes, ranging from particles as fine as flour to blocks as big as houses, becomes part of the glacier's load.

The second major erosional process is **abrasion**

(Figure 12.8). As the ice and its load of rock fragments slide over bedrock, they function as a kind of "sandpaper" to smooth and polish the surface below. The pulverized rock produced by the glacial "grist mill" is appropriately called **rock flour**. So much rock flour may be produced that meltwater streams flowing out of a glacier often have the grayish appearance of skimmed milk and offer visible evidence of the grinding power of ice. When the ice at the bottom of a glacier contains large fragments of rock, long scratches and grooves called **glacial striations** may be cut into the bedrock (Figure 12.8). These linear grooves provide clues as to the direction of ice flow. By mapping the striations over large areas, patterns of glacial flow can often be reconstructed. On the other hand, not all abrasive action produces striations. The rock surfaces over which the glacier moves may also become highly polished by the ice and its load of finer particles. The broad expanses of smoothly polished granite in Yosemite National Park, California, provide an excellent example.

The erosional effects of valley glaciers and ice sheets are quite different. A visitor to a glaciated mountain region is likely to see a sharp and angular topography. The reason is that as alpine glaciers move downvalley, they tend to accentuate the irregu-

BOX 12.2

Surging Glaciers

The copious snows that fall in the mountains of southeastern Alaska and adjacent Canada nurture an extensive system of valley glaciers. Because these glaciers are relatively accessible and exhibit a diversity of behaviors, they are of great interest to scientists. The ice lobes are used as natural laboratories for the study of glacial processes and climatic change.

Although many of these glaciers are retreating, others are advancing, and a few exhibit the sudden movements called *surges*. Hubbard Glacier is one example. From about 1900, when records of its movements were begun, until 1986, the ice within Hubbard Glacier moved forward at a relatively steady rate of 100 meters or so per year. Since ablation did not entirely offset this movement, the front of the ice gradually advanced. Then unexpectedly, during the winter of 1986, Hubbard Glacier experienced a period of surging. At times the ice moved as fast as 14 meters per day. A tributary also surged forward at rates exceeding 30 meters per day.

Certainly the most studied glacier exhibiting a surging-type movement is the 24-kilometer-long Variegated Glacier northwest of Juneau, Alaska (Figure 12.B). Unlike Hubbard Glacier, Variegated Glacier is known to surge at regular intervals. Since 1906, surges have occurred about once every 17 to 20 years. The most recent and thoroughly investigated event took place in 1982–1983. During this surge the ice advanced at rates as high as 54 meters (180 feet) per day.

It is not yet clear whether the mechanism that triggers these rapid movements is the same for each surging-type glacier. However, researchers studying the Variegated Glacier have determined that the surges of this ice mass take the form of a rapid increase in basal sliding that is caused by increases in water pressure beneath the ice. The increased water pressure at the base of the glacier acts to reduce friction between the underlying bedrock and the moving ice. The pressure buildup, in turn, is related to changes in the system of passageways that conduct water along the glacier's bed and deliver it as outflow to the terminus. Geologists study the surge phenomenon not only because it is an unsolved problem but also because an understanding of the mechanism that triggers surges may have wider significance. This broader interest exists,

> . . . because of the possibilities that glacier surges may impinge upon works of man and that surging of the Antarctic ice sheet may be a factor in the initiation of ice ages and in cyclic variations of sea level.*

*Barclay Kamb et al., "Glacier Surge Mechanism: 1982–1983 Surge of Variegated Glacier, Alaska," *Science* 227 (February 1985): 469.

larities of the mountain landscape by creating steeper canyon walls and making bold peaks even more jagged. By contrast, continental ice sheets generally override the terrain and hence tend to subdue rather than accentuate the irregularities they encounter.

Finally, it should be pointed out that, as is the case with other agents of erosion, the rate of glacial erosion is highly variable. This differential erosion by ice is largely controlled by four factors: (1) rate of glacial movement; (2) thickness of the ice; (3) shape, abundance, and hardness of the rock fragments contained in the ice at the base of the glacier; and (4) the erodibility of the surface beneath the glacier. Variations in any or all of these factors from time to time and/or from place to place mean that the features, effects, and degree of landscape modification in glaciated regions can vary greatly.

LANDFORMS CREATED BY GLACIAL EROSION

Although the erosional potential of ice sheets is enormous, landforms carved by these huge ice masses usually do not inspire the same degree of

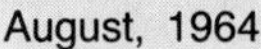

August, 1964

August, 1965

FIGURE 12.B
The surge of Variegated Glacier, a valley glacier near Yakutat, Alaska, is captured in these two aerial photographs taken one year apart. During a surge, ice velocities in Variegated Glacier are 20 to 50 times greater than during a quiescent phase. (Photos by Austin Post, U.S. Geological Survey)

FIGURE 12.9
Prior to glaciation, a mountain valley is typically narrow and V-shaped. During glaciation, an alpine glacier widens, deepens, and straightens the valley, creating a U-shaped glacial trough. The series of lakes are termed pater noster lakes. This valley is in Glacier National Park, Montana. (Photo by John Montagne)

wonderment and awe as do the erosional features created by valley glaciers. In regions where the erosional effects of continental ice sheets are significant, glacially scoured surfaces and subdued terrain are the rule. By contrast, erosion by valley glaciers accentuates an already mountainous topography. Much of the rugged mountain scenery so celebrated for its majestic beauty is the product of glacial erosion.

Glaciated Valleys

A hike up a glaciated valley will reveal a number of striking ice-created features. The valley itself is often a dramatic sight. Rather than creating their own valleys, glaciers take the path of least resistance by following the courses of pre-existing stream valleys. Prior to glaciation mountain valleys are characteristically narrow and V-shaped because streams are well above base level and are therefore downcutting. However, during glaciation these narrow valleys undergo a transformation as the glacier widens and deepens them, creating a U-shaped **glacial trough** (Figure 12.9). In addition to producing a broader and deeper valley, the glacier also straightens the valley. As ice flows around sharp curves, its great erosional force removes the spurs of land that extend into the valley. The results of this activity are triangular-shaped cliffs called **truncated spurs**.

Since the intensity of glacial erosion depends in part upon the thickness of the ice, main (trunk) glaciers cut their valleys deeper than do their smaller tributary glaciers. Thus, after the glaciers have receded, the valleys of feeder glaciers stand above the main glacial trough and are termed **hanging valleys**. Rivers flowing through hanging valleys may produce spectacular waterfalls, such as those in Yosemite National Park (Figure 12.10).

As hikers walk up a glacial trough, they may pass a series of bedrock depressions on the valley floor that were probably formed by plucking and scoured by the abrasive force of the ice. If these depressions are filled with water, they are called **pater noster lakes**, a reference to the fact that, from high above or on a map, they resemble a string of beads (see Figure 12.9).

At the head of a glacial valley is a very characteristic and often imposing feature associated with an alpine glacier—a **cirque**. As Figure 12.11 illustrates, these bowl-shaped depressions have precipitous walls on three sides but are open on the downvalley side. The cirque represents the focal point of the glacier's growth, that is, the area of snow accumulation and ice formation. Although the origin of cirques is still not totally clear, they are believed to begin as irregularities in the mountainside that are subsequently enlarged by frost wedging and plucking along the sides and bottom of the glacier. After the glacier has melted away, the cirque basin is often occupied by a small lake called a **tarn** (Figure 12.12).

Sometimes when two glaciers exist on opposite sides of a divide, one flowing away from the other, the common headwall (back wall) between the cirques is largely eliminated as plucking and frost action enlarge each cirque. When this occurs, the two glacial troughs intersect to create a gap or pass from one valley into the other. Such a feature is termed a **col**. Some important and well-known mountain passes that are cols include St. Gotthard Pass in the Swiss Alps, Tioga Pass in the Sierra Nevada, and Berthoud Pass in the Rocky Mountains.

Before leaving the topic of glacial troughs and their associated features, one more rather well-

FIGURE 12.10
Bridalveil Falls in Yosemite National Park cascades from a hanging valley into the glacial trough below. (Photo by E. J. Tarbuck)

FIGURE 12.11
A cirque in Alaska's Alsex Range. These bowl-shaped depressions are found at the heads of glacial valleys and represent a focal point of ice formation. (Photo by Bruce F. Molnia, courtesy of Terraphotographics/BPS)

known feature should be discussed—fiords. **Fiords** are deep, often spectacular, steep-sided inlets of the sea that are present at high latitudes where mountains are adjacent to the ocean (Figure 12.13). Norway, British Columbia, Greenland, New Zealand, Chile, and Alaska all have coastlines characterized by fiords. They represent drowned glacial troughs that were submerged as the ice left the valley and sea level rose following the Ice Age. The depths of fiords may exceed 1000 meters. However, the great depths of these flooded troughs is only partly explained by the post–Ice Age rise in sea level. Unlike the situation governing the downward erosional work of rivers, sea level does not act as base level for glaciers. As a consequence, glaciers are capable of eroding their beds far below the surface of the sea. For example, a 300-meter thick alpine glacier can carve its valley floor more than 250 meters below sea level before downward erosion ceases and the ice begins to float.

Arêtes and Horns

A visit to the Alps, the Northern Rockies, or many other scenic mountain landscapes carved by valley glaciers would reveal not only glacial troughs, cirques, pater noster lakes, and the other related features just discussed. In addition, a visitor would likely see sinuous, sharp-edged ridges called **arêtes** and sharp, pyramid-like peaks called **horns** projecting above the surroundings. Both features can originate from the same basic process, the enlargement of

FIGURE 12.12
This portion of the Rocky Mountains in Montana has clearly been eroded by valley glaciers. The smallest lake is called a tarn. A tarn is a lake that occupies a cirque. (Photo by Michael Collier)

FIGURE 12.13
Like other fiords, Muir Inlet, Alaska, is a drowned glacial trough. (Photo by Bruce F. Molnia, courtesy of Terraphotographics/BPS)

cirques produced by plucking and frost action. In the case of the spires of rock called horns, a group of cirques around a single high mountain are responsible. As the cirques enlarge and converge, an isolated horn is produced. Certainly the most famous example is the classic Matterhorn in the Swiss Alps (Figure 12.14). Arêtes can be formed in a similar manner except that the cirques are not clustered around a

FIGURE 12.14
Horns are sharp, pyramid-like peaks that are fashioned by alpine glaciers. This example is the famous Matterhorn in the Swiss Alps. (Photo by E. J. Tarbuck)

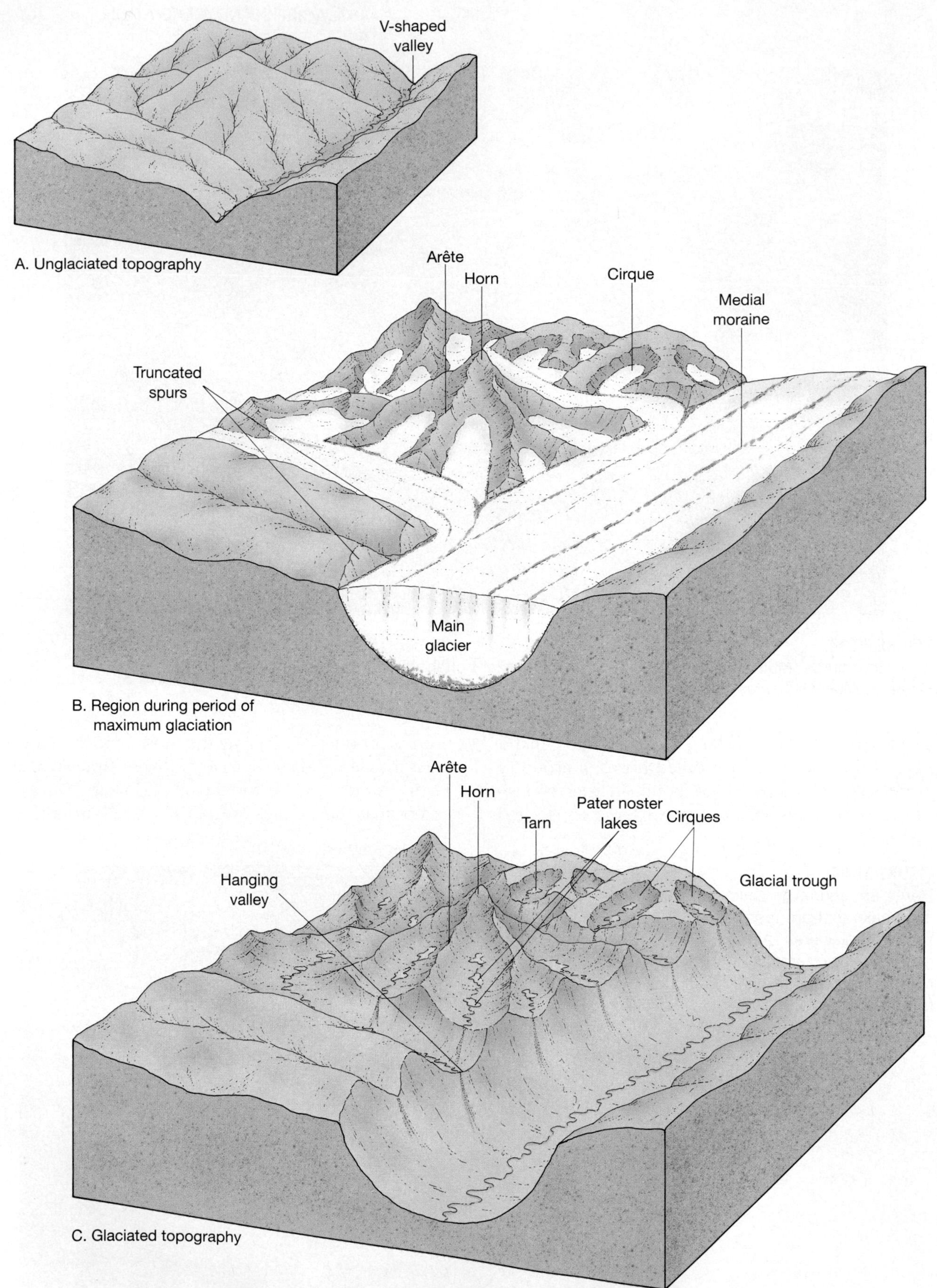

FIGURE 12.15
Erosional landforms created by alpine glaciers.

point but rather exist on opposite sides of a divide. As the cirques grow, the divide separating them is reduced to a very narrow knife-like partition. An arête, however, may also be created in another way. When two glaciers occupy parallel valleys, an arête can form when the divide separating the moving tongues of ice is progressively narrowed as the glaciers scour and widen their valleys.

The landforms created by glacial erosion in an alpine environment are summarized in Figure 12.15.

Roches Moutonnées

In many glaciated landscapes, but most frequently where continental ice sheets modified the terrain, the ice carves small streamlined hills from protruding bedrock knobs. These asymmetrical knobs of bedrock, called **roches moutonnées**, are formed when glacial abrasion smoothes the gentle slope facing the oncoming ice sheet and plucking steepens the opposite side as the ice rides over the knob (Figure 12.16). When roches moutonnées are present, they can be used to indicate the direction of glacial flow, because the gentler slope is on the side from which the ice advanced.

GLACIAL DEPOSITS

As we have seen, glaciers are capable of acquiring and transporting a huge load of debris as they slowly and relentlessly advance across the land. Of course, these materials must be deposited when the ice eventually melts. Thus in regions of deposition, glacial sediment can play a truly significant role in shaping the physical landscape. For example, in many areas once covered by the continental ice sheets of the recent Ice Age, the bedrock is rarely exposed because glacial deposits that are tens or even hundreds of meters thick completely mantle the terrain. The general effect of these deposits is to reduce the local relief and thus level the topography. Indeed, many of the rural country scenes that are familiar to many of us—rocky pastures in New England, wheat fields in the Dakotas, rolling farmland in the Midwest—result directly from glacial deposition.

Long before the theory of an extensive Ice Age was ever proposed, much of the soil and rock debris covering portions of Europe was recognized as coming from somewhere else. At the time, these "foreign" materials were believed to have been "drifted" into their present positions by floating ice during an ancient flood. As a consequence, the term *drift* was applied to this sediment. Although rooted in an incorrect concept, this term was so well established by the time the true glacial origin of the debris became widely recognized that it remained in the basic glacial vocabulary. Today the word **drift** is an all-embracing term for sediments of glacial origin, no matter how, where, or in what shape they were deposited.

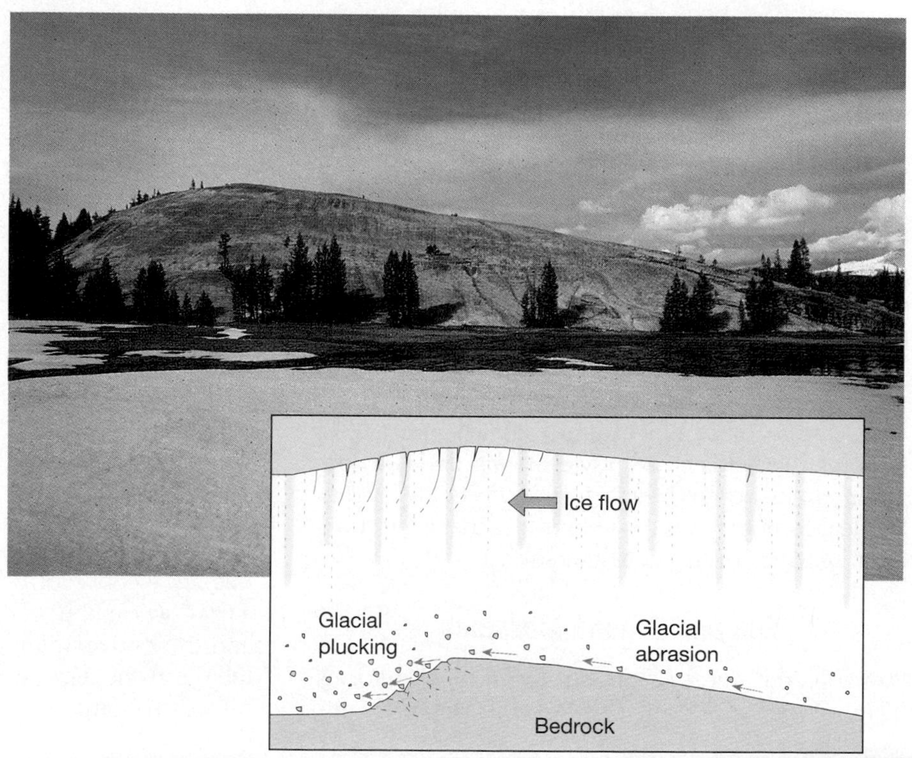

FIGURE 12.16
Roche moutonnée, in Yosemite National Park, California. The gentle slope was abraded and the steep side was plucked. The ice moved from right to left. (Photo by E. J. Tarbuck)

One of the features that distinguishes drift from sediments laid down by other erosional agents is that glacial deposits consist primarily of mechanically weathered rock debris that underwent little or no chemical weathering prior to deposition. Thus, minerals that are notably prone to chemical decomposition, such as hornblende and the plagioclase feldspars, are often conspicuous components in glacial sediments.

Glacial drift is divided by geologists into two distinct types: (1) materials deposited directly by the glacier, which are known as **till**; and (2) sediments laid down by glacial meltwater, called **stratified drift**.

LANDFORMS MADE OF TILL

Till is deposited as glacial ice melts and drops its load of rock fragments. Unlike moving water and wind, ice cannot sort the sediment it carries; therefore, deposits of till are characteristically unsorted mixtures of many particle sizes (Figure 12.17). A close examination of this sediment would reveal that many of the pieces are striated and polished as the result of being dragged along by the glacier. Such pieces help distinguish till from other deposits that may also consist of a mixture of different sediment sizes, such as the debris from a mudflow or a rockslide.

FIGURE 12.17
Glacial till is an unsorted mixture of many different sediment sizes. (Photo by E. J. Tarbuck)

Boulders found in the till or lying free on the surface are called **glacial erratics**, if they are different from the bedrock below. Of course, this means that they must have been derived from a source outside the area wherein they are found. Although the locality of origin for most erratics is unknown, the origin of some can be determined. In many cases evidence shows that the boulders were transported as far as 500 kilometers from their source area and, in a few instances, more than 1000 kilometers. Therefore, by studying glacial erratics as well as the mineral composition of the remaining till, geologists are sometimes able to trace the path of a lobe of ice. In portions of New England, as well as other areas, erratics dot pastures and farm fields. In fact, in some places these large rock fragments were cleared from fields and piled to make fences and walls (Figure 12.18). Keeping the fields clear, however, was and is an ongoing chore because each spring the fields have to be cleared of newly exposed erratics that wintertime frost heaving has lifted to the surface.

End and Ground Moraines

Probably the most common term for landforms made of glacial deposits is *moraine*. Originally this term was used by French peasants when referring to the ridges and embankments of debris found near the margins of glaciers in the French Alps. Today, however, moraine has a broader meaning, because it is applied to a number of landforms, all of which are composed primarily of till.

An **end moraine** is a ridge of till that forms at the terminus of both alpine and continental glaciers. These relatively common landforms are deposited when a state of equilibrium is attained between ablation and ice accumulation. That is, the end moraine forms when the ice is melting and evaporating near the end of the glacier at a rate equal to the forward advance of the glacier from its region of nourishment. Although the terminus of the glacier is now stationary, the ice continues to flow forward, delivering a continuous supply of sediment in the same manner a conveyor-belt delivers goods to the end of a production line. As the ice melts, the till is dropped and the end moraine grows. Therefore, the longer the ice front remains stable, the larger the ridge of till will become.

Eventually the time comes when ablation exceeds nourishment. At this point the front of the glacier will begin to recede in the direction from which it originally advanced. However, as the ice front retreats, the conveyor belt action of the glacier continues to provide fresh supplies of till to the terminus. In this manner a large quantity of till is deposited as the ice melts away, creating a rock-strewn, undulating plain. This gently rolling layer of till laid down as the ice front recedes is termed **ground moraine**. Ground moraine has a leveling effect, filling in low spots and clogging old stream channels, often leading to a derangement of the existing drainage system. In areas where this layer of till is still relatively fresh, such as the northern Great Lakes region, poorly drained swampy land is quite common.

Periodically, a glacier will retreat to a point where ablation and nourishment once again balance. When this happens the ice front stabilizes and a new end moraine is created.

The pattern of end moraine formation and ground moraine deposition may be repeated many times before the glacier has completely vanished. Such a pattern is illustrated by Figure 12.19. It should be pointed out that the outermost end moraine marks the limit of the glacial advance. Because of its special status, this end moraine is also called the **terminal moraine**. On the other hand, the end moraines that were created as the ice front occasionally stabilized during retreat are termed **recessional moraines**. Note that both terminal and recessional moraines are essentially alike; the only difference between them is their relative positions.

End moraines deposited by the last major stage of Ice Age glaciation are prominent features in many parts of the Midwest and Northeast. In Wisconsin, the wooded, hilly terrain of the Kettle Moraine near Milwaukee is a particularly picturesque example. Certainly a well-known example in the Northeast is Long Island. This linear strip of glacial sediment that extends northeastward from New York City is part of an end moraine complex that stretches from eastern

FIGURE 12.18
Land cleared of glacial erratics which were then piled atop one another to build this stone wall in Wisconsin. (Photo by E. J. Tarbuck)

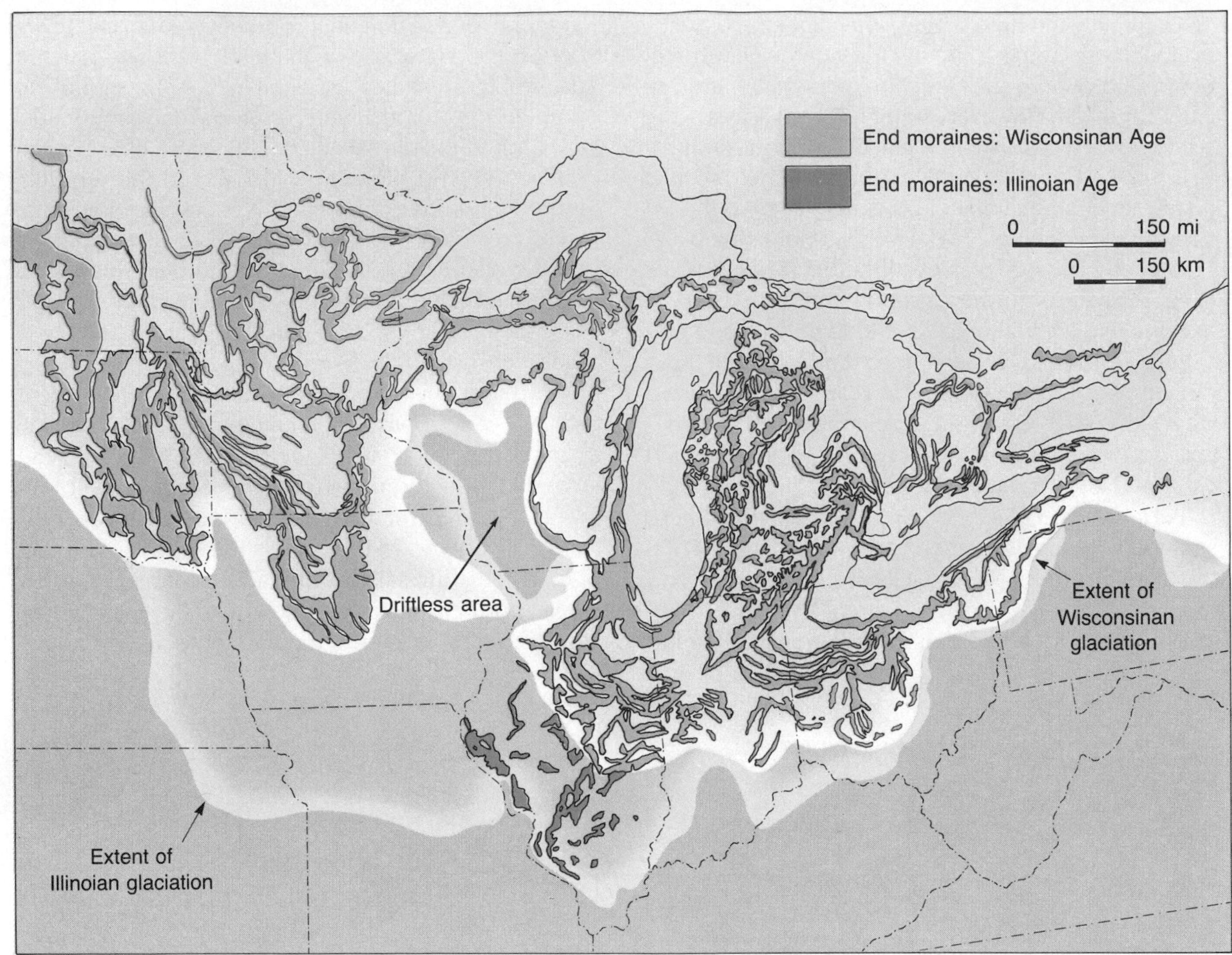

FIGURE 12.19
End moraines of the Great Lakes region. Those deposited during the most recent (Wisconsinan) stage are most prominent.

Pennsylvania to Cape Cod, Massachusetts. The end moraines that make up Long Island represent materials that were deposited by a continental ice sheet in the relatively shallow waters off the coast and built up many meters above sea level. Long Island Sound, the narrow body of water separating the island and the mainland, was not built up as much by glacial deposition and was therefore subsequently flooded by the rising sea following the Ice Age.

Lateral and Medial Moraines

Alpine glaciers produce two types of moraines that occur exclusively in mountain valleys. The first of these is called a **lateral moraine** (Figure 12.20). As we learned earlier, when an alpine glacier moves downvalley, the ice erodes the sides of the valley with great efficiency. In addition, large quantities of debris are added to the glacier's surface as rubble falls or slides from higher up on the valley walls and collects on the edges of the moving ice. When the ice eventually melts, this accumulation of debris is dropped next to the valley walls. These ridges of till paralleling the sides of the valley constitute the lateral moraines.

The second type of moraine that is unique to alpine glaciers is the **medial moraine**. Medial moraines are created when two alpine glaciers coalesce to form a single ice stream. The till that was once carried along the edges of each glacier joins to form a single dark stripe of debris within the newly enlarged glacier. The creation of these dark stripes within the ice stream is one obvious proof that glacial ice moves, because the moraine could not form if the ice did not flow downvalley. It is quite common to see several medial moraines within a single large

FIGURE 12.20
A well-developed lateral moraine deposited by the shrinking Athabaska Glacier in the Canadian Rockies. Note the highway to the left of the lateral moraine for scale. (Photo by James E. Patterson)

FIGURE 12.21
Drumlins, such as this one in upstate New York, are depositional features associated with continental ice sheets. (Courtesy of Ward's Natural Science Establishment, Inc., Rochester, N.Y.)

alpine glacier, because a streak will form whenever a tributary glacier joins the main valley.

Drumlins

Moraines are not the only landforms deposited by glaciers. In some areas that were once covered by continental ice sheets, a special variety of glacial landscape exists—one characterized by smooth, elongate, parallel hills called **drumlins** (Figure 12.21). Certainly one of the best-known drumlins is Bunker Hill in Boston, the site of the famous Revolutionary War battle in 1775. An examination of Bunker Hill or other less famous drumlins would reveal that they are streamlined asymmetrical hills composed largely of till. They range in height from about 15 to 50 meters and may be up to one kilometer long. The steep side of the hill faces the direction from which the ice advanced, while the gentler, longer slope points in the direction the ice moved. Drumlins are

BOX 12.3

Medial Moraines and Mineral Exploration

The study of medial moraines may prove useful to exploration geologists who wish to determine the mineral resource potential of actively glaciated areas (Figure 12.C). In high mountainous terrains that contain extensive valley glaciers, mineral exploration is inhibited because it is difficult to acquire suitable material for geochemical analysis. Not only is sampling difficult, time-consuming, and expensive, but many exposures of bedrock are inaccessible, even by helicopter. A new research technique assesses the mineral resource possibilities of such rugged areas by sampling debris from medial moraines. This is possible because each medial moraine contains a representative sample of the bedrock composing the valley wall in which its debris originated. Since glacial ice moves by laminar flow, there is no mixing of the material in medial moraines for most of the length of the glacier. After sampling and analyzing the material from many medial moraines, geologists can use the data to determine the nature of the bedrock in the valleys where each moraine formed.

> Each medial moraine can be traced, using aerial photos, back to its bedrock source. Therefore, if the field examination reveals visible mineralization, or if the geochemical analysis reveals . . . high metal contents, its bedrock location can be pinpointed accurately and more studies done.*

Although this method is not a substitute for conventional bedrock mapping, it does provide a useful tool for reconnaissance-level studies. According to the researchers who developed the techniques, it will allow ". . . the exploration geologist . . . a means for rapid, accurate evaluation of resource potential of high, mountainous glaciated regions."†

*George C. Stephens et al., "Geochemical Research Program Underway," *Geotimes* 30 (July 1985): 16.

†George C. Stephens, p. 17.

FIGURE 12.C
Medial moraines form when the lateral moraines of merging valley glaciers join. Sampling medial moraines is one method of evaluating the resource potential in high mountainous regions containing extensive valley glacier systems. St. Elias Mountains, Yukon Territory. (Photo by Warren Hamilton, U.S. Geological Survey)

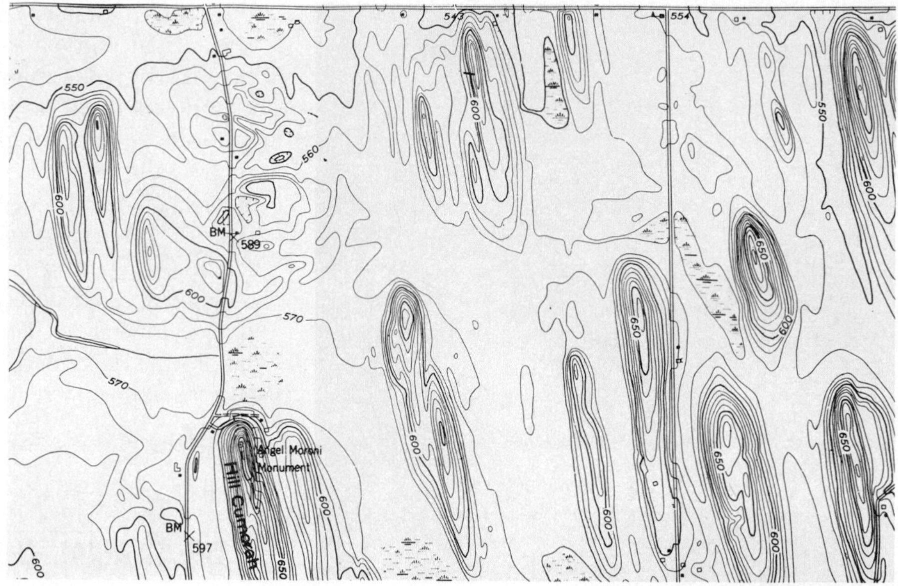

FIGURE 12.22
Portion of a drumlin field shown on the Palmyra, New York, 7.5 minute topographic map. North is at the top. The drumlins are steepest on the north side, indicating that the ice advanced from this direction.

not found as isolated landforms, but rather occur in clusters called *drumlin fields* (Figure 12.22). One such cluster, east of Rochester, New York, is estimated to contain about 10,000 drumlins. Although drumlin formation is not fully understood, their streamlined shape would indicate that they were molded in the zone of plastic flow within an active glacier. It is believed that many drumlins originate when glaciers advance over previously deposited drift and reshape the material.

LANDFORMS MADE OF STRATIFIED DRIFT

As the name implies, stratified drift is sorted according to the size and weight of the particles. Since ice is not capable of such sorting activity, these materials are not deposited directly by the glacier as till is, but rather reflect the sorting action of glacial meltwater. Accumulations of stratified drift often consist largely of sand and gravel, that is, bed load material, because the finer rock flour remains suspended and therefore is commonly carried far from the glacier by the meltwater streams. An indication that stratified drift consists primarily of sand and gravel is reflected in the fact that in many areas these deposits are actively mined as a source of aggregate for road work and other construction projects.

Outwash Plains and Valley Trains

At the same time that an end moraine is forming, water from the melting glacier cascades over the till, sweeping some of it out in front of the growing ridge of unsorted debris. Meltwater generally emerges from the ice in rapidly moving streams that are often choked with suspended material and carry a substantial bed load as well. As the water leaves the glacier, it moves onto the relatively flat surface beyond and rapidly loses velocity. As a consequence, much of its bed load is dropped and the meltwater begins weaving a complex pattern of braided channels (see Figure 10.16). In this way a broad, ramplike surface composed of stratified drift is built adjacent to the downstream edge of most end moraines. When the feature is formed in association with an ice sheet, it is termed an **outwash plain**, and when largely confined to a mountain valley, it is usually referred to as a **valley train**.

Often outwash plains and valley trains are pockmarked with basins or depressions known as **kettles** (Figure 12.23). Kettles also occur in deposits of till. Kettles are formed when a block of stagnant ice becomes wholly or partly buried in drift and ultimately melts, leaving a pit in the glacial sediment. Although most kettles do not exceed 2 kilometers in diameter, some with diameters exceeding 10 kilometers occur in Minnesota. Likewise, the typical depth of most kettles is less than 10 meters, although the vertical dimensions of some approach 50 meters. In many cases, water eventually fills the depression and forms a pond or lake. One well-known example is Walden Pond near Concord, Massachusetts. It is here that Henry David Thoreau lived alone for two years in the 1840s and about which he wrote *Walden,* his classic of American literature.

Ice-Contact Deposits

When the wasting terminus of a glacier shrinks to a critical point, flow virtually stops and the ice be-

FIGURE 12.23
These ponds occupy depressions called kettles. A kettle forms when a block of ice that was buried in drift melts and leaves a pit. (Photo by Bruce F. Molnia, courtesy of Terraphotographics/BPS)

comes stagnant. With time, meltwater flowing over, within, and at the base of the motionless ice deposits stratified drift. Then, as the supporting ice melts away, the stratified sediment is left behind in the form of hills, terraces, and ridges. Such accumulations are collectively termed **ice-contact deposits** and are classified according to their shapes.

When the ice-contact stratified drift is in the form of a mound or steep-sided hill, it is called a **kame** (Figure 12.24). Some kames represent bodies of sediment deposited by meltwater in openings within or depressions on top of the ice. Others originate as deltas or fans built outward from the ice by meltwater streams. Later, when the stagnant ice melts away, these various accumulations of sediment collapse to form isolated, irregular mounds.

When glacial ice occupies a valley, terraces known as **kame terraces** may be built along the sides of the valley. These features are commonly narrow masses of stratified drift laid down between the glacier and the side of the valley by streams that drop debris along the margins of the shrinking ice mass.

Finally, on some glacial landscapes long, narrow, sinuous ridges composed largely of sand and gravel are present. Some ridges are more than 100 meters high with lengths in excess of 100 kilometers. The dimensions of many others, however, are far less spectacular. Known as **eskers,** these ridges were deposited by meltwater rivers flowing in confined channels within, on top of, and beneath a mass of motionless, stagnant glacial ice (Figure 12.25). While many sediment sizes were carried by the torrents of meltwater in the ice-banked channels, only the coarser material could settle out of the turbulent stream.

Figure 12.26 depicts a hypothetical area during and following glaciation. It illustrates many of the landforms described in the preceding sections that may be found in regions affected by continental ice sheets.

THE GLACIAL THEORY AND THE ICE AGE

At various points in the preceding pages mention was made of the Ice Age, a time when ice sheets and alpine glaciers were far more extensive than they are today. As was noted earlier, there was a time when the most popular explanation for what we now know to be glacial deposits was that the materials had been drifted in by means of icebergs or perhaps simply swept across the landscape by a catastrophic flood. What convinced geologists that an extensive ice age was responsible for these deposits as well as for many other features?

In 1821 a Swiss engineer, Ignaz Venetz, presented a paper suggesting that glacial features occurred at considerable distances from the existing margins of

FIGURE 12.24
Bodies of ice-contact stratified drift are classified according to their shape. When it is in the form of a steep-sided hill, the feature is called a kame. This kame, known as Garriety Mountain, is in southern Wisconsin. (Photo by T. S. Laudon)

A.

FIGURE 12.25
The retreat of the glacier reveals an esker, the sinuous ridge of sand and gravel in the center of the photograph. The esker consists of materials deposited by meltwater in a channel flowing beneath the stagnant ice. Where the stream emptied into a small standing body of water, a delta was formed. (Photo by Bradford Washburn)

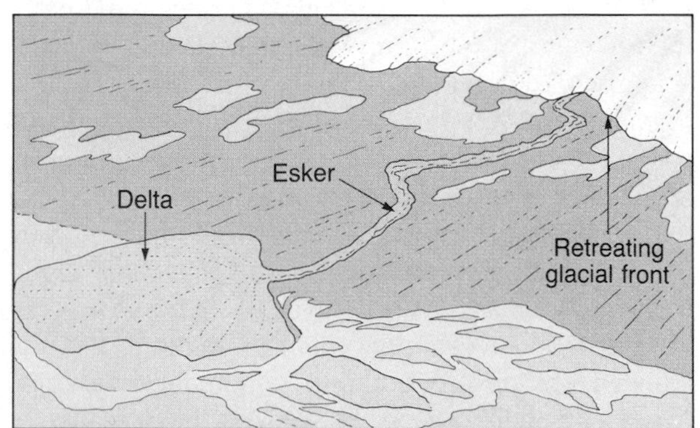

B.

glaciers in the Alps. The inference therefore was that the glaciers had once occupied positions farther downvalley. In 1836, Louis Agassiz (Figure 12.27), another Swiss scientist, who doubted the proposal of widespread glacial activity put forth by Venetz and later by Jean de Charpentier, set out to prove that the idea was not valid. Instead, his fieldwork in the Alps convinced him of the merits of his colleagues' hypothesis. A year later Agassiz authored the theory of a great ice age that had had extensive and far-reaching effects—an idea that was to give Agassiz widespread fame.

The proof of the glacial theory proposed by Agassiz and others constitutes a classic example of apply-

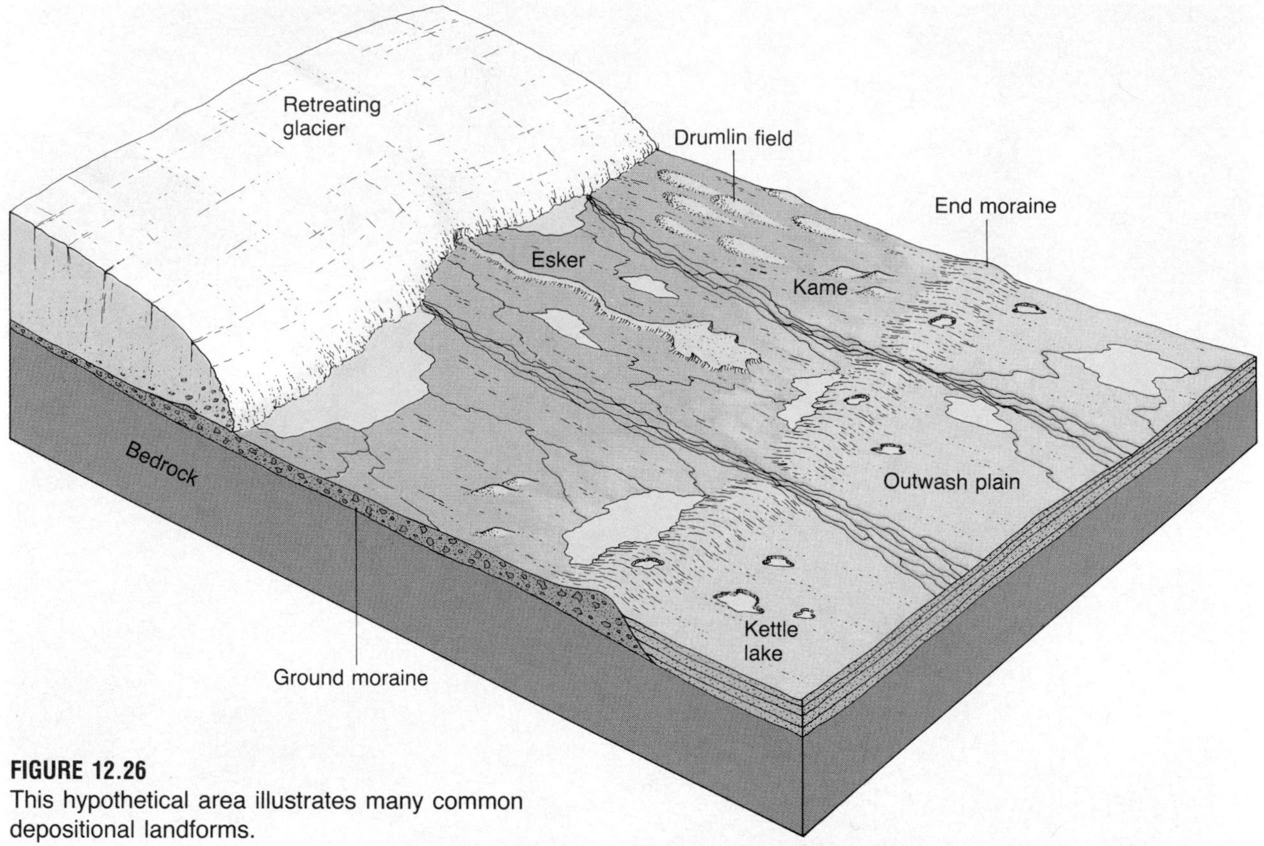

FIGURE 12.26
This hypothetical area illustrates many common depositional landforms.

ing the principle of uniformitarianism. Realizing that certain features are produced by no other known process but glacial action, they were able to begin reconstructing the extent of now-vanished ice sheets based on the presence of features and deposits found far beyond the margins of present-day glaciers. In this manner the development and verification of the glacial theory continued during the nineteenth century, and through the efforts of many scientists, a knowledge of the nature and extent of former ice sheets became clear.

By the turn of the twentieth century, geologists had largely determined the areal extent of the Ice Age glaciation. Further, during the course of their investigations they had discovered that many glaciated regions had not one layer of drift but several. Moreover, close examination of these older deposits showed well-developed zones of chemical weathering and soil formation as well as the remains of plants that require warm temperatures. The evidence was clear; there had not been just one glacial advance but several, each separated by extended periods when climates were as warm as or warmer than at present. The Ice Age then had not simply been a time when the ice advanced over the land, lingered for a while, and then receded. Rather, the period was

FIGURE 12.27
Louis Agassiz (1807–1873) played a major role in the development of glacial theory. (Courtesy of Harvard University Archives)

FIGURE 12.28
At their maximum extent, Pleistocene glaciers covered about 10 million square kilometers (4 million square miles) of North America.

a very complex event characterized by a number of advances and withdrawals of glacial ice.

By early in the twentieth century a four-fold division of the Ice Age had been established for both North America and Europe. The divisions were based largely on studies of glacial deposits. In North America each of the four major stages was named for the midwestern state where deposits of that stage were well exposed and/or were first studied. These are, in order of occurrence, the Nebraskan, Kansan, Illinoian, and Wisconsinan. These traditional divisions remained in place until relatively recently when it was learned that sediment cores from the ocean floor contain a much more complete record of climate change during the Ice Age.* Unlike the glacial record on land, which is punctuated by many unconformities, sea-floor sediments provide an uninterrupted record of climatic cycles for this period. Studies of these sea-floor sediments showed that glacial/interglacial cycles had occurred about every 100,000 years. About twenty such cycles of cooling and warming were identified for the span we call the Ice Age.

During the glacial age, ice left its imprint on almost 30 percent of the earth's land area, including about 10 million square kilometers of North America (Figure 12.28), 5 million square kilometers of Europe, and 4 million square kilometers of Siberia. The amount of glacial ice in the Northern Hemisphere was roughly twice that of the Southern Hemisphere. The primary reason is that the southern polar ice could not spread far beyond the margins of Antarctica. By contrast, North America and Eurasia provided great expanses of land for the spread of ice sheets.

Today we know that the Ice Age began between two and three million years ago. This means that most of the major glacial stages occurred during a division of the geologic time scale called the **Pleistocene epoch**. Although the Pleistocene is commonly used as a synonym for the Ice Age, note that this epoch does not encompass all of the last glacial period. The Antarctic ice sheet, for example, is believed to have formed at least 14 million years ago, and, in fact, might be much older.

*For more on this topic, see Box 6.2, "Sea-Floor Sediments and Climatic Change."

SOME INDIRECT EFFECTS OF ICE-AGE GLACIERS

In addition to the massive erosional and depositional work carried on by Pleistocene glaciers, the ice sheets had other, sometimes profound, effects upon the landscape. For example, as the ice advanced and retreated, animals and plants were forced to migrate. This led to stresses that some organisms could not tolerate. Hence, a number of different plants and animals became extinct. Furthermore, many present-day stream courses bear little resemblance to their preglacial routes. The Missouri River once flowed northward toward Hudson Bay, while the Mississippi River followed a path through central Illinois and the head of the Ohio River reached only as far as Indiana. Other rivers that today carry only a trickle of water but nevertheless occupy broad channels are testimony to the fact that they once carried torrents of glacial meltwater.

In areas that were centers of ice accumulation, such as Scandinavia and the Canadian Shield, the land has been slowly rising over the past several thousand years. Uplifting of almost 300 meters has occurred in the Hudson Bay region. This, too, is the result of the continental ice sheets. But how can glacial ice cause crustal movement? We now understand that the land is rising because the added weight of the three-kilometer-thick mass of ice caused downwarping of the earth's crust. Following the removal of this immense load, the crust has been adjusting by gradually rebounding upward ever since. (Figure 12.29).*

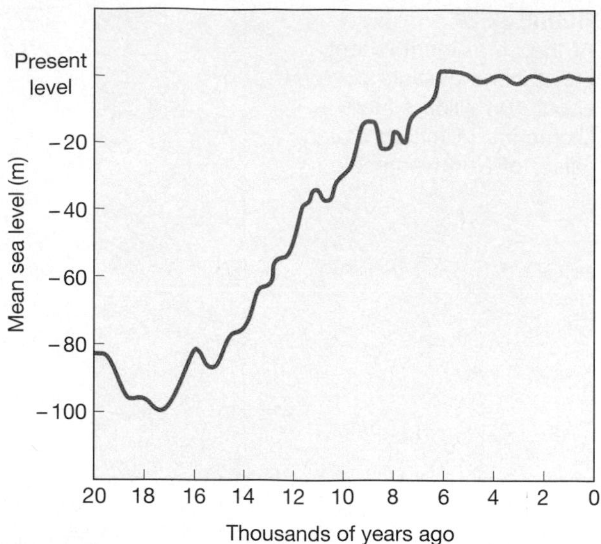

FIGURE 12.30
Changing sea level during the past 20,000 years. The lowest level shown on the graph represents the time about 18,000 years ago when the most recent ice advance was at a maximum.

Certainly one of the most interesting and perhaps dramatic effects of the ice age was the fall and rise of sea level that accompanied the advance and retreat of the glaciers. In Box 12.1 it was pointed out that sea level would rise by an estimated 60 or 70 meters if the water locked up in the Antarctic Ice Sheet were to melt completely. Such an occurrence would flood many densely populated coastal areas. Although the total volume of glacial ice today is great, exceeding 25 million cubic kilometers, during the Ice Age the volume of glacial ice amounted to about 70 million cubic kilometers, or 45 million cubic kilometers more than at present. Since we know that the snow from which glaciers are made ultimately comes from the evaporation of ocean water, the growth of ice sheets must have caused a worldwide drop in sea level (Figure 12.30). Indeed, estimates suggest that sea level was as much as 100 meters lower than today. Thus, land that is presently flooded by the oceans was dry. The Atlantic Coast of the United States lay more than 100 kilometers to the east of New York City; France and Britain were joined where the famous English Channel is today; Alaska and Siberia were connected across the Bering Strait; and

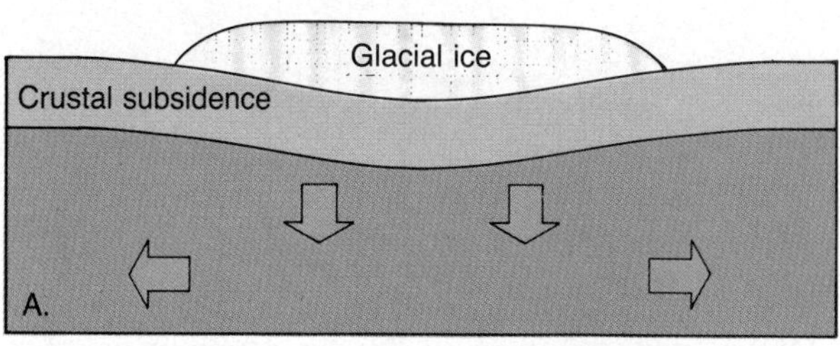

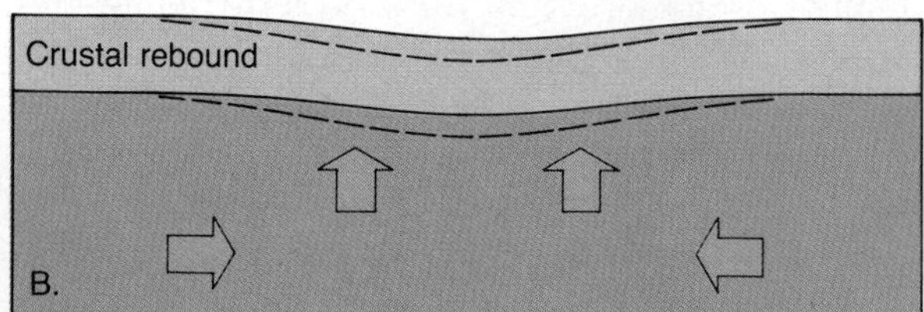

FIGURE 12.29
Simplified illustration showing crustal subsidence and rebound resulting from the addition and removal of continental ice sheets. **A.** In northern Canada and Scandinavia, where there was the greatest accumulation of glacial ice, the added weight caused downwarping of the crust. **B.** Ever since the ice melted, there has been gradual uplift or rebound of the crust.

*For a more complete discussion of this concept, termed *isostatic adjustment,* see the section "Isostasy" in Chapter 20.

Southeast Asia was tied by dry land to the islands of Indonesia.

While the formation and growth of ice sheets was an obvious response to significant changes in climate, the existence of the glaciers themselves triggered important climatic changes in the regions beyond their margins. In arid and semiarid areas on all of the continents, temperatures and thus evaporation rates were lower, and at the same time moderate precipitation totals were experienced. This cooler, wetter climate resulted in the formation of many lakes called **pluvial lakes**, from the Latin term *pluvia* meaning *rain*. In North America, the greatest concentration of pluvial lakes occurred in the vast Basin and Range region of Nevada and Utah (Figure 12.31). By far the largest of the lakes in this region was Lake Bonneville. With maximum depths exceeding 300 meters and an area of 50,000 square kilometers, Lake Bonneville was nearly the same size as present-day Lake Michigan. As the ice sheets waned, the climate again grew more arid, and the lake levels lowered in response. Although most of the lakes completely disappeared, a few small remnants of Lake Bonneville remain, the Great Salt Lake being the largest and best known.

CAUSES OF GLACIATION

A great deal is known about glaciers and glaciation. Much has been learned about glacier formation and movement, the extent of glaciers past and present, and the features created by glaciers, both erosional and depositional. However, a widely accepted theory for the causes of glacial ages has not yet been established. Although nearly 150 years have elapsed since Louis Agassiz proposed his theory of a great ice age, no complete agreement exists as to the causes of such events.

While widespread glaciation has been a rare occurrence in the earth's history, the ice age that encompassed the Pleistocene epoch is not the only glacial period for which a record exists. Other earlier glaciations are indicated by deposits called **tillite**, a sedimentary rock formed when glacial till is lithified. Such deposits, found in strata of several different ages, usually contain striated rock fragments, and some overlie grooved and polished bedrock surfaces or are associated with sandstones and conglomerates that show features of outwash deposits. Two Precambrian glacial episodes have been identified in the geologic record, the first approximately two billion years ago and the second about 600 million years ago. Further, a well-documented record of an earlier glacial age is found in late Paleozoic rocks that are about 250 million years old and which exist on several landmasses.

Any theory that attempts to explain the causes of glacial ages must successfully answer two basic questions. The first question is: What causes the onset of glacial conditions? In order for continental ice sheets to have formed, average temperatures must have been somewhat lower than at present and perhaps substantially lower than throughout much of geologic time. Thus, a successful theory would have to account for the gradual cooling that finally leads to glacial conditions. The second question that requires an answer is: What caused the alternation of glacial and interglacial stages that have been documented for the Pleistocene epoch? While the first question deals with long-term trends in temperature that occur on a scale of millions of years, this second question relates to much shorter-term changes.

Although the literature of science contains a vast array of hypotheses relating to the possible causes of glacial periods, we will discuss only a few major ideas in an effort to summarize current thought.

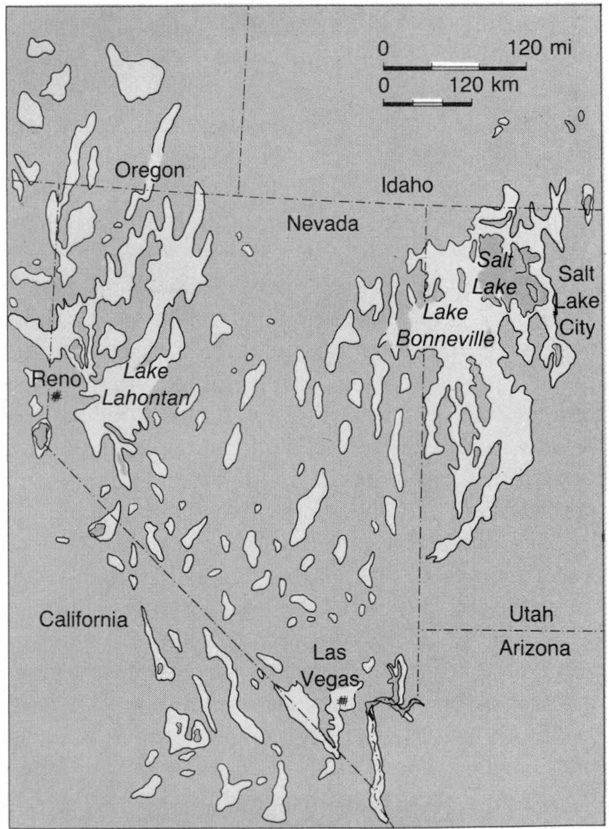

FIGURE 12.31
Pluvial lakes of the western United States. (After R. F. Flint, *Glacial and Quaternary Geology,* New York: John Wiley & Sons.)

Probably the most attractive proposal for explaining the fact that extensive glaciations have occurred only a few times in the geologic past comes from the theory of plate tectonics.* Not only does this theory provide geologists with explanations about many previously misunderstood processes and features, but it also provides a possible explanation for some hitherto unexplainable climatic changes, including the onset of glacial conditions. Since glaciers can form only on land, we know that landmasses must exist somewhere in the higher latitudes before an ice age can commence. Many believe that ice ages have occurred only when the earth's shifting crustal plates have carried the continents from tropical latitudes to more poleward positions.

Glacial features in present-day Africa, Australia, South America, and India indicate that these regions, which are now tropical or subtropical, experienced an ice age near the end of the Paleozoic era, about 250 million years ago. However, there is no evidence that ice sheets existed during this same period in what are today the higher latitudes of North America and Eurasia. For many years this puzzled scientists. Was the climate in these relatively tropical latitudes once like it is today in Greenland and Antarctica? Why did glaciers not form in North America and Eurasia? Until the plate tectonics theory was formulated, there had been no reasonable explanation. Today scientists realize that the areas containing these ancient glacial features were joined together as a single supercontinent located at latitudes far to the south of their present positions. Later, this landmass broke apart, and its pieces, each moving on a different plate, drifted toward their present locations (Figure 12.32). It is now understood that during the geologic past, plate movements accounted for many dramatic climatic changes as landmasses shifted in relation to one another and moved to different latitudinal positions. Changes in oceanic circulation also must have occurred, altering the transport of heat and moisture and consequently the climate as well. Since the rate of plate movement is very slow—on the order of a few centimeters per year—appreciable changes in the positions of the continents occur only over great spans of geologic time. Thus, climatic changes brought about by shifting plates are extremely gradual and happen on a scale of millions of years.

Since climatic changes brought about by moving plates are extremely gradual, the plate tectonics theory cannot be used to explain the alternation between glacial and interglacial climates that occurred during the Pleistocene epoch. Therefore we must look to some other triggering mechanism that may cause climatic change on a scale of thousands rather than millions of years. Today many scientists believe or strongly suspect that the climatic oscillations that characterized the Pleistocene may be linked to variations in the earth's orbit (Figure 12.33). This hypothesis was first developed and strongly advocated by the Yugoslavian scientist Milutin Milankovitch and is based on the premise that variations in incoming solar radiation are a principal factor in controlling the climate of the earth. Milankovitch formulated a comprehensive mathematical model based on the following elements:

1. Variations in the shape *(eccentricity)* of the earth's orbit about the sun;
2. Changes in *obliquity;* that is, changes in the angle that the axis makes with the plane of the earth's orbit; and
3. The wobbling of the earth's axis, called *precession.*

Using these factors, Milankovitch calculated variations in the receipt of solar energy and the corresponding surface temperature of earth back into time in an attempt to correlate these changes with the climatic fluctuations of the Pleistocene. In explaining climatic changes that result from these three variables, it should be noted that they cause little or no variation in the total amount of solar energy reaching the ground. Instead, their impact is felt because they change the degree of contrast between the seasons. Somewhat milder winters in the middle to high latitudes means greater snowfall totals, while cooler summers would bring a reduction in snowmelt.

Over the years the astronomical theory of Milankovitch has been widely accepted, then largely rejected, and now, in light of recent investigations, has once again gained significant support. Among the studies that have added credibility to the astronomical theory is one in which deep-sea sediments containing certain climatically sensitive microorganisms were analyzed to establish a chronology of temperature changes going back nearly one-half million years.* This time scale of climatic change was then compared to astronomical calculations of eccentricity, obliquity, and precession to determine if a correlation did indeed exist. Although the study was very involved and mathematically complex, the conclusions were straightforward. The authors found that major variations in climate over the past several hun-

*A brief overview of the theory appears in Chapter 1, and a more extensive discussion may be found in Chapter 18.

*J. D. Hays, John Imbrie, and N. J. Shackelton, "Variations in the Earth's Orbit: Pacemaker of the Ice Ages," *Science* 194 (1976): 1121–32.

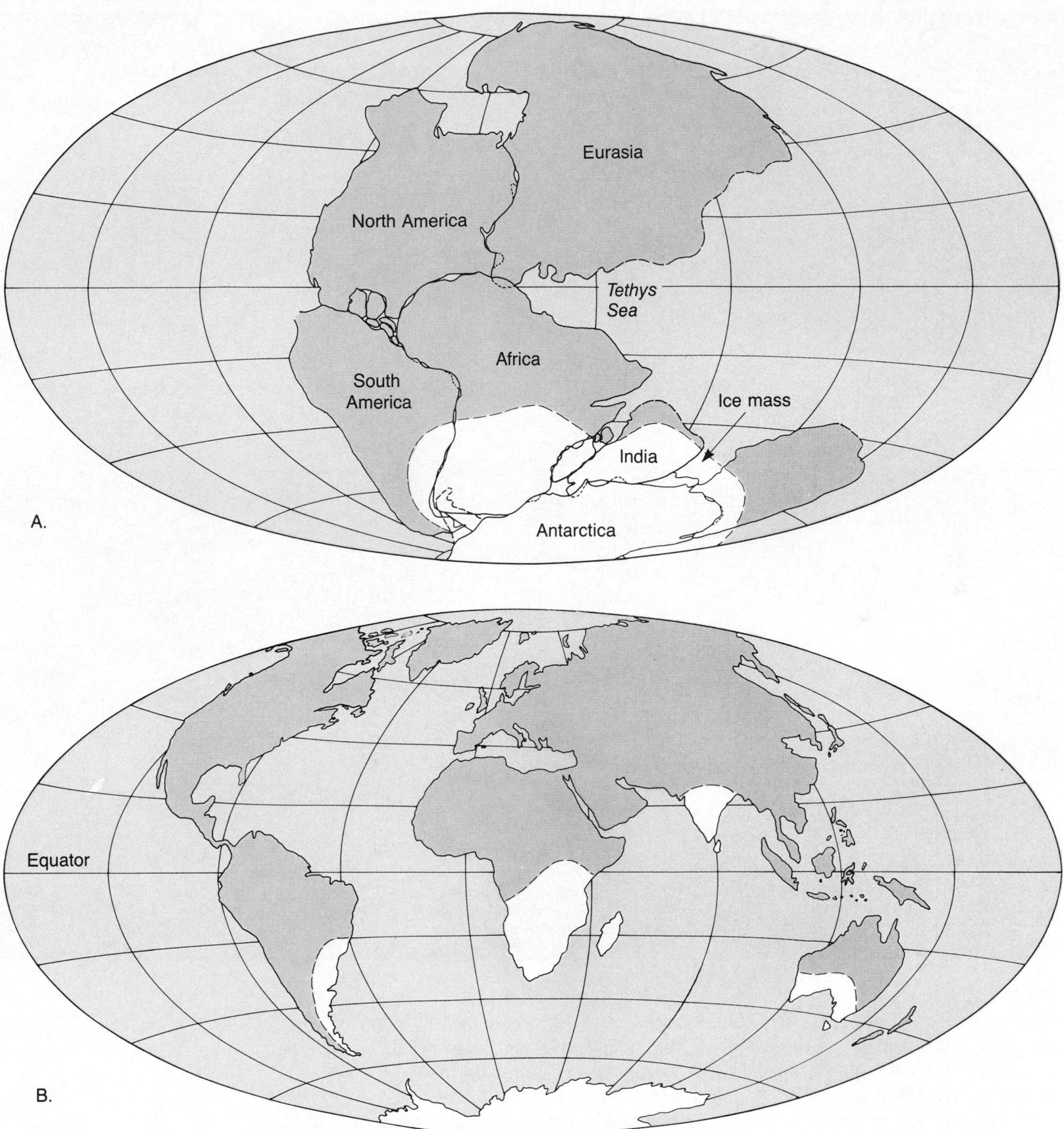

FIGURE 12.32
A. The supercontinent Pangaea showing the area covered by glacial ice 300 million years ago. **B.** The continents as they are today. The shading outlines areas where evidence of the old ice sheets exists. (After R. F. Flint and B. J. Skinner, *Physical Geology,* 2nd ed., p. 418, New York: Wiley, 1977)

dred thousand years were closely associated with changes in the geometry of the earth's orbit; that is, cycles of climatic change were shown to correspond closely with the periods of obliquity, precession, and orbital eccentricity. More specifically, they stated: "It is concluded that changes in the earth's orbital geometry are the fundamental cause of the succession of Quaternary ice ages."*

*J. D. Hays et al., p. 1131. The term *Quaternary* refers to the period on the geologic time scale that encompasses the last 1.6 million years.

FIGURE 12.33
This glacier—McBride Glacier, Glacier Bay, Alaska—is presently shrinking and receding upvalley. Its meltwater is contributing to a rise in sea level. During the Ice Age, glaciers of all types advanced and retreated many times. Today many scientists support the idea that the climatic changes that triggered the growth and shrinkage of glaciers during the Pleistocene were linked to cyclical changes in the earth's orbit. (Photo by Breck Kent)

Let us briefly summarize the ideas that were just described. The theory of plate tectonics provides us with an explanation for the widely spaced and nonperiodic onset of glacial conditions at various times in the geologic past, while the astronomical theory proposed by Milankovitch and recently supported by the work of J. D. Hays and his colleagues furnishes an explanation for the alternating glacial and interglacial episodes of the Pleistocene.

In conclusion, it should be emphasized at this point that the ideas just discussed do not represent the only possible explanations for glacial ages. Although interesting and attractive, these proposals are certainly not without critics; nor are they the only possibilities currently under study. Other factors may and, in fact, probably do enter into the picture.

REVIEW QUESTIONS

1. What is a glacier? Under what circumstances does glacial ice form?
2. Each statement below refers to a particular type of glacier. Name the type of glacier.
   (a) The term *continental* is often used to describe this type of glacier.
   (b) This type of glacier is also called an *alpine glacier.*
   (c) This is a stream of ice leading from the margin of an ice sheet through the mountains to the sea.
   (d) This is a glacier formed when one or more valley glaciers spreads out at the base of a steep mountain front.
   (e) Greenland is the only example in the Northern Hemisphere.
3. Where are glaciers found today? What percentage of the earth's land area do they cover? How does this compare to the area covered by glaciers during the Pleistocene?
4. Describe glacial flow. In a valley glacier does all of the ice move at the same rate? Explain.
5. Why do crevasses form in the upper portion of a glacier but not below 50 meters?
6. Under what circumstances will the front of a glacier advance? Retreat? Remain stationary?
7. Describe the processes of glacial erosion.
8. How does a glaciated mountain valley differ from a mountain valley that was not glaciated?
9. List and describe the erosional features you might expect to see in an area where valley glaciers exist or have recently existed.
10. What is glacial drift? What is the difference between till and stratified drift? What general effect do glacial deposits have on the landscape?
11. List the four basic moraine types. What do all moraines have in common? What is the significance of terminal and recessional moraines?
12. What type of moraine may be used to evaluate the mineral resource potential in regions occupied by valley glaciers? (See Box 12.3.)
13. How do kettles form?
14. What direction was the ice sheet moving that affected the area shown in Figure 12.22? Explain how you were able to determine this.
15. What are ice-contact deposits? Distinguish between kames and eskers.
16. The development of the glacial theory is a good example of applying the principle of uniformitarianism. Explain briefly.
17. In North America four major stages of glaciation have been recognized. List them in the order they occurred.
18. During the Pleistocene epoch the amount of glacial ice in the Northern Hemisphere was about twice as great as in the Southern Hemisphere. Briefly explain why this was the case.
19. List three indirect effects of Ice Age glaciers.
20. How might plate tectonics help explain the cause of ice ages? Can plate tectonics explain the alternation between glacial and interglacial climates during the Pleistocene?

KEY TERMS

ablation (p. 295)
abrasion (p. 295)
alpine glacier (p. 289)
arête (p. 300)
basal slip (p. 292)
calving (p. 294)
cirque (p. 299)
col (p. 299)
crevasse (p. 293)
drift (p. 303)
drumlin (p. 308)
end moraine (p. 304)
esker (p. 310)
fiord (p. 300)
firn (p. 292)
glacial erratic (p. 304)
glacial striations (p. 296)
glacial trough (p. 298)
glacier (p. 288)
ground moraine (p. 305)
hanging valley (p. 299)
horn (p. 300)
ice cap (p. 291)
ice-contact deposit (p. 310)
ice sheet (p. 290)
ice shelf (p. 291)
kame (p. 310)
kame terrace (p. 310)
kettle (p. 309)
lateral moraine (p. 306)
medial moraine (p. 306)
outlet glacier (p. 291)
outwash plain (p. 309)
pater noster lakes (p. 299)
piedmont glacier (p. 292)
plastic flow (p. 292)
Pleistocene epoch (p. 313)
plucking (p. 295)
pluvial lake (p. 315)
recessional moraine (p. 305)
roche moutonnée (p. 303)
rock flour (p. 296)
snowline (p. 294)
stratified drift (p. 304)
surge (p. 294)
tarn (p. 299)
terminal moraine (p. 305)
till (p. 304)
tillite (p. 315)
truncated spur (p. 298)
valley glacier (p. 289)
valley train (p. 309)
zone of accumulation (p. 294)
zone of fracture (p. 293)

13

Deserts and Winds

Opposite: Wind has created ripple marks on these sand deposits in Arizona's Monument Valley. (Photo by Tom Till)
(Top photo by Michael Collier)

The word *desert* literally means *deserted* or *unoccupied.* Many desert areas are not truly deserted and indeed, in some cases, many people live there. Nevertheless, the world's dry regions are probably the least familiar land areas on earth outside of the polar realm.

Desert landscapes frequently appear stark. Their profiles are not softened by a carpet of soil and abundant plant life. Instead, barren rocky outcrops with steep, angular slopes are common. At some places the rocks are tinted orange and red. At others they are gray and brown and streaked with black. For many visitors desert scenery exhibits a striking beauty; to others, the terrain seems bleak. No matter which feeling is elicited, it is clear that deserts are very different from the more humid places where most people live. As we shall see, arid regions are not dominated by a single geologic process. Rather, the effects of tectonic forces, running water, and wind are all apparent. Because these processes combine in different ways from place to place, the appearance of desert landscapes varies a great deal as well (Figure 13.1).

DISTRIBUTION AND CAUSES OF DRY LANDS

The dry regions of the world encompass about 42 million square kilometers, nearly 30 percent of the earth's land surface. No other climatic group covers so large a land area. Within these water-deficient regions, two climatic types are commonly recognized: **desert**, or arid, and **steppe**, or semiarid. The two categories have many features in common; their differences are primarily a matter of degree. The steppe is a marginal and more humid variant of the desert and represents a transition zone that surrounds the desert and separates it from bordering humid climates. The world map showing the distribution of desert and steppe regions reveals that dry lands are

FIGURE 13.1
A rocky hillside in the Sonoran Desert west of Tucson, Arizona. The appearance of desert landscapes varies a great deal from place to place. (Photo by Michael Collier)

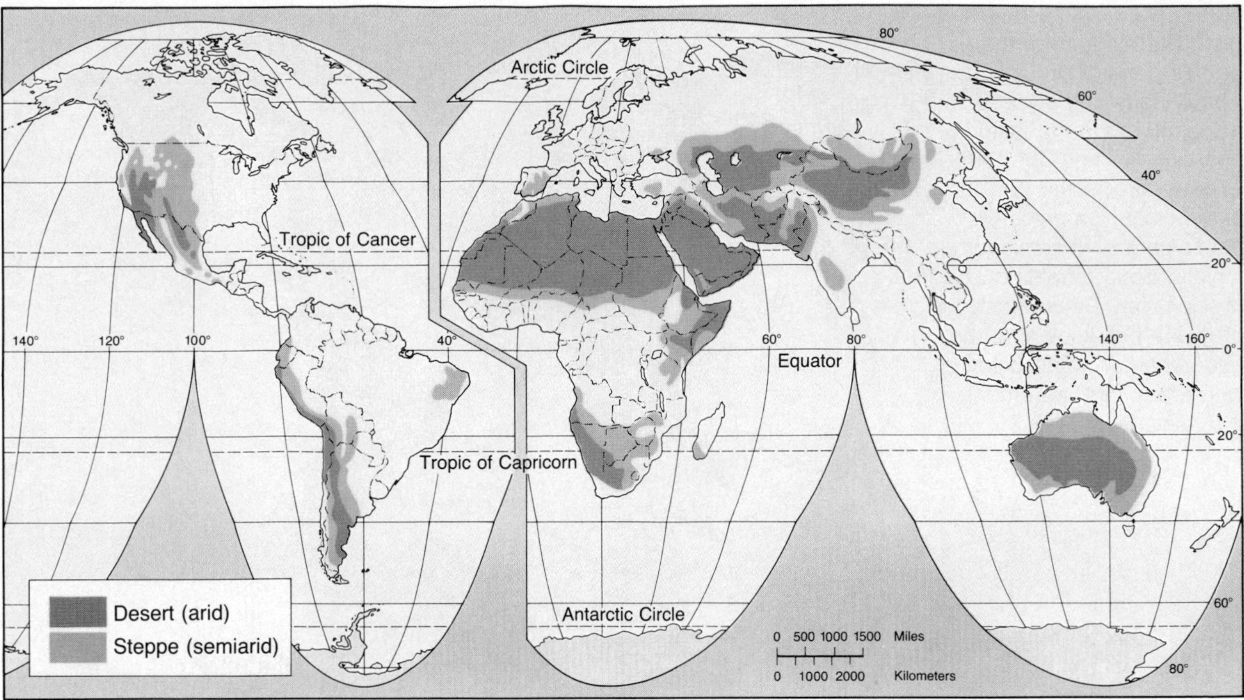

FIGURE 13.2
Arid and semiarid climates cover about 30 percent of the earth's land surface. No other climatic group covers so large an area.

concentrated in the subtropics and in the middle latitudes (Figure 13.2).

The heart of the low-latitude dry climates lies in the vicinities of the Tropics of Cancer and Capricorn. A glance at Figure 13.2 shows a virtually unbroken desert environment stretching for more than 9300 kilometers from the Atlantic coast of North Africa to the dry lands of northwestern India. In addition to this single great expanse, the Northern Hemisphere contains another much smaller area of tropical desert and steppe in northern Mexico and of the south-western United States. In the Southern Hemisphere, dry climates dominate Australia. Almost 40 percent of the continent is desert, and much of the remainder is steppe. In addition, arid and semiarid areas are found in southern Africa and make a limited appearance in coastal Chile and Peru. The existence of this dry subtropical realm is primarily the result of the prevailing global distribution of air pressure and winds. Figure 13.3, an idealized diagram of the earth's general circulation, helps visualize the relationship. Heated air in the pressure belt known as the *equatorial low* rises to great heights (usually between 15 and 20 kilometers) and then spreads out. As the upper-level flow reaches 20° to 30° latitude, north or south, it sinks toward the surface. Air that rises through the atmosphere expands and cools, a process that leads to the development of clouds and precipitation. For this reason, the areas under the influence of the equatorial low are among the rainiest on earth. Just the opposite is true for the regions in the vicinity of 30° north and south latitude, where high pressure predominates. Here, in the zones known as the *subtropical highs,* air is subsiding. When air sinks, it is compressed and warmed. Such conditions preclude cloud formation and precipitation. Therefore clear skies, a maximum of sunshine, and drought characterize places that are strongly influenced by high pressure. Indeed, the earth's low-latitude deserts and steppes coincide with the subtropical high-pressure belts.

Unlike their low-latitude counterparts, middle-latitude deserts and steppes are not controlled by the subsiding air masses associated with high pressure. Instead, these dry lands exist principally because of their positions in the deep interiors of large landmasses far removed from the oceans, which are the ultimate source of moisture for cloud formation and precipitation. In addition, the presence of high mountains across the paths of prevailing winds further acts to separate these areas from water-bearing, maritime air masses. As prevailing winds meet mountain barriers, the air is forced to ascend. Recall that when air rises it expands and cools, a process that can produce clouds and precipitation. The windward sides of mountains, therefore, are often charac-

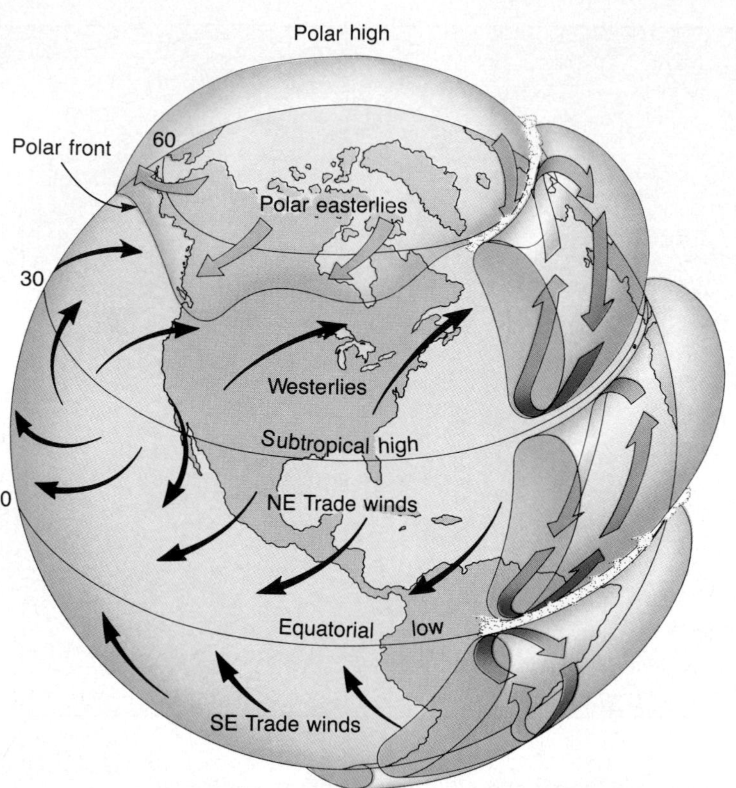

FIGURE 13.3
Idealized diagram of the earth's general circulation. The deserts and steppes that are centered in the latitude between 20° and 30° north and south coincide with the subtropical high-pressure belts. Here dry, subsiding air inhibits cloud formation and precipitation. By contrast, the pressure belt known as the equatorial low is associated with areas that are among the rainiest on earth.

terized by relatively high precipitation totals. By contrast, the leeward sides of mountains are usually much drier (Figure 13.4). This situation exists because air reaching the leeward side has lost much of its moisture and, if the air descends, it is compressed and warmed, making cloud formation even less likely. The dry region that results is often referred to as a **rainshadow desert**. Because many middle-latitude deserts occupy sites on the leeward sides of mountains they can also be classified as rainshadow deserts. In North America the Coast Ranges, Sierra Nevada, and Cascades are the foremost mountain barriers, whereas in Asia, the great Himalayan chain prevents the summertime monsoon flow of moist Indian Ocean air from reaching into the interior. Because the Southern Hemisphere lacks extensive land areas in the middle latitudes, only a small area of desert and steppe are found in this latitude range,

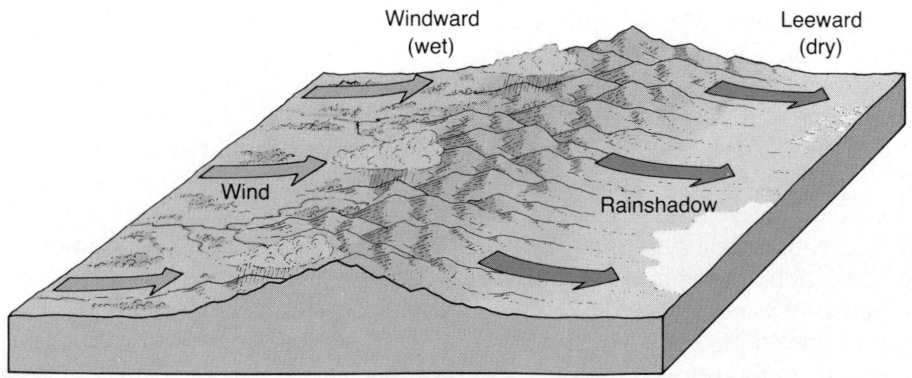

FIGURE 13.4
Many deserts in the middle latitudes are rainshadow deserts. As moving air meets a mountain barrier, it is forced to rise. Clouds and precipitation on the windward side often result. Air descending the leeward side is much drier. The mountains effectively cut the leeward side off from the source of moisture, producing a rainshadow desert.

existing primarily near the southern tip of South America in the rainshadow of the towering Andes. In the case of middle-latitude deserts, we have an example of the impact of tectonic processes on climate. Rainshadow deserts exist by virtue of the mountains produced when plates collide. Without such mountain-building episodes, wetter climates would prevail where many dry regions exist today.

BOX 13.1

What Is Meant by "Dry"?

Albuquerque, New Mexico, in the southwestern United States, receives an average of 20.7 centimeters (8.07 inches) of rainfall annually. As you might expect, because Albuquerque's precipitation total is modest, the station is classified as a desert when the commonly used Köppen climate classification is applied. The Russian city of Verkhoyansk is a remote station located near the Arctic Circle in Siberia. The yearly precipitation total there averages 15.5 centimeters (6.05 inches), about 5 centimeters less than Albuquerque's. Although Verkhoyansk receives less precipitation than Albuquerque, its classification is that of a humid climate. How can this occur?

We all recognize that deserts are dry places, but just what is meant by the term *dry?* That is, how much rain defines the boundary between humid and dry regions? Sometimes it is arbitrarily defined by a single rainfall figure, for example, 25 centimeters (10 inches) per year of precipitation. However, the concept of dryness is a relative one that refers to any situation in which a water deficiency exists. Hence, climatologists define *dry climate* as one in which yearly precipitation is not as great as the potential loss of water by evaporation. Dryness then not only is related to annual rainfall totals but is also a function of evaporation which, in turn, is closely dependent upon temperature. As temperatures climb, potential evaporation also increases. Fifteen to twenty-five centimeters of precipitation may be sufficient to support coniferous forests in northern Scandinavia or Siberia, where evaporation into the cool, humid air is slight and a surplus of water remains in the soil. However, the same amount of rain falling on New Mexico or Iran supports only a sparse vegetative cover because evaporation into the hot, dry air is great. So clearly no specific amount of precipitation can serve as a universal boundary for dry climates.

To establish the boundary between dry and humid climates, the widely used Köppen classification uses formulas that involve three variables: (1) average annual precipitation, (2) average annual temperature, and (3) seasonal distribution of precipitation. The use of average annual temperature reflects its importance as an index of evaporation. The amount of rainfall defining the humid-dry boundary will be larger where mean annual temperatures are high and smaller where temperatures are low. The use of seasonal precipitation distribution as a variable is also related to this idea. If rain is concentrated in the warmest months, loss to evaporation is greater than if the precipitation is concentrated in the cooler months.

Table 13.A summarizes the precipitation amounts that divide dry and humid climates. Notice that a station with an annual mean of 20°C (68°F) and a summer rainfall maximum of 68 centimeters (26.5 inches) is classified as dry. If the rain falls primarily in winter, however, the station must receive only 40 centimeters (15.6 inches) or more to be considered humid. If the precipitation is more evenly distributed, the figure defining the humid-dry boundary is between the other two.

TABLE 13.A
Average annual precipitation defining the boundary between dry and humid climates.

Average Annual Temp. (°C)	Winter Rainfall Maximum (centimeters)	Even Distribution (centimeters)	Summer Rainfall Maximum (centimeters)
5	10	24	38
10	20	34	48
15	30	44	58
20	40	54	68
25	50	64	78
30	60	74	88

GEOLOGIC PROCESSES IN ARID CLIMATES

The angular hills, the sheer canyon walls, and the pebble- or sand-covered surface of the desert contrast sharply with the rounded hills and curving slopes of more humid places. Indeed, to a visitor from a humid region, a desert landscape may seem to have been shaped by forces different from those operating in well-watered areas. However, although the contrasts may be striking, they are not a reflection of different processes but merely the differing effects of the same processes operating under contrasting climatic conditions.

In humid regions relatively fine-textured soils support an almost continuous cover of vegetation that mantles the surface. Here the slopes and rock edges are rounded. Such a landscape reflects the strong influence of chemical weathering in a humid climate. By contrast, much of the weathered debris in deserts consists of unaltered rock and mineral fragments—the result of mechanical weathering processes. In dry lands rock weathering of any type is

A.

B.

FIGURE 13.5
A. Most of the time, desert stream channels are dry.
B. An ephemeral stream shortly after a heavy shower. Although such floods are short-lived, large amounts of erosion occur. (Photos by James E. Patterson)

FIGURE 13.6
This desert thunderstorm lasted just half an hour. There are often many weeks, months, or occasionally even years separating periods of rain in the desert. When rains do occur, they are often heavy and of relatively short duration. Because the rainfall intensity is high, all of the water cannot soak in, and rapid runoff results. (Photo by Stephen Trimble)

greatly reduced because of the lack of moisture and the scarcity of organic acids from decaying plants. Chemical weathering, however, is not completely lacking in deserts. Over long spans of time clays and thin soils do form and many iron-bearing silicate minerals oxidize, producing the rust-colored stain found tinting some desert landscapes.

Although permanent streams are common in humid regions, practically all desert streams are dry most of the time (Figure 13.5A). Desert streams are said to be **ephemeral**, which means that they carry water only in response to specific episodes of rainfall. An average ephemeral stream may flow only a few days or perhaps just a few hours during the year. In some years the channel may carry no water at all. This fact is often obvious even to the casual observer who, while traveling, notices the number of bridges with no streams beneath them or the number of dips in the road where dry channels cross. However, when the rare heavy showers do come, so much rain falls in such a short time that all of it cannot soak in (Figure 13.6). Since the vegetative cover is sparse, runoff is largely unhindered and consequently rapid, often creating flash floods along valley floors (Figure 13.5B). Such floods, however, are quite unlike floods in humid regions. A flood on a river like the Mississippi may take several days to reach its crest and then to subside again, whereas desert floods arrive suddenly and likewise subside in a short time. Because much of the surface material is not anchored by vegetation, the amount of erosional work that occurs during one of these short-lived events is impressive. In the dry western United States a number of different names are used for ephemeral streams. Two of the most common are *wash* and *arroyo*. In other parts of the world, a dry desert stream may be called a *wadi* (Arabia and North Africa), a *donga* (South America), or a *nullah* (India).

Unlike the drainage in humid regions, stream courses in arid regions are seldom well integrated. That is, desert streams lack an extensive system of tributaries. In fact, a basic characteristic of deserts is that most of the streams that originate in them are small and die out before reaching the sea. Because the water table is usually far below the surface, few desert streams can draw upon it. Without a steady supply of water, evaporation and infiltration soon deplete the stream. The few permanent streams that do cross arid regions, such as the Colorado and Nile rivers, originate outside the desert, often in well-watered mountains. Here the water supply must be great to compensate for the losses occurring as the stream crosses the desert. For example, after the Nile leaves the lakes and mountains of central Africa that are its source, it traverses almost 3000 kilometers of the Sahara without the contribution of a single tributary. By contrast, in humid regions the discharge of a river becomes larger as it flows downstream because tributaries and groundwater contribute additional water along the way.

It should be emphasized that running water, although an infrequent occurrence, nevertheless does most of the erosional work in deserts. This is contrary to a commonly held belief that wind is the most important erosional agent sculpturing desert landscapes. Although wind erosion is more significant in

BOX 13.2

Common Misconceptions about Deserts

Deserts are hot, lifeless, sand-covered landscapes shaped largely by the force of wind. The preceding statement summarizes the image of arid regions that many people hold, especially those living in more humid places. Is it an accurate view? The answer is no. Although there are clearly elements of reality in such an impression, it is a generalization that contains a number of misconceptions (Figure 13.A).

One common fallacy about deserts is that they are lifeless or almost lifeless. Although reduced in amount and different in character, plant and animal life are usually present. Desert plants may differ widely from one part of the world to another, but all have one characteristic in common: they have developed adaptations that make them highly tolerant of drought. Many have waxy leaves, stems, or branches or a thickened cuticle (outermost protective layer) to reduce water loss. Others have very small leaves or no leaves at all. Also, the roots of some species often extend to great depths in order to tap the moisture found there, whereas others produce a shallow but widespread root system that enables them to absorb great amounts of moisture quickly from the infrequent desert downpours. Often the stems of these plants are thickened by a spongy tissue that can store enough water to sustain the plant until the next rainfall comes. Thus, although widely dispersed and providing little ground cover, plants of many kinds flourish in the desert.

A second widely held belief about the world's dry lands is that they are always hot. This fact seems to be reinforced by commonly quoted temperature statistics. The highest accepted temperature record for the United States as well as the entire Western Hemisphere is 57°C (134°F). This long-standing record was set at Death Valley, California, on July 10, 1913. The nearly 59°C (137°F) reading at Azizia, Libya, in North Africa's Sahara Desert on September 13, 1922, is the world record. Despite these remarkably high figures, cold temperatures are also experienced in desert regions. For example, the average daily minimum in January at Phoenix, Arizona, is 1.7°C (35°F), just barely above freezing. At Ulan Bator in Mongolia's Gobi Desert, the average *high* temperature on January days is only -19°C (-2°F)! Dry climates are found from the tropics poleward to the high middle latitudes. Consequently, although tropical deserts lack a cold season, deserts in the middle latitudes do experience seasonal temperature changes.

The last two commonly held misconceptions about the world's deserts are more geologic than climatic. One mistaken assumption is that they consist of mile after mile of drifting sand. It is true that sand accumulations do exist in some areas and may be striking features, but they represent only a small percentage of the total desert area. For example, in the Sahara, the world's largest desert, accumulations of sand cover only one-tenth of its area. The sandiest of all deserts is the Arabian, one-third of which consists of sand. The final mistaken assumption is the seemingly logical idea that wind is the most important agent of erosion in deserts. Although wind is relatively more significant in dry areas than anywhere else, most erosional landforms in deserts are created by running water. When the rains come, they usually take the form of thunderstorms. Because the heavy rain associated with these storms cannot all soak in, rapid runoff results. Without a thick vegetative cover to protect the ground, erosion is great.

dry areas than elsewhere, most desert landforms are nevertheless carved by running water. As we shall see shortly, the main role of wind is in the transportation and deposition of sediment, creating and shaping the ridges and mounds we call dunes.

TRANSPORTATION OF SEDIMENT BY WIND

Moving air, like moving water, is turbulent and able to pick up loose debris and transport it to other locations. Just as in a stream, the velocity of wind increases with height above the surface. Also like a stream, wind transports fine particles in suspension while heavier ones are carried as bed load. However, the transport of sediment by wind differs from that of running water in two significant ways. First, wind has a low density compared to water; thus it is not capable of picking up and transporting coarse materials. Second, because wind is not confined to channels, it can spread sediment over large areas, as well as high into the atmosphere.

FIGURE 13.A
Desert plants surrounded by snow-covered ground in dry southern Utah. As this scene demonstrates, deserts are not necessarily hot, lifeless, dune-covered expanses. (Photo by Stephen Trimble)

Bed Load

The **bed load** carried by wind consists of sand grains. Observations in the field and experiments using wind tunnels indicate that windblown sand moves by skipping and bouncing along the surface—a process termed **saltation.** The term is not a reference to salt, but instead derives from the Latin word meaning "to jump." The movement of sand grains begins when wind reaches a velocity sufficient to overcome the inertia of the resting particles. At first, the sand rolls along the surface. Upon striking another grain, one or both of the grains may jump into the air. Once in the air, the sand is carried forward by the wind until gravity pulls the grain back toward the surface. When the sand hits the surface, it either bounces back into the air or dislodges other grains which then jump upward. In this manner a chain reaction is established, filling the air near the ground with saltating sand grains in a short period of time (Figure 13.7).

Bouncing sand grains never travel far from the surface. Even when winds are very strong, the height

FIGURE 13.7
A cloud of saltating sand grains moving up the gentle slope of a dune. (Photo by Stephen Trimble)

of the saltating sand seldom exceeds one meter and under less extreme conditions is usually confined to heights no greater than one-half meter. Some sand grains are too large to be thrown into the air by impact from other particles. When this is the case, the energy provided by the impact of the smaller saltating grains drives the larger grains forward. Estimates indicate that between 20 and 25 percent of the sand transported in a sandstorm is moved in this way.

Suspended Load

Unlike sand, dust can be swept high into the atmosphere by the wind. Since dust is often composed of rather flat particles that have large surface areas when compared to the weight of the particle, it is relatively easy for turbulent air to counterbalance the pull of gravity and keep these fine particles suspended for extended periods. Although both silt and

FIGURE 13.8
Winds are capable of transporting enormous quantities of fine material. Dust storms such as this one occur when the ground is dry and not protected by vegetation. In some agricultural regions the removal of topsoil by wind is a serious problem. (Photo by John S. Shelton)

clay can be carried in suspension, silt commonly makes up the bulk of the **suspended load** because the reduced level of chemical weathering in deserts produces only small amounts of clay.

Fine particles are easily carried by the wind, but they are not easily acquired by the turbulent air. The reason is that the wind velocity is practically zero within a very thin layer close to the ground. Thus, the wind cannot lift the sediment by itself. Instead, the dust must be ejected or spattered into the moving current of the air by bouncing sand grains or other disturbances. This idea is illustrated nicely by a dry, unpaved country road on a windy day. Left undisturbed, little dust is raised by the wind. However, as a car or truck moves over the road, the previously smooth layer of silt is disturbed, creating a thick cloud of dust.

Although the suspended load is usually deposited relatively near its source, high winds are capable of carrying large quantities of dust great distances (Figure 13.8). In the 1930s, silt picked up in Kansas was transported to New England and beyond into the North Atlantic. Similarly, dust blown from the Sahara has been traced as far as the West Indies.

WIND EROSION

Compared to running water and moving ice, wind is a relatively insignificant erosional agent. Recall that even in deserts, few major erosional landforms are created by the wind. Although wind erosion is not restricted to arid and semiarid regions, it does its most effective work in these areas. In humid places moisture binds particles together and vegetation anchors the soil so that wind erosion is negligible. For wind to be effective, dryness and scanty vegetation are important prerequisites. When such circumstances exist, wind may pick up, transport, and deposit great quantities of fine sediment. During the 1930s parts of the Great Plains experienced great dust storms. The plowing under of the natural vegetative cover for farming, followed by severe drought, made the land ripe for wind erosion, and led to the area being labeled the Dust Bowl.*

One way that winds erode is by **deflation**, the lifting and removal of loose material. Although the effects of deflation are sometimes difficult to notice because the entire surface is being lowered at the same time, they can be significant (Figure 13.9). In portions of the 1930s Dust Bowl, vast areas of land were lowered by as much as one meter in only a few years.

*For more information, see Box 5.2 "Dust Bowl: Soil Erosion in the Great Plains."

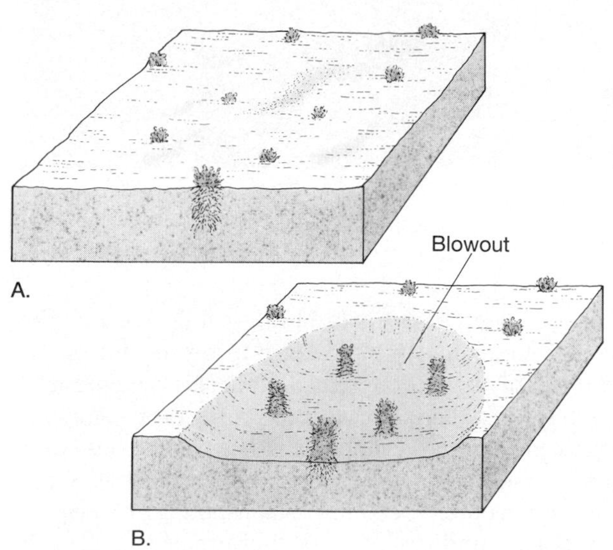

FIGURE 13.9
Formation of a blowout. **A.** Area prior to deflation. **B.** Area after deflation has created a shallow depression. **C.** This 1.5-meter-high mound of soil that was anchored by vegetation shows the level of the land prior to the formation of the blowout. (Photo by E. J. Tarbuck)

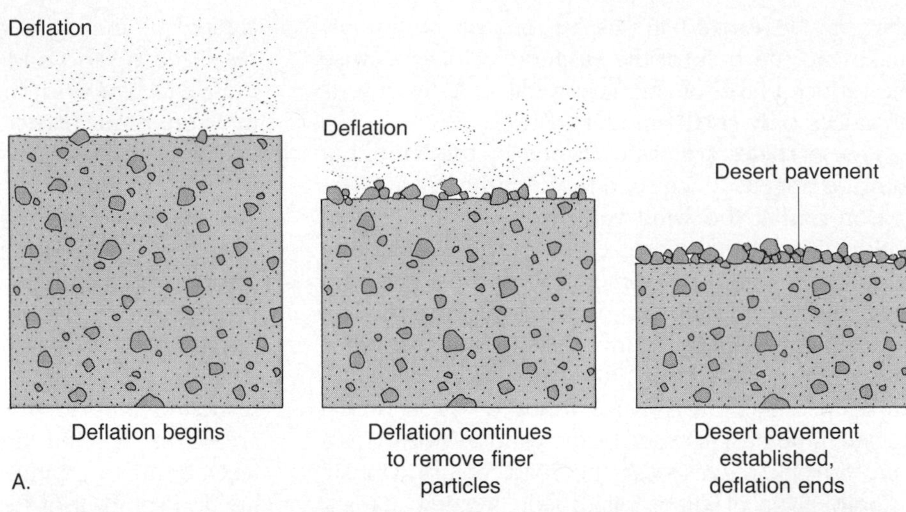

FIGURE 13.10
A. Formation of desert pavement. As these cross sections illustrate, coarse particles gradually become concentrated into a tightly packed layer as deflation lowers the surface by removing sand and silt. **B.** If left undisturbed, desert pavement such as this will protect the surface from further deflation. (Photo by Peter Kresan)

The most noticeable results of deflation in some places are shallow depressions which are quite appropriately called **blowouts** (Figure 13.9). In the Great Plains region, from Texas north to Montana, thousands of blowouts are visible on the landscape. They range in size from small dimples less than one meter deep and three meters wide to depressions that approach 50 meters in depth and several kilometers across. The factor that controls the depths of these basins (that is, acts as base level) is the local water table. When blowouts are lowered to the water table, damp ground and vegetation prevent further deflation.

In portions of many deserts the surface is characterized by a closely packed layer of coarse pebbles and cobbles that are too large to be moved by the wind. This stony veneer, called **desert pavement**, is created as deflation lowers the surface by removing sand and silt until eventually only a continuous cover of coarse particles remains (Figure 13.10). Once desert pavement becomes established, a process that may take hundreds of years, the surface is effectively protected from further deflation if left undisturbed. However, since the layer is only one or two stones thick, the passage of vehicles or animals can dislodge the pavement and expose the fine-grained material below. If this happens, the surface is once again subject to deflation.

Like glaciers and streams, wind erodes by **abrasion.** In dry regions as well as along some beaches, windblown sand cuts and polishes exposed rock surfaces. However, abrasion is often credited for accomplishments beyond its actual capabilities. Such features as balanced rocks that stand high atop narrow pedestals, and intricate detailing on tall pinnacles, are not the results of abrasion. Since sand seldom travels more than a meter above the surface, the wind's sandblasting effect is obviously quite limited in vertical extent. Abrasion by windblown sand, however, does create interestingly shaped stones called **ventifacts.** The side of the stone exposed to the prevailing wind is abraded, leaving it polished, pitted, and with sharp edges. If the wind is not consistently from one direction, or if the pebble becomes reoriented, it may have several faceted surfaces.

WIND DEPOSITS

Although wind is relatively unimportant as a producer of erosional landforms, wind deposits are significant features in some regions. Accumulations of windblown sediment are particularly conspicuous landscape elements in the world's dry lands and along many sandy coasts. Wind deposits are of two distinctive types: (1) mounds and ridges of sand from the wind's bed load and (2) extensive blankets of silt that once were carried in suspension.

Sand Deposits

As is the case with running water, wind drops its load of sediment when velocity falls and the energy available for transport diminishes. Thus sand begins to accumulate wherever an obstruction across the path of the wind slows the movement of the air. Unlike many deposits of silt, which form blanket-like layers over large areas, winds commonly deposit sand in mounds or ridges called **dunes** (Figure 13.11).

As moving air encounters an object, such as a clump of vegetation or a rock, the wind sweeps around and over it, leaving a shadow of slower-moving air behind the obstacle as well as a smaller zone of quieter air just in front of the obstacle. Some

FIGURE 13.11
Sand sliding down the steep face of a dune, Coral Pink Sand Dunes, Utah. (Photo by Michael Collier)

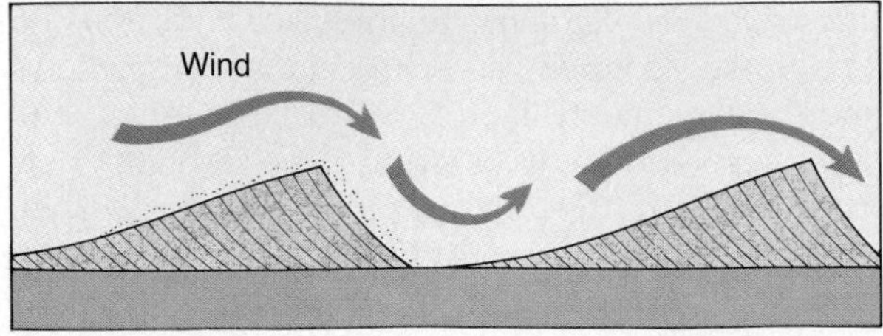

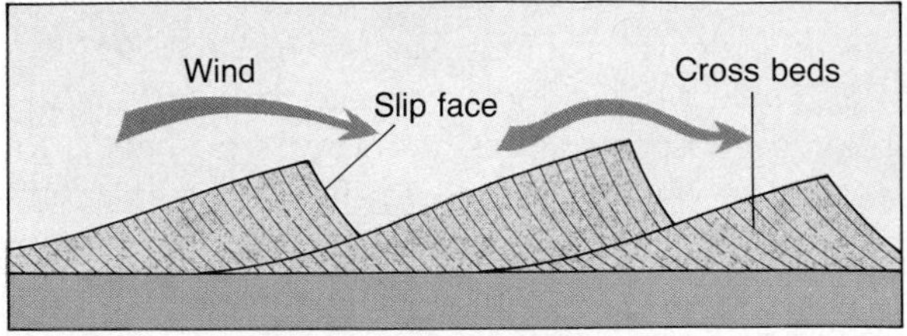

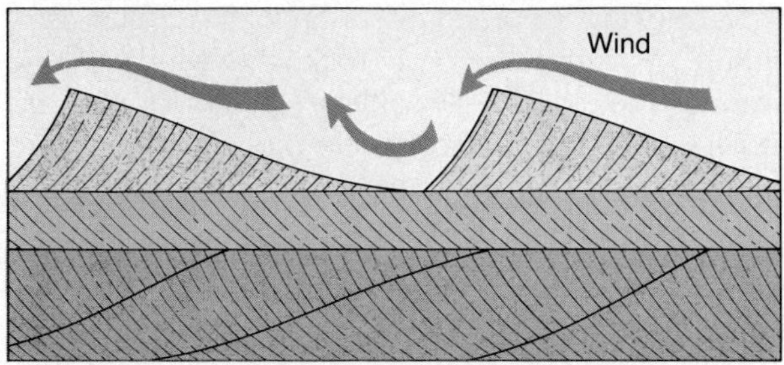

FIGURE 13.12
Dunes commonly have an asymmetrical shape. The steeper leeward side is called the slip face. Sand grains deposited on the slip face at the angle of repose create the cross bedding of the dunes. A complex pattern develops in response to changes in prevailing winds.

of the saltating sand grains moving with the wind come to rest in these wind shadows. As the accumulation of sand continues, it becomes a more imposing barrier to the wind and thus a more efficient trap for even more sand. If there is a sufficient supply of sand and the wind blows steadily for a long enough time, the mound of sand grows into a dune.

A profile of a dune shows an asymmetrical shape with the leeward slope being steep and the windward slope more gently inclined (Figure 13.12). Sand moves up the gentle slope on the windward side by saltation. Just beyond the crest of the dune, where the wind velocity is reduced, the sand accumulates. As more sand collects, the slope steepens and eventually some of it slides or slumps under the pull of gravity. In this way the leeward slope of the dune, called the **slip face**, maintains an angle of about 34 degrees, the angle of repose for loose dry sand (see Figure 13.11).* Continued sand accumulation coupled with periodic slides down the slip face results in the slow migration of the dune in the direction of air movement.

For some areas, moving sand is troublesome. In portions of the Middle East, valuable oil rigs must be protected from encroaching dunes. In some cases fences are built sufficiently upwind of the dunes to stop their migration. As sand continues to collect, however, the fences must be built higher. In Kuwait, protective fences extend for almost 10 kilometers around one important oil field. Migrating dunes can also pose a problem to the construction and maintenance of highways and railroads that cross sandy desert regions. For example, in order to keep a portion of Highway 95 near Winnemucca, Nevada, open to traffic, sand must be taken away about three times per year. Each time, between 1500 and 4000 cubic meters of sand are removed. Attempts at stabilizing the dunes by planting different varieties of grasses have been unsuccessful because the meager rainfall cannot support the plants.

As sand is deposited on the slip face, layers form which are inclined in the direction the wind is blowing. These sloping layers are called **cross beds** (Figure 13.12). When the dunes are eventually buried under other layers of sediment and become part of the sedimentary rock record, their asymmetrical shape is destroyed, but the cross beds remain (Figure 13.13). By studying the orientation of these beds, geologists can determine the average direction of ancient winds. This information, together with other data, is then used in reconstructing climates of the geologic past. This knowledge of past climatic conditions, in turn, aids in determining earlier positions of the earth's moving lithospheric plates.

Types of Sand Dunes

Dunes are not just random heaps of sediment. Rather, they are accumulations that usually assume patterns that are surprisingly consistent (Figure 13.14). Addressing this point, a leading early investigator of dunes, the British engineer R. A. Bagnold, observed: "Instead of finding chaos and disorder, the observer never fails to be amazed at a simplicity of form, an exactitude of repetition, and a geometric order. . . . " A broad assortment of dune forms exists; so to simplify and provide some order, several major types are recognized. Of course, there are gradations

*Recall from Chapter 9 that the angle of repose is the steepest angle at which loose material remains stable.

among different forms as well as irregularly shaped dunes that do not fit easily into any category. Several factors influence the form and size that dunes ultimately assume. These include wind direction and velocity, availability of sand, and the amount of vegetation present.

Barchan Dunes Solitary sand dunes shaped like crescents and with their tips pointing downwind are called **barchan dunes** (Figures 13.14A and 13.15). These dunes form where supplies of sand are limited and the surface is relatively flat, hard, and lacking vegetation. They migrate slowly with the wind at a rate of up to 15 meters per year. Their size is usually modest with the largest barchans reaching heights of about 30 meters while the maximum spread between their horns approaches 300 meters. When the wind direction is nearly constant, the crescent form of these dunes is nearly symmetrical. However, when the wind direction is not perfectly fixed, one tip becomes longer than the other.

FIGURE 13.13
Cross beds are an obvious characteristic of the Navajo Sandstone, here exposed in Zion National Park, Utah. When dunes are buried and become part of the sedimentary record, the cross-bedded structure is preserved. (Photo by E. J. Tarbuck)

Transverse Dunes In regions where the prevailing winds are steady, sand is plentiful, and vegetation is sparse or absent, the dunes form a series of long ridges that are separated by troughs and oriented at right angles to the prevailing wind. Because of this orientation, they are termed **transverse dunes** (Figure 13.14B). Typically, many coastal dunes are of this type. In addition, transverse dunes are common in many arid regions where the extensive surface of wavy sand is sometimes called a *sand sea.* In some parts of the Sahara and Arabian deserts, transverse dunes reach heights of 200 meters, are 1 to 3 kilometers across, and can extend for distances of 100 kilometers or more.

There is a relatively common dune form that is intermediate between isolated barchans and extensive waves of transverse dunes. Such dunes, called **barchanoid dunes,** form scalloped rows of sand oriented at right angles to the wind (Figure 13.14C). The rows resemble a series of barchans that have been positioned side by side. Visitors exploring the gypsum dunes at White Sands National Monument, New Mexico, will recognize this form (Figure 13.16).

Longitudinal Dunes **Longitudinal dunes** are long ridges of sand that form more or less parallel to the prevailing wind and where sand supplies are limited (Figure 13.14D). Apparently the prevailing wind direction must vary somewhat, but still remain in the same quadrant of the compass. Although the smaller types are only three or four meters high and several tens of meters long, in some large deserts longitudinal dunes can reach great size. For example, in portions of North Africa, Arabia, and central Australia, these dunes may approach a height of 100 meters and extend for distances of more than 100 kilometers (62 miles).

Parabolic Dunes Unlike the other dunes that have been described thus far, **parabolic dunes** form where vegetation partially covers the sand. The shape of these dunes resembles the shape of bar-

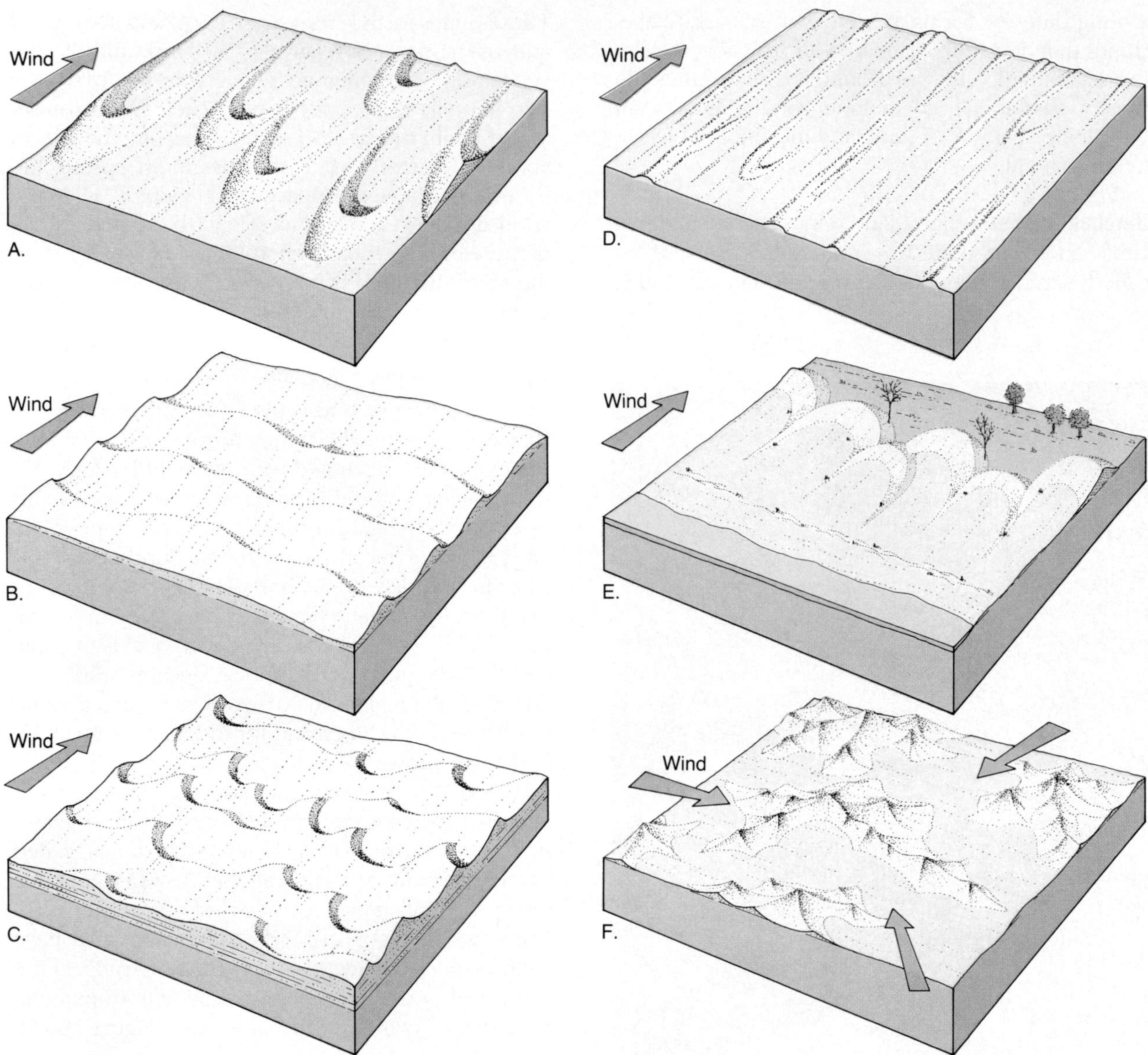

FIGURE 13.14
Sand dune types. **A.** Barchan dunes. **B.** Transverse dunes. **C.** Barchanoid dunes. **D.** Longitudinal dunes. **E.** Parabolic dunes. **F.** Star dunes.

chans except that their tips point into the wind rather than downwind (Figure 13.14E). Parabolic dunes often form along coasts where there are strong onshore winds and abundant sand. If the sand's sparse vegetative cover is disturbed at some spot, deflation creates a blowout. Sand is then transported out of the depression and deposited as a curved rim which grows higher as deflation enlarges the blowout.

Star Dunes. Confined largely to parts of the Sahara and Arabian deserts, **star dunes** are isolated hills of sand that exhibit a complex form (Figure 13.14F). Their name is derived from the fact that the bases of these dunes resemble multipointed stars. Usually three or four sharp-crested ridges diverge from a central high point that in some cases may approach a height of 90 meters. As their form suggests, star dunes develop where wind directions are variable.

Loess

In some parts of the world the surface topography is mantled with deposits of windblown silt. Over periods of perhaps thousands of years dust storms de-

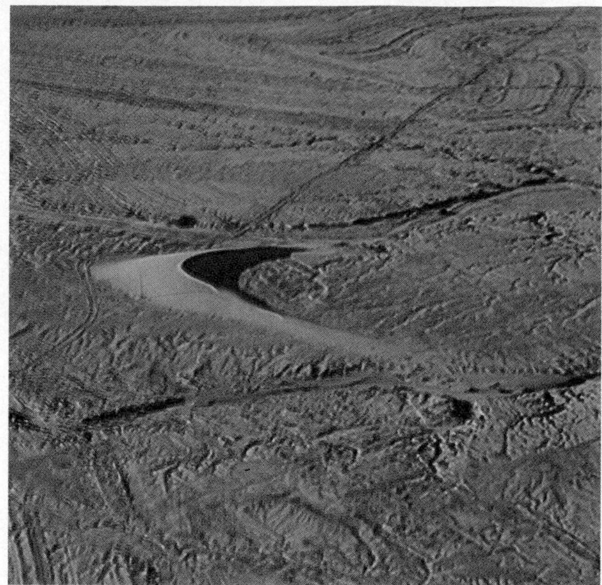

FIGURE 13.15
Aerial view of a solitary barchan. The gentle slope is on the side from which the wind is blowing. (Photo by John S. Shelton)

posited this material, which is called **loess.** As can be seen in Figure 13.17, when loess is breached by streams or road cuts, it tends to maintain vertical cliffs and lacks any visible layers. The distribution of loess indicates that there are two primary sources for this sediment: deserts and glacial outwash deposits. The thickest and most extensive deposits of loess in the world occur in western and northern China, where accumulations of 30 meters are not uncommon and thicknesses of more than 100 meters have been measured. It is this fine, buff-colored sediment that gives the Yellow River (Hwang Ho) and the adjacent Yellow Sea their names. The sources of China's 800,000 square kilometers of loess are the extensive desert basins of central Asia.

In the United States, deposits of loess are significant in many areas, including South Dakota, Nebraska, Iowa, Missouri, and Illinois as well as portions of the Columbia Plateau in the Pacific Northwest. The correlation between the distribution of loess and important farming regions in the Midwest and eastern Washington is not just a coincidence, because soils derived from this wind-deposited sediment are among the most fertile in the world. Unlike the deposits in China, the loess in the United States, as well as in Europe, is an indirect product of glaciation, for its source was deposits of stratified drift. During the retreat of the glacial ice, many river valleys were choked with sediment that was provided by meltwater. Strong westerly winds sweeping across the barren floodplains picked up the finer sediment and dropped it as a blanket on the east sides of the valleys. Such an origin is confirmed by the fact that loess deposits are thickest and coarsest on the lee side of such major glacier drainage outlets as the Mississippi and Illinois rivers and rapidly thin with increasing distance from the valleys. Furthermore, the angular mechanically weathered particles composing the loess are essentially the same as the rock flour produced by the grinding action of glaciers.

FIGURE 13.16
Barchanoid dunes represent a type that is intermediate between isolated barchans on the one hand and extensive transverse dunes on the other. The gypsum dunes at White Sands National Monument, New Mexico, are an example. (Photo by E. J. Tarbuck)

FIGURE 13.17
This vertical loess bluff near the Mississippi River in southern Illinois is about 3 meters high. (Photo by James E. Patterson)

BOX 13.3

Desertification

On nearly any list of major environmental issues facing the world, one is likely to find reference to the problem of *desertification.* The term by itself simply implies the expansion of desertlike conditions into nondesert areas. Such transformations can result from natural processes that act gradually over extended spans of time. However, the use of the term in recent years has been restricted primarily to situations in which there is a relatively rapid alteration of the land to desertlike conditions as the result of human activities.

Desertification commonly takes place on the margins of deserts. The advancement of desertlike conditions into areas that were previously productive is not a process in which the borders of a desert gradually expand in a uniform manner. Rather, degeneration into desert usually occurs as a patchy transformation of dry but habitable land into dry land that is uninhabitable. It is primarily the product of inappropriate land use and is aided and accelerated by drought. The process may be halted during wet years, only to advance rapidly during succeeding dry years.

On marginal land used for crops, natural vegetation is cleared. During periods of drought, crops fail and the unprotected soil is exposed to the forces of erosion. Gullying of slopes and accumulations of sediment in stream channels are visible signs on the landscape, as are the clouds of dust created as topsoil is removed by the wind. Where crops are not grown, the raising of livestock leads to degradation of the land. Although the modest vegetation associated with marginal lands may be adequate to maintain local wildlife, it cannot support the intensive grazing of large domesticated herds. Overgrazing reduces or eliminates plant cover. When the vegetative cover is destroyed beyond the minimum required for protection of the soil against erosion, the destruction becomes irreversible.

Desertification first received worldwide attention when drought struck a region in Africa called the *Sahel* in the late 1960s (Figure 13.B). During that period and others since then, the people in this vast expanse lying south of the Sahara Desert have suffered from malnutrition and death by starvation. Livestock herds have been decimated, and the loss of productive land has been great. Hundreds of thousands of people have

THE EVOLUTION OF A DESERT LANDSCAPE

Since arid regions typically lack permanent streams, they are characterized as having **interior drainage**, that is, a discontinuous pattern of intermittent streams that do not flow out of the desert to the ocean. In the United States, the dry Basin and Range region provides an excellent example. The region includes southern Oregon, all of Nevada, western Utah, southeastern California, as well as southern Arizona and New Mexico. The name Basin and Range is an apt description for this almost 800,000 square kilometer region, since it is characterized by more than 200 relatively small mountain ranges which rise 900–1500 meters above the basins that separate them. In this region, as in others like it around the world, erosion is carried out for the most part without reference to the ocean (ultimate base level) because drainage is in the form of local interior systems. Even in areas where permanent streams flow to the ocean, few tributaries exist, and thus only a relatively narrow strip of land adjacent to the stream has sea level as the ultimate level of land reduction.

The block models shown in Figure 13.18 depict the stages of landscape evolution in a mountainous desert such as the Basin and Range region. During and following the uplift of the mountains, running water begins carving the elevated mass and depositing large quantities of debris in the basin. During this early stage, relief is greatest, for as erosion lowers the mountains and sediment fills the basins, elevation differences diminish.

When the occasional torrents of water produced by sporadic rains move down the mountain canyons, they are heavily loaded with sediment. Emerging from the confines of the canyon, the runoff spreads over the gentler slopes at the base of the mountains and quickly loses velocity. Consequently most of its load is dumped within a short distance. The result is a cone of debris at the mouth of a canyon known as an **alluvial fan** (Figure 13.19). Since the coarsest material is dropped first, the head of the fan is steepest, having a slope of perhaps 10 to 15 degrees. Moving down the fan, the size of the sediment and the steep-

been forced to migrate. As agricultural lands shrink, people must rely on smaller areas for food production. This, in turn, places greater stress on the environment and accelerates the desertification process.

Although human suffering associated with desertification is most serious along the southern margins of the Sahara, the problem is by no means confined to the Sahel. The problem also exists in other parts of Africa as well as on every other continent except Antarctica. Each year millions of acres are lost beyond practical hope for reclamation. Recurrent droughts may seem to be the obvious reason for desertification, but the stresses placed by people on a tenuous environment with fragile soils is the chief cause.

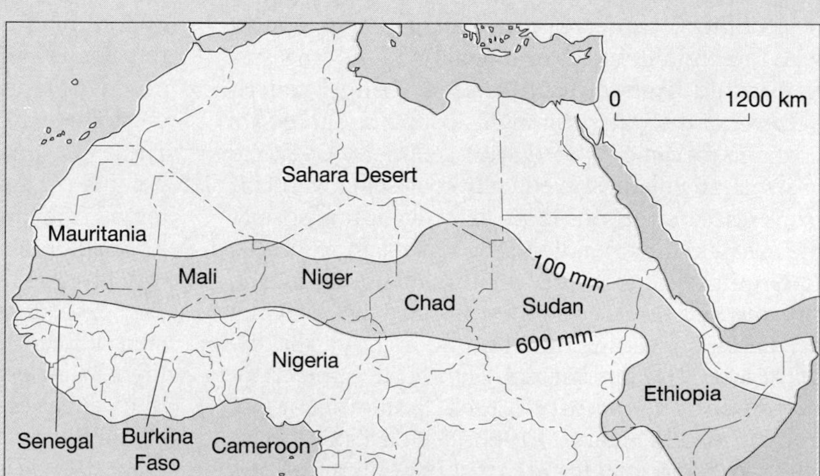

FIGURE 13.B
Desertification is most serious along the southern margin of the Sahara in a region known as the Sahel. The lines defining the approximate boundaries of the Sahel represent average annual rainfall in millimeters.

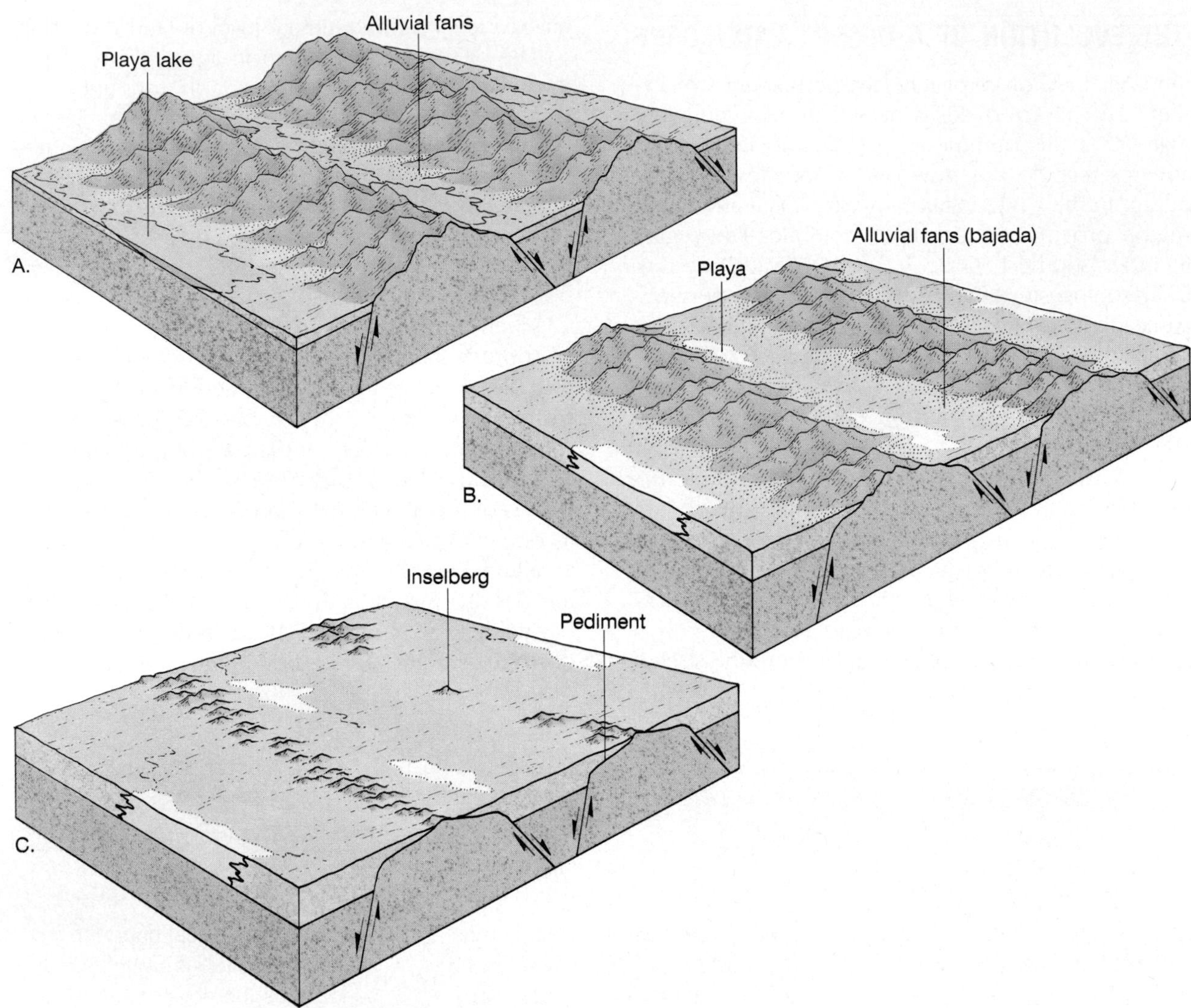

FIGURE 13.18
Stages of landscape evolution in a mountainous desert. As erosion of the mountains and deposition in the basins continue, relief diminishes.
A. Early stage. **B.** Middle stage. **C.** Late stage.

ness of the slope decrease and merge imperceptibly with the basin floor. An examination of the fan's surface would likely reveal a braided channel pattern because of the water shifting its course as successive channels became choked with sediment. Over the years, a fan enlarges, eventually coalescing with fans from adjacent canyons to produce an apron of sediment called a **bajada** along the mountain front.

On the rare occasions when rainfall is abundant, streams may flow across the bajada to the center of the basin, converting the basin floor into a shallow **playa lake**. The intermittent lake in the upper right portion of the Furnace Creek, California, topographic map in Figure 13.19B is such a feature. Playa lakes are temporary features that last only a few days or at best a few weeks before evaporation and infiltration remove the water. The dry, flat lake bed that remains is called a **playa**. Playas are typically composed of fine silts and clays, and occasionally encrusted with salts precipitated during evaporation. These precipitated salts may be unusual. A case in point is the sodium borate (better known as borax) mined from ancient playa lake deposits in Death Valley, California.

With time the mountain front is worn back and a sloping bedrock platform, called a **pediment**, is created adjacent to the steep mountain front. A pediment is an erosional surface, usually covered by a thin veneer of debris, that is formed by the action of running water. Just how the water carves the pediment, however, is unclear and still a matter for debate.

A.

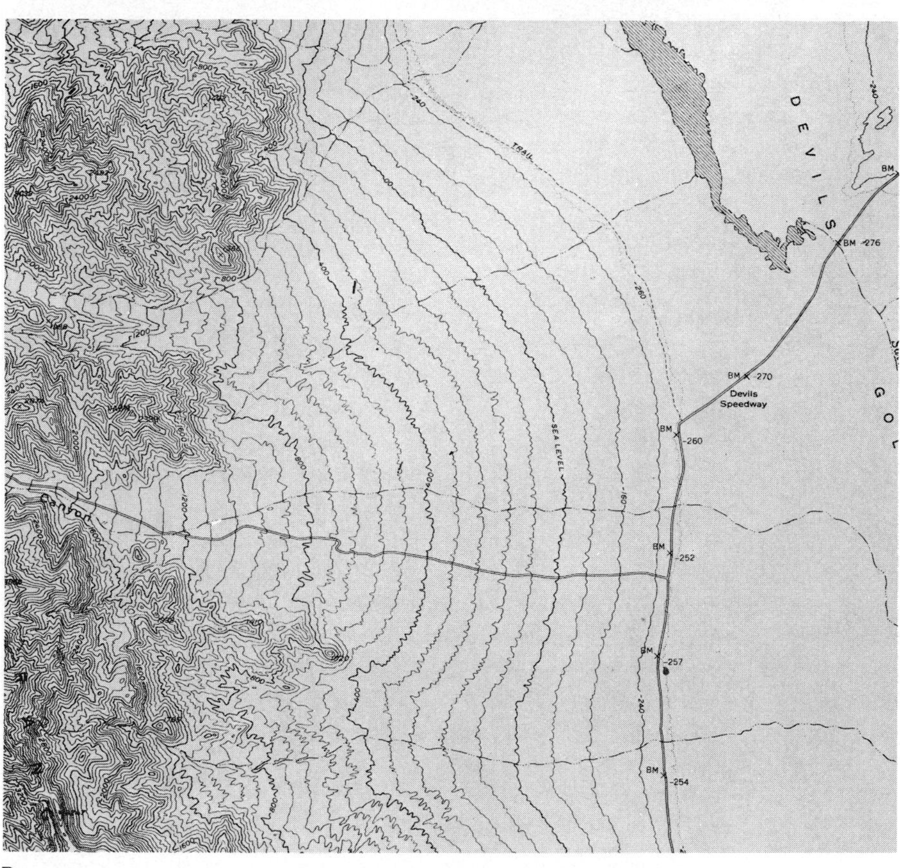

B.

FIGURE 13.19
A. Aerial view of alluvial fans in Death Valley, California. The size of the fan depends upon the size of the drainage basin. As the fans grow, they eventually coalesce to form a bajada. (Photo by John S. Shelton). **B.** A portion of the Furnace Creek, California, topographic map showing excellent alluvial fan development and playa lake. (Courtesy of the U.S. Geological Survey)

With the ongoing dissection of the mountain mass into an intricate series of valleys and sharp divides as well as the accompanying sedimentation, the local relief continues to diminish. After more time passes, the steady retreat of the mountain front enlarges the pediment. Eventually this pediment growth results in nearly the entire mountain mass being consumed. Thus, by the late stages of erosion, the mountain areas are reduced to a few large bedrock knobs projecting above the surrounding pediment and sediment-filled basin. These isolated erosional remnants on an old-age desert landscape are called **inselbergs**, a German word meaning "island mountains."*

Each of the stages of landscape evolution in an arid climate depicted in Figure 13.18 can be observed in the Basin and Range region. Recently uplifted mountains in an early stage of erosion are found in southern Oregon and northern Nevada. Death Valley, California, and southern Nevada fit into the more advanced middle stage, whereas the late stage, with its inselbergs and extensive pediments, can be seen in southern Arizona.

*For more on inselbergs, see Box 5.1, "Ayers Rock: An Example of Differential Weathering."

REVIEW QUESTIONS

1. How extensive are the desert and steppe regions of the earth?
2. What is the primary cause of subtropical deserts? Of middle-latitude deserts?
3. In which hemisphere (Northern or Southern) are middle-latitude deserts most common?
4. Why is the amount of precipitation that is used to determine whether a place has a dry climate or a humid climate a variable figure? (See Box 13.1.)
5. List four common misconceptions about deserts. (See Box 13.2.)
6. Why is rock weathering reduced in deserts?
7. As a permanent stream such as the Nile River crosses a desert, does discharge increase or decrease? How does this compare to a river in a humid region?
8. What is the most important erosional agent in deserts?
9. Describe the way in which wind transports sand. During very strong winds, how high above the surface can sand be carried?
10. Why is wind erosion relatively more important in arid regions than in humid areas?
11. What factor limits the depths of blowouts?
12. How do sand dunes migrate?
13. List three factors that influence the form and size of a sand dune.
14. Six major dune types are recognized. Indicate which type of dune is associated with each of the statements below.
    (a) Dunes whose tips point into the wind.
    (b) Long sand ridges oriented at right angles to the wind.
    (c) Dunes that often form along coasts where strong winds create a blowout.
    (d) Solitary dunes whose tips point downwind.
    (e) Long sand ridges that are oriented more or less parallel to the prevailing wind.
    (f) An isolated dune consisting of three or four sharp-crested ridges diverging from a central high point.
    (g) Scalloped rows of sand oriented at right angles to the wind.
15. Although sand dunes are the best-known wind deposits, accumulations of loess are very significant in some parts of the world. What is loess? Where are such deposits found? What are the origins of this sediment?
16. Why is sea level (ultimate base level) not a significant factor influencing erosion in desert regions?
17. Describe the features and characteristics associated with each of the stages in the evolution of a mountainous desert. Where in the United States can these stages be observed?
18. What term refers to the process by which desertlike conditions expand into areas that were previously productive? Is this strictly a natural process? (See Box 13.3.)

KEY TERMS

abrasion (p. 335)
alluvial fan (p. 341)
bajada (p. 342)
barchan dune (p. 337)
barchanoid dune (p. 337)
bed load (p. 331)
blowout (p. 334)
cross beds (p. 336)
deflation (p. 333)
desert (p. 324)
desert pavement (p. 334)
dune (p. 335)
ephemeral stream (p. 329)
inselberg (p. 344)
interior drainage (p. 341)
loess (p. 339)
longitudinal dune (p. 337)
parabolic dune (p. 337)
pediment (p. 342)
playa (p. 342)
playa lake (p. 342)
rainshadow desert (p. 326)
saltation (p. 331)
slip face (p. 336)
star dune (p. 338)
steppe (p. 324)
suspended load (p. 333)
transverse dune (p. 337)
ventifact (p. 335)

14 Shorelines

Opposite: Waves crashing against the rugged coast at Hawaii Volcanoes National Park. (Photo by Steven Trimble) (Top photo by E. J. Tarbuck)

The waters of the ocean are constantly in motion. The restless nature of the water is more noticeable along the shore—the dynamic interface between land and sea. Here we can observe the rhythmic rise and fall of the tides and see waves constantly rolling in and breaking. Sometimes the waves are low and gentle. At other times, they pound the shore with an awesome fury.

Although it may not be readily apparent to the occasional visitor, the shoreline is constantly being

A.

FIGURE 14.1
A. This satellite image includes the familiar outline of Cape Cod. Boston is in the upper left corner. The two large islands off the south shore of Cape Cod are Martha's Vineyard (left) and Nantucket (right). Although the work of waves constantly modifies this coastal landscape, shoreline processes are not responsible for creating it. Rather, the present size and shape of Cape Cod result from the positioning of moraines and other glacial materials deposited during the Pleistocene epoch. (Photo courtesy of Earth Satellite Corporation) **B.** High-altitude image of the Point Reyes area north of San Francisco, California. The 5.5-kilometer-long south-facing cliffs at Point Reyes (lower left corner) are exposed to the full force of the waves from the Pacific Ocean. Nevertheless, this promontory retreats slowly because the bedrock from which it is formed is very resistant. (Photo courtesy of USDA-ASCS)

shaped and modified by the moving ocean waters. Two well-known areas exhibiting the effects of this activity are Cape Cod, Massachusetts, and Point Reyes, California (Figure 14.1). At Cape Cod, waves striking portions of the eastern shore are causing cliffs consisting of poorly consolidated glacial sediments to retreat at rates approaching 1 meter per year. By contrast, the durable bedrock cliffs at Point Reyes are retreating much more slowly. At both locations wave activity is moving sediment along the shore and building narrow sand bars that protrude into and across some bays. However, the nature of present-day shorelines is not just the result of the relentless attack of the land by the sea. Indeed, the shore is a complex zone whose unique character is the result of many geologic processes. For example, practically all coastal areas were affected by the worldwide rise in sea level that accompanied the melting of glaciers at the close of the Pleistocene epoch. As the sea edged landward, the shoreline became superimposed upon landscapes that had been shaped by such processes as stream erosion, glaciation, volcanic activity, and the forces of mountain building.

WAVES

Wind-generated waves provide most of the energy that shapes and modifies shorelines. Where the land and sea meet, waves that may have traveled unimpeded for hundreds or thousands of kilometers suddenly encounter a barrier that will not allow them to advance farther. Stated another way, the shore is the location where a practically irresistible force confronts an almost immovable object. The conflict that results is never-ending and sometimes dramatic.

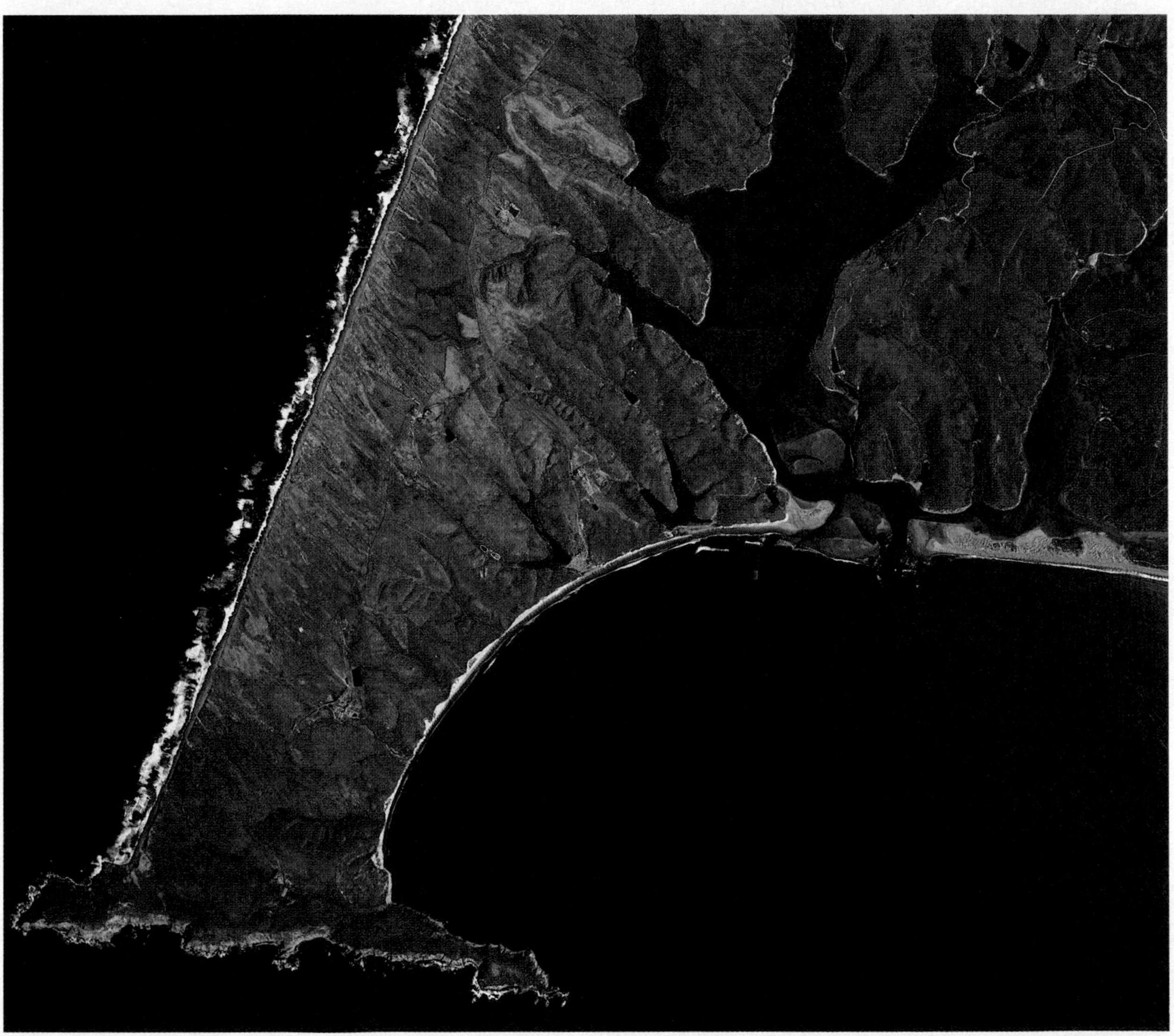

B.

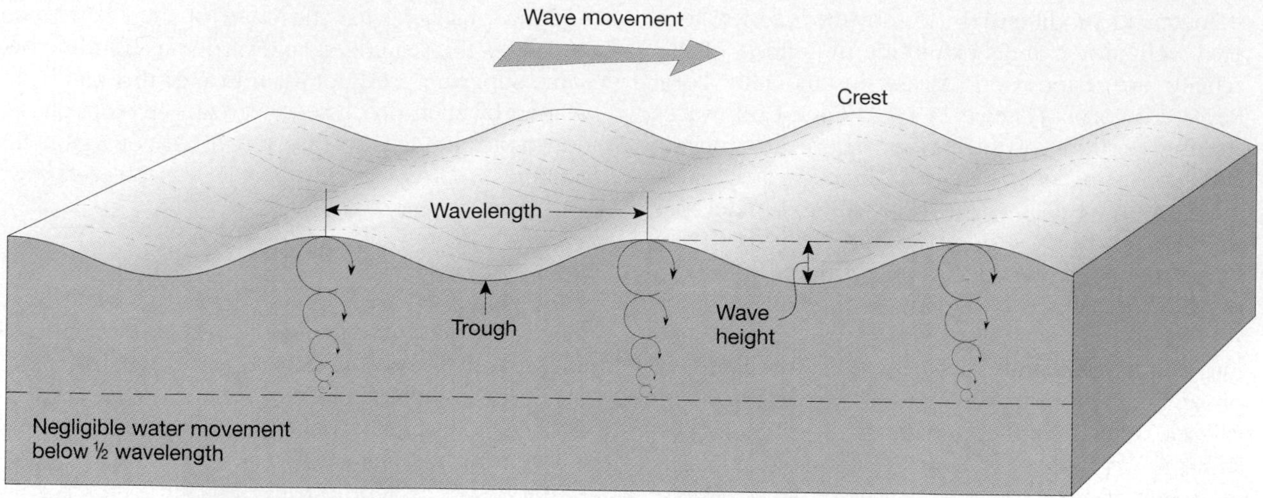

FIGURE 14.2
This diagram illustrates the basic parts of a wave as well as the movement of water particles with the passage of the wave. Negligible water movement occurs below a depth equal to one-half the wavelength (the level of the dashed line).

The undulations of the water surface, called waves, derive their energy and motion from the wind. If a breeze of less that 3 kilometers (2 miles) per hour starts to blow across still water, small wavelets appear almost instantly. When the breeze dies, the ripples disappear as suddenly as they formed. However, if the wind exceeds 3 kilometers per hour, more-stable waves gradually form and progress with the wind.

All waves are described in terms of the characteristics illustrated in Figure 14.2. The tops of the waves are the *crests,* which are separated by *troughs.* The vertical distance between trough and crest is called the **wave height**, and the horizontal distance separating successive crests is the **wavelength.** The **wave period** is the time interval between the passage of successive crests at a stationary point. The height, length, and period that are eventually achieved by a wave depend upon three factors: (1) the wind speed; (2) the length of time the wind has blown; and (3) the **fetch**, or distance that the wind has traveled across open water. As the quantity of energy trans-

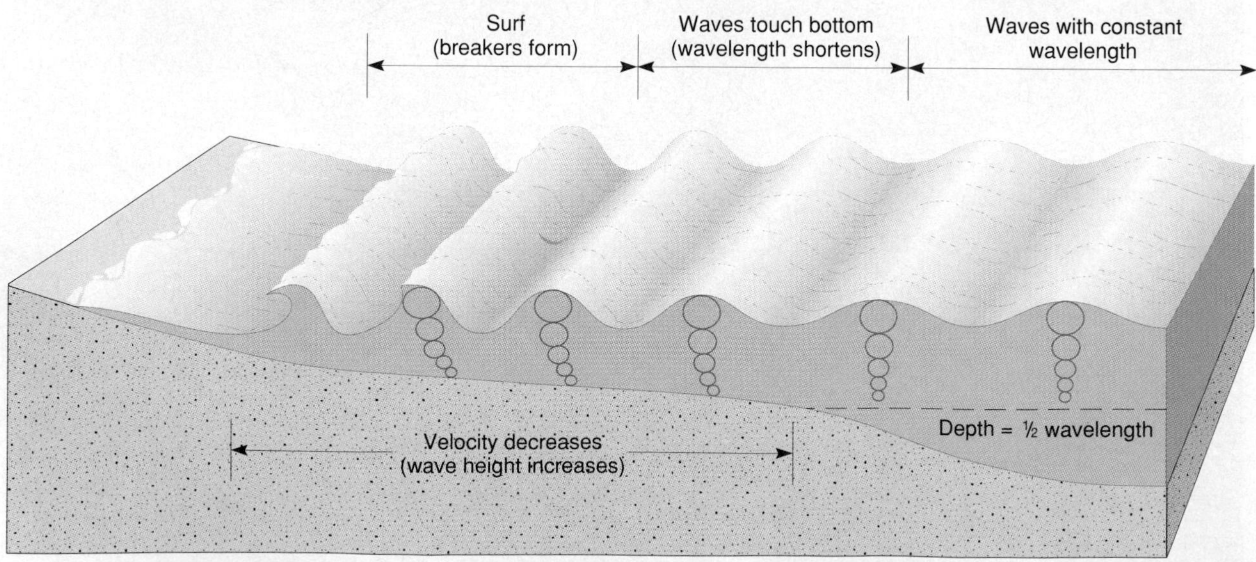

FIGURE 14.3
Changes that occur when a wave moves onto shore.

ferred from the wind to the water increases, the height and steepness of the waves increase as well. Eventually a critical point is reached and ocean breakers called *whitecaps* form.

For a particular wind speed, there is a maximum fetch and duration of wind beyond which waves will no longer increase in size. When the maximum fetch and duration are reached for a given wind velocity, the waves are said to be "fully developed." The reason that waves can grow no further is that they are losing as much energy through the breaking of whitecaps as they are receiving from the wind.

When wind stops or changes direction or if waves leave the stormy area where they were created, they continue on without relation to local winds. The waves also undergo a gradual change to *swells*, which are lower and longer, and may carry a storm's energy to distant shores. Because many independent wave systems exist at the same time, the sea surface acquires a complex, irregular pattern. Hence the sea waves we watch from the shore are usually a mixture of swells from faraway storms and waves created by local winds.

An important point to remember is that in the open sea the motion of the wave is different from the motion of the water particles within it. It is the wave form that moves forward, not the water itself. Each water particle moves in a circular path during the passage of a wave (Figure 14.2). As a wave passes, a water particle returns almost to its original position. The circular orbits followed by the water particles at the surface have a diameter equal to the wave height. When water is part of the wave crest, it moves in the same direction as the advancing wave form. When it is in the trough, the waver moves in the opposite direction. This is demonstrated by observing the behavior of a floating cork as a wave passes. The cork merely seems to bob up and down and sway slightly to and fro without advancing appreciably from its original position.* Because of this, waves in the open sea are called **waves of oscillation**. The energy contributed by the wind to the water is transmitted not only along the surface of the sea but downward as well. However, beneath the surface the circular motion rapidly diminishes until, at a depth equal to about one-half the wavelength, the movement of water particles becomes negligible. This is shown by the rapidly diminishing diameters of water-particle orbits in Figure 14.2.

As long as a wave is in deep water, it is unaffected by water depth. However, when a wave approaches

*The wind does drag the water forward slightly, causing the surface circulation of the oceans.

FIGURE 14.4
When waves break against the shore, the force of the water can be powerful and the erosional work that is accomplished can be great. (Photo by E. J. Tarbuck)

the shore, the water becomes shallower and influences wave behavior. The wave begins to "feel bottom" at a water depth equal to about one-half its wavelength. Such depths interfere with water movement at the base of the wave and slow its advance. As the wave continues to move toward the shore, the slightly faster seaward waves catch up, decreasing the wavelength. As the speed and length of the wave diminish, the wave steadily grows higher. Finally a critical point is reached when the steep wave front is unable to support the wave, and it collapses, or *breaks* (Figure 14.3). What had been a wave of oscillation now becomes a **wave of translation** in which the water advances up the shore. The turbulent water created by breaking waves is called **surf**. On the landward margin of the surf zone the turbulent sheet of water from collapsing breakers, called *swash*, moves up the slope of the beach. When the energy of the swash has been expended, the water flows back down the beach toward the surf zone as *backwash*.

WAVE EROSION

During periods of calm weather wave action is minimal. However, just as streams do most of their work during floods, so too do waves accomplish most of their work during storms. The impact of high, storm-induced waves against the shore can be awesome in its violence (Figure 14.4). Each breaking wave may hurl thousands of tons of water against the land, sometimes causing the earth to literally tremble. The pressures exerted by Atlantic waves, for example, average nearly 10,000 kilograms per square meter (more than 2000 pounds per square foot) in winter. During storms the force is even greater. During one such storm a 1350-ton portion of a steel and concrete breakwater was ripped from the rest of the structure and moved to a useless position toward the shore at Wick Bay, Scotland. Five years later the 2600-ton unit that replaced the first met a similar fate. There are many such stories that demonstrate the great force of breaking waves. It is no wonder then that cracks and crevices are quickly opened in cliffs, seawalls, breakwaters, and anything else that is subjected to these enormous shocks. Water is forced into every opening, causing air in the cracks to become highly compressed by the thrust of crashing waves. When the wave subsides, the air expands rapidly, dislodging rock fragments and enlarging and extending preexisting fractures.

In addition to the erosion caused by wave impact and pressure, **abrasion**, the sawing and grinding action of the water armed with rock fragments, is also important. In fact, abrasion is probably more intense in the surf zone that in any other environment. Smooth, rounded stones and pebbles along the shore are obvious reminders of the grinding action of rock against rock in the surf zone. Further, such fragments are used as "tools" by the waves as they cut horizontally into the land (Figure 14.5).

Along shorelines composed of unconsolidated material rather than hard rock, the rate of erosion by breaking waves can be extraordinary. In parts of Britain, where waves have the easy task of eroding glacial deposits of sand, gravel, and clay, the coast has been worn back 3 to 5 kilometers since Roman times, sweeping away many villages and ancient landmarks. A similar retreat may be seen along the cliffs of Cape Cod, which in places are retreating at a rate of up to 1 meter per year.

FIGURE 14.5
Cliff undercut by wave erosion along the Oregon coast. (Photo by E. J. Tarbuck)

FIGURE 14.6
Wave bending around the end of a beach at Stinson Beach, California. (Photo by James E. Patterson)

WAVE REFRACTION AND LONGSHORE TRANSPORT

The bending of waves, called **wave refraction**, is an important factor when shoreline processes are considered (Figure 14.6). It is significant because it affects the distribution of energy along the shore, thus strongly influencing where and to what degree erosion, sediment transport, and deposition will take place.

Waves seldom approach the shore straight on. Rather, most waves move toward the shore at an angle. However, when they reach the shallow water of a smoothly sloping bottom, they are bent and tend to become parallel to the shore. Such bending occurs because the part of the wave nearest the shore touches bottom and slows down first, whereas the end that is in deeper water continues forward at its regular speed. The net result is a wave front that may approach nearly parallel to the shore regardless of the original direction of the wave.

Due to refraction, wave impact is concentrated against the sides and ends of headlands projecting into the water, while wave attack is weakened in bays. This differential wave attack along irregular coastlines is illustrated in Figure 14.7. Since the waves reach the shallow water in front of the headland sooner than they do in adjacent bays, they are

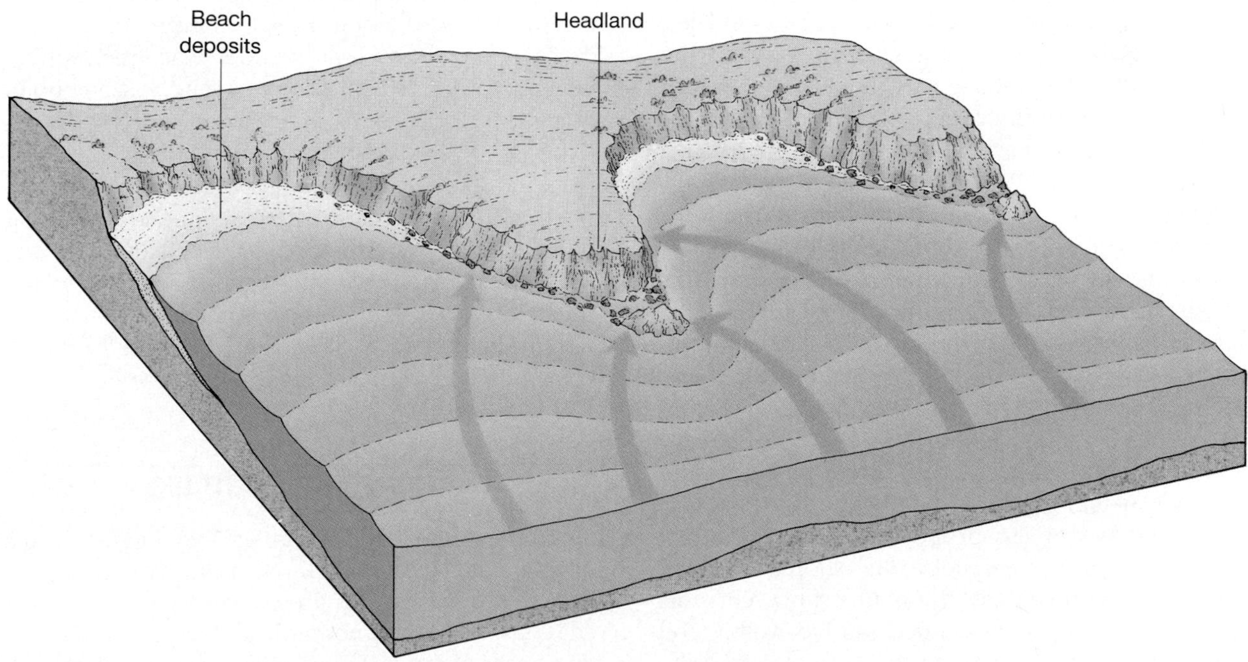

FIGURE 14.7
Because of wave refraction, the greatest erosional power is concentrated on the headlands. In the bays, the force of the waves is much weaker.

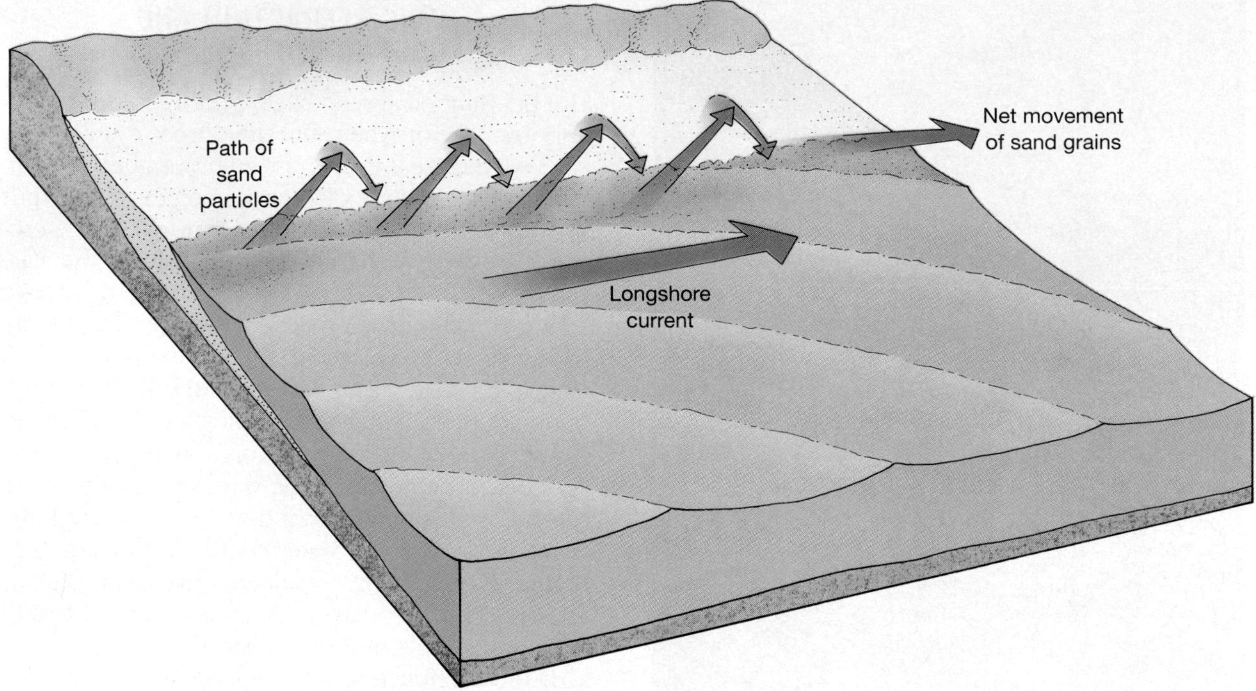

FIGURE 14.8
Beach drift and longshore currents are created by obliquely breaking waves. These processes transport large quantities of material along the beach and in the surf zone.

bent more nearly parallel to the protruding land and strike it from all three sides. By contrast, refraction in the bays causes waves to diverge and expend less energy. In these zones of weakened wave activity, sediments can accumulate and form sandy beaches. Over a long period, erosion of the headlands and deposition in the bays will straighten an irregular shoreline.

Although waves are refracted, most still reach the shore at some angle, however slight. Consequently, the uprush of water from each breaking wave (the swash) is oblique. However, the backwash is in the direction of the slope of the beach. The effect of this pattern of water movement is to transport particles of sediment in a zigzag pattern along the beach (Figure 14.8). This movement is called **beach drift**, and it can transport sand and pebbles hundreds or even thousands of meters each day.

Oblique waves also produce currents within the surf zone that flow parallel to the shore. Since the water here is turbulent, these **longshore currents** easily move the fine suspended sand as well as roll larger sand and gravel along the bottom. When the sediment transported by longshore currents is added to the quantity moved by beach drift, the total amount can be very large. At Sandy Hook, New Jersey, for example, the quantity of sand transported along the shore over a 48-year period averaged almost 750,000 tons per year. For a 10-year period at Oxnard, California, more than 1.5 million tons of sediment moved along the shore each year.

There should be little wonder that beaches have been characterized as "rivers of sand." At any point along a beach there is likely to be more sediment that was derived elsewhere than material eroded from the shore area immediately behind it. It is also worth noting that much of the sediment composing beaches is not wave-eroded debris. Rather, in many areas sediment-laden rivers that discharge into the ocean are the major sources of material. Hence, if it were not for beach drift and longshore currents, many beaches would be nearly sandless.

SHORELINE FEATURES

Along the rugged and irregular New England coast or along the steep shorelines of the West Coast, the effects of wave erosion are often easy to see. **Wave-cut cliffs**, as their name implies, originate by the cutting action of the surf against the base of coastal land. As erosion progresses, rocks overhanging the notch at the base of the cliff crumble into the surf, and the cliff retreats (see Figure 14.5). A relatively flat, benchlike surface, the **wave-cut platform**, is left behind by the receding cliff (Figure 14.9). The platform broad-

FIGURE 14.9
Elevated wave-cut platform along the California coast near San Francisco. A new platform is being created at the base of the cliff. (Photo by John S. Shelton)

FIGURE 14.10
Sea arch and sea stack along the coast of Iceland. (Photo by Bruce F. Molnia, courtesy of Terraphotographics/BPS)

ens as wave attack continues. Some of the debris produced by the breaking waves remains along the water's edge as part of the beach, while the remainder is transported farther seaward.

Headlands that extend into the sea are vigorously attacked by waves because of refraction. The surf erodes the rock selectively, wearing away the softer or more highly fractured rock at the fastest rate. At first, sea caves may form. When two caves on opposite sides of a headland unite, a **sea arch** results. Finally the arch falls in, leaving an isolated remnant, or **sea stack**, on the wave-cut platform (Figure 14.10). Eventually it too will be consumed by the action of the waves.

Where beach drift and longshore currents are active, several features related to the movement of sediments along the shore may develop. **Spits** are elongated ridges of sand that project from the land into the mouth of an adjacent bay (Figure 14.11). Often the end in the water hooks landward in response to wave-generated currents. The term **baymouth bar** is applied to a sand bar that completely crosses a bay, sealing it off from the open ocean (Figure 14.11). Such a feature tends to form across bays where currents are weak, allowing a spit to extend to the other side. A **tombolo**, a ridge of sand that connects an island to the mainland or to another island, forms in much the same manner as a spit.

The Atlantic and Gulf Coast Plains are relatively flat and slope gently seaward. The shore zone is characterized by **barrier islands**. These low ridges of sand parallel the coast at distances from 3 to 30 kilometers offshore. From Cape Cod, Massachusetts, to Padre Island, Texas, nearly 300 barrier islands rim the coast (Figure 14.12). Most are from 1 to 5 kilometers wide and between 15 and 30 kilometers long. The highest topographic features are sand dunes, which usually reach heights of 5 to 10 meters; in a few areas, unvegetated dunes are more than 30 meters high. The lagoons separating these narrow islands from the

FIGURE 14.11
High-altitude image of a well-developed spit and baymouth bar along the coast of Martha's Vineyard, Massachusetts. (Photo courtesy of USDA-ASCS)

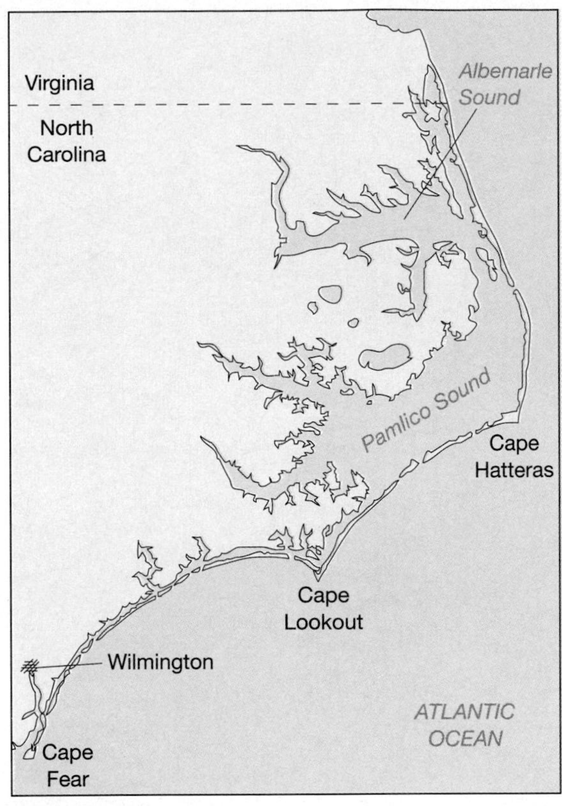

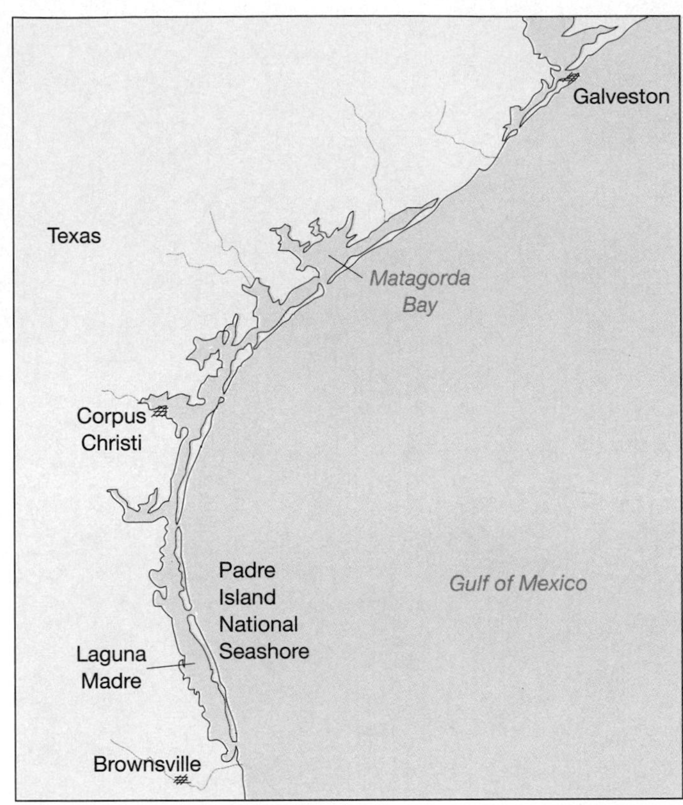

FIGURE 14.12
The islands of North Carolina's Outer Banks (left) and those along the south Texas coast (right) are some of the best examples of the nearly 300 barrier islands that rim the Atlantic and Gulf coasts.

shore represent zones of relatively quiet water that allow small craft traveling between New York and northern Florida to avoid the rough waters of the North Atlantic.

How barrier islands originate is still not certain. They possibly form in three or more ways. Some are thought to have originated as spits that were subsequently severed from the mainland by wave erosion or by the general rise in sea level following the last episode of glaciation. Some barrier islands may have been created when turbulent waters in the line of breakers heaped up sand that was scoured from the bottom. Since these sand barriers rise above normal sea level, the piling of sand likely resulted from the work of storm waves at high tide. Finally, some studies suggest that barrier islands may be former sand dune ridges that originated along the shore during the last glacial period, when sea level was lower. When the ice sheets melted, sea level rose and flooded the area behind the beach-dune complex.

There is little question that a shoreline soon undergoes modification regardless of its initial configuration. At first most coastlines are irregular, although the degree of and reason for the irregularity may vary considerably from place to place. Along a coastline that is characterized by varied geology, the pounding surf may at first increase its irregularity because the waves will erode the weaker rocks more easily than the stronger ones. However, it is commonly agreed that if a shoreline remains stable, marine erosion and deposition will eventually produce a more regular coast. Figure 14.13 illustrates the evolution of an initially irregular coast. As waves erode the headlands, creating cliffs and a wave-cut platform, sediment is carried along the shore. Some material is deposited in the bays, while other debris is formed into spits and baymouth bars. At the same time rivers fill the bays with sediment. Ultimately a smooth coast results.

SHORELINE EROSION PROBLEMS

Compared with other natural hazards such as earthquakes, volcanic eruptions, and landslides, shoreline erosion is often perceived to be a more continuous and predictable process that appears to cause rela-

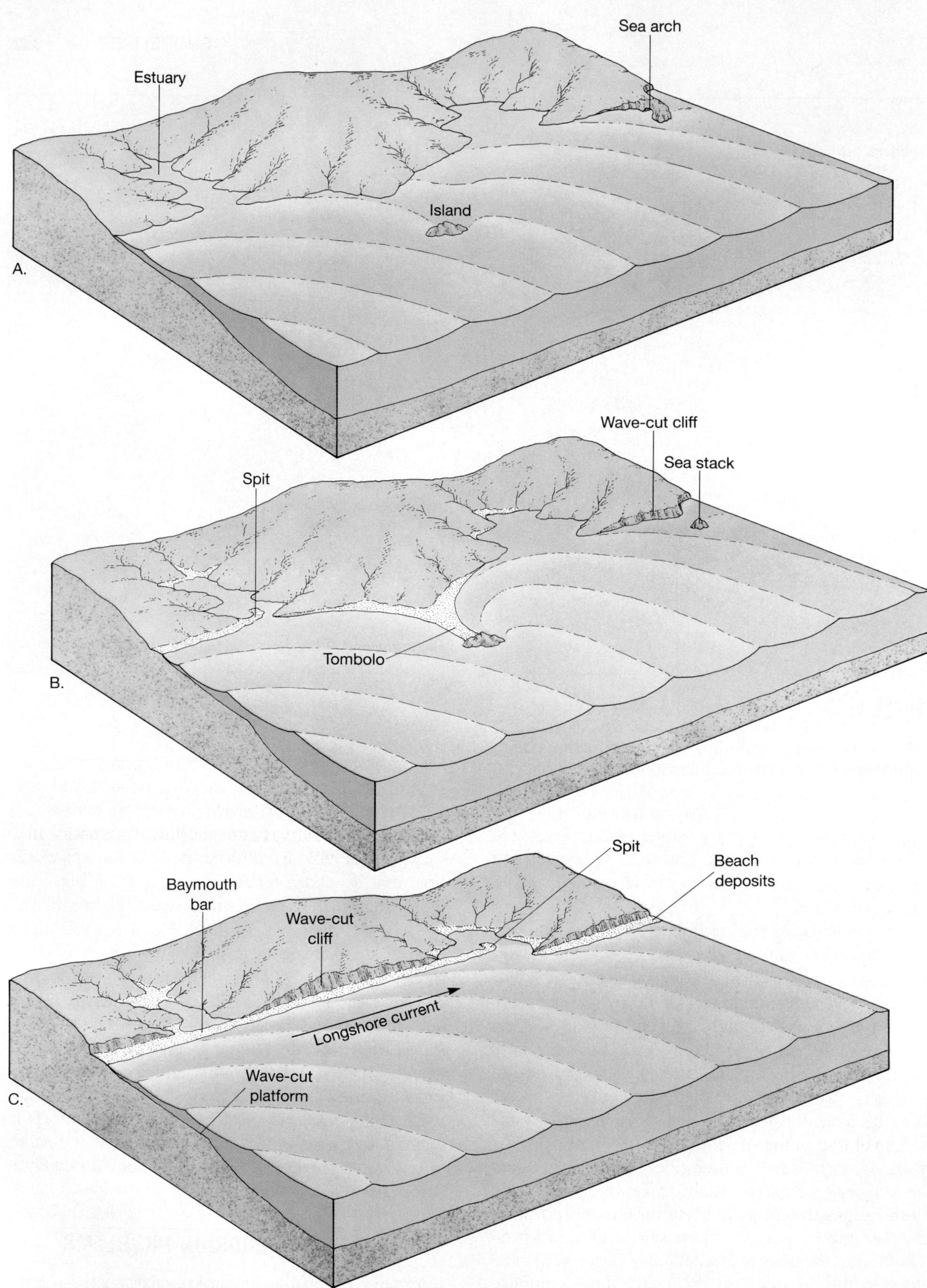

FIGURE 14.13
These diagrams illustrate the changes that can take place through time along an initially irregular coastline that remains relatively stable. The diagrams also serve to illustrate many of the features described in the section on shoreline features.

FIGURE 14.14
This abandoned cottage was once thought to be relatively safe from wave attack. However, as the shoreline retreated, the cottage was exposed to the full force of the sea. (Photo by Kenneth Hasson)

tively modest damage to limited areas. This is not always true. The shoreline is a dynamic place that can change rapidly in response to natural forces. Exceptional storms are capable of eroding beaches and cliffs at rates that are far in excess of the long-term average. Such bursts of accelerated erosion not only have a significant impact on the natural evolution of a coast, but can also have a profound impact on people who reside in the coastal zone. Erosion along our coasts has caused and will continue to cause significant property damage (Figure 14.14). Large sums are spent annually not only to repair damage, but also to prevent or control erosion. Already a problem at many sites, shoreline erosion is certain to become an increasingly serious problem as extensive development in coastal areas continues.

Although the same processes cause change along every coast, not all coasts respond in the same way. Interactions among different processes and the relative importance of each process depend upon local factors. The factors include: (1) the proximity of a coast to sediment-laden rivers; (2) the degree of tectonic activity; (3) the topography and composition of the land; (4) prevailing winds and weather patterns; and (5) the configuration of the coastline and nearshore areas.

The shoreline along the Pacific Coast of the United States is strikingly different from that characterizing the Atlantic and Gulf coast regions. Some of the differences are related to plate tectonics. The West Coast represents the leading edge of the North American plate, and because of this, it experiences active uplift and deformation. By contrast, the East Coast is a tectonically quiet region that is far from any active plate margin. Because of this basic geological difference, the nature of shoreline erosion problems along America's opposite coasts is different.

The discussion that follows will focus first on the problems faced along the East Coast of the United States, especially its vulnerable barrier islands. This will be contrasted with a look at the equally serious erosion problems occurring at many West Coast locations.

Gulf and Atlantic Coasts

During this century growing affluence and increasing demands for recreation have brought unprecedented development to many coastal areas. Much of this development has occurred on barrier islands. Typically barrier islands, also termed *barrier beaches* or *coastal barriers*, consist of a wide beach that is backed by dunes and separated from the mainland by marshy lagoons. The broad expanses of sand and exposure to the ocean have made barrier islands exceedingly attractive sites for development (Figure 14.15). Unfortunately, development has taken place more rapidly than our understanding of barrier island dynamics.

Because barrier islands face the open ocean, they receive the full force of major storms that strike the coast. When a storm occurs, the barriers absorb the energy of the waves primarily through the move-

BOX 14.1

Coasts: The Conflict of People and Nature

People often treat the coast as if it were a stable platform on which structures can be safely built. This attitude inevitably leads to conflicts between people and nature. The waves from a single major storm can move large quantities of coastal sediment and can flood vast areas in a matter of hours.*

"The sea won this round." These words were uttered by President George Bush, November 3, 1991, as he surveyed the devastating damage to his vacation home in Kennebunkport, Maine, following a fierce Atlantic storm that ravaged a large portion of the East Coast for two days. The President vowed to rebuild the turn-of-the-century structure.

Conflicts such as the one experienced by President Bush have always occurred along coasts. However, the increasing desirability and accessibility of coasts as places to live and work have increased the frequency and intensity of these conflicts over the past 50 years. In 1990, 50 percent of the United States population lived within 75 kilometers (45 miles) of a coast. This number is projected to increase to 75 percent by the year 2010. The concentration of such large numbers of people near the shoreline means that hurricanes and other large storms not only place millions at risk but can result in extraordinary property damage as well. The graph in Figure 14.A clearly shows this trend.

Hurricane Hugo illustrates what can occur when a powerful storm strikes a developed coast. Between September 11 and 22, 1989, Hurricane Hugo tracked across the tropical Atlantic and the eastern margin of the Caribbean before making landfall in the mainland United States. Prior to slamming into the South Carolina coast near Charleston (Figure 14.B), Hugo was responsible for serious destruction on the islands of Puerto Rico and St. Croix. In addition to destroying homes and property, the storm moved vast quantities of sand. On Puerto Rico, sand was removed from the beach and deposited behind the dunes outside the immediate coastal environment. Coral colonies were choked by sediment, and many were destroyed. A large offshore sand deposit (about 90,000,000 cubic meters) was leveled by storm waves and dispersed over the ocean floor.

When the storm came ashore along the South Carolina coast, the storm surge washed over some barrier islands, demolishing seawalls, roads, and structures of all kinds (Figure 14.C). Where barrier islands were narrow and lacked dunes, they were breached by the storm-driven waters, and inlets were created. The damage Hugo caused along South Carolina's barrier islands and shore communities clearly shows the importance of good construction and the natural protection provided by a wide beach and high sand dunes. Folly Beach, south of Charleston, was on the weaker (south) side of the hurricane, but its beaches were already severely narrowed by long-term erosion; homeowners had dumped boulders and concrete rubble on the beach for protection. The rocks and rubble proved useless during Hugo. The storm surge (over 3.5 meters at Folly Beach) overtopped the structure, caused major damage to beachfront houses, and totally swept away a popular seafood restaurant. Isle of Palms and Sullivans Island, north of Charleston, suffered substantial damage, but the effects were somewhat lessened by a wide beach and dunes. Houses that survived were built to withstand high winds and flooding, in contrast to homes built to lower standards that were totally destroyed.

Hugo was the strongest hurricane to strike the continental United States since Hurricane Camille, 20 years earlier. The total property loss resulting from Hugo was estimated to be $10 billion, with more than $7 billion of that in the continental United States alone. The loss of life included 28 on Caribbean islands and 21 in the continental United States. Although storm warnings have kept the loss of life from coastal storms at a low figure, the National Weather Service is concerned that the high population increases in coastal areas could set the stage for a major hurricane disaster. Unfortunately, the rate of improvement in achieving forecast lead times is not keeping pace with the time requirements for evacuating the ever-increasing coastal population.

*This discussion is based on one in S. J. Williams et al., *Coasts in Crisis,* U.S. Geological Survey Circular 1075, (Washington, D.C., U.S. Government Printing Office, 1991).

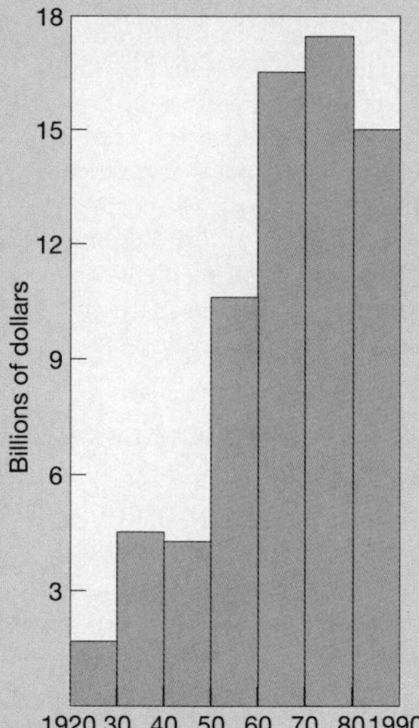

FIGURE 14.A
Property losses from hurricanes in the continental United States by decades (1920–1990). The extraordinary rise in the amount of property damage reflects the rapid population increase in coastal areas. (From National Hurricane Center/NOAA)

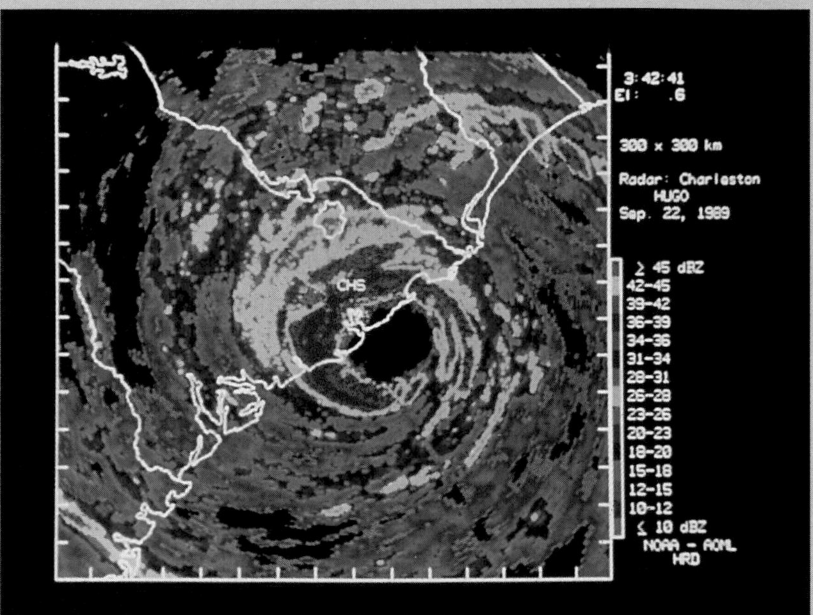

FIGURE 14.B
Radar image of Hurricane Hugo as the storm struck the South Carolina coast. The storm had sustained winds of 215 kilometers (135 miles) per hour and a maximum storm surge of about 6 meters. Barrier islands here are generally less than 3 meters above sea level. Colors show rainfall intensity. Red is the most intense and blue is less intense. No rain is falling in the black areas. The rainless "hole" in the center is the eye. The most intense part of the storm, known as the eye wall, surrounds the eye. (Courtesy of Peter Dodge, NOAA, Hurricane Research Division)

FIGURE 14.C
The damage caused by Hurricane Hugo in September, 1989, to some of South Carolina's barrier islands underscores the need for knowledge of beach and storm processes in order to protect lives and property. (Photo by S. J. Williams, U.S. Geological Survey)

FIGURE 14.15
This high-altitude infrared image shows a portion of an urbanized barrier island off the New Jersey coast. The problems and hazards associated with constructing buildings on barrier islands remain. It is just as unsafe to erect a building on shifting sand today as it was a century ago. In fact, considering the high population density on some islands, the potential for a disaster is even greater. (Photo courtesy of USDA-ASCS)

ment of sand. Frank Lowenstein describes this process and the dilemma that results:

> Waves may move sand from the beach to offshore areas or, conversely, into the dunes; they may erode the dunes, depositing sand onto the beach or carrying it out to sea; or they may carry sand from the beach and the dunes into the marshes behind the barrier, a process known as overwash. The common factor is movement. Just as a flexible reed may survive a wind that destroys an oak tree, so the barriers survive hurricanes and nor'easters not through unyielding strength but by giving before the storm.
>
> This picture changes when a barrier is developed for homes or as a resort. Storm waves that previously rushed harmlessly through gaps between the dunes now encounter buildings and roadways. Moreover, since the dynamic nature of the barriers is readily perceived only during storms, homeowners tend to attribute damage to a particular storm, rather than to the basic mobility of coastal barriers. With their homes or investments at stake, local residents are more likely to seek to hold the sand in place and the waves at bay than to admit that development was improperly placed to begin with.*

Protecting property from storm-induced waves as well as from the ongoing movement of sand by longshore currents has been a significant concern in developed coastal areas for many years. Attempts at controlling the dynamic beach environment include building such artificial structures as jetties, breakwaters, groins, and seawalls. Such interference can create many new problems and result in unwanted changes that are difficult and expensive to correct.

From relatively early in America's history a principal goal in coastal areas was the development and maintenance of harbors. In many cases, this involved the construction of jetty systems. **Jetties** are usually built in pairs and extend into the ocean at the entrances to rivers and harbors. With the flow of water confined to a narrow zone, the ebb and flow caused by the rise and fall of the tides keep the sand in motion and prevent deposition in the channel. However, as illustrated in Figure 14.16, the jetty may act as a dam against which the longshore current and beach drift deposit sand. At the same time, wave activity removes sand on the other side. Since the other side is not receiving any new sand, there is soon no beach at all.

To maintain or widen beaches that are losing sand, groins are sometimes constructed. A **groin** is a barrier built at a right angle to the beach for the purpose of trapping sand that is moving parallel to the shore (see Figure 14.16). The result is an irregular but wider beach. These structures often do their job so effectively that the longshore current beyond the groin is sand deficient. As a result, the current removes sand from the beach on the leeward side of the groin. Such a situation is illustrated in Figure 14.17. To offset this effect, property owners downcurrent from the structure may erect groins on their property. In this manner, the number of groins multiplies. An example of such proliferation is the shore-

*"Beaches or Bedrooms—The Choice as Sea Level Rises," *Oceanus* 28 (No. 3, Fall 1985): 22.

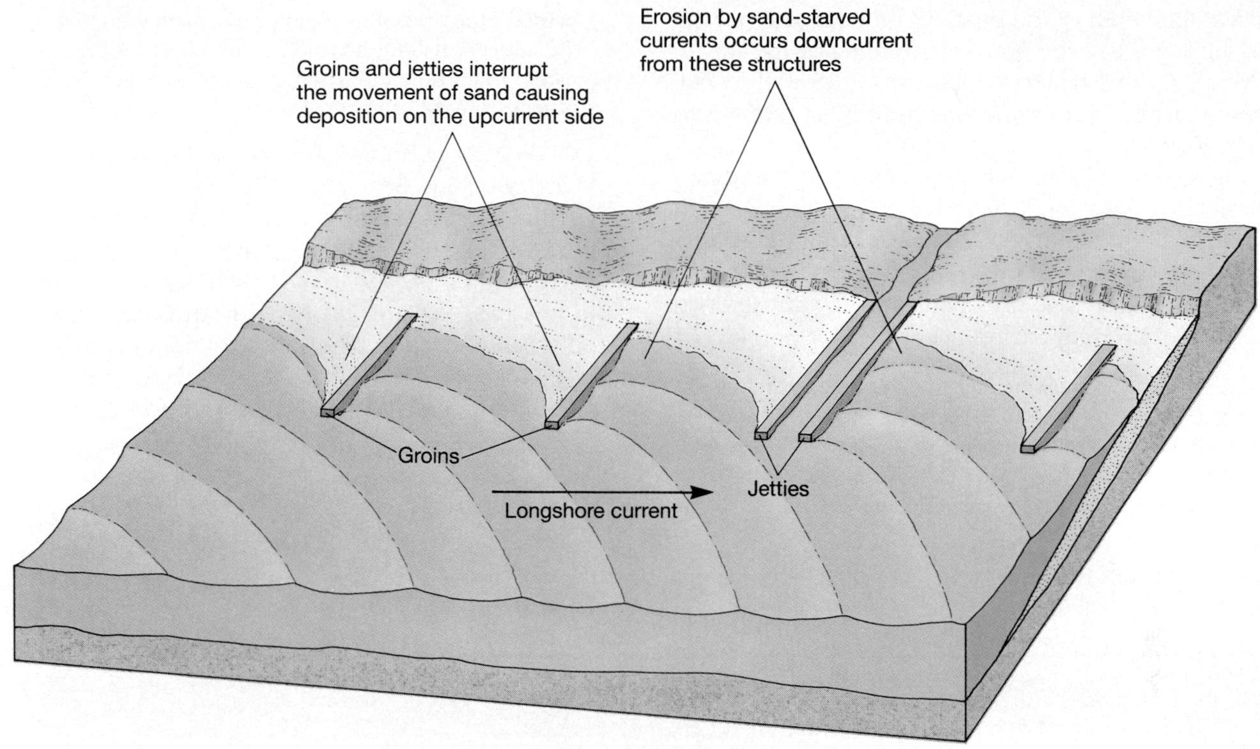

FIGURE 14.16
Jetties and groins interrupt the movement of sand by beach drift and longshore currents. Beach erosion often results downcurrent from the site of the structure.

line of New Jersey, where more than three hundred such structures have been built. Since it has been shown that groins often do not provide a satisfactory solution, they are no longer the preferred method of keeping beach erosion in check.

In some coastal areas a **breakwater** may be constructed parallel to the shoreline. The purpose of such a structure is to protect boats from the force of large breaking waves by creating a quiet water zone near the shore. However, when this is done, the reduced wave activity along the shore behind the structure may allow sand to accumulate. If this happens, the marina will eventually fill with sand while the downstream beach erodes and retreats. At Santa Monica, California, where the building of a breakwater created such a problem, the city had to install a dredge to remove sand from the protected quiet water zone and deposit it down the beach where longshore currents and beach drift could recirculate the sand (Figure 14.18).

As development has moved ever closer to the beach, seawalls represent yet another structure that is built to defend property from the force of breaking waves. **Seawalls** are simply massive barriers intended to prevent waves from reaching the areas behind the wall. Waves expend much of their energy as they move across an open beach. Seawalls cut this process short by reflecting the force of unspent waves seaward. As a consequence, the beach to the seaward side of the seawall experiences significant erosion and may, in some instances, be eliminated entirely.

FIGURE 14.17
At Manasquan, New Jersey, a groin has trapped sand on the upcurrent side leaving the downcurrent side starved of sand. (Photo by S. J. Williams, U.S. Geological Survey)

Once the width of the beach is reduced, the seawall is subjected to even greater pounding by the waves. Eventually this battering will cause the wall to fail, and a larger, more expensive wall must be built to take its place.

In recent years, the wisdom of building temporary protective structures has been questioned with greater frequency. The feelings of many coastal scientists are expressed in the following excerpt from a position paper that grew out of a conference on America's Eroding Shoreline:

> It is now clear that halting the receding shoreline with protective structures benefits only a few and seriously degrades or destroys the natural beach and the value it holds for the majority. Protective structures divert the ocean's energy temporarily from private properties, but usually refocus that energy on the adjacent natural beaches. Many interrupt the natural sand flow in coastal currents, robbing many beaches of vital sand replacement.*

Of the various approaches to stabilizing the sands of barrier islands, **beach nourishment** is now considered the most acceptable method. As the term implies, this practice simply involves the addition of large quantities of sand to the beach system. Building the beaches seaward improves both beach quality and storm protection. The source of the sand may be the bottom of a nearby lagoon or inland dunes. In some cases, sand is trucked in and added to the

*"Strategy for Beach Preservation Proposed," *Geotimes* 30 (No. 12, December 1985): 15.

BOX 14.2

Ocean City: An Urbanized Barrier Island

Conflicts between human activities and natural systems frequently occur when the two come together in a coastal environment. One striking example of such conflict has occurred at Ocean City, Maryland.*

Ocean City has been a popular beach resort for more than a century. During the 1920s, several large hotels and a boardwalk were built to accommodate visitors. Development was modest but steady until the early 1950s, when a dramatic boom in construction began that lasted nearly 30 years. Early concerns about the coastal environment were raised in the late 1970s and led to federal and state laws to limit dredging and filling of wetlands.

The resort is built on the southern end of Fenwick Island, one of the chain of barrier islands stretching from New York to Florida (Figure 14.D). Ocean City Inlet, which connects the quiet bay on the mainland side of the island with the Atlantic Ocean at the southern end of the island, was opened during the great hurricane of 1933. To maintain the inlet as a navigation channel, two stone jetties were constructed by the U.S. Army Corps of Engineers shortly after the storm. The jetties have stabilized the inlet, but they have drastically altered the sand-transport process near the inlet. The net longshore drift at Ocean City is southerly; it has produced a wide beach at Ocean City north of the jetty, but Assateague Island, south of the inlet, has been starved of sediment. The result is a westerly offset of more than 500 meters in the once-straight barrier island (Figure 14.E).

The most damaging storm to hit Ocean City within historic times was the Ash Wednesday northeaster of early March, 1962. It caused severe erosion and flooding along much of the middle Atlantic Coast. For two days, over five high-tide cycles, all of Fenwick Island except the highest dune areas was repeatedly washed over by storm waves superimposed on the 2-meter-high storm surge. Property damage to Ocean City alone was estimated at $7.5 million. Given the dense development of the island over the last 20 years, damage from a similar storm would today be hundreds of millions of dollars.

*This discussion is based on one in S. J. Williams et al., *Coasts in Crisis*, U.S. Geological Survey Circular 1075, (Washington, D.C., U.S. Government Printing Office, 1991).

beach. In other instances, the sand is added at an upstream location to be distributed down the coast by wave activity. Beach nourishment, however, is not a permanent solution to the problem of shrinking beaches. A case in point is Virginia Beach, Virginia, one of the largest seaside resort communities on the Atlantic Coast. Each year since 1952 this community has added the equivalent of 30,000 dump trucks' worth of sand along the shore in order to maintain just a thin strip of beach. The price is about $1.5 million per year. When beach nourishment was used to renew 24 kilometers of Miami Beach, the cost was $64 million. Furthermore, in some instances, beach nourishment can lead to unwanted environmental effects. For example, beach replenishment at Waikiki Beach, Hawaii, involved replacing coarse calcareous sand with softer, muddier calcareous sand. Destruction of the soft beach sand by breaking waves increased the water's turbidity and killed offshore coral reefs. At Miami Beach, where quartz sand was replaced by calcareous sand, the increased turbidity damaged local coral communities.

Beach nourishment appears to be an economically viable long-range solution to the beach preservation problem only in areas where there are dense development, large supplies of sand, relatively low wave energy, and reconcilable environmental issues. Unfortunately, few areas possess all these attributes.

So far two basic responses to shoreline erosion problems have been considered: (1) the building of structures such as seawalls to hold the shoreline in place and (2) the addition of sand to replenish eroding beaches. However, a third option is also available and that is to relocate buildings away from the beach. For some areas the costs of adding new sand to the beaches and/or building stronger and stronger sea-

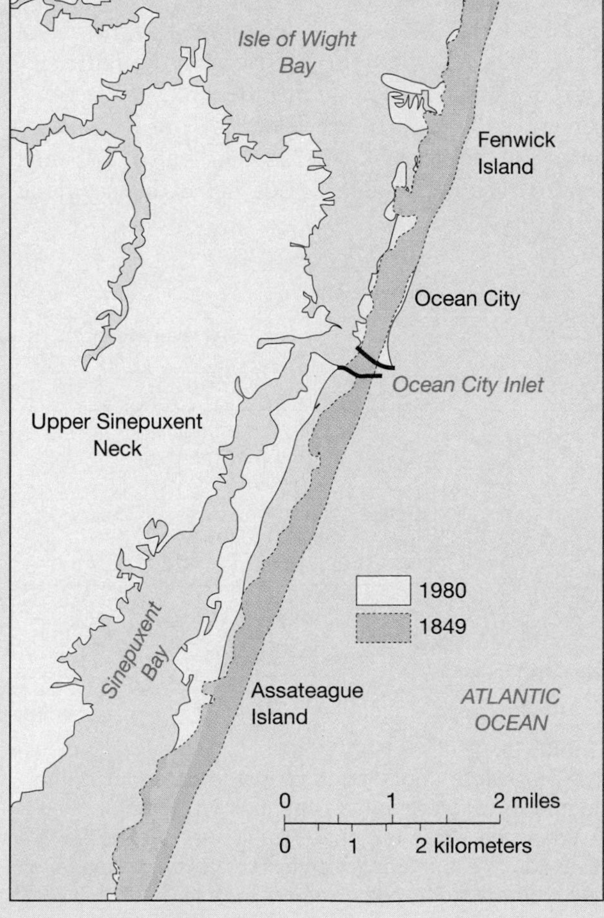

FIGURE 14.D
Shoreline changes in the Fenwick Island–Assateague Island region between 1849 and 1980 show the dramatic effects of the two large jetties at Ocean City Inlet on the natural sediment-transport processes along the coast. (After U.S. Geological Survey Circular 1075)

FIGURE 14.E
Oblique view of Ocean City Inlet (about 1980) showing the 500-meter offset between Fenwick Island (background) and Assateague Island (foreground) due to sediment starvation as a result of the construction of jetties in the 1930s. (Photo by S. J. Williams, U.S. Geological Survey)

A.

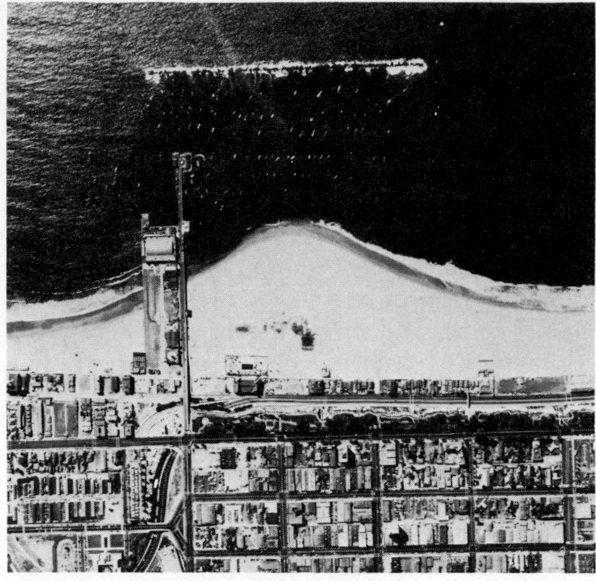

B.

FIGURE 14.18
A. The shoreline at Santa Monica pier as it appeared in 1931. **B.** The same area in 1949. The construction of the breakwater disrupted longshore transport and caused the seaward growth of the beach. (Photos courtesy of Fairchild Aerial Photography Collection, Whittier College)

walls may exceed the value of the property to be saved. When this point is reached, abandonment and relocation is the alternative.

Sea Level Is Rising

The shifting, dynamic nature of barrier islands and the ineffectiveness of most shoreline protection measures are now relatively well-established facts. Unfortunately, recent research, which indicates that sea level is rising, has compounded this already distressing situation. Studies indicate that sea level has risen between 10 and 15 centimeters over the past century. Furthermore, some investigators predict an accelerated sea-level rise in the years to come—as much as 30 centimeters or more by the year 2025. Although such a vertical change may seem modest, many coastal geologists believe that any given rise in sea level along the gently sloping Atlantic and Gulf coasts will cause from 10 to 1000 times as much horizontal shoreline retreat (Figure 14.19).

The belief that sea level will continue to rise in the coming decades is linked to the results of climatic studies that predict a global warming trend. Such predictions are based upon the now well-established fact that the carbon dioxide (CO_2) content of the atmosphere has been rising at an accelerating rate for more than a century. The CO_2 is added primarily as a by-product of the combustion of ever-increasing quantities of fossil fuels. If we assume that the use of fossil fuels will continue to rise at projected rates, current estimates indicate that the atmosphere's CO_2 content will grow by an additional 40 percent or more by sometime in the second half of the next century. The importance of CO_2 lies in the fact that it

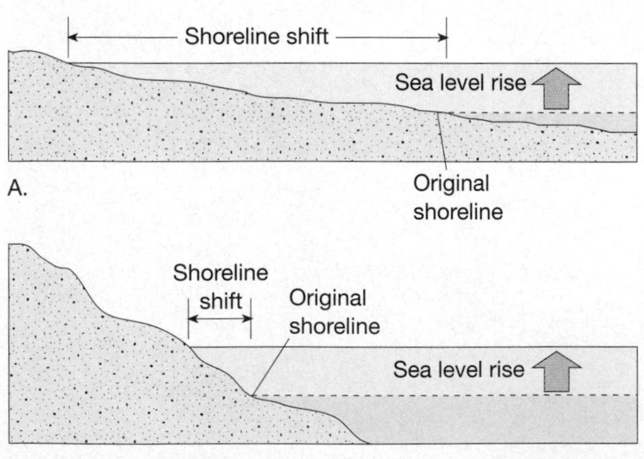

FIGURE 14.19
The slope of a shoreline is critical to determining the degree to which sea-level changes will affect it. **A** When the slope is gentle, small changes in sea level cause a substantial shift. **B.** The same sea-level rise along a steep coast results in only a small shoreline shift.

traps a portion of the radiation emitted by the earth and thereby keeps the air near the earth's surface warmer than it would be without CO_2. In other words, CO_2 is an important heat-absorbing gas; it follows logically that an increase in the air's CO_2 content should lead to higher atmospheric temperatures.

Carbon dioxide is not the only gas contributing to an upward trend in global temperatures. Atmospheric scientists have come to realize that the industrial and agricultural activities of people are causing a buildup of certain trace gases that may also play a significant role. The trace gases that appear to be most important are methane (CH_4), nitrous oxide (N_2O), and certain types of chlorofluorocarbons. These gases absorb wavelengths of outgoing earth radiation that would otherwise escape into space. Although individually their impact is small, taken together the effects of these trace gases may be as great as those of CO_2 in warming the earth.

How is a warmer atmosphere related to a global rise in sea level? First, higher temperatures can cause glacial ice to melt. About one-half of the 10- to 15-centimeter rise in sea level over the past century is attributed to the melting of small glaciers and ice sheets. Second, a warmer atmosphere causes an increase in ocean volume through thermal expansion. That is, higher air temperatures raise the temperature of the upper layers of the ocean. This, in turn, causes the water to expand and sea level to rise. It is also believed that a warmer ocean may spur storm development. Of course, an increase in storm activity would compound an already serious problem in many coastal areas.

Since rising sea level is a gradual phenomenon, it may be overlooked by coastal residents as a significant contributor to shoreline erosion problems. Rather the blame is assigned to other forces, especially storm activity. Although a given storm may be the immediate cause, the magnitude of its destruction may result from the relatively small sea-level rise that allowed the storm's power to cross a much greater land area.

Pacific Coast

In contrast to the broad, gently sloping coastal plains of the East, much of the Pacific Coast is characterized by relatively narrow beaches that are backed by steep cliffs and mountain ranges. Recall that America's western margin is a more rugged and tectonically active region than the East. Since uplift continues, the apparent rise in sea level in the West is not so readily apparent. Nevertheless, like the shoreline erosion problems facing the East's barrier islands, West Coast difficulties also stem largely from the alteration of a natural system by people.

A major problem facing the Pacific shoreline, and especially portions of southern California, is a significant narrowing of many beaches. The bulk of the sand on many of these beaches is supplied by rivers that transport it from the mountainous regions to the coast. Over the years this natural flow of material to the coast has been interrupted by dams built for irrigation and flood control. The reservoirs effectively trap the sand that would otherwise nourish the beach environment. When the beaches were wider, they served to protect the cliffs behind them from the force of storm waves. Now, however, the waves move across the narrowed beaches without losing much energy and cause more rapid erosion of the sea cliffs.

Although the retreat of the cliffs provides material to replace some of the sand impounded behind dams, it also endangers homes and roads built on the bluffs. In addition, development atop the cliffs aggravates the problem. Urbanization increases runoff which, if not carefully controlled, can result in serious bluff erosion. Watering lawns and gardens adds significant quantities of water to the slope. This water percolates downward toward the base of the cliff, where it may emerge in small seeps. This action reduces the slope's stability and facilitates mass wasting.

Shoreline erosion along the Pacific Coast varies considerably from one year to the next, largely because of the sporadic occurrence of storms. As a consequence, when the infrequent but serious episodes of erosion occur, the damage is often blamed on the unusual storms and not on coastal development or the sediment-trapping dams that may be great distances away. If, as predicted, sea level rises at an increasing rate in the years to come, increased shoreline erosion and sea cliff retreat should be expected along many parts of the Pacific Coast.

EMERGENT AND SUBMERGENT COASTS

The great variety of present-day shorelines suggests that they are complex areas. Indeed, to understand the nature of any particular coastal area, many factors must be considered, including rock types, size and direction of waves, frequency of storms, tidal range, and submarine profile. Moreover, recent tectonic events and changes in sea level must also be taken into account. These many variables make shoreline classification difficult.

One way that many geologists classify coasts is based upon changes that have occurred with respect

FIGURE 14.20
Satellite image of a portion of the East Coast showing Chesapeake Bay, an estuary created when the lower portion of a river valley was submerged by the rise in sea level that followed the end of the Ice Age. (Photo courtesy of Earth Satellite Corporation)

to sea level. This commonly used but incomplete classification divides coasts into the two categories of emergent and submergent. **Emergent coasts** develop either because an area has been uplifted or as a result of a drop in sea level. Conversely, **submergent coasts** are created when sea level rises or the land adjacent to the sea subsides.

In some areas the coast is clearly emergent because rising land or a falling water level exposes wave-cut cliffs and platforms above sea level. Excellent examples include portions of coastal California where uplift has occurred in the recent geological past. The elevated wave-cut platform shown in Figure 14.9 illustrates this situation. In the case of the Palos Verdes Hills, south of Los Angeles, seven different terrace levels exist, indicating seven episodes of uplift. The ever-persistent sea is now cutting a new platform at the base of the cliff. If uplift follows, it too will become an elevated marine terrace.

Other examples of emergent coasts include regions that were once buried beneath great ice sheets. When glaciers were present, their weight depressed the crust; when the ice melted, the crust began to gradually spring back. Consequently, prehistoric shoreline features may now be found high above sea level. The Hudson Bay region of Canada is such an area, portions of which are still rising at a rate of more than one centimeter per year.

In contrast to the preceding examples, other coastal areas show definite signs of submergence. The shoreline of a coast that has been submerged in the relatively recent past is often highly irregular because the sea typically floods the lower reaches of river valleys. The ridges separating the valleys, however, remain above sea level and project into the sea as headlands. These drowned river mouths, which are often called **estuaries**, characterize many coasts today. Along the Atlantic Coast, the Chesapeake and Delaware bays are examples of large estuaries created by submergence (Figure 14.20). The picturesque coast of Maine, particularly in the vicinity of Acadia National Park, is another excellent example of an area that was flooded by the post-glacial rise in sea level and transformed into a highly irregular coastline.

It should be kept in mind that most coasts have complicated geologic histories. With respect to sea level, many have at various times emerged and then submerged. Each time they may retain some of the features created during the previous situation.

TIDES

Tides are periodic changes in the elevation of the ocean surface at a specific location. Their rhythmic rise and fall along coastlines have been known since antiquity, and other than waves, they are the easiest ocean movements to observe (Figure 14.21). Although known for centuries, tides were not explained satisfactorily until Sir Isaac Newton applied the law of gravitation to them. Newton showed that there is a mutual attractive force between two bodies, and that since oceans are free to move, they are deformed by this force. Hence tides result from the gravitational attraction exerted upon the earth by the moon, and to a lesser extent by the sun.

To illustrate how tides are produced, we will assume that the earth is a rotating sphere covered to a uniform depth with water (Figure 14.22A). It is easy to see how the moon's gravitational force can cause the water to bulge on the side of the earth nearest the moon. In addition, however, an equally large tidal bulge is produced on the side of the earth di-

FIGURE 14.21
High tide and low tide on Nova Scotia's Minas Basin in the Bay of Fundy. (Photos courtesy of Nova Scotia Department of Tourism)

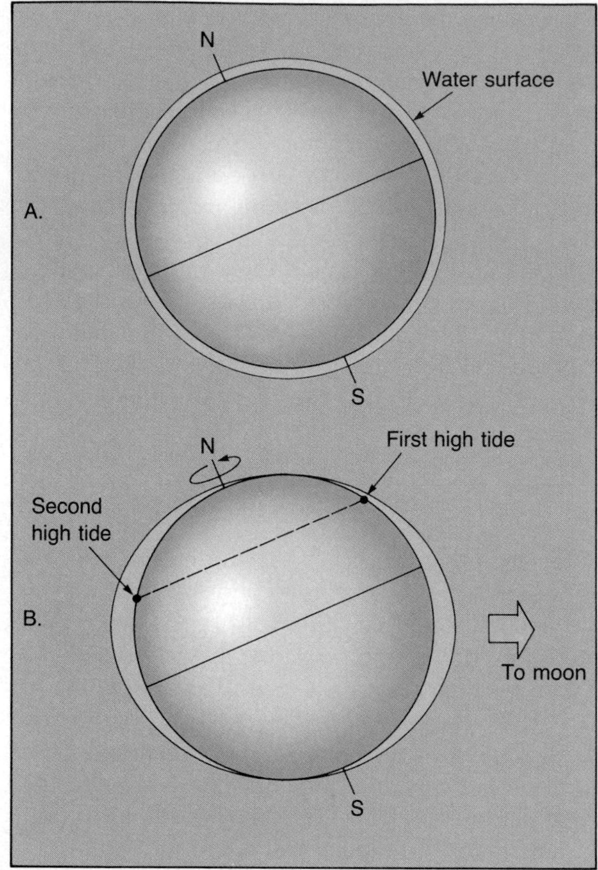

FIGURE 14.22
Tides on an earth that is covered to a uniform depth with water. Depending upon the moon's position, tidal bulges may be inclined to the equator. In this situation, an observer will experience two unequal high tides.

rectly opposite the moon. Both tidal bulges are caused, as Newton discovered, by the pull of gravity, a force that is inversely proportional to the square of the distance between two objects. In this case, the objects involved are the moon and the earth. Because the force of gravity decreases as distance increases, the moon's gravitational pull on the earth is slightly greater on the near side of the earth than on the far side. The result of this differential pulling is to stretch (elongate) the earth. Although the solid earth is stretched by the moon's gravitational pull, the amount of elongation is very slight. However, the world's oceans, which are mobile, are deformed quite dramatically by this effect to produce the two opposing tidal bulges.

Since the position of the moon changes only slightly in a single day, it is the tidal bulges that remain in place while the earth rotates beneath them. Therefore the earth will carry an observer at any given place alternately into areas of deeper and shallower water. When he is carried into the regions of deeper water, the tide rises, and as he is carried away, the tide falls. Therefore, during one day the observer would experience two high tides and two low tides. In addition to the earth rotating, the tidal bulges also move as the moon revolves around the earth about every 29 days. As a result, the tides, like the time of moonrise, occur about 50 minutes later each day. After about 29 days one cycle is complete and a new one begins.

There may be an inequality between the high tides during a given day. Depending upon the position of the moon, the tidal bulges may be inclined to the equator as in Figure 14.22B. This figure illustrates that the first high tide experienced by an observer in the Northern Hemisphere is considerably lower than the high tide half a day later. On the other hand, a Southern Hemisphere observer would experience the opposite effect.

The sun also influences the tides, but because it is so far away, the effect is considerably less than that of the moon. In fact, the tide-generating potential of the sun is slightly less than half that of the moon. At the times of new and full moons, the sun and moon are aligned and their forces are added together (Figure 14.23A). Accordingly, the two tide-producing bodies cause higher tidal bulges (high tides) and lower tidal troughs (low tides). These are called the **spring tides**. Spring tides create the largest daily tidal range, that is, the largest variation between high and low tides. Conversely, at times of the first and third quarters of the moon, the gravitational forces of the moon and sun on the earth are at right angles, and each partially offsets the influence of the other (Figure 14.23B). As a result, the daily tidal range is less. These are the **neap tides**.

Although the discussion thus far explains the basic causes and patterns of tides, these theoretical considerations cannot be used to predict either the height or the time of actual tides at a particular place. The shape of coastlines and the configuration of ocean basins greatly influence the tide. Consequently, tides at various places respond differently to the tide-producing forces. This being the case, the nature of the tide at any location can be determined most accurately by actual observation. The predictions in tidal tables and the tidal data on nautical charts are based upon such observations.

Tidal Currents

Tidal current is the term used to denote the horizontal flow of water accompanying the rise and fall of the tide. These water movements induced by tidal forces can be important in some coastal areas. Tidal cur-

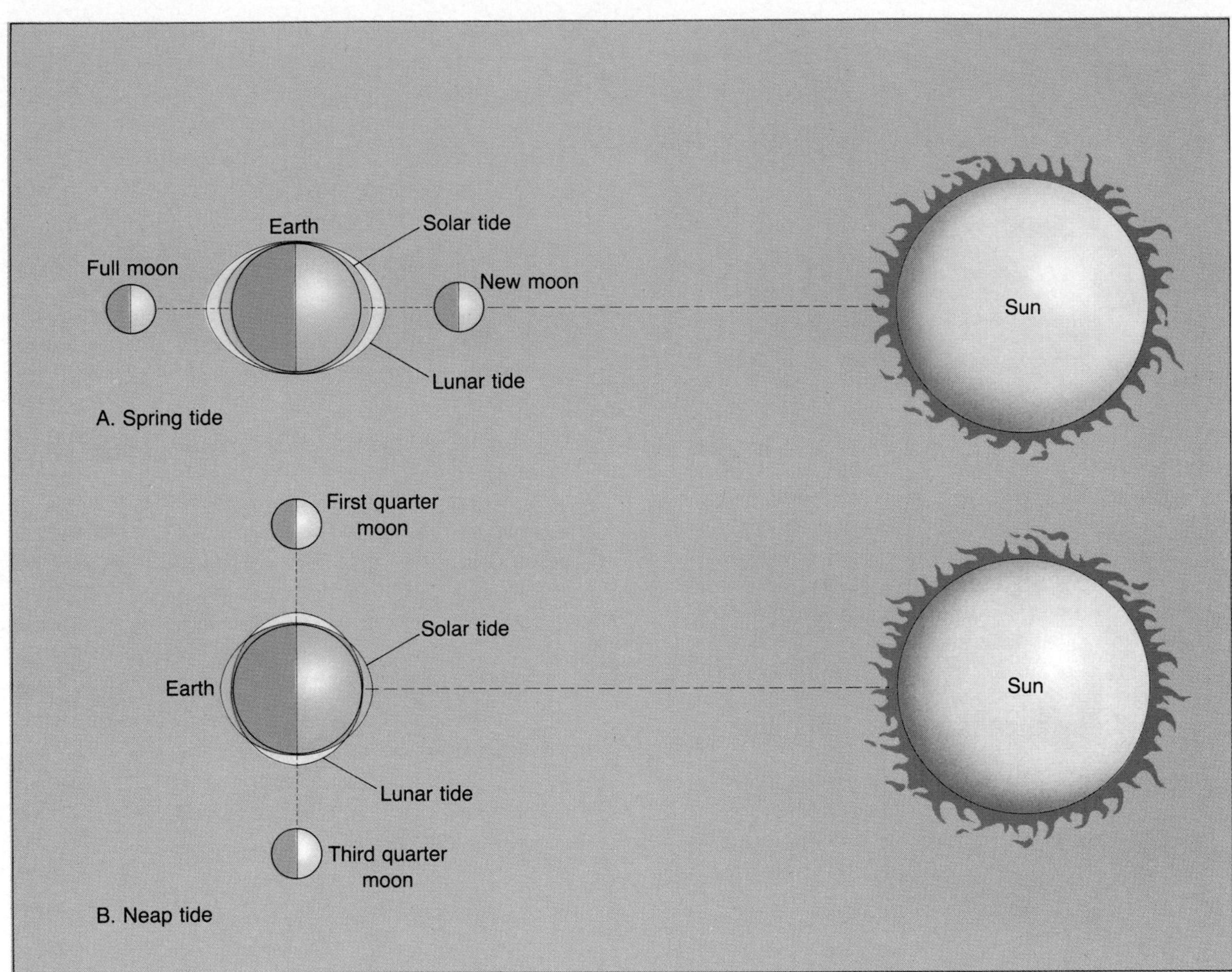

FIGURE 14.23
Relationship of the moon and sun to the earth during **A.** spring tides and **B.** neap tides. (The Tasa Collection: Shorelines. Published by Merrill/Macmillan Publishing Co., Columbus, OH, Copyright © 1986, by Tasa Graphic Arts, Inc. All rights reserved)

rents flow in one direction during a portion of the tidal cycle and reverse their flow during the remainder. Tidal currents that advance into the coastal zone as the tide rises are called **flood currents.** Those that move seaward as the level of the tide falls are called **ebb currents.** Periods of little or no current, called *slack water,* separate flood and ebb. The areas affected by these alternating tidal currents are **tidal flats.** Depending upon the nature of the coastal zone, tidal flats vary in size from narrow strips lying seaward of the beach to extensive zones that may extend for several kilometers.

Although tidal currents are not important in the open sea, they can be rapid in bays, river estuaries, straits, and other narrow places. Off the coast of Brittany, for example, tidal currents which accompany a high tide of 12 meters (40 feet) may attain a speed of 20 kilometers (12 miles) per hour. While tidal currents are not generally believed to be major agents of erosion and sediment transport, notable exceptions occur where tides move through narrow inlets. Here they constantly scour the small entrances to many good harbors that would otherwise be blocked.

Sometimes deposits called **tidal deltas** are created by tidal currents (Figure 14.24). They may develop either as *flood deltas* landward of an inlet or as *ebb deltas* on the seaward side of an inlet. Because wave activity and longshore currents are reduced on the sheltered landward side, flood deltas are more common and usually more prominent. They form after the tidal current moves rapidly through an inlet. As the current emerges into more open waters from the narrow passage, it slows and deposits its load of sediment.

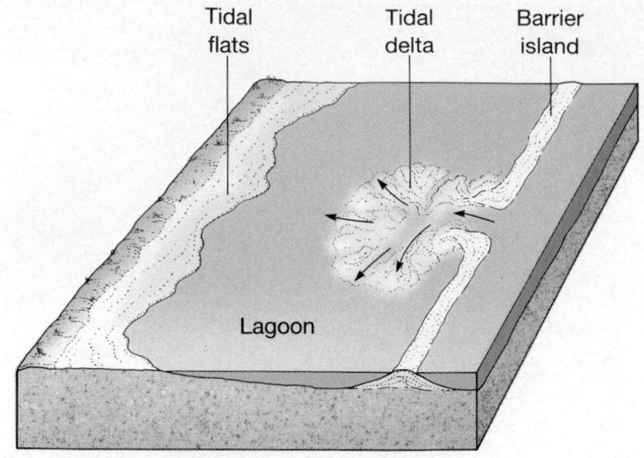

FIGURE 14.24
Because this tidal delta is forming in the relatively quiet waters on the landward side of a barrier island, it is termed a flood delta. As a rapidly moving tidal current emerges from the inlet, it slows and deposits sediment. The shapes of tidal deltas range from irregular to typical delta shaped.

Tides and the Earth's Rotation

The tides, by friction against the floor of the ocean basins, act as weak brakes that are steadily slowing the earth's rotation. The rate of slowing, however, is not great. Astronomers who have precisely measured the length of the day over the past 300 years, have found that the day is increasing by 0.002 second per century. Although this may seen inconsequential, over millions of years this small effect will become very large. Eventually, thousands of millions of years into the future, rotation will cease and the earth will no longer have alternating days and nights.

If the earth's rotation is slowing, the length of each day must have been shorter and the number of days per year must have been greater in the geologic past. One method used to investigate this phenomenon involves the microscopic examination of shells of certain invertebrates. Clams and corals, as well as other organisms, grow a microscopically thin layer of new shell material each day. By studying the daily growth rings of some well-preserved fossil specimens, we can ascertain the number of days in a year. Studies using this ingenious technique indicate that early in the Cambrian period, about 570 million years ago, the length of the day was only 21 hours. Since the length of the year, which is determined by the earth's revolution about the sun, does not change, the Cambrian year contained 424 twenty-one hour days. By late Devonian time, about 365 million years ago, a year consisted of about 410 days, and as the Permian period opened, 286 million years ago, there were 390 days in a year.

REVIEW QUESTIONS

1. List three factors that determine the height, length, and period of a wave.
2. Describe the motion of a water particle as a wave passes (see Figure 14.2).
3. Explain what happens when a wave breaks.
4. Describe two ways in which waves cause erosion.
5. What is wave refraction? What is the effect of this process along irregular coastlines (see Figure 14.7)?
6. Why are beaches often called "rivers of sand"?
7. Describe the formation of the following features: wave-cut cliff, wave-cut platform, sea stack, spit, baymouth bar, and tombolo.
8. Discuss three possible ways in which barrier islands originate.
9. For what purpose is a groin built? Why might the building of one groin lead to the building of others?
10. How did the building of jetties at Ocean City, Maryland, affect the areas north and south of the inlet? (See Box 14.2).
11. How might a seawall lead to increased beach erosion?
12. Of the various approaches used to combat beach erosion, which is viewed as the most acceptable? List two drawbacks of this method.
13. What is the basis for the predictions that global air temperatures will rise? How can a warmer atmosphere lead to a rise in sea level?
14. Relate the damming of rivers to the shrinking of beaches at many locations along the West Coast of the United States. Why do narrower beaches lead to accelerated sea cliff retreat?
15. What observable features would lead you to classify a coastal area as emergent?
16. Are estuaries associated with submergent or emergent coasts? Why?
17. Discuss the origin of ocean tides.
18. Explain why an observer can experience two unequal high tides during one day (see Figure 14.21).
19. How does the sun influence tides?
20. Distinguish between flood current and ebb current.
21. How have tides affected the earth's rotation? How did geologists substantiate this idea?

KEY TERMS

abrasion (p. 352)
barrier island (p. 356)
baymouth bar (p. 356)
beach drift (p. 354)
beach nourishment (p. 364)
breakwater (p. 363)
ebb current (p. 371)
emergent coast (p. 369)
estuary (p. 369)
fetch (p. 350)
flood current (p. 371)
groin (p. 362)
jetty (p. 362)
longshore current (p. 354)
neap tide (p. 370)
sea arch (p. 356)
sea stack (p. 356)
seawall (p. 363)
spit (p. 356)
spring tide (p. 370)
submergent coast (p. 369)
surf (p. 352)
tidal current (p. 370)
tidal delta (p. 371)
tidal flats (p. 371)
tide (p. 369)
tombolo (p. 356)
wave-cut cliff (p. 354)
wave-cut platform (p. 354)
wave height (p. 350)
wavelength (p. 350)
wave of oscillation (p. 351)
wave of translation (p. 352)
wave period (p. 350)
wave refraction (p. 353)

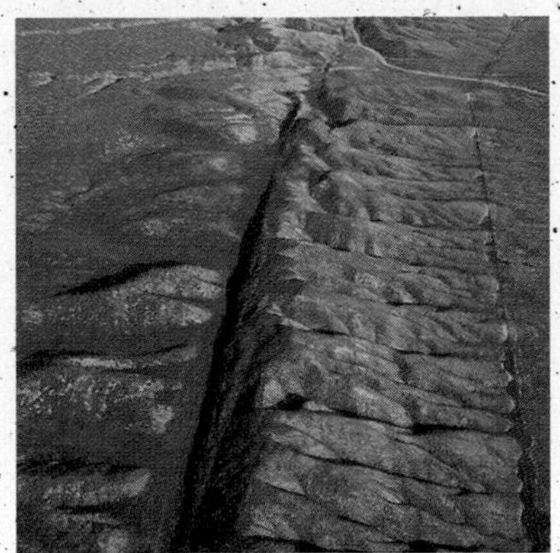

15
Crustal Deformation

DEFORMATION

TYPES OF DEFORMATION

STRIKE AND DIP

FOLDS
Types of Folds ■ Domes and Basins

FAULTS
Dip-Slip Faults ■ Strike-Slip Faults

JOINTS

Opposite: Aerial photo of the San Andreas fault as it crosses the Carrizo Plain about 160 kilometers north of Los Angeles.
(Photo by David Parker, Photo Researchers, Inc.)
(Top photo by R. E. Wallace, U.S. Geological Survey)

The earth is a dynamic planet. In the preceding chapters, we learned that weathering, mass wasting, and erosion continually sculpture the physical landscape. In addition, tectonic forces deform rocks in the earth's crust. Evidence for the enormous forces that operate within the earth includes highly fractured and folded rocks. In the Canadian Rockies, for example, some rock units have been thrust over others in a nearly horizontal manner for hundreds of kilometers. On a smaller scale, crustal movements of a few meters occur along faults during large earthquakes. In addition, rifting and extension of the crust produce elongated depressions and create ocean basins.

The results of many diverse tectonic activities are strikingly apparent in the earth's major mountain belts (Figure 15.1). Here rocks containing fossils of marine organisms may be found thousands of meters above sea level, and massive rock units are intensely folded as if they were made of putty.

In this chapter we will examine the forces that deform rock, as well as the structures that result. The basic geologic structures associated with deformation are folds, faults, joints, rock cleavage, and foliation. Since rock cleavage and foliation were examined in Chapter 7, this chpater will be devoted to the remaining rock structures. It is worth mentioning that structural geologists are always looking at the finished products of deformation. By studying the orientation of folds and faults, geologists can often determine the original geologic setting and the nature and direction of the forces that produced these rock structures. The complex events that make up geologic history are thereby unraveled.

In addition to their importance in interpreting the geologic past, rock structures are economically sig-

FIGURE 15.1
Uplifted sedimentary strata in the Canadian Rockies near Lake Louise, Alberta, Canada. (Photo by E. J. Tarbuck)

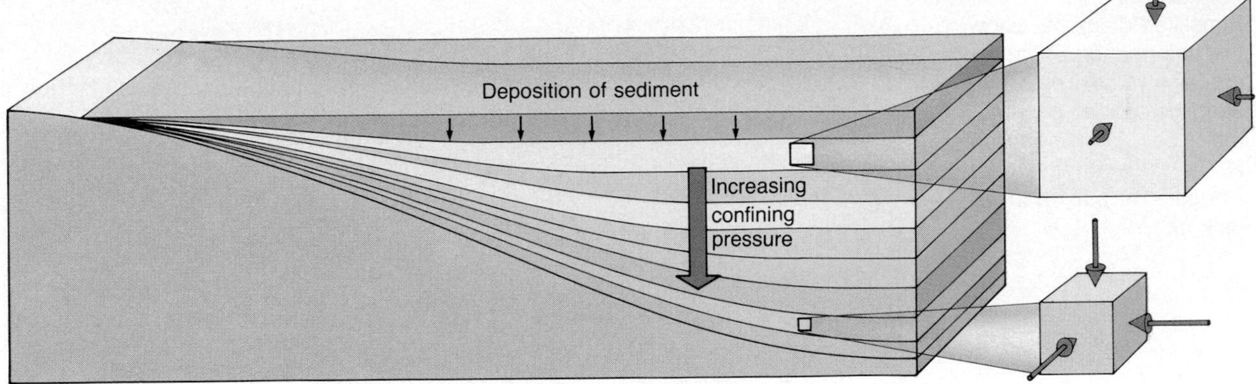

FIGURE 15.2
In depositional environments rock layers are buried under additional layers of sediment. As confining pressure increases, rocks deform by decreasing in volume.

nificant as well. For example, most occurrences of oil and natural gas are associated with geologic structures that act to trap these important fluids in reservoirs of one type or another (see Chapter 21). Furthermore, hydrothermal mineralization that occurs along rock fractures is a major source of metallic ores. Moreover, the orientation and characteristics of rock structures must be considered when selecting sites for major construction projects such as bridges, hydroelectric dams, and nuclear power plants. In short, a working knowledge of these features is essential to our modern way of life.

DEFORMATION

Deformation is a general term that refers to all changes in volume or shape of a rock body. Among the forces that deform rock is **confining pressure**, which like air pressure is uniform in all directions. Confining pressure is caused by the load of the overlying rocks. For example, in basins where rapid deposition occurs, rock is buried deeper and deeper as additional layers of sediment accumulate. The result is a reduction in volume and a more compact rock (Figure 15.2). Confining pressure is also important because it affects the way rocks behave when deformed by directional forces. In near-surface environments, where the confining pressure and temperature are low, rocks are described as *brittle* because they will fracture when deformed. However, at great depths, where confining pressures are high, rocks become *ductile* and flow rather than fracture when a directional force is applied (Figure 15.3).

Besides confining pressure, other geological forces deform rocks by causing changes in shape. One such situation is along plate margins. Recall from Chapter 1 that the lithosphere consists of plates that move relative to one another. As plates interact along their boundaries, stresses are produced that

FIGURE 15.3
Rocks exhibiting ductile behavior during deformation. These rocks were deformed at great depth and were subsequently exposed along Pipe Creek in Grand Canyon National Park. (Photo by Peter Kresan)

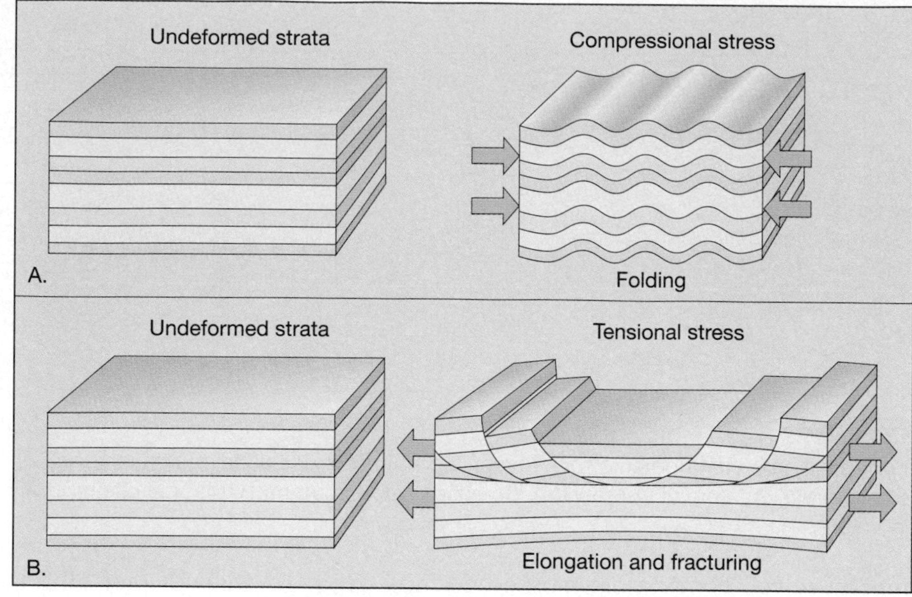

FIGURE 15.4
Simplified diagrams showing the deformation of flat-lying strata. **A.** Compressional stresses tend to shorten a rock body, often by folding. **B.** Tensional stresses act to elongate, or pull apart, a rock unit.

cause deformation. **Stress** is a force that acts on a rock unit to change its shape and/or its volume. When stresses are acting to shorten a rock body, they are known as **compressional stresses** (Figure 15.4A). Conversely, when stresses are acting in opposing directions, they tend to elongate, or pull apart, the rock unit and are known as **tensional stresses** (Figure 15.4B). In addition, directional stress can cause a rock to **shear**. Shearing is similar to the slippages that occur between individual playing cards when a deck

FIGURE 15.5
Deformed sedimentary strata exposed in a road cut near Palmdale, California. In addition to the obvious folding, light-colored beds are offset along a fault located about one-third of the way in from the right-hand edge of the photograph. (Photo by E. J. Tarbuck)

is held between your hands and the top of the deck is moved relative to the bottom. In near-surface environments, shearing occurs when relatively brittle rock is broken into thin slabs that are forced to slide past one another. By contrast, rock located at great depths is warmer and behaves in a ductile fashion during deformation. This accounts for its ability to flow and bend into intricate folds when subjected to shearing.

TYPES OF DEFORMATION

When rocks are subjected to stresses greater than their own strength, they begin to deform, usually by folding or fracturing (Figure 15.5). It is easy to visualize how rocks break, but how are large rock units bent into intricate folds without being appreciably broken during the process? In an attempt to answer this question, structural geologists used the laboratory. Here rocks were subjected to stresses under conditions that simulated those existing at various depths within the crust.

Although each rock type deforms somewhat differently, the general characteristics of rock deformation were determined from these experiments. Geologists discovered that when stress is applied, rocks first respond by deforming elastically. Changes resulting from **elastic deformation** are reversible; that is, like a rubber band, the rock will return to nearly its original size and shape when the stress is removed. However, once the elastic limit is surpassed, rocks either deform plastically or fracture. **Plastic deformation** results in permanent changes, that is, the size and shape of a rock unit are altered through folding and flowing. Laboratory experiments confirmed the speculation that, at high temperatures and pressures, most rocks deform plastically once their elastic limit is surpassed (Figure 15.6). Rocks tested under surface conditions also deform elastically, but once they exceed their elastic limit, most behave like a brittle solid and fracture. As we shall see in the next chapter, the energy for most earthquakes comes from stored elastic energy that is released as rock ruptures and snaps back to its original shape.

In addition to the environment, the mineral composition of rocks greatly influences how individual masses will respond to deformation. For example, some rocks are very strong and tend to fail by brittle fracture, whereas others are weak and generally deform by ductile flow. Rocks that are weak, and thus most likely to behave in a ductile manner when sub-

FIGURE 15.6
A marble cylinder deformed in the laboratory by applying thousands of pounds of load from above. Each sample was deformed in an environment that duplicated the confining pressure found at various depths. Notice that when the confining pressure was low, the sample deformed by brittle fracture, whereas when the confining pressure was high, the sample deformed plastically. (Photo courtesy of M. S. Paterson, Australian National University)

jected to stress, include rock salt, gypsum, marble, and shale. By comparison, quartzite, granite, and gneiss are strong and brittle. In a near-surface environment, strong, brittle rocks will fail by fracture when subjected to stresses that exceed their strength.

One factor that researchers are unable to duplicate in the laboratory is geologic time. We know that if stress is applied quickly, as with a hammer, rocks tend to fracture. On the other hand, these same materials may deform plastically if stress is applied over an extended period. For example, marble benches have been known to sag under their own weight over a period of a hundred years or so, In nature, small forces applied over long time periods surely play an important role in the deformation of rock.

To summarize, three factors determine how rocks will behave when subjected to stresses that exceed their strength. First, the environment strongly influences how a rock will deform. In near-surface environments, where temperatures and pressures are

A.

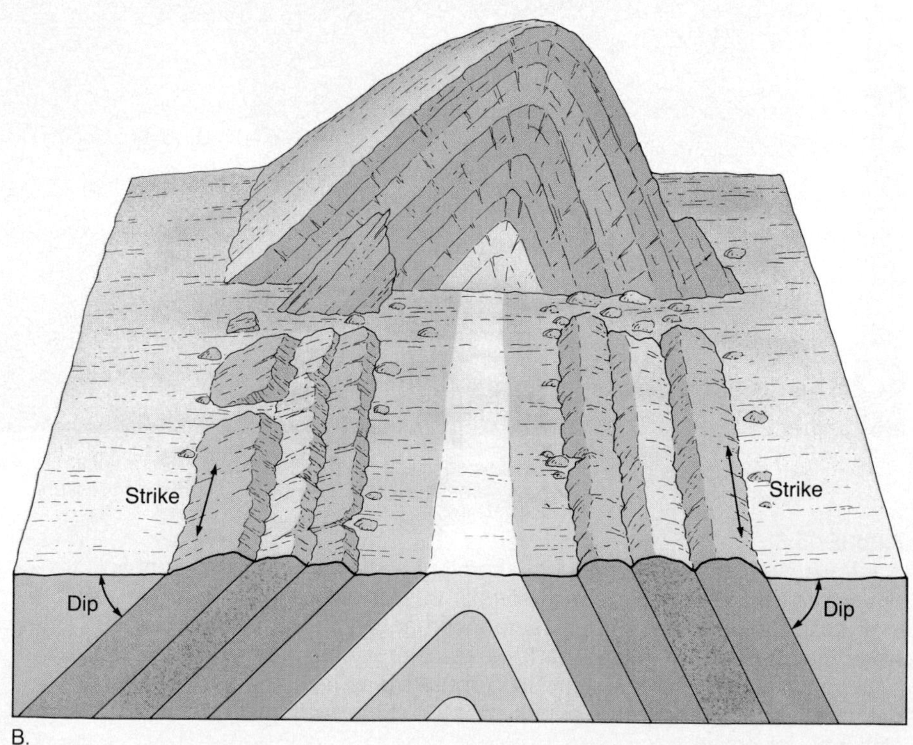

B.

FIGURE 15.7
A. Photo of sedimentary strata sharply folded into an asymmetrical anticline and attacked by erosion. **B.** Drawing of this fold to illustrate the concept of strike and dip. (Reproduced by permission of the Director, Institute of Geological Sciences (NERC); NERC Copyright reserved/Crown Copyright reserved)

low, most rocks exhibit brittle fracture. However, in the high-temperature, high-pressure regimes that exist deep in the crust the same rocks will deform by ductile flow. A second factor relates to the strength of the materials. Some rocks such as gypsum are very weak and are more likely to flow under conditions that would cause stronger rocks to fail by brittle fracture. Third, time plays a major role in rock deformation. Rocks that would fracture when stress is applied rapidly are known to flow when stress is applied gradually over a long time span.

The processes of deformation generate features at many different scales. At one extreme are the earth's major mountain systems. At the other extreme, highly localized stresses create minor fractures in bedrock. All of these phenomena, from the largest folds in the Alps to the smallest fractures in a slab of rock, are referred to as **rock structures**. Before beginning our discussion of rock structures, we need to familiarize ourselves with the way geologists describe them.

STRIKE AND DIP

When conducting a study of a region, a geologist attempts to identify and describe the dominant structures. In many cases a structure is so large that only a small portion is visible from any particular vantage point. In other situations most of the bedrock is concealed by vegetation or by recent sedimentation. Consequently, the reconstruction must be done using data gathered from a limited number of *outcrops,* which are sites where bedrock is exposed at the surface. Despite such difficulties, a number of mapping techniques enable geologists to reconstruct the orientation and shape of the existing structures. In recent years, this work has been aided by advances in aerial photography and satellite imagery.

Geologic mapping is most easily accomplished in areas where sedimentary strata are exposed. Because sediments are usually deposited in horizontal layers, inclined strata indicate that a period of deformation occurred following deposition. Two measurements used to establish the orientation of deformed sedimentary beds (or fault surfaces) are strike and dip. **Strike** is the trend, or direction, of the strata, whereas **dip** is the angle of inclination of the bedding surface. Perhaps the easiest way to understand these measurements is to examine sedimentary strata that are outcropping in an otherwise flat terrain (Figure 15.7). Strike is defined as the direction of the line produced by the intersection of the surface represented by the inclined strata with a horizontal surface, which in this example is the surface of the land. Dip, on the other hand, is defined as the angle of maximum inclination of a bed measured in a direction perpendicular to the strike. In the field, geologists measure the strike (trend) and dip (inclination) of sedimentary rocks at as many outcrops as practical. These data are then plotted on a topographic map or an aerial photograph along with a color-coded description of the rock. From the orientation of the strata, an inferred orientation and shape of the structure can be established as shown in Figure 15.8.

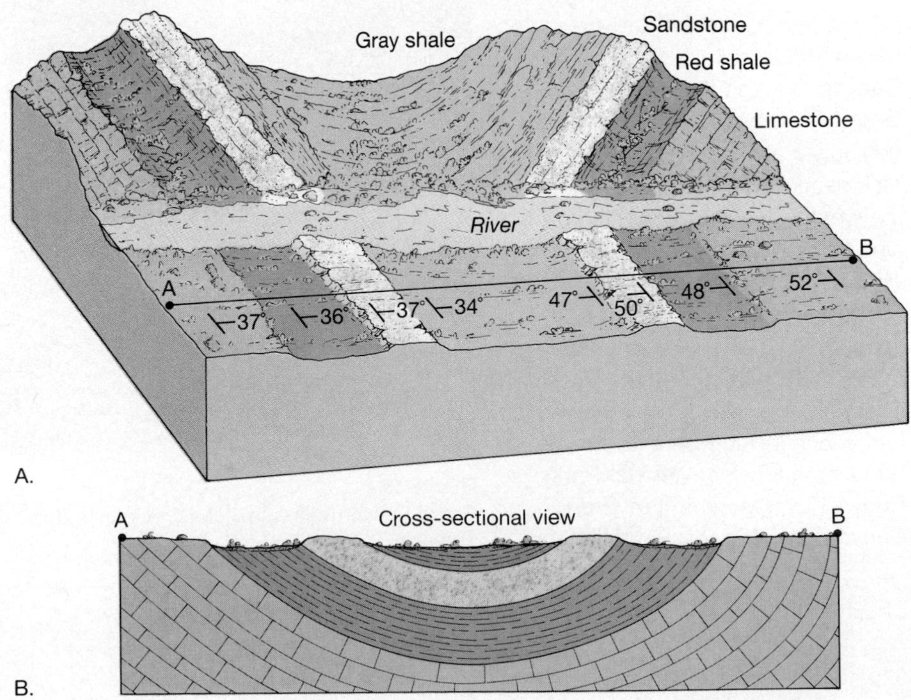

FIGURE 15.8
By establishing the strike and dip of outcropping sedimentary beds (Part A), geologists can infer the orientation of the structure below ground (Part B).

BOX 15.1

Naming Local Rock Units

One of the primary goals of geology is reconstructing the earth's long and complex history through the systematic study of rocks. In most areas, exposures of rocks are not continuous over great distances. Consequently, the study of rock layers must be done locally and then correlated with data from adjoining areas to produce a larger and more complete picture. The first step in the effort to unravel past geologic events is that of describing and mapping rock units exposed in local outcrops.

Describing anything as complex as a thick sequence of rocks requires subdividing the layers into units of manageable size. The most basic rock division is called a *formation,* which is simply a distinctive series of strata that originated through the same geologic processes. More precisely, a formation is a mappable rock unit that has definite boundaries (or contacts) and certain obvious characteristics by which it may be traced from place to place and distinguished from other rock units.

Figure 15.A shows several named formations that are exposed in the walls of the Grand Canyon. Just as the layers in the Grand Canyon were subdivided, geologists attempt to subdivide rock sequences found throughout the world into these non-overlapping units known as formations. Those who have had the opportunity to travel to some of the national parks in the West may already be familiar with the names of certain formations. Well-known formations include the Navajo Sandstone in Zion National Park, Redwall Limestone in the Grand Canyon, the Entrada Sandstone in Arches National Park, and the Wasatch Formation in Bryce Canyon National Park.

Although formations can consist of igneous or metamorphic rocks, the vast majority of established formations are sedimentary rocks. A formation may be relatively thin and composed of a single rock type, for example, a 1-meter-thick layer of limestone. At the other extreme, formations can be thousands of meters thick and consist of an interbedded sequence of rock types such as sandstones and shales. The most important condition to be met when establishing a formation is that it constitutes a unit of rock produced by uniform or uniformly alternating conditions.

In most regions of the world, the name of each formation consists of two parts, for example, the Oswego Sandstone and the Carmel Formation. The first part of the name is generally taken from a geologic structure or a locality where the formation is clearly and completely exposed. For example, the expansive Morrison Formation is well exposed at Morrison, Colorado. As a result, this particular exposure is known as the *type locality.* Ideally, the second part of the name indicates the dominant rock type as exemplified by such names as the Dakota Sandstone, the Kaibab Limestone, and the Burgess Shale. When no single rock type dominates, the term *formation* is used, as, for example, the well-known Chinle Formation, exposed in Arizona's Petrified Forest National Park.

In summary, describing and naming formations is an important first step in the process of organizing and simplifying the study and analysis of earth history.

FIGURE 15.A
Grand Canyon with a few of its rock units (formations) named. (Photo by E. J. Tarbuck)

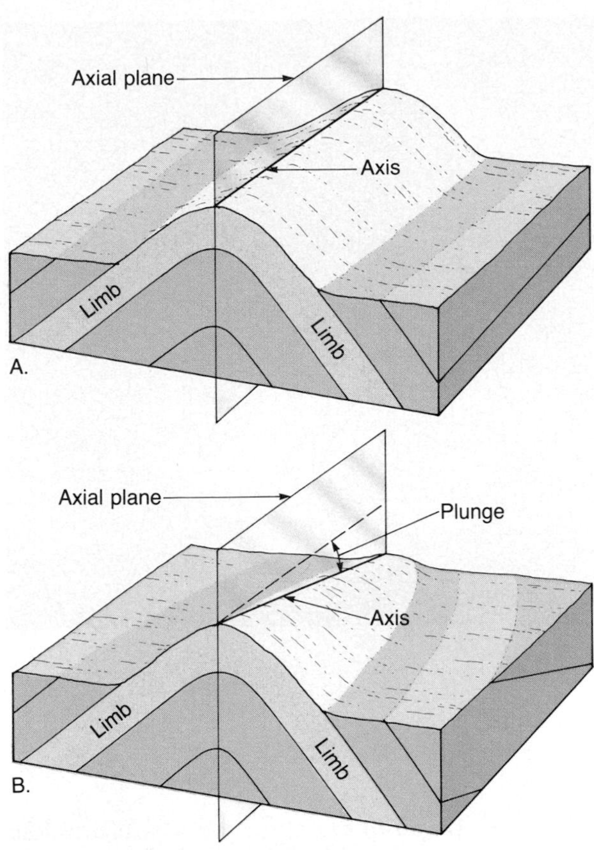

FIGURE 15.9
Idealized sketches illustrating the features associated with symmetrical folds. The axis of the fold in Part A is horizontal, whereas the axis of the fold in Part B is plunging.

Using this information, the geologist can reconstruct the pre-erosional structures and is also better able to interpret the region's geologic history.

FOLDS

During mountain building, flat-lying sedimentary and volcanic rocks are often bent into a series of wavelike undulations called **folds.** Folds in sedimentary strata are much like those that would form if you were to hold the ends of a sheet of paper and then push them together. In nature, folds come in a wide variety of sizes and configurations. Some folds are broad flexures in which rock units hundreds of meters thick have been slightly warped. Others are very tight, microscopic structures found in metamorphic rocks. Size differences notwithstanding, most folds are the result of compressional stresses that result in the shortening and thickening of the crust. Occasionally, folds are found singly, but most often they occur as a series of undulations.

To aid our understanding of folds and folding, we need to become familar with the terminology used to name the parts of a fold. As shown in Figure 15.9, the two sides of a fold are called *limbs.* A line drawn along the points of maximum curvature of each layer is termed the *axis* of the fold. In some folds, as Figure 15.9A illustrates, the axis is oriented horizontally. However, in more complex folding, the fold axis is often inclined at an angle known as the *plunge* (Figure 15.9B). Further, the *axial plane* is an imaginary surface that divides a fold as symmetrically as possible.

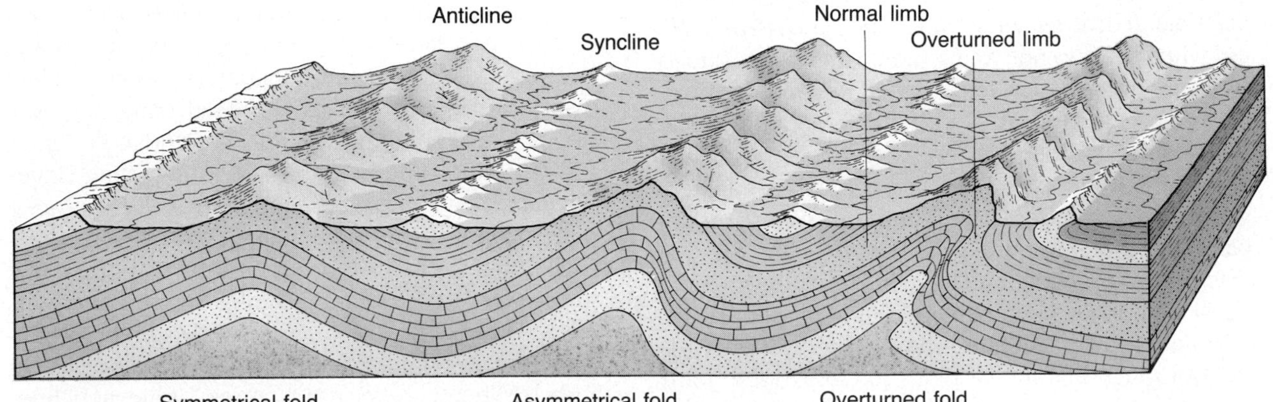

FIGURE 15.10
Block diagram of principal types of folded strata. The upfolded or arched structures are anticlines. The downfolds or troughs are synclines. Notice that the limb of an anticline is also the limb of the adjacent syncline.

FIGURE 15.11
Recumbent fold in Precambrian rocks of the Umanak area, Greenland. (Photo by T. C. R. Pulvertaft, Geological Survey of Greenland)

Types of Folds

The two most common types of folds are anticlines and synclines (Figure 15.10). An **anticline** is most commonly formed by the upfolding, or arching, of rock layers.* Anticlines are sometimes spectacularly displayed where highways have been cut through deformed strata. Often found in association with anticlines are downfolds, or troughs, called **synclines**. Notice in Figure 15.10 that the limb of an anticline is also a limb on the adjacent syncline. Depending on their orientation, these basic folds are described as *symmetrical* when the limbs on either side of the axial plane diverge at the same angle and *asymmetrical* when they do not. An asymmetrical fold is said to be *overturned* if one limb is tilted beyond the vertical (Figure 15.10). An overturned fold can also "lie on its side" so that a plane extending through the axis of the fold would have a horizontal orientation. These *recumbent* folds are common in some mountainous regions (Figure 15.11). In the Alps there is evidence that certain deformed strata have been shoved up to 50 kilometers over the adjacent rocks.

Folds do not continue forever; rather, their ends die out much like the wrinkles in cloth. Some folds are said to be *plunging* since the axis of the fold penetrates into the ground (Figure 15.12). Figure 15.13 shows an example of a plunging anticline and the pattern produced when erosion removes the upper layers of the structure and exposes its interior. Note that the outcrop pattern of an anticline points in the direction it is plunging, whereas the opposite is true for a syncline. A good example of the kind of topography that results when erosional forces attack folded sedimentary strata is found in the Valley and Ridge Province of the Appalachians (see Figure 20.20).

It is important to point out that ridges are not necessarily associated with anticlines; nor are valleys always related to synclines. Rather, ridges and valleys result because of differential weathering and erosion. For example, in the Valley and Ridge Province, resistant sandstone beds remain as imposing ridges separated by valleys cut into more easily eroded shale or limestone beds.

Although most folds are caused by compressional stresses that squeeze and crumble strata, some folds are a consequence of vertical displacement. **Monoclines**, broad flexures found on the Colorado Plateau and elsewhere, are such structures. Unlike anticlines and synclines, which have two limbs, monoclines have just one. These folds are thought to result from nearly vertical faulting in deep-lying basement rocks as shown in Figure 15.14. Whereas the rigid basement complex responded to vertical stress by frac-

*By strict definition, an anticline is a structure in which the oldest strata are found in the center. This most typically occurs when strata are upfolded. Further, a syncline is strictly defined as a structure in which the youngest strata are found in the center. This occurs most commonly when strata are downfolded.

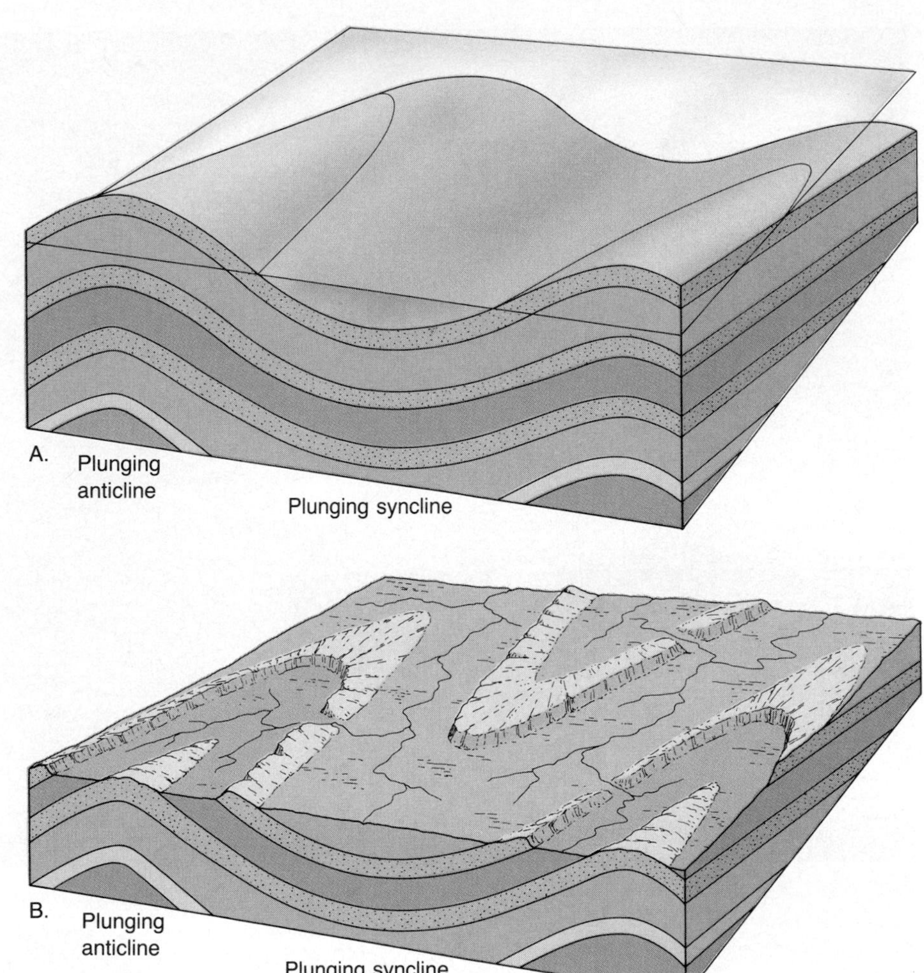

FIGURE 15.12
Plunging folds **A.** Idealized view of plunging folds in which a planar surface has been added to represent an imaginary horizontal surface. **B.** View of plunging folds as they might appear after extensive erosion. Notice that in a plunging anticline the outcrop pattern "points" in the direction of the plunge, while the opposite is true of plunging synclines.

turing, the relatively flexible sedimentary strata above were deformed by folding.

Domes and Basins

Broad upwarps in basement rock may deform the overlying cover of sedimentary strata and generate large folds. When this upwarping produces a circular or elongated structure, the feature is called a **dome** (Figure 15.15A). Downwarped structures having a similar shape are termed **basins** (Figure 15.15B). The Black Hills of western South Dakota is a large domed structure. Here erosion has stripped away the highest portions of the unwarped sedimentary beds, exposing older igneous and metamorphic rocks in the center (Figure 15.16). Remnants of these once-continuous sedimentary layers are visible flanking the crystalline core of this mountain range. The more resistant strata are easy to identify because differential erosion has left them outcropping as prominent angular ridges called **hogbacks**. Since hogbacks can form whenever resistant strata are steeply inclined, they are also associated with other types of folds.

FIGURE 15.13
Sheep Mountain, a plunging anticline. Note that erosion has cut the flanking sedimentary beds into low ridges that make a "V" pointing in the direction of plunge. (Photo by John S. Shelton)

A.

Several large basins exist in the United States. The basins of Michigan and Illinois have very gently sloping beds similar to saucers. Because large basins contain sedimentary beds sloping at such low angles, they are usually identified by the age of the rocks composing them. The youngest rocks are found near the center and the oldest rocks are at the flanks. This is just the opposite order of a domed structure, such as the Black Hills, where the oldest rocks form the core.

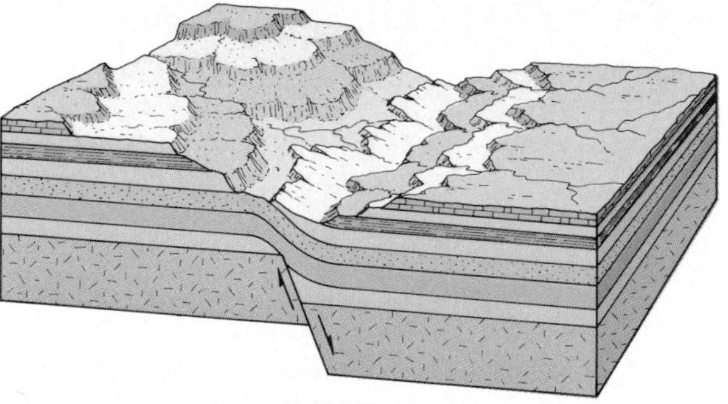

B.

FIGURE 15.14
Monocline. **A.** The Waterpocket monocline, southern Utah. (Photo by Michael Collier) **B.** Monocline consisting of bent sedimentary beds that were deformed by faulting in the bedrock below.

FAULTS

Faults are fractures in the crust along which appreciable displacement has taken place. Occasionally, small faults can be recognized in road cuts where sedimentary beds have been offset a few meters as shown in Figure 15.17. Faults of this scale usually occur as single discrete breaks. By contrast, large faults, like the San Andreas fault in California, have displacements of hundreds of kilometers and consist of many interconnecting fault surfaces. These so-called *fault zones* can be several kilometers wide and are often easier to identify from high-altitude photographs than at ground level (see chapter-opening photo).

Sudden movements along faults are the cause of most earthquakes. However, many faults are inactive and thus are remnants of past deformation. Along active faults, rock is often broken and pulverized as crustal blocks on opposite sides of a fault grind past

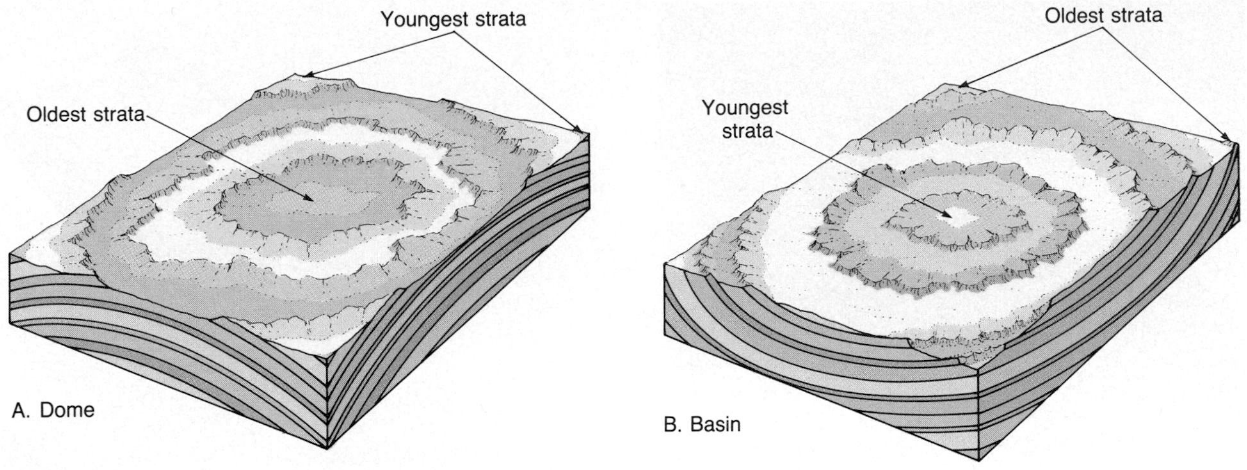

FIGURE 15.15
Gentle upwarping and downwarping of crustal rocks produce domes and basins. Erosion of these structures results in an outcrop pattern, which is roughly circular or elongated.

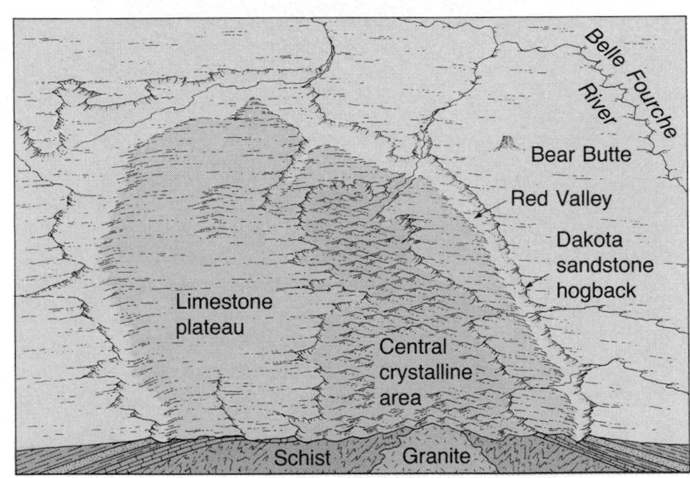

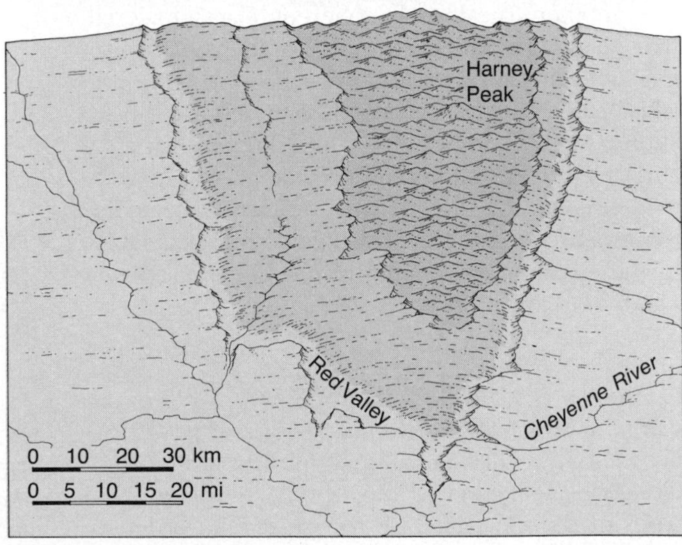

FIGURE 15.16
The Black Hills of South Dakota, an example of a domal structure in which the resistant igneous and metamorphic central core has been exposed by erosion. (After Arthur N. Strahler, *Introduction to Physical Geography,* 3rd ed., New York: John Wiley & Sons, 1973. Reprinted by permission)

FIGURE 15.17
Faulting caused the vertical displacement of these sedimentary beds in southern Nevada. Arrows show relative motion of rock units. (Photo by E. J. Tarbuck)

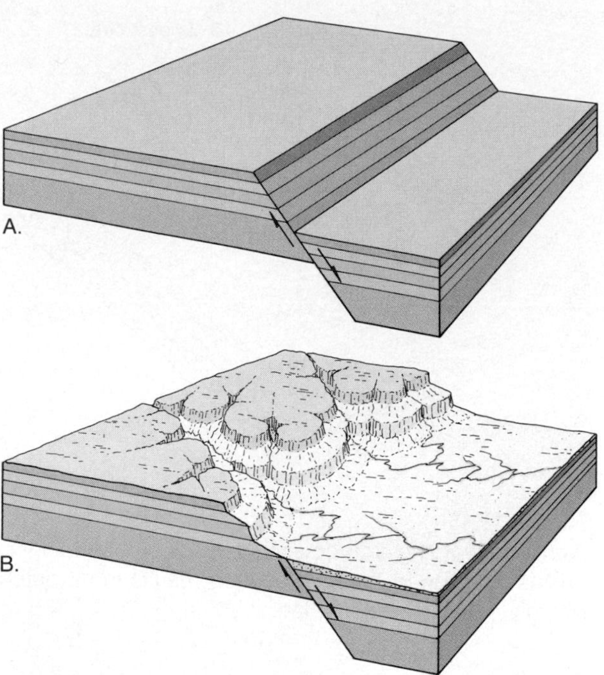

FIGURE 15.18
Block diagrams of a normal fault. **A.** The relative movement of displaced blocks. **B.** How erosion would alter the upfaulted block.

one another. The loosely coherent, clayey material that results from this activity is called *fault gouge.* On some fault surfaces, the rocks become highly polished and striated, or grooved, as the crustal blocks slide past one another. These polished and striated surfaces, called *slickensides,* provide geologists with evidence for the direction of the most recent displacement along the fault. Geologists classify faults by these relative movements, which can be predominantly horizontal, vertical, or oblique.

Dip-Slip Faults

Faults in which the movement is primarily vertical are called **dip-slip faults** since the displacement is along the inclination, or dip, of the fault plane. Two types of dip-slip faults are recognized. In order to distinguish between the two types, it has become

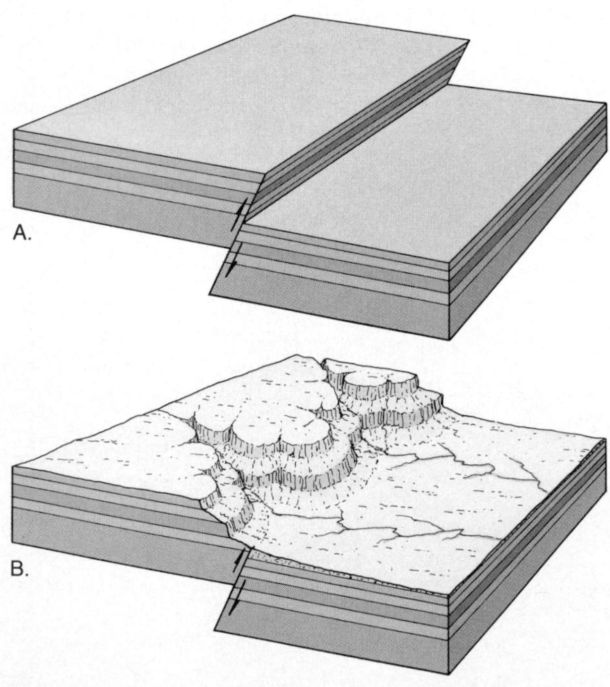

FIGURE 15.19
Block diagrams of a reverse fault. **A.** The relative movement of displaced blocks. **B.** How erosion would alter the upfaulted block.

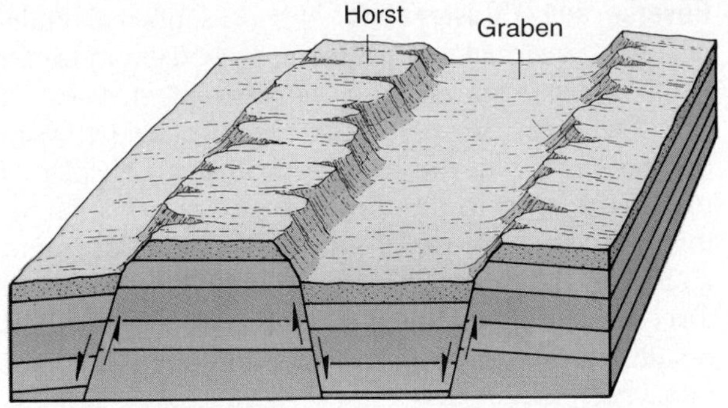

FIGURE 15.20
Diagrammatic sketch of downfaulted block (graben) and upfaulted block (horst).

common practice to call the rock immediately above the fault surface the *hanging wall* and to call the rock below the *footwall.* This nomenclature arose from prospectors and miners who excavated shafts along fault zones because they are frequently sites of ore deposition. During these operations, the miners would walk on the rocks below the fault trace (the footwall) and hang their lanterns on the rocks above (the hanging wall). Dip-slip faults are classified as **normal faults** when the hanging wall moves down relative to the footwall (Figure 15.18). Conversely, **reverse faults** occur when the hanging wall moves up relative to the footwall (Figure 15.19). Reverse faults having a very low angle are also referred to as **thrust faults**.

Normal Faults Fault motion provides the geologist with a method of determining the nature of the forces at work within the earth. Normal faults indicate the existence of tensional stresses that pull the crust apart. This "pulling apart" can be accomplished either by uplifting that causes the surfaces to stretch and break or by horizontal forces that actually rip the crust apart. Normal faulting occurs at spreading centers where plate divergence is prevalent. Here, a central block called a **graben** is bounded by normal faults and drops as the plates separate (Figure 15.20). These grabens produce an elongated valley bounded by upfaulted structures called **horsts**. The Great Rift Valley of East Africa is made up of several large grabens, above which tilted horsts produce a linear mountainous topography. This valley, nearly 6000 kilometers (3600 miles) long, contains the excavations sites of some of the earliest human fossils. Other rift valleys include the Rhine Valley in Germany and the valley of the Dead Sea in the Middle East.

In the western United States, normal faults are also associated with structures called **fault-block mountains**. Excellent examples of fault-block mountains are found in the Basin and Range Province, a region that encompasses Nevada and portions of the surrounding states (Figure 15.21). Here the crust has been elongated and broken to create more than 200 relatively small mountain ranges. Averaging about 80 kilometers in length, the ranges rise 900-1500 meters above the adjacent down-faulted basins. Other exam-

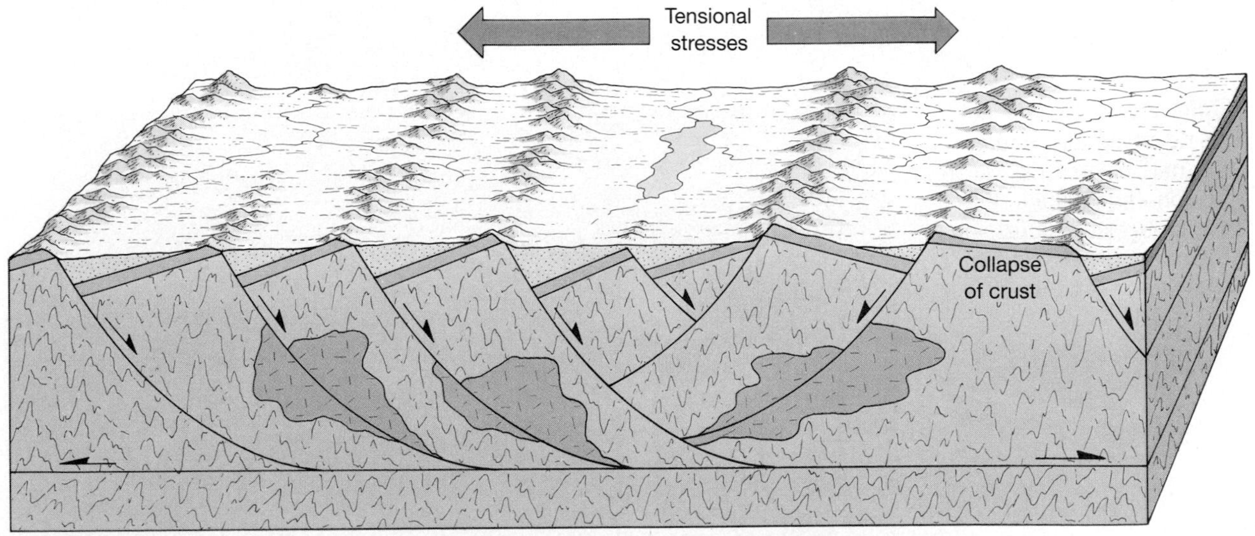

FIGURE 15.21
Normal faulting in the Basin and Range Province. Here tensional stresses have elongated and fractured the crust into numerous blocks. Movement along these fractures has tilted the blocks, producing parallel mountain ranges called fault-block mountains.

FIGURE 15.22
Fault scarp with a displacement of nearly 3 meters located in central Idaho. (Photo courtesy of Glenn Embree and Edmund Williams)

ples of fault-block mountains include the Teton Range of Wyoming and the Sierra Nevada of California. Both are faulted along their eastern flanks, which were uplifted as the blocks tilted downward to the west. These precipitous mountain fronts were produced over a period of 5-10 million years by many irregularly spaced episodes of faulting. Each event was responsible for just a few meters of displacement. Vertical displacements along faults often produce long, low cliffs called **fault scarps.** Fault scarps such as the one shown in Figure 15.22 are produced by displacements that occur during an earthquake. A strong earthquake in central Idaho in 1983 produced this 3-meter scarp.

Reverse and Thrust Faults Whereas normal faults occur in tensional environments, reverse and thrust faults result from strong compressional stresses. In these settings, crustal blocks are displaced toward one another, with the hanging wall being displaced upward relative to the footwall. The regions where this activity is most pronounced are convergent zones where plates are colliding. Compressional forces generally produce folds as well as faults and result in a thickening and shortening of the material involved.

In mountainous regions such as the Alps, the American Cordillera, and the Appalachians, thrust faults have displaced strata as far as 50 kilometers over adjacent rock units. The result of this large-scale movement is that older strata end up overlying younger rocks. The photo of Nevada's Keystone Overthrust in Figure 15.23 illustrates this phenomenon. Here 500-million-year-old dark-colored limestone has been thrust on top of 150-million-year-old light-colored sandstone.

A classic site of thrust faulting occurs in Glacier National Park (Figure 15.24). Here the mountain peaks that provide the park's majestic scenery have been carved from Precambrian rocks that were displaced over much younger Cretaceous strata. At the eastern edge of Glacier National Park there is an outlying peak called Chief Mountain. This structure is an isolated remnant of the thrust sheet that was severed by the erosional forces of glacial ice and running water. An isolated block, such as Chief Mountain, is called a **klippe**. In many instances, thrust faults form in conjunction with large, broad folds.

FIGURE 15.23
The Keystone Overthrust. Here dark-colored Cambrian limestone has been thrust over light-colored Jurassic sandstone. (Photo by John S. Shelton)

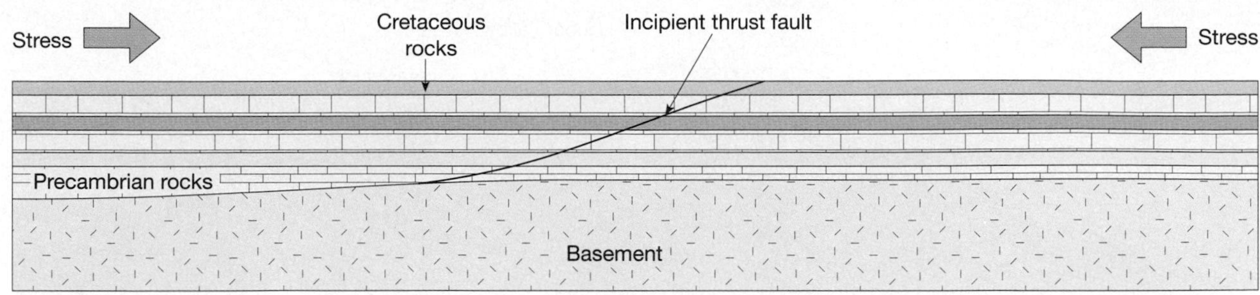

A. Initial conditions

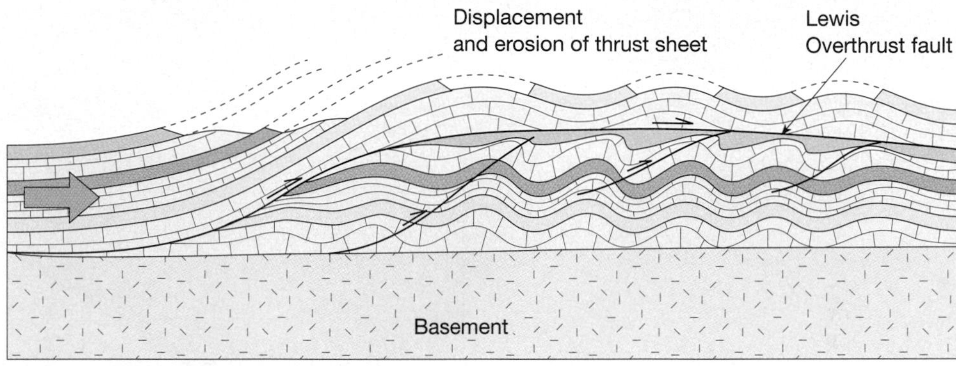

B. Displacement and erosion

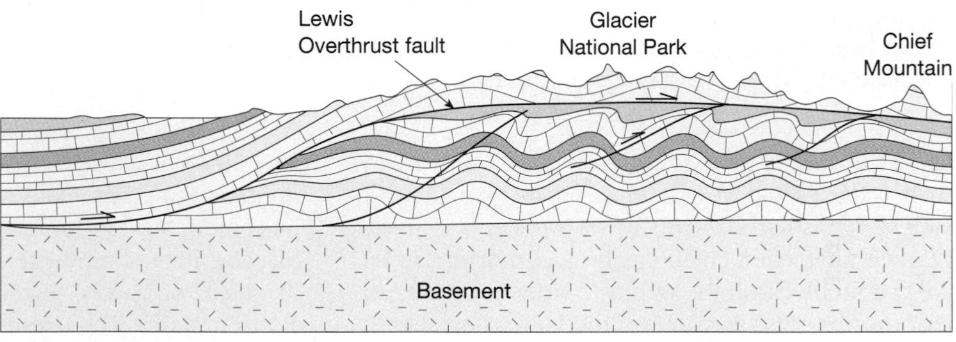

C. Present day

FIGURE 15.24
Idealized development of Lewis Overthrust fault. **A.** Geologic setting prior to deformation. **B.** Large-scale movement along a thrust fault displaced Precambrian rock over Cretaceous strata in the region of Glacier National Park. **C.** Erosion by glacial ice and running water sculptured the thrust sheet into a majestic landscape and isolated a remnant of the thrust sheet called Chief Mountain.

Strike-Slip Faults

Faults in which the dominant displacement is horizontal and parallel to the strike of the fault surface are called **strike-slip faults**. Because of their large size and linear nature, many strike-slip faults produce a trace that is visible over a great distance (see chapter-opening photo). Rather than a single fracture along which movement takes place, large strike-slip faults consist of a zone of roughly parallel fractures up to several kilometers in width. The most recent movement, however, is often along a strand only a few meters wide which may offset features such as stream channels (Figure 15.25). Furthermore, crushed and broken rocks produced during faulting are more easily eroded, producing linear valleys or troughs that often mark the locations of strike-slip faults.

The earliest scientific records of strike-slip faulting were made following surface ruptures that produced large earthquakes. One of the most noteworthy of these was the great San Francisco earthquake

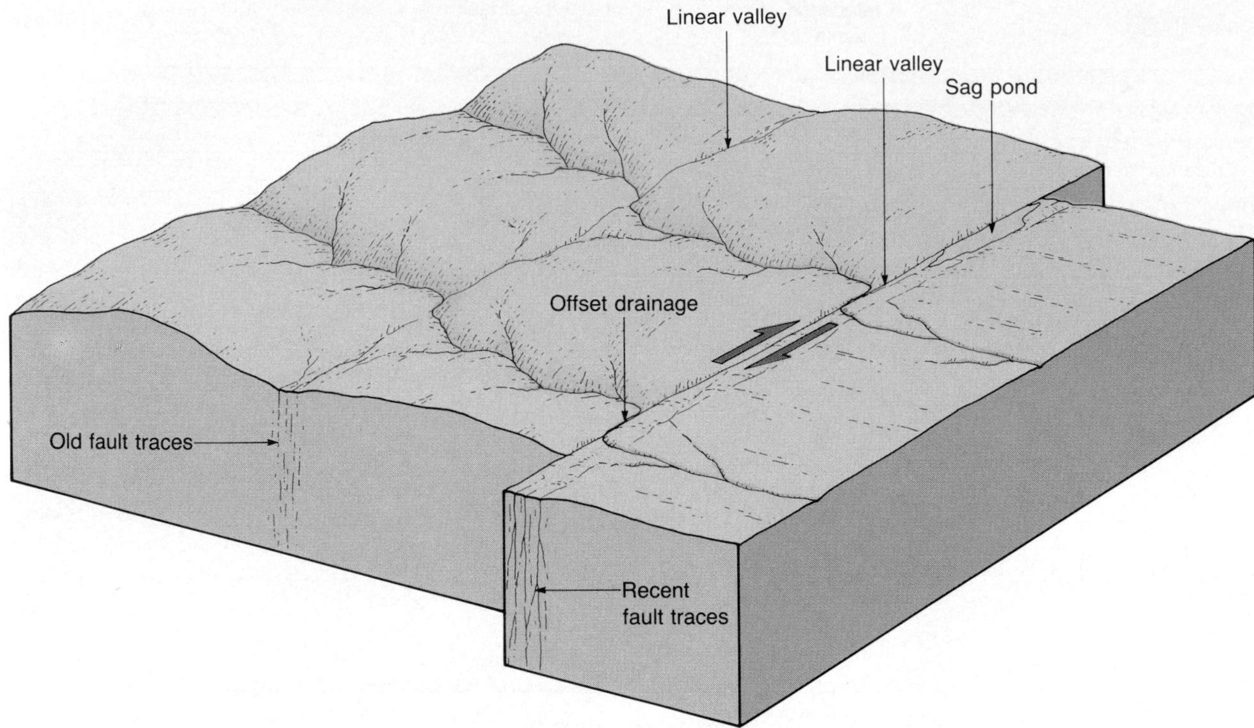

FIGURE 15.25
Block diagram illustrating the features associated with strike-slip faults. (Modified after R. L. Wesson and others)

of 1906. During this strong earthquake, structures such as fences that were built across the San Andreas fault were displaced as much as 4.7 meters (15 feet). Because the movement along the San Andreas causes the crustal block on the opposite side of the fault to move to the right as you face the fault, it is called a *right-lateral* strike-slip fault. The Great Glen fault in Scotland is a well-known example of a *left-lateral* strike-slip fault, which exhibits the opposite sense of displacement. The total displacement along the Great Glen fault is estimated to exceed 100 kilometers (60 miles). Also associated with this fault trace are numerous lakes, including Loch Ness, the home of the legendary monster.

Many major strike-slip faults cut through the lithosphere and accommodate motion between two large crustal plates. This special kind of strike-slip fault is called a **transform fault**. Numerous transform faults cut the oceanic lithosphere and link spreading oceanic ridges. Others accommodate displacement between continental plates which move horizontally with respect to each other. One of the best-known transform faults is the San Andreas fault in California. This plate-bounding fault can be traced for about 950 kilometers (600 miles) from the Gulf of California to a point along the Pacific Coast north of San Francisco, where it heads out to sea. Since its formation about 29 million years ago, displacement along the San Andreas fault has exceeded 330 kilometers. This movement has accommodated the northward displacement of southwestern California and the Baja Peninsula of Mexico in relation to the remainder of North America. The nature of these important structures will be discussed in more detail in Chapter 18.

Strike-slip faults and dip-slip faults are on the opposite ends of a spectrum of fault structures. Faults that exhibit a combination of dip-slip and strike-slip movements are called **oblique-slip faults**.

JOINTS

Among the most common rock structures are fractures called joints. Unlike faults, **joints** are fractures along which no appreciable displacement has occurred. Although some joints have a random orientation, most occur in roughly parallel groups (see Figure 5.7).

We have already considered two types of joints. Earlier we learned that *columnar joints* form when igneous rocks cool and develop shrinkage fractures that produce elongated, pillarlike columns (Figure 15.26). Also recall that sheeting produces a pattern of gently curved joints that develop more or less paral-

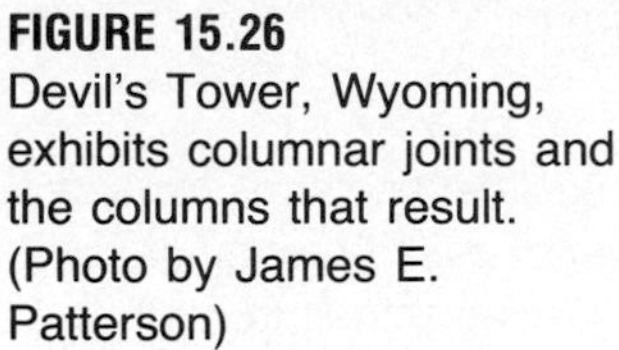

FIGURE 15.26
Devil's Tower, Wyoming, exhibits columnar joints and the columns that result. (Photo by James E. Patterson)

lel to the surface of large exposed igneous bodies such as batholiths. Here the jointing is thought to result from the gradual expansion that occurs when erosion removes the overlying load.

In contrast to the situations just described, most joints are produced when rocks are deformed, particularly by the tensional and shearing stresses associated with crustal movements. For example, when folding occurs, rocks situated at the axes of the folds are elongated and pulled apart to produce tensional joints. Extensive joint patterns can also develop in response to relatively subtle and often barely perceptible regional upwarping and downwarping of the crust. It should be pointed out that, in many cases, the cause for jointing at a particular locale is not readily apparent.

Many rocks are broken by two or even three sets of intersecting joints that slice the rock into numerous regularly shaped blocks. These joint sets often exert a strong influence on other geologic processes. For example, chemical weathering tends to be concentrated along joints, and in many areas, groundwater movement and the resulting solution activity in soluble rocks is controlled by the joint pattern (Figure 15.27). Moreover, a system of joints can influence the direction that stream courses follow. The rectangular drainage pattern described in Chapter 10 is such a case.

Joints may also be significant from an economic standpoint. Some of the world's largest and most important mineral deposits were emplaced along joint systems. Hydrothermal solutions, which are basically mineralized fluids, can migrate into fractured host rocks and precipitate economically important amounts of copper, silver, gold, zinc, lead, and uranium. Further, the construction of engineering projects, including highways and dams, presents certain risks in areas of highly jointed rocks. On June

FIGURE 15.27
Chemical weathering is enhanced along joints in granitic rocks near the top of Lembert Dome, Yosemite National Park. (Photo by E. J. Tarbuck)

BOX 15.2

The San Andreas Fault System

The San Andreas, the best-known and largest fault system in North America, first attracted wide attention after the great 1906 San Francisco earthquake and fire. Following this devastating event, geologic studies demonstrated that a displacement of as much as 5 meters along the fault was responsible for the earthquake. It is now known that this dramatic event is just one of many thousands of earthquakes that have resulted from repeated movements along the San Andreas throughout its 29-million-year history.

Where is the San Andreas fault system located? As shown in Figure 15.B, it trends in a northwesterly direction for nearly 1300 kilometers (780 miles) through much of western California. At its southern end, the San Andreas connects with a spreading center located in the Gulf of California. In the north, the fault enters the Pacific Ocean at Point Arena, where it is thought to continue its northwesterly trend, eventually joining the Mendocino fracture zone. In the central section, the San Andreas is relatively simple and straight. However, at its two extremities, several branches spread from the main trace, so that in some areas the fault zone exceeds 100 kilometers (60 miles) in width.

With the advent of the theory of plate tectonics, geologists have begun to realize the significance of this great fault system. The San Andreas fault is a transform boundary separating two crustal plates that move very slowly over the earth's surface. The Pacific plate, located to the west, moves northwestward relative to the North American plate, causing earthquakes along the fault (Table 15.A).

Over much of its extent, a linear trough reveals the presence of the San Andreas fault. When the system is viewed from the air, linear scars, offset stream channels, and elongated ponds mark the trace in a most striking manner. On the ground, however, surface expressions of the fault are much more difficult to detect. Some of the most distinctive landforms include long, straight escarpments, narrow ridges, and sag ponds formed by settling of blocks within the fault zone. Further, many stream channels characteristically bend sharply to the right where they cross the fault (Figure 15.C).

The San Andreas is undoubtedly the most studied of any fault system in the world. Although many questions remain unanswered, geologists have learned that each fault segment exhibits somewhat different behavior. Some portions of the San Andreas exhibit a slow creep with little noticeable seismic activity. Other segments regularly slip, producing small earthquakes, while still other segments seem to store elastic energy for hundreds of years and rupture in great earthquakes. This knowledge is useful when assigning the potential earthquake hazard to a given segment of the fault zone.

Because of the great length and complexity of the San Andreas fault, it is more appropriately re-

TABLE 15.A
Major earthquakes on the San Andreas fault system.

Date	Location	Magnitude	Remarks
1812	Wrightwood, CA	7	Church at San Juan Capistrano collapsed, killing 40 worshipers.
1812	Santa Barbara channel	7	Churches and other buildings wrecked in and around Santa Barbara.
1838	San Francisco peninsula	7	At one time thought to have been comparable to the great earthquake of 1906.
1857	Fort Tejon, CA	8.25	One of the greatest U.S. earthquakes. Occurred near Los Angeles, then a city of 4000.
1868	Hayward, CA	7	Rupture of the Hayward fault caused extensive damage in San Francisco Bay area.
1906	San Francisco, CA	8.25	The great San Francisco earthquake. As much as 80 percent of the damage caused by fire.
1940	Imperial Valley	7.1	Sixty-five-kilometer displacement on the newly discovered Imperial fault.
1952	Kern County	7.7	Rupture of the White Wolf fault. Largest earthquake in California since 1906. Sixty million dollars in damages and 12 people killed.
1971	San Fernando Valley	6.5	One-half billion dollars in damage and 58 lives claimed.
1989	Santa Cruz Mountains	7.1	Loma Prieta earthquake. Six billion dollars in damages, 62 lives lost, and 3757 people injured.

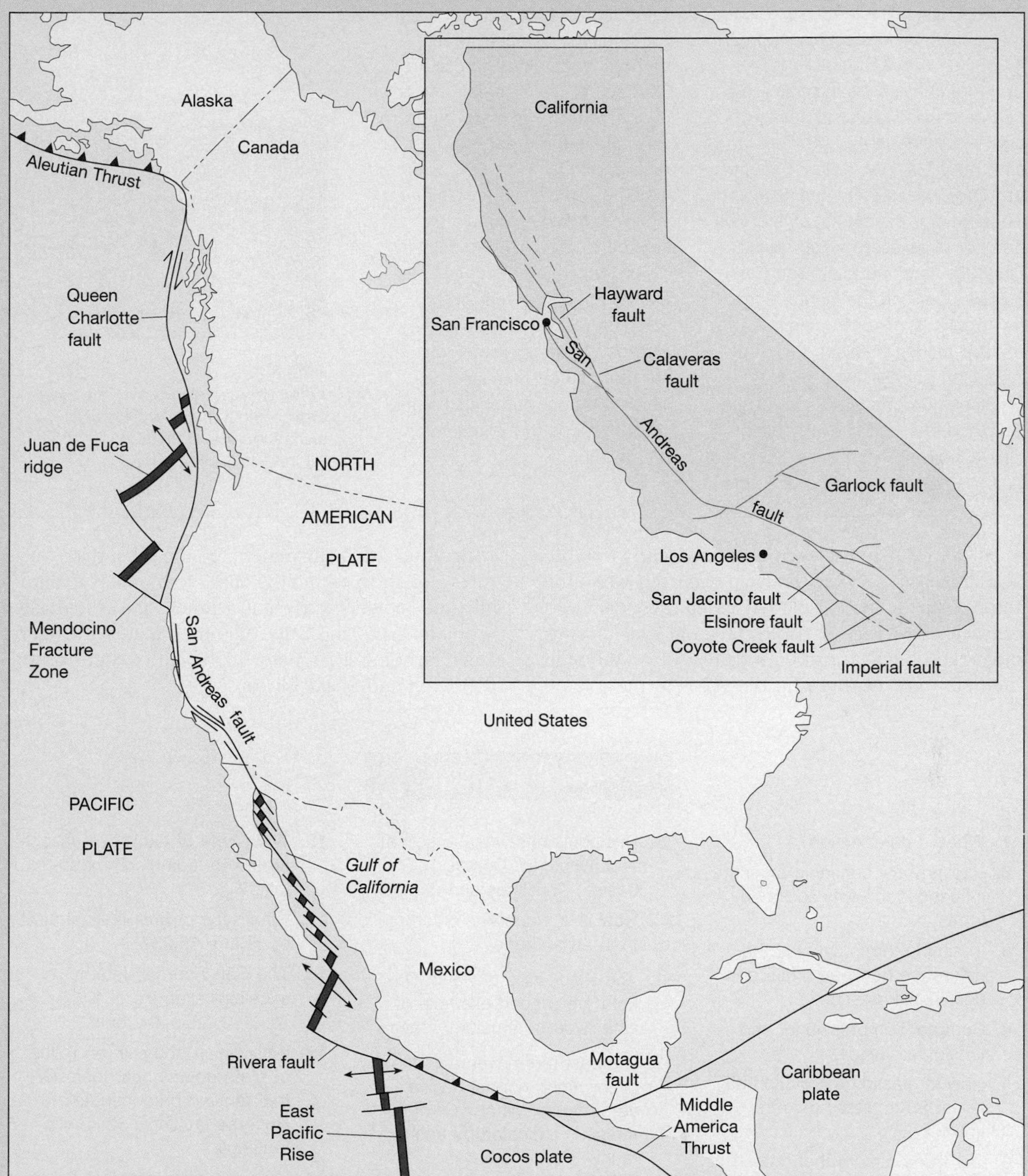

FIGURE 15.B
Map showing the extent of the San Andreas fault system. Insert shows only a few of the many splinter faults that are part of this great fault system.

ferred to as a "fault system." This major fault system consists primarily of San Andreas fault and several major branches including the Hayward and Calaveras faults of central California and the San Jacinto and Elsinore faults of southern California (see Figure 15.B). These major segments, plus a vast number of smaller faults that include the Imperial fault, San Fernando fault, and the Santa Monica fault, collectively accommodate the relative motion between the North American and Pacific plates.

Blocks on opposite sides of the San Andreas fault move horizontally in opposite directions, such that if a person stood on one side of the fault, the block on the opposite side would appear to move to the right when slippage occurred. This type of displacement is known as *right-lateral strike-slip* by geologists. Since the great San Francisco earthquake of 1906, when as much as 5 meters of displacement occurred, geologists have attempted to establish the cumulative displacement along this fault over its 29-million-year history. By matching rock units across the fault, geologists have determined that the total accumulated displacement from earthquakes and creep exceeds 560 kilometers (340 miles).

FIGURE 15.C
Aerial view showing the displacement of stream channels along the trace of the San Andreas fault. (Photo by John S. Shelton)

5, 1976, fourteen lives were lost and nearly one billion dollars in property damage occurred when the Teton Dam in Idaho failed. This earthen dam was constructed of very erodible clays and silts and was situated on highly fractured volcanic rocks. Although attempts were made to fill the voids in the jointed rock, water gradually penetrated the subsurface fractures and undermined the dam's foundation. Eventually, the moving water cut a tunnel into the easily erodible clays and silts. Within minutes the dam failed, sending a 20-meter high wall of water down the Teton and Snake rivers.

REVIEW QUESTIONS

1. What is deformation?
2. Explain how confining pressure influences the way rocks deform.
3. In simple terms, what is the difference between brittle and ductile deformation?
4. Contrast compressional and tensional stresses.
5. How is elastic deformation different from plastic deformation?
6. List three factors that determine how rocks will behave when exposed to stresses that exceed their strength. Briefly explain the role of each.
7. What is an outcrop?
8. What two measurements are used to establish the orientation of deformed strata? Distinguish between them.
9. Distinguish between anticlines and synclines. Domes and basins. Anticlines and domes.
10. How is a monocline different from an anticline?
11. The Black Hills of South Dakota are a good example of what type of structural feature?
12. Contrast the movements that occur along normal and reverse faults. What type of stress is indicated by each fault?
13. Is the fault shown in Figure 15.17 a normal or a reverse fault?
14. What type of fault is shown in the chapter-opening photo?
15. Describe a horst and a graben. Explain how a graben valley forms and name one.
16. What type of faults are associated with fault-block mountains?
17. What type of fault is illustrated by Figure 15.23?
18. The San Andreas fault is an excellent example of a ______________ fault.
19. With which of the three types of plate boundaries does normal faulting predominate? Reverse faulting? Strike-slip faulting?
20. How are joints different from faults?

KEY TERMS

anticline (p. 384)
basin (p. 385)
compressional stress (p. 378)
confining pressure (p. 377)
deformation (p. 377)
dip (p. 381)
dip-slip fault (p. 388)
dome (p. 385)
elastic deformation (p. 379)
fault (p. 386)
fault-block mountain (p. 389)
fault scarp (p. 390)
fold (p. 383)
graben (p. 389)
hogback (p. 385)
horst (p. 389)
joint (p. 392)
klippe (p. 390)
monocline (p. 384)
normal fault (p. 389)
oblique-slip fault (p. 392)
plastic deformation (p. 379)
reverse fault (p. 389)
rock structure (p. 381)
shear (p. 378)
stress (p. 378)
strike (p. 381)
strike-slip fault (p. 391)
syncline (p. 384)
tensional stress (p. 378)
thrust fault (p. 389)
transform fault (p. 392)

16
Earthquakes

Opposite: Collapsed upper deck of Highway 880 caused by the 1989 Loma Prieta earthquake. (Photo by Howard G. Wilshire, U.S. Geological Survey) (Top photo by R. W. Wallace, U.S. Geological Survey)

On October 17, 1989, at 5:04 P.M. Pacific Daylight Time, strong tremors shook the San Francisco Bay area. Millions of Americans and others around the world were just getting ready to watch the World Series, but instead saw their television sets go black as the shock hit Candlestick Park. Although the Loma Prieta earthquake was centered in a remote section of the Santa Cruz Mountains, about 16 kilometers north of the city of Santa Cruz, major damage occurred in the Marina District of San Francisco 100 kilometers to the north (Figure 16.1A). Here, as many as 60 row houses were so badly damaged that they had to be demolished. Although wood frame structures often survive earthquakes with little or no structural damage, many of these homes were built over garages that were supported only by thin wooden columns. Simply, there were no walls on the lower levels to resist the horizontal stresses.

The most tragic result of the violent shaking was the collapse of some double-decked sections of Interstate 880, also known as the Nimitz Freeway (Figure 16.1B). The ground motions caused the upper deck to sway, shattering the concrete support columns along a mile-long section of the freeway. The upper deck then collapsed onto the lower roadway, flattening cars as if they were aluminum beverage cans. Other roadways that were damaged during this earthquake included a 50-foot section of the upper deck of the Bay Bridge, which is a major artery connecting the cities of Oakland and San Francisco. The vibrations caused cars on the bridge to bounce up and down vigorously. A motorcyclist on the upper deck described how the roadway bulged and rippled toward him: "It was like bumper cars—only you could die. . . !" Fortunately, only one motorist on the bridge was killed.

The Loma Prieta earthquake lasted just 15 seconds and occurred along the northern segment of the San Andreas fault. This active fault zone and associated faults, including the Hayward fault, which runs through Oakland, and the San Jacinto fault, near San Bernardino, extend northward from southern California for over 1000 kilometers. It is along this fault system that two great sections of the earth, the North American plate and the Pacific plate, grind past each other at the rate of a few centimeters per year. Along

FIGURE 16.1
Damage to structures caused by the October 17, 1989, Loma Prieta earthquake. Damage in the Marina district of San Francisco. (Photo by Charles E. Meyers, U.S. Geological Survey)

BOX 16.1

Severe Earthquakes

Earthquakes are among nature's most destructive forces. Many people in the United States are familiar with the Loma Prieta earthquake that struck the San Francisco Bay area during the 1989 World Series. Although this event was dramatic and destructive, earthquakes can be much worse. Two of the most damaging earthquakes of the 1980s occurred in Armenia and Mexico.

On December 7, 1988, a violent earthquake jolted northwestern Armenia, devastating several towns and killing an estimated 25,000 persons. Poor construction practices were blamed for the high death toll in this rugged region located near the Caucasus Mountains. To quote a seismologist who was asked about this destructive event, "Earthquakes don't kill people, buildings do." Many of the taller structures in the region were made of precast concrete slabs that were held together by metal hooks. A San Francisco structural engineer studying the destruction stated, "There was very little reinforcing to tie these buildings together. The buildings basically came apart the way they were put together." In addition, many of the village homes had thick roofs made of mud and rock, which were deadly when they collapsed. Nearly the entire town of Spitak, which was located near the epicenter, was leveled.

This destructive earthquake registered 6.9 on the Richter scale and was centered near the Armenian-Turkish border. A few minutes after the main shock struck, a large aftershock of magnitude 5.8 collapsed many structures that had been weakened by the main tremor. The cause of this devastating earthquake was the sudden movement along a fault that had not been previously identified. This newly discovered fault is believed to have an origin similar to that of the many other faults that slice through the region. They are all associated with the collision of the Arabian plate with the Eurasian plate.

Although the Armenian earthquake was one of the century's worst in terms of deaths, the magnitude of the shock was much less than that of the earthquake that struck Mexico City in 1985. The latter quake, which registered 8.1 on the Richter scale, was centered along the Pacific Coast nearly 400 kilometers (250 miles) from Mexico City. In less than two minutes, the quake battered downtown Mexico City, causing 412 buildings to collapse and damaging over 7000 others (Figure 16.A). Fortunately, the earthquake struck during the morning rush hour when businesses and schools were not fully occupied. Nevertheless, casualties in Mexico City included more than 9500 deaths, with 30,000 injured and 50,000 left homeless.

The vibrations were felt as far north as Houston, Texas, where skyscrapers swayed. Ironically, coastal towns such as Ixtapa, which were closer to the epicenter than Mexico City, suffered less during the event. The fact that the central district of Mexico City is built on the soft, moist sediments of an ancient lake bed contributed to the destruction. According to a researcher at the California Institute of Technology, this earthquake caused the lake bed to vibrate "like a bowl of jelly when the seismic waves came through the ground beneath it."

Although the destruction to downtown Mexico City was staggering, the damage was very restricted. Out of a total of more than 800,000 structures located in the city, only a few hundred buildings collapsed. Further, only one percent of the city was heavily damaged. This is not the first earthquake to batter Mexico City, and it will not be the last. During this century, California has had five earthquakes with Richter magnitudes greater than 7, while Mexico has had over 40, many of which have brought great human suffering.

FIGURE 16.A
This building collapsed during the September 19, 1985, Mexican earthquake. (Photo by James L. Beck)

much of this fault zone the rocks on either side tend to remain locked, resisting the overall motion. In time the stress builds to a point where the strength of the rocks is exceeded and the plates slide past each other, releasing the stored energy in short bursts. The result is an earthquake.

In spite of the huge economic loss in the Bay Area, it is remarkable that the losses were not even greater. Just a year earlier, a somewhat weaker earthquake killed an estimated 25,000 people in Soviet Armenia. Unquestionably, the efforts to upgrade structures to conform to the building codes of California helped minimize what could have been a catastrophic event.

Although some comfort can be taken from the fact that most buildings in the Bay Area held up well during this event, it is also clear that this was not the long-feared "Big One." Whereas the Loma Prieta had a magnitude of 7.1 on the Richter scale, the 1985 Mexico City earthquake, with a magnitude of 8.1, involved the release of more than 30 times as much energy (see Box 16.1). The latter quake was centered along the Pacific Coast nearly 400 kilometers (250 miles) from Mexico City. In less than two minutes the quake battered downtown Mexico City, causing 412 buildings to collapse and killing an estimated 9500 persons. The vibrations from this tremor were felt as far north as Houston, Texas, where skyscrapers swayed. The fact that the central district of Mexico City is built on the soft, moist sediment of an ancient lake bed contributed to the destruction. According to a researcher at the California Institute of Technology, this earthquake caused the lake bed to vibrate "like a bowl of jelly when the seismic waves came through the ground beneath it."

It is estimated that over 30,000 earthquakes, strong enough to be felt, occur worldwide annually. Fortunately most of these are minor tremors and do very little damage. Generally only about 75 significant earthquakes take place each year, and many of these occur in remote regions. However, occasionally a large earthquake occurs near a large population center. When such an event takes place, it is among the most destructive natural forces on earth. The shaking of the ground coupled with the liquefaction of some soils wreaks havoc on buildings. In addition, when a quake occurs in a populated area,

FIGURE 16.2
San Francisco in flames after the 1906 earthquake. (Reproduced from the collection of the Library of Congress)

power and gas lines are often ruptured, causing numerous fires. In the 1906 San Francisco earthquake, much of the damage was caused by fires which ran unchecked when broken water mains left firefighters with only trickles of water (Figure 16.2).

WHAT IS AN EARTHQUAKE?

An **earthquake** is the vibration of the earth produced by the rapid release of energy. This energy radiates in all directions from its source, or **focus**, in the form of waves analogous to those produced when a stone is dropped into a calm pond. Just as the impact of the stone sets water waves in motion, an earthquake generates seismic waves that radiate throughout the earth. Even though the energy dissipates rapidly with increasing distance from the focus, instruments located throughout the world record the event.

The tremendous energy released by atomic explosions or by volcanic eruptions can produce an earthquake; but these events are usually weak and infrequent. What mechanism does produce a destructive earthquake? Ample evidence exists that the earth is not a static planet. Numerous ancient wave-cut benches can be found many meters above the level of the highest tides, which indicates crustal uplifting. Other regions exhibit evidence of extensive subsidence. In addition to these vertical displacements, offsets in fence lines, roads, and other structures indicate that horizontal movement is also prevalent (Figure 16.3). These movements are usually associated with large fractures in the earth called **faults.** Most of the motion along faults can be satisfactorily explained by the plate tectonics theory. The plate model proposes that large slabs of the earth are continually in motion. These mobile plates interact with neighboring plates, straining and deforming the rocks at their edges. It is along faults associated with plate boundaries that most earthquakes occur. Furthermore, earthquakes are repetitive; that is, as soon as one earthquake is over, the continuous motion of the plates begins to add strain to the rocks until they fail again.

The actual mechanism of earthquake generation eluded geologists until H. F. Reid of Johns Hopkins University conducted a study following the great 1906 San Francisco earthquake. This earthquake was accompanied by displacements of several meters along the northern portion of the San Andreas fault, a 1300-kilometer (780-mile)-long fracture that runs northward through southern California. This large fault zone separates two great sections of the earth, the North American plate and the Pacific plate. Field investigations determined that during this single earthquake the Pacific plate slid as much as 4.7 meters (15 feet) in a northward direction past the adjacent North American plate.

Using land surveys conducted several years apart, Reid discovered that during the 50 years prior to the 1906 earthquake the land at distant points on both sides of the fault showed a relative displacement of slightly more than 3 meters (10 feet). The mechanism of earthquake formation which Reid deduced from this information is illustrated in Figure 16.4. Tectonic forces ever so slowly deform the crustal rocks on both sides of the fault as illustrated by the bent features. Under these conditions, rocks are bending and storing elastic energy, much like a wooden stick would if bent. Eventually, the forces

FIGURE 16.3
This fence was offset 2.5 meters (8.5 feet) during the 1906 San Francisco earthquake. (Photo by G. K. Gilbert, U.S. Geological Survey)

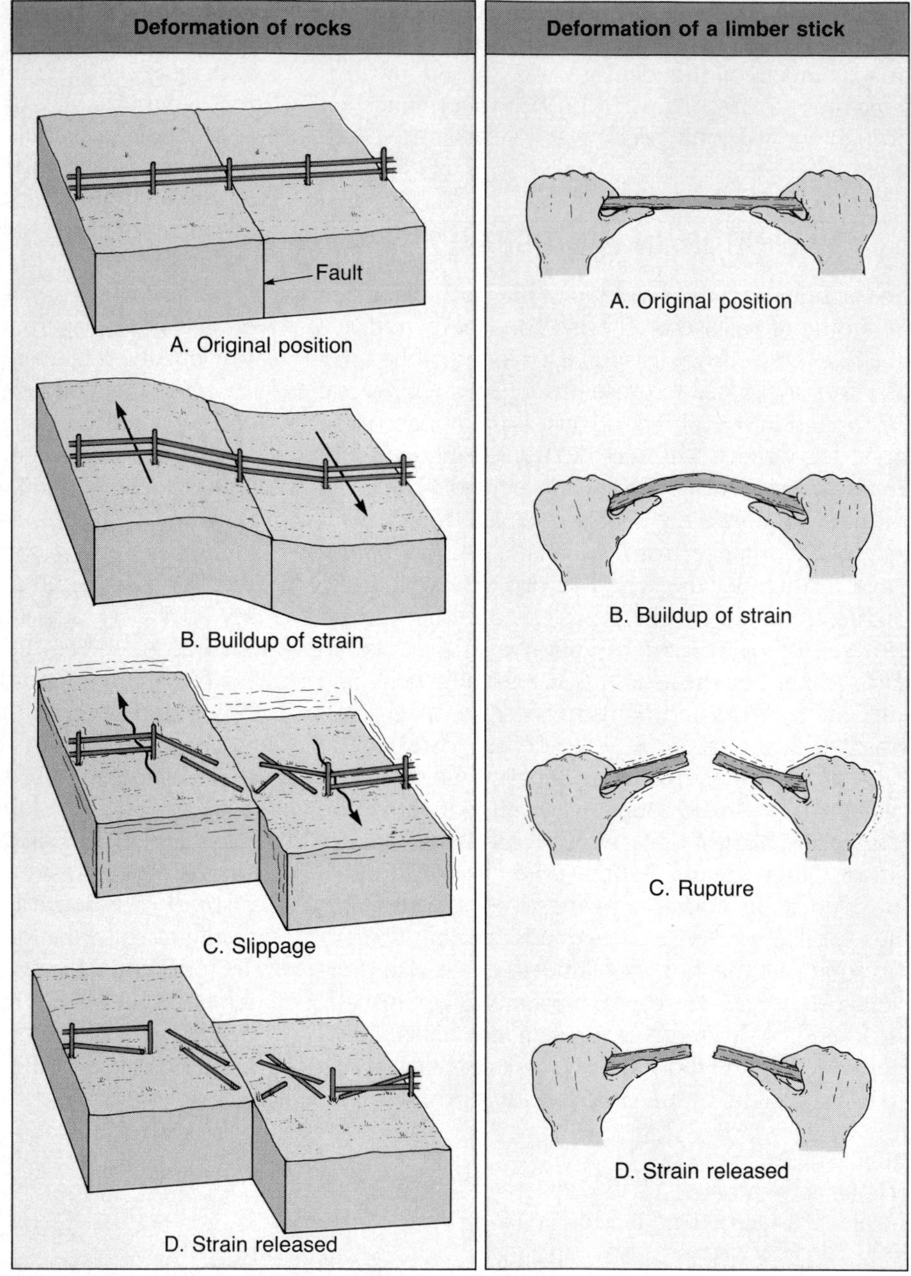

FIGURE 16.4
Elastic rebound. As rock is deformed, it bends, storing elastic energy. Once the rock is strained beyond its breaking point, it ruptures, releasing the stored-up energy in the form of earthquake waves. (Modified after Foster, *Physical Geology,* 4th edition, Columbus, Ohio: Merrill/ Macmillan 1983)

holding the rocks together are overcome. As slippage occurs at the weakest point (the focus), displacement will exert farther along the fault where additional slippage will occur until most of the built-up strain is released. This slippage allows the deformed rock to "snap back." The vibrations we know as an earthquake occur as the rock elastically returns to its original shape. The "springing back" of the rock was termed **elastic rebound** by Reid, since the rock behaves elastically, much like a stretched rubber band does when it is released. In summary, most earthquakes are produced by the rapid release of elastic energy stored in rock that has been subjected to great differential stress. Once the strength of the rock is exceeded, it suddenly ruptures, which results in the vibrations of an earthquake. Earthquakes also occur along existing faults when the frictional forces on the fault surfaces are overcome.

The intense vibrations of the 1906 San Francisco earthquake lasted about 40 seconds. Although most of the displacement along the fault occurred in this rather short period, additional movements and adjustments in the rocks occurred for several days following the main quake. The adjustments that follow a major earthquake often generate smaller earthquakes called **aftershocks.** Although aftershocks are usually much weaker than the main earthquake, they can sometimes cause significant destruction to already badly weakened structures. This occurred, for example, during the 1988 Armenian earthquake when a large aftershock of magnitude 5.8 collapsed many structures that had been weakened by the main tremor. In addition, several small earthquakes, called **foreshocks** often precede a major earthquake by several days or in some cases by as much as several years. Monitoring of these foreshocks has been used as a means of predicting a forthcoming major earthquake. We will consider the topic of earthquake prediction in a later section of this chapter.

The tectonic forces which created the strain that was eventually released during the 1906 San Francisco earthquake are still active. Currently, laser beams* are used for establishing the relative motion between the opposite sides of this fault. These measurements have revealed a displacement of from 2 to 5 centimeters per year. Although this rate of movement seems slow, it is indeed appreciable on a geologic time scale. In 30 million years this rate of displacement is sufficient to slide the western portion of California northward so that Los Angeles, on the Pacific plate, would be adjacent to San Francisco on the North American plate. More importantly in the short term, a displacement of just 2 centimeters per year produces 2 meters of offset every 100 years. Consequently, the 4 meters of displacement produced during the 1906 San Francisco earthquake should occur at least every 200 years along this segment of the fault zone.

*Laser beams are used in very precise surveying instruments because of their incredibly accurate straight line qualities.

The San Andreas is undoubtedly the most studied fault system in the world. Over the years investigations have shown that displacement occurs along discrete segments that are 100 to 200 kilometers long. Further, each fault segment behaves somewhat differently from the others. Some portions of the San Andreas exhibit a slow, gradual displacement known as *creep,* which occurs with little noticeable seismic activity. Other segments regularly slip, producing small earthquakes, while still other segments remain locked and store elastic energy for hundreds of years before rupturing in great earthquakes. The latter process is described as *stick-slip* motion, since the fault exhibits alternating periods of locked behavior followed by sudden slippage. It is estimated that great earthquakes should occur about every 50 to 200 years along those sections of the San Andreas fault that exhibit stick-slip motion. This knowledge is useful when assigning a potential earthquake risk to a given segment of the fault zone.

Not all movement along faults is horizontal. Vertical displacement, in which one side is lifted higher in relation to the other, is also common. Figure 16.5 shows a *fault scarp* (cliff) produced during the 1964 Good Friday earthquake in Alaska. Further, many earthquakes occur at such great depths that no displacement is evident at the surface.

SEISMOLOGY

The study of earthquake waves, **seismology**, dates back to attempts made by the Chinese almost 2000 years ago to determine the direction from which these waves originated. The seismic instrument used

FIGURE 16.5
Fault scarp that resulted from vertical displacement along a fault during the Madison Canyon earthquake in Montana. (Photo by Stephen Trimble)

FIGURE 16.6
Ancient Chinese seismograph. During an earth tremor, the dragons located in the direction of the main vibrations would drop a ball into the mouths of the frogs below.

by the Chinese was a large hollow jar that probably contained a mass suspended from the top (Figure 16.6). This suspended mass (similar to a clock pendulum) was connected in some fashion to the jaws of several dragon figurines that encircled the container. The jaws of each dragon held a metal ball. When earthquake waves reached the instrument, the relative motion between the suspended mass and the jar would dislodge some of the metal balls into the waiting mouths of frogs directly below. The Chinese were probably aware that the first strong ground motion from an earthquake is directional, and when it is strong enough, all poorly supported items will topple over in the same direction. Apparently the Chinese used this fact plus the position of the dislodged ball to detect the direction to an earthquake's source. However, the complex motion of seismic waves makes it unlikely that the actual direction to an earthquake was very often determined.

In principle at least, modern **seismographs**, instruments that record seismic waves, are not unlike the device used by the early Chinese. Seismographs have a mass freely suspended from a support that is attached to the ground (Figure 16.7). When the vibration from a distant earthquake reaches the instrument, the **inertia*** of the mass keeps it relatively stationary, while the earth and support move. The movement of the earth in relation to the stationary mass is recorded on a rotating drum or magnetic tape.

Earthquakes cause both vertical and horizontal ground motion; therefore, more than one type of seismograph is needed. The instrument shown in Figure 16.7 is designed so that the mass is permitted

*Inertia: Simply stated, objects at rest tend to stay at rest and objects in motion tend to remain in motion unless either is acted upon by an outside force. You probably have experienced this phenomenon when you tried to quickly stop your automobile and your body continued to move forward.

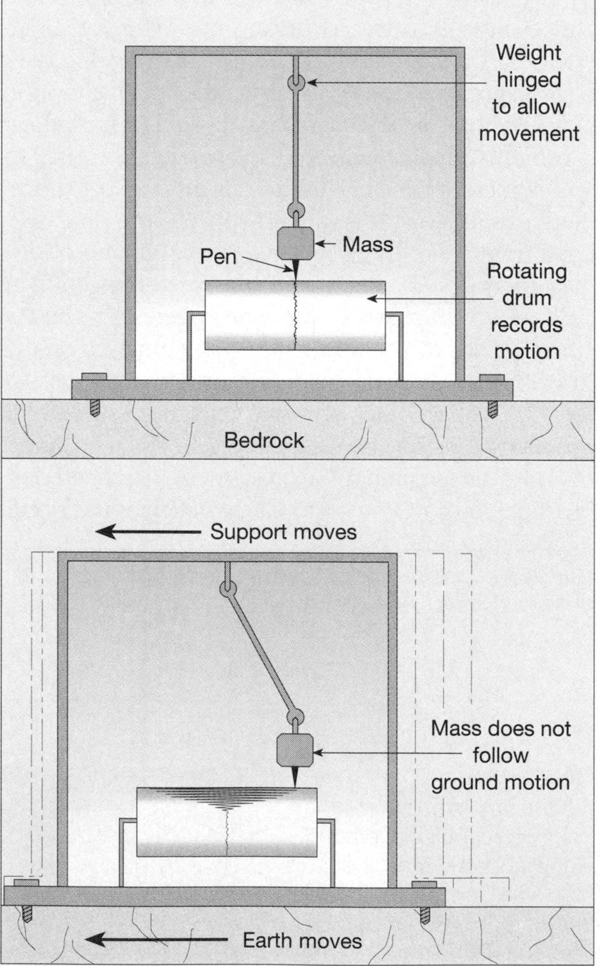

FIGURE 16.7
Principle of the seismograph. The inertia of the suspended mass tends to keep it motionless, while the recording drum, which is anchored to bedrock, vibrates in response to seismic waves. Thus, the stationary mass provides a reference point from which to measure the amount of displacement occurring as the seismic wave passes through the ground below.

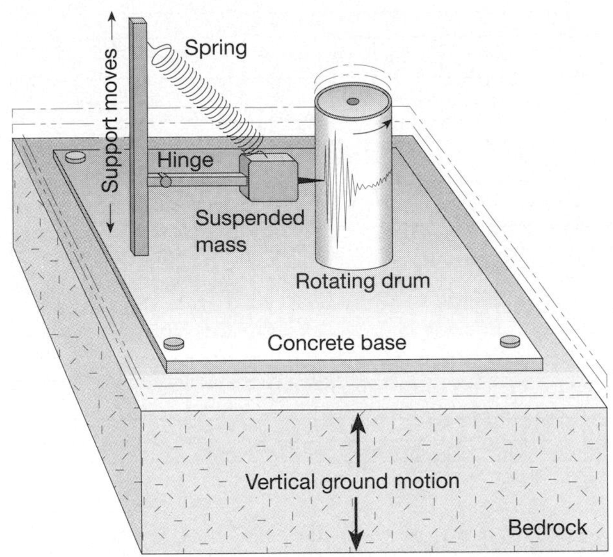

FIGURE 16.8
Seismograph designed to record vertical ground motion.

to swing from side-to-side and thus it detects horizontal ground motion. Usually two horizontal seismographs are employed, one oriented north-south and the other placed with an east-west orientation. Vertical ground motion can be detected if the mass is suspended from a spring as shown in Figure 16.8.

In order to detect very weak earthquakes, or a great earthquake that occurred in another part of the world, seismic instruments are designed to magnify the actual ground motion. Conversely, other instruments are designed to withstand the violent shaking that occurs very near the earthquake source.

The records obtained from seismographs, called **seismograms**, provide a great deal of information concerning the behavior of seismic waves. Simply stated, seismic waves are elastic energy that radiates out in all directions from the focus. The propagation (transmission) of this energy can be compared to the shaking of gelatin in a bowl which results as some is spooned out. Whereas the gelatin will have one mode of vibration, seismograms reveal that two main groups of seismic waves are generated by the slippage of a rock mass. One of these wave types travels

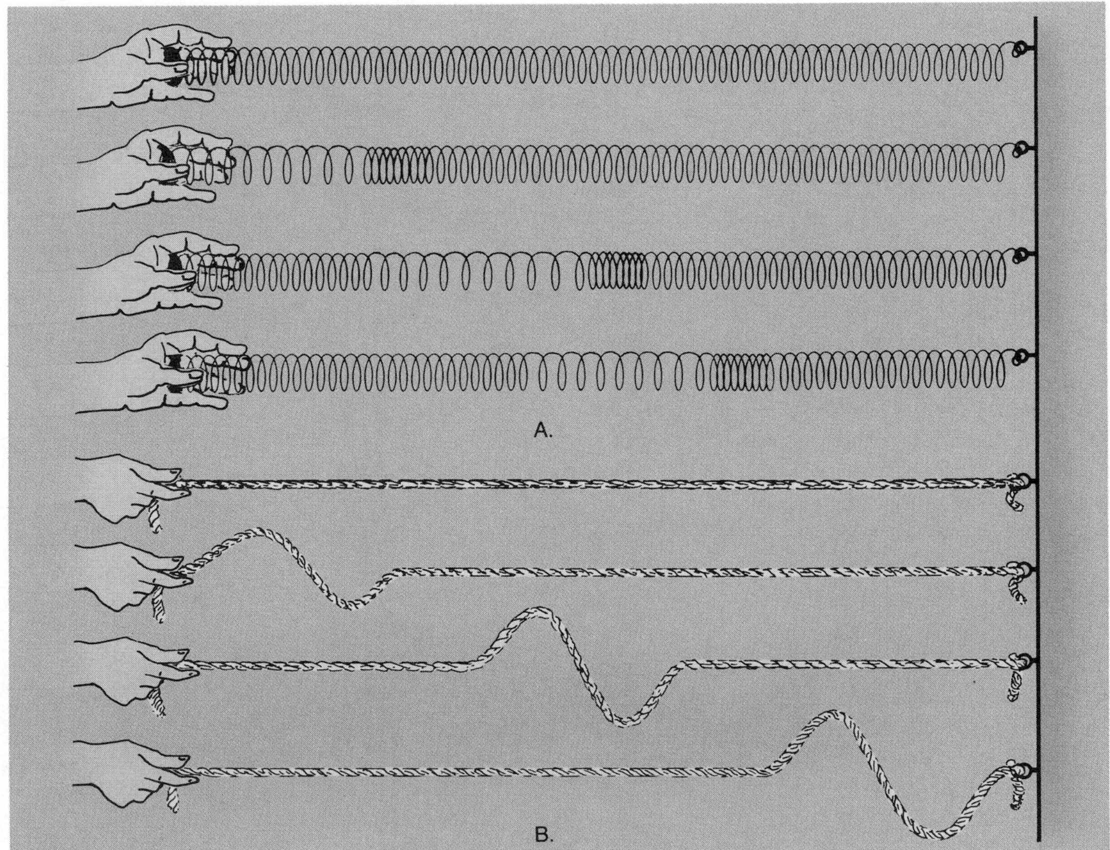

FIGURE 16.9
Types of seismic waves and their characteristic motion. **A.** P waves cause the particles in the material to vibrate back and forth in the same direction as the waves move. **B.** S waves cause particles to oscillate at right angles to the direction of wave motion.

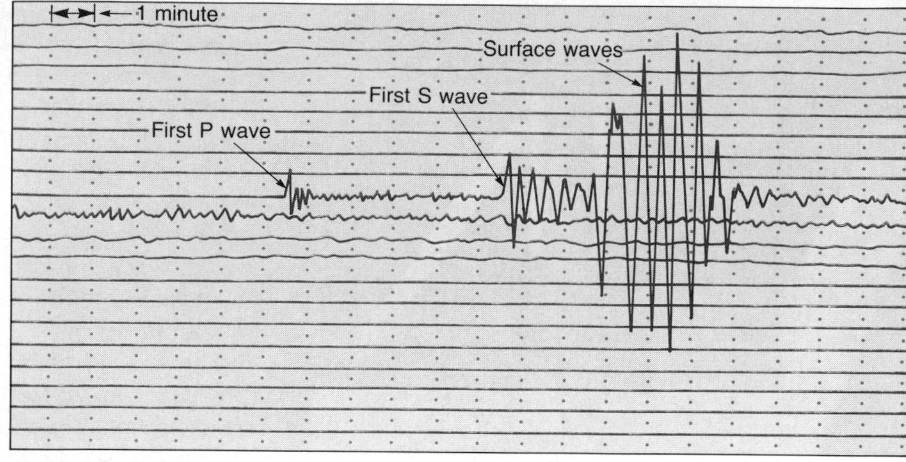

FIGURE 16.10
Typical seismic record. Note the time interval between the arrival of the first P waves and the arrival of the first S waves.

along the outer part of the earth. These are called **surface waves.** Others travel through the earth's interior and are called **body waves.** Body waves are further divided into two types called **primary** or **P waves** and **secondary** or **S waves.**

Body waves are divided on the basis of their mode of propagation through intervening material. P waves push (compress) and pull (dilate) rocks in the direction the wave is traveling (Figure 16.9A). This wave motion is analogous to that generated by human vocal cords as they move air to create sound. Solids, liquids, and gases resist a change in volume when compressed and will elastically spring back once the force is removed. Therefore, P waves, which are compressional waves, can travel through all these materials. S waves, on the other hand, "shake" the particles at right angles to their direction of travel. This can be illustrated by tying one end of a rope to a post and shaking the other end as shown in Figure 16.9B. Unlike P waves, which change the volume of the intervening material, S waves change only the shape of the material that transmits them. Since fluids (gases and liquids) do not resist changes in shape, they will not transmit S waves.

The motion of surface waves is somewhat more complex. As surface waves travel along the ground, they cause it and anything resting upon it to move in a manner similar to the way ocean swells toss a ship about. In addition to the up-and-down motion generated, surface waves also have a side-to-side motion which is similar to an S wave oriented in a horizontal plane. This latter motion is particularly damaging to the foundations of structures.

By observing a "typical" seismic record as shown in Figure 16.10, we find that some of the differences among these seismic waves became apparent. P waves arrive at the recording station before S waves, which themselves arrive before the surface waves. This is a consequence of their relative velocities. For purposes of illustration, the velocity of P waves through granite within the crust is about 6 km/sec, whereas S waves under the same conditions will travel at 3.6 km/sec. Differences in density and elastic properties of the transmitting material greatly influence the velocities of these waves. In water, for example, the velocity of P waves is 1.5 km/sec, while S waves are not transmitted through this medium. However, in any solid material P waves travel about 1.7 times faster than S waves, and surface waves can be expected to travel 0.9 times the velocity of the S waves.

In addition to velocity differences, also notice in Figure 16.10 that the height, or more correctly, the amplitude, of these wave types varies. The S waves have a slightly greater amplitude than the P waves,

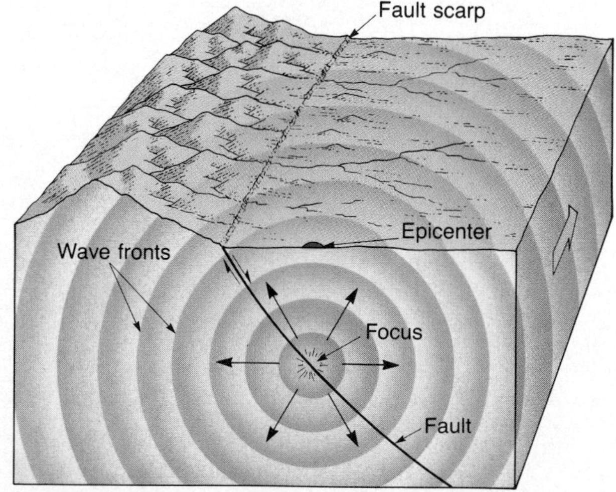

FIGURE 16.11
Earthquake focus and epicenter. The focus is the zone within the earth where the initial displacement occurs. The epicenter is the surface location directly above the focus.

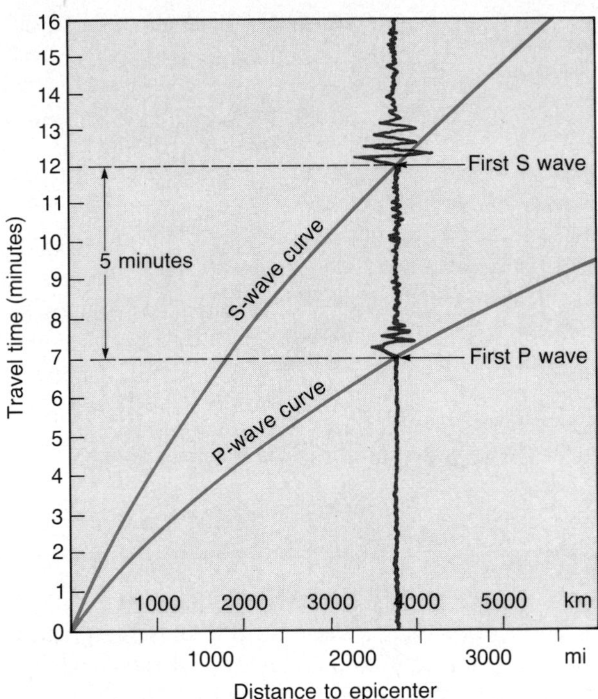

FIGURE 16.12
A travel-time graph is used to determine the distance to the epicenter. The difference in arrival times of the first P and S waves in the example is 5 minutes. Thus, the epicenter is roughly 3800 kilometers (2350 miles) away.

while the surface waves, which cause the greatest destruction, exhibit an even greater amplitude. Because surface waves are confined to a narrow region near the surface and are not spread throughout the earth as P and S waves are, they retain their maximum amplitude longer. Surface waves also have longer periods (time interval between crests); therefore they are often referred to as **long waves**, or **L waves**.

As we shall see, seismic waves are useful in determining the location and magnitude of earthquakes. In addition, seismic waves provide a tool for probing the earth's interior.

LOCATING THE SOURCE OF AN EARTHQUAKE

Recall that the focus is the place within the earth where the earthquake waves originate (Figure 16.11). The **epicenter** is the location on the surface directly above the focus. For shallow earthquakes the difference in velocities of P and S waves provides a method for determining the distance to an earthquake. The principle used is analogous to a race between two autos, one faster than the other. The greater the distance of the race, the greater will be the difference in the arrival times at the finish line.

FIGURE 16.13
Earthquake epicenter is located using the distances obtained from three seismic stations.

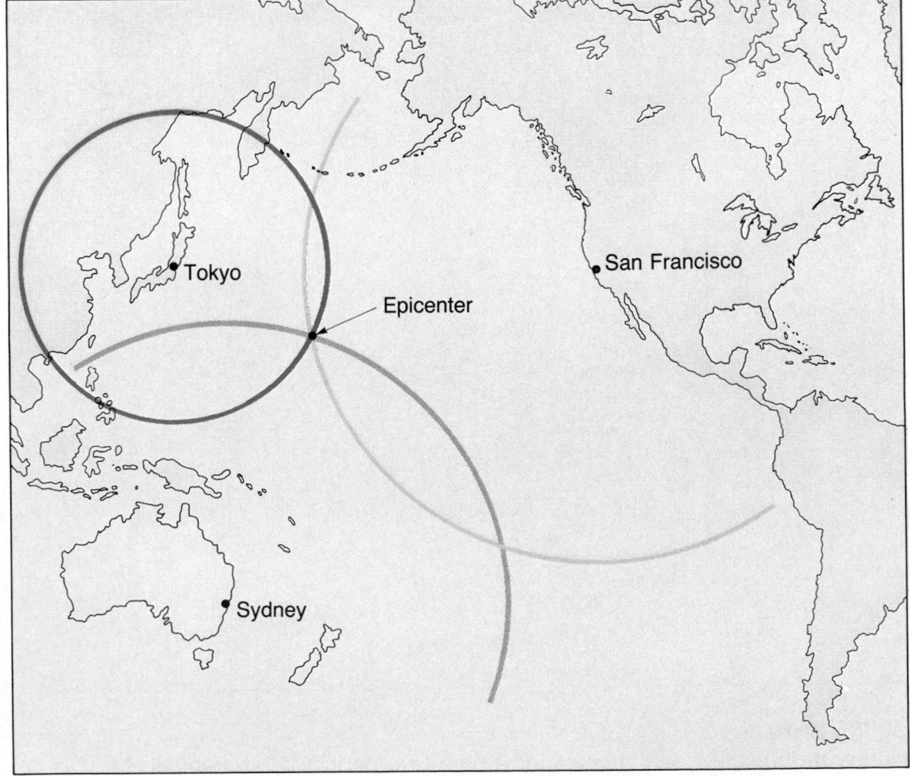

Therefore, the greater the interval between the arrival of the first P wave and the first S wave, the greater the distance to the earthquake.

A system for locating earthquake epicenters was developed through the use of seismograms from earthquakes whose epicenters could be easily pinpointed from physical evidence. From these seismograms, travel-time graphs as shown in Figure 16.12 were constructed. The first travel-time graphs were refined when seismograms from nuclear explosions became available, since the location and time of detonation were precisely known.

Using the sample seismogram in Figure 16.10 and the travel-time curves in Figure 16.12, we can determine the distance separating the recording station and the earthquake. This is accomplished by establishing the time interval between the arrival of the first P wave and the first S wave and then finding the place on the travel-time graphs which exhibits an equivalent time spread between the P and S wave curves. From this information we can determine that this earthquake occurred 3800 kilometers from the recording instrument. Although the distance to an earthquake is established in this manner, its location could be in any direction from the observer. As shown in Figure 16.13, the precise location can be found only when the distance is known from three or more seismic stations. By drawing circles representing the epicenter distance for each of these observatories, seismologists can establish an accurate location.

The study of earthquakes was greatly bolstered during the 1960s through efforts to discriminate between underground nuclear explosions and natural earthquakes. The United States established a worldwide network of over 100 seismic stations coordinated through Golden, Colorado. The largest of these, located in Billings, Montana, consists of an array of 525 instruments grouped in 21 clusters covering a region 200 km in diameter. Using data from this array, seismologists employing high-speed computers are able to locate an epicenter by a trial-and-error technique.

EARTHQUAKE BELTS

About 95 percent of the energy released by earthquakes is concentrated in a few relatively narrow zones that wind around the globe (Figure 16.14). The greatest energy is released along a path located near the outer edge of the Pacific Ocean known as the *circum-Pacific belt.* Included in this zone are regions of great seismic activity such as Japan, the Philippines, Chile, and numerous volcanic island chains, as exemplified by the Aleutian Islands. Another

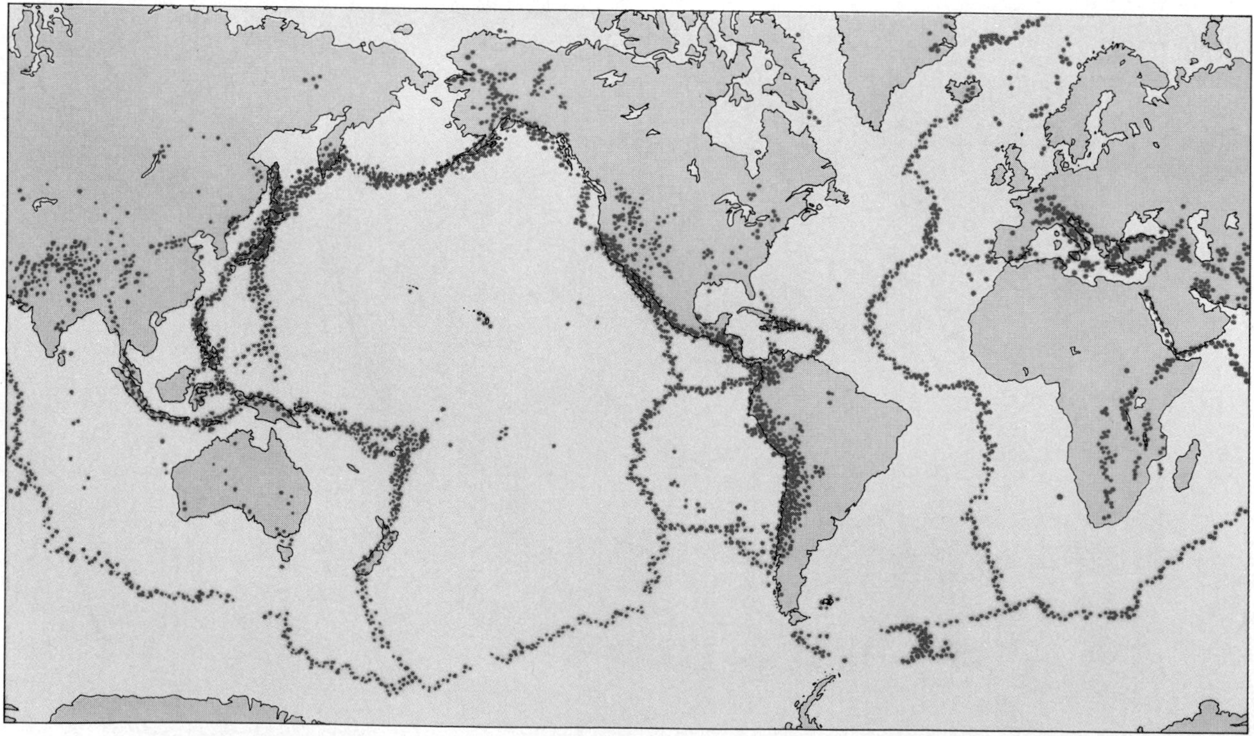

FIGURE 16.14
World distribution of earthquakes for a nine-year period. (Data from NOAA)

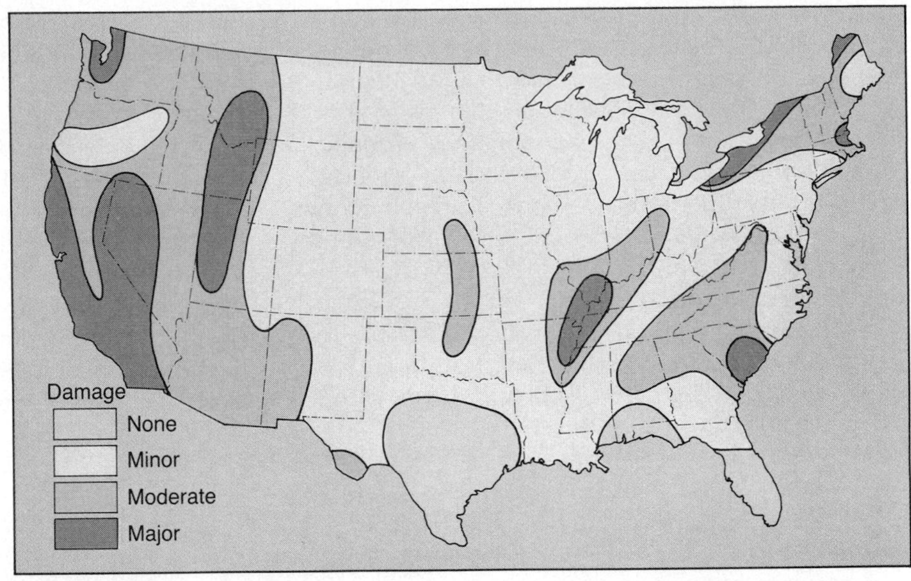

A.

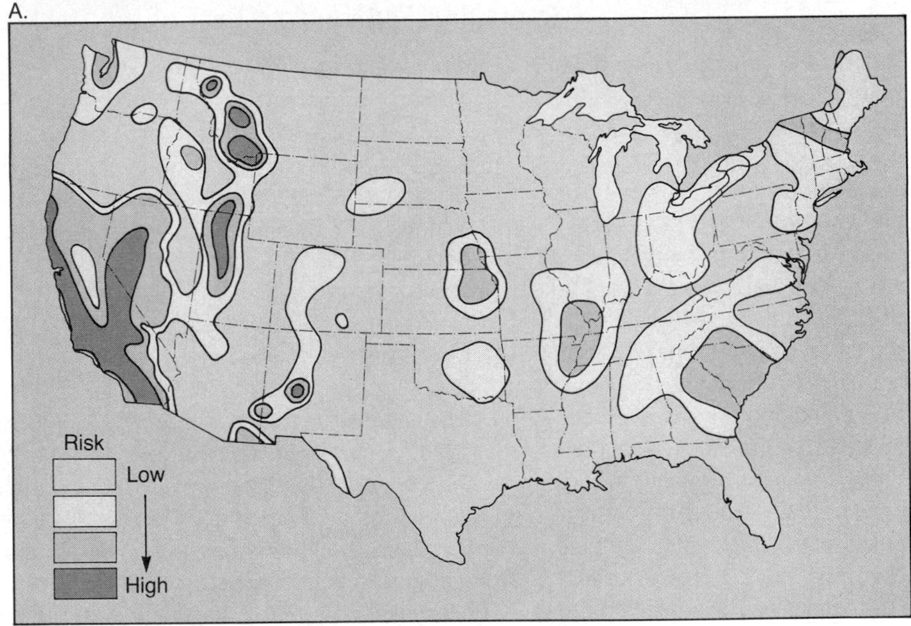

B.

FIGURE 16.15
Two types of earthquake risk maps. Map A shows the relative risk of seismic destruction based on where earthquakes have occurred in the past. Map B is based on the probability that a certain level of ground motion will occur during an earthquake over a 50-year period. (From Environmental Science Service Administration)

major concentration of strong seismic activity runs through the mountainous regions that flank the Mediterranean Sea and continues through Iran and on past the Himalayan complex. Figure 16.14 indicates that yet another continuous belt extends for thousands of kilometers through the world's oceans. This zone coincides with the oceanic ridge system, which is an area of frequent but low-intensity seismic activity.

The areas of the United States included in the circum-Pacific belt lie adjacent to the San Andreas fault and along the western coastal regions of Alaska, including the Aleutian Islands. In addition to these high-risk areas, other sections of the United States are regarded as regions of relatively high earthquake probability (Figure 16.15). One of these regions extends from southern Illinois southward along the Mississippi River (see Box 16.2). This region is the site of three strong shocks that devastated the town New Madrid, Missouri, in 1811–1812. The drainage of the Mississippi was also altered by these quakes, and Tennessee's Reelfoot Lake was enlarged. Although the total amount of structural damage caused by the New Madrid earthquakes was slight, it should be remembered that in the early 1800s this was a sparsely populated frontier region. A similar earthquake in this region today would be truly catastrophic.

Another strong earthquake centered away from the active circum-Pacific belt occurred August 31, 1886, in Charleston, South Carolina. The event, which spanned one minute, caused 60 deaths, numerous injuries, and great economic loss within a radius of 200 kilometers (120 miles) of Charleston. Within eight minutes, the effects of the vibrations were felt as far away as Chicago, Illinois, and St. Louis, Missouri, where strong vibrations on upper floors of buildings caused people to rush outdoors. According to reports made shortly after the earthquake, not a single building in Charleston escaped some damage, and only a few buildings avoided serious damage.

The causes of earthquakes in relatively stable regions such as Charleston and New Madrid are not easily explained. Both of these locations are far from the active tectonic regions near plate boundaries where stress builds between moving sections of lithosphere. Nevertheless, a hypothesis has been proposed to explain the seismic activity in the New Madrid region. This proposal is based on recent findings that the Mississippi embayment is underlain by crust weakened by an ancient rift. Although the faults that produced this rift have been inactive for millions of years, seismic studies indicate that they have been remobilized. The stress responsible for the displacement during the 1811–1812 events is thought to be compressional with an east-west orientation. This agrees with measurements made by the

BOX 16.2

Damaging Earthquakes East of the Rockies

Whenever the topic of earthquake destruction is raised, places that are located near plate boundaries, such as California and Japan, usually come to mind. However, people who do not live near earthquake-prone plate boundaries are not necessarily immune from these destructive events. Recent estimates made by a team of seismologists indicate that the probability of a damaging earthquake occurring east of the Rocky Mountains during the next 30 years is roughly two-thirds as likely as an earthquake with comparable damage occurring in California. Like all earthquake risk assessments, this prediction is based in part on the geographic distribution and average rate of earthquake occurrences in these regions.

At least six major earthquakes have occurred in the central and eastern United States since colonial times. Three of these had estimated Richter magnitudes of 7.5, 7.3, and 7.8, and they were centered near the Mississippi River Valley in southeastern Missouri. Occurring on December 16, 1811, January 23, 1812, and February 7, 1812, these earthquakes, plus numerous smaller tremors, destroyed the town of New Madrid, Missouri, triggered massive landslides, and caused damage over a six-state area. The course of the Mississippi River was altered, and Tennessee's Reelfoot Lake was enlarged. The distances over which these earthquakes were felt are truly remarkable. Chimneys were reported downed in Cincinnati, Ohio, and Richmond, Virginia, while Boston residents, located 1770 kilometers (1100 miles) to the northeast, felt the tremor. Although the total amount of destruction caused by the New Madrid earthquake was slight compared to that caused by the Loma Prieta earthquake of 1989, it should be remembered that in the early 1800s the Midwest was a sparsely populated region. Memphis, Tennessee, which is located near the epicenter, had not yet been established, and St. Louis was but a small frontier town of a few thousand inhabitants. Furthermore, damaging earthquakes that occurred in Aurora, Illinois (1909), and Valentine, Texas (1931), remind us that many areas in the central United States are vulnerable.

The greatest historical earthquake in the eastern states occurred August 31, 1886, in Charleston, South Carolina. The event, which spanned one minute, caused 60 deaths, numerous injuries, and great economic loss within a radius of 200 kilometers (120 miles) of Charleston. Within eight minutes, effects were felt as far away as Chicago, Illinois, and St. Louis, Missouri, where strong vibrations shook the upper floors of buildings, causing people to rush outdoors. In Charleston alone over one hundred buildings were destroyed, and 90 percent of the remaining structures were damaged. It was difficult to find a chimney that was still standing. (Figure 16.B).

Numerous other strong earthquakes have been recorded in the central and eastern United States. New England and adjacent areas have experienced sizable shocks since colonial times. The first reported earthquake in the Northeast took place in Plymouth, Massachusetts, in 1683, and was followed in 1755 by the destructive Cambridge, Massachusetts, earthquake. Moreover, since records have been kept, New York State alone has experienced over 300

U.S. Geological Survey which show that most of the mid-continent is under compressional stress. It has been suggested that the stresses generated by the interaction of lithospheric pressure plates is transmitted to the interior of the plate. Thus, the source of stress for mid-plate earthquakes appears to be the same forces that cause plate motions and the earthquakes associated with plate boundaries.

EARTHQUAKE DEPTHS

Evidence from seismic records reveals that earthquakes originate at depths ranging from 5 to nearly 700 kilometers. In a somewhat arbitrary fashion, earthquake foci have been classified by their depth of occurrence. Those with points of origin within 70 kilometers of the surface are referred to as *shallow,* while those generated between 70 and 300 km are considered *intermediate,* and those with a focus greater than 300 km are classified as *deep.* About 90 percent of all earthquakes occur at depths of less than 100 km, and nearly all very damaging earthquakes appear to originate at shallow depths. For example, the 1906 San Francisco earthquake involved movement within the upper 15 km of the earth's crust, whereas the 1964 Alaskan earthquake had a focal depth of 33 km. Seismic data reveal that

earthquakes large enough to be felt.

Earthquakes in the central and eastern United States occur far less frequently than in California. Yet history indicates that the East is vulnerable. Further, the shocks that have occurred east of the Rockies have generally produced structural damage over a larger area than their counterparts of similar magnitude in California. The reason is that the underlying bedrock in the central and eastern United States is older and more rigid. As a result, seismic waves are able to travel greater distances with less attenuation of intensity than in the western United States. It is estimated that for similar earthquakes, the region of maximum ground motion in the East may be up to ten times larger than in the West. Consequently, the higher rate of earthquake occurrence in the western United States is balanced, at least partially, by the fact that earthquakes occurring in the central and eastern United States are capable of producing damage over a larger area.

Despite the recent geologic history of the surrounding region, Memphis, Tennessee, the largest population center in the area of the New Madrid earthquake, does not have adequate provisions regarding earthquakes in its building code. Further, because Memphis is located on unconsolidated floodplain deposits, buildings are more susceptible to damage than similar structures built on bedrock. It has been estimated that if an earthquake the size of the 1811–1812 New Madrid event were to strike in the next decade, it would result in casualties in the thousands and damages in tens of billions of dollars.

FIGURE 16.B
Damage to Charleston, South Carolina, during the August 31, 1886, earthquake. Damage ranged from toppled chimneys and broken plaster to total collapse. (Photo courtesy of U.S. Geological Survey)

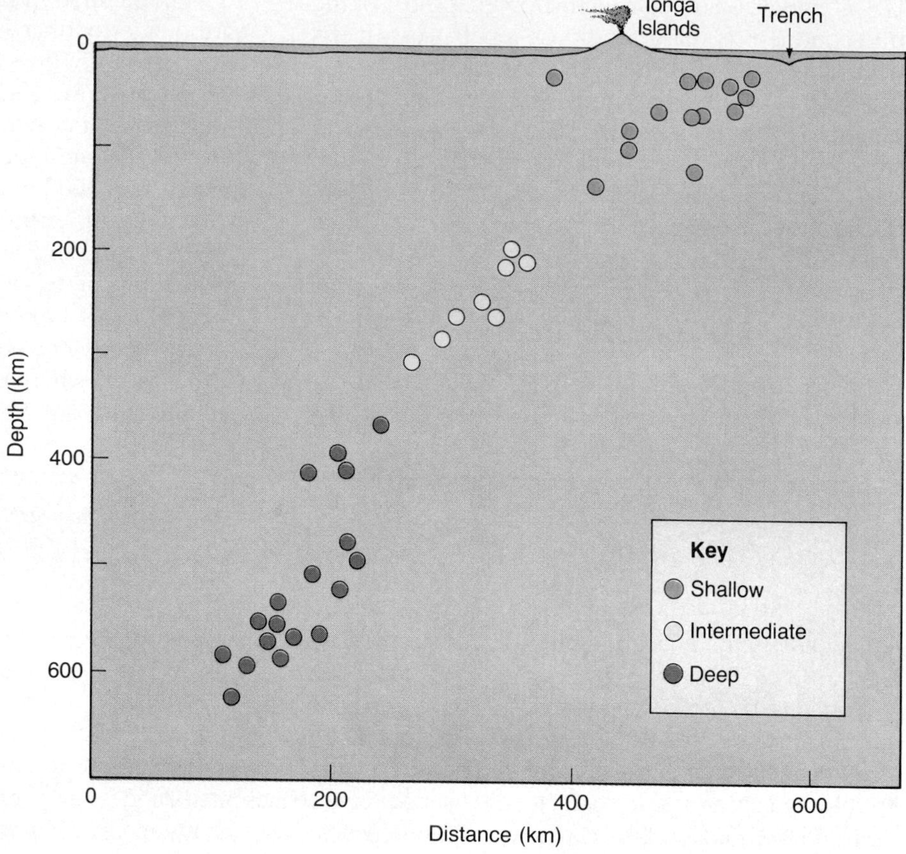

FIGURE 16.16
Distribution of earthquake foci in 1965 in the vicinity of the Tonga Islands. (Data from B. Isacks, J. Oliver, and L. R. Sykes)

while shallow-focus earthquakes have been recorded with Richter magnitudes of 8.6, the strongest intermediate-depth quakes have had values below 7.5, and deep-focus earthquakes have not exceeded 6.9 in magnitude.

When earthquake data were plotted according to geographic location and depth, several interesting observations were noted. Rather than a random mixture of shallow and deep earthquakes, some very definite distribution patterns emerged. Earthquakes generated along the oceanic ridge system always have a shallow focus and none is very strong. Further, it was noted that almost all deep-focus earthquakes occurred in the circum-Pacific belt, particularly in regions situated landward of deep-ocean trenches. In a study conducted in the southwestern Pacific near the Tonga trench by H. Benioff, it was discovered that focal depths increased with distance from the trench as shown in Figure 16.16. These seismic regions, called **Benioff zones*** after the man who extensively studied them, are oriented about 35 to nearly 90 degrees to the surface. Why should earthquakes be oriented along a narrow zone which plunges almost 700 kilometers into the earth's interior? We will consider this question further in Chapter 18.

*These zones are also called Wadati-Benioff zones to recognize the Japanese seismologist who first presented convincing evidence for the existence of deep-focus earthquakes.

EARTHQUAKE INTENSITY AND MAGNITUDE

Earthquake **intensity** is a measure of the effects of an earthquake at a particular locale. It is important to note that earthquake intensity depends not only on the strength of the earthquake but on other factors as well. These include the distance from the epicenter, the nature of the surface materials, and building design. Recall that the modest 6.9 Richter magnitude Armenian earthquake was very destructive mainly because of poor construction practices. Moreover, the nature of the surface materials contributed significantly to the destruction in central Mexico City during the 1985 quake. Early attempts to establish the intensity of an earthquake relied heavily on descriptions of the event. There was an obvious problem related to this method—people's accounts varied widely, making an accurate classification of a quake's intensity difficult. Then in 1902 a fairly reliable scale based on the amount of damage caused to various types of structures was developed by Giuseppe Mer-

TABLE 16.1
Modified Mercalli intensity scale.

I	Not felt except by a very few under specially favorable circumstances.
II	Felt only by a few persons at rest, especially on upper floors of buildings.
III	Felt quite noticeably indoors, especially on upper floors of buildings, but many people do not recognize it as an earthquake.
IV	During the day felt indoors by many, outdoors by few. Sensation like heavy truck striking building.
V	Felt by nearly everyone, many awakened. Disturbances of trees, poles, and other tall objects sometimes noticed.
VI	Felt by all; many frightened and run outdoors. Some heavy furniture moved; few instances of fallen plaster or damaged chimneys. Damage slight.
VII	Everybody runs outdoors. Damage negligible in buildings of good design and construction; slight to moderate in well-built ordinary structures; considerable in poorly built or badly designed structures.
VIII	Damage slight in specially designed structures; considerable in ordinary substantial buildings with partial collapse; great in poorly built structures. (Fall of chimneys, factory stacks, columns, monuments, walls.)
IX	Damage considerable in specially designed structures. Buildings shifted off foundations. Ground cracked conspicuously.
X	Some well-built wooden structures destroyed. Most masonry and frame structures destroyed. Ground badly cracked.
XI	Few, if any (masonry) structures remain standing. Bridges destroyed. Broad fissures in ground.
XII	Damage total. Waves seen on ground surfaces. Objects thrown upward into air.

SOURCE: U.S.Coast and Geodetic Survey.

calli. A modified form of the **Mercalli intensity scale** is presently used by the U.S. Coast and Geodetic Survey (Table 16.1).

By contrast, earthquake **magnitude** is a measure of the strength of an earthquake, or the amount of energy released during the event. Ideally, the magnitude of an earthquake can be determined from the amount of material that slides along the fault and the distance it is displaced. Even in an ideal setting such as that of the 1906 San Francisco earthquake, where the fault trace is visible and displacement can be measured from physical evidence, this method can provide only a crude estimate of the forces involved. In most earthquakes, the fault does not penetrate the surface; therefore the amount of displacement cannot be measured directly. In 1935, Charles Richter of the California Institute of Technology attempted to rank the earthquakes of southern California into groups of large, medium, and small magnitude. The system he developed determines earthquake magnitudes from the motions measured by seismic instruments.

Today a refined **Richter scale** is used worldwide to describe earthquake magnitude. Using Richter's scale, the magnitude is determined by measuring the amplitude of the largest wave recorded on the seismogram (Figure 16.17). Although seismographs greatly magnify the ground motion, large-magnitude earthquakes will cause the recording pen to be displaced farther than will small-magnitude earthquakes. In order for seismic stations worldwide to obtain the same magnitude for a given earthquake, adjustments must be made for the weakening of the seismic waves as they move from the focus, and for the sensitivity of the recording instrument. Richter established 100 kilometers as the standard distance and the Wood-Anderson instrument as the standard recording device.

The largest earthquakes ever recorded have Richter magnitudes near 8.6. These great shocks released approximately 10^{26} ergs of energy—roughly equivalent to the detonation of 1 billion tons of TNT. Apparently, earthquakes having a Richter magnitude greater than 9 do not occur. Conversely, earthquakes with a Richter magnitude of less than 2.0 are usually not felt by humans. With the advent of more sensitive instruments, tremors of a magnitude of <2 have been recorded. Table 16.2 shows how earthquake magnitudes and their effects are related.

As we have seen, earthquakes vary enormously in strength. Furthermore, great earthquakes produce traces having wave amplitudes that are thousands of times larger than those generated by weak tremors. To accommodate this wide variation, Richter used a logarithmic scale to express magnitude. On this scale a tenfold increase in wave amplitude corresponds to an increase of one on the magnitude scale. Thus, the amplitude of the largest surface wave for a 5-magnitude earthquake is 10 times greater than the wave amplitude produced by an earthquake having a magnitude of 4. Further, each unit of magnitude increase on the Richter scale equates to roughly a 32-fold increase in the energy released. Thus, an earthquake with a magnitude of 6.5 releases 32 times more energy than one with a magnitude of 5.5, and roughly 1000 times more energy than a 4.5-magnitude quake. A major earthquake with a magnitude of 8.5 releases millions of times more energy than the smallest earthquakes felt by humans. This should dispel the notion that a moderate earthquake decreases the

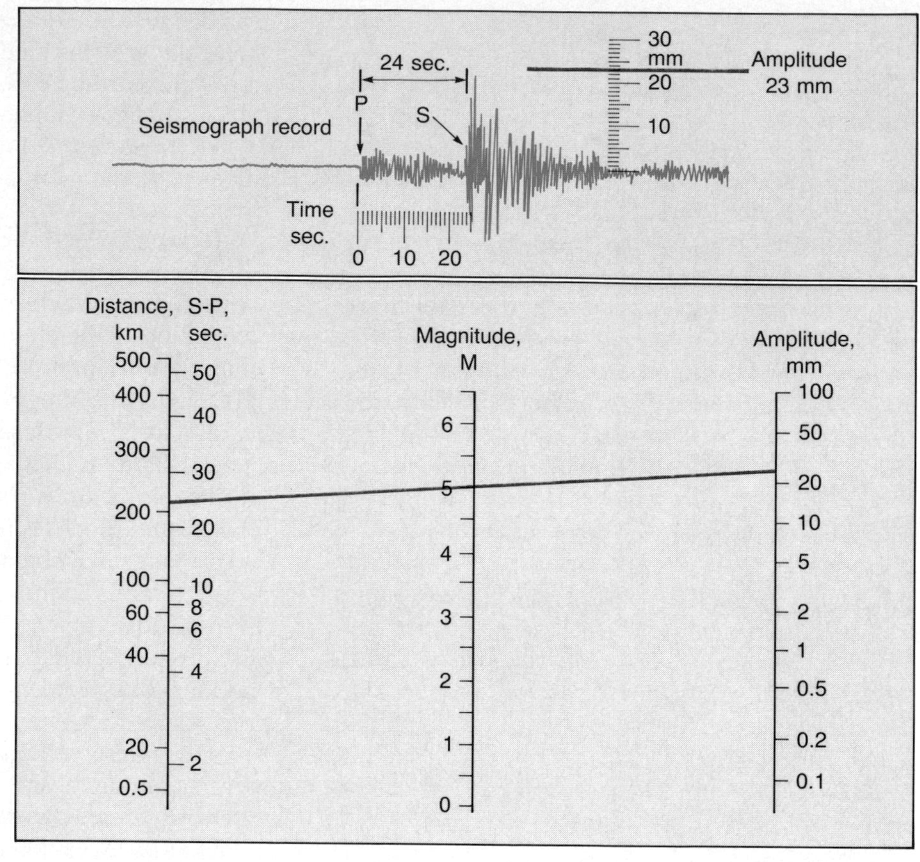

FIGURE 16.17
Illustration showing how the Richter magnitude of an earthquake can be determined graphically using a seismograph record from a Wood-Anderson instrument. (Data from California Institute of Technology)

chances for the occurrence of a major quake in the same region. Thousands of moderate tremors would be needed to release the amount of energy equal to one "great" earthquake. Some of the world's major earthquakes and their corresponding Richter magnitudes are listed in Table 16.3.

Recent studies have shown that the Richter scale does not adequately differentiate those earthquakes which have very high magnitudes. Since all of the very strongest quakes have nearly equal wave amplitudes, the Richter scale becomes saturated at this level. Consequently, methods to extend the Richter scale so that it can accurately measure the magnitudes of the largest earthquakes have been pro-

TABLE 16.2
Earthquake magnitudes and expected world incidence.

Richter Magnitudes	Earthquake Effects	Estimated Number per Year
<2.0	Generally not felt, but recorded.	600,000
2.0–2.9	Potentially perceptible.	300,000
3.0–3.9	Felt by some.	49,000
4.0–4.9	Felt by most.	6200
5.0–5.9	Damaging shocks.	800
6.0–6.9	Destructive in populous regions.	266
7.0–7.9	Major earthquakes. Inflict serious damage.	18
≥8.0	Great earthquakes. Produce total destruction to communities near epicenter.	1.4

SOURCE: *Earthquake Information Bulletin* and others.

TABLE 16.3
Some notable earthquakes.

Year	Location	Deaths (est.)	Magnitude	Comments
1290	Chihli (Hopei), China	100,000		
1556	Shensi, China	830,000		Possibly the greatest natural disaster.
1737	Calcutta, India	300,000		
1755	Lisbon, Portugal	70,000		Tsunami damage extensive.
*1811–1812	New Madrid, Missouri	Few		Three major earthquakes.
*1886	Charleston, South Carolina	60		Greatest historical earthquake in the eastern United States.
*1906	San Francisco, California	1500	8.1–8.2	Fires caused extensive damage.
1908	Messina, Italy	120,000		
1920	Kansu, China	180,000		
1923	Tokyo, Japan	143,000	7.9	Fire caused extensive destruction.
1960	Southern Chile	5700	8.5–8.6	Possibly the largest-magnitude earthquake ever recorded.
*1964	Alaska	131	8.3–8.4	
1970	Peru	66,000	7.8	Great rockslide.
*1971	San Fernando, California	65	6.5	Damage exceeded $1 billion.
1975	Liaoning Province, China	Few	7.5	First major earthquake to be predicted.
1976	Tangshan, China	240,000	7.6	Not predicted.
1985	Mexico City	9500	8.1	Major damage occurred 400 km from epicenter.
1988	Soviet Armenia	25,000	6.9	Poor construction practices contributed to destruction.
*1989	San Francisco Bay area	62	7.1	Damages exceeded $6 billion.
1990	Northwestern Iran	50,000	7.3	Landslides and poor construction practices caused great damage.

*U.S. earthquakes.
SOURCE: U.S. National Oceanic and Atmospheric Administration.

posed. One method analyzes very long-period seismic waves for this purpose. On this extended scale, the 1906 San Francisco earthquake with a surface wave magnitude of 8.3 would be demoted to 7.9, whereas the 1964 Alaskan earthquake with an 8.4–8.6 magnitude would be increased to 9.2. Using this system, the strongest earthquake on record is the 1960 Chilean earthquake with a magnitude of 9.5

EARTHQUAKE DESTRUCTION

The most violent earthquake to jar North America this century—the Good Friday Alaskan earthquake—occurred at 5:36 P.M. on March 27, 1964. Felt throughout that state, the earthquake had a magnitude of 8.4–8.6 on the Richter scale and reportedly lasted 3 to 4 minutes. This brief event left 131 persons dead, thousands homeless, and the economy of the state badly disrupted. Had the schools and business districts been open, the toll surely would have been higher. Within 24 hours of the initial shock, 28 aftershocks were recorded, 10 of which exceeded a magnitude of 6 on the Richter scale. The location of the epicenter and the towns that were hardest hit by the quake are shown in Figure 16.18.

Many factors determine the amount of destruction that will accompany an earthquake. The most obvious of these are the magnitude of the earthquake and the proximity of the quake to a populated area. Fortunately, most earthquakes are small and occur in remote regions of the earth. However, about 20 major earthquakes are reported annually, one or two of which can be catastrophic.

Ordinarily during an earthquake, the region within 20 to 50 kilometers (12.5 to 30 miles) of the

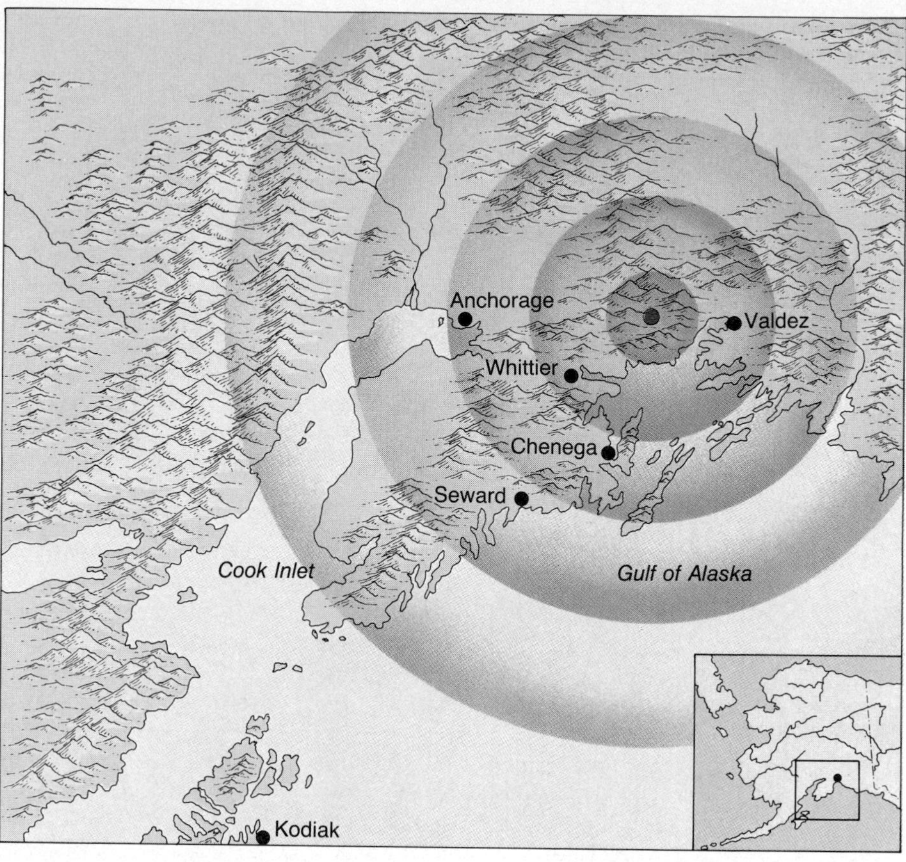

FIGURE 16.18
Region most affected by the Good Friday earthquake of 1964. Note the location of the epicenter (red dot). (After U.S. Geological Survey)

fault will experience roughly the same degree of ground shaking, but beyond this limit the vibration deteriorates rapidly. Occasionally, during earthquakes that occur in the stable continental interior, such as the New Madrid earthquake of 1811, the area of influence can be much larger. The epicenter of this earthquake was located directly south of Cairo, Illinois, and the vibrations were felt from the Gulf of Mexico to Canada, and from the Rockies to the Atlantic seaboard.

Destruction Caused by Seismic Vibrations

The 1964 Alaskan earthquake provided geologists with new insights into the role of ground shaking as a destructive force. As the energy released by an earthquake travels along the earth's surface, it causes the ground to vibrate in a complex manner by moving up and down as well as from side to side. The amount of structural damage attributable to the vibrations depends on several factors, including (1) the intensity and duration of the vibrations; (2) the nature of the material upon which the structure rests; and (3) the design of the structure.

All of the multistory structures in Anchorage were damaged by the vibrations; the more flexible wood frame residential buildings fared best. However, many homes were destroyed when the ground failed. A striking example of how construction variations affect earthquake damage is shown in Figure 16.19. We can see in this photo that the steel-frame building on the left withstood the vibrations, whereas the relatively rigid concrete structure was badly damaged.

Most of the large structures in Anchorage were damaged even though they were built to conform to the earthquake provisions of the Uniform Building Code of California. Perhaps some of that destruction can be attributed to the unusually long duration of this earthquake, which was estimated at 3 to 4 minutes. Most earthquakes consist of tremors that last less than one minute. For example, the San Francisco earthquake of 1906 was felt for about 40 seconds, whereas the strong vibrations of the 1989 Loma Prieta earthquake lasted less than 15 seconds.

Although the region within 20 to 50 kilometers of the epicenter will experience about the same intensity of ground shaking, the destruction varies considerably within this area. This difference is mainly attributable to the nature of the ground on which the structures are built. Soft sediments, for example, generally amplify the vibration more than solid bedrock. Thus, the buildings located in Anchorage, which were situated on unconsolidated sediments, experienced heavy structural damage. By contrast, most of the town of Whittier, although located much nearer to the epicenter than Anchorage, rests on a firm foundation of granite and hence suffered much less damage from the seismic vibrations. However, Whittier was damaged by a seismic sea wave.

The 1985 Mexican earthquake gave seismologists and engineers a vivid reminder of what had been learned following the 1964 Alaskan earthquake. The Mexican coast, where the earthquake was centered, experienced unusually mild tremors despite the strength of the quake. As expected, the seismic waves became progressively weaker with increasing distance from the epicenter. However, in the central section of Mexico City, nearly 400 kilometers from the source, the vibrations intensified to five times that experienced in outlying districts. Much of this amplified ground motion can be attributed to soft sediments, remnants of an ancient lake bed, that underlie portions of the city.

To understand what happened in Mexico City, recall that as seismic waves pass through the earth, they cause the intervening material to vibrate much as a tuning fork vibrates when it is struck. Although most objects can be forced to vibrate over a wide range of frequencies, each has a natural period of vibration that is preferred. Different earth materials, like different-length tuning forks, also have different natural periods of vibration.*

*To demonstrate the natural period of vibration of an object, hold a ruler over the edge of a desk so that most of it is not supported by the desk. Start it vibrating and notice the noise it makes. By changing the length of the unsupported portion of the ruler, the natural period of vibration will change accordingly.

FIGURE 16.19
Damage caused to the five-story J. C. Penney Co. building, Anchorage, Alaska. Very little structural damage was incurred by the adjacent building. (Courtesy of NOAA)

FIGURE 16.20
During the 1985 Mexican earthquake, multistory buildings swayed back and forth as much as one meter. Many, including the hotel shown here, collapsed or were seriously damaged. (Photo by James L. Beck)

The key to ground motion amplification is to match the natural period of ground vibration with that of the seismic waves. A common example of this phenomenon occurs when a parent pushes a child on a swing. When the parent periodically pushes the child in rhythm with the frequency of the swing, the child moves back and forth in a greater and greater arc. By chance, the column of sediment beneath Mexico City had a natural period of vibration of about two seconds, matching that of the strongest seismic waves. Thus, when the seismic waves began shaking the soft sediments, a resonance developed which greatly increased the amplitude of the vibrations. This amplified motion began throwing the ground back and forth about 40 centimeters every two seconds for a period of nearly two minutes. That shaking was too intense for many of the poorly designed buildings in the city. In addition, intermediate-height buildings (5 to 15 stories) sway back and forth with a period of about two seconds. Thus, resonance also developed between these buildings and the ground, and most of the building failures were in this height range (Figure 16.20).

In areas where unconsolidated materials are saturated with water, earthquakes can generate a phenomenon known as **liquefaction.** Under these conditions, what had been a stable soil turns into a fluid that is not capable of supporting buildings or other structures (Figure 16.21). As a result, underground objects such as storage tanks and sewer lines may literally float toward the surface, while buildings and other structures may settle and collapse. In San Francisco's Marina District, foundations failed and geysers of sand and water shot from the ground, indicating that liquefaction occurred during the Loma Prieta earthquake of 1989 (Figure 16.22).

Poor construction practices were blamed for the deaths of an estimated 25,000 persons in the 1988

FIGURE 16.21
Effects of liquefaction. This tilted building rests on unconsolidated sediment that imitated quicksand during the 1985 Mexican earthquake. (Photo by James L. Beck)

earthquake that jolted Soviet Armenia. To quote a seismologist who was asked about this destructive event, "Earthquakes don't kill people, buildings do." Many of the taller structures in the region were made of precast concrete slabs that were held together by metal hooks. A San Francisco structural engineer studying the destruction stated, "There was very little reinforcing to tie these buildings together. The buildings basically came apart the way they were put together." In addition, many of the village homes had thick roofs made of mud and rock, which were deadly when they collapsed. Nearly the entire town of Spitak, which was located adjacent to the fault break, was leveled. Not far away, in northwestern Iran, poor building construction was also responsible for a large share of the approximately 50,000 deaths that resulted from an earthquake in June, 1990. The earthquake had a Richter magnitude of at least 7.3. When it struck, thousands of mud brick and concrete buildings were reduced to rubble. In addition to the large death toll, an estimated one-half million people were left homeless.

The effects of great earthquakes may be felt thousands of kilometers from their source. Ground motion may generate *seiches,* the rhythmic sloshing of water in lakes, reservoirs, and enclosed basins such as the Gulf of Mexico. The 1964 Alaskan earthquake, for example, generated 2-meter waves off the coast of Texas, which damaged small craft while much smaller waves were noticed in swimming pools in both Texas and Louisiana. Seiches can be particularly dangerous when they occur in reservoirs retained by earthen dams. These waves have been known to slosh over reservoir walls and weaken the structure, thereby endangering the lives of those downstream. Another response to these large-scale vibrations is the fluctuation of water levels in wells. During this earthquake, numerous wells along a belt from South Dakota to Georgia were affected. The largest reported fluctuation in water level was 3.5 meters.

FIGURE 16.22
This "mud volcano" was one of several produced by the Loma Prieta earthquake of 1989. These structures were produced when geysers of sand and water shot from the ground, an indication that liquefaction occurred. (Photo by Richard Hilton, courtesy of Dennis Fox)

Tsunamis

Most of the deaths associated with the 1964 Alaskan quake were caused by **seismic sea waves,** or **tsunamis.*** These destructive waves have popularly been called "tidal waves." However, this name is not accurate since these waves are not generated by the tidal effect of the moon or sun.

Most tsunamis result from the vertical displacement of the ocean floor during an earthquake as illustrated in Figure 16.23. Once formed, a tsunami resembles the ripples formed when a pebble is dropped into a pond. In contrast to ripples, tsunamis advance at speeds between 500 and 950 kilometers per hour. Despite this striking characteristic, a tsunami in the open ocean can pass undetected because its height is usually less than one meter and the distance between wave crests ranges from 100 to 700 kilometers. However, upon entering shallower coastal waters, these destructive waves are slowed down and the water begins to pile up to heights that occasionally exceed 30 meters (Figure 16.23). As the crest of a tsunami approaches the shore, it appears as a rapid rise in sea level with a turbulent and chaotic surface.

Usually the first warning of an approaching tsunami is a rather rapid withdrawal of water away from beaches. Residents of coastal areas have learned to heed this warning and move to higher ground. About 5 to 30 minutes later the retreat of water is followed by a surge capable of extending hundreds of meters inland. In a successive fashion, each surge is followed by rapid oceanward retreat of the water.

*Seismic sea waves were given the name *tsunami* by the Japanese who have suffered a great deal from them. The term *tsunami* is now used worldwide.

Tsunamis are able to traverse large stretches of the ocean before their energy is totally dissipated. The tsunami generated by the 1960 Chilean earthquake, in addition to completely destroying villages along an 800-kilometer stretch of coastal South America, traveled 17,000 kilometers across the Pacific to Japan. Here, about 22 hours after the quake, considerable destruction was inflicted upon southern coastal villages of Honshu, the major island of Japan. For several days afterward, tidal gauges located in Hilo, Hawaii, were able to detect these diminishing waves as they bounced about the Pacific.

The tsunami generated in the 1964 Alaskan earthquake inflicted heavy damage to the communities in the vicinity of the Gulf of Alaska, completely destroying the town of Chenega. Kodiak was also heavily damaged and most of its fishing fleet destroyed when a seismic sea wave carried many vessels into the business district (Figure 16.24). The deaths of 107 persons have been attributed to this tsunami. By contrast, only 9 persons died in Anchorage as a direct result of the vibrations. Tsunami damage following the Alaskan earthquake extended along much of the west coast of North America, and in spite of a 1-hour warning, 12 persons perished in Crescent City, California, where all of the deaths and most of the destruction were caused by the fifth wave. The first wave crested about 4 meters (13 feet) above low tide

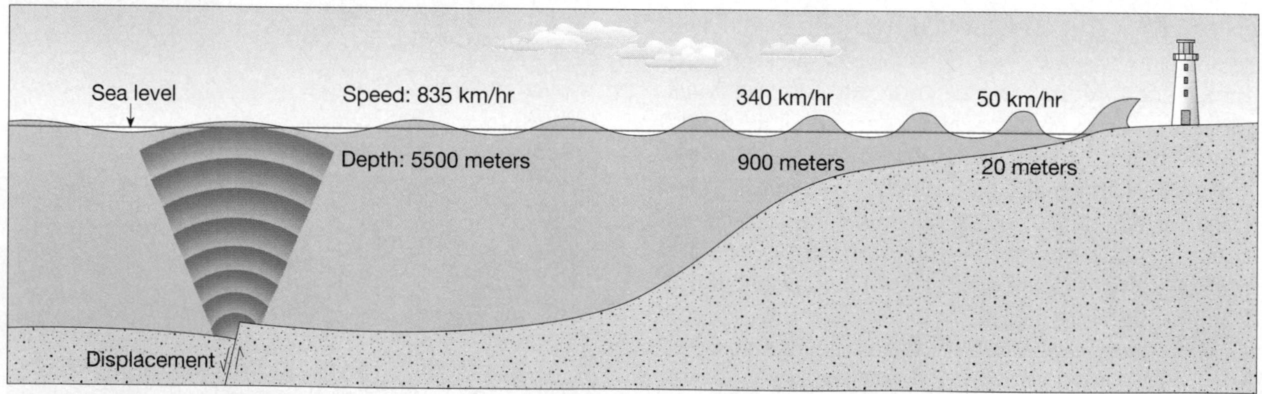

FIGURE 16.23
Schematic drawing of a tsunami generated by displacement of the ocean floor. The speed of a wave column correlates with ocean depth. As shown, waves moving in deep water advance at speeds in excess of 800 kilometers per hour. Speed gradually slows to 50 kilometers per hour at depths of 20 meters. Decreasing depth slows the movement of the wave column. As waves slow in shallow water, they grow in height until they topple and rush onto shore with tremendous force. The size and spacing of these swells are not to scale.

FIGURE 16.24
A tsunami washed this fishing fleet into the heart of the village of Kodiak, Alaska. (Photo by W. R. Hansen, U.S. Geological Survey)

and was followed by three progressively smaller waves. Believing that the tsunami had ceased, people returned to the shore, only to be met by the fifth and most devastating wave, which, superimposed upon high tide, crested about 6 meters higher than the level of low tide.

Although most tsunamis are generated by earthquakes, a volcanic eruption in the ocean can gener-

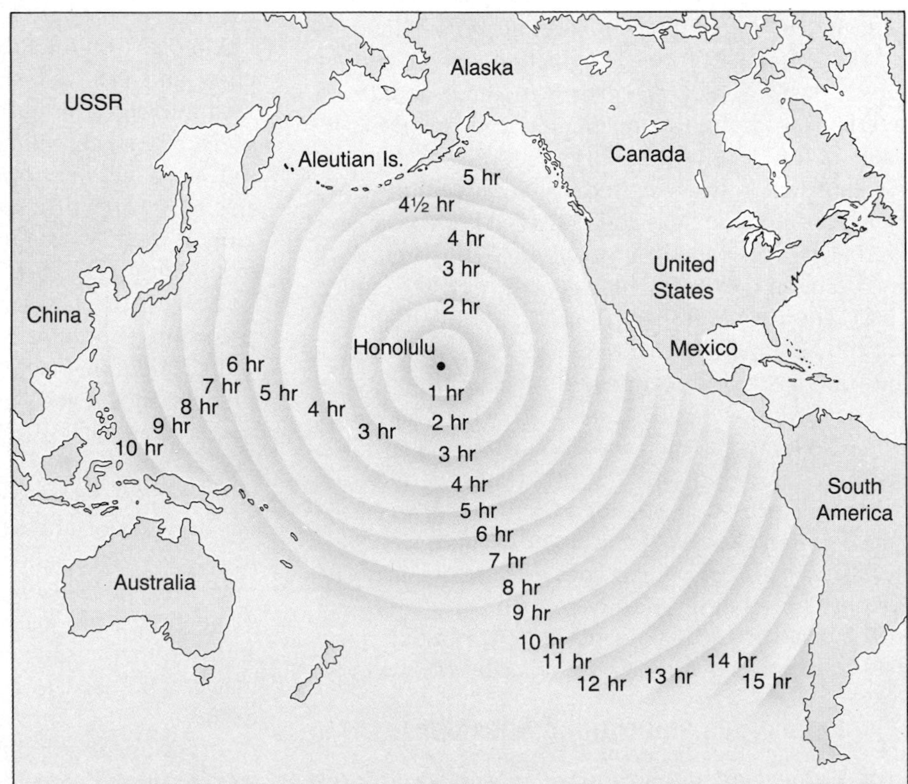

FIGURE 16.25
Tsunami travel times to Honolulu, Hawaii, from locations throughout the Pacific. (From NOAA)

ate this destructive phenomenon as well. For example, the 1883 volcanic explosion of Krakatoa generated a tsunami that drowned an estimated 36,000 coastal residents of Java and Sumatra.

In 1946, a large tsunami struck the Hawaiian Islands without warning. A wave more than 15 meters high left several coastal villages in shambles. This destruction motivated the United States Coast and Geodetic Survey to establish a tsunami warning system for the coastal areas of the Pacific. From seismic observatories throughout the area, warnings of large earthquakes are reported to the Tsunami Warning Center near Honolulu. Using tidal gauges, a determination is made as to whether a tsunami has been formed. Within an hour a warning is issued. Although tsunamis travel very rapidly, there is sufficient time to evacuate all but the region nearest the epicenter (Figure 16.25). For example, a tsunami generated near the Aleutian Islands would take 5 hours to reach Hawaii, and one generated near the coast of Chile would travel 15 hours before reaching Hawaii. Fortunately, most earthquakes do not generate tsunamis. On the average, only about 1.5 destructive tsunamis are generated worldwide each year. Of these, only about one every ten years is catastrophic.

Fire

Fire caused only minor damage during the 1989 Loma Prieta earthquake, but sometimes it is the most destructive force. The 1906 earthquake centered near the city of San Francisco reminds us of the formidable threat of fire. The central city contained mostly large, older wooden structures and brick buildings. Although extensive damage was done to many of the unreinforced brick buildings, the greatest destruction was caused by innumerable fires which started when gas and electrical lines were severed. The fires raged out of control for three days and devastated over 500 blocks of the city (see Figure 16.2). The problem was compounded by the initial ground shaking which broke the city's water lines into hundreds of unconnected pieces.

The fire was finally contained when buildings were dynamited along a wide boulevard to provide a fire break. Although only a few deaths were attributed to the fires, that is not always the case. An earthquake that rocked Japan in 1923 triggered an estimated 250 fires, which devastated the city of Yokohama and destroyed more than half the homes in Tokyo. Over 100,000 deaths were attributed to the fires which were driven by unusually high winds.

Landslides and Ground Subsidence

In the 1964 Alaskan earthquake, it was not ground vibrations directly but landslides and ground subsidence triggered by the vibrations that caused the greatest damage to structures. At Valdez and Seward the violent shaking caused deltaic materials to experience liquefaction; the subsequent slumping carried both water-fronts away. Because of the threat of recurrence, the entire town of Valdez was relocated about 7 kilometers away in a region of stable ground. The destruction at Valdez was compounded by the tragic loss of 31 lives. While waiting for an incoming vessel, the 31 persons died when the dock slid into the sea.

Most of the damage in the city of Anchorage was also attributed to landslides caused by the shaking and lurching ground. Many homes were destroyed in Turnagain Heights when a layer of clay lost its strength and carried over 200 acres of land toward the ocean (Figure 16.26). A portion of this spectacular landslide was left in its natural condition as a reminder of this destructive event. The site was appropriately named "Earthquake Park." Downtown Anchorage was also disrupted as sections of the main business district dropped by as much as 3 meters (10 feet).

EARTHQUAKE PREDICTION AND CONTROL

The vibrations that shook the San Francisco Bay Area during rush hour on October 17, 1989, inflicted 62 deaths and about $6 billion in damages, all from an earthquake that lasted less than 15 seconds and had a rating of 7.1 on the Richter scale. Although San Francisco and Oakland suffered considerable damage, seismologists warn that quakes of comparable or greater strength will someday be centered beneath the more heavily populated parts of the San Francisco Bay Area. These earthquakes will be far less kind to the Bay Area unless ways to reduce the risks from earthquakes are forthcoming.

Because the risks are high, substantial research programs are underway in Japan, the United States, China, and the Soviet Union. These projects strive to identify those geologic phenomena that precede major earthquakes. In Japan, for example, researchers have established a complex seismic network extending 200 kilometers (125 miles) out into the ocean. Here on the ocean floor where background noise is slight, the Japanese plan to monitor microearthquakes, which are known to precede some great earthquakes. Seismologists hope that by studying this seismic activity, they will find a pattern that can be used to accurately predict forthcoming tremors.

In California, uplift or subsidence of the land, changes in the movements along a fault zone, and a period of seismic quiescence often followed by re-

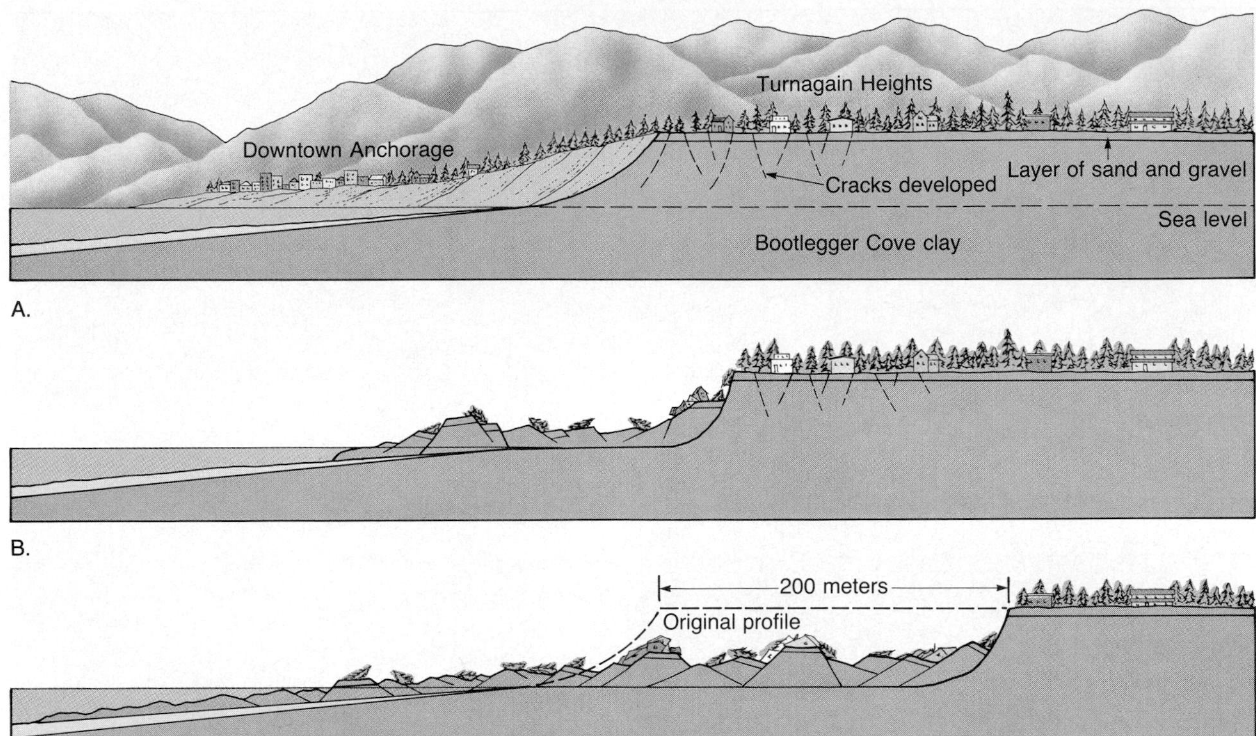

FIGURE 16.26
Turnagain Heights slide caused by the 1964 Alaskan earthquake. **A.** Vibrations from the earthquake caused cracks to appear near the edge of the bluff. **B.** Within seconds blocks of land began to slide toward the sea upon a weak layer of clay. **C.** In less than five minutes, as much as 200 meters of the Turnagain Heights bluff area had been destroyed. **D.** Photo of a small portion of the Turnagain Heights slide. (Photo courtesy of U.S. Geological Survey)

newed activity have been found to precede some earthquakes. It therefore seems reasonable that earthquakes may be predicted by continually monitoring ground tilt, fault movement, and seismic activity. Perhaps the most ambitious earthquake prediction experiment of this type is underway along a segment of the San Andreas fault near the town of Parkfield in central California. Here, earthquakes of moderate intensity have occurred on a regular basis about once every 22 years since 1857. The most recent rupture was a 5.6-magnitude quake that occurred in 1966. With the next event already overdue, the U.S. Geological Survey has established an elaborate monitoring network. Included are creepmeters, tiltmeters, and bore hole strain meters that are used to measure the accumulation and release of strain. Moreover, 70 seismometers of various designs have been installed to record foreshocks as well as the main event. Finally, a network of distance-measuring devices that employ lasers measures movement across the fault. The object is to identify ground displacements that may precede a sizable rupture.

Although no reliable method of short-range prediction has yet been devised, a few successful predictions have reportedly been made. Most notable was the prediction that foretold the 1975 earthquake in the Lianoning Province of China. Here, for the first and only time, seismologists succeeded in forecasting a large earthquake that was about to destroy a major city. The evacuation of an estimated 3 million residents from unreinforced masonry structures saved tens of thousands of lives. Western observers confirmed Chinese reports that almost 90 percent of the structures in the city of Haicheng were heavily damaged. Although this event was predicted months earlier, swarms of small earthquakes that preceded

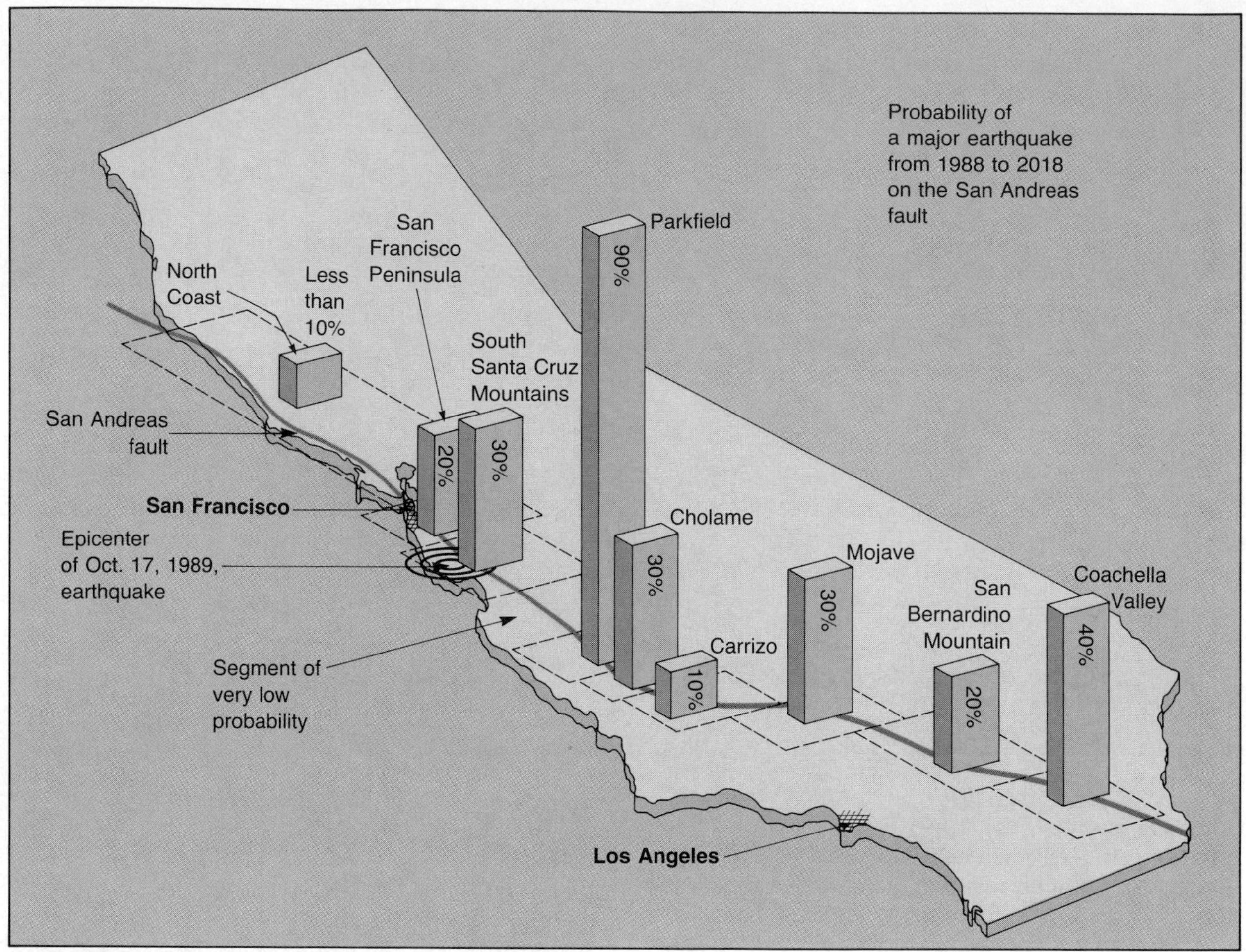

FIGURE 16.27
Probability of a major earthquake from 1988 to 2018 on the San Andreas fault.

this earthquake undoubtedly aided the prediction and also prompted the citizenry to heed the warning.

Unfortunately, the Chinese were unable to pinpoint the date of the great Tangshan earthquake of 1976. Their long-range warning of a forthcoming earthquake was not sufficiently precise to prevent the loss of an estimated 240,000 lives. The Chinese have also issued false alarms. In a province near Hong Kong, people left their dwellings for over a month, but no earthquake followed. The debate that would precede an order to evacuate a large city in the United States, such as Los Angeles or San Francisco, would be considerable. The cost of evacuating millions of people, to say nothing of lost work time, and the innumerable other problems associated with an evacuation would have to be weighed against the earthquake's probability.

What constitutes success or failure for earthquake predictions? Most experts agree that a successful earthquake prediction must specify the geographic area involved, the expected magnitude, and the probable time of occurrence. As you might expect, researchers have found that establishing a time for the event is the most elusive factor. This is particularly true if the forecasts attempt to establish the time of a particular quake within the relatively narrow period of a few days, or at most a few weeks. On the other hand, substantial progress has been made in producing long-range earthquake predictions.

Long-term forecasts are based on the premise that earthquakes are repetitive. In other words, as soon as one earthquake is over, the continuing motion of the earth's plates begins to add strain to the rocks until they fail again. This fact has led seismologists to conclude that by studying the history of earthquakes, they can predict their occurrences. One study conducted by the U.S. Geological Survey gives the probability of a rupture along various segments of the San Andreas fault for the 30 years from 1988 to 2018 (Figure 16.27). From this study, the October 17, 1989,

BOX 16.3

Long-Term Earthquake Predictions

Most long-term earthquake forecasts are based on the premise that earthquakes are cyclical events. In other words, strain is continually building along active faults until rupture occurs; immediately thereafter, the accumulation of strain begins anew. With this concept in mind, a group of seismologists plotted the distribution of rupture zones associated with great earthquakes that have occurred in the seismically active regions of the Pacific basin. The maps revealed that individual rupture zones tended to occur adjacent to one another without appreciable overlap, thereby tracing out a plate boundary. Recall that most earthquakes are generated along plate boundaries by the relative motions of large crustal blocks. Because plates are in constant motion, the researchers predicted that over a span of one or two centuries, major earthquakes would occur along each segment of the Pacific plate boundary. When the researchers studied historical records, they discovered that some rupture zones had not produced a great earthquake in more than a century. These quiet zones, called *seismic gaps,* were identified as probable sites for major earthquakes in the next few decades (Figure 16.C). In the 20 years since the original studies were conducted, some of these gaps have ruptured. Included in this group is the zone that produced the earthquake which devastated portions of Mexico City in September, 1985. Hence, some long-range predictions have already been fulfilled.

Recently, another method of long-term prediction, known as *paleoseismology,* has been implemented. This technique involves the study of layered deposits that were offset by prehistoric seismic disturbances. To date, the most complete investigation that used this method focused on a segment of the San Andreas fault about 50 kilometers northeast of Los Angeles. Here the drainage of Pallet Creek has been repeatedly disturbed by successive ruptures along the fault zone. Ditches excavated across the creek bed have exposed sediments that had apparently been displaced by nine large earthquakes over a period of 1400 years. From these data it was determined that a great earthquake occurs here on an average of once every 140 to 150 years. The last major event occurred along this segment of the San Andreas fault in 1857. Thus, roughly 130 years have elapsed. If earthquakes are truly cyclic, a major event in southern California seems imminent. Such information led the U.S. Geological Survey to predict that there is a 50 percent probability that an earthquake of magnitude 8.3 will occur along the southern San Andreas fault within the next 30 years.

It now appears that the best prospects for making reasonably accurate long-term predictions involve forecasting earthquake magnitudes and locations, but with times of occurrence stated in terms of years or perhaps even decades. Although they will not be as precise as we would like, such forecasts will nevertheless be important because the information can be used to implement appropriate building codes and land-use planning.

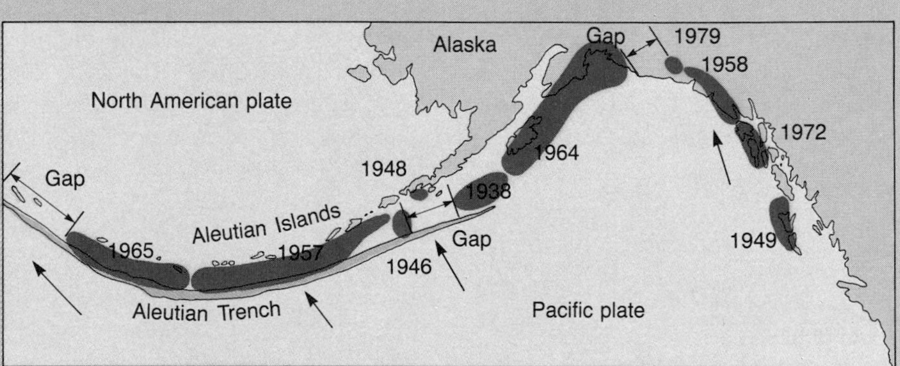

FIGURE 16.C
The distribution of rupture areas of large, shallow earthquakes from 1930 to 1979 along the southwestern coast of Alaska and the Aleutian Islands. The three seismic gaps denote the most likely locations for the next large earthquakes along this plate boundary. (After J. C. Savage et al., U.S. Geological Survey)

Loma Prieta earthquake in the Santa Cruz Mountains was given a 30 percent probability of producing a 6.5-magnitude earthquake during this time period. The region given the highest probability (90 percent) of generating a quake is the Parkfield section. This area has been called the "Old Faithful" of earthquake zones because activity here has been very regular for many years. One section of the San Andreas fault between Parkfield and the Santa Cruz Mountains is given a very low probability of generating an earthquake. This area has experienced very little seismic activity in historical times and is thought to exhibit a slow, continual movement known as *creep.*

A significant concern for many Californians is the occurrence of the next "Big One." Although seismologists are not willing to make a formal prediction, most point to southern California as the most likely location. The reason is that southern California has not had a major earthquake since 1857.

In addition to the attention given to California, new concerns are being brought to the remainder of the nation as well. For example, Charleston, South Carolina, was hit by a major earthquake in 1886 and could be a site for renewed activity. In addition, three major earthquakes struck the New Madrid, Missouri, region in 1811–1812. Although this area, near the confluence of the Mississippi and Ohio rivers, was sparsely populated in the early nineteenth century, this is no longer the case. This portion of the Mississippi Valley now has significant agricultural and urban development with Memphis, Tennessee, being the major city in the region. A 1985 federal study concluded that a 7.6-magnitude earthquake in this area could cause an estimated 2500 deaths, collapse 3000 structures, cause $25 billion in damages, and displace a quarter of a million people in Memphis alone.

The actual control of earthquakes is another matter altogether. The discovery that humans have inadvertently triggered earthquakes has given earth scientists some encouragement. The most convincing evidence that people can initiate earthquakes came between 1962 and 1966, when studies of the seismic activity at the Rocky Mountain Arsenal near Denver were conducted. For a period of 80 years prior to 1962, the U.S. Coast and Geodetic Survey reported no significant earthquake activity in the Denver region. In 1962 the arsenal began disposing wastes from its chemical warfare production into a well over 3600 meters deep. During the period of fluid waste injection, from April, 1962, to September, 1965, about 700 microearthquakes were reported, 75 intense enough to be felt. The injection of water under pressure is believed to have "lubricated" the fault, which had been building up strain over the years. This lubricating effect is not one of making rocks along the fault zone slippery. Rather, the water exerts an outward force that is directed perpendicular to the fault line. The induced outward force opposes the natural inward force caused by the weight of rock piled above. When the injection was halted for about a year, a marked drop in seismic activity was also detected. When pumping resumed, the frequency of tremors increased markedly.

Other earthquakes triggered by human activity have occurred in regions adjacent to large reservoirs such as Lake Mead on the Arizona-Nevada border. Since Lake Mead was filled in 1936, hundreds of small tremors have been recorded. They are thought to have been caused by the added weight of the lake, and perhaps aided by the "lubricating" effect of water seeping into the rock below. Another large reservoir in India is believed to be responsible for triggering a disastrous earthquake in which 200 persons were killed. Underground nuclear explosions have also been responsible for initiating numerous small aftershocks, although none has been as great as the explosion itself.

The hope of many scientists is that we may someday be able to reduce the threat of earthquakes by triggering numerous small earthquakes using fluid injections or nuclear explosions. Such methods would slowly and continually release the elastic strain that might otherwise build up and be released as a high-magnitude earthquake. Recall, however, that many thousands of minor tremors are required to equal the energy released by one strong earthquake. This fact coupled with the inaccessibility of many fault zones makes the possibility of earthquake control not nearly as feasible as it first appears.

REVIEW QUESTIONS

1. What is an earthquake? Under what circumstances do earthquakes occur?
2. How are faults, foci, and epicenters associated?
3. Earthquakes occur only in the rigid lithosphere, not in the plastic asthenosphere. Using the elastic rebound idea, explain this phenomenon.
4. Faults that are experiencing no active creep may be considered "safe." Rebut or defend this statement.
5. Describe the principle of a seismograph.
6. Contrast the motion produced by P waves with the movements created by S waves.
7. P waves move through solids, liquids, and gases, whereas S waves move only through solids. Explain.
8. Which type of seismic wave causes the greatest destruction to buildings?
9. Using Figure 16.12, determine the distance between an earthquake and a seismic station if the first S wave arrives 3 minutes after the first P wave.
10. Most strong earthquakes occur in a zone on the globe known as the ______________.
11. Deep-focus earthquakes occur several hundred kilometers below what prominent feature of the deep-ocean floor?
12. How does earthquake magnitude differ from earthquake intensity?
13. For each increase of one on the Richter scale, wave amplitude increases _____ times.
14. An earthquake measuring 7 on the Richter scale releases about _____ times more energy than an earthquake with a magnitude of 6.
15. List three factors that affect the amount of destruction caused by seismic vibrations.
16. What factor contributed most to the extensive damage that occurred in the central portion of Mexico City during the 1985 earthquake?
17. The 1988 Armenian earthquake had a Richter magnitude of 6.9, far less than the magnitude of the great quakes in Alaska in 1964 and San Francisco in 1906. Nevertheless, the loss of life was greater in the Armenian event. Why?
18. In addition to the destruction created directly by seismic vibrations, list three other types of destruction associated with earthquakes.

KEY TERMS

aftershock (p. 405)
Benioff zone (p. 414)
body wave (p. 408)
earthquake (p. 403)
elastic rebound (p. 404)
epicenter (p. 409)
fault (p. 403)
focus (p. 403)
foreshock (p. 405)
inertia (p. 406)
intensity (p. 414)
liquefaction (p. 420)
long (L) waves (p. 409)
magnitude (p. 415)
Mercalli intensity scale (p. 415)
primary (P) waves (p. 408)
Richter scale (p. 415)
secondary (S) waves (p. 408)
seismic sea wave (p. 422)
seismogram (p. 407)
seismograph (p. 406)
seismology (p. 405)
surface wave (p. 408)
tsunami (p. 422)

17
The Earth's Interior

Opposite: Attempts to probe the earth's interior via drilling have rarely exceeded depths of 10 kilometers. (Photo by Lowell Georgia, Photo Researchers, Inc.) (Top photo courtesy of Gulf Oil)

Although the earth's interior lies just below us, its accessibility to direct observation is very limited. Wells drilled into the crust in search for oil, gas, and other natural resources have generally been confined to the upper 7 kilometers. Even the Kola Well, a super-deep research well located in a remote northern outpost of Russia, has penetrated to a depth of only 13 kilometers. Although volcanic activity can be considered as a window into the earth's interior because materials are brought up from below, this activity allows only a glimpse of the outer 200 kilometers of our planet—only a small fraction of the earth's 6370-kilometer radius.

Fortunately geologists have learned a great deal about the earth's composition through space exploration, by high-pressure laboratory experiments and from samples of the solar system (meteorites) that frequently collide with the earth. More importantly, many clues to the physical conditions inside our planet have been obtained through the study of seismic waves generated by earthquakes and nuclear explosions. As seismic waves pass through the earth, they carry information about the materials through which they were transmitted to the surface. Hence, when carefully analyzed, seismic records provide an X-raylike picture of the earth's interior.

PROBING THE EARTH'S INTERIOR

Much of our knowledge of the earth's interior comes from the study of P (compressional) and S (shear) waves that penetrate the earth and emerge at some distant point. Simply stated, the technique involves accurately measuring the time required for seismic waves to travel from an earthquake or nuclear explosion to a seismographic station. Since the time required for P and S waves to travel through the earth depends upon the properties of the rock materials encountered, seismologists search for variations in travel times that cannot be accounted for simply by differences in the distances traveled. These variations correspond to changes in rock properties.

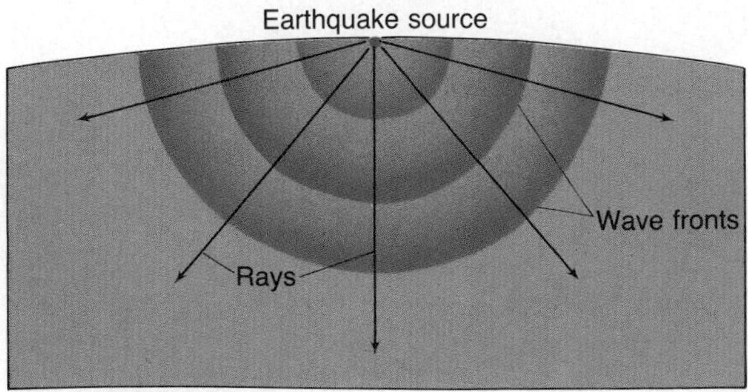

FIGURE 17.1
Seismic waves travel in all directions from an earthquake source (focus). The direction of this motion can also be shown as rays, lines drawn perpendicular to the wave fronts.

One major problem is that to obtain accurate travel times, seismologists must establish the exact location and time of the seismic event. For earthquakes, this information can be obtained only from the seismic waves themselves, which makes measurement somewhat uncertain. Nuclear test explosions, on the other hand, have the advantage over earthquakes in that the precise time and location are known. However, despite the limitations of studying seismic waves generated by earthquakes, seismologists during the first half of the century were able to detect the major layers of the earth. Yet, not until the early 1960s, when nuclear testing was in its heyday and arrays consisting of hundreds of sensitive seismographs were deployed, were the fine structures of the earth first established with certainty.

The Nature of Seismic Waves

To examine the earth's composition and structure, we must first study some of the basic properties of wave transmission, or propagation. As stated in the preceding chapter, seismic waves travel out from their source in all directions as a wave. For purposes of description, the common practice is to consider the paths taken by these waves as rays, or lines drawn perpendicular to the wave front as shown in Figure 17.1. Significant characteristics of seismic waves include the following:

1. The velocity of seismic waves depends on the density and elasticity of the intervening material. Seismic waves travel most rapidly in rigid materials which elastically spring back to their original shapes when the stress is removed. For instance, crystalline rock transmits seismic waves more rapidly than a layer of unconsolidated material.
2. Within a given layer the speed of seismic waves generally increases with depth because pressure increases and squeezes the rock into a more compact elastic material.
3. Compressional waves (P waves), which oscillate back and forth in the same direction as their direction of motion, travel through solids as well as liquids because when compressed these materials behave elastically; that is, they spring back to their original shape when the stress is removed (Figure 17.2A). Shear waves (S waves), which oscillate at right angles to their direction of motion, cannot travel through liquids because, unlike solids, liquids have no shear strength (Figure 17.2B). That

FIGURE 17.2
The transmission of P and S waves through a solid. **A.** The passage of P waves causes the intervening material to suffer alternate compressions and expansions. **B.** The passage of S waves causes a change in shape without changing the volume of the material. Because liquids behave elastically when compressed (they spring back when the stress is removed), they will transmit P waves. However, since liquids do not resist changes in shape, S waves cannot be transmitted through liquids. (From O. M. Phillips, *The Heart of the Earth,* San Francisco: Freeman, Cooper and Co., 1968).

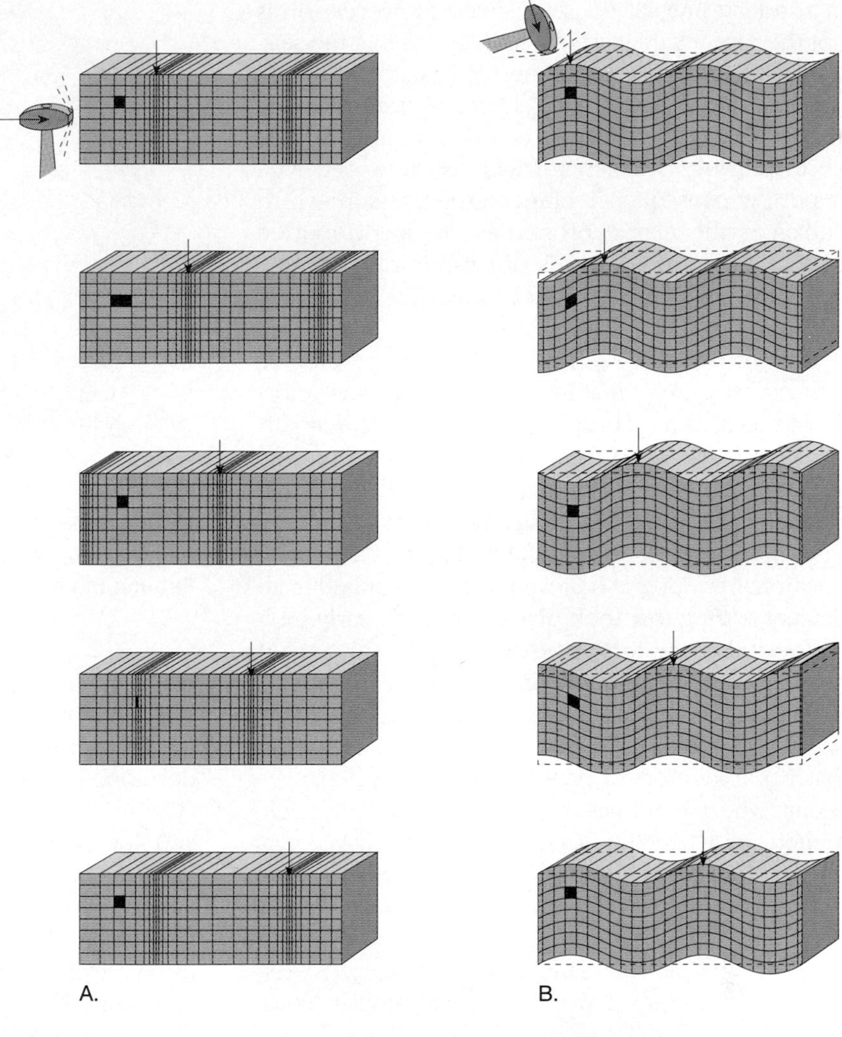

is, when liquids are subjected to forces that act to change their shapes, they simply flow.

4. In all materials, P waves travel faster than S waves.
5. When seismic waves pass from one material to another, the path of the wave is refracted (bent).* In addition, some of the energy is reflected from the **discontinuity** (the boundary between the two dissimilar materials). This is similar to what happens to light when it passes from air into water.

Thus, depending on the nature of the layers through which they pass, seismic waves are speeded up or slowed down, refracted (change direction), or reflected. These observable changes in seismic wave motions enable seismologists to probe the earth's interior.

If the earth were a perfectly homogeneous body, seismic waves would spread through the earth in all directions as shown in Figure 17.3. Seismic waves

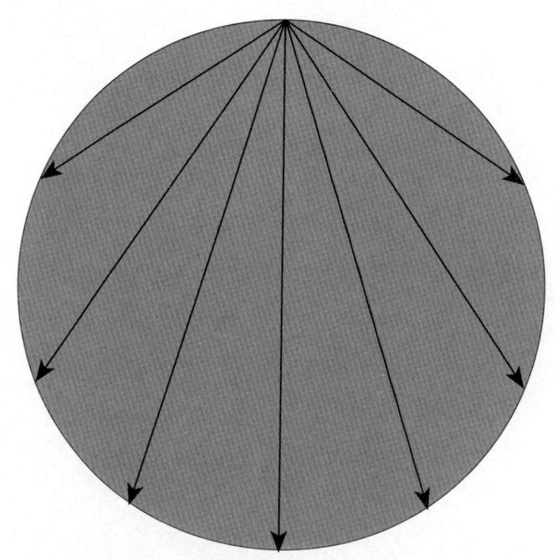

FIGURE 17.3
Seismic waves would travel through a hypothetical planet with uniform properties along linear paths and at constant velocities.

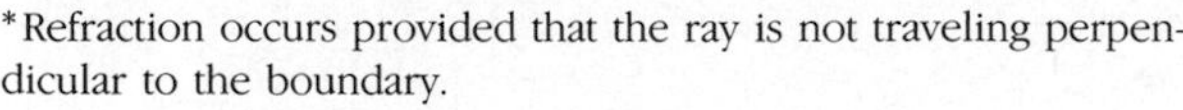

*Refraction occurs provided that the ray is not traveling perpendicular to the boundary.

traveling through such an ideal planet would travel in a straight line at a constant speed. However, this is not the case for the earth. It so happens that the seismic waves that reach seismographic stations located farther from an earthquake travel at faster average speeds than those that are recorded at more nearby observatories. This general increase in speed with depth is a consequence of increased pressure which enhances the elastic properties of deeply buried rock. As a result, the paths of seismic rays through the earth are refracted (bent) in the manner shown in Figure 17.4.

As more sensitive seismographs were developed, it became apparent that in addition to gradual velocity changes, rather abrupt changes also occur at particular depths. Since these discontinuities were detected worldwide, seismologists concluded that the earth must be composed of distinct layers or shells of varying composition or structure (Figure 17.5). Compositional layering is believed to have resulted from density sorting that took place during an early molten period in the earth's history. During this period heavy substances sank while lighter components floated upward. Structural layering, on the other hand, represents material of the same composition that has undergone a phase change. Phase changes occur when rock has melted or almost entirely melted, or when the atoms in minerals have rearranged themselves into tighter crystalline structures in response to the enormous pressures existing at great depths.

Seismological data gathered from numerous seismographic stations have been continuously compiled and analyzed for many years. From this information, seismologists have, over the past 80 years, developed a detailed picture of the earth's interior (Figure 17.6). This model is continually rechecked and fine-tuned as more data become available and as new seismic techniques are employed. Furthermore, laboratory studies that experimentally determine the properties of various earth materials under the extreme environments found deep in the earth add to this body of knowledge.

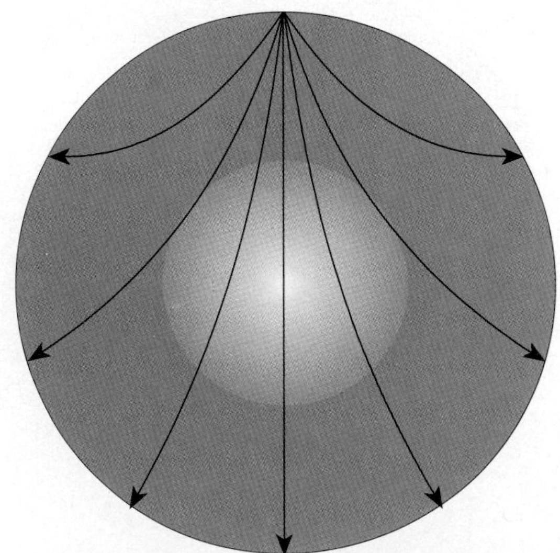

FIGURE 17.4
Wave paths through a planet where velocity increases with depth.

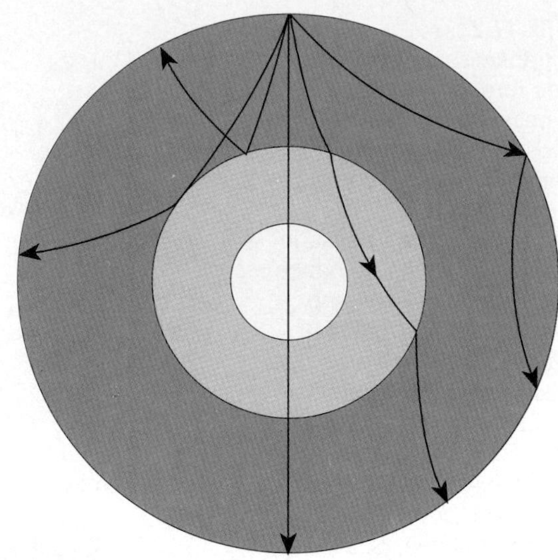

FIGURE 17.5
A few of many possible paths seismic rays take through the earth.

Based upon this seismological data, the earth has been divided into four major layers: (1) the **crust**, a very thin outer layer; (2) the **mantle**, a rocky layer located below the crust and having a maximum thickness of 2885 kilometers; (3) the **outer core**, a layer about 2270 kilometers thick which exhibits characteristics of a mobile liquid; and (4) the **inner core**, a solid metallic sphere, about 1216 kilometers in radius. As we shall see, within these layers some notable features have also been discovered.

Discovering the Earth's Structure

In 1909, a pioneering Yugoslavian seismologist, Andrija Mohorovičić, presented the first convincing evidence for layering within the earth.* The boundary he discovered separates crustal rocks from rocks of different composition in the underlying mantle and was named the **Mohorovičić discontinuity** in his

*Discontinuities in the earth had been predicted by earlier researchers, but their arguments for a central core were inconclusive.

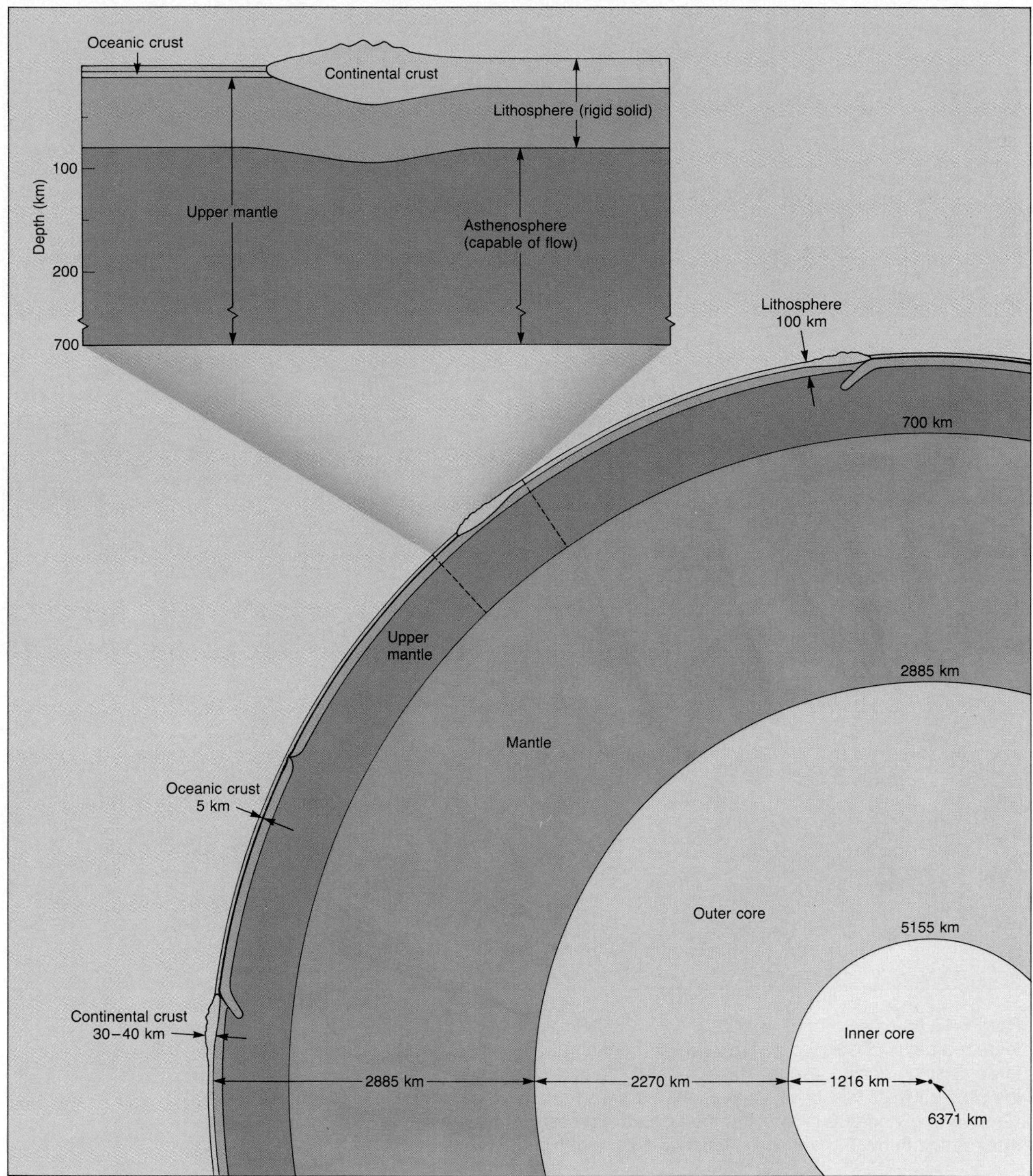

FIGURE 17.6
Cross-sectional view of the earth showing the internal structure.

honor. For reasons that are obvious, the name for this boundary was quickly shortened to **Moho**.

By carefully examining the seismograms of shallow earthquakes, Mohorovičić found that seismographic stations located more than 200 kilometers from an earthquake obtained appreciably faster average travel velocities for P waves than stations located nearer the quake. In particular, P waves that reached the closer stations first had velocities that averaged about 6 kilometers per second. By contrast, the seismic energy recorded at more distant stations traveled at speeds that approached 8 kilometers per sec-

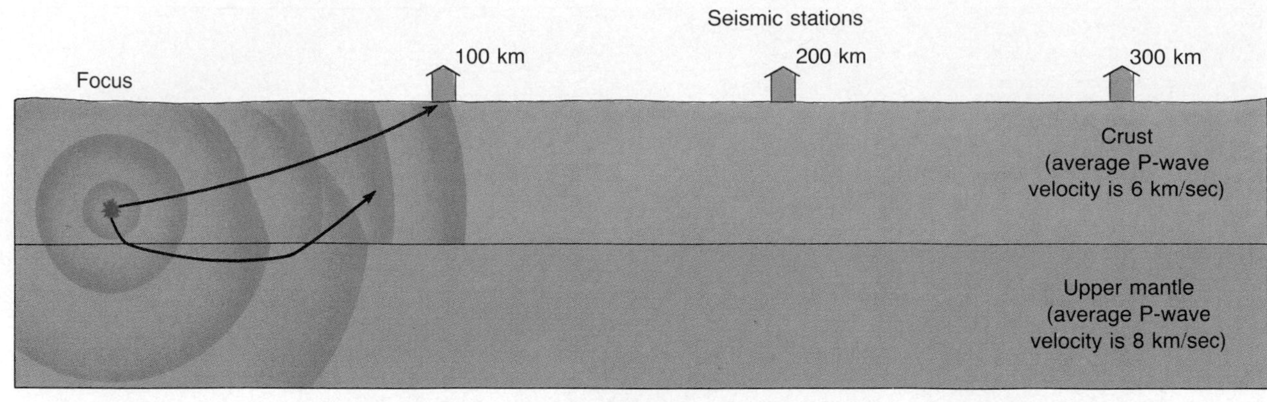

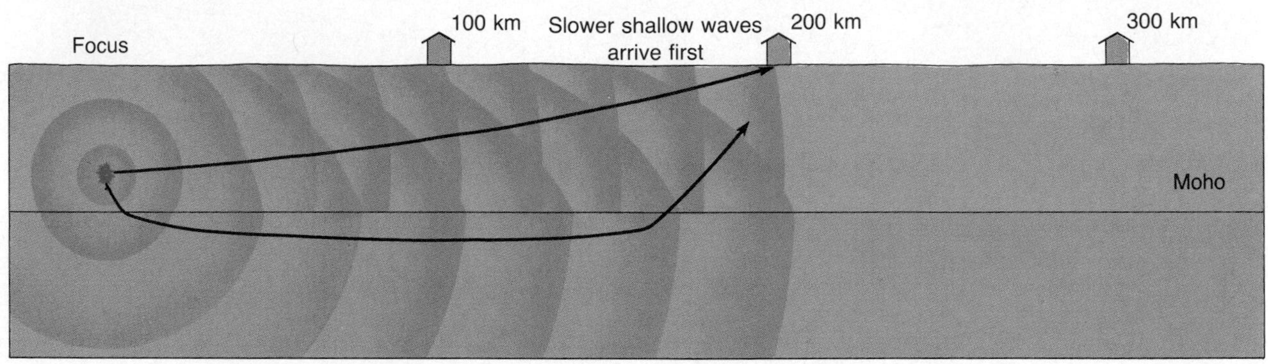

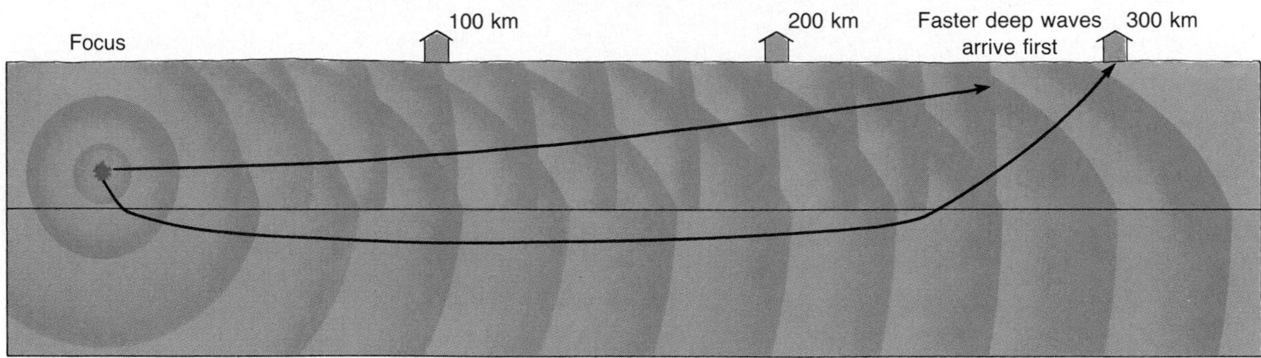

FIGURE 17.7
Idealized paths of seismic waves traveling from an earthquake focus to three seismographic stations. The two nearest recording stations receive the slower waves first because the waves traveled a shorter distance. However, beyond 200 kilometers, the first waves received are the waves that passed through the mantle, which is a zone of higher velocity.

ond. This abrupt jump in velocity did not fit the general pattern that had been observed previously. From these data, Mohorovičić concluded that below 30 kilometers there exists a layer with physical properties markedly different from those of the earth's outer shell.

Figure 17.7 illustrates how Mohorovičić came to this important conclusion. Notice that the first ray to reach the seismographic station located 100 kilometers from the epicenter traveled the shortest route directly through the crust. However, at the seismographic station that was 300 kilometers from the epicenter, the first P wave entered and traveled in the mantle, a zone of higher velocity. Thus, although this wave traveled a greater distance, it reached the recording instrument sooner than any of the direct rays, because a large portion of its journey was through a region having a different composition.

This principle is analogous to a driver taking a bypass route around a large city during rush hour. Although this alternate route is longer, it may be faster.

A few years later another major boundary was discovered by the German seismologist Beno Gutenberg. This discovery was based primarily on the observation that P waves diminish and eventually die out completely about 105 degrees from an earthquake. Then, about 140 degrees away, the P waves reappear, but about two minutes later than would be expected based on the distance traveled. This belt where direct seismic waves are absent is about 35 degrees wide and has been named the **shadow zone*** (Figure 17.8). Gutenberg realized that the shadow zone could be explained if the earth contained a core composed of material unlike the overlying mantle and had a radius of 3420 kilometers. The core must somehow hinder the transmission of P waves in a manner similar to the light rays blocked by an opaque object which casts a shadow. However, rather than actually stopping the P waves, the shadow zone is produced by the bending of P waves which enter the core as shown in Figure 17.8.

*As more sensitive instruments were developed, weak and delayed P waves that enter this zone via reflection were detected.

It was further learned that S waves could not propagate through the core; therefore, geologists concluded that at least a portion of this region is liquid (Figure 17.9). This conclusion was further supported by the observation that P wave velocities suddenly decrease about 40 percent as they enter the core. Since melting would reduce the elasticity of rock, all evidence points to the existence of a liquid layer below the rocky mantle.

In 1936, the last major subdivision of the earth's interior was predicted by the discovery of seismic waves believed to be reflected from a boundary within the core. Hence, a core within a core was discovered. The actual size of the inner core was not accurately calculated until the early 1960s when underground nuclear tests were conducted in Nevada. Because the precise location and time of the explosions were known, echoes from seismic waves that bounced off the inner core provided an accurate means of determining its size (Figure 17.10). From

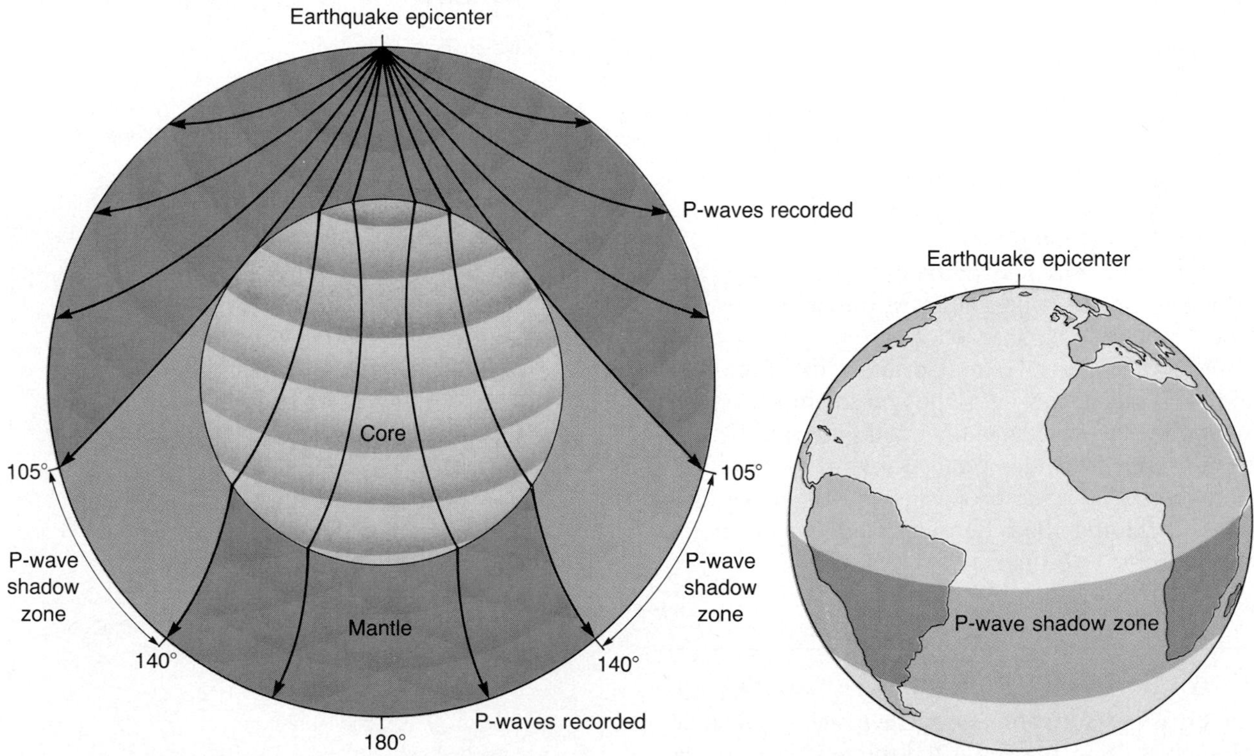

A. Cross-sectional view of P-wave shadow zone

B. Global view of P-wave shadow zone

FIGURE 17.8
The abrupt change in physical properties at the mantle-core boundary causes the wave paths to bend sharply. This abrupt change in wave direction results in a shadow zone for P waves between about 105 and 140 degrees.

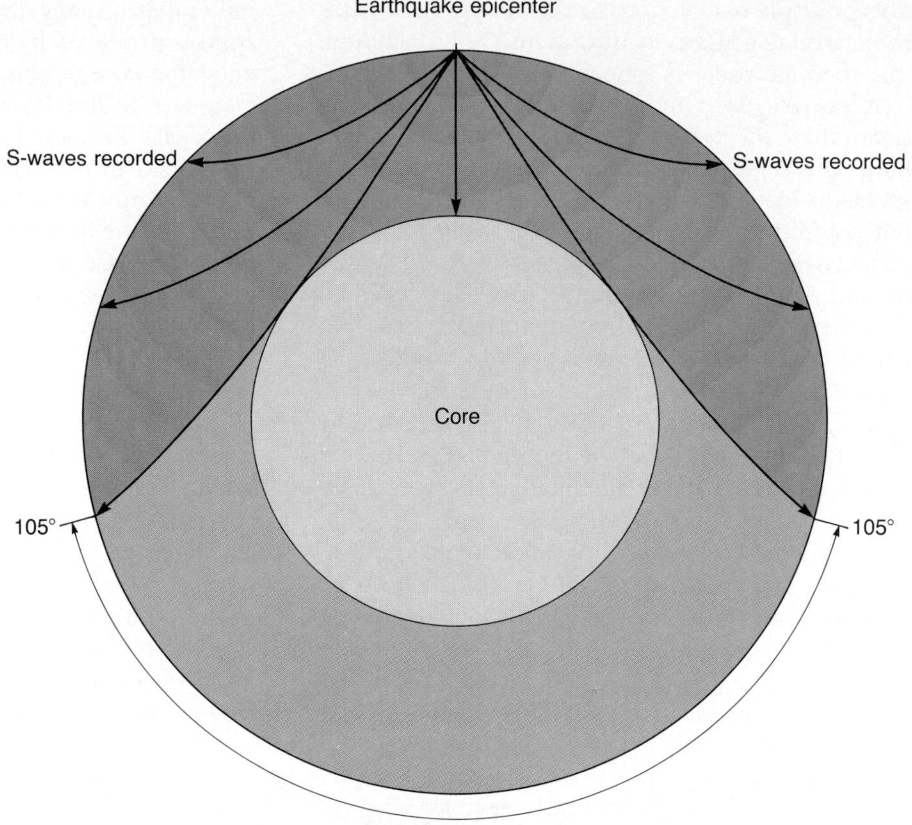

FIGURE 17.9
View of the earth's interior showing the paths of S waves. Any location more than 105 degrees from the earthquake epicenter will not receive direct S waves, since the outer core will not transmit them.

these data and subsequent studies, the inner core was found to have a radius of about 1216 kilometers. Furthermore, P waves passing through the inner core have appreciably faster velocities than those penetrating the outer core exclusively. The apparent increase in the elasticity of the inner core material is evidence for the solid nature of this innermost region.

Over the past 25 years, advances in seismology and rock mechanics have allowed for much refinement of the gross view of the earth's interior that has been presented to this point. Some of these refinements as well as other properties of these major divisions, including their densities and compositions, will be considered next.

THE CRUST

The crust of the earth is on the average less than 20 kilometers thick, making it the thinnest layer so far distinguished (see Figure 17.6). However, along this eggshell-thin layer great variations in rock composition and thickness exist. Whereas the crustal rocks of the continental masses are roughly 35 kilometers thick, the oceanic crust is much thinner, averaging only 5 kilometers. In a few exceptionally prominent

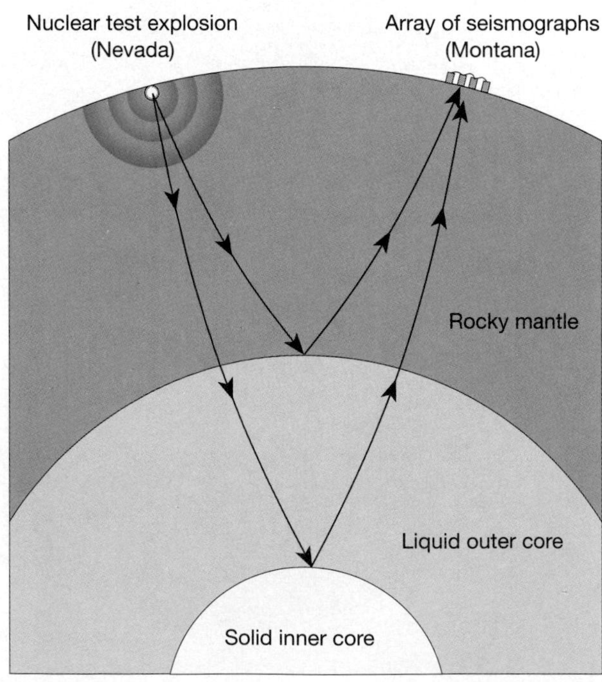

FIGURE 17.10
Travel times of seismic waves generated from nuclear test explosions were used to accurately measure the depth of the inner core. An array of seismographs located in Montana detected the "echoes" that bounced back from the boundary of the inner core.

mountainous regions, the crust obtains its greatest thickness, exceeding 60 kilometers. By contrast, in the stable continental interior, its thickness is closer to 30 kilometers.

The discovery that crustal rocks of the deep-ocean basins are compositionally different from those of the continental masses was first established through studies of seismic velocities. P-wave travel times indicate that velocities of 6 kilometers per second are typical for continental rocks, whereas velocities of 7 kilometers per second are recorded for the oceanic crust. Laboratory experiments were designed to determine which earth materials could produce seismic velocities most like those recorded for these rocky layers. From these experiments, as well as from direct observations, the average composition of continental rocks has been compared with that of the igneous rock granite. Like granite, the continental crust is believed to be enriched in the elements potassium, sodium, and silicon and to have an average density about 2.8 times that of water. Although numerous granitic intrusions and equivalent metamorphic rocks such as gneiss can be found, large outpourings of basalt and volcanic chains composed of andesitic rocks are also abundant. Consequently, indications are that the continental crust's average composition is more similar to rocks of intermediate composition, such as andesite and diorite, than that of "true" granite.

Until recently, geologists could only speculate on the composition of the deep-oceanic crust, which lies beneath 4 kilometers of seawater as well as hundreds of meters of sediments. With the development of deep-sea drilling ships, the recovery of core samples from the ocean floor became possible. As predicted, the samples obtained were predominantly basaltic; indeed, they were different from the rocks that compose the continents. Recall that volcanic eruptions of basaltic material are known to have generated the islands located within the deep-ocean basins.

THE MANTLE

Over 80 percent of the earth's volume is contained within the mantle, a 2885-kilometer thick shell of rock extending from the base of the crust (Moho) to the liquid outer core. Our knowledge of the mantle's composition comes from experimental data, as well as from the examination of material intruded into the crust from below. In particular, the rocks composing kimberlite pipes, in which diamonds are often found, are thought to have originated at depths approaching 200 kilometers, well within the mantle. These kimberlite deposits are composed of peridotite, a rock that contains iron and magnesium-rich silicate minerals, mainly olivine and pyroxene, plus lesser amounts of garnet. Further, because S waves are readily propagated through the mantle, we conclude that it behaves as an elastic solid. Thus, the mantle is described as a solid rocky layer, the upper portion of which has the same composition as the rock peridotite.

As might be expected, this simple picture of the mantle is far from complete. Any working model of the mantle must explain the temperature distribution calculated for this layer. Whereas the crust has a large increase in temperature with depth, this same trend does not continue downward into the mantle. Rather, the temperature increase with depth in the mantle is apparently much more gradual. This means that the mantle has an effective method of transmitting heat outward. If heat were transmitted through the mantle by conduction, as occurs in the crust, the lower mantle would, out of necessity, be hundreds of times hotter than the outer mantle since the conduction of heat through rock is very slow. Consequently, most geologists conclude that some form of mass transport (convection) of hot rock must exist within the mantle. This being the case, the rock of the mantle, where temperatures and pressures are extreme, must be capable of flow.

If this is true, how does the rocky mantle transmit S waves, which can travel only through solids, and at the same time flow like a fluid? This apparent contradiction can be resolved if the material behaves like a solid under certain conditions and like a fluid under other conditions. Geologists generally describe material of this type as exhibiting *plastic* behavior. This means that when the material encounters short-lived stresses, such as those produced by seismic waves, the material behaves like an elastic solid. However, in response to long-term stresses, this same rocky material will flow. This also explains why S waves can penetrate the mantle, yet at the same time, this layer is not able to store elastic energy like a brittle solid and is thus incapable of generating earthquakes. This apparently unusual phenomenon is not restricted to mantle rocks. The manmade substances Silly Putty and some taffy candies also exhibit plastic behavior. When struck with a hammer these materials shatter like a brittle solid. However, when slowly pulled apart they flow plastically. From this example do not get the idea that the mantle is composed of soft, putty-like material. Rather, it is composed of hot, solid rock, which under extreme pressures unknown on the surface of the earth, exhibits the ability to flow.

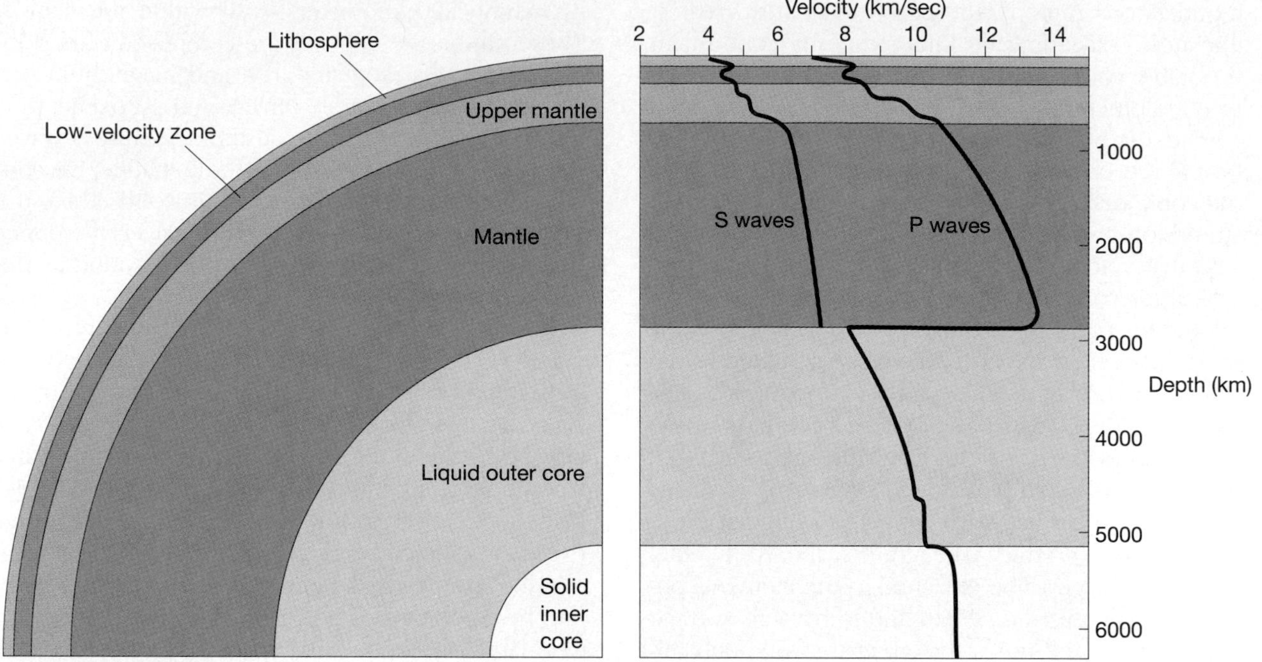

FIGURE 17.11
Variations in P and S wave velocities with depth. Abrupt changes in average wave velocities delineate the major features of the earth's interior. At a depth of about 100 kilometers, the sharp decrease in wave velocity corresponds to the top of the low-velocity zone. Two other bends in the velocity curves occur in the upper mantle at depths of about 400 and 700 kilometers. These variations are thought to be caused by minerals that have undergone phase changes, rather than resulting from compositional differences. The abrupt decrease in P-wave velocity and the absence of S waves at 2885 kilometers marks the core-mantle boundary. The liquid outer core will not transmit S waves, and within this layer the propagation of P waves is slowed. As the P waves enter the solid inner core, their velocity once again increases. (Data from Bruce A. Bolt)

FIGURE 17.12
Respective locations of the asthenosphere and lithosphere.

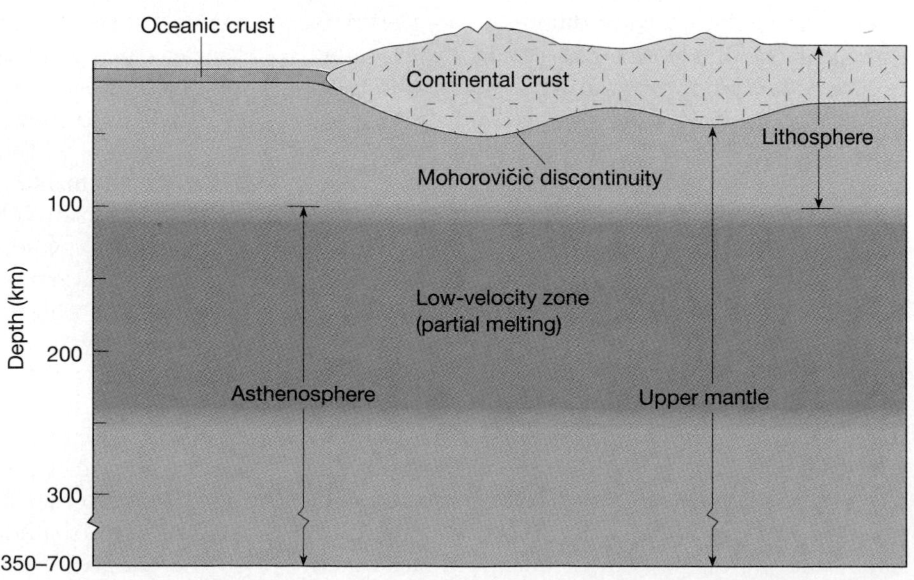

More recent efforts to probe the upper mantle have confirmed earlier speculations that finer divisions also exist. One of the most significant of these subdivisions is a region located between the depths of 100 and 250 kilometers called the **low-velocity zone.** When penetrating this zone, P and S waves show a marked decrease in velocity (Figure 17.11). The most probable explanation for the observed slowing of seismic energy is that this zone contains a small percentage of melt. Although found below the oceanic crust and portions of the continents, this low-velocity zone does not encircle the earth. It is notably absent, for example, below the older shield areas of the continents.

The discovery of the low-velocity zone supports a proposal made earlier that a zone of weak rock exists below 100 kilometers (Figure 17.12). This region, called the **asthenosphere**, is located between depths of 100 and 350 kilometers and may extend down to 700 kilometers. The asthenosphere incorporates the low-velocity zone. However, unlike this zone of partially melted rock which is absent under portions of older continental crust, the asthenosphere is thought to be a global feature of the upper mantle. As can be seen in Figure 17.13, this weak layer exists because the rock at this level is nearer its melting temperature than the rock above or below it. Thus, like red-hot iron, the rock within this zone is easily deformed. The discovery of the asthenosphere was an important contribution to the theory of plate tectonics because this weak layer facilitates the motion of the rigid layer above.

Situated above the asthenosphere is the cool brittle layer about 100 kilometers thick called the **lithosphere** (Figure 17.12). Actually the lithosphere includes the entire crust as well as the uppermost mantle and is defined as that layer of the earth cool enough to behave like a brittle solid.

At the depth of about 400 kilometers a relatively abrupt increase in seismic velocity has been detected (Figure 17.11). While the velocity increase at the crust-mantle boundary is thought to represent a change in composition, the increase at the 400-kilometer level is believed to be the result of a phase change. A phase change occurs when the crystalline structure of a mineral changes in response to changes in temperature and/or pressure. Laboratory studies show that the mineral olivine, $(Mg, Fe)_2SiO_4$, which is one of the main constituents in the rock peridotite, will collapse to a more compact, high-pressure mineral (spinel) at the pressures experienced at this depth. This structural change to a denser crystal form could explain the increased seismic velocities observed.

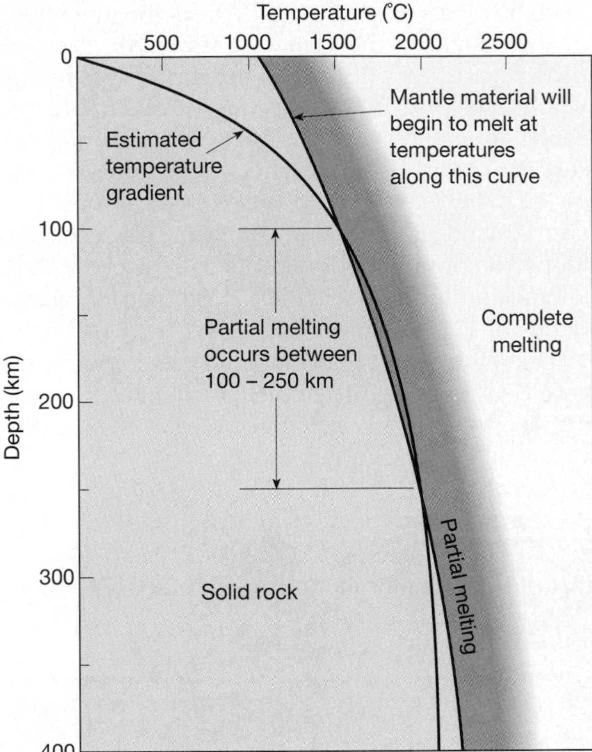

FIGURE 17.13
Relationship between the temperature gradient (gradual increase in temperature with depth) and the melting-temperature (minimum temperature required to melt rock) for mantle rocks. Notice that in the zone between 100 and 250 kilometers the temperature of mantle materials slightly exceeds the temperature at which melting begins. This is thought to account for the existence of the zone of partial melting and hence the low-velocity zone and weak nature of the asthenosphere.

Another boundary has been detected from variations in seismic velocity at a depth of 670 kilometers (Figure 17.11). At this depth the mineral spinel is believed to undergo a transformation to the mineral perovskite $(Mg, Fe)SiO_3$. Since perovskite is thought to be pervasive in the lower mantle, it is perhaps the most abundant mineral in the earth.

THE CORE

Like the other layers discussed thus far, the fact that the earth contained a central core was established from seismological data. Having a radius of 3486 kilometers, this dense sphere inside the earth is larger than the planet Mars. Extending from the inner edge of the mantle to the center of the earth, the core constitutes about one-sixth of the earth's volume and nearly one-third of its total mass. Pressures at the center are millions of times greater than the air pressure at the earth's surface, and temperatures are estimated to be between 4000° and 5000°C. As more precise seismic data became available, the core was found to consist of a liquid outer layer about 2270 kilometers thick, and a solid inner portion with a radius of 1216 kilometers.

One of the most interesting characteristics of the core is its great density. At the core-mantle boundary the density is nearly ten times greater than that of water, and at the center the density is 13.5 times greater than that of water. Even under the extreme pressures at those depths, the common silicate minerals found in the crust with densities 2.6 to 3.5 times that of water could not be compacted enough to account for the great densities calculated for the core. Consequently, attempts were undertaken to determine the earth material that could account for this property.

Surprisingly enough, meteorites provided an important clue to the earth's internal composition. Since meteorites are part of the solar system, they are assumed to be representative samples of the material from which the earth originally accreted. Their composition ranges from metallic types made primarily of iron and nickel to stony meteorites composed of rocky substances that closely resemble the rock peridotite. Because the earth's crust contains a much smaller percentage of iron than is dictated by the relative abundance of iron in the debris of the solar system, geologists concluded that the interior of the earth must be enriched in this heavy material. Further, iron is the only abundant substance found in the solar system that exhibits the proper density.

BOX 17.1

The Earth's Magnetic Field

Anyone who has used a compass to find direction knows that the earth's magnetic field has a north pole and a south pole. In many respects our planet's magnetic field resembles that produced by a simple bar magnet. Invisible lines of force pass through the earth and out into space while extending from one pole to the other (Figure 17.A). A compass needle, itself a small magnet free to move about, becomes aligned with these lines of force and points toward the magnetic poles. It should be noted that the earth's magnetic poles do not coincide exactly with the geographic poles. The north magnetic pole is located in northeastern Canada, near Hudson Bay, while the south magnetic pole is located near Antarctica in the Indian Ocean south of Australia.

In the early 1960s, geophysicists learned that the earth's magnetic field periodically (every million years or so) reverses polarity; that is, the north magnetic pole becomes the south magnetic pole, and vice versa. The cause of these reversals is apparently linked to the fact that the earth's magnetic field experiences long-term fluctuations in intensity. Recent calculations indicate that the magnetic field has weakened by about 5 percent over the past century. If this trend continues for another 1500 years, the earth's magnetic field will become very weak or even nonexistent. It has been suggested that the decline in magnetic intensity is related to changes in the convective flow in the core. In a similar manner, magnetic reversals may be triggered when something disturbs the main convection pattern of the fluid core. After a reversal takes place, the flow is reestablished and builds a magnetic field with opposite polarity. Magnetic reversals are not unique to the earth. The sun's magnetic field regularly reverses polarity, having an average period of about 22 years. These solar reversals are closely tied to the well-known 11-year sunspot cycle.

When the earth's magnetic field was first described in 1600, it was thought to originate from permanently magnetized materials located deep within the earth. We have since learned that, except for the upper crust, the planet is much too hot for magnetic materials to retain their magnetism. Furthermore, permanently magnetized materials are not known to vary their intensity in a manner that would account for the observed long-term waxing and waning of the earth's magnetic field.

It is not yet known exactly how the earth's magnetic field is produced. Nevertheless, most investigators agree that the gradual flow of molten iron in the outer core is an important part of the process. The most widely accepted view proposes that the core behaves like a self-sustaining *dynamo,* a device that converts mechanical energy into magnetic energy. The

Although the core is predominantly iron, it cannot be pure iron. Experiments indicate that the density of pure iron under the extreme pressures of the core is about 10 percent higher than the density that was actually established. This being the case, the suggestion has been made that the core must also contain some lighter elements which alloy with iron and lower its density. This idea is supported by the fact that the best estimates of core temperatures are below the melting environment for pure iron. Thus, if the outer core were pure iron, it would have long ago crystallized, a condition that contradicts seismological data. A liquid outer core can be explained by the addition of lighter elements which, when mixed with iron, lower its melting point. The elements most likely to alloy with iron and account for the core's observed density and the liquid state of the outer core are sulfur and oxygen. However, other substances, including nickel, silicon, and carbon, are undoubtedly present in minor amounts.

Although the existence of a metallic central core is well established, efforts to explain the core's origin are more speculative. The most widely accepted scenario suggests that the core formed early in the earth's history from what was originally a relatively homogeneous body. During the period of accretion the entire earth was heated by the energy released by the infalling material. Sometime late in this period of growth, the earth's internal temperature was sufficiently high to mobilize the accumulated material. Blobs of heavy iron-rich material collected and sank toward the center. Simultaneously, lighter substances may have floated upward to generate the mantle, and possibly portions of the crust as well. In a short time, geologically speaking, the earth took on a layered configuration perhaps not much different from what we find today.

How can we explain the existence of a liquid outer core when the inner core, which must be hotter, is solid? Most probably in its formative stage the

driving forces of this system are the earth's rotation and the unequal distribution of heat in the earth's interior which propels the highly conductive molten iron in the outer core. As the iron churns in the outer core, it interacts with the earth's magnetic field. This interaction generates an electric current, just as moving a wire past a magnet creates a current in the wire. Once established, the electric current produces a magnetic field that reinforces the earth's magnetic field. As long as the flow of molten iron around the core continues, electric currents will be produced and the earth's magnetic field will be sustained. Thus, the earth's dynamo is self-perpetuating. After having been started by a small external magnetic field, the dynamo generates its own field without an external supply of magnetism.

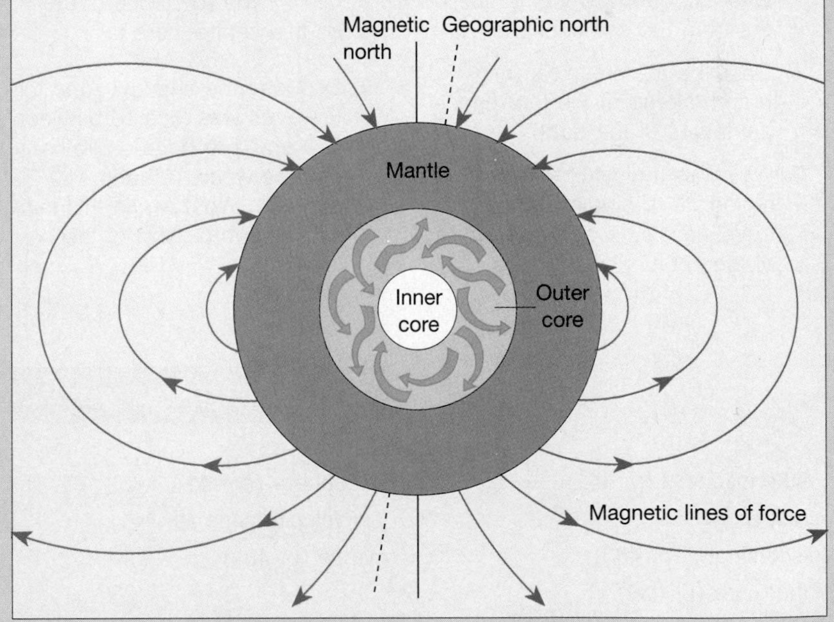

FIGURE 17.A
The earth's magnetic field is thought to be generated by the vigorous mixing of molten iron alloy in the liquid outer core.

entire core was liquid. Further, this liquid iron alloy was in a state of vigorous mixing. However, during the last 3.5 billion years, the material of the core has been slowly segregating. As the core cooled, a portion of the iron components gradually migrated downward while some of the lighter components floated upward toward the outer edge of the core. The sinking iron-rich components, depleted of the lighter elements which act to depress the melting point, began to solidify. The downward migration and crystallization of the heavier material release gravitational energy and the heat of fusion which drives the currents in the remaining liquid shell above.

Our picture of the core with its solid inner sphere surrounded by a mobile liquid shell is further supported by the existence of the earth's magnetic field, which behaves as if a large bar magnet were situated deep within the earth. However, we know that the source of the magnetic field cannot be permanently magnetized material, because the earth's interior is too hot for any material to retain its magnetism. The most widely accepted mechanism explaining the earth's magnetic field requires that the core be made of a material that conducts electricity, such as iron, and one that is mobile enough to circulate. (See Box 17.1.) Both of these conditions are met by the model of the earth's core that was established on the basis of seismological data.

REVIEW QUESTIONS

1. List the major differences between P and S waves.
2. How does the boundary between the crust and mantle (Moho) differ from the boundaries that occur at depths of about 400 and 700 kilometers?
3. Describe the lithosphere. In what important way is it different from the asthenosphere?
4. Describe the chemical (mineral) makeup of the four principal layers of the earth.
5. Why was it difficult for seismologists to obtain precise travel-time data before the turn of the century?
6. Describe the method first used to accurately measure the size of the inner core.
7. How were the first samples (in place) of the deep-ocean floor obtained?
8. What evidence did Gutenberg use for the existence of the earth's central core?
9. Suppose the shadow zone for P waves was located between 120 and 160 degrees, rather than between 105 and 140 degrees. What would this indicate about the size of the core?
10. Explain why the asthenosphere is able to flow like a fluid yet has the ability to transmit S waves which cannot travel through fluids.
11. Why are meteorites considered important clues to the composition of the earth's interior?
12. What evidence is provided by seismology to indicate that the outer core is liquid? What other evidence exists for a molten outer core?
13. Why is it possible for the outer core to be molten when the inner core (which has a higher temperature) is in the solid state?

KEY TERMS

asthenosphere (p. 441)
crust (p. 434)
discontinuity (p. 433)
inner core (p. 434)
lithosphere (p. 441)
low-velocity zone (p. 441)
mantle (p. 434)
Mohorovičić discontinuity, or Moho (pp. 434–435)
outer core (p. 434)
shadow zone (p. 437)

18
Plate Tectonics

Opposite: The Gulf of Suez (left) and the Gulf of Aqaba (right) are major rift zones caused by movement of the Arabian Peninsula from Africa. (Photo courtesy of Earth Satellite Corporation) (Top photo LANDSAT image courtesy of Phillips Petroleum Company, Exploration Projects Section)

Early in this century geologic thought about the age of the ocean basins was dominated by a belief in their antiquity. Moreover, most geologists accepted the geographic permanency of the oceans and continents. Mountains were believed to result from the contraction of the earth caused by gradual cooling from a once-molten state. As the interior cooled and contracted, the earth's solid outer skin was deformed by folding to fit the shrinking planet. Mountains were therefore regarded as analogous to the wrinkles on a dried-out piece of fruit. This model of the earth's tectonic* processes, however inadequate, was firmly entrenched in the geologic thought of the time. Even changes in sea level, evident from the record of marine fossils found deep in the continental interiors, were explained using the model of a gradually contracting earth. As the earth's solid outer shell was deformed, some regions subsided and were inundated by the sea, while other areas emerged as dry land.

During the last few decades spectacular developments have taken place in the earth sciences. Due to the vast accumulation of new data, our ideas about the structure and workings of the earth have changed dramatically. Earth scientists now realize that the positions of landmasses are not fixed. Rather, the continents gradually migrate across the globe. The splitting of continental blocks has resulted in the formation of new ocean basins, while older segments of the sea floor are continually being recycled in areas where we find deep-ocean trenches. Further, because of this movement, once disjointed segments of continental material have collided and formed the earth's great mountain ranges. In short, a revolutionary new model of the earth's tectonic processes has emerged in marked contrast to what was accepted just a few decades ago.

This profound reversal of scientific opinion has been appropriately described as a scientific revolution. Like other scientific revolutions, an appreciable length of time elapsed between the idea's inception and its general acceptance. The revolution began in the early part of the twentieth century as a relatively straightforward proposal that the continents drifted about the face of the earth. After many years of heated debate, the idea of drifting continents was rejected by the vast majority of earth scientists as being improbable.

The concept of a mobile earth was particularly distasteful to North American geologists, perhaps because much of the supporting evidence was gathered from the southern continents, with which most North American geologists were essentially unfamiliar. This fact is evidenced by the meager amount of material concerning continental drift in the scientific literature in the United States between 1930 and 1950. However, during the 1950s and 1960s new evidence began to rekindle interest in this abandoned proposal. By 1968 these new developments led to the unfolding of a far more encompassing theory than continental drift—a theory known as plate tectonics.

In this chapter we will examine the events that led to this dramatic reversal of scientific opinion in an attempt to provide some insight into how science works. We will briefly trace the developments that took place from the inception of the concept of continental drift through the general acceptance of the theory of plate tectonics. The evidence gathered to support the concept of a mobile earth will also be provided.

***Tectonics* refers to the deformation of earth's crust and results in the formation of structural features such as mountains.

CONTINENTAL DRIFT: AN IDEA BEFORE ITS TIME

The idea that continents, particularly South America and Africa, fit together like pieces of a jigsaw puzzle originated with improved world maps. However, little significance was given this idea until 1915, when Alfred Wegener,* a German meteorologist and geophysicist, published an expanded version of a 1912 lecture in his book *The Origin of Continents and Oceans.* In this monograph, Wegener set forth the basic outline of his radical hypothesis of **continental drift.** One of his major tenets suggested that a supercontinent called **Pangaea** (meaning "all land") once existed (Figure 18.1). He further hypothesized that about 200 million years ago this supercontinent began breaking into smaller continents, which then "drifted" to their present positions. Wegener and others who advocated this position collected substantial evidence to support these claims. The fit of South America and Africa, ancient climatic similarities, fossil evidence, and rock structures all seemed to support the idea that these now separate landmasses were once joined.

Fit of the Continents

Like a few others before him, Wegener first suspected that the continents might have been joined

*Wegener's ideas were actually preceded by those of an American geologist, F. B. Taylor, who in 1910 published a paper on continental drift. Taylor's paper provided little corroborating evidence for continental drift, which may have been the reason that it had a relatively small impact on the scientific community.

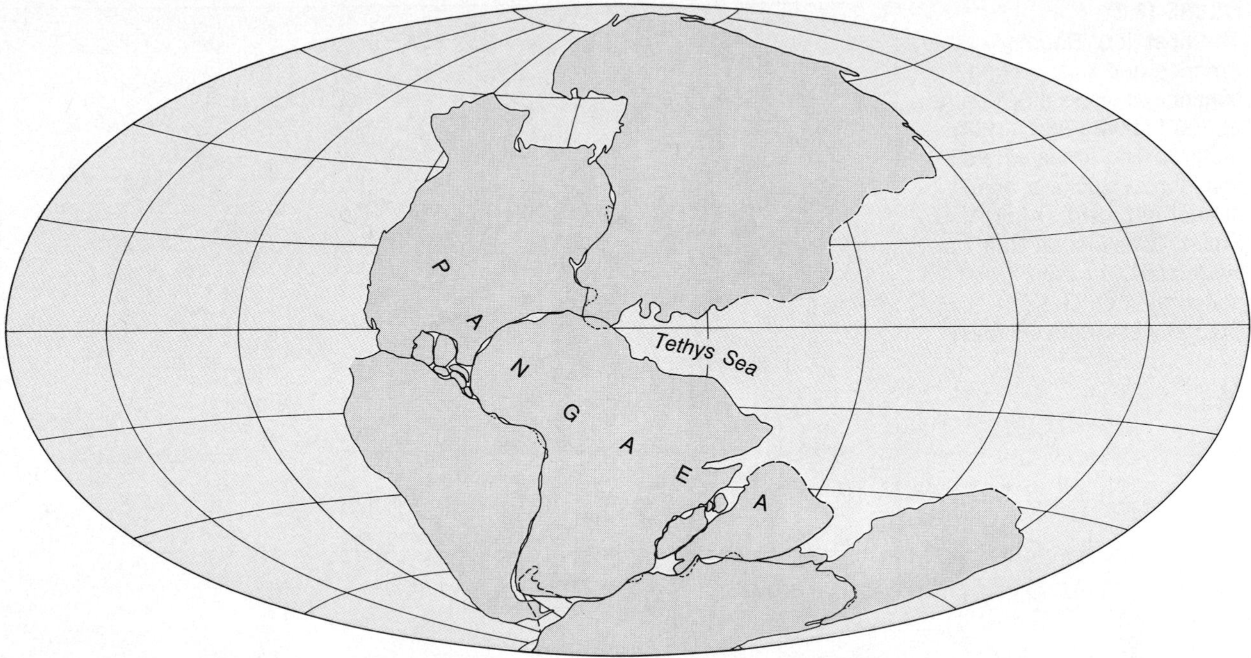

FIGURE 18.1
Reconstruction of Pangaea as it is thought to have appeared 200 million years ago. (After R. S. Deitz and J. C. Holden. *Journal of Geophysical Research* 75: 4943. Copyright by American Geophysical Union)

when he noticed the remarkable similarity between the coastlines on opposite sides of the South Atlantic. However, his use of present-day shorelines to make a fit of the continents was challenged immediately by other earth scientists. These opponents correctly argued that shorelines are continually modified by erosional processes and even if continental displacement had taken place, a good fit today would be unlikely. Furthermore, abundant fossil evidence exists that indicates most of the world's land areas have experienced periods of either uplift or subsidence in the recent geologic past. This would have markedly altered the position of the global coastlines. Wegener appeared to be aware of these problems, and, in fact, his original jigsaw fit of the continents was only crude.

A much better approximation of the outer boundary of the continents is the seaward margin of the continental shelf. Today the continental shelf's edge lies several hundred meters below sea level. In the early 1960s, Sir Edward Bullard and two associates produced a map with the aid of computers that attempted to fit the continents at a depth of about 900 meters. The remarkable fit that was obtained is shown in Figure 18.2. Although the continents overlap in a few places, these are regions where streams have deposited large quantities of sediment, thus enlarging the continents. The overall fit obtained by Bullard and his associates was better than even the supporters of continental drift suspected it would be.

Fossil Evidence

Although Wegener was intrigued by the remarkable similarities of the shorelines on opposite sides of the Atlantic, he at first thought the idea of a mobile earth improbable. Not until he came across an article citing fossil evidence for the existence of a land bridge connecting South America and Africa did he begin to take his own idea seriously. Through a search of the literature, Wegener learned that most paleontologists were in agreement that some type of land connection was needed to explain the existence of identical fossils on the widely separated landmasses. This requirement was particularly true for late Paleozoic and early Mesozoic life forms.

To add credibility to his argument for the existence of Pangaea, Wegener cited documented cases of several fossil organisms that had been found on different landmasses but which could not have crossed the vast oceans presently separating the continents. Of particular interest were organisms that were restricted in geographical distribution, but which nevertheless appeared in two or more areas that are presently separated by major barriers. The classic example is *Mesosaurus,* a presumably aquatic,

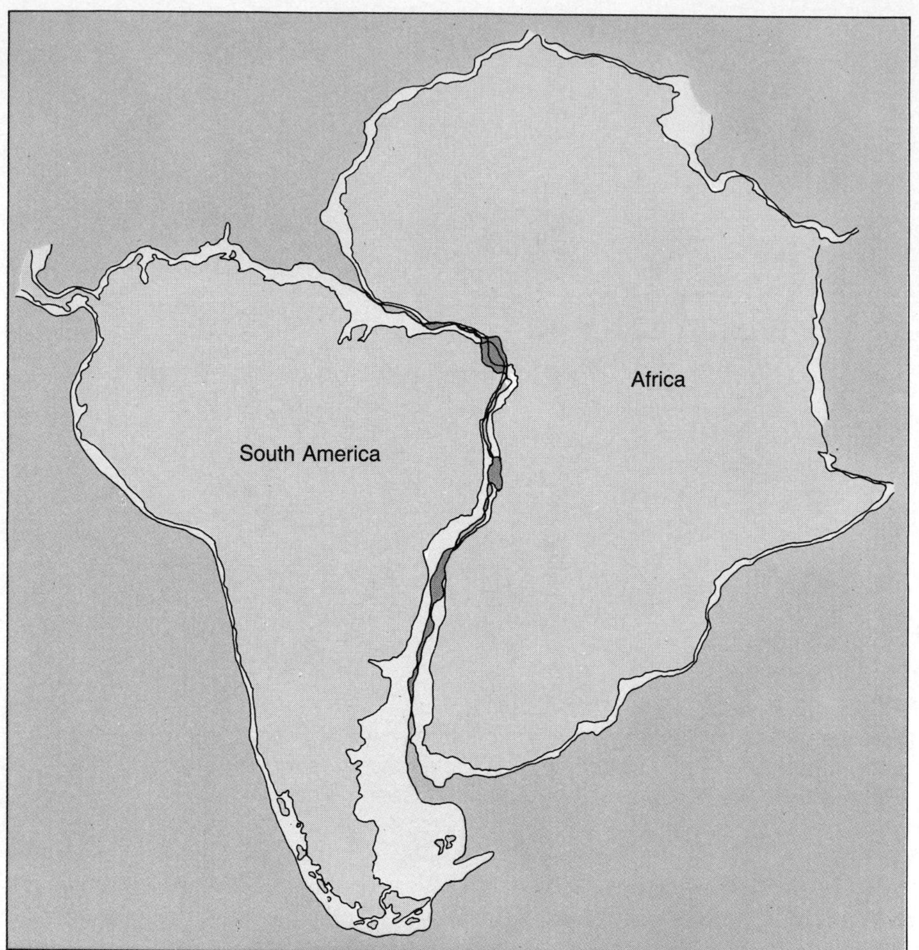

FIGURE 18.2
The best fit of South America and Africa along the continental slope at a depth of 500 fathoms (about 900 meters). The areas where continental blocks overlap appear in brown. (After A. G. Smith. "Continental Drift." In *Understanding the Earth,* edited by I. G. Gass. Courtesy of Artemis Press)

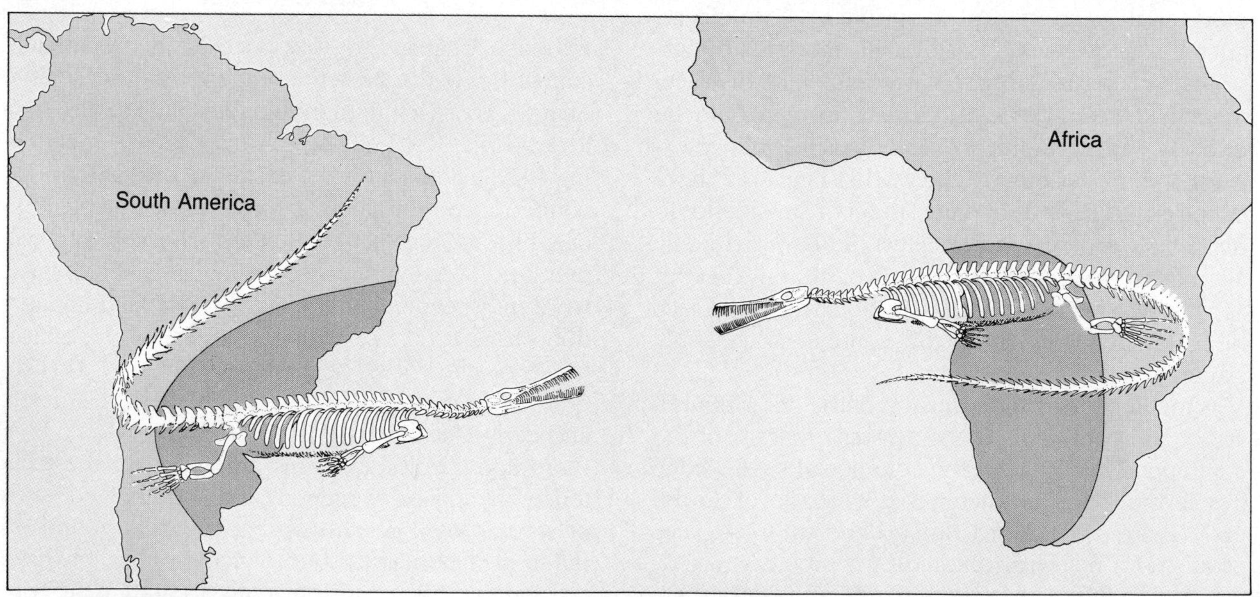

FIGURE 18.3
Fossils of *Mesosaurus* have been found on both sides of the South Atlantic and nowhere else in the world. Fossil remains of this and other organisms found on the continents of Africa and South America appear to link these landmasses during the late Paleozoic and early Mesozoic eras.

snaggle-toothed reptile whose fossil remains are known to be limited to eastern South America and southern Africa (Figure 18.3). If *Mesosaurus* had been able to swim well enough to cross a vast ocean, its remains should be widely distributed. Since this was not the case, Wegener argued that South America and Africa must have been joined.

Wegener also cited the distribution of the fossil fern *Glossopteris* as evidence for the existence of Pangaea. This plant, identified by its large seeds that could not be blown very far, was known to be widely dispersed among Africa, Australia, India, and South America during the late Paleozoic era. Later, fossil remains of *Glossopteris* were discovered in Antarctica as well. Wegener knew that these seed ferns and associated flora grew only in a subpolar climate; therefore, he concluded that these landmasses must have been joined, since they presently include climatic regions that are too diverse to support such flora. For Wegener, fossils proved without question that a supercontinent had existed.

In his book, Wegener also cited the distribution of present-day organisms as evidence to support the concept of drifting continents. For example, modern organisms with similar ancestries clearly had to evolve in isolation during the last few tens of millions of years. Most obvious of these are the Australian marsupials, which have a direct fossil link to the marsupial opossums found in the Americas.

How could these fossil flora and fauna be so similar in places separated by thousands of kilometers of open ocean? The idea of land bridges was the most widely accepted solution to the problem of migration (Figure 18.4). We know, for example, that during the recent glacial period the lowering of sea level allowed animals to cross the narrow Bering Straits between Asia and North America. Was it possible then that land bridges once connected Africa and South America? We are now quite certain that land bridges of this magnitude did not exist, for their remnants should still lie below sea level, but are nowhere to be found.

Rock Type and Structural Similarities

Anyone who has worked a picture puzzle knows that in addition to the pieces fitting together, the picture must be continuous as well. The picture that must match in the "Continental Drift Puzzle" is represented by the rock types and mountain belts found on the continents. If the continents were once together, the rocks found in a particular region on one continent should closely match in age and type with those found in adjacent positions on the matching continent. For example, a good correlation between rocks found in northwestern Africa was made with rocks in eastern Brazil. Recent re-examination of this early evidence has supported Wegener's claim. In both regions, 550 million-year-old rocks lie adjacent to rocks dated at more than 2 billion years in such a manner that the line separating them is continuous when the two continents are brought together.

Further evidence to support the concept of continental drift comes from several mountainous belts

FIGURE 18.4
These sketches by John Holden illustrate various explanations for the occurrence of similar species on landmasses that are presently separated by vast oceans. (Reprinted with permission of John Holden)

that appear to terminate at one coastline only to reappear again on a landmass across the ocean. For instance, the mountain belt that includes the Appalachians trends northeastward through the eastern United States and disappears off the coast of Newfoundland. Mountains of comparable age and structure are found in Greenland and Northern Europe (Figure 18.5B). When these landmasses are reassembled as in Figure 18.5A, the mountain chains form a nearly continuous belt. Numerous other rock structures exist that appear to have formed at the same time and were subsequently split apart.

Wegener was very satisfied that the similarities in rock structure on both sides of the Atlantic linked these landmasses. In fact, he was too zealous with this evidence and incorrectly suggested that glacial

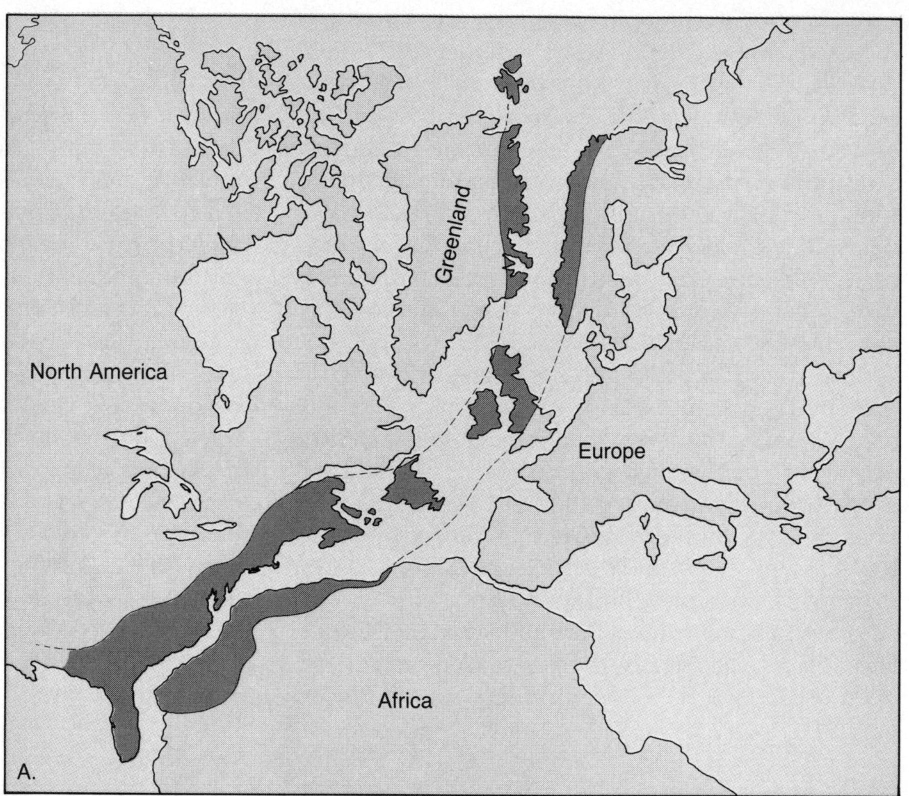

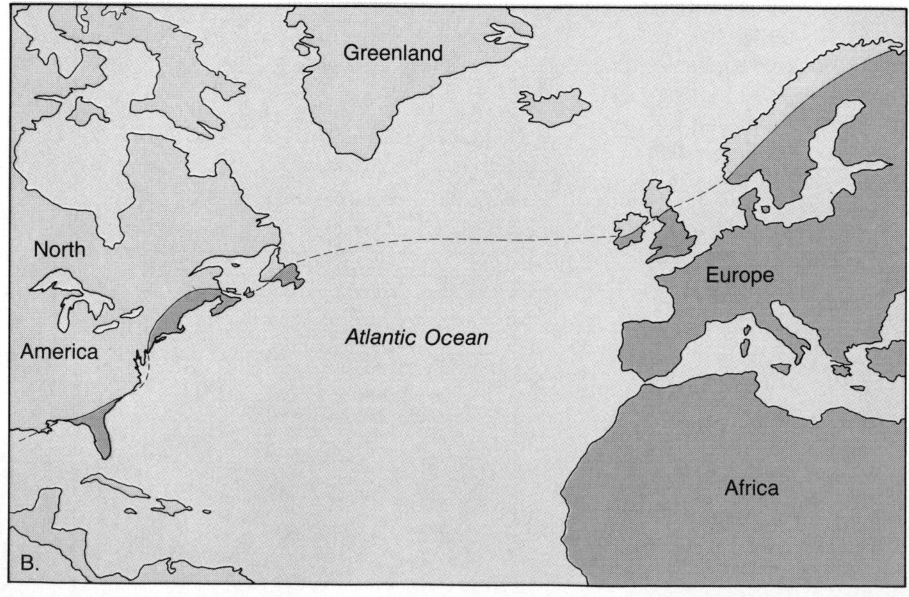

FIGURE 18.5
Matching structural belts across the North Atlantic. **A.** When the North Atlantic landmasses are positioned in their pre-drift locations, ancient mountain chains form a nearly continuous belt. These folded rocks include the Appalachians, which trend northeastward into Newfoundland and match up with structures of similar age in Greenland, Ireland, Great Britain, and Norway. **B.** The folded belts shown in Part A were formed as the landmasses collided during the formation of Pangaea. Later, when the continents broke apart, generating the modern Atlantic, the line of separation did not exactly correspond to the margins of the accreted landmasses. In particular, part of ancient North America remained attached to Europe, forming Scotland, Ireland, and Norway. Further, a portion of northeastern North America and much of Florida is believed to have once been a part of the "Euro-African" continent. (After A. Hallam and others)

moraines in North America matched up with those of Northern Europe. In his own words, "It is just as if we were to refit the torn pieces of a newspaper by matching their edges and then check whether the lines of print run smoothly across. If they do, there is nothing left but to conclude that the pieces were in fact joined in this way."*

Paleoclimatic Evidence

Since Alfred Wegener was a meteorologist by profession, he was keenly interested in obtaining paleoclimatic (ancient climate) data in support of continental drift. His efforts in this area were rewarded when he found evidence for apparently dramatic climatic changes. For instance, glacial deposits indicated that near the end of the Paleozoic era (between 220 and 300 million years ago), ice sheets covered extensive areas of the Southern Hemisphere. Layers of glacial till were found at the same stratigraphic position in southern Africa and South America, as well as in India and Australia. Below these beds of glacial debris lay striated and grooved bedrock. In some locations the striations and grooves indicated that the ice had moved from what is now the sea onto land (Figure 18.6). Much of the land area containing evidence of this late Paleozoic glaciation presently lies within 30 degrees of the equator in a subtropical or tropical climate.

Could the earth have gone through a period sufficiently cold to have generated extensive continental glaciers in what is presently a tropical region? Wegener rejected this explanation because during the late Paleozoic, large swamps existed in the Northern Hemisphere. These swamps with their lush vegetation eventually became the major coal fields of the eastern United States, Europe, and Siberia.

Fossils from these coal fields indicate that the tree ferns which produced the coal deposits had large fronds that are indicative of a tropical setting. Furthermore, the tree trunks lacked growth rings, a characteristic of tropical plants caused by minimal seasonal fluctuations in temperature. Wegener believed that a better explanation for these paleoclimatic regimes is provided when the landmasses are fitted together as a supercontinent with South Africa centered over the South Pole. This would account for the conditions necessary to generate extensive expanses of glacial ice over much of the Southern Hemisphere. At the same time, this geography would place the northern landmasses nearer the tropics and account for their vast coal deposits. Wegener was so convinced that his explanation was correct that he wrote, "This evidence is so compelling that by comparison all other criteria must take a back seat."*

How does a glacier develop in hot, arid Australia? How do land animals migrate across wide expanses of open water? As compelling as this evidence may have been, fifty years passed before most of the scientific community would accept it and the logical conclusions to which it led.

THE GREAT DEBATE

Wegener's proposal did not attract much open criticism until 1924 when his book was translated into English, French, Spanish, and Russian. From this time on, until his death in 1930, his drift hypothesis encountered a great deal of hostile criticism. To quote the respected American geologist R. T. Chamberlin, "Wegener's hypothesis in general is of the foot-loose type, in that it takes considerable liberty with our globe, and is less bound by restrictions or tied down by awkward, ugly facts than most of its rival theories. Its appeal seems to lie in the fact that it plays a game in which there are few restrictive rules and no sharply drawn code of conduct." W. B. Scott, former president of the American Philosophical Society, expressed the prevalent American view of continental drift in fewer words when he described the theory, as "utter damned rot!"

One of the main objections to Wegener's hypothesis stemmed from his inability to provide a mechanism for continental drift. Wegener proposed two possible energy sources for drift. One of these, the tidal influence of the moon, was presumed by Wegener to be strong enough to give the continents a westward motion. However, the prominent physicist Harold Jeffreys quickly countered with the argument that tidal friction of the magnitude needed to displace the continents would bring the earth's rotation to a halt in a matter of a few years. Further, Wegener suggested that the larger and sturdier continents broke through the oceanic crust, much like ice breakers cut through ice. However, no evidence existed to suggest that the ocean floor was weak enough to permit passage of the continents without themselves being appreciably deformed in the process. By 1929 criticisms of Wegener's ideas were pouring in from all areas of the scientific community.

*Alfred Wegener, *The Origin of Continents and Oceans.* Translated from the 4th revised German edition of 1929 by J. Birman. (London: Methuen, 1966)

*Alfred Wegener, *The Origin of Continents and Oceans.* Translated from the 4th revised German edition of 1929 by J. Birman. (London: Methuen, 1966)

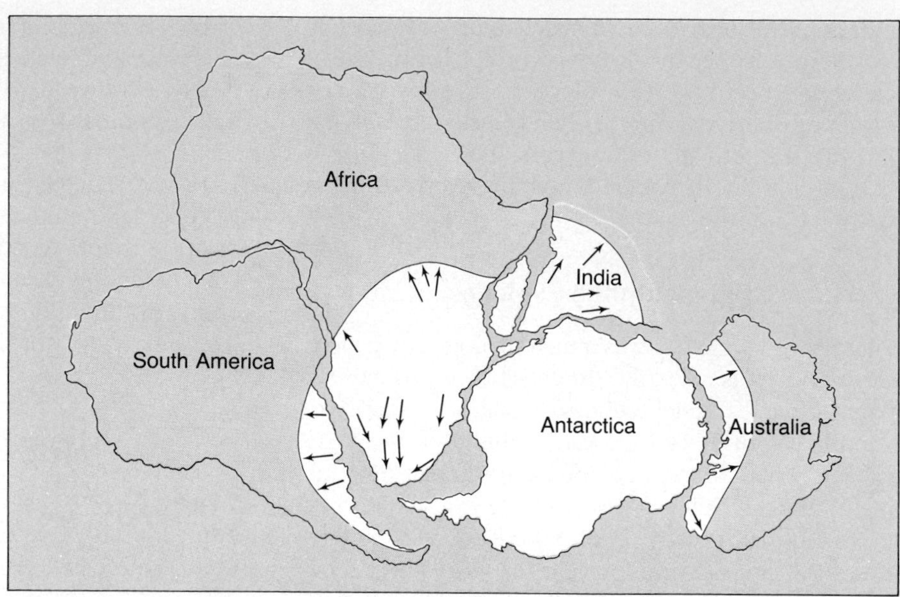

A.

B.

FIGURE 18.6
A. Direction of ice movement in the southern supercontinent called Gondwanaland by the founders of the continental drift concept. **B.** Glacial striations in the bedrock of Hallet Cove, South Australia, indicate direction of ice movement. (Photo by W. B. Hamilton, U.S. Geological Survey)

Despite these affronts, Wegener wrote the fourth and final edition of his book, maintaining his basic hypothesis and adding supporting evidence.

In 1930, Wegener made his third and final trip to the Greenland ice sheet (Figure 18.7). Although the primary focus of this expedition was to study the harsh winter weather on the ice-covered island, he planned to test his continental drift hypothesis as well. Wegener felt that by precisely establishing the locations of specific points and then measuring their changes over a period of years, he could demonstrate the westward drift of Greenland with respect to Europe. In November, 1930, while returning from Eismitte (an experimental station located in the center of Greenland), Wegener perished along with a companion. His intriguing idea, however, did not die with him.

Why was Wegener not able to overturn the established view of his day? Although his hypothesis was correct in principle, it also contained many incorrect details. For example, the continents do not break through the ocean floor, and tidal energy is not the driving mechanism for continental displacement. In order for any scientific viewpoint to gain universal

FIGURE 18.7
Alfred Wegener shown waiting out the 1912–1913 Arctic winter during an expedition to Greenland, where he made a 1200-kilometer traverse across the widest part of the island's ice sheet. (Photo courtesy of Bildarchiv Preussischer Kulturbesitz, Berlin)

acceptance, supporting evidence from all realms of science must be found. This same idea was stated very well by Wegener himself in response to his critics, when he said, "Scientists still do not appear to understand sufficiently that all earth sciences must contribute evidence toward unveiling the state of our planet in earlier times, and the truth of the matter can only be reached by combining all this evidence."* Wegener's great contribution to our understanding of the earth notwithstanding, not *all* of the evidence supported the continental drift hypothesis as he had proposed it. Therefore, Wegener himself answered the very question he must have asked many times: "Why do they reject my proposal?"

Although most of Wegener's contemporaries opposed his views, even to the point of openly ridiculing them, a few considered his ideas plausible. Among the most notable of this latter group were the eminent South African geologist Alexander du Toit and the well-known Scottish geologist Arthur Holmes. In 1937, du Toit published *Our Wandering Continents,* in which he eliminated some of Wegener's weaker points and added a wealth of new evidence in support of this revolutionary idea. Arthur Holmes contributed to the cause by proposing a plausible driving mechanism for continental drift. In Holmes' book *Physical Geology,* he suggested that convection currents operating within the mantle were responsible for propelling the continents across the globe. Although even to this day geologists are not in agreement on the nature of the driving mechanism for continental drift, the concept proposed by Holmes is still one of the most appealing.

For these few geologists who continued the search, the concept of continents in motion evidently provided enough excitement to hold their interest. Others undoubtedly viewed continental drift as a solution to previously unexplainable observations.

CONTINENTAL DRIFT AND PALEOMAGNETISM

Very little new light was shed on the continental drift hypothesis between the time of Wegener's death in 1930 and the early 1950s. Little was known about the land beneath the sea, which makes up over 70 percent of the earth's surface and was in fact the key to unraveling the secrets of our planet. Perhaps the initial impetus for the renewed interest in continental drift came from rock magnetism, a comparatively new field of study.

Early workers studying rock magnetism set out to investigate ancient changes in the earth's magnetic field in hopes of better understanding the nature of the present-day magnetic field. Anyone who has used a compass to find direction knows that the earth's magnetic field has a north pole and a south pole. These magnetic poles align closely, but not exactly, with the respective geographic poles. In many re-

*Alfred Wegener, *The Origin of Continents and Oceans.* Translated from the 4th revised German edition of 1929 by J. Birman. (London: Methuen, 1966)

spects the earth's magnetic field is very much like that produced by a simple bar magnet. Invisible lines of force pass through the earth and extend from one pole to the other (Figure 18.8). A compass needle, itself a small magnet free to move about, becomes aligned with these lines of force and thus points toward the magnetic poles.

The technique used to study ancient magnetic fields relies on the fact that certain rocks contain minerals that serve as fossil compasses. These iron-rich minerals such as magnetite are abundant, for example, in lava flows of basaltic composition. When heated above a certain temperature called the **Curie point**, these magnetic minerals lose their magnetism. However, when these iron-rich grains cool below their Curie point (about 580°C), they become magnetized in the direction parallel to the existing magnetic field. Once the minerals solidify, the magnetism they possess will remain "frozen" in this position. In this regard, they behave much like a compass needle inasmuch as they "point" toward the existing magnetic poles. If the rock is moved or the magnetic pole changes position, the rock magnetism will, in most instances, retain its original alignment. Rocks formed thousands or millions of years ago thus "remember" the location of the magnetic poles at the time of their formation and are said to possess fossil magnetism, or **paleomagnetism**.

One important aspect of rock magnetism is that the magnetized minerals not only indicate the direction to the poles (like a compass), but they also provide a means of determining the latitude of their origin. To envision how latitude can be established

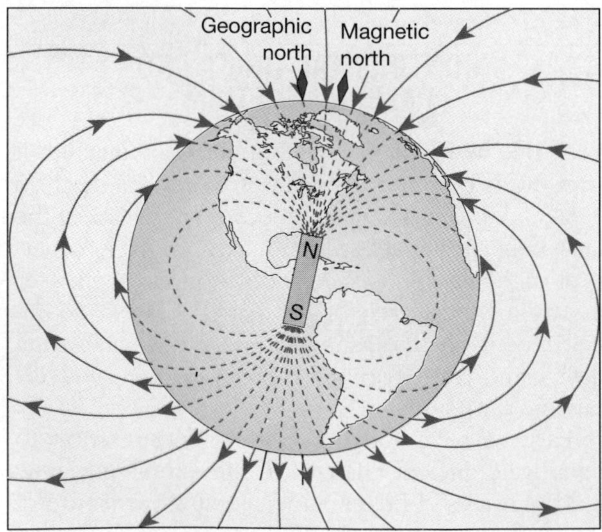

FIGURE 18.8
The earth's magnetic field consists of lines of force much like those a giant bar magnet would produce if placed at the center of the earth.

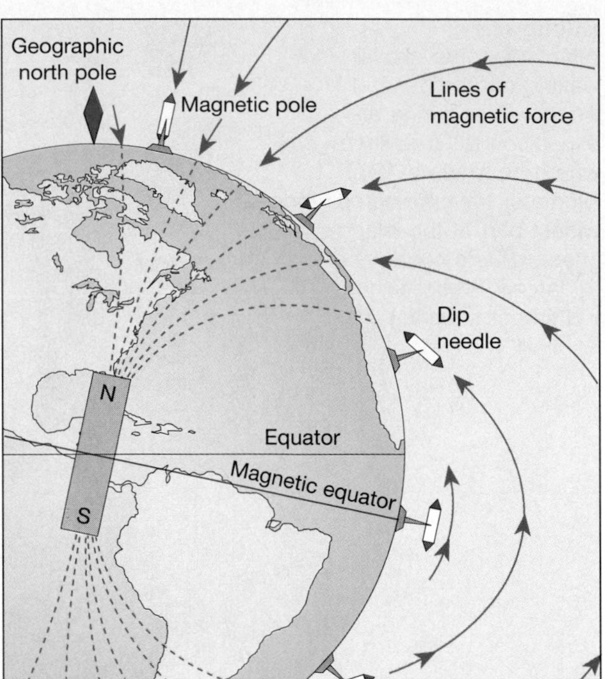

FIGURE 18.9
The earth's magnetic field causes a dip needle (compass oriented in a vertical plane) to align with the lines of magnetic force. The dip angle decreases uniformly from 90 degrees at the magnetic poles to 0 degrees at the magnetic equator. Consequently, the distance to the magnetic poles can be determined from the dip angle.

from paleomagnetism, imagine a compass needle mounted in a vertical plane rather than horizontally like an ordinary compass. As shown in Figure 18.9, when this modified compass (dip needle) is located over the north magnetic pole, it will point straight down. However, as this dip needle is moved closer to the equator, the angle of dip is reduced until it becomes horizontal at the equator. Thus, from the dip needle's angle of inclination, one can determine the latitude. In a similar manner, the orientation of the paleomagnetism in rocks indicates the latitude of the rock at the time it became magnetized.

A study conducted in Europe by S. K. Runcorn and his associates during the 1950s led to an interesting discovery. The magnetic alignment in the iron-rich minerals in lava flows of different ages was found to vary widely. A plot of the apparent position of the magnetic north pole revealed that during the past 500 million years the position of the pole had gradually wandered from a location near Hawaii northward through eastern Siberia and finally to its present location (Figure 18.10). This was clear evidence that either the magnetic poles had migrated through time, an idea known as *polar wandering,* or the continents had drifted.

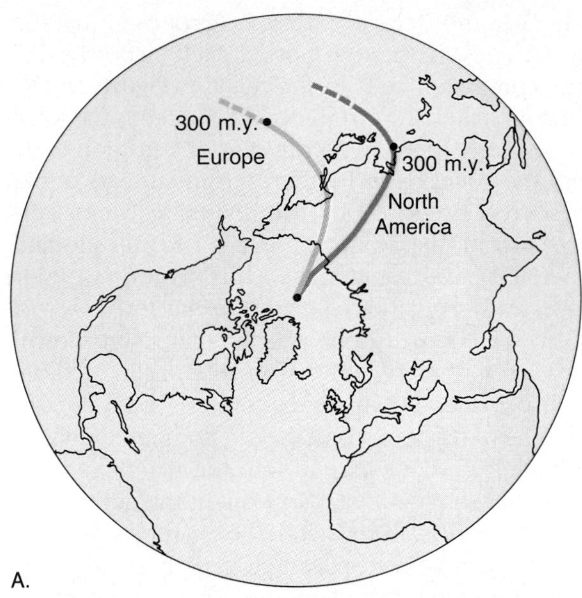

A.

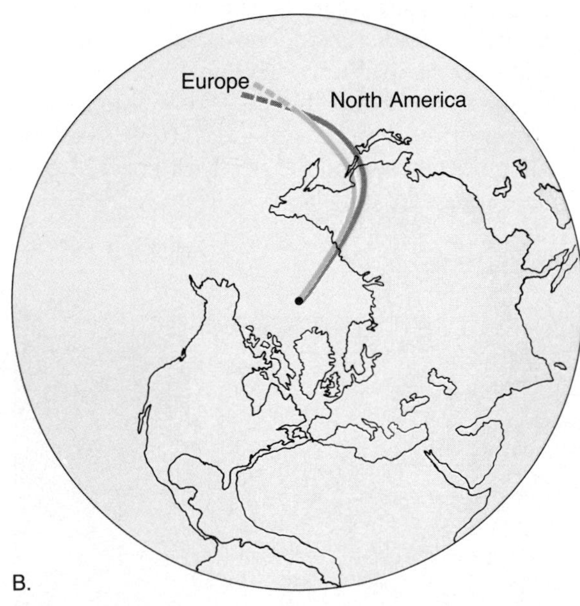

B.

FIGURE 18.10
Simplified apparent polar-wandering paths as established from North American and European paleomagnetic data. **A.** The more westerly path determined from North American data is thought to have been caused by the westward drift of North America by about 24 degrees from Europe. **B.** The positions of the wandering paths when the landmasses are reassembled in their pre-drift locations.

Although the magnetic poles are known to move, studies of the magnetic field indicate that the average positions of the magnetic poles correspond closely to the positions of the geographic poles. This is consistent with our knowledge of the earth's magnetic field, which is generated in part by the rotation of the earth about its axis. If the geographic poles do not wander appreciably, which we believe is true, neither can the magnetic poles. Therefore, a more acceptable explanation for the apparent polar wandering was provided by the continental drift hypothesis. If the magnetic poles remain stationary, their apparent movement was produced by the drifting of the continents.

The later idea was further supported by comparing the latitude of Europe as determined from rock magnetism with evidence from paleoclimatic studies. In particular, during the period when coal-producing swamps covered much of Europe, paleomagnetic evidence places Europe near the equator—a fact consistent with the tropical environment indicated by the coal deposits.*

Further evidence for continental drift came a few years later when a polar-wandering curve was constructed for North America (Figure 18.10A). To nearly everyone's surprise the curves for North America and Europe had similar paths, except that they were separated by about 30 degrees longitude. When these rocks solidified, could there have been two magnetic north poles which migrated parallel to each other? This is very unlikely. The differences in these migration paths, however, can be reconciled if the two presently separated continents are placed next to one another, as we now believe they were prior to opening of the Atlantic Ocean (Figure 18.10B).

Although these new data rekindled interest in continental drift, they by no means caused a major swing in opinion. For one thing, the techniques used in extracting paleomagnetic data were relatively new and untested. Furthermore, rock magnetism tends to weaken with time, and rocks can also obtain a secondary magnetization. Despite these problems and other conflicting evidence, some researchers were convinced that continental drift had indeed occurred. A new era had begun.

A SCIENTIFIC REVOLUTION BEGINS

During the 1950s and 1960s great technological strides permitted extensive and detailed mapping of the ocean floor. From this work came the discovery of a global oceanic ridge system. Examination of the Mid-Atlantic Ridge reveals a trend that parallels the continental margins on both sides of the Atlantic (see Figure 19.12). Also of importance was the discovery

*Fossil trees found in this coal lack tree rings (growth rings), a characteristic of tropical vegetation.

of a central rift valley extending for the length of the Mid-Atlantic Ridge, an indication that great tensional forces were at work. In addition, high heat flow and some volcanism were found to characterize the oceanic ridge system.

In other parts of the ocean additional discoveries were being made. Earthquake studies conducted in the vicinity of the deep-ocean trenches suggested activity was occurring at great depths beneath the ocean. Flat-topped seamounts hundreds of meters below sea level showed signs of formerly being islands. Of equal importance, dredging of the oceanic crust was unable to bring up rock that was older than Mesozoic age. Could the ocean floors actually be geologically young features?

Sea-Floor Spreading

In the early 1960s, all of these newly discovered facts were put together by Harry Hess of Princeton University into a hypothesis called **sea-floor spreading.** Hess was so lacking in confirmed data that he presented his paper as an "essay in geopoetry." Unlike its forerunner, continental drift, which essentially neglects the ocean basins, sea-floor spreading is centered on the activity beyond our direct view.

In Hess' now classic paper, he proposed that the ocean ridges are located above upwelling portions of large convection cells in the mantle (Figure 18.11). As rising material from the mantle spreads laterally, sea floor is carried in a conveyor-belt fashion away from the ridge crest. Further, tensional tears at the ridge crest produced by the diverging lateral currents provide pathways for magma to intrude and generate new oceanic crust. Thus, as the sea floor moves away from the ridge crest, newly formed crust replaces it. Hess further proposed that the downward limbs of these convection cells are located beneath the deep-ocean trenches. Here, according to Hess, the older portions of the sea floor are gradually consumed as they descend into the mantle. As one researcher summarized, "No wonder the ocean floor was young—it was constantly being renewed!"

With the sea-floor spreading hypothesis in place, Harry Hess had initiated another phase of this scientific revolution. The conclusive evidence to support his ideas came a few years later from the work of a young English graduate student, Fred Vine, and his supervisor, D. H. Matthews. The greatness of Vine's and Matthews' work was that they were able to connect two previously unrelated ideas: Hess' sea-floor spreading hypothesis and the newly discovered geomagnetic reversals.

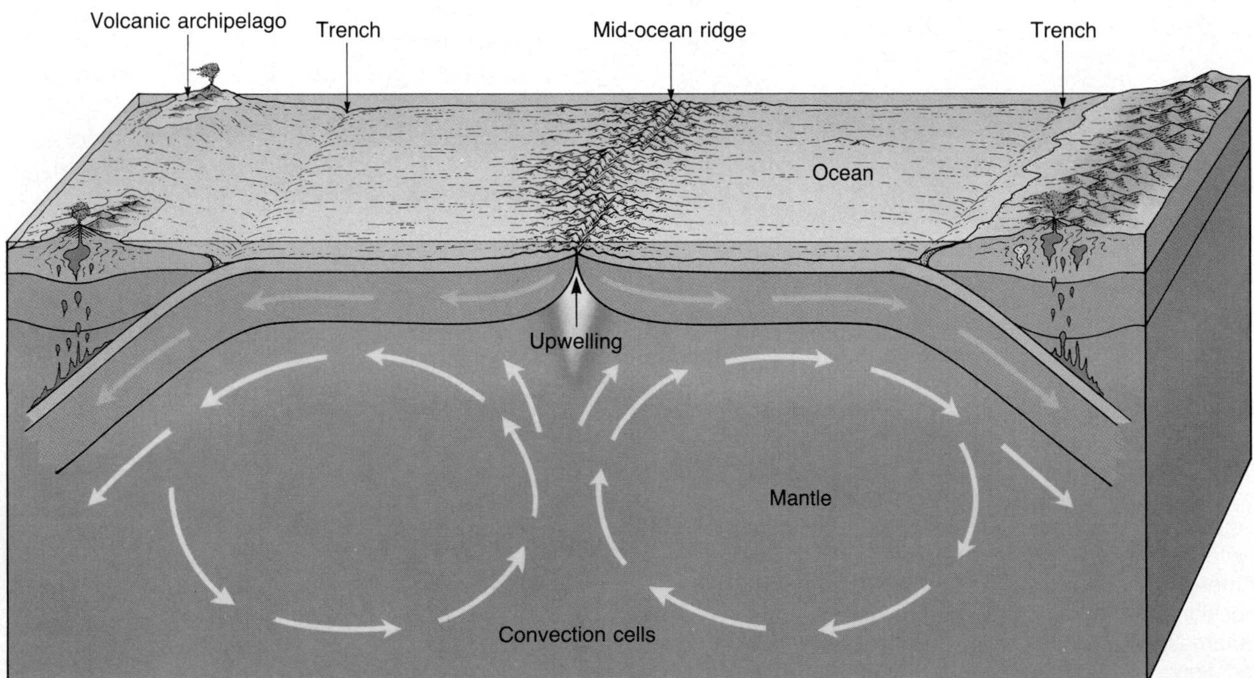

FIGURE 18.11
Sea-floor spreading. Harry Hess proposed that upwelling of mantle material along the mid-ocean ridge system created new sea floor. The convective motion of mantle material carries the sea floor in a conveyor-belt fashion to the deep-ocean trenches, where the sea floor descends into the mantle.

Geomagnetic Reversals

About the time that Hess formulated his ideas, geophysicists had begun to accept the fact that the earth's magnetic field periodically reverses polarity; that is, the north magnetic pole becomes the south magnetic pole, and vice versa. (See Box 17.1) A rock solidifying during one of the periods of reverse polarity will be magnetized with the polarity opposite that of rocks being formed today. When rocks exhibit the same magnetism as the present magnetic field, they are said to possess **normal polarity**, while those rocks exhibiting the opposite magnetism are said to have **reverse polarity**. Evidence for magnetic reversals was obtained from lavas and sediments from around the world. Once the concept of magnetic reversals was confirmed, researchers set out to establish a time scale for polarity reversals. There are many areas where volcanic activity has occurred sporadically for periods of millions of years. The task was to measure the directions of paleomagnetism in numerous lava flows of various ages (Figure 18.12). These data were collected from several places and were used to determine the dates when the polarity of the earth's magnetic field changed. Figure 18.13A shows the time scale of the polarity reversals established for the last few million years.

A significant relationship between magnetic reversals and the sea-floor spreading hypothesis was developed from data obtained when very sensitive instruments called **magnetometers** were towed by research vessels across a segment of the ocean floor located off the west coast of the United States. Here workers from the Scripps Institution of Oceanography discovered alternating stripes of high- and low-intensity magnetism that trended in roughly a north-south direction. This relatively simple pattern of magnetic variation defied explanation until 1963, when Fred Vine and D. H. Matthews* tied the discovery of the high- and low-intensity stripes to Hess' concept of sea-floor spreading. Vine and Matthews suggested that the stripes of high-intensity magnetism are regions where the paleomagnetism of the ocean crust is of the normal type. Consequently, these positively magnetized rocks enhance the existing magnetic field. Conversely, the low-intensity stripes represent regions where the ocean crust is polarized in the reverse direction and, therefore, weaken the existing magnetic field. But how do parallel stripes of normally and reversely magnetized rock become distributed across the ocean floor?

*This idea was also put forth a few months earlier by L. W. Morely, but his paper was rejected for publication because of its highly speculative nature.

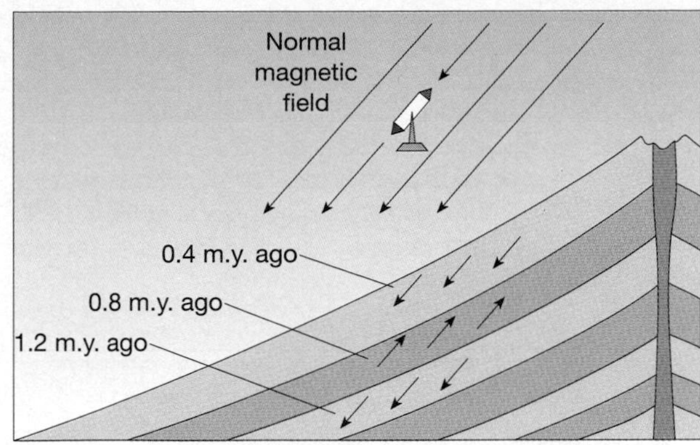

FIGURE 18.12
Schematic illustration of paleomagnetism preserved in lava flows of various ages. Data such as these from various locales were used to establish the time scale of polarity reversals shown in Figure 18.13A.

Vine and Matthews reasoned that as new basalt was added to the ocean floor at the oceanic ridges, it would be magnetized according to the existing magnetic field. Since the new rock is added in approximately equal amounts to both trailing edges of the spreading oceanic floor, we should expect strips of equal size and polarity to parallel both sides of the ocean ridges (Figure 18.13B). This explanation of the alternating strips of normal and reverse polarity, which lay as mirror images across the ocean ridges, was the strongest evidence so far presented in support of the concept of sea-floor spreading.

A short time later, Vine's and Matthews' proposal was supported by a study group from Columbia University's Lamont-Doherty Geological Observatory. Towing magnetometers across a segment of the Reykjanes Ridge lying south of Iceland, researchers found that the magnetic variations there were indeed symmetrical with the ridge crest (Figure 18.14). By 1968, magnetic variations having similar patterns were identified paralleling most oceanic ridges.

Now that the dates of the most recent magnetic reversals have been established, the rate at which spreading occurs at the various ridges can be determined accurately. In the Pacific Ocean, for example, the magnetic stripes are much wider for corresponding time intervals than those of the Atlantic Ocean. Hence, we conclude that a faster spreading rate exists for the spreading center of the Pacific as compared to the Atlantic. When we apply absolute dates to these magnetic events, we find that the spreading rate for the North Atlantic Ridge is only 1 or 2 centimeters per year.* The rate is somewhat faster for the

*Note that each side of the ridge spreads at this rate.

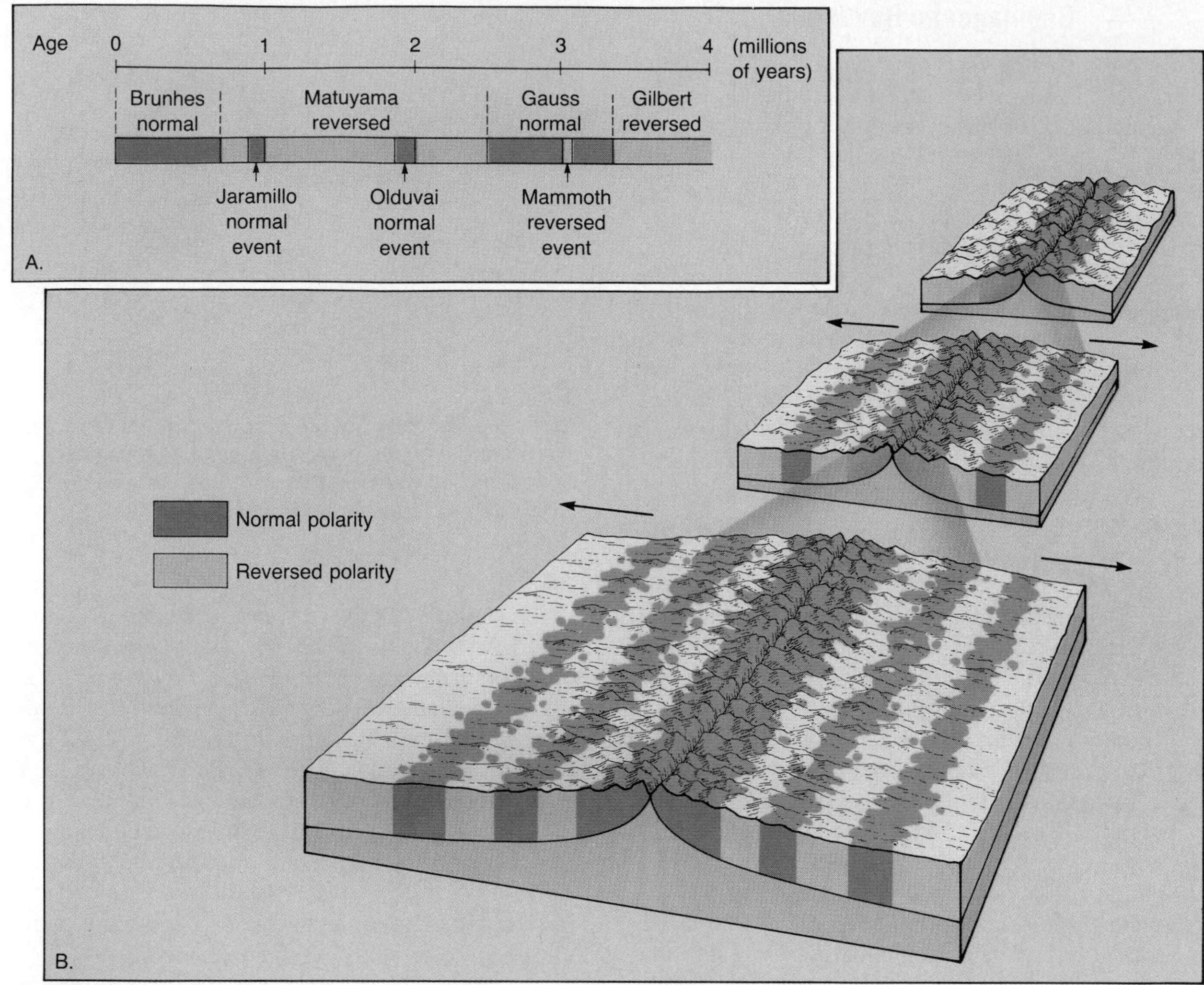

FIGURE 18.13
Magnetic evidence in support of the sea-floor spreading hypothesis. **A.** Time scale of the earth's magnetic field in the recent past. This time scale was developed by establishing the magnetic polarity for volcanic lavas of known ages. **B.** New sea floor records the polarity of the magnetic field at the time it formed. Hence it behaves much like a tape recorder, as it records each reversal of the earth's magnetic field. (Data from Allan Cox and G. B. Dalrymple)

South Atlantic. The spreading rates for the East Pacific Rise generally range between 3 and 8 centimeters per year, with a maximum rate of about 10 centimeters per year in one segment. Thus, not only had Vine and Matthews discovered a magnetic tape recorder that detailed changes in the earth's magnetic field, but this recorder could also be used to determine the rate of sea-floor spreading.

There is now general agreement that paleomagnetism was the most convincing evidence set forth in support of the concepts of continental drift and sea-floor spreading. By 1968 geologists began reversing their stand on this issue in a manner not unlike a magnetic reversal. The tide of scientific opinion had indeed switched in favor of a mobile earth.

PLATE TECTONICS: A MODERN VERSION OF AN OLD IDEA

By 1968, the concepts of continental drift and sea-floor spreading were united into a much more encompassing theory known as plate tectonics.

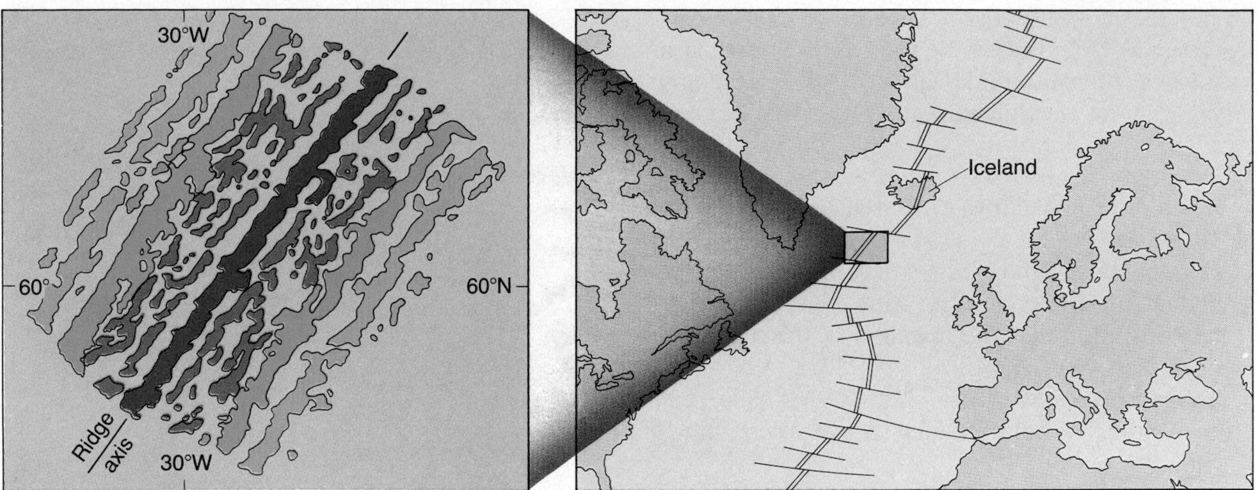

FIGURE 18.14
The symmetrical pattern of magnetic variation found across the Reykjanes Ridge just southwest of Iceland. The colored stripes are zones where the basaltic lavas have normal magnetization. The areas between the colored stripes represent zones of reverse magnetization. (After J. R. Heirtzler, S. Le Pichon, and J. G. Baron, *Deep-Sea Res.* 13 (1966): 247. Reprinted with permission from Pergamon Press, Ltd.)

The implications of plate tectonics are so far-reaching that this theory can be considered the framework from which to view most other geologic processes.

The theory of plate tectonics states that the outer, rigid lithosphere consists of several individual segments, called **plates.** So far, a dozen or so large and numerous smaller plates have been identified (Figure 18.15). Of these, the largest is the Pacific plate, which is located mostly within the ocean proper, except for a small sliver of North America that includes southwestern California and the Baja Peninsula. Notice from Figure 18.15 that all of the other large plates contain both continental and oceanic crust—a major departure from the continental drift theory, which proposed that the continents moved through, not with, the ocean floor. Many of the smaller plates, on the other hand, consist exclusively of oceanic material, as, for example, the Nazca plate located off the west coast of South America. Although not clearly defined in Figure 18.15, one small plate that roughly coincides with Turkey is located exclusively within a continent.

The lithosphere is the earth's rigid outer shell. It consists of a number of irregularly shaped plates that vary in thickness. Plates are thinnest in the oceans, where their thicknesses vary from as little as 10 kilometers at the ocean ridges to 100 kilometers in the deep-ocean basins. By contrast, continental blocks are generally 100 to 150 kilometers thick but may extend to 250 kilometers below the stable continental shields. Located below the lithosphere is the hotter and weaker zone known as the *asthenosphere.* The weak nature of the rock within the asthenosphere facilitates motion in the earth's rigid outer shell.

One of the main tenets of the plate tectonics theory is that plates are assumed to be rigid. Therefore, two places on the same plate are not in motion relative to each other. For example, as the plates move, the distance between New York and Denver, located on the same plate, remains unchanged, while the distance between New York and London, which are located on different plates, is constantly changing. Since each plate moves as a distinct unit, all major interactions between plates occur along plate boundaries. Thus, most of the earth's seismic activity, volcanism, and mountain building occur along these dynamic margins.

PLATE BOUNDARIES

For some time now, tectonic activity has been known to be concentrated along plate boundaries, such as the so-called *Ring of Fire* that encircles the Pacific. Thus, the first approximations of plate margins relied on the distribution of earthquake and volcanic activity. Later work indicated the existence of three distinct types of plate boundaries, each differentiated by the movement it exhibits (Figure 18.16). These are:

1. **Divergent boundaries**—where plates move apart, resulting in upwelling of material from the mantle to create new sea floor.
2. **Convergent boundaries**—where plates move together, causing one of the slabs of lithosphere to be consumed into the mantle as it descends beneath an overriding plate.
3. **Transform boundaries**—where plates slide past each other without creating or destroying lithosphere.

In the following sections we will briefly summarize the nature of these three types of plate boundaries. Then, in the next two chapters, the role of these plate margins in sea-floor spreading and mountain building will be considered in more detail.

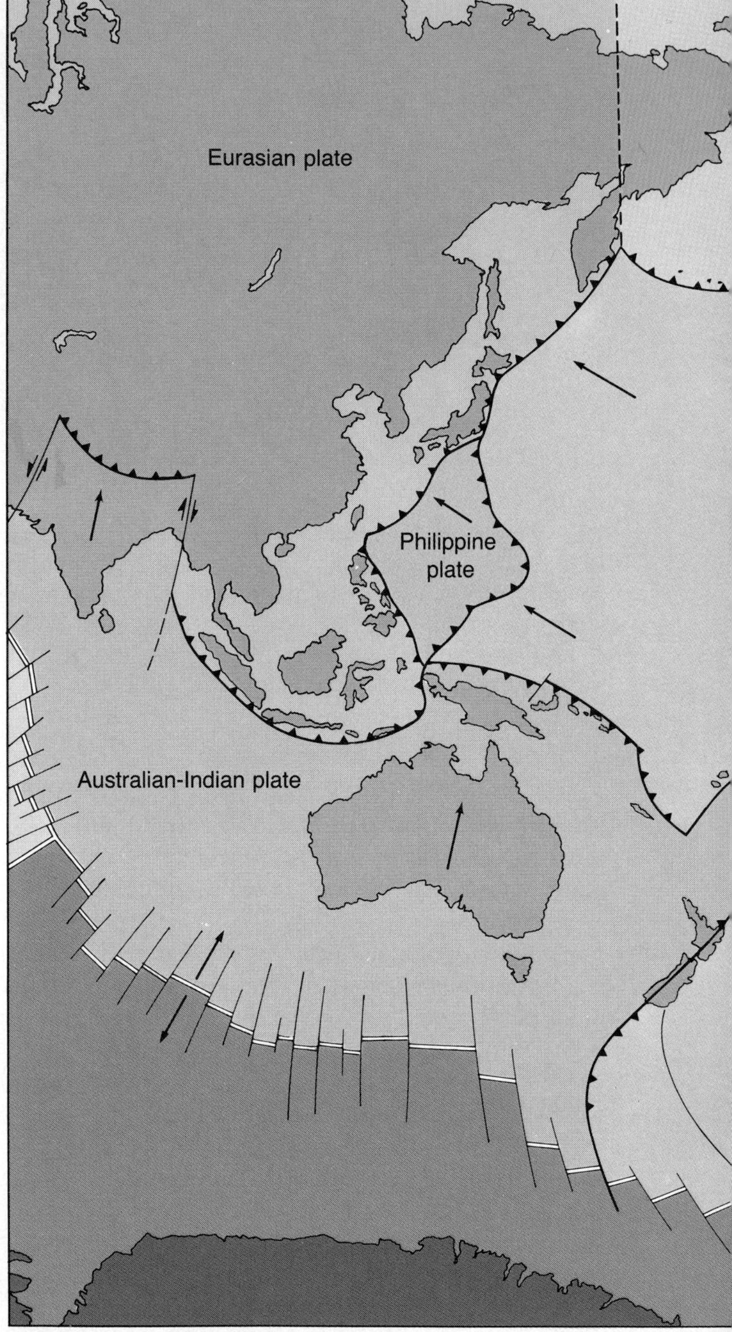

FIGURE 18.15
Mosaic of rigid plates that constitute the earth's outer shell. **A.** Divergent boundary. **B.** Convergent boundary. **C.** Transform boundary. (After W. B. Hamilton, U.S. Geological Survey)

Divergent Boundaries

Most divergent boundaries, where plate spreading occurs, are situated at the crests of oceanic ridges. Here, as the plates move away from the ridge axis, the fractures created are immediately filled with molten rock that oozes up from the hot asthenosphere. This material cools slowly to produce new slivers of sea floor. In a continuous manner, successive separations and injections of magma add new oceanic crust (lithosphere) between the diverging plates. As noted earlier, this mechanism is called sea-floor spreading and produced the floor of the Atlantic Ocean during the past 165 million years.

Not all spreading centers have been in existence as long as the Mid-Atlantic Ridge, and not all are found in the middle of large oceans. The Red Sea is believed to be the site of a recently formed divergent boundary. Here the Arabian Peninsula separated from Africa and began to move toward the northeast. Consequently, the Red Sea is providing oceanographers with a view of how the Atlantic Ocean may have looked in its infancy. Another narrow, linear sea produced by sea-floor spreading in the recent geologic past is the Gulf of California.

When a spreading center develops within a continent, the landmass may split into smaller segments as Wegener had proposed for the breakup of Pangaea. The fragmentation of a continent is thought to be associated with the upward movement of hot rock from below. The effect of this activity is to upwarp the crust directly above the hot rising plume. The crustal stretching associated with the doming generates numerous tensional cracks as shown in Figure 18.17A). As the plates move from the area of upwelling, the broken slabs are displaced downward, creat-

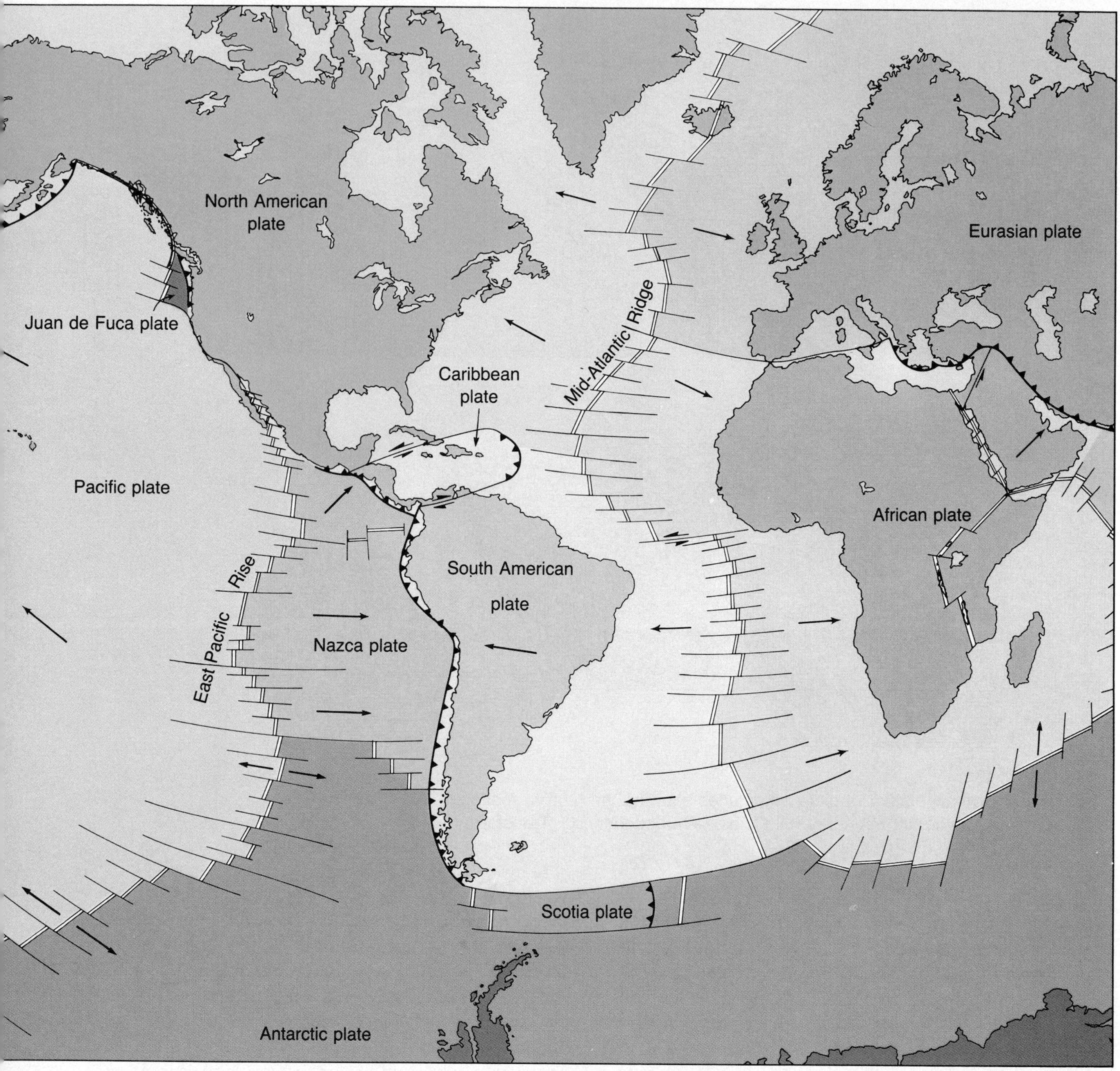

North American plate
Eurasian plate
Juan de Fuca plate
Mid-Atlantic Ridge
Caribbean plate
Pacific plate
African plate
South American plate
East Pacific Rise
Nazca plate
Scotia plate
Antarctic plate

A. Divergent plate boundaries
B. Convergent plate boundaries
C. Transform plate boundaries

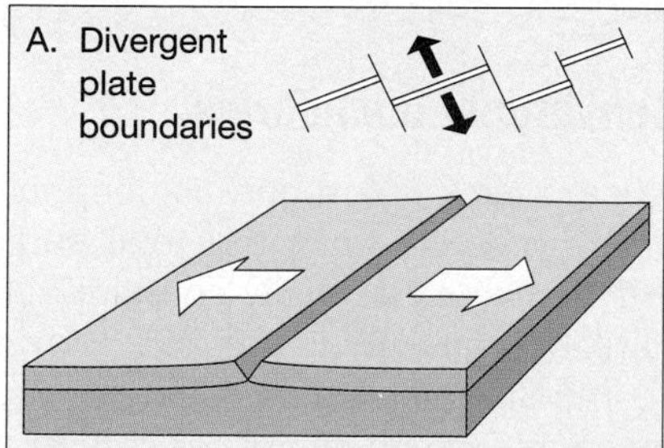

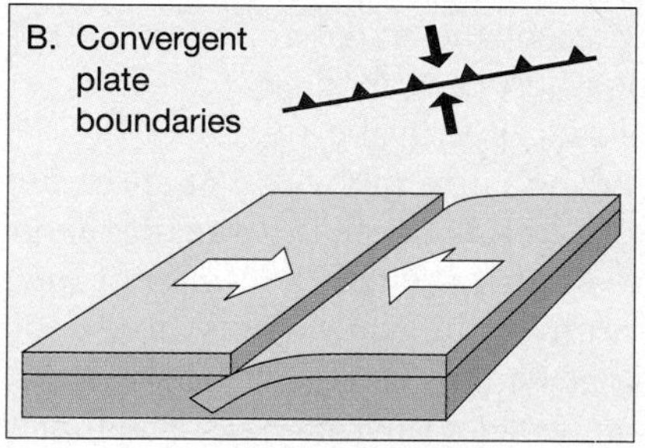

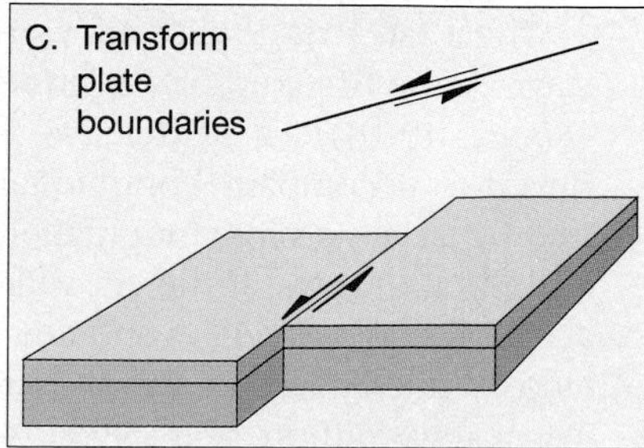

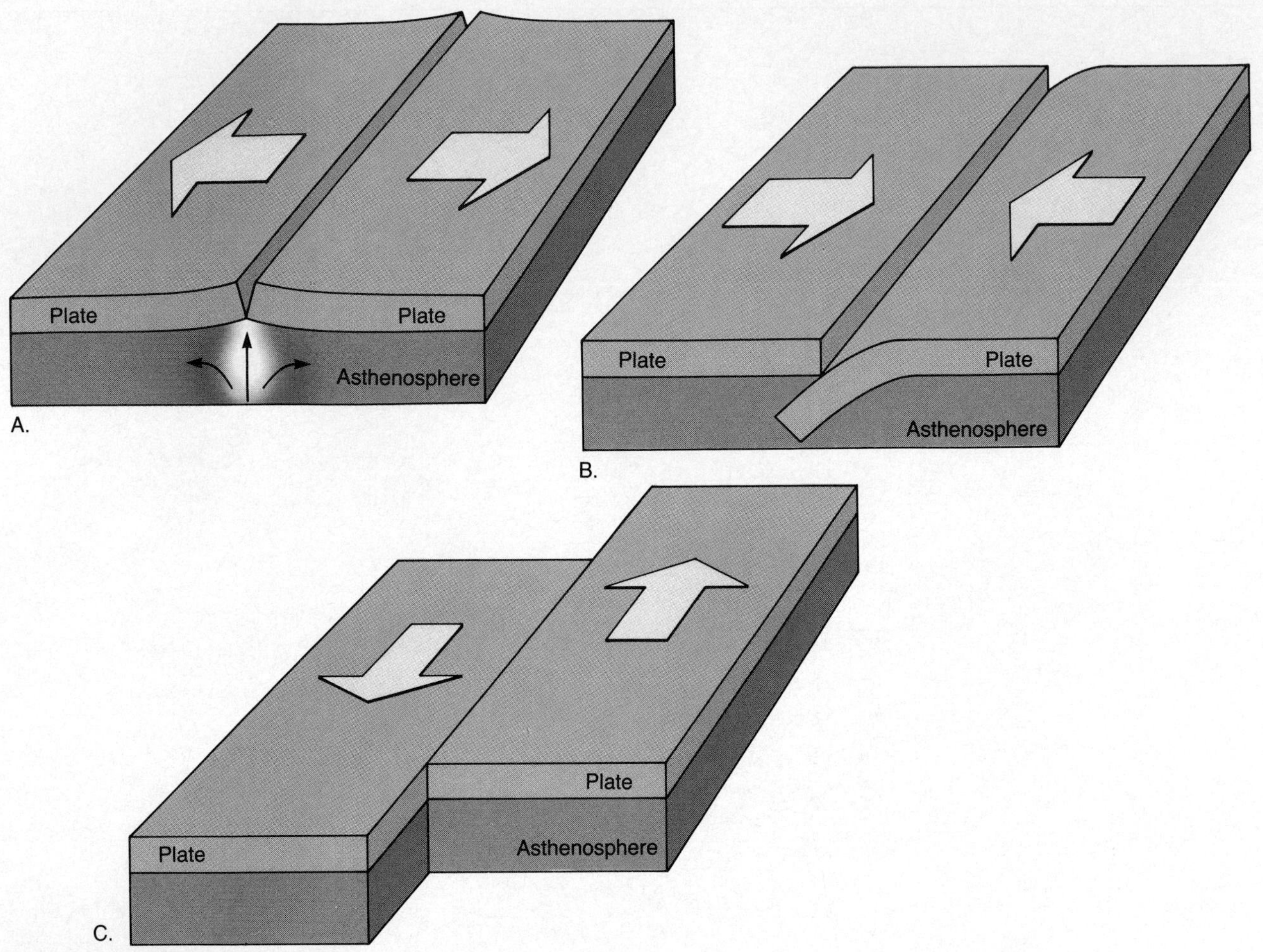

FIGURE 18.16
Schematic representation of plate boundaries showing the relative motion of plates. **A.** Divergent boundary. **B.** Convergent boundary. **C.** Transform boundary.

ing downfaulted valleys called **rifts** or **rift valleys** (Figure 18.17B). As the spreading continues, the rift valley will lengthen and deepen, eventually extending out into the ocean. At this point the valley will become a narrow linear sea with an outlet to the ocean similar to the Red Sea today (Figure 18.17C). The zone of rifting will remain the site of igneous activity, continually generating new sea floor in an ever-expanding ocean basin (Figure 18.17D).

The East African rift valleys represent the initial stage in the breakup of a continent as just described (Figure 18.18). The extensive volcanic activity believed to accompany continental rifting is exemplified by large volcanic mountains such as Kilimanjaro and Mount Kenya. If the rift valleys in Africa remain active, East Africa will eventually part from the mainland in much the same way the Arabian Peninsula did just a few million years ago. However, not all rift valleys develop into full-fledged spreading centers. Running through the central United States is an aborted rift zone extending from Lake Superior to Kansas. This once-active rift valley is filled with rock that was extruded onto the crust more than one billion years ago. Why one rift valley continues to develop while others are abandoned is not yet known.

Convergent Boundaries

At spreading centers new lithosphere is continually being generated; however, since the total surface area of the earth remains essentially constant, lithosphere must also be consumed. The zone of plate convergence is the site where lithosphere is reabsorbed into the mantle. When two plates collide, the leading edge of one is bent downward, allowing it to descend beneath the other. The typical angle of descent ranges from 35 to nearly 90 degrees from the surface.

Although all convergent zones are basically similar, the nature of plate collisions is influenced greatly by the type of crustal material involved. Convergence can occur between two oceanic plates, one oceanic and one continental plate, or two continental plates, as shown in Figure 18.19. Whenever the leading edge of a plate capped with continental crust converges with oceanic crust, the less dense continental material remains "floating," while the denser oceanic slab sinks into the asthenosphere. The region where an oceanic plate descends into the asthenosphere because of convergence is called a **subduction zone**. As the oceanic plate slides beneath the overriding plate, the oceanic plate bends, thereby producing a **deep-ocean trench** adjacent to the zone of subduction (Figure 18.19A). Trenches formed in this manner

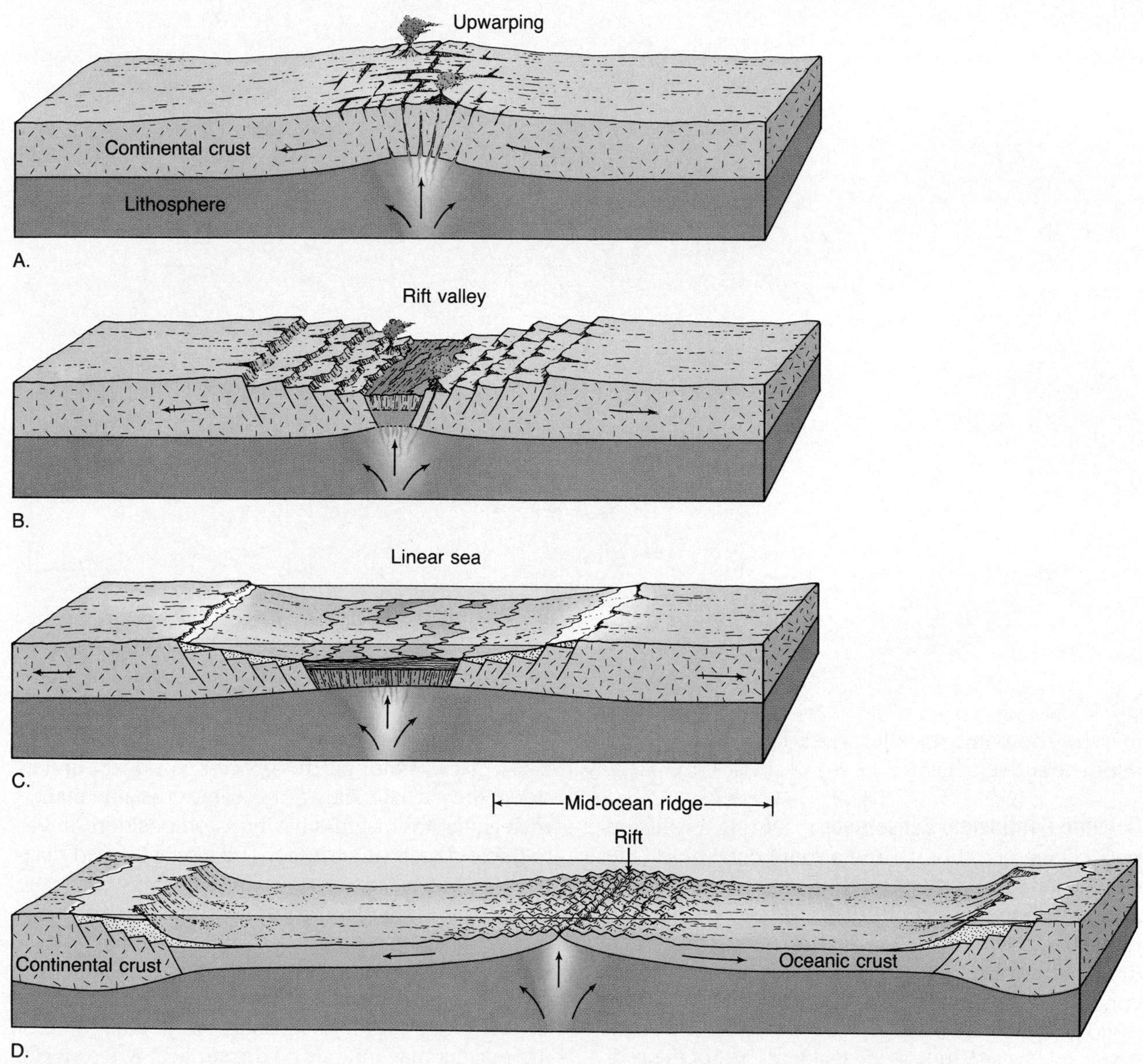

FIGURE 18.17
A. Rising magma upwarps the crust, causing numerous cracks in the rigid lithosphere. **B.** As the crust is pulled apart, large slabs of rock sink, generating a rift zone. **C.** Further spreading generates a narrow sea. **D.** Eventually, an expansive ocean basin and ridge system are created.

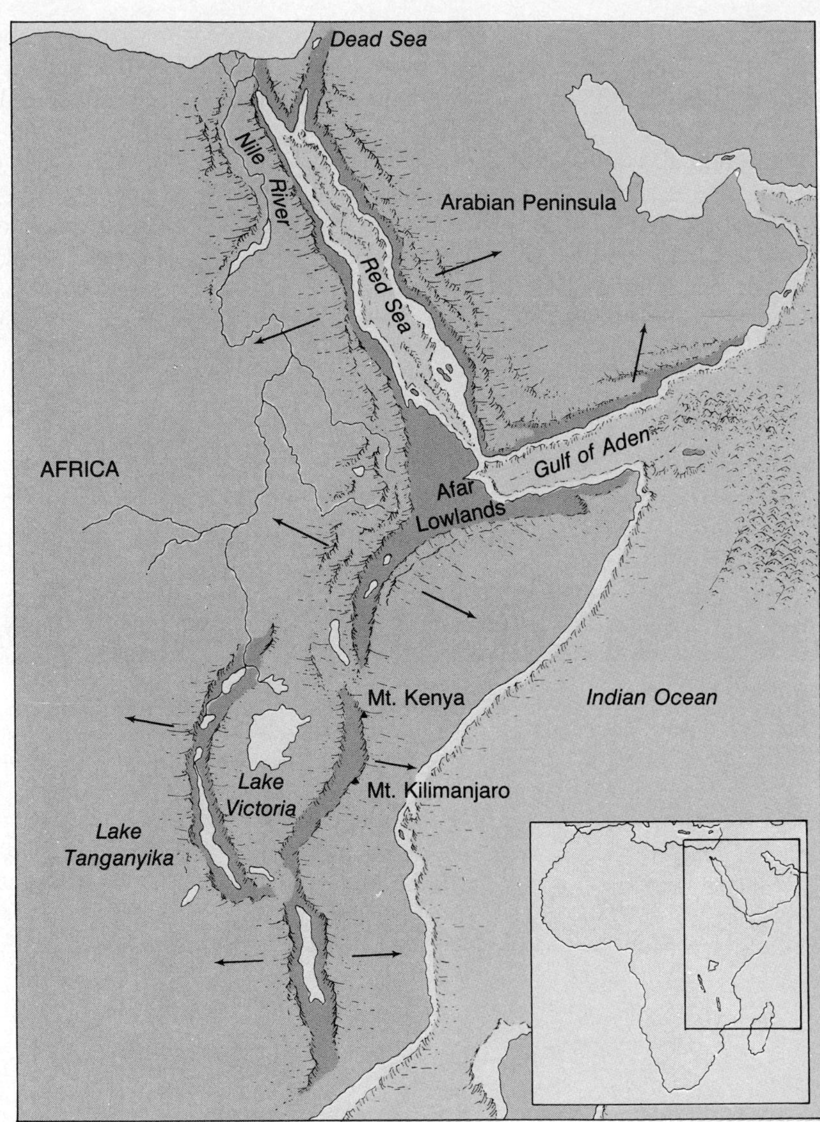

FIGURE 18.18
East African rift valleys and associated features.

may be thousands of kilometers long and 8 to 11 kilometers deep (Figure 18.20).

Oceanic-Continental Convergence During a collision between an oceanic slab and a continental block, the oceanic crust is bent, permitting its descent into the asthenosphere (Figure 18.19A). As the oceanic slab descends, some of the soft sediments carried upon the sinking plate are scraped off by the overriding continental block. Studies conducted in the coastal regions of western Mexico, where the Cocos plate is being subducted, indicate that at least half of the sediment carried on the descending plate can be removed in this manner. Therefore, this process contributes to the already substantial accumulation of sediment deposited along the continental margins.

When the descending oceanic plate reaches a depth of about 100 kilometers, partial melting of the water-rich ocean crust and the overlying mantle takes place. Although the process is poorly understood, the partial melting of oceanic crust and mantle rock generates magmas having a composition similar to that of basalt or andesite. The newly formed magmas created in this manner are less dense than the rocks of the mantle. Consequently, when sufficient quantities have accumulated, the molten rock will slowly buoy upward. Most of the rising magma will be emplaced in the overlying continental crust where it will cool and crystallize at depth. Some of the magma may migrate to the surface, where it can give rise to numerous and occasionally explosive volcanic eruptions. The volcanic portions of the Andes Mountains are believed to have been produced by such activity when the Nazca plate melted as it descended beneath the continent of South America (see Figure 18.15). The frequent earth-

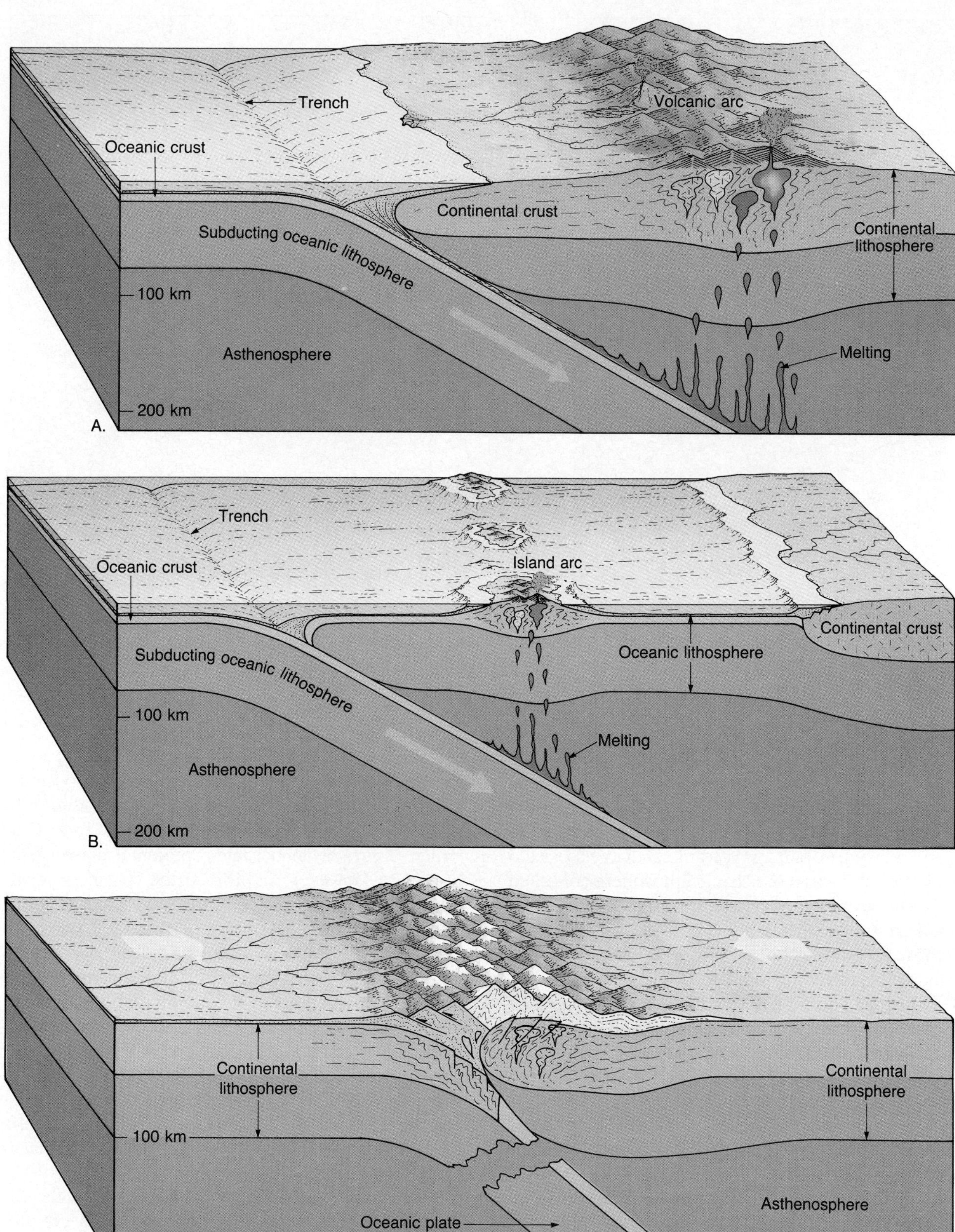

FIGURE 18.19
Zones of plate convergence. **A.** Oceanic-continental. **B.** Oceanic-oceanic. **C.** Continental-continental.

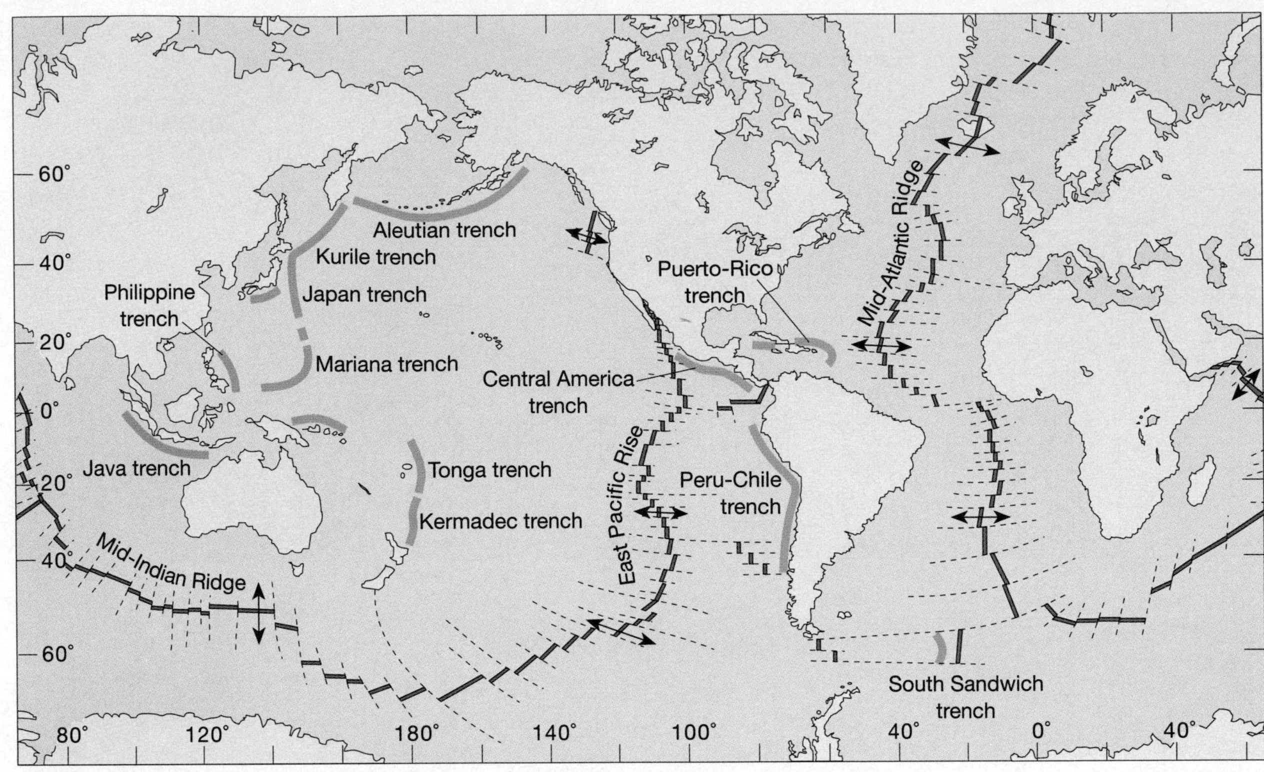

FIGURE 18.20
Distribution of the world's oceanic trenches, ridge system, and transform faults. Where transform faults offset ridge segments, they permit the ridge to change direction (curve) as can be seen in the Atlantic Ocean.

quakes occurring within the Andes testify to the activity beyond our view.

Mountains such as the Andes that are believed to be produced in part by volcanic activity associated with the subduction of oceanic lithosphere are called **volcanic arcs**. Two volcanic arcs are located in the western United States. One of these, the Cascade Range, is composed of several well-known volcanic mountains, including Mounts Rainier, Shasta, and St. Helens (see Figure 4.32). As the recent eruptions of Mount St. Helens testify, the Cascade Range is still quite active. The magma here arises from the melting of a small remaining segment of the Juan de Fuca plate. Since the rate of subduction at this plate is slow, the volcanoes of the Cascades are thought to be magma deficient, which partly accounts for their rather sporadic activity. The second volcanic arc is the Sierra Nevada, in which Yosemite National Park is located. The Sierra Nevada system is much older and has been inactive for several million years as evidenced by the absence of volcanic cones. Here erosion has stripped away most of the obvious traces of volcanic activity and left exposed the large, crystallized magma chambers that once fed lofty volcanoes.

Oceanic-Oceanic Convergence When two oceanic slabs converge, one descends beneath the other, initiating volcanic activity in a manner similar to that which occurs at an oceanic-continental convergent boundary. However, in this case, the volcanoes form on the ocean floor rather than on land (Figure 18.19B). If this volcanic activity is sustained, dry land will eventually emerge from the ocean depths. In early stages of development, this newly formed land consists of a chain of small volcanic islands called an **island arc**. The Aleutian, Mariana, and Tonga islands exemplify such features. Island arcs such as these are generally located a few hundred kilometers from an ocean trench where active subduction of the lithosphere is occurring. Adjacent to the island arcs just mentioned are the Aleutian trench, the Mariana trench, and the Tonga trench.

Over an extended period, numerous episodes of volcanic activity build large volcanic piles on the ocean floor. This volcanic activity, plus the buoyancy of the intrusive igneous rock emplaced within the crust below, gradually increases the size and elevation of the developing arc. This growth, in turn, increases the amount of eroded sediments added to the sea floor. Some of these sediments reach the trench and are deformed and metamorphosed by the compressional forces exerted by the two converging plates. The result of these diverse activities is the development of a mature island arc composed of a complex system of volcanic rocks, folded and metamorphosed sedimentary rocks, and intrusive igneous rocks. Examples of mature island arc systems are the Alaskan Peninsula, the Philippines, and Japan.

Continental-Continental Convergence When two plates carrying continental crust converge, neither plate will subduct beneath the other because of the low density and thus the buoyant nature of continental rocks. The result is a collision between the two continental blocks (Figure 18.19C). Such a collision is believed to have occurred when the once-separated continent of India "rammed" into Asia and produced the Himalayas, perhaps the most spectacular mountain range on earth (Figure 18.21). During this collision, the continental crust buckled, fractured, and was generally shortened. In addition to the Himalayas, several other major mountain systems, including the Alps, Appalachians, and Urals, are thought to have formed during continental collisions.

Prior to a continental collision, the landmasses involved are separated by an ocean basin. As the continental blocks converge, the intervening sea floor is subducted beneath one of the plates. The partial melting of the descending ocean slab and adjacent mantle rocks generates a volcanic arc. Depending on the location of the subduction zone, the volcanic arc could develop on either of the converging landmasses, or if the subduction zone developed at an appreciable distance into the ocean, an island arc would form. In any case, erosion of the newly formed volcanic arc would add large quantities of

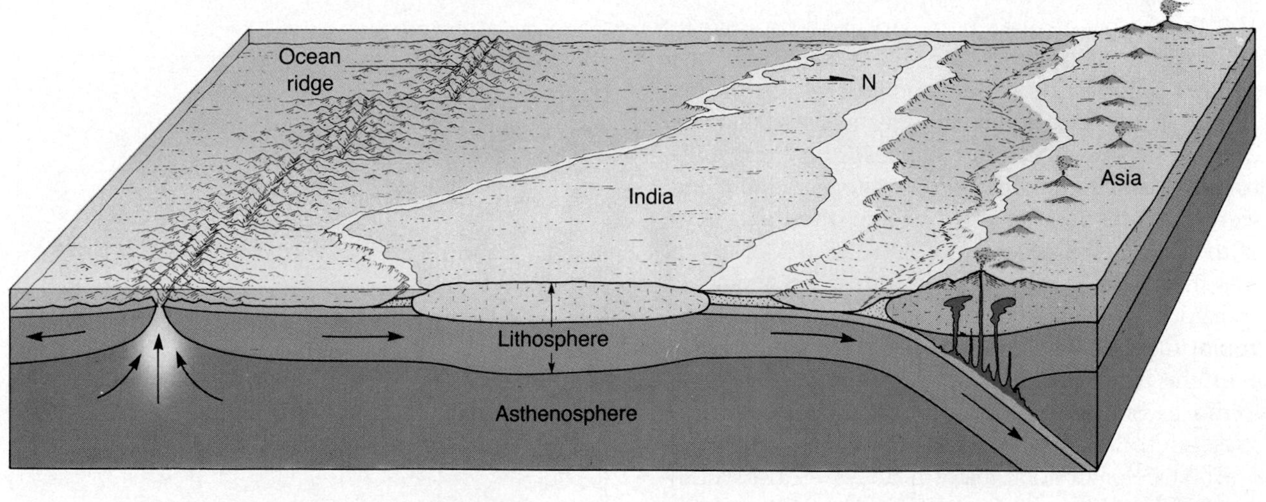

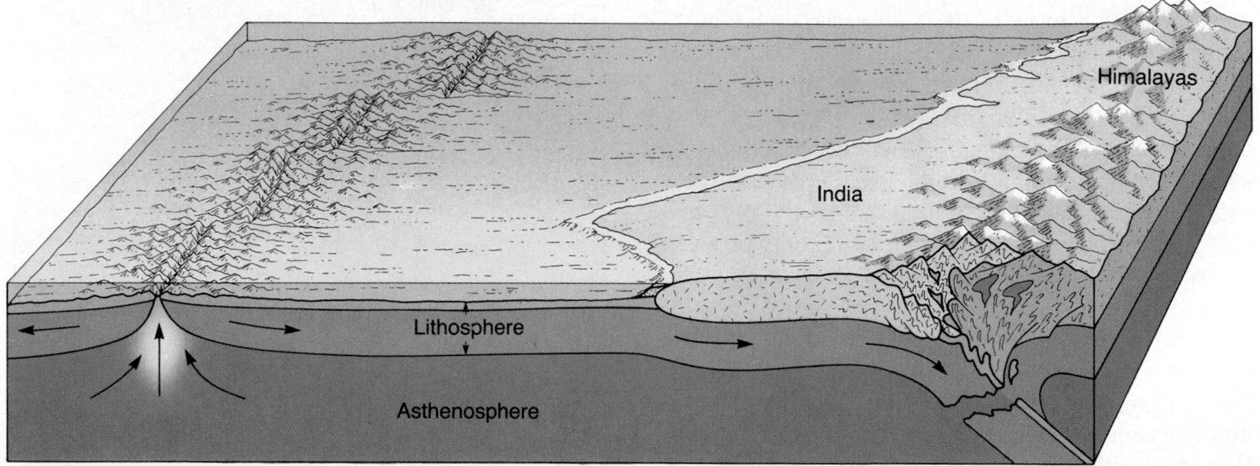

FIGURE 18.21
The collision of India and Asia about 45 million years ago produced the majestic Himalayas.

sediment to the already sediment-laden continental margins. Eventually, as the intervening sea floor was consumed, these continental masses would collide, thereby squeezing, folding, and generally deforming the sediments as if they were placed in a gigantic vise. The result would be the formation of a new mountain range composed of deformed sedimentary rocks and fragments of the volcanic arc.

After continents collide, the descending oceanic material may separate from the continental block and continue moving downward. However, because of its buoyancy, continental lithosphere cannot be carried very far into the mantle. In the case of the Himalayas, the leading edge of the Indian plate was forced partially under Asia, generating an unusually great thickness of continental lithosphere. This accumulation accounts, in part, for the high elevation of the Himalayas and the Tibetan Plateau to the north.

Transform Boundaries

The third type of plate boundary is the transform fault, which is located where plates slide past one another without the production of crust, as occurs along oceanic ridges, or without the destruction of crust, as occurs at oceanic trenches. Transform faults roughly parallel the direction of plate movement and were first identified where they join offset segments of the oceanic ridge system (Figure 18.20). At first it was erroneously assumed that the ridge segments were originally aligned and had been offset by horizontal displacement along these large faults. However, the relative motion along these fault zones was found to be in the opposite direction required to produce the offsets observed.

The true nature of transform faults was discovered in 1965 by J. Tuzo Wilson of the University of Toronto. Wilson suggested that these large fractures connected the global active belts into a continuous network that divides the earth's outer shell into several rigid plates. Thus, Wilson became the first to suggest that the earth was made of individual plates, while at the same time identifying the zones along which relative motion between the plates is made possible. In this latter role, transform faults provide the means by which the oceanic crust created at the ridge crests can be transported to its site of destruction, the deep-ocean trenches. Figure 18.22 illustrates this role. Notice that the Juan de Fuca plate moves in a southeasterly direction, eventually being subducted under the west coast of the United States. The southern end of this relatively small plate is bounded by the Mendocino transform fault. This transform fault boundary connects an active spreading center to a subduction zone. Therefore, the fault facilitates the movement of the crustal material created at the ridge crest to its destination beneath the North American continent.

Wilson called these special faults *transform faults* because the relative motion of the plates can be changed, or transformed, along them. As we saw in the preceding example, divergence occurring at a spreading center can be transformed into convergence at a subduction zone. Since transform faults connect convergent and divergent boundaries in various combinations, other changes in relative plate motions are possible along transform faults.

Most commonly, transform faults join two ridge segments. When a transform fault connects two spreading centers as shown in Figure 18.23, the newly created sea floor is moving in opposite directions only in the region between the two ridges. Thus, the only active part of the fault lies between the two offset ridge segments. This active zone is also a site of frequent, but generally weak, seismic activity.

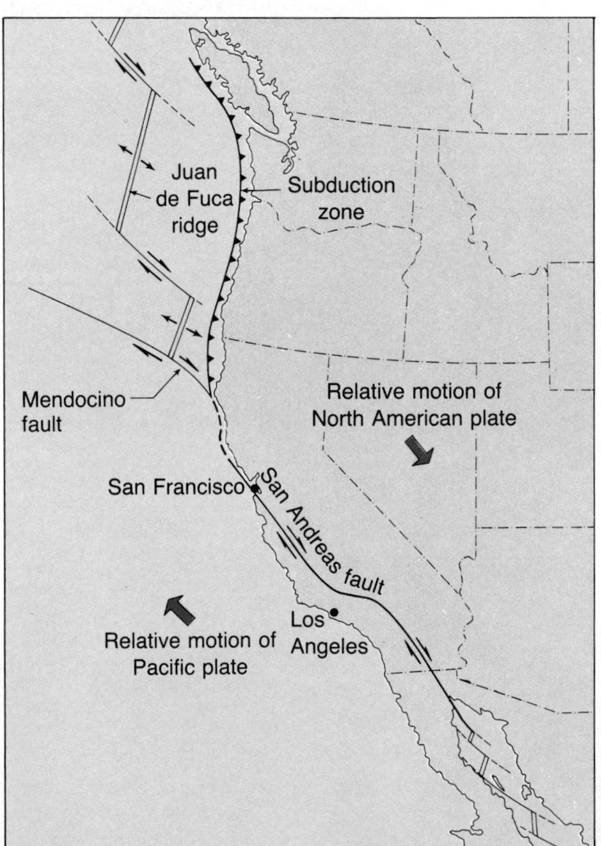

FIGURE 18.22
The role of transform faults in permitting relative motion between adjacent plates. The Mendocino transform fault permits sea floor generated at the Juan de Fuca ridge to move southeastward past the Pacific plate.

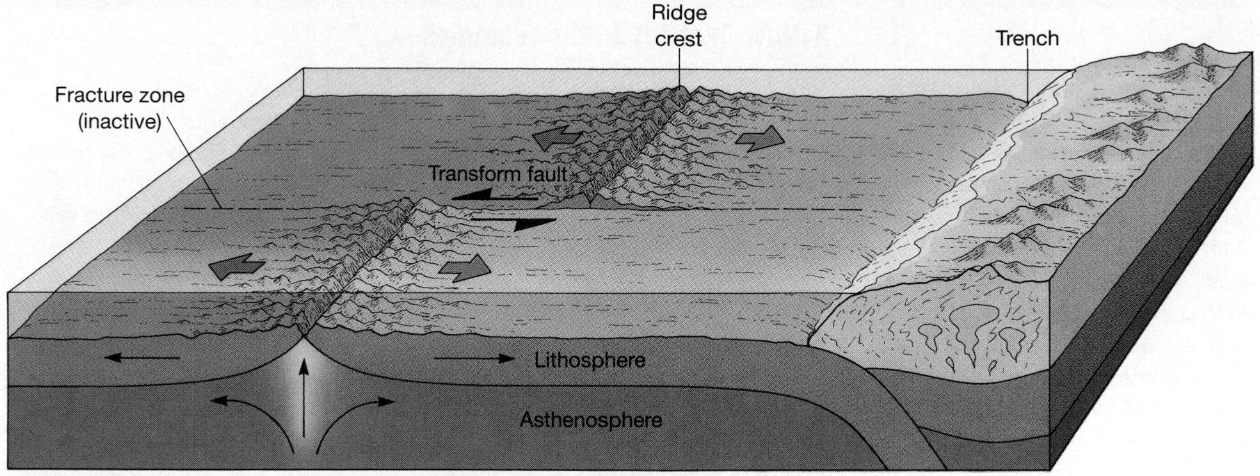

FIGURE 18.23
Transform faults often connect offset segments of an oceanic ridge. Note that the lithosphere is moving in opposite directions in the region between the two ridges.

Although most transform faults are located within the ocean basins, a few, including California's San Andreas fault, cut through continental crust (Figure 18.22). Along the San Andreas fault, the Pacific plate is moving toward the northwest, past the North American plate. If this movement continues, that part of California west of the fault zone, including the Baja Peninsula, will become an island off the west coast of the United States and Canada, and it could eventually reach Alaska. However, the more immediate concerns are the earthquakes triggered by movements along this fault system.

TESTING THE MODEL

With the birth of the plate tectonics model, researchers from all of the earth sciences began testing this proposal. Some of the evidence supporting continental drift and sea-floor spreading has already been presented in this chapter. In addition, some of the evidence that was instrumental in solidifying the support for this new concept follows. It should be pointed out that much of this evidence was not new; rather, it was new interpretations of old data that swayed the tide of opinion. Further, some of the data were compiled to refute rather than support global tectonics. As one researcher said, "My observations are not compatible with sea-floor spreading, and I shall prepare a critical demonstration that this is so and thus demolish this nutty idea and we can all get back to work." However, he, like many others, found that his results were indeed compatible with this new theory. The revolution had produced a new model from which to view all tectonic processes operating on the earth.

Plate Tectonics and Earthquakes

By 1968, the basic outline of global tectonics was firmly established. In this same year, three seismologists at Lamont-Doherty Geological Observatory published papers demonstrating how successfully the new plate tectonics model accounted for the global distribution of earthquakes (Figure 18.24). In particular, these scientists were able to account for the close association between deep-focus earthquakes and trench-volcanic arc systems. Furthermore, the absence of deep-focus earthquakes along the oceanic ridge system was also shown to be consistent with the new theory.

Note the close association between plate boundaries and earthquakes by comparing the distribution of earthquakes shown in Figure 18.24 with the map of plate boundaries in Figure 18.15. In trench regions, where dense slabs of lithosphere plunge into the mantle, this association is especially striking. When the depths of earthquake foci and their locations within the trench systems are plotted, an interesting pattern emerges. Figure 18.25, which shows the distribution of earthquakes in the vicinity of the Japan trench, is an example. Here most shallow-focus earthquakes occur within, or adjacent to, the trench, whereas intermediate- and deep-focus earthquakes occur toward the mainland. A similar distribution pattern exists along the western margin of South

BOX 18.1

A New Test for Plate Tectonics

Until the late 1980s the evidence supporting the theory of plate tectonics was acquired from the study of geologic phenomena such as volcanoes, earthquakes, and sea-floor sediments. Recently, however, it became possible to test the theory directly. Specifically, scientists are now able to confirm the fact that the plates shift in relation to one another in the way that the plate tectonics theory predicts.

The new evidence comes from two different techniques which allow distances between widely separated points on the earth's surface to be measured with unprecedented accuracy. Called *Satellite Laser Ranging* (SLR) and *Very Long Baseline Interferometry* (VLBI), these methods can detect the motion of any one site with respect to another at a level of better than 1 centimeter per year. Thus, for the first time, scientists can directly measure the relative motions of the earth's plates. Further, because these techniques are quite different, scientists use them to cross-check one against the other by comparing measurements made for the same sites.

The Satellite Laser Ranging system employs ground-based stations that bounce laser pulses off satellites whose orbital positions are well established. Precise timing of the round-trip travel of these pulses allows scientists to calculate the precise locations of the ground-based stations. By monitoring these stations over time, researchers can establish the relative motions of the sites.

FIGURE 18.A
Radio telescopes like these located at Socorro, New Mexico, are used to accurately determine the distance between two distant sites. Data collected by repeated measurements have detected relative plate motions of from 1 to 12 centimeters per year between various sites worldwide. (Photo by Geoff Chester)

The Very Long Baseline Interferometry system uses large radio telescopes to record signals from very distant quasars (Figure 18.A). Since quasars (quasi-stellar objects) lie billions of light years from the earth, they act as stationary reference points. The millisecond differences in the arrival time of the same signal at different earthbound observational sites provides a means of establishing the distance between receivers.

Confirming data from these two techniques leave little doubt that real plate motion has been detected. Calculations show that Hawaii is moving in a northwesterly direction and approaching Japan at a rate of 8.3 centimeters per year. Moreover, a site located in Maryland is retreating from one in England at a rate of about 1.7 centimeters per year. This rate is roughly equal to the 2.2 centimeters per year of sea-floor spreading that was established from paleomagnetic evidence.

America where the Nazca plate is being subducted beneath the continent.

In the plate tectonics model, deep-ocean trenches are produced where cold, dense slabs of oceanic lithosphere plunge into the mantle (Figure 18.26). Shallow-focus earthquakes are produced as the descending plate interacts with the overriding lithosphere. As the slab descends farther into the asthenosphere, deeper-focus earthquakes are generated. Since the earthquakes occur within the rigid subducting plate rather than in the "plastic" mantle, they provide a method for tracking the plate's descent. Recall that the zones of inclined seismic activity that extend from the trench into the mantle are called *Benioff zones* after an American seismologist who conducted extensive studies on the distribution of earthquake foci. Very few earthquakes have been recorded below 690 kilometers, possibly because the slab has been heated sufficiently to lose its rigidity.

The cause of intermediate- and deep-focus earthquakes occurring between 70 and 690 kilometers has been a long-standing problem in geology. At depths below about 70 kilometers, rocks are expected to deform by ductile flow rather than by brittle fracture or frictional sliding. With the development of the plate tectonics theory, researchers were provided a new avenue to explore this old problem. Experimental evidence indicates that brittle fracturing within a cold descending slab may occur to depths as great as 300 kilometers. Thus, intermediate-focus earthquakes could be generated in a manner similar to shallow quakes, that is, through the release of elastic

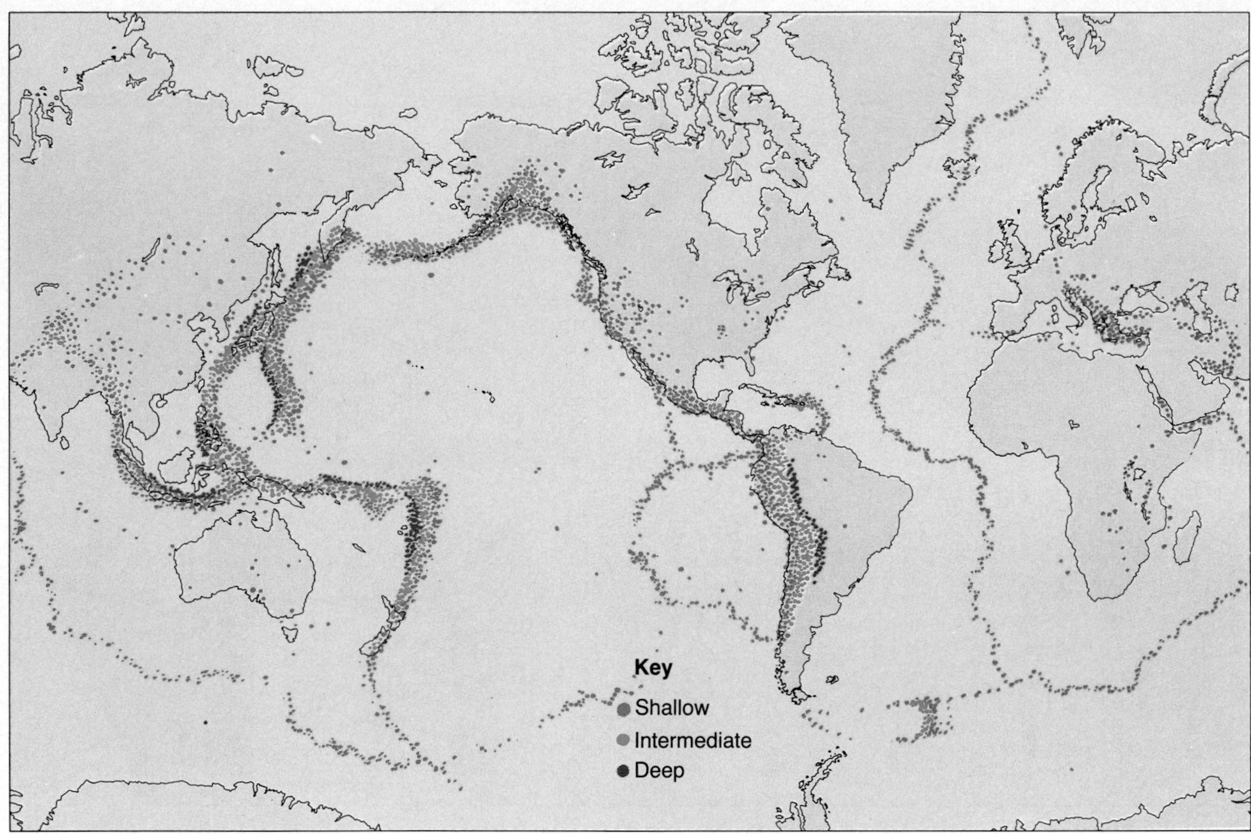

FIGURE 18.24
Distribution of shallow-, intermediate-, and deep-focus earthquakes. (Data from NOAA)

energy along a fault surface. But what causes deep-focus earthquakes?

The mechanism responsible for earthquakes that occur between 300 and 690 kilometers remains elusive. One hypothesis that has gained substantial support is based on evidence that during subduction, increasing pressures cause some minerals, such as olivine, to go through a phase change. A phase change produces a mineral having a more compact crystalline structure. It has been suggested that when this transformation occurs in the coldest plate interiors, it causes a type of high-pressure faulting, which in turn produces a deep-focus earthquake. To test this and other proposals, researchers are duplicating mantle conditions by using specially designed anvils that can apply enormous pressures to minute rock samples between two diamonds. The samples are heated by shining a laser through one of the diamonds, and sensors are used to detect acoustical events thought to be analogous to earthquakes. Although these studies are preliminary, they do indicate that at high pressures and temperatures, minerals can fail through slippage in the manner predicted.

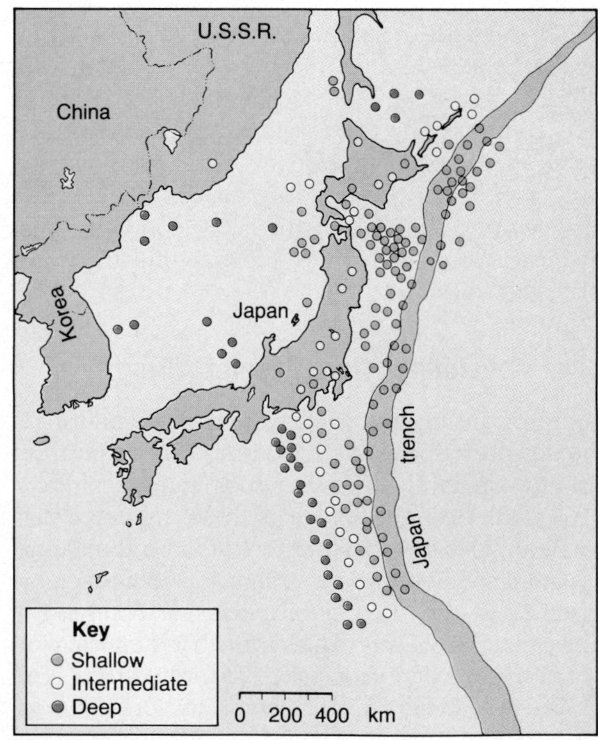

FIGURE 18.25
Distribution of earthquake foci in the vicinity of the Japan trench. (Data from NOAA)

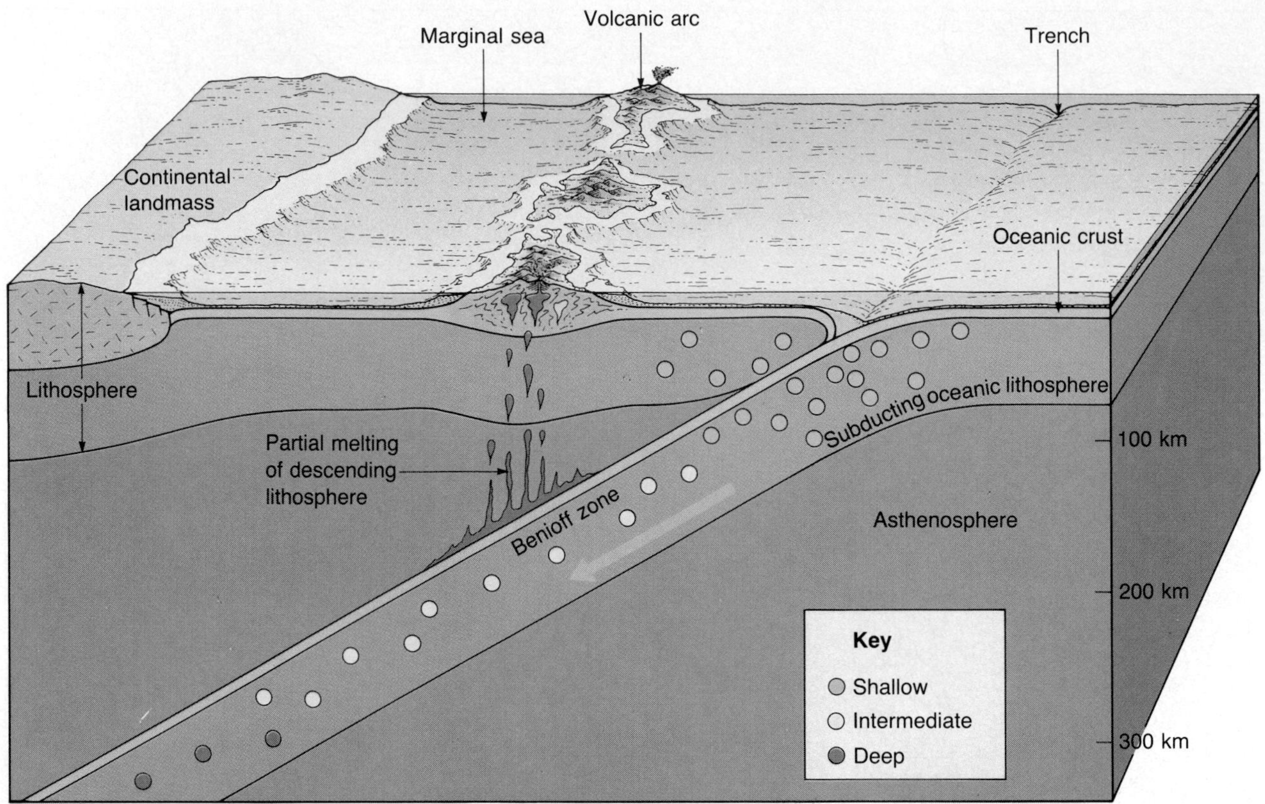

FIGURE 18.26
Relationship between the descending plate and depth of earthquake foci.

Although the exact cause of deep-focus earthquakes is still debated, their close association with subduction zones is well documented. Since subduction zones are the only regions where cold crustal rocks are forced to great depths, these should be the only sites of deep-focus earthquakes. Indeed, the absence of deep-focus earthquakes along divergent and transform boundaries supports the theory of plate tectonics.

Evidence from Ocean Drilling

Some of the most convincing evidence confirming the sea-floor spreading hypothesis has come from drilling directly into ocean-floor sediment. From 1968 until 1983, the source of these important data was the Deep Sea Drilling Project, an international program sponsored by several major oceanographic institutions and the National Science Foundation. The primary goal was to gather firsthand information about the age and processes that formed the ocean basins. Researchers felt that the predictions concerning sea-floor spreading that were based on paleomagnetic data could best be confirmed by the direct sampling of sediments from the floor of the deep-ocean basins. To accomplish this, a new drilling ship, the *Glomar Challenger,* was built. This ship represented a significant technological breakthrough, because it was capable of lowering drill pipe thousands of meters to the ocean floor and then drilling hundreds of meters into the sediments and underlying basaltic crust.

Operations began in August, 1968, and shortly thereafter important evidence was gathered in the South Atlantic. At several sites holes were drilled through the entire thickness of sediments to the basaltic rock below. An important objective was to gather samples of sediment from just above the igneous crust as a means of dating the sea floor at each site.* Since sedimentation begins immediately after the oceanic crust forms, remains of microorganisms found in the oldest sediments (that is, those resting directly on the basalt) can be used to date the ocean floor at that site. When the oldest sediment from each drill site was plotted against its distance from the ridge crest, the plot revealed that the age of the sediment increased with increasing distance from the ridge. This finding was in agreement with the sea-floor spreading hypothesis which predicted that

*Radiometric dates of the ocean crust itself are unreliable because of the alteration of basalt by seawater.

the youngest oceanic crust is to be found at the ridge crest and that the oldest oceanic crust flanks the continental margins. Further, the rate of sea-floor spreading determined from the ages of sediments was identical to the rate previously estimated from magnetic evidence. Subsequent drilling in the Pacific Ocean verified these findings. These excellent correlations were a striking confirmation of sea-floor spreading.

The data from the Deep Sea Drilling Project also reinforced the idea that the ocean basins are geologically youthful because no sediment with an age in excess of 160 million years was found. By comparison, some continental crust has been dated at 3.9 billion years.

The thickness of ocean-floor sediments provided additional verification of sea-floor spreading. Drill cores from the *Glomar Challenger* revealed that sediments are almost entirely absent on the ridge crest and that the sediment thickens with increasing distance from the ridge. Since the ridge crest is younger than the areas farther away from it, this pattern of sediment distribution should be expected if the sea-floor spreading hypothesis is correct. Furthermore, measurements in the open ocean have shown that sediment accumulates at a rate of about 1 centimeter per 1000 years. Therefore, if the ocean floor were an ancient feature, sediments would be many kilometers thick. However, data from hundreds of drilling sites indicate that the greatest thickness of sediment in the deep-ocean basins is only a few hundred meters, equivalent to intervals of only a few tens of millions of years. Thus, here is yet another fact that strongly suggests the ocean floor is indeed a young geologic feature.

The Deep Sea Drilling Project provided an enormous quantity of basic information about the history of the oceans and confirmed many important aspects of the plate tectonics theory. To summarize, the sea-floor spreading hypothesis was upheld when deep-ocean core samples showed that the age of the oldest sediments and the thickness of sediments increase with increasing distance from the ridge crest. Moreover, these data on the thickness and ages of sediments strongly support the idea of geologically youthful ocean basins.

During its 15 years of operation the *Glomar Challenger* logged more than 600,000 kilometers on 96 voyages across every ocean. The drilling of 1092 holes yielded more than 96 kilometers (60 miles) of invaluable core samples. Although the Deep Sea Drilling Project ended and the *Glomar Challenger* was retired, the important work of sampling the floors of the world's ocean basins continues. The Ocean Drilling Project has succeeded the Deep Sea Drilling Project and, like its predecessor, it is a major international program. The more technologically advanced drilling ship, the *JOIDES Resolution,* now continues the work of the *Glomar Challenger.** The *JOIDES Resolution* can drill in water depths as great as 8100 meters (27,000 feet) and contains onboard laboratories equipped with the largest and most varied array of seagoing scientific research equipment in the world (Figure 18.27). The work of the Ocean Drilling Project is to continue until at least 1995.

Hot Spots

Mapping of seamounts in the Pacific revealed a chain of volcanic structures extending from the Hawaiian Islands to Midway Island and then continuing northward toward the Aleutian trench (Figure 18.28). Potassium-argon dating of numerous volcanoes in this chain revealed an increase in age with an increase in distance from Hawaii. Suiko Seamount, which is located near the Aleutian trench, is 65 million years old, Midway Island is 27 million years old, and the island of Hawaii rose from the sea less than 1 million years ago (Figure 18.28).

Researchers have proposed that a rising plume of mantle material is located below the island of Hawaii. Melting of this hot rock as it enters the low-pressure environment near the surface generates a volcanic area or **hot spot**. Presumably, as the Pacific plate moved over the hot spot, successive volcanic structures emerged. The age of each volcano indicates the time when it was situated over the relatively stationary mantle plumes. Kauai is the oldest of the large islands in the Hawaiian chain. Five million years ago, when it was positioned over the hot spot, Kauai was the only Hawaiian Island in existence (Figure 18.28). Visible evidence of the age of Kauai can be seen by examining the extinct volcanoes which have been eroded into jagged peaks and vast canyons. By contrast, the south slopes of the island of Hawaii consist of fresh lava flows, and two of Hawaii's volcanoes, Mauna Loa and Kilauea, remain active. Recent evidence indicates that a new volcanic pile, named Loihi Seamount, is forming on the ocean floor about 35 kilometers off the southeast coast of Hawaii. Geologically speaking, it should not be long before another tropical island will be added to the Hawaiian chain.

Two island groups parallel the Hawaiian Island–Emperor Seamount chain. One chain consists of the Tuamotu and Line islands, and the other includes the Austral, Gilbert, and Marshall islands. In each case,

*JOIDES stands for Joint Oceanographic Institutions for Deep Earth Sampling.

BOX 18.2

Hot Spots and Flood Basalts

Massive accumulations of basaltic lava occur worldwide. One of the largest of these is the Deccan Traps, a thick sequence of flat-lying basalt flows covering nearly 500,000 square kilometers of west central India. When the Deccan Traps were produced 66 million years ago, nearly 2 million cubic kilometers of lava were extruded in less than 1 million years. Some scientists have proposed that this event may have substantially altered the earth's climate which in turn led to the extinction of many life forms, including the dinosaurs (see Box 8.2). Although considerably smaller than the Deccan Traps, the area covered by the Columbia River basalts in the northwestern United States consists of a lava plain which in places is 2 to 3 kilometers thick. Here, individual fissure eruptions extruded enormous volumes of very fluid basaltic lava that greatly exceed the amounts generated by any contemporary volcanic process. Further, because these lava flows were fluid enough to travel up to 150 kilometers from their source, they became known as *flood basalts.* Figure 18.B shows the distribution of the flood basalt provinces that have formed in the last 250 million years. Although most of the known provinces are located on land, geologists believe that many of the large suboceanic plateaus, such as Ontong Java, are similar in structure.

Until recently, the processes that generated these huge piles of volcanic rock were not well understood. Geologists were particularly puzzled by the fact that basalt plateaus formed over geologically short time spans of from 1 to 2 million years. No currently operating volcanic process extrudes lava at the rate needed to produce

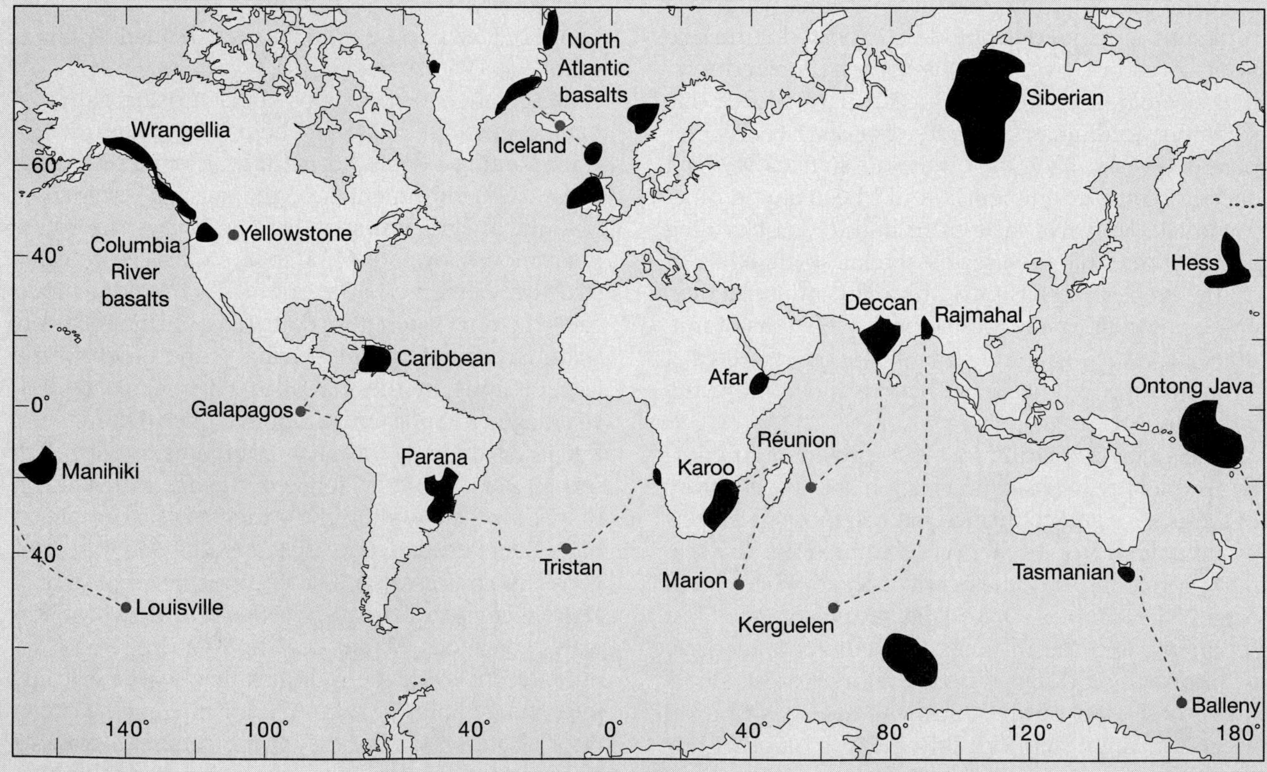

FIGURE 18.B
Global distribution of flood basalt provinces (shown in black) and associated hot spots (shown as red dots). The red dashed lines are hot spot tracks, which are revealed as a line of volcanic structures on the floor of the ocean. (After Robert A. Duncan)

these structures in such a short time frame.

It now appears that the largest basalt plateaus mark the onset of volcanic activity at newly forming hot spots. Recall that hot spots arise from plumes of partially molten rock that are believed to originate in the deep mantle. Because hot spots remain relatively stationary, they leave a trail of volcanic structures on the surface of the lithospheric plates passing above. A classic example is the 5000-kilometer-long Hawaiian-Emperor seamount chain in the North Pacific.

It has been suggested that hot spots originate when a mass of warm rock located at the core-mantle boundary gradually rises through the mantle. Laboratory experiments predict that these ascending plumes of hot rock will develop a mushroom-shaped head and a long, narrow tail (Figure 18.C). Upon reaching the comparatively low pressure environment at the base of the cooler lithosphere, the hot plume flattens and melts, producing enormous quantities of basaltic magma. It is this event that generates the magma to feed the fissure eruptions that form the flood basalt provinces. Subsequent volcanism associated with the plume's tail produces a chain of volcanic structures that connects the basalt plateau with an active hot spot (Figure 18.C). Since hot spots may persist for 120 million years or more, the trail of volcanic structures can be thousands of kilometers long. Sixty-six million years ago, when the Deccan Traps formed, India was situated over the Reunion hot spot (see Figure 18.B). Since that time, the Indian subcontinent drifted northward to collide with Asia. A trail of volcanic islands and seamounts connects the flood basalts of the Deccan Traps to the volcanic island of Reunion.

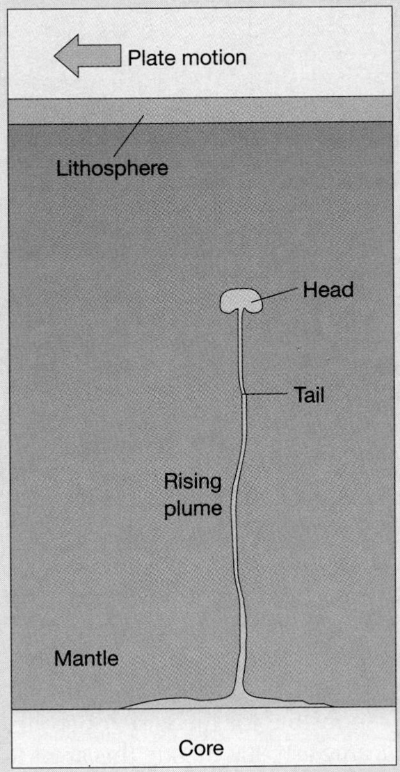

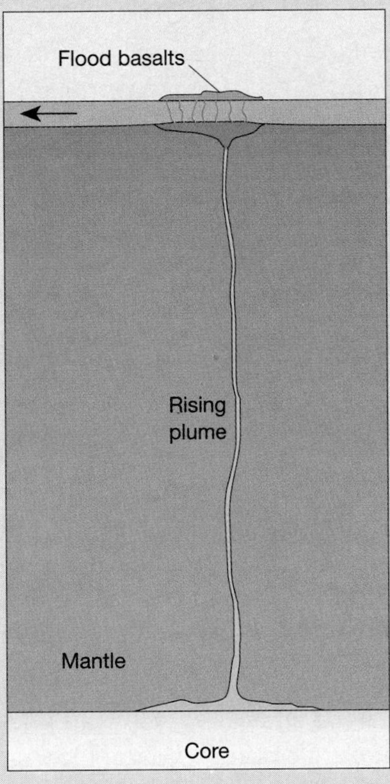

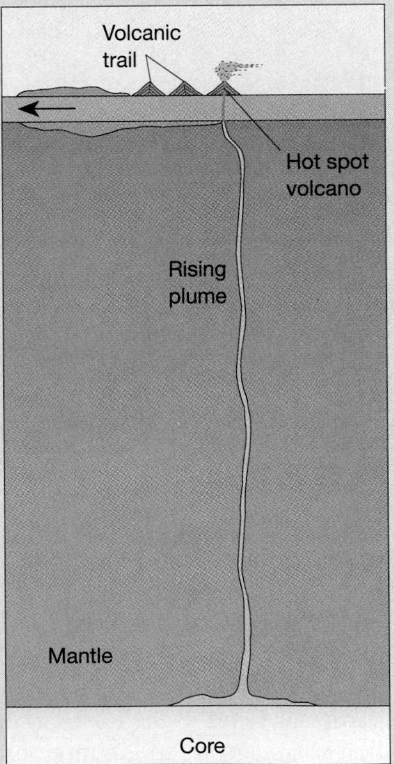

FIGURE 18.C
Development of a flood basalt province and a volcanic trail from a rising mantle plume.

FIGURE 18.27
The *JOIDES Resolution,* the drilling ship of the Ocean Drilling Program. This modern drilling ship has replaced the *Glomar Challenger* in the important work of sampling the floors of the world's oceans. (Photo courtesy of Ocean Drilling Program)

the most recent volcanic activity has occurred at the southeastern end of the chain, and the islands get progressively older to the northwest. Thus, like the Hawaiian Island–Emperor Seamount chain, these volcanic structures apparently formed by the same motion of the Pacific plate over fixed mantle plumes. Not only does this evidence support the fact that the plates do indeed move relative to the earth's interior, but also the hot spot "tracks" provide a frame of reference for tracing the direction of the plate motion. Notice, for example, in Figure 18.28 that the Hawaiian Island–Emperor Seamount chain bends. This particular bend in the trace occurred about 40 million years ago when the motion of the Pacific plate changed from nearly due north to a northwesterly path. Similarly, hot spots found on the floor of the Atlantic have increased our understanding of the migration of landmasses following the breakup of Pangaea.

Although the existence of mantle plumes is well documented, their exact role in plate tectonics is not altogether clear. Some researchers suggest that mantle plumes originate deep in the mantle, perhaps at the core-mantle boundary. Here a region of abnormally high temperatures produces a rising plume of rock that initiates hot spot volcanism at the earth's surface. Most evidence indicates that hot spots remain relatively stationary. Of the 50 to 120 hot spots believed to exist, about a dozen, or so, are located near divergent plate boundaries. A hot spot beneath Iceland is thought to be responsible for the unusually large accumulation of lava found in that portion of the Mid-Atlantic Ridge. Another hot spot is believed to be located beneath Yellowstone National Park and may be responsible for the large outpourings of lava and volcanic ash that mantle this area. If the Yellowstone region was indeed modified by hot spot volcanism, there is good reason to expect additional activity in the future.

THE DRIVING MECHANISM

The plate tectonics theory describes plate motion and the effects of this motion. Therefore, acceptance

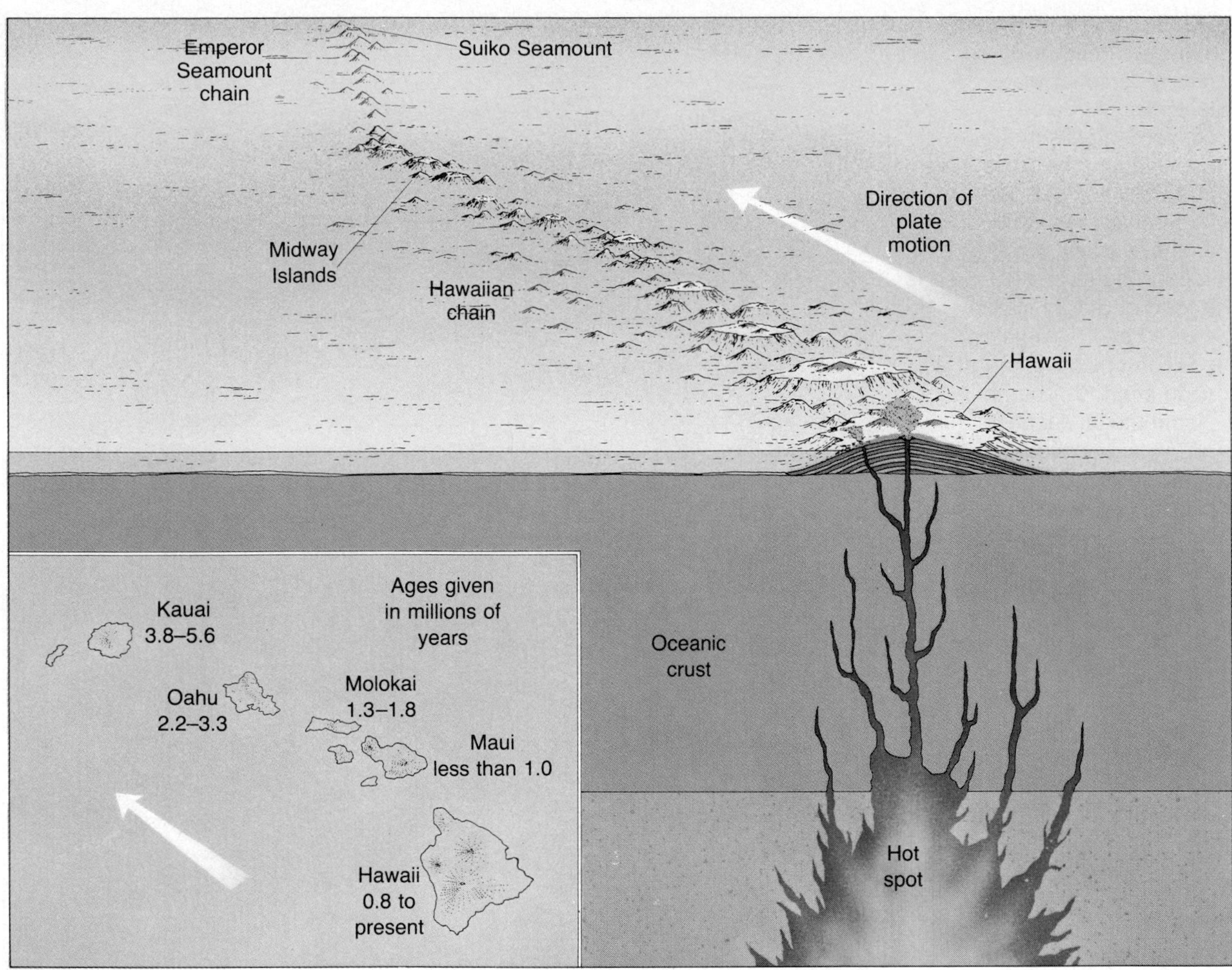

FIGURE 18.28
The chain of islands and seamounts that extends from Hawaii to the Aleutian trench results from the movement of the Pacific plate over an apparently stationary hot spot. Radiometric dating of the Hawaiian Islands shows that the volcanic activity decreases in age toward the island of Hawaii.

does not rely on a knowledge of the force or forces moving the plates. This is fortunate, since none of the driving mechanisms yet proposed can account for all major facets of plate motion. Nevertheless, it is clear that the unequal distribution of heat within the earth is the underlying driving force for plate movement.

One of the first models used to explain the movements of plates was originally proposed by the eminent English geologist Arthur Holmes as a possible driving mechanism for continental drift. Adapted to plate tectonics, this hypothesis suggested that large convection currents drive plate motion (Figure 18.29A). The warm, less dense material of the mantle rises very slowly in the regions of oceanic ridges. As the material spreads laterally, it drags the lithosphere along. Eventually, the material cools and begins to sink back into the mantle, where it is reheated. Partly because of its simplicity, this proposal had wide appeal. However, researchers employing modern research techniques have learned that the flow of material in the mantle is far more complex than that exhibited by simple convection cells. Furthermore, there is considerable debate as to whether mantle circulation is confined to the upper 700 kilometers or involves the whole mantle.

Many other mechanisms that may contribute to or influence plate motion have been suggested. One relies on the fact that as a newly formed slab of oceanic crust moves away from the ridge crest, it gradually cools and becomes denser. Eventually, the cold oceanic slab becomes denser than the asthenosphere and begins to descend. When this occurs, the

FIGURE 18.29
Proposed models of the driving force for plate tectonics. **A.** Large convection cells in the mantle carry the lithosphere in a conveyor-belt fashion. **B.** Slab-pull results from negative buoyancy of a subducting slab. Slab-push is a form of gravity sliding caused by the elevated position of lithosphere at a ridge crest. **C.** The hot plume model suggests that all upward convection is confined to a few narrow plumes, while the downward limbs of these convection cells are the cold, dense subducting oceanic plates.

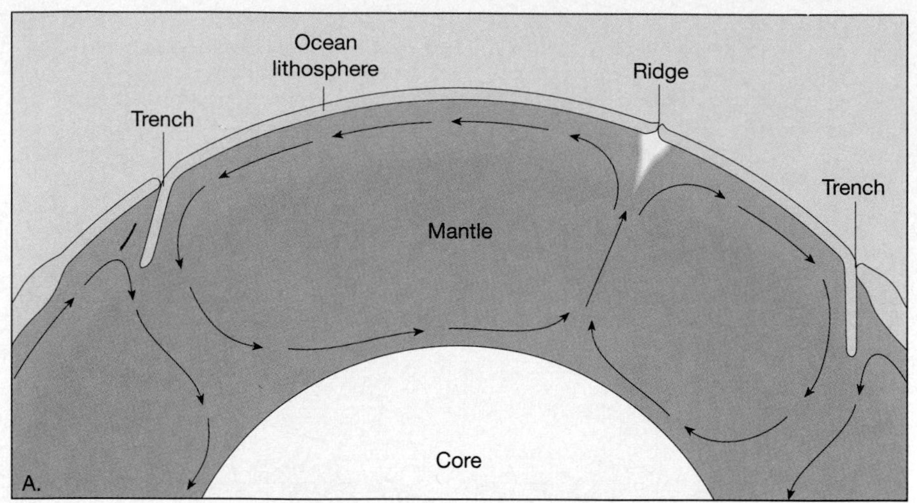

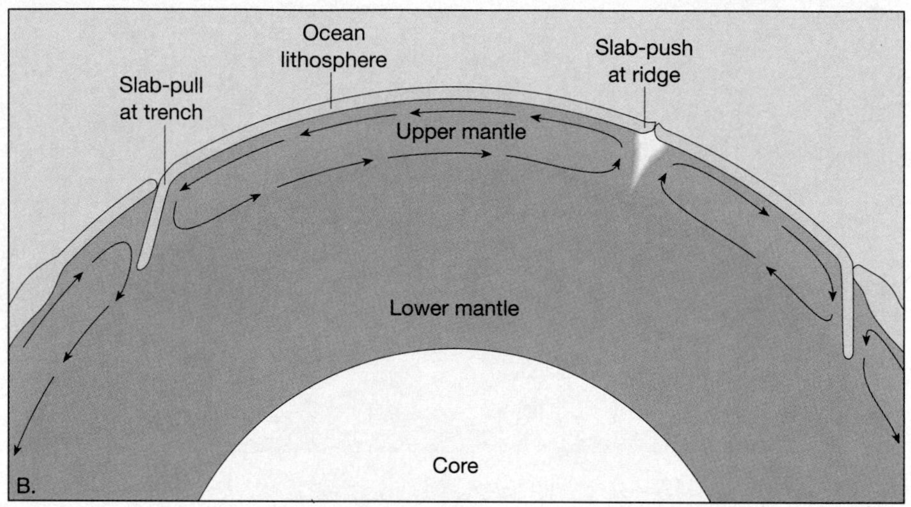

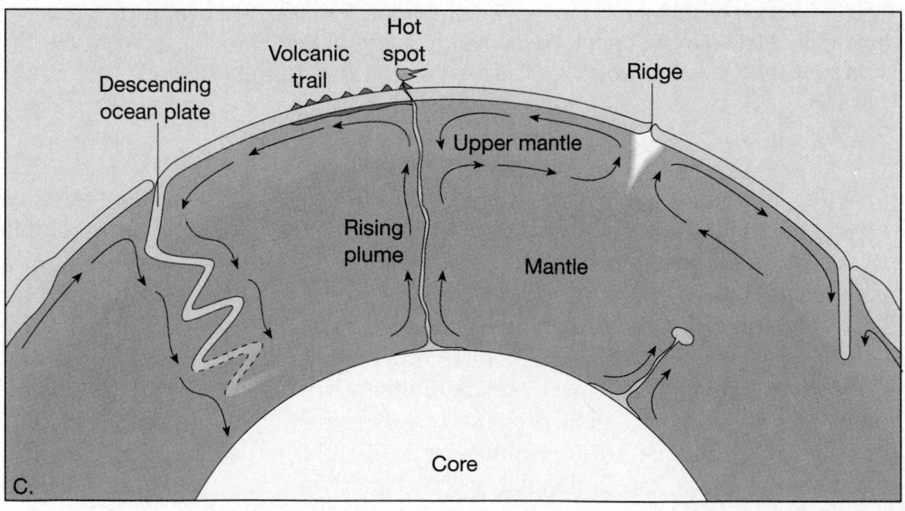

BOX 18.3

Convective Flow in the Mantle

Convective flow in the mantle—in which warm, less dense rocks rise and cooler, denser material sinks—is the most important process operating in the earth's interior. Thermal convection is the driving force that propels the rigid lithospheric plates across the globe, and it ultimately generates the earth's mountain belts and worldwide earthquake and volcanic activity. Furthermore, the upward flow of hot material from the core-mantle boundary is responsible for transferring most of the heat that is lost from the earth's interior. Driving this flow is heat given off by the decay of radioactive elements within the mantle, as well as heat transferred by conduction from the core to the mantle.

Recently, some new techniques have become available that may significantly enhance our knowledge of convective flow in the mantle. One analytical tool, called *seismic tomography,* is similar in principle to CAT scanning (computer-aided tomography) which is used in medical diagnoses. Whereas CAT scanning uses X rays to penetrate the human body, information on the earth's

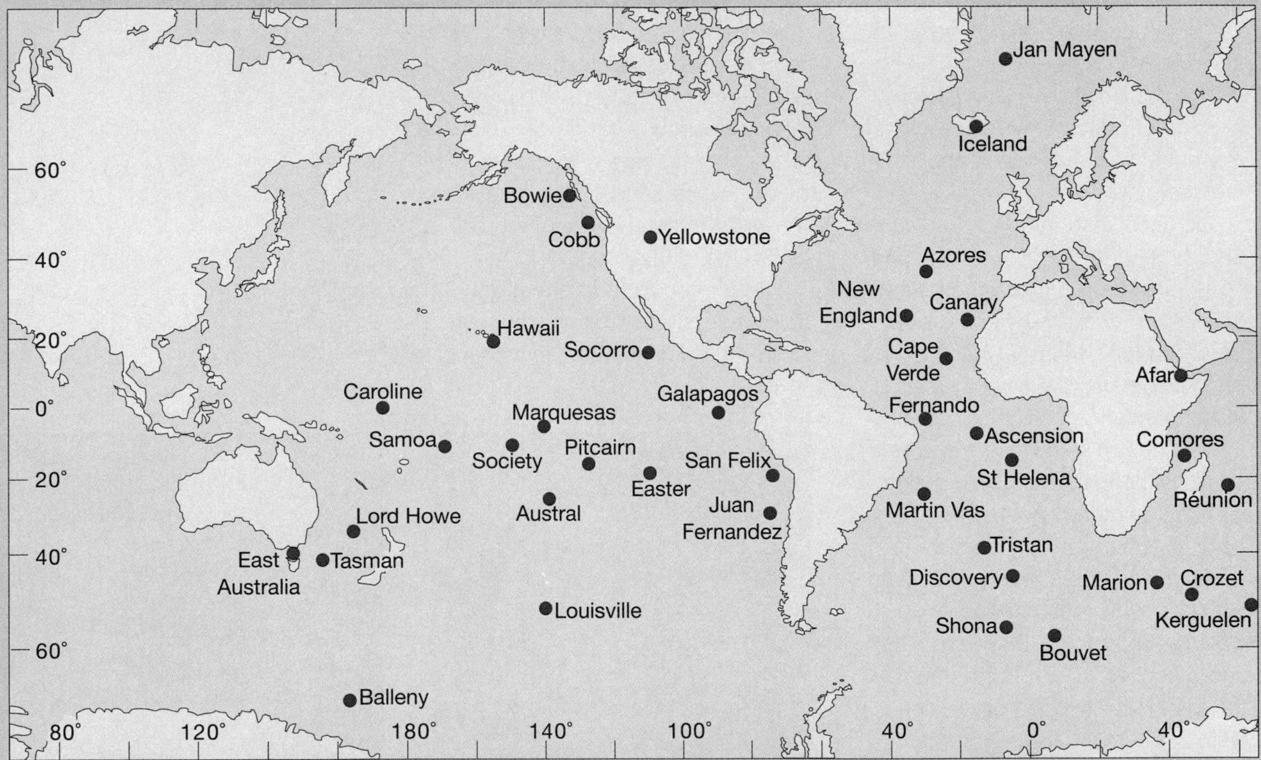

FIGURE 18.D
The global distribution of major hot spots. Hot spots are not randomly distributed but are concentrated in two large groups. One group occupies the south central Pacific. The other large group, which is elongated in a north-south direction, is centered beneath Africa and includes much of the eastern Atlantic. These hot spots are thought to be connected with large-scale convective upwelling in the mantle. (Data from Robert A. Duncan)

dense sinking slab pulls the trailing lithosphere along. This hypothesis is similar to another model which suggests that the elevated position of an oceanic ridge could cause the lithosphere to slide under the influence of gravity. However, some ridge systems are subdued, which would reduce the effectiveness of the *slab-push* model. Further, some ocean basins, notably the Atlantic, lack subduction zones; thus, the *slab-pull* mechanism cannot adequately explain the spreading occurring at all ridges. Never-

interior is obtained from seismic waves triggered by earthquakes. Like CAT scanning, seismic tomography uses computers to combine data to build a three-dimensional image of the object that the waves have traversed. Recall that the velocities of seismic waves are strongly influenced by the properties of the transmitting materials. In tomographic studies, the information from many criss-crossing waves is combined to map regions of "slow" and "fast" seismic velocity. In general, regions of slow seismic velocity are associated with hot, upwelling rock, whereas regions of fast seismic velocity represent areas in which cool rock is descending.

Early seismic tomography studies reveal that the flow in the mantle is far more complex than simple convection cells, in which hot material gradually rises at ridge crests and cold material slowly descends at subduction zones. It appears that upwelling is confined to a few large cylindrical structures, one of which is centered beneath west central Africa and another located beneath the south central Pacific. Embedded in these zones of upwelling are most of the earth's hot spots (Figure 18.D). Furthermore, these studies show downwelling beneath convergent boundaries where plates are being subducted. This flow appears to extend to the lower mantle, but other interpretations of the data are possible. Although seismic tomography has provided new information about convective flow in the mantle, its usefulness has been limited by an inadequate number of digital seismic stations. To overcome this limitation, a project is underway to improve quality by increasing the number of seismic stations carrying out this kind of research.

Another innovative technique called *numerical modeling* has been employed to simulate thermal convection in the mantle. Simply, this method uses high-speed computers to solve mathematical equations that describe the dynamics of mantlelike fluids. Because of a number of uncertainties, including precise knowledge of the mode of mantle heating, a range of conditions are simulated. The results of these studies can be graphically represented as shown in Figure 18.E. One study concludes that downwelling occurs in sheetlike structures, supporting seismic evidence that descending lithospheric slabs are an integral part of mantle circulation. Furthermore, large cylindrical mantle plumes were found to be the main mechanism of upwelling in the mantle. These simulations also verify that mid-ocean ridges are not a consequence of active upwelling from deep in the mantle. Rather, upwelling beneath ridge crests is a shallow and passive response to fracturing of the plates under the pull of sinking slabs.

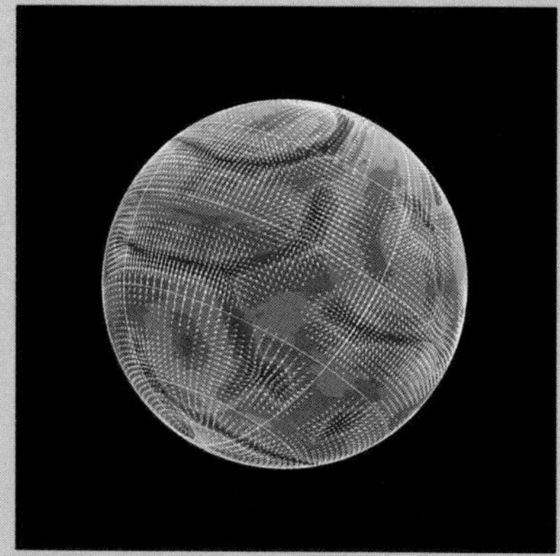

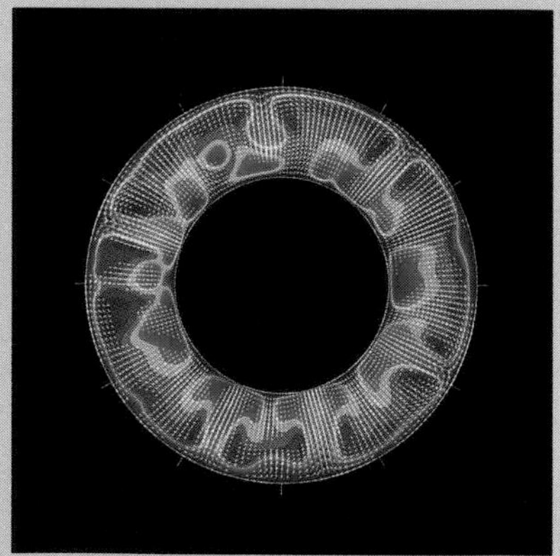

FIGURE 18.E
Cross-sectional view of numerically simulated thermal convection in the mantle. Red and yellow lines indicate hot upwelling currents, while blue lines depict regions of cold downwelling currents. (Courtesy of D. Bercovici, G. Schubert, and G. A. Glatzmaier)

theless, the slab-push and slab-pull phenomena appear to be active in some ridge-trench systems (Figure 18.29B).

One version of the thermal convection model suggests that relatively narrow, hot plumes of rock contribute to plate motion (Figure 18.29C). These hot plumes are presumed to extend upward from the vicinity of the mantle-core boundary. Upon reaching the lithosphere, they spread laterally and facilitate the plate motion away from the zone of upwelling. These mantle plumes reveal themselves as long-lived volcanic areas (hot spots) in such places as Iceland. A dozen or so hot spots have been identified along ridge systems where they may contribute to plate divergence. Recall, however, that many hot spots, including the one which generated the Hawaiian Islands, are not located in ridge areas.

In another version of the hot plume model, all upward convection is confined to a few large cylindrical structures. Embedded in these large zones of upwelling are most of the earth's hot spots (see Box 18.3). The downward limbs of these convection cells are the cold, dense subducting lithospheric plates. Moreover, advocates of this view suggest that subducting slabs may descend all the way to the core-mantle boundary. However, convincing seismic evidence for the existence of these sheetlike structures below 700 kilometers is lacking.

Although there is still much to be learned about the mechanisms that cause plates to move, some facts are clear. The unequal distribution of heat in the earth generates some type of thermal convection in the mantle which ultimately drives plate motion. Whether the upwelling is mainly in the form of rising limbs of convection currents or cylindrical plumes of various sizes and shapes is yet to be determined. Studies do show, however, that the oceanic ridges are not closely aligned with the active upwelling that originates deep in the mantle. Except for hot spots, upwelling beneath ridges appears to be a shallow feature, responding to the tearing of the lithosphere under the pull of the descending slabs. Furthermore, the descending lithospheric plates are active components of downwelling, and they serve to transport cold material into the mantle.

REVIEW QUESTIONS

1. What first led scientists such as Alfred Wegener to suspect that the continents were once joined?
2. What was Pangaea?
3. List the evidence that Wegener and his followers gathered to support the continental drift hypothesis.
4. Early in this century, what was the prevailing view of how land animals migrated across vast expanses of ocean?
5. Briefly explain why the recent acceptance of plate tectonics has been described as a scientific "revolution."
6. How does evidence for a late Paleozoic glaciation in the Southern Hemisphere support the continental drift hypothesis?
7. Explain how paleomagnetism can be used to establish the latitude of a specific place at some distant time.
8. What is meant by sea-floor spreading? Who is credited with formulating the concept of sea-floor spreading?
9. Describe how Fred Vine and D. H. Matthews related the sea-floor spreading hypothesis to magnetic reversals.
10. On what basis were plate boundaries first established?
11. Where is lithosphere being formed? Consumed? Why must the production and destruction of the lithosphere be going on at about the same rate?
12. Why is the oceanic portion of a lithospheric plate subducted while the continental portion is not?
13. In what ways may the origin of the Japanese Islands be considered similar to the formation of the Andes Mountains? How do they differ?
14. Differentiate between transform faults and the two other types of plate boundaries.
15. Some people predict that California will sink into the ocean. Is this idea consistent with the concept of plate tectonics?
16. Applying the idea that hot spots remain fixed, in what direction was the Pacific plate moving while the Emperor Seamounts were being produced? (See Figure 18.28) While the Hawaiian Seamounts were being produced?
17. With what type of plate boundary are the following places or features associated (be as specific as possible): Himalayas, Aleutian Islands, Red Sea, Andes Mountains, San Andreas fault, Iceland, Japan, Mount St. Helens?

KEY TERMS

continental drift (p. 448)
convergent boundary (p. 462)
Curie point (p. 456)
deep-ocean trench (p. 465)
divergent boundary (p. 462)
hot spot (p. 475)
island arc (p. 468)
magnetometer (p. 459)
normal polarity (p. 459)
paleomagnetism (p. 456)
Pangaea (p. 448)
plate (p. 461)
reverse polarity (p. 459)
rift, or rift valley (p. 464)
sea-floor spreading (p. 458)
subduction zone (p. 465)
transform boundary (p. 462)
volcanic arc (p. 468)

19

The Ocean Floor and Its Evolution

CONTINENTAL MARGINS

SUBMARINE CANYONS AND TURBIDITY CURRENTS

FEATURES OF THE DEEP-OCEAN BASIN

Deep-Ocean Trenches ■ Abyssal Plains ■ Seamounts

CORAL REEFS AND ATOLLS

SEA-FLOOR SEDIMENTS

MID-OCEAN RIDGES

THE OCEAN FLOOR AND SEA-FLOOR SPREADING

OPENING AND CLOSING OF THE OCEAN BASINS

PANGAEA: BEFORE AND AFTER

Breakup of Pangaea ■ Before Pangaea ■ A Look into the Future

Opposite: View of the open ocean at sunset, Roads End, Oregon coast. (Photo by Craig Tuttle Photography, The Stock Market)
(Top photo by E. J. Tarbuck)

If all water were drained from the ocean basins, the exposed surface would not have a quiet, subdued topography as was once thought. Rather, a great diversity of features, including towering mountain chains and deep canyons as well as flat plains, would be found. The scenery would be just as varied as that on the continents.

The ocean's vast expanse first became apparent through voyages of discovery in the fifteenth and sixteenth centuries. An understanding of the ocean floor's varied topography did not unfold until much later with the historic $3\frac{1}{2}$-year voyage of the H.M.S. *Challenger* (Figure 19.1). From December 1872 to May 1876, the *Challenger* expedition made the first, and still perhaps most comprehensive, study of the global ocean ever attempted by one agency. The 110,000-kilometer (68,000-mile) trip took the ship and its crew of scientists to every ocean except the Arctic. Throughout the voyage they sampled the depth of the water by laboriously lowering a weighted line overboard. Not many years later the knowledge gained by the *Challenger* of the ocean's great depths and varied topography was further expanded with the laying of transatlantic cables. However, as long as ocean depth had to be measured with weighted lines, our knowledge of the sea floor remained slight. Then, in the 1920s a technological breakthrough occurred with the invention of electronic depth-sounding equipment (**echo sounder**).

The echo sounder works by transmitting sound waves toward the ocean bottom (Figure 19.2A). A delicate receiver intercepts the echo reflected from the bottom, and a clock precisely measures the time interval to fractions of a second. By knowing the velocity of the sound waves in water (about 1500 meters per second) and the time required for the energy pulse to reach the ocean floor, scientists can establish the depth. The depths determined from continuous monitoring of these echos are normally plotted so that a profile of the ocean floor is produced (Figure 19.2B). Since the invention of the echo sounder, millions of kilometers of continuous sonic-depth determinations have provided a more complete and detailed view of the ocean floor.

Oceanographers studying the topography of the oceans have delineated three major units: the continental margins, the deep-ocean basins, and the mid-ocean ridges. The map in Figure 19.3 outlines these provinces for the North Atlantic, and the profile at the bottom illustrates the varied topography. Such profiles usually have their vertical dimension exaggerated many times—40 times in this case—to make topographic features more conspicuous. Because of this, the slopes shown in the sea-floor profile appear to be much steeper than they actually are.

CONTINENTAL MARGINS

The features comprising the **continental margin** include the continental shelf, the continental slope, and the continental rise (Figure 19.4). The first of these parts, the **continental shelf**, is a gently sloping submerged surface extending from the shoreline toward the deep-ocean basin. Since it is underlain by continental-type crust, it is clearly a flooded extension of the continents. The continental shelf varies greatly in width. Almost nonexistent along some con-

FIGURE 19.1
The H.M.S. *Challenger.* (From C. W. Thomson and Sir John Murray, *Report on the Scientific Results of the Voyage of the H.M.S. Challenger,* Vol. 1. Great Britain: Challenger Office, 1895, Plate 1)

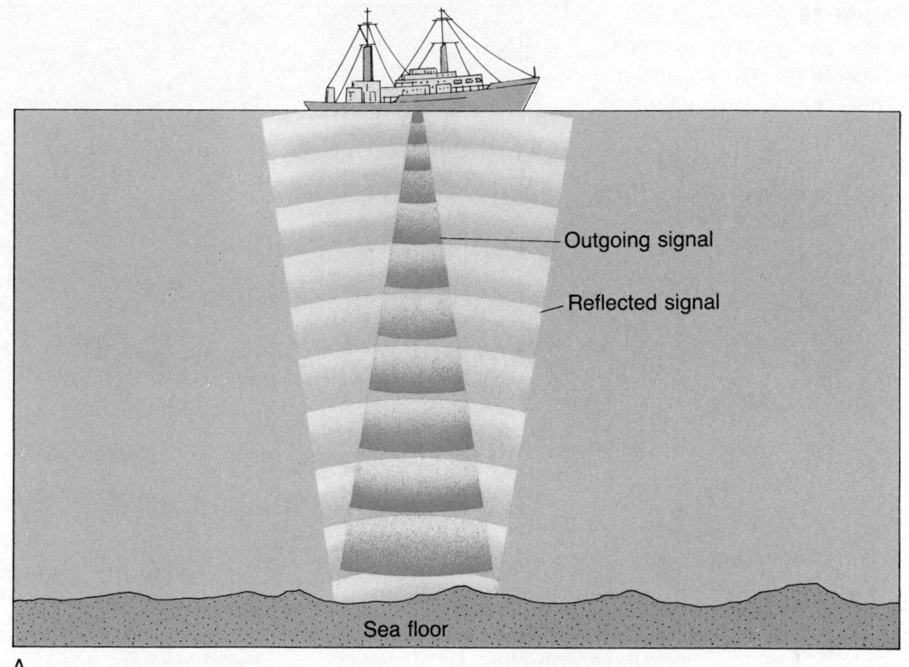

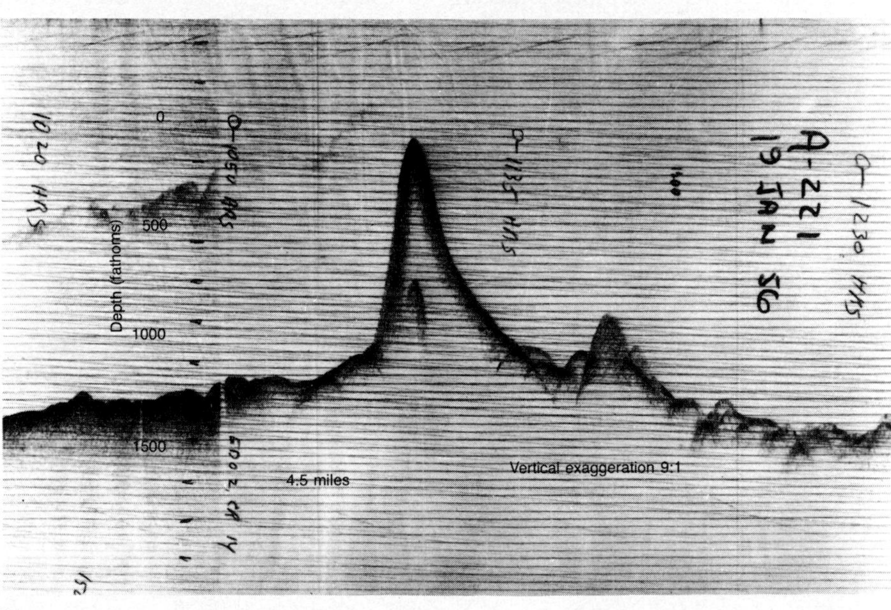

FIGURE 19.2
A. An echo sounder determines the water depth by measuring the time interval required for a sonic wave to travel from a ship to the sea floor and back. The speed of sound in water is 1500 m/sec. Therefore, depth = $\frac{1}{2}$(1500 m/sec × echo travel time). **B.** Sea-floor profile made by an echo sounder. (Courtesy of Woods Hole Oceanographic Institution)

tinents, the shelf may extend seaward as far as 1500 kilometers along others. On the average, the continental shelf is about 80 kilometers wide and 130 meters deep at the seaward edge. The average inclination of the continental shelf is less than one-tenth of one degree, a drop of less than 2 meters per kilometer. The slope is so slight that it would appear to an observer to be a horizontal surface.

Although the continental shelf is relatively featureless, it is not completely smooth. Some continental shelf areas are mantled by extensive glacial deposits and are thus quite rugged. The most profound features are long valleys running from the coastline into deeper waters. Many of these valleys are the seaward extensions of river valleys on the adjacent landmass. Such valleys appear to have been excavated during the Pleistocene epoch (Ice Age). During this time great quantities of water were tied up in vast ice sheets on the continents. This caused sea level to drop by 100 meters or more, exposing large areas of the continental shelves (see Figure 12.A). Because of this drop in sea level, rivers extended their courses,

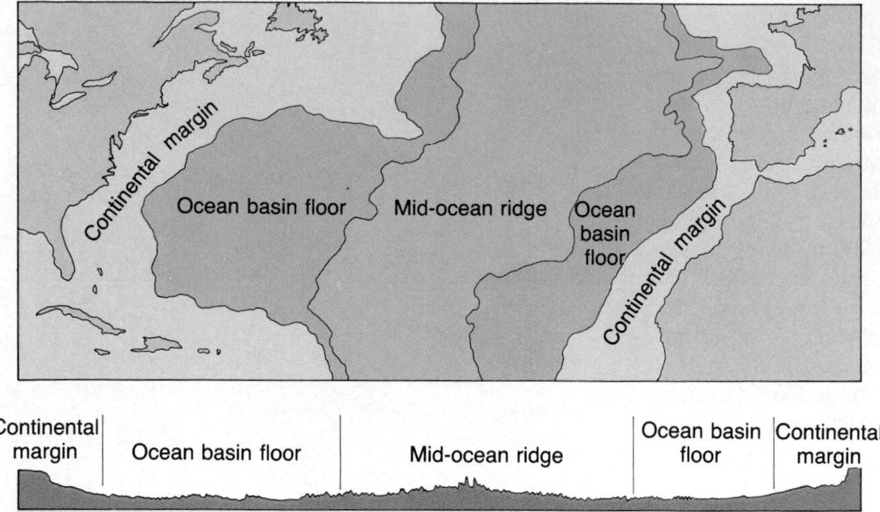

FIGURE 19.3
Major topographic divisions of the North Atlantic and a profile from New England to the coast of North Africa. (After B. C. Heezen, M. Tharp, and M. Ewing. "The Floors of the Oceans," *Geological Society of America Special Paper* 65, p. 16)

and land-dwelling plants and animals inhabited the newly exposed portions of the continents. Dredging off the east coast of North America has produced the remains of numerous land dwellers, including mammoths, mastodons, and horses, adding to the evidence that portions of the continental shelves were once above sea level.

Marking the seaward edge of the continental shelf is the **continental slope.** This feature is characterized by a steep gradient as compared with the shelf, and it marks the boundary between continental crust and oceanic crust. Although the inclination of the continental slope varies greatly from place to place, the average is about 5 degrees and in places may exceed 25 degrees.

In regions where trenches do not exist, the steep continental slope merges into a more gradual incline known as the **continental rise.** Here the gradient lessens to between 4 and 8 meters per kilometer. Whereas the width of the continental slope averages about 20 kilometers, the continental rise may extend for hundreds of kilometers into the deep-ocean basin. This feature consists of a thick accumulation of sediment that moved downslope from the continental shelf to the deep-ocean floor. Although continental rises are relatively featureless, their surfaces are occasionally interrupted by submarine canyons or by submarine volcanoes that have not yet been completely buried by the sediments.

Some continental shelves, such as those along the east coast of the United States, are wide and consist of thick accumulations of shallow-water sediments. These sediments are frequently several kilometers thick and are interbedded with limestones that

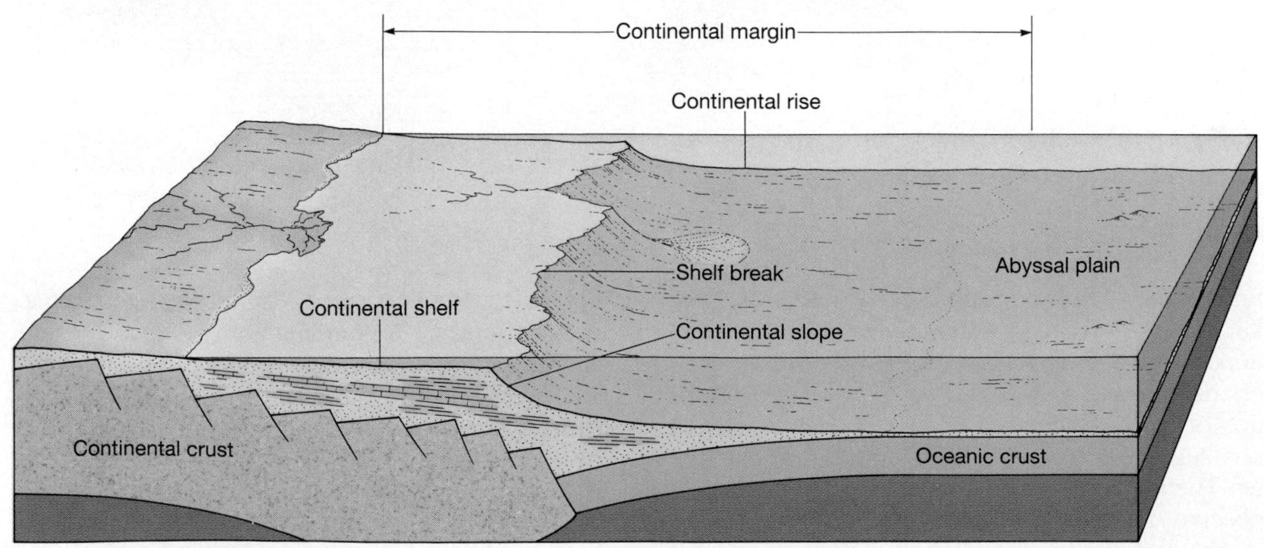

FIGURE 19.4
Schematic view showing the provinces of the continental margin.

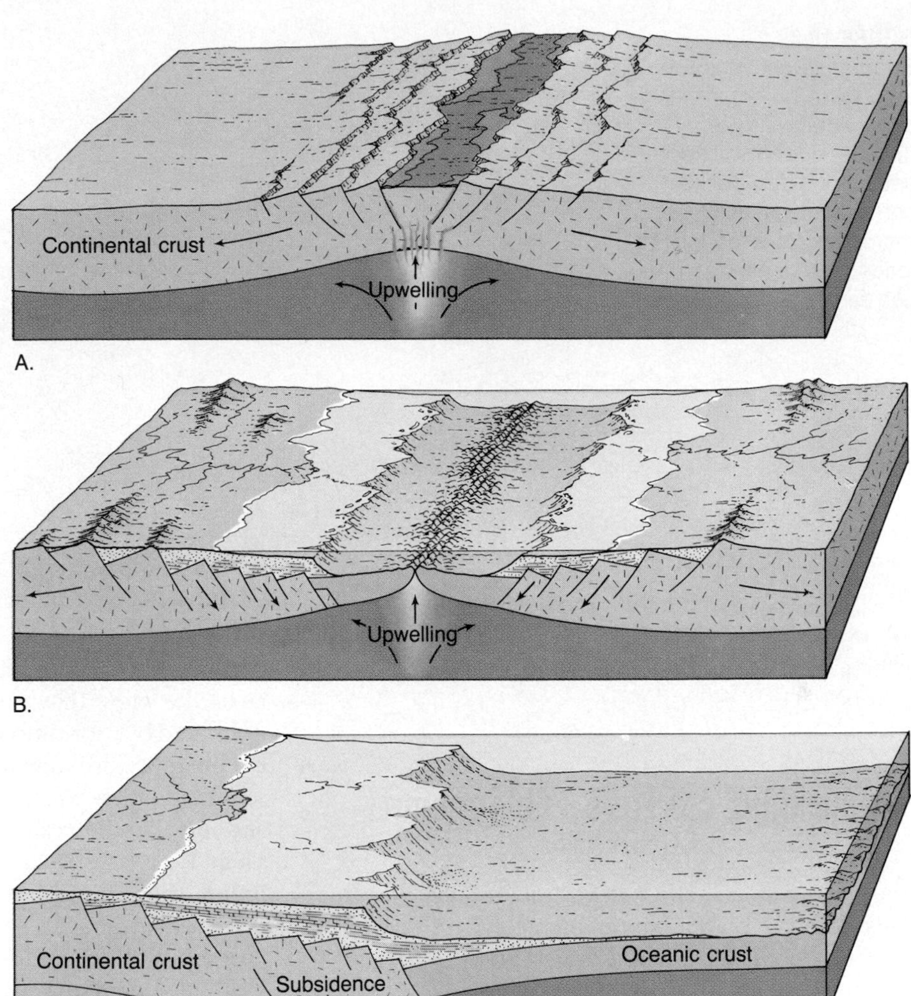

FIGURE 19.5
Stages in the breakup of a continental landmass. **A.** Upwelling of mantle rock causes fragmentation of the crust above. **B.** As the landmass moves away from the zone of upwelling, cooling results in the subsidence of the continental margins. **C.** Material derived from the adjacent highlands accumulates into thick wedges of sediment.

formed during earlier periods of coral reef building, a process that occurs only in shallow water. This evidence led researchers to conclude that these thick accumulations of sediment are produced along a gradually subsiding continental margin.

This explanation fits very well with our knowledge of how new continental margins are produced during the breakup of a continental landmass. Figure 19.5 illustrates the stages in the process. Notice that upwelling of mantle material causes doming, which stretches and fractures the crust. As sea-floor spreading progresses, the stretched and fragmented crust is wedged away from the zone of upwelling where gradual cooling leads to shrinkage and hence subsidence. Sediments carried from adjacent highlands begin to accumulate on the young continental margin. This additional load is believed to contribute to the subsidence.

Passive continental margins of this type are found around the Atlantic Ocean on the trailing edges of the continents. The accretion of material on the passive margins of the continents results in a gradual increase in the size of the landmass. In addition, such vast accumulations of sediments have an important role in mountain building as we shall see in Chapter 20.

Along some mountainous coasts the continental slope descends abruptly into deep-ocean trenches found between the continent and ocean basin. In such cases, the shelf is very narrow or does not exist at all. The side of the trench and the continental slope are essentially the same feature and grade into the adjacent mountains which tower thousands of meters above sea level. These narrow continental margins are located primarily around the Pacific Ocean in areas where the leading edge of a continent is overrunning oceanic lithosphere. Here the stress between the converging plates results in deformation of the continental margin. An example of this activity is found along the west coast of South America. Here the vertical distance from the peaks of the Andes Mountains to the floor of the deep Peru-Chile

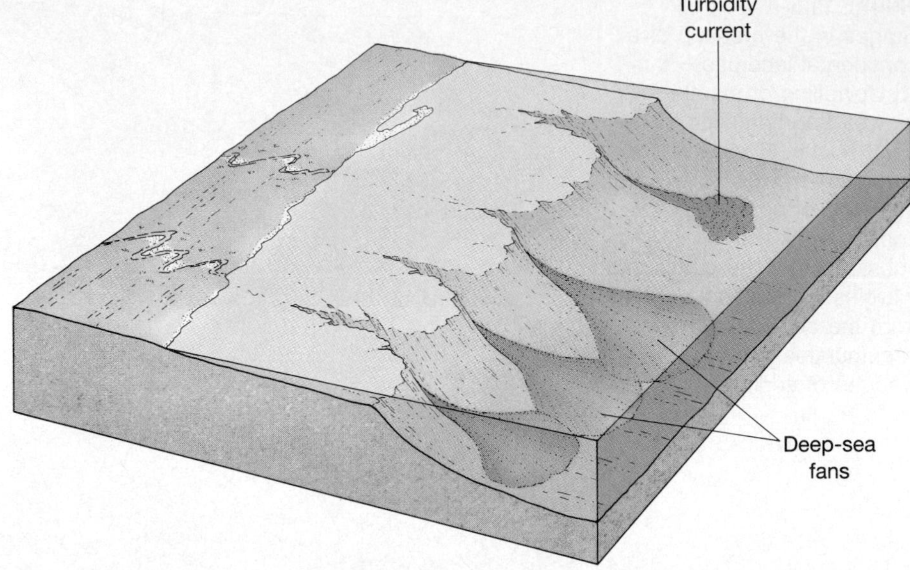

FIGURE 19.6
Submarine canyons are thought to be cut into the continental margins by turbidity currents. These sediment-laden density currents eventually lose momentum and deposit their loads of sediment as deep-sea fans.

trench bordering the continent exceeds 12,000 meters.

SUBMARINE CANYONS AND TURBIDITY CURRENTS

Deep, steep-sided valleys known as **submarine canyons** originate on the continental slope and may extend to depths of 3 kilometers (Figure 19.6). Although some of these canyons appear to be the seaward extensions of river valleys such as the Hudson Valley, many others do not line up in this manner. Furthermore, since these canyons extend to depths far below the maximum lowering of sea level during the Ice Age, we cannot attribute their formation to stream erosion. These features must be created by some process that operates far below the ocean surface. Most available information seems to favor the view that submarine canyons have been excavated by turbidity currents (Figure 19.6). **Turbidity currents** are downslope movements of dense, sediment-laden water. They are created when sand and mud on the continental shelf and slope are dislodged, perhaps by an earthquake, and are thrown into suspension. Since such mud-choked water is denser than normal sea water, it flows downslope, eroding and accumulating more sediment (Figure 19.6). The erosional work repeatedly carried on by these muddy torrents is thought to be the major force in the excavation of most submarine canyons.

Turbidity currents usually originate along the continental slope and continue across the continental rise, still cutting channels. Eventually they lose momentum and come to rest along the bottom of the ocean basin (Figure 19.7). As these currents slow, the

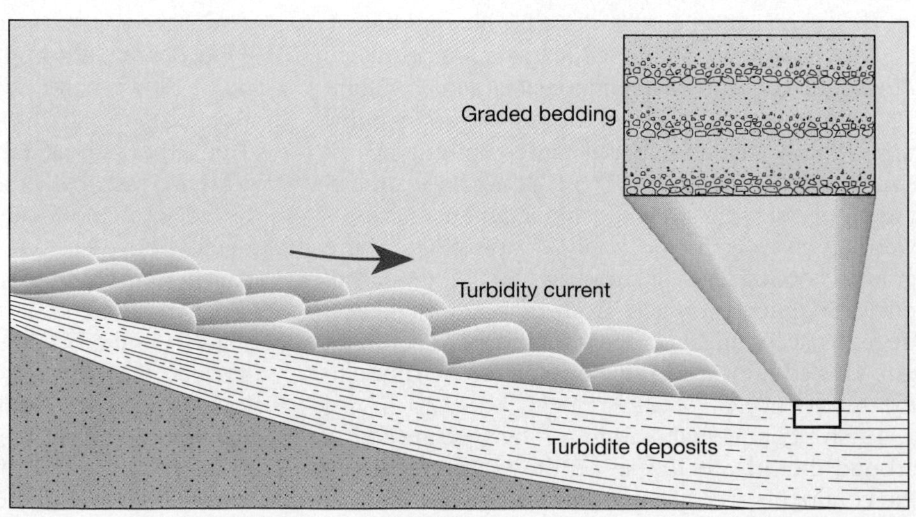

FIGURE 19.7
Turbidity currents move downslope, eroding the continental margin to produce submarine canyons. The deposits, called turbidites, are characterized by a decrease in sediment grain size from bottom to top, a phenomenon known as graded bedding.

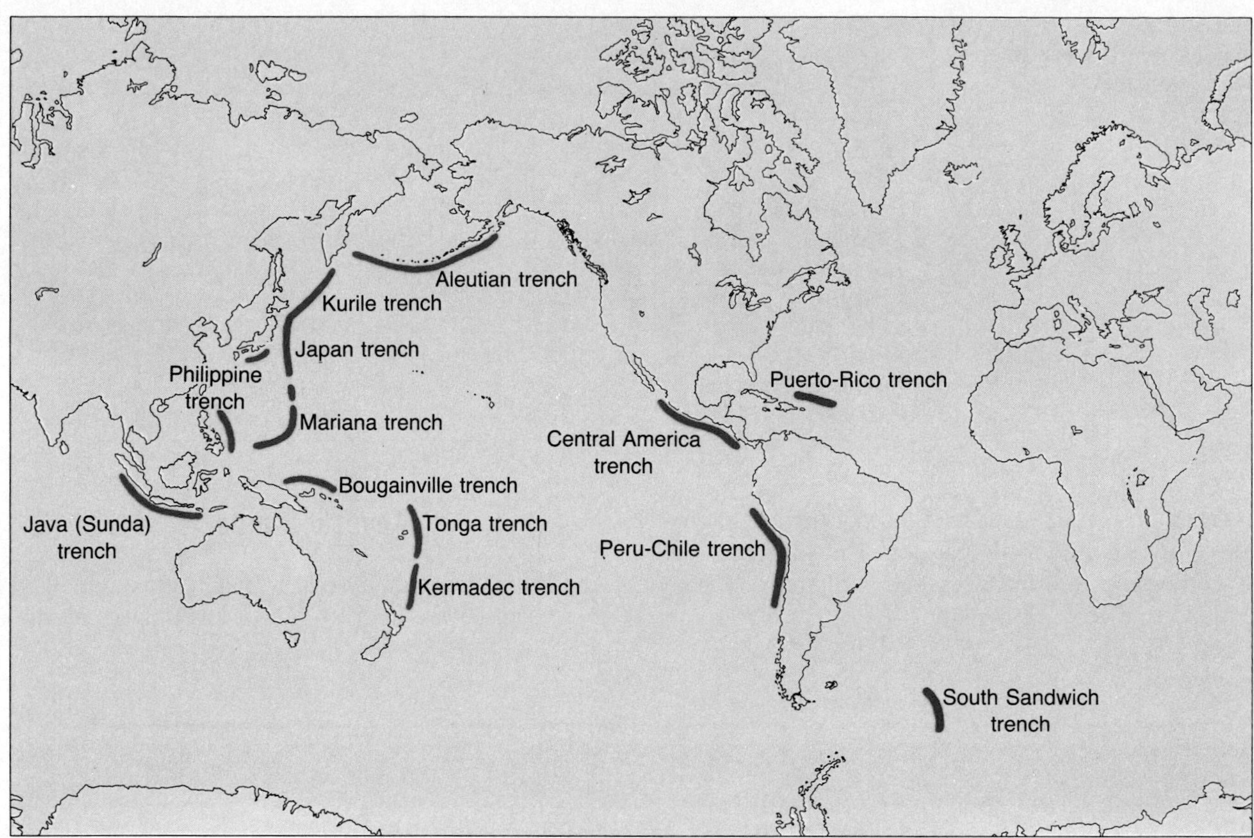

FIGURE 19.8
Distribution of the world's major oceanic trenches.

suspended sediments begin to settle out. First, the coarser sand is dropped, followed by successively finer deposits of silt and then clay. Consequently, these deposits, called **turbidites**, are characterized by a decrease in sediment grain size from bottom to top, a phenomenon known as **graded bedding** (Figure 19.7).

Although there is still more to be learned about the complex workings of turbidity currents, it has been well established that they are an important mechanism of sediment transport in the ocean. By the action of turbidity currents, submarine canyons are created and sediments are carried to the deep-ocean floor.

FEATURES OF THE DEEP-OCEAN BASIN

Between the continental margin and the oceanic ridge system lies the **deep-ocean basin**. The size of this region—almost 30 percent of the earth's surface—is roughly comparable to the percentage of the surface that projects above the sea level as land. Here we find remarkably flat regions known as abyssal plains, steep-sided volcanic peaks called seamounts, and deep-ocean trenches, which are extremely deep linear depressions in the ocean floor.

Deep-Ocean Trenches

Deep-ocean **trenches** are long, relatively narrow features that form the deepest parts of the ocean. Most trenches are located in the Pacific Ocean where some approach or exceed 10,000 meters in depth, and at least a portion of one, the Challenger Deep in the Mariana trench, is more than 11,000 meters below sea level (Figure 19.8). Table 19.1 presents dimensional characteristics of some of the larger trenches.

Although deep-ocean trenches represent only a very small portion of the area of the ocean floor, they are nevertheless significant geologic features. Trenches are the sites where moving crustal plates plunge back into the mantle. In addition to the earthquakes created as one plate descends beneath another, volcanic activity is also associated with trench regions. Thus, trenches in the open ocean are often paralleled by volcanic island arcs. Furthermore, volcanic mountains, such as those making up a portion

TABLE 19.1
Dimensions of some deep-ocean trenches.

Trench	Depth (kilometers)	Average Width (kilometers)	Length (kilometers)
Aleutian	7.7	50	3700
Japan	8.4	100	800
Java	7.5	80	4500
Kurile-Kamchatka	10.5	120	2200
Mariana	11.0	70	2550
Central America	6.7	40	2800
Peru-Chile	8.1	100	5900
Philippine	10.5	60	1400
Puerto Rico	8.4	120	1550
South Sandwich	8.4	90	1450
Tonga	10.8	55	1400

of the Andes, are located parallel to trenches that lie adjacent to continental margins. The melting of a descending plate produces the molten rock that leads to this volcanic activity.

Abyssal Plains

Abyssal plains are incredibly flat features; in fact, these regions are likely the most level places on the

BOX 19.1

Evidence for the Existence of Turbidity Currents

For many years the existence of turbidity currents in the ocean was a matter of considerable debate among marine geologists. Not until the 1950s did the speculation begin to subside. Two lines of evidence helped establish turbidity currents as important mechanisms of submarine erosion and sediment transportation. The first important evidence came from records of a rather severe earthquake that took place off the coast of Newfoundland in 1929 and resulted in the breakage of 13 transatlantic telephone and telegraph cables. At the time, it was presumed that the tremor had caused the multiple breaks. However, when the data were analyzed, it appeared that this was not the case. After plotting the locations of the breaks on a map, researchers saw that all the breaks had occurred along the steep continental slope and the gentler continental rise. Since the time of each break was known from information provided by automatic recorders, a pattern of what had happened could be deduced. The breaks high up on the continental slope took place first, almost concurrently with the earthquake. The other breaks happened in succession, the last occurring 13 hours later, some 720 kilometers (450 miles) from the source of the quake (Figure 19.A). The breaks downslope had obviously taken place too long after the tremor to have been caused by the shock of the earthquake. The existence of a turbidity current, triggered by the quake, thus appeared as a plausible alternative. As the avalanche of sediment-choked water raced downslope, it snapped the cables in its path. Investigators calculated that the current reached speeds approaching 80 kilometers (50 miles) per hour on the continental slope and about 24 kilometers (15 miles) per hour on the more gently sloping continental rise.

A second compelling line of evidence relating turbidity currents to submarine erosion and transportation of sediment came from the examination of deep-sea sediment samples. These cores show that extensive graded beds of sand, silt, and clay exist in the quiet waters of the deep ocean. Some samples also include fragments of plants and animals that live only in the shallower waters of the continental shelves. No mechanism other than turbidity currents has been identified to satisfactorily explain the existence of these deposits.

More recently, turbidity currents have been measured directly. In one study, instruments were deployed in Bute Inlet, a deep fiord along the coast of British Columbia in western Canada. The inlet served as a natural laboratory for the study of turbidity current dynamics. The year-long program found turbidity currents to be com-

earth. The abyssal plain found off the coast of Argentina, for example, has less than 3 meters of relief over a distance exceeding 1300 kilometers. The monotonous topography of abyssal plains will occasionally be interrupted by the protruding summit of a buried volcanic structure.

By employing seismic profilers, instruments whose signals penetrate far below the ocean floor, researchers have determined that abyssal plains owe their relatively featureless topography to thick accumulations of sediment that have buried an otherwise rugged ocean floor. The nature of the sediment indicates that these plains consist primarily of sediments transported far out to sea by turbidity currents.

Abyssal plains are found in all the oceans. However, since the Atlantic Ocean has fewer trenches to act as traps for the sediments carried down the continental slope, it has more extensive abyssal plains than the Pacific.

Seamounts

Dotting the ocean floors are isolated volcanic peaks called **seamounts** that may rise hundreds of meters above the surrounding topography. Although these steep-sided conical peaks have been discovered in all the oceans, the greatest number have been identified in the Pacific.

Many of these undersea volcanoes form near oceanic ridges, regions of sea-floor spreading. If a volcano grows rapidly, it may emerge as an island. Examples of volcanic islands in the Atlantic include the Azores, Ascension, Tristan da Cunha, and St. Helena.

While they exist as islands, some of these volcanoes are eroded to near sea level by running water and wave action. Over a span of millions of years the islands gradually sink as the moving plate slowly carries them from the oceanic ridge area. These submerged, flat-topped seamounts are called **guyots**. In

mon events. Although the majority were low-velocity flows that covered limited distances, occasional faster-moving, large-scale currents were also measured. These larger events moved significant amounts of sand down the gently sloping floor of the fiord for distances of 40 to 50 kilometers. The dense currents of sediment in this investigation were linked to submarine landslides in a delta at the head of the fiord.

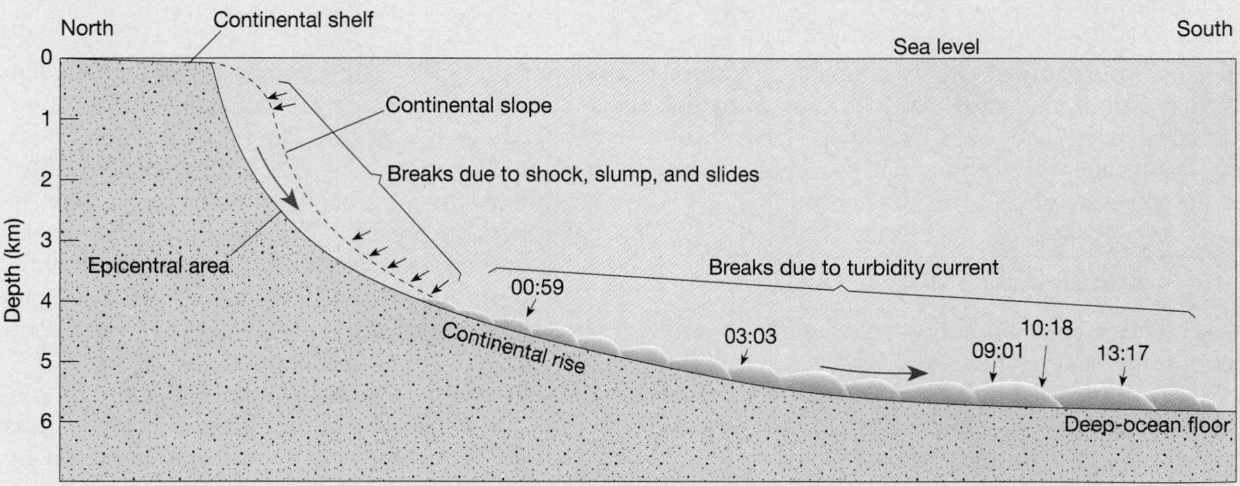

FIGURE 19.A
Profile of the sea floor showing the events of the November 18, 1929, earthquake off the coast of Newfoundland. The arrows point to cable breaks; the numbers show times of breaks in hours and minutes after the earthquake. Vertical scale is greatly exaggerated. (After B. C. Heezen and M. Ewing, "Turbidity Currents and Submarine Slump and the 1929 Grand Banks Earthquake," *American Journal of Science* 250: 867)

FIGURE 19.9
View from space of a group of atolls in the Pacific Ocean. (Photo courtesy of NASA)

other instances, guyots may be remnants of eroded volcanic islands that were formed away from the ridge crest, possibly by hot spot activity. Here subsidence occurs after the volcanic activity ceases and the sea floor cools and contracts.

CORAL REEFS AND ATOLLS

Coral reefs are among the most picturesque features found in the ocean. They are constructed primarily from the calcareous (calcite-rich) skeletal remains and secretions of corals and certain algae. The term *coral reef* is somewhat misleading in that it makes no mention of the skeletons of many small animals and plants found inside the branching framework built by the corals; nor does it reveal that limy secretions of algae help bind the entire structure together.

Coral reefs are confined largely to the warm, clear waters of the Pacific and Indian oceans, although a few occur elsewhere. Reef-building corals grow best in waters with an average annual temperature of about 24°C (75°F). They can survive neither sudden temperature changes nor prolonged exposure to temperatures below 18°C (64°F). In addition, these reef-builders require clear sunlit water. Consequently, the limiting depth of active reef growth is only about 45 meters. Clear blue waters such as those in the Bahamas support active reef building.

In 1831 the naturalist Charles Darwin set out aboard the British ship H.M.S. *Beagle* on its famous five-year expedition that circumnavigated the globe. One outcome of Darwin's studies was the development of a theory on the formation of coral islands, called **atolls.** As Figure 19.9 illustrates, atolls consist of a nearly continuous ring of coral reef surrounding a central lagoon. Darwin's theory explained what seemed to be a paradox; that is, how can corals, which require warm, shallow, sunlit water no deeper than a few tens of meters to live, create structures that reach thousands of meters to the floor of the ocean? Commenting on this in his book *The Voyage of the Beagle,* Darwin states:

> . . . from the fact of the reef-building corals not living at great depths, it is absolutely certain that throughout these vast areas, wherever there is now an atoll, a foundation must have originally existed within a depth of from 20 to 30 fathoms* from the surface.

The essence of Darwin's theory was that coral reefs form on the flanks of sinking volcanic islands. As the island slowly sinks, the corals continue to build the reef complex upward (Figure 19.10).

> For as mountain after mountain, and island after island slowly sank beneath the water, fresh bases would be successively afforded for the growth of the corals.

Thus atolls, like guyots, are thought to owe their existence to the gradual sinking of oceanic crust. In succeeding years there were numerous challenges to Darwin's theory. These arguments were not completely put to rest until after World War II when the United States made extensive studies of two atolls (Eniwetok and Bikini) that became sites for testing atomic bombs. Drilling operations at these atolls revealed that volcanic rock did indeed underlie the thick coral reef structure. This finding was a striking confirmation of Darwin's theory.

*One fathom equals 1.8 meters (6 feet), the approximate distance from fingertip to fingertip of a person with outstretched arms.

SEA-FLOOR SEDIMENTS

Except for a few areas, such as near the crests of mid-ocean ridges, the ocean floor is mantled with sediment. Part of this material has been deposited by turbidity currents, and the rest has slowly settled to the bottom from above. The thickness of this carpet of debris varies greatly. In some trenches, which act as traps for sediments originating on the continental

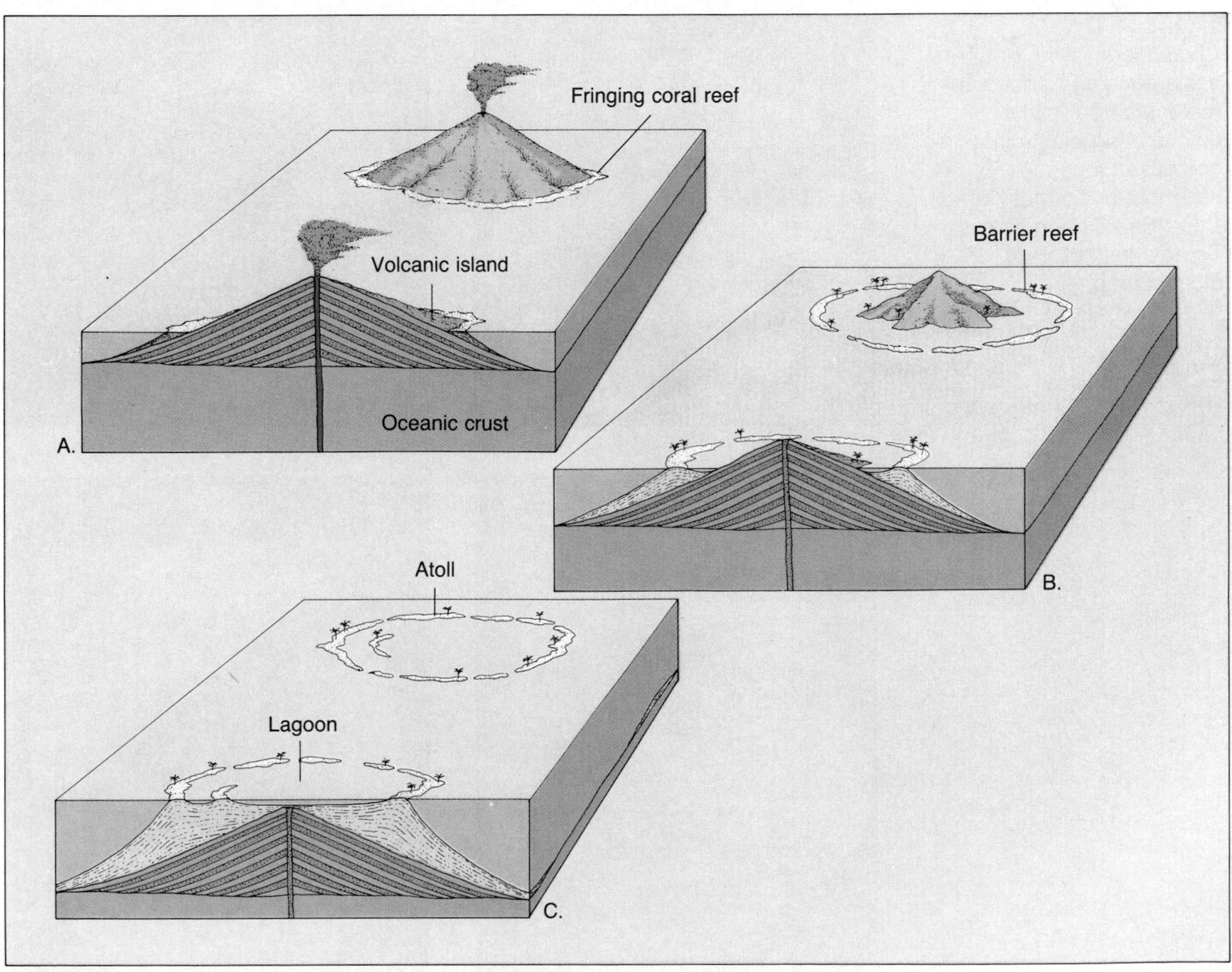

FIGURE 19.10
Formation of a coral atoll due to the gradual sinking of oceanic crust.

margin, accumulations may exceed 9 kilometers. In general, however, sediment accumulations are considerably less. In the Pacific Ocean, uncompacted sediment measures about 600 meters or less, while on the floor of the Atlantic, the thickness varies from 500 to 1000 meters.

Although deposits of sand-sized particles are found on the deep-ocean floor, mud is the most common sediment covering this region. Muds also predominate on the continental shelf and slopes, but the sediments in these areas are coarser overall because of greater quantities of sand. Sampling has revealed that sand deposits are most prevalent as beach deposits along the shore. However, in some cases coarse sediment, which we normally expect to be deposited near the shore, occurs in irregular patches at greater depths near the seaward limits of the continental shelves. While some sand may have been deposited by strong localized currents capable of moving coarse sediment far from shore, much of it appears to be the result of sand deposition on ancient beaches. Such beaches evidently formed during the Ice Age, when sea level was much lower than today.

Sea-floor sediments can be classified according to their origin into three broad categories: (1) terrigenous ("derived from the land"); (2) biogenous ("derived from organisms"); and (3) hydrogenous ("derived from water"). Although each category is discussed separately, remember that all sea-floor sediments are mixtures. No body of sediment comes from a single source.

Terrigenous sediment consists primarily of mineral grains that were weathered from continental rocks and transported to the ocean. The sand-sized particles settle near shore. However, since the very smallest particles take years to settle to the ocean floor, they may be carried for thousands of kilome-

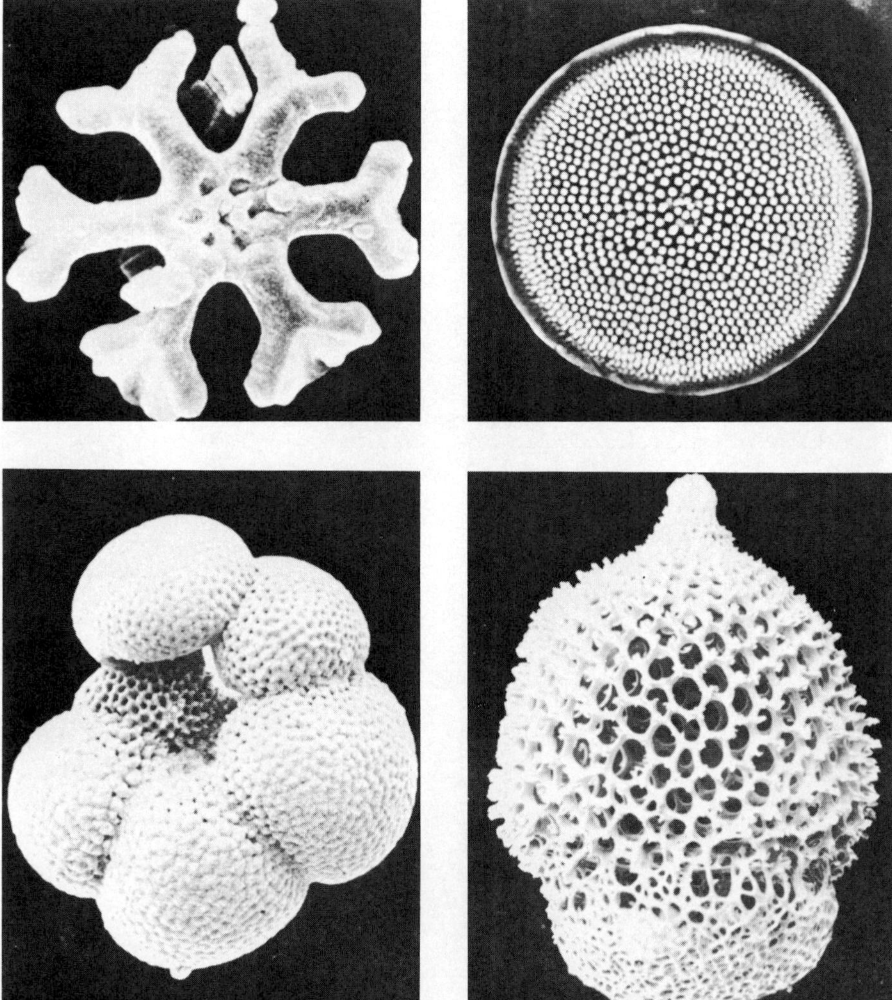

FIGURE 19.11
Photomicrographs of typical calcareous and siliceous materials from minute animals and plants that compose biogenous sediments. (Courtesy of Deep Sea Drilling Project, Scripps Institution of Oceanography)

ters by ocean currents. As a consequence, virtually every part of the ocean receives some terrigenous sediment. However, the rate at which this sediment accumulates on the deep-ocean floor is indeed very slow. From 5000 to 50,000 years are necessary for a 1-centimeter layer to form. Conversely, on the continental margins near the mouths of large rivers, terrigenous sediment accumulates rapidly. In the Gulf of Mexico, for example, the sediment has reached a depth of many kilometers.

Since fine particles remain suspended in the water for a very long time, there is ample opportunity for chemical reactions to occur. Because of this, the colors of the deep-sea sediments are often red or brown. This results when iron on the particle or in the water reacts with dissolved oxygen in the water and produces a coating of iron oxide (rust).

Biogenous sediment consists of shells and skeletons of marine animals and plants (Figure 19.11). This debris is produced mostly by microscopic organisms living in the sunlit waters near the ocean surface. The remains constantly "rain" down upon the sea floor.

The most common biogenous sediments are known as *calcareous* ($CaCO_3$) *oozes,* and as their name implies, they have the consistency of thick mud. These sediments are produced by organisms that inhabit warm surface waters. When calcareous hard parts slowly sink through a cool layer of water, they begin to dissolve. This results because cold seawater is rich in carbon dioxide and is thus more acidic than warm water. In seawater deeper than about 4500 meters (15,000 feet), calcareous shells will completely dissolve before they reach the bottom. Consequently, calcareous ooze does not accumulate in the deep-ocean basins.

Other biogenous sediments include *siliceous* (SiO_2) *oozes* and phosphate-rich materials. The former are composed primarily of opaline skeletons of diatoms (single-celled algae) and radiolarians (single-celled animals), while the latter are derived from the bones, teeth, and scales of fish and other marine organisms.

Hydrogenous sediment consists of minerals that crystallize directly from seawater through various chemical reactions. For example, some limestones are formed when calcium carbonate precipitates directly from the water; however, most limestone is composed of biogenous sediment.

One of the principal examples of hydrogenous sediment, and one of the most important sediments on the ocean floor in terms of economic potential, are **manganese nodules**. These rounded blackish lumps are composed of a complex mixture of minerals that form very slowly on the floors of the ocean basins. (See Box 19.2.)

MID-OCEAN RIDGES

Our knowledge of **mid-ocean ridges**, the sites of seafloor spreading, comes from soundings taken of the ocean floor, core samples obtained from deep-sea drilling, visual inspection using deep-diving submersibles, and even first-hand inspection of slices of ocean floor that have been shoved up onto dry land. Ocean ridge systems are characterized by an elevated position, extensive faulting, and numerous volcanic structures which have developed upon the newly formed crust (Figure 19.12).

Ocean ridges are found in all of the major oceans and represent more than 20 percent of the earth's surface. They are certainly the most prominent topographic features in the oceans, for they form a continuous mountain range that extends for about 65,000 kilometers (40,000 miles) in a manner similar to the seam on a baseball. Although ocean ridges stand high above the adjacent deep-ocean basins, they are much different from the mountains found on the continents. Rather than consisting of thick sequences of folded and faulted sedimentary rocks, oceanic ridges consist of layer upon layer of basaltic rocks that have been faulted and uplifted. The term *ridge* may also be misleading since these features are not narrow but have widths of from 500 to 5000 kilometers and, in places, may occupy as much as one-half of the total area of the ocean floor.

Despite their large size, most of the accretions of new sea floor occur along a narrow region centered on the ridge crest. These active **rift zones** are characterized by frequent but generally weak earthquakes and a rate of heat flow that is greater than that of other crustal segments. Here vertical displacement of large slabs of oceanic crust caused by faulting and the growth of volcanic piles contribute to the characteristically rugged topography of the oceanic ridge system. Further, the rocks along the ridge axis appear very fresh and are nearly void of sediment. Away from the ridge axis the topography becomes more subdued and the thickness of the sediments as well as the depth of the water increases. Gradually the ridge system grades into the flat, sediment-laden abyssal plains of the deep-ocean basin.

FIGURE 19.12
The topography of the earth's solid surface is shown on the following two pages. (Copyright © by Marie Tharp)

CANADA
ABYSSAL
PLAIN
ELLESMERE
ISLAND
APTEV SEA
NEW SIBERIAN
ISLANDS
MELVILLE ISLAND
DEVON I.
BANKS I.
BAFFIN
BAY
WRANGEL
ISLAND
CHUKCHI SEA
VICTORIA ISLAND
BAFFIN
ISLAND
BROOKS RANGE
ALASKA
BERING
STRAIT
VERKHOYANSK MTS.
KOLYMSKOYE MTS.
BERING SEA
HUDSON
BAY
STANOVOY MTS.
OKHOTSK
SEA
KAMCHATKA
BERING ABYSSAL PLAIN
NORTH AMERICA
LABRADOR
ALEUTIAN TRENCH
ROCKY MOUNTAINS
ALEUTIAN
ABYSSAL PLAIN
MT. SHASTA
Chicago
New York
Washington
Montreal
San Francisco
MURRAY FRACTURE ZONE
MOLOKAI FRACTURE ZONE
CLARION FRACTURE ZONE
CLIPPERTON FRACTURE ZONE
New Orleans
CHRISTMAS I.
NEW GUINEA
Darwin
Bogota
Quito
AMAZON
SOUTH AM
SIMPSON
DESERT
AUSTRALIA
NULLARBOR PLAIN
LORD HOWE RISE
Sydney
Melbourne
TASMANIA
NEW ZEALAND
MT. COOK
JUAN
FERNANDEZ I.
Santiago
Buenos Aires
FALKLAND
ISLANDS
RELLINGSHAUSEN ABYSSAL PLAIN
ANTARCTICA

GREENLAND
CONTINENTAL SHELF
BARENTS SEA
KARA SEA
NOVAYA ZEMLYA
TAYMYR PENINSULA
Murmansk
VORING PLATEAU
ICELAND
FAEROE ISLANDS
ROCKALL BANK
NORTH SEA
Oslo
Helsinki
Leningrad
Moscow
Copenhagen
Glasgow
Dublin
London
Amsterdam
Berlin
Warsaw
EUROPA
URAL MOUNTAINS
VOLGA
SIBERIAN PLATEAU
ASIA
LAKE BAYKAL
Irkutsk
YABLONOVYY MTS
Ulan Bator
GOBI DESERT
Paris
Vienna
CARPATHIAN MTS
Bern
ALPS
DANUBE
PYRENEES
Madrid
Lisbon
Rome
CAUCASUS
CASPIAN SEA
LAKE BALKHASH
Tashkent
KARA KOUM
TIEN SHAN MTS
Peking
FLEMISH CAP
GRAND BANKS
GIBBS FRACTURE ZONE
Ankara
Teheran
HINDU KUSH
TIBET PLATEAU
Rabat
ATLAS MOUNTAINS
Tunis
Tripoli
Jerusalem
Bagdad
Cairo
CANARY ISLANDS
New Delhi
THAR DESERT
HIMALAYAS
GANGES
Shanghai
SAHARA
AHAGGAR
TIBESTI
NILE
ARABIA
Rhyad
Karachi
Dacca
Hanoi
Canton
Bombay
INDUS CONE
GANGES CONE
CAPE VERDE ISLANDS
Dakar
NIGER
Niamey
LAKE CHAD
N'Djamena
Khartoum
Aden
Addis Ababa
ANDAMAN ISLANDS
Rangoon
Bangkok
Saigon
AFRICA
Abidjan
Accra
Lagos
MT. CAMEROON
Douala
CONGO
WESTERN RIFT VALLEY
LAKE VICTORIA
Nairobi
Kinshasa
Luanda
CONGO CONE
Singapore
BORNEO
JAVA TRENCH
SUMATRA
Djakarta
JAVA
MATO GROSSO
Brasilia
Lusaka
ZAMBEZI
Salisbury
ANGOLA ABYSSAL PLAIN
MADAGASCAR
Rio de Janeiro
SANTOS PLATEAU
WALFISCH RIDGE
KALAHARI DESERT
Pretoria
ORANGE
Cape Town
Perth
ARGENTINE RISE
ARGENTINE ABYSSAL PLAIN
DIAMANTINA FRACTURE ZONE
CROZET ISLANDS
KERGUELEN ISLANDS
LENA GUYOT
KERGUELEN
WILKES ABYSSAL PLAIN
FALKLAND PLATEAU
ENDERBY ABYSSAL PLAIN
CONTINENTAL RISE
WEDDELL ABYSSAL PLAIN
MAUD RISE
WILKES LAND
ENDERBY LAND

BOX 19.2

Resources from the Deep-Ocean Floor

The demand for mineral resources continues to climb in response to a rapidly growing world population and the desire of people everywhere to have a higher standard of living. As demand rises and technology improves, the floor of the ocean may become an important source of some valuable minerals. Today oil, gas, titanium, tin, gold, and diamonds are being recovered from relatively shallow coastal waters. In the future the deep-ocean basins may also become sites of mineral production. Exploration in this environment has already begun. The effort has resulted in the discovery of substantial deposits of manganese nodules as well as rich accumulations of metallic sulfides.

Deposits of manganese nodules that may have significant economic potential are found in many parts of the deep-ocean basins. Despite existing technological and political problems, it appears as though the mining of these deposits may take place in the relatively near future. In fact, at least one company has attempted to file a claim on a deposit of manganese nodules covering a vast area of the Pacific Ocean floor.

Manganese nodules are rounded, dark lumps that are composed of a mixture of minerals (Figure 19.B). The major component is manganese dioxide (MnO_2), which constitutes about 30 percent of a typical nodule. Iron, in the form of iron oxide (Fe_2O_3), is next in abundance and makes up roughly 20 percent of each nodule. Although manganese and iron are important materials in their own right, the interest in manganese nodules as a potential resource lies in the fact that other, more valuable metals may be enriched in them. Nodules may contain significant quantities of copper, nickel, and cobalt.

Although nodules are widely distributed in the deep-ocean basins, not all regions are equally promising sites for development. Potential mining locations must have abundant nodules (more than 5 kilograms per square meter) that contain the economically optimum mix of cobalt, copper, and nickel. Sites meeting these criteria are relatively limited. Furthermore, before such areas prove to be valuable commercial sources for these metals, the logistics of extracting nodules from the floors of the deep-ocean basins, and the political ramifications, must be worked out.

A second potential source of mineral resources on the deep-ocean floor are massive sulfide deposits. Exploration of a series of deep basins along the axis of the Red Sea has led to the discovery of the richest submarine metallic sulfide deposits yet known. The largest deposit found to date ranks with the largest deposits on land. It is estimated to contain 100 million metric tons of potential ore, consisting of 29 percent iron, over 3 percent zinc, 1 percent copper, and substantial concentrations of gold and silver. Located above these deposits are pools of metallic brines that have salinities two to three times normal values and temperatures that exceed 36°C (97°F). Hydrothermal convection systems involving these reactive brines are believed to extract metallic ions from the underlying crustal rocks and deposit them as metallic sulfides on the sea floor. Despite the fact that these deposits are located at depths of 2000 meters or more, mining operations are possible in the future as world demand for metals increases.

FIGURE 19.B
Manganese nodules photographed at a depth of 2909 fathoms (5323 meters) beneath the *Robert Conrad* south of Tahiti. (Photo courtesy of Lawrence Sullivan, Lamont-Doherty Geological Observatory)

During sea-floor spreading new material is added about equally to the two diverging plates; hence, we would expect new ocean floor to grow symmetrically about a centrally located ridge. Indeed, the ridge systems of the Atlantic and Indian oceans are located near the middle of these water bodies and as a consequence are named mid-ocean ridges. However, the East Pacific Rise is situated far from the center of the Pacific Ocean. Despite uniform spreading along the East Pacific Rise, much of the Pacific basin that once lay east of this spreading center has been overridden by the westward migration of the American plate.

Partly because of its accessibility to both American and European scientists, the Mid-Atlantic Ridge has been studied more thoroughly than other ridge sys-

FIGURE 19.13
A photograph taken from the *Alvin* during project FAMOUS shows lava extrusions in the rift valley of the Mid-Atlantic Ridge. Large toothpastelike extrusions such as this were common features. A mechanical arm is sampling an adjacent blisterlike extrusion. (Photo courtesy of Woods Hole Oceanographic Institution)

tems. The Mid-Atlantic Ridge is a gigantic submerged mountain range standing 2500-3000 meters above the adjacent floor of the deep-ocean basins. In a few places, such as Iceland, the ridge has actually grown above sea level. Throughout most of its length, however, the divergent plate boundary lies 2500 meters below sea level. Another prominent feature of the Mid-Atlantic Ridge is a deep linear valley extending along the ridge axis. In places this rift valley is deeper than the Grand Canyon of the Colorado and two or three times as wide. The name *rift valley* has been applied to this feature because it is so strikingly similar to continental rift valleys such as the East African rift valleys. An examination of Figure 19.12 reveals that this central rift is broken into sections that are offset by transform faults.

THE OCEAN FLOOR AND SEA-FLOOR SPREADING

The concept of sea-floor spreading was formulated in the early 1960s by Harry H. Hess of Princeton University. Later, using deep-diving submersibles, geologists were able to support Hess' thesis that sea-floor spreading occurs along relatively narrow zones located at the crests of ocean ridges. (See Box 19.3.) Along the East Pacific Rise the active zones of sea-floor formation appear to be only about a kilometer wide, whereas along the Mid-Atlantic Ridge these zones may extend for tens of kilometers. As the plates move apart, magma intrudes into the newly created fracture zone and generates new sections of oceanic crust (Figure 19.13). This apparently unending process generates new lithosphere that moves from the ridge crest in a conveyor-belt fashion.

As various segments of the oceanic ridge system were studied in detail, numerous differences came to light. For example, the East Pacific Rise has a relatively fast rate of spreading that averages about 6 centimeters per year and reaches a maximum of 10 centimeters per year along a section of the ridge located near Easter Island (Figure 19.14). By contrast, the spreading rate in the North Atlantic is much less, averaging about 2 centimeters per year. Apparently the rate of spreading strongly influences the appearance of the ridge system. The slow spreading along the Mid-Atlantic Ridge is believed to contribute to its very rugged topography and large central rift valley. By contrast, the rapid spreading of the East Pacific Rise is thought to account for its more subdued topography and the lack of a well-defined rift valley through much of its length. Despite these differences, all ridge systems are thought to generate new sea floor in a similar manner.

Although most oceanic crust forms out of view, far below sea level, geologists have been able to examine the structure of the ocean floor firsthand. In such locations as Newfoundland, Cyprus, and California, slivers of oceanic crust have been elevated high above sea level. From these outcrops it appears that the ocean floor consists of three distinct layers (Figure 19.15A). The upper layer is composed mainly of pillow basalts. The middle layer is made up of numerous interconnected dikes called **sheeted dikes**. Finally, the lower layer is made up of gabbro, the coarse-grained equivalent of basalt, that crystallized at depth. This sequence of rocks is called an **ophiolite complex** (Figure 19.15A). From studies of various ophiolite complexes and related data, geologists have pieced together a scenario for the formation of the ocean floor.

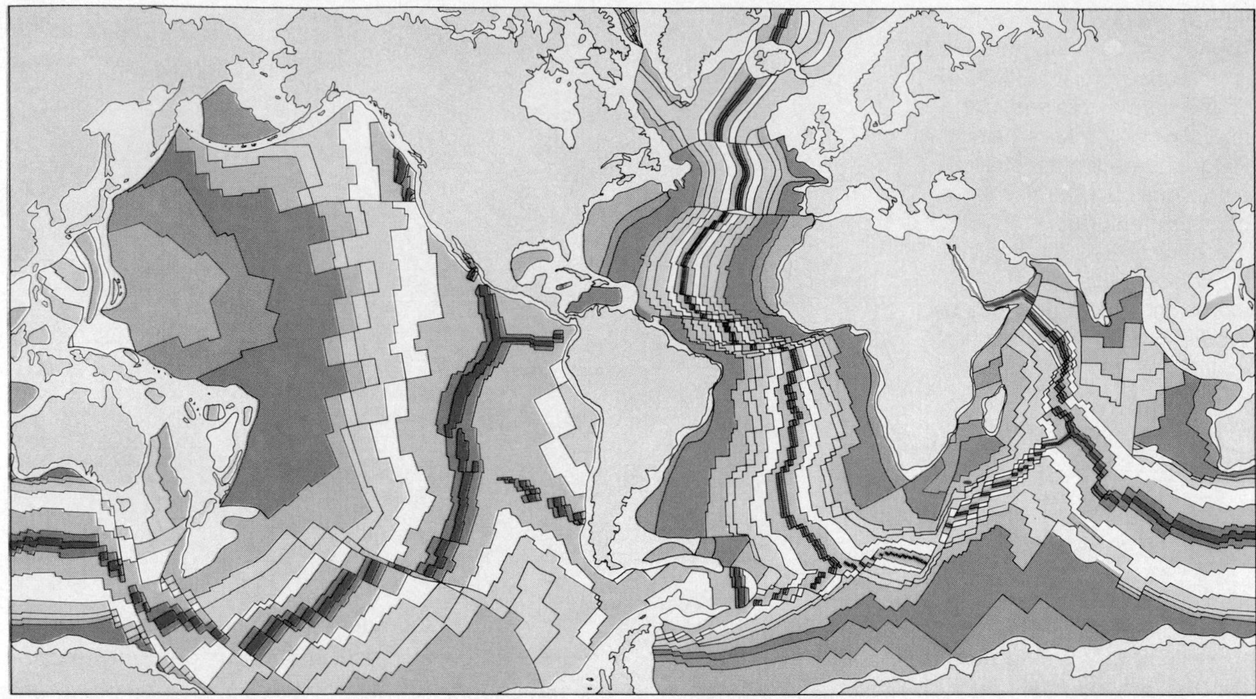

A.

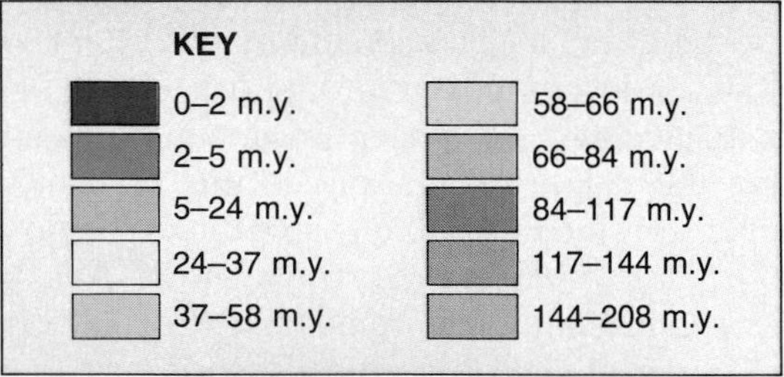

B.

FIGURE 19.14
The relative age of oceanic crust beneath deep-sea deposits. Notice that the youngest rocks (bright red areas) are found along the oceanic ridge crests and the oldest oceanic crust (brown areas) is located adjacent to the continents. When you observe the Atlantic basin, a symmetrical pattern centered on the Mid-Atlantic Ridge crest becomes apparent. This pattern verifies the fact that sea-floor spreading generates new oceanic crust equally on both sides of a spreading center. Further, compare the widths of the yellow stripes in the Pacific basin with those in the South Atlantic. Because these stripes were produced during the same time period, this comparison verifies that the rate of sea-floor spreading was faster in the Pacific than in the South Atlantic. (After *The Bedrock Geology of the World,* by R. L. Larson et al. Copyright © by W. H. Freeman)

The magma that migrates upward to create new ocean floor originates from partially melted peridotite in the asthenosphere. In the region of the rift zone, this magma source may lie no more than 35 kilometers below the sea floor. Being molten and less dense than the surrounding solid rock, the magma gradually moves upward where it is believed to enter large reservoirs located only a few kilometers below the ridge crest (Figure 19.15B). As the ocean floor is pulled or pushed apart, numerous fractures develop in the crust, permitting this molten rock to migrate upward. During each eruptive phase, the initial flows are thought to be quite fluid and spread over the rift zone in broad, thin sheets. As new lava flows are added to the ocean floor, each is cut by fractures that allow additional lava to migrate upward and form overlying layers. Later in each eruptive cycle, as the magma in the shallow reservoir cools and thickens, shorter flows with a more characteristic pillow form occur. Recall that pillow lava has the appearance of large, elongate sand bags stacked one atop another (Figure 19.16). Depending upon the rate of flow, the thick pillow lavas may build into volcano-sized mounds. These mounds will eventually be cut off from their supply of magma and be carried away from the ridge crest by sea-floor spreading. The magma that does not flow upward will crystallize at depth to generate thick units of coarse-grained gabbro. This lowest rock unit forms as crystallization takes place along the walls and floor of the magma chamber. In this manner the processes at work at the ridge systems are producing the entire sequence of rocks found in the ophiolite complex.

Because newly formed sections of the ocean floor

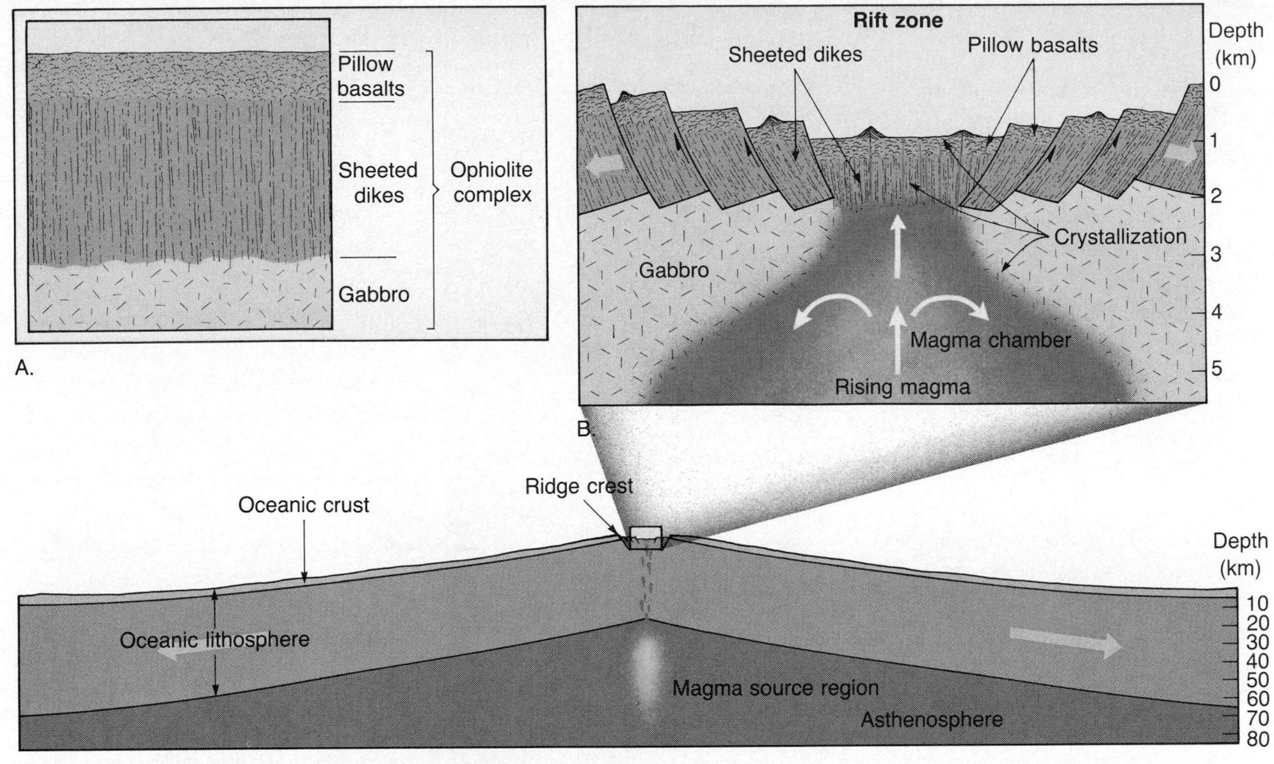

FIGURE 19.15
A. The structure of oceanic crust is thought to be equivalent to the ophiolite complexes that have been discovered elevated above sea level in such places as California and Newfoundland. **B.** The formation of the three units of an ophiolite complex in the rift zone of an oceanic ridge. **C.** Diagram illustrating the site where new ocean crust is generated.

FIGURE 19.16
Ancient pillow lava at Trinity Bay, Newfoundland. (Photo courtesy of the Geological Survey of Canada, photo no. 152581)

BOX 19.3

A Close-up View of the Ocean Floor

Much has been (and continues to be) learned about the floor of the ocean from echo sounders and other remote sensing equipment, as well as from drilling and sampling from surface ships. However, oceanographers in the 1970s became aware that direct manned observation was essential to bring about a clearer understanding of many deep-sea phenomena. What was needed was a firsthand view of the previously unseen world below.

Today, the names and accomplishments of deep-diving, manned submersibles such as the *Alvin* are common knowledge among oceanographers (Figure 19.C). Manned submersibles are now extending the coverage provided by traditional oceanographic research vessels by allowing scientists to investigate the fine-scale features that previously eluded detection.

One of the pioneering research projects that used deep-diving submersibles was a cooperative venture called Project FAMOUS (French-American Mid-Ocean Undersea Study). In 1974, after three years of preliminary surveying and study by surface ships, two French submersibles and one American vessel made a total of 44 dives to the floor of the Atlantic. The primary purpose of the project was to examine the structure of the rift valley in the Mid-Atlantic Ridge. The data collected by the three vessels proved invaluable and led to more realistic explanations of how the spreading process works in creating new ocean floor.

Later, dives were made by the *Alvin* to a spreading center at 20°N

FIGURE 19.C
The deep-diving submersible *Alvin* is 7.6 meters long, weighs 16 tons, has a cruising speed of 1 knot, and can reach depths as great as 4000 meters. A pilot and two scientific observers are along during a normal 6- to 10-hour dive. (Photo courtesy of Woods Hole Oceanographic Institution)

latitude on the East Pacific Rise near the mouth of the Gulf of California. Here, in addition to gathering large quantities of basic data, the scientists aboard the *Alvin* discovered the existence of spectacular, geyserlike hot springs. They witnessed 2- to 5-meter-high chimneylike structures spewing dark, mineral-rich, hot (350°C–400°C) water (Figure 19.D). As the heated solutions hit the surrounding 2°C seawater, sulfides of copper, iron, and zinc precipitated immediately, forming mounds of minerals around the steaming vents. In addition to viewing firsthand the formation of massive sulfide deposits, the scientists aboard the *Alvin* found communities of exotic, bottom-dwelling animals living near cooler (20°C) hot springs. The discovery of an animal community thriving more than 3 kilometers below the surface where no light can reach was totally unexpected. An analysis of the sulfur-rich vent water, as well as the stomach contents of some animals, revealed that the base of the food chain was sulfur-oxidizing bacteria.

Dives such as the ones briefly highlighted here have demonstrated the value of deep-diving, manned submersibles in detecting and studying the fine-scale features of the ocean floor. These vessels now appear to occupy a permanent and important place as tools in oceanographic research.

FIGURE 19.D
A black smoker spewing hot, mineral-rich seawater along the East Pacific Rise. As heated solutions meet cold seawater, sulfides of copper, iron, and zinc precipitate immediately, forming mounds of minerals around these vents. (Photo by Dudley Foster, Woods Hole Oceanographic Institution)

are warm, they are also rather buoyant. This buoyancy is thought to cause large blocks to shear from the sea floor and be elevated. Since the rate of spreading along the Mid-Atlantic Ridge is relatively slow, the displacement of oceanic slabs is more pronounced there than along faster-spreading centers such as the East Pacific Rise. Thus, along the Mid-Atlantic Ridge, uplifted sections form nearly vertical walls that border the central rift zone. As sea-floor spreading continues, the earlier-formed blocks are wedged away from the ridge axis and replaced by more recently formed segments of ocean crust. This process contributes to the imposing height of the Mid-Atlantic Ridge as well as to its topography.

The primary reason for the elevated position of a ridge system is the fact that newly created oceanic crust is hot, and therefore it occupies more volume than cooler rocks of the deep-ocean basin. As the young lithosphere travels away from the spreading center, it gradually cools and contracts. This thermal contraction accounts in part for the greater ocean depths that exist away from the ridge. Almost 100 million years must pass before cooling and contraction cease completely. By this time, rock that was once a part of a majestic ocean mountain system is located in the deep-ocean basin, where it is mantled by thick accumulations of sediment.

OPENING AND CLOSING OF THE OCEAN BASINS

A great deal of evidence has been gathered to support the fact that Wegener's vast continent of Pangaea began to break apart about 200 million years ago. An important consequence of this episode of continental rifting was the creation of a "new" ocean basin, the Atlantic. The breakup of Pangaea and the formation of the Atlantic Ocean basin apparently occurred over a span of nearly 150 million years, with the last phase, the separation of Greenland and Eurasia, beginning only about 50 million years ago.

Although continental rifting is well documented, the question that remains open to debate is, What causes a continent to break apart? We have already considered the role of convection currents as a possible driving mechanism for plate motion (see Chapter 18). It seems reasonable to assume that the slow movement of mantle material could initiate continental rifting. However, the shape of the rifted continental margins and the large number of hot spots located along ridge crests led some geologists to a different conclusion. They have proposed that hot spots initiate continental fragmentation.

Recall that **hot spots** are plumes of molten rock that are believed to rise from deep within the mantle. Hot spots are generally characterized by large outpourings of basaltic lava for relatively long periods of time. Worldwide, as many as 120 isolated volcanic

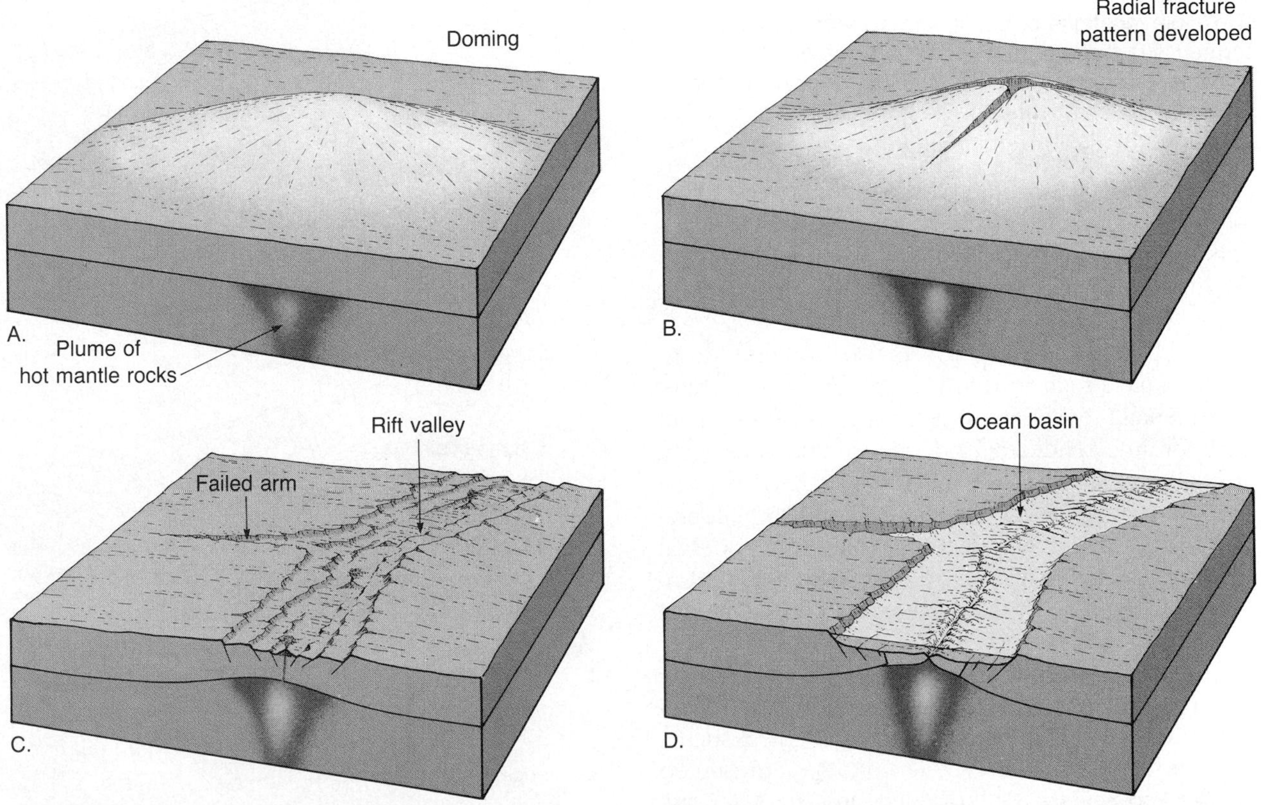

FIGURE 19.17
Diagrammatic representation of the doming and fragmentation of continental crust above an upwelling hot spot. (After Burke and Wilson)

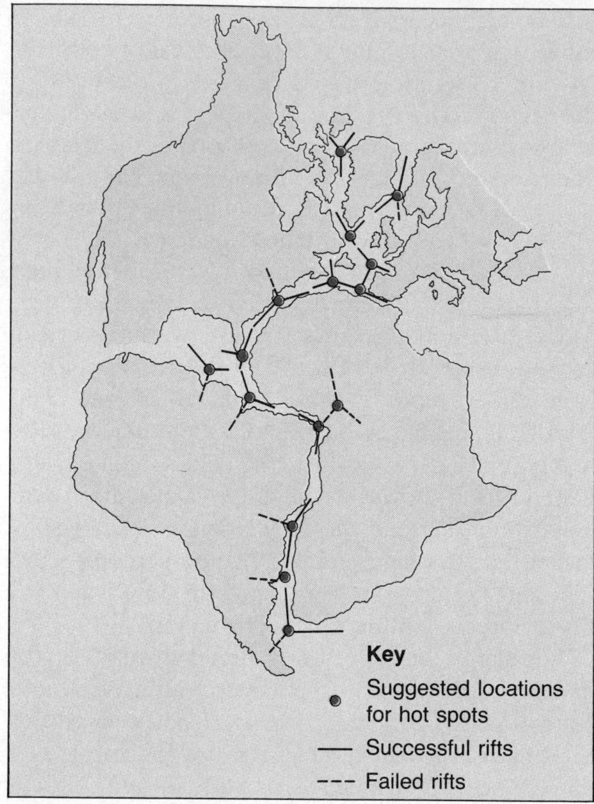

FIGURE 19.18
Possible locations of hot spots and associated three-armed rifts that account for the shapes of the continents surrounding the Atlantic. In most instances two of the arms opened while the third failed. (After Burke and Wilson)

sites have been attributed to hot spot activity. Since hot spots appear to remain nearly stationary over extended time periods, they form volcanic trails, such as the Hawaiian chain, upon the moving oceanic plates above.

The Canadian geologist J. Tuzo Wilson and his associates have suggested that when a thick segment of continental lithosphere remains stationary over a hot spot for an extended period, the conditions are right for continental rifting. Initially, upwelling of material from below generates a dome, roughly 200 kilometers in diameter, within the overlying continental crust as shown in Figure 19.17A. As the dome enlarges, it fractures with a characteristic three-armed pattern (Figure 19.17B). Rifting continues along two of the arms, resulting in the development of a new ocean basin, while the third arm often fails to develop further (Figure 19.17C). An example of such a three-armed rift system is believed to be represented by the Red Sea, the Gulf of Aden, and the Afar Lowlands (see Figure 18.18). Here the arm extending from the Afar Lowlands into the interior of Africa is the failed arm. The two active arms have subsequently generated long, narrow seas.

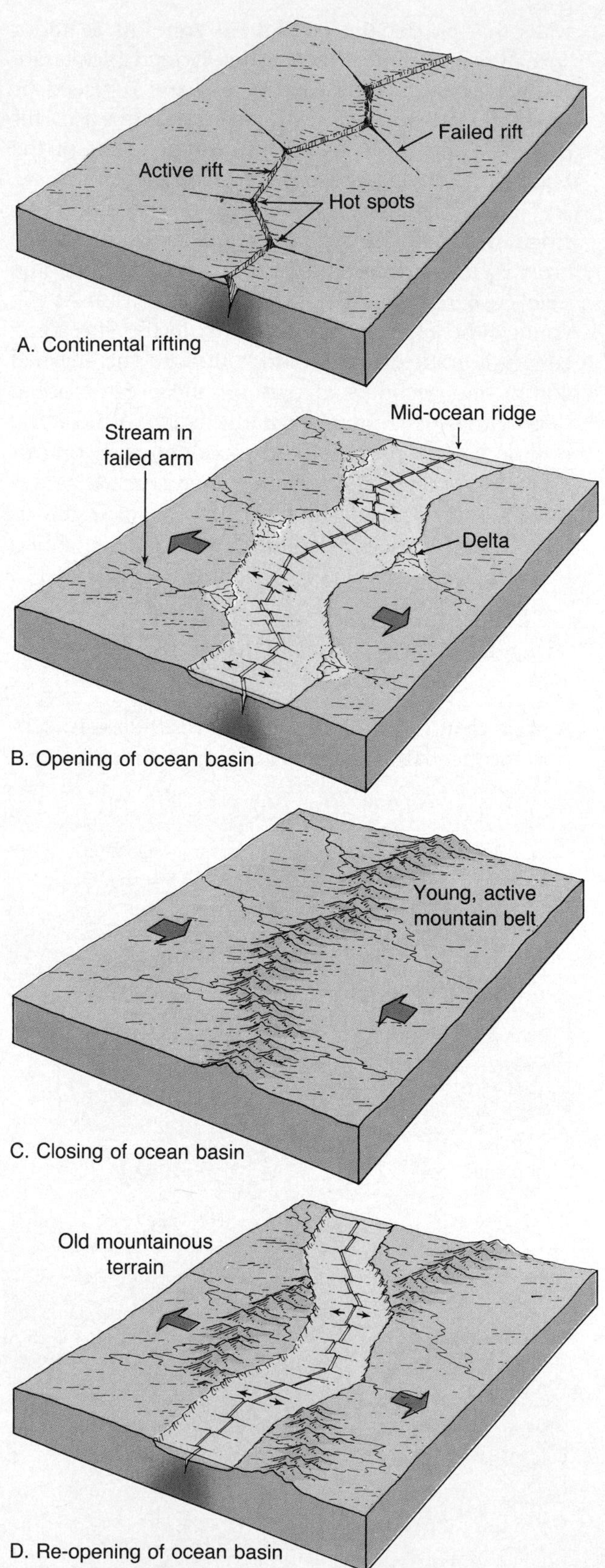

FIGURE 19.19
The Wilson cycle. Diagrammatic representation of the generation of mountainous terrains through the opening and closing of ocean basins.

Professor Wilson suggested that about 20 hot spots guided the fracturing of Pangaea. Figure 19.18 shows the proposed locations of several hot spots and the associated fractures that could account for the shape of the continental margins that presently border the Atlantic. Notice that in some situations all three arms opened, whereas in others, one of the arms failed. Many of these failed rifts extend into the continental interior, where they are represented by deep, narrow, sediment-filled troughs. Some of the world's major river systems, including the Niger and the Amazon, apparently occupy segments of these failed rifts.

In 1966, Wilson also proposed that several times in the geologic past the continents joined to form a supercontinent, which later broke apart. The term **Wilson cycle** is now applied to the cyclic processes that are responsible for the rifting of continents to form ocean basins and the subsequent closing of ocean basins to produce supercontinents (Figure 19.19). In their modern form, Wilson cycles begin with a stationary supercontinent composed of thick continental crust. Because continental crust is a poor conductor of heat, it acts like a thick blanket retarding the outward flow of heat from the mantle. Thus, a buildup of heat causes the supercontinent to bulge upward and eventually break apart. Upwelling of hot material between the rifted continental fragments produces a new ocean floor. This activity also increases the rate of heat flow from the mantle. As the ocean basin grows in size, the sea floor cools and becomes denser. After perhaps 200 million years, the oldest part of the ocean floor becomes dense enough to sink into the mantle. The subduction of the oceanic crust begins the inward flight of the continents, which eventually collide to once again form a supercontinent. The time period of each complete cycle, from breakup to reassembly of the continents, is thought to take about 500 million years.

Although the Wilson cycle has not yet gained general acceptance, it does seem to explain the evolution of the continental margins surrounding the Atlantic. In particular, the closing of the proto-Atlantic is believed to have caused a collision between North America and Africa and Europe that resulted in the formation of the Appalachian mountain belt. The breakup that followed this event is once again dispersing these continental fragments around the globe.

When an ocean closes and reopens, the zone of fragmentation may not occur at the same location as the suture where the landmasses were joined. During the closing of the proto-Atlantic about 400 million years ago, the suture formed along a mountainous belt extending from Alabama to the British Isles and Norway. However, when the Atlantic began to reopen about 200 million years ago, the split occurred along a somewhat different trend. Thus the size and shape of continental fragments appear to change through time. In the next section we will examine further the concept of "Wilson cycles" as it relates to the formation and breakup of Pangaea.

PANGAEA: BEFORE AND AFTER

Robert Dietz and John Holden have rather precisely projected the gross details of the migrations of individual continents over the past 500 million years. By extrapolating plate motion back in time using such evidence as the orientation of volcanic structures left behind on moving plates, the distribution and movements of transform faults, and paleomagnetism, Dietz and Holden were able to reconstruct Pangaea (see Figure 18.1). The use of radiometric dating helped them establish the time frame for the formation and eventual breakup of Pangaea, and the relatively stationary positions of hot spots through time helped to fix the locations of the continents.

Breakup of Pangaea

The fragmentation of Pangaea began about 200 million years ago. Figure 19.20 illustrates the breakup and subsequent paths taken by the landmasses involved. As we can readily see in Figure 19.20A, two major rifts initiated the breakup. The rift zone between North America and Africa generated numerous outpourings of Jurassic-age basalts which are presently visible along the eastern seaboard of the United States. Radiometric dating of these basalts indicates that rifting occurred between 200 and 165 million years ago. This date can be used as the birth date of this section of the North Atlantic. The rift that formed in the southern landmass of Gondwanaland developed a Y-shaped fracture which sent India on a northward journey and simultaneously separated South America-Africa from Australia-Antarctica.

Figure 19.20B illustrates the position of the continents 135 million years ago, about the time Africa and South America began splitting apart to form the South Atlantic. India can be seen halfway into its journey to Asia, while the southern portion of the North Atlantic has widened greatly. By the end of the Cretaceous period, about 65 million years ago, Madagascar had separated from Africa, and the South Atlantic had emerged as a full-fledged ocean (Figure 19.20C). At this juncture, India had drifted over a hot spot that generated numerous fluid basalt flows across a region in western India now called the Deccan Plateau. These lava flows are very similar to those that make up the Columbia Plateau in the Pacific Northwest.

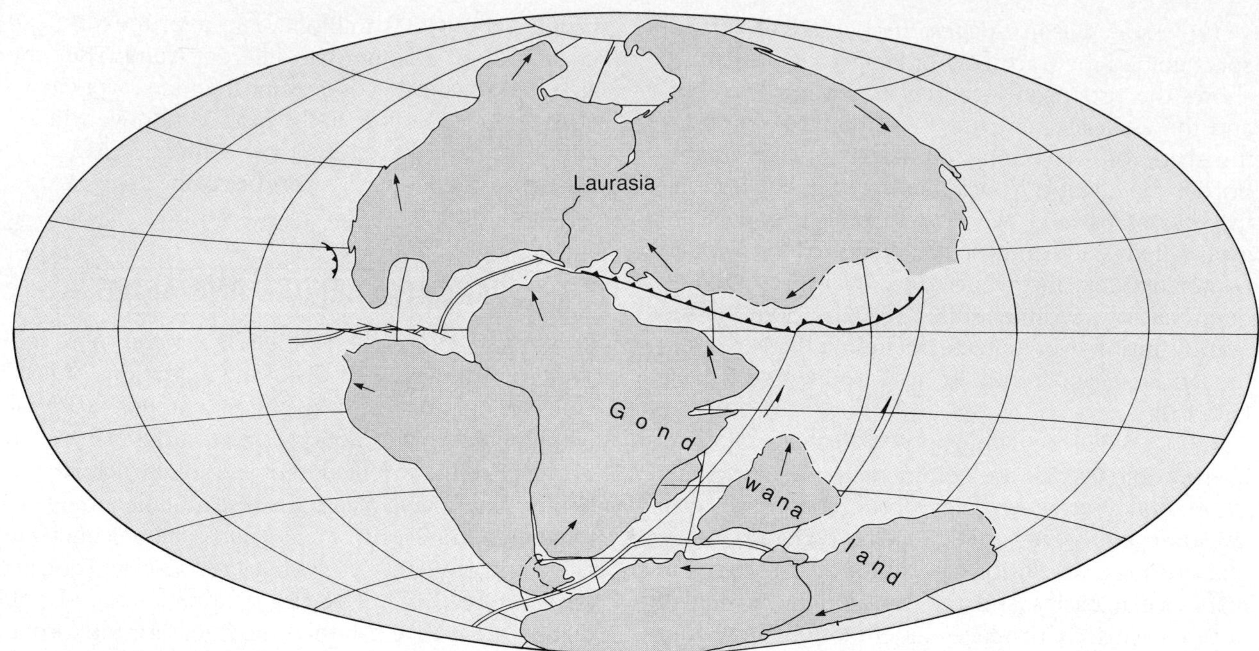

A. 180 million years ago (Triassic Period)

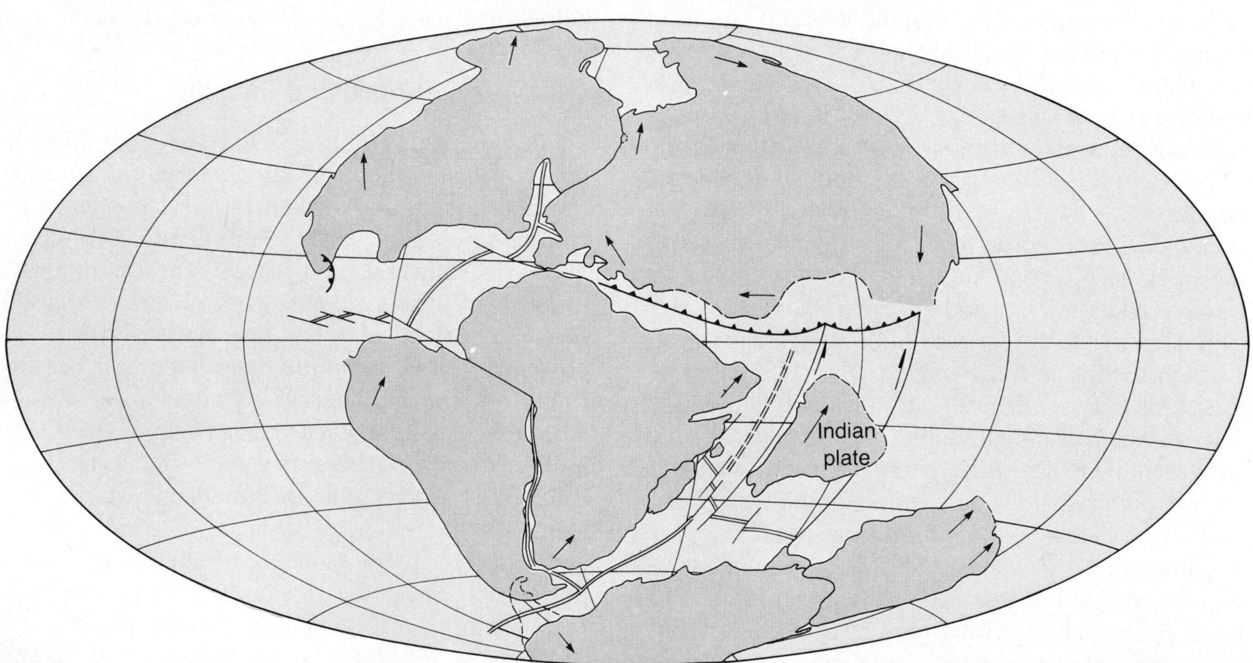

B. 135 million years ago (Jurassic Period)

FIGURE 19.20A.,B.
Several views of the breakup of Pangaea over a period of 200 million years according to Dietz and Holden. (Robert S. Dietz and John C. Holden, *Journal of Geophysical Research* 75: 4939–56. 1970 Copyright by American Geophysical Union)

The current map (Figure 19.20D) shows India in contact with Asia, an event that began about 45 million years ago and created the highest mountains on earth, the Himalayas, along with the Tibetan Highlands. It is interesting to note that the average height of Tibet is 5000 meters, higher than any spot in the contiguous United States. India's continued northward migration is believed to cause the numerous and often destructive earthquakes which plague that part of the world.

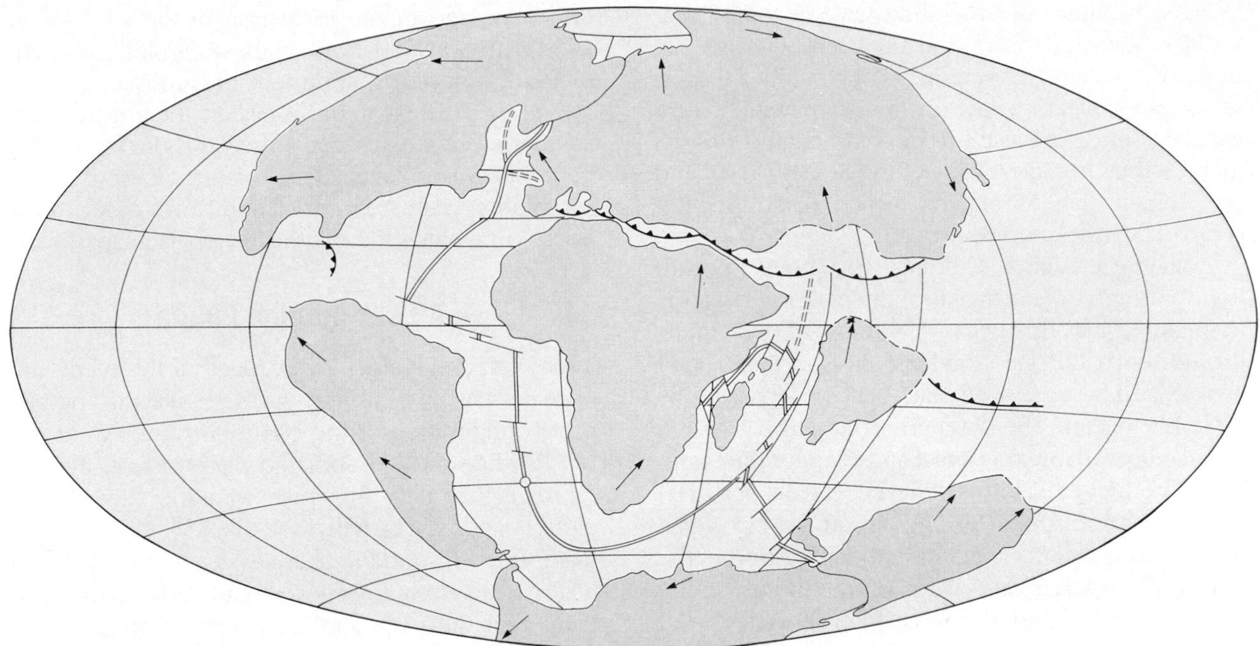

C. 65 million years ago (Late Cretaceous)

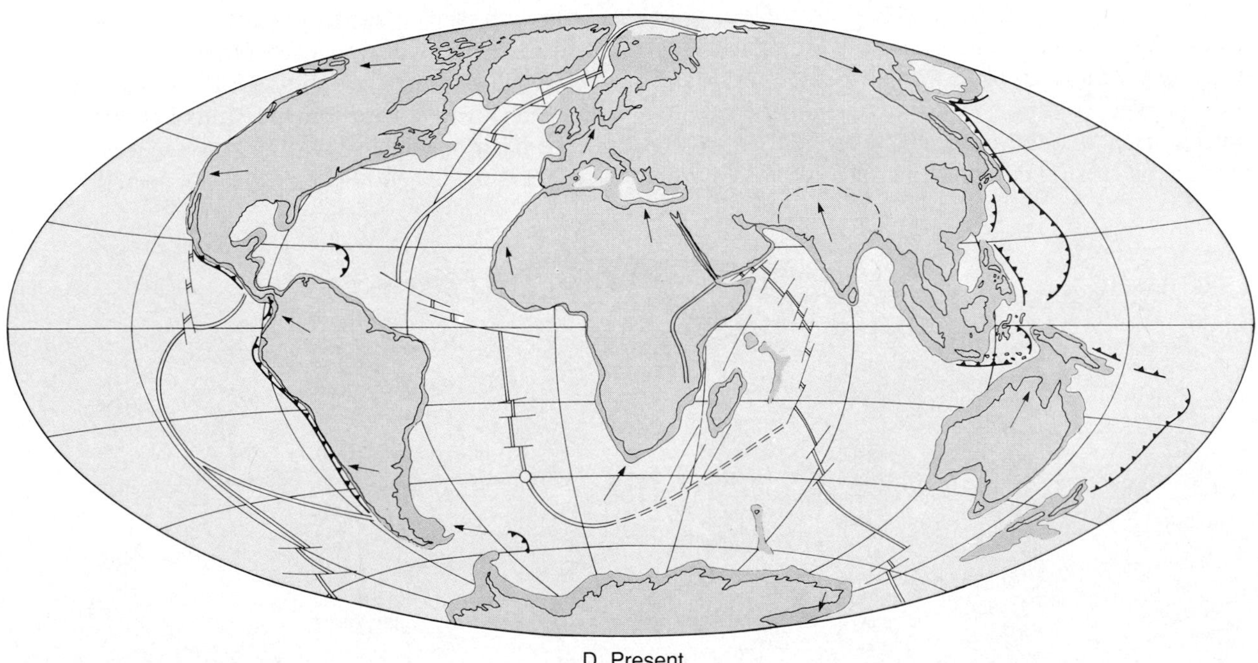

D. Present

FIGURE 19.20C.,D.

By comparing Figures 19.20C and 19.20D, we can see that the separation of Greenland from Eurasia was a recent event in geologic history. Also notice the recent formation of the Baja Peninsula along with the Gulf of California. This event is thought to have occurred less than 10 million years ago.

Before Pangaea

Prior to the formation of Pangaea, the landmasses had probably gone through several episodes of fragmentation similar to what we see happening today. Also like today, these ancient continents moved away from each other only to collide again at some other location. During the period between 500 and 225 million years ago, the fragments of an earlier dispersal began collecting to form the continent of Pangaea. Evidence of these earlier continental collisions include the Ural Mountains of the Soviet Union and the Appalachian Mountains, which flank the east coast of North America.

Available evidence indicates that about 500 million years ago the northern continent of Laurasia was fragmented into three major sections—North America, northern Europe (southern Europe was part of Africa), and Siberia—with each section separated by a sizable ocean. The southern continent of Gondwanaland probably was intact and lay near the South Pole. The first collision is believed to have occurred as North America and Europe closed the pre-North Atlantic. This activity resulted in the formation of the northern Appalachians. Parts of the floor of the former ocean can be seen today high above sea level in Nova Scotia. The sliver of eastern Canada and the United States which lies seaward of the zone of this collision is truly a gift from Europe. It is also thought that before North America and Europe collided, part of Scotland, Ireland, and Norway were attached to the North American plate. While North America and Europe were joining, Siberia was closing the gap between itself and Europe, which lay farther to the west. This closing culminated about 300 to 350 million years ago in the formation of the Ural Mountains. The consolidation of these landmasses completed the northern continent of Laurasia.

During the next 50 million years the northern and southern landmasses converged, producing the supercontinent of Pangaea. At this time (about 250 to 300 million years ago) Africa and North America collided to produce the southern Appalachians.

A Look into the Future

After Dietz and Holden drew together the events that over the past 500 million years resulted in the current configuration of the continents, they went one step further and extrapolated plate motion into the future. Figure 19.21 illustrates what they envision the earth's landmasses will look like 50 million years from now. Important changes are seen in Africa, where a new sea emerges as East Africa parts company with the mainland. In North America we see that the Baja Peninsula and the portion of southern California that lies west of the San Andreas fault have slid past the North American plate. If this northward migration takes place as predicted, Los Angeles and San Francisco will pass each other.

In another part of the world, Africa will have moved slowly toward Europe, initiating perhaps the next major mountain-building stage on our dynamic planet. Australia and New Guinea are seen on a colli-

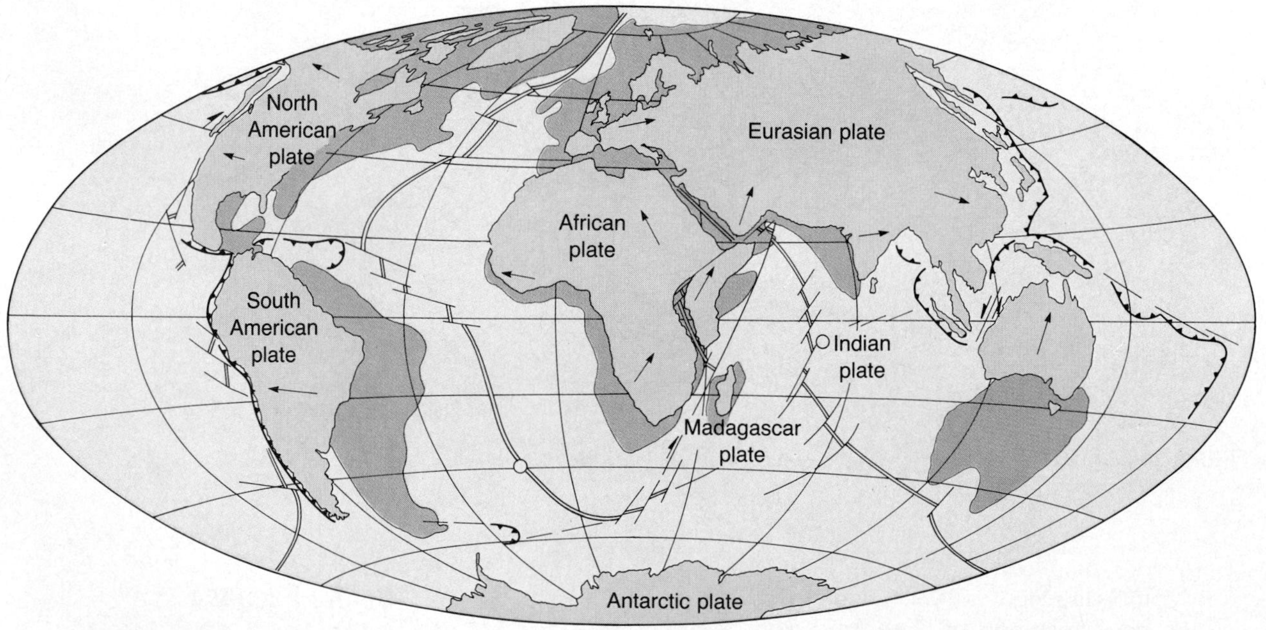

FIGURE 19.21
The world as it may look 50 million years from now. (From "The Breakup of Pangaea," Robert S. Dietz and John C. Holden. Copyright 1970 by Scientific American, Inc. All rights reserved)

sion course with Asia, and North and South America are once again separating. These projections into the future, although interesting, must be viewed with caution since many assumptions must be correct for these events to unfold as just described. Nevertheless, similar types of changes in the shapes and positions of the continents will undoubtedly occur for many millions of years to come.

REVIEW QUESTIONS

1. Assuming that the average speed of sound waves in water is 1500 meters per second, determine the water depth if the signal sent out by an echo sounder requires 6 seconds to strike bottom and return to the recorder (see Figure 19.2A).
2. List the three major features that comprise the continental margin. Which of these features is considered a flooded extension of the continent? Which has the steepest slope?
3. How does the continental margin along the west coast of South America differ from the continental margin along the east coast of North America?
4. Defend or rebut the following statement: "Most of the submarine canyons found on the continental slope and rise were formed during the Ice Age when rivers extended their valleys seaward."
5. What are turbidites? What is meant by the term *graded bedding?*
6. Discuss the evidence that helped confirm the existence of turbidity currents in the ocean.
7. Why are abyssal plains more extensive on the floor of the Atlantic than on the floor of the Pacific?
8. What is an atoll? Describe Darwin's theory on the origin of atolls. Was the theory ever confirmed?
9. Differentiate among the three basic types of sea-floor sediment.
10. If you were to examine recently deposited biogenous sediment taken from a depth in excess of 4500 meters (15,000 feet), would it more likely be rich in calcareous materials or siliceous materials? Explain.
11. How are mid-ocean ridges and deep-ocean trenches related to sea-floor spreading?
12. What is the primary reason for the elevated position of the oceanic ridge system?
13. Briefly describe J. Tuzo Wilson's proposal for continental rifting. If Wilson's idea is correct, how does it affect previous notions about convection currents?
14. Describe the continental collisions that created the Appalachian and Ural mountains.
15. Presently Los Angeles is south of San Francisco. However, millions of years from now, their positions may be reversed; that is, Los Angeles may be north of San Francisco. Explain why this could occur.

KEY TERMS

abyssal plain (p. 494)
atoll (p. 496)
biogenous sediment (p. 499)
continental margin (p. 488)
continental rise (p. 490)
continental shelf (p. 488)
continental slope (p. 490)
deep-ocean basin (p. 493)
echo sounder (p. 488)
graded bedding (p. 493)
guyot (p. 495)
hot spot (p. 507)
hydrogenous sediment (p. 499)
manganese nodule (p. 499)
mid-ocean ridge (p. 499)
ophiolite complex (p. 503)
rift zone (p. 499)
seamount (p. 495)
sheeted dike (p. 503)
submarine canyon (p. 492)
terrigenous sediment (p. 498)
trench (p. 493)
turbidite (p. 493)
turbidity current (p. 492)
Wilson cycle (p. 509)

20
Mountain Building and the Evolution of Continents

Opposite: The Great Trango Tower in the Karakoram Himalayas, Pakistan. (Photo by Galen Rowell) (Top photo by James E. Patterson)

FIGURE 20.1
Sunrise in the Himalayas, Nepal. (Photo by Stephen Trimble)

Mountains are often spectacular features that rise several hundred meters or more above the surrounding terrain. Some occur as single isolated masses; the volcanic cone Kilimanjaro, for example, stands almost 6000 meters (20,000 feet) above sea level, overlooking the expansive grasslands of East Africa. Other peaks are parts of extensive mountain belts, such as the American Cordillera, which runs continuously from the tip of South America through Alaska. Chains such as the Himalayas consist of youthful, towering peaks that are still rising, whereas others, including the Appalachian Mountains in the eastern United States, are much older and nearly worn down (Figure 20.1).

All major mountain systems show evidence of enormous horizontal forces that have folded, faulted, and generally deformed large sections of the earth's crust. Although the processes of folding and faulting have contributed to the majestic appearance of mountains, much of the credit for their beauty must be given to the work of running water and glacial ice, which sculpture these uplifted masses in an unending effort to lower them to sea level.

Geologists have come to realize that, in addition to providing spectacular scenery, mountains play a significant role in the evolution of continental crust. In particular, most geologists agree that the continents have gradually grown larger by the addition of linear mountainous terrains to their flanks. Examples of this growth include North America's Appalachian Mountains as well as the Andes of South America. This hypothesis predicts that nearly all continental areas once stood as mountains and were subsequently lowered to their present elevations by erosion. We will return to this idea later in the chapter, but first let us consider the nature of mountain belts and the forces that elevate them above adjacent lowlands.

MOUNTAIN BELTS

Like other people, geologists have been inspired more by the earth's mountains than by any other landform. Through extensive scientific exploration over the last one hundred and fifty years, much has been learned about the internal processes that generate these often spectacular terrains. The name for the processes that collectively produce a mountain system is **orogenesis**, from the Greek *oros* ("mountain") and *genesis* ("to come into being"). The rocks comprising mountains provide striking visual evidence of the enormous compressional forces that have deformed large sections of the earth's crust and subsequently elevated them to their present positions. Although folding is often the most conspicuous sign of these forces, thrust faulting, metamorphism, and igneous activity are always present in varying degrees.

When geologists speak of mountain-building processes, they are generally referring to the major mountain chains shown in Figure 20.2. Included in this group are the Alps, Urals, Himalayas, Appalachians, and the American Cordillera. Mountain belts are found on every continent, extending as linear chains for hundreds or even thousands of kilometers. Typically, mountain chains consist of numerous *mountain ranges* that show evidence of being formed during the same mountain-building episode.

This chapter deals with the most important result of crustal deformation—the earth's major mountain belts. Yet it is worth noting that some regions exhibit mountainous topography that is produced without appreciable crustal deformation. For example, plateaus, which are areas of high-standing rocks that are essentially horizontal, can be deeply dissected by erosional forces into rugged, mountainlike landscapes. Although these highlands resemble mountains topographically, they lack the structures associated with orogenesis. The opposite situation also exists. For instance, the Piedmont section of the eastern Appalachians exhibits topography that is nearly as subdued as that seen in the Great Plains. Yet, because this region is composed of deformed metamorphic rocks, it is clearly part of the Appalachian Mountains.

ISOSTASY AND CRUSTAL UPLIFT

Two major questions have been paramount to the understanding of the processes that build mountains.

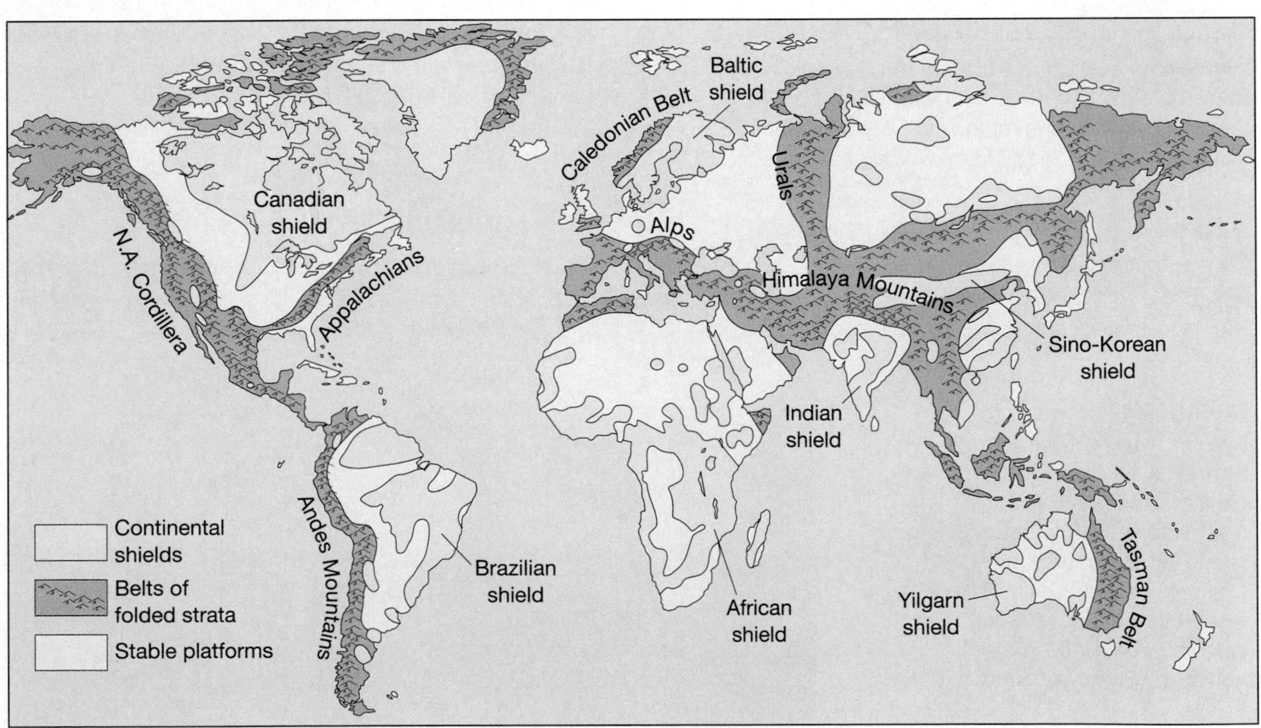

FIGURE 20.2
The continental shields of the world are composed largely of metamorphic rocks. In addition, the deformed portions of many mountain belts are also metamorphic. The light tan area shown on this map is the stable continental interior, which generally consists of undeformed sedimentary beds that overlie metamorphic and igneous basement rocks. (Modified after *Earth History and Plate Tectonics,* 2nd ed., by Carl K. Seyfert and Leslie A. Sirkin. New York: Harper & Row, 1979)

FIGURE 20.3
Remaining columns of the ancient Roman temple of Serapis, Pozzuoli, Italy, in 1836. Clam borings 6 meters above sea level indicate former submergence. (Charles Lyell, *Principles of Geology,* 10th ed., 1867)

First, what is the origin of the enormous horizontal forces that deform crustal rocks during orogenesis? Second, what keeps mountains elevated above the surrounding lowlands?

Evidence for Crustal Uplift

The fossilized shells of marine organisms are often found in mountain regions, an indication that the sedimentary rocks composing the peaks were once below sea level. This is convincing evidence that some drastic changes occurred between the time these animals died and the time when their fossilized remains were discovered. Evidence for crustal uplift such as this is common in the geologic record and is even present in the historical record. For example, Figure 20.3 shows three columns remaining from a Roman temple. The columns have clam borings to a height of about 6 meters (20 feet), indicating that the land upon which the temple was built submerged and was later partially uplifted. These elevated clam borings might also be explained by a recent change in sea level; however, a similar change is not recorded at any other location for that same time period. Further evidence for crustal uplift can be found along the coastline of the western United States. When a coastal area remains undisturbed for an extended period, wave action cuts a gentle sloping platform. In parts of California, ancient wave-cut platforms can now be found as terraces, hundreds of meters above sea level (Figure 20.4). Each terrace represents a period when that area was at sea level. Unfortunately, the reasons for uplift are not always as easy to determine as the evidence for the movements.

Do Mountains Have Roots?

One of the major advances in determining the structure of mountains occurred in the 1840s when Sir George Everest (after whom Mount Everest is

FIGURE 20.4
Former wave-cut platforms now exist as a series of elevated terraces on the west side of San Clemente Island off the southern California coast. Once at sea level, the highest terraces are now about 400 meters above it. (Photo by John S. Shelton)

named) conducted the first topographical survey in India. During this survey the distance between the towns of Kalianpur and Kaliana, located south of the Himalayan range, was measured using two different methods. One method employed the conventional surveying technique of triangulation, and the other method determined the distance astronomically. Although the two techniques should have given similar results, the astronomical calculations placed these towns nearly 150 meters closer to each other than did the triangulation survey.

The discrepancy was attributed to the gravitational attraction exerted by the massive Himalayas on the plumb bob used for leveling the astronomical instrument. (A plumb bob is a metal weight suspended by a cord, used to determine a vertical orientation.) It was suggested that the deflection of the plumb bob would be greater at Kaliana than at Kalianpur because it is closer to the mountains (Figure 20.5).

A few years later J.H. Pratt estimated the mass of the Himalayas and calculated the error that should have been caused by the gravitational influence of the mountains. To his surprise, Pratt discovered that the mountains should have produced an error three times larger than was actually observed. Simply stated, the mountains were not "pulling their weight." It was as if they had a hollow central core.

A hypothesis to explain the apparent "missing" mass was developed by George Airy. Airy suggested that the earth's lighter crustal rocks float on the denser, more plastic mantle. Further, he correctly argued that the crust must be thicker under mountains beneath the adjacent lowlands. In other words, mountainous terrains are supported by light crustal material that extends as "roots" into the denser mantle (Figure 20.5). This phenomenon is exhibited by icebergs, which are buoyed up by the weight of the displaced water. If the Himalayas do have roots of light crustal rocks that extend far beneath them, then these mountains would exert a smaller gravitational attraction that Pratt had calculated. Hence, Airy's model explained why the plumb bob was deflected much less than expected.

Seismological and gravitational studies have confirmed the existence of crustal roots under some mountain ranges. The thickness of continental crust is normally about 35 kilometers, but crustal thicknesses exceeding 70 kilometers have been determined for these mountain chains. Elevated regions that are supported primarily by the buoyant nature of crustal roots are the Andes Mountains, the Sierra Nevada, and the Tibetan Plateau.

Isostasy

We know that the force of gravity must play an important role in determining the elevation of the land. The concept of a floating crust in gravitational balance, as Airy proposed, is called **isostasy**. Perhaps the easiest way to grasp the concept of isostasy is to envision a series of wooden blocks of different heights floating in water (Figure 20.6). Note that the thicker wooden blocks float higher in water than the thinner blocks. In a similar manner mountain belts stand high above the surface and also extend farther into the supporting material below. Now visualize what would happen if another small block of wood were placed atop one of the blocks in Figure 20.6. The combined block would sink until a new isostatic balance was reached. However, the top of the combined block would actually be higher than before and the bottom would be lower. This process of establishing a new level of equilibrium is called **isostatic adjustment**.

If the concept of isostasy is correct, we should expect that when weight is added to the crust, the crust will respond by subsiding, and that when weight is removed there will be uplifting. (Visualize

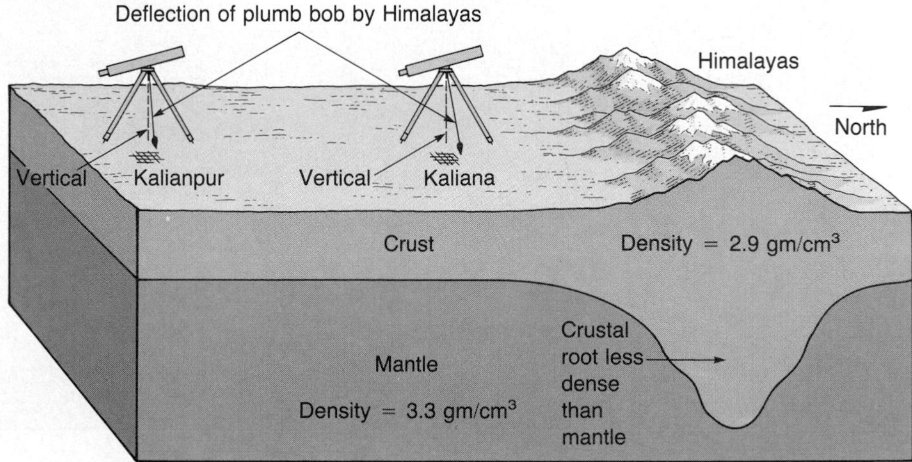

FIGURE 20.5
During the first survey of India an error in measurement occurred because the plumb bob on an instrument was deflected by the massive Himalayas. Later work by George Airy predicted that the mountains have roots of light crustal rock. Airy's model explained why the plumb bob was deflected much less than expected.

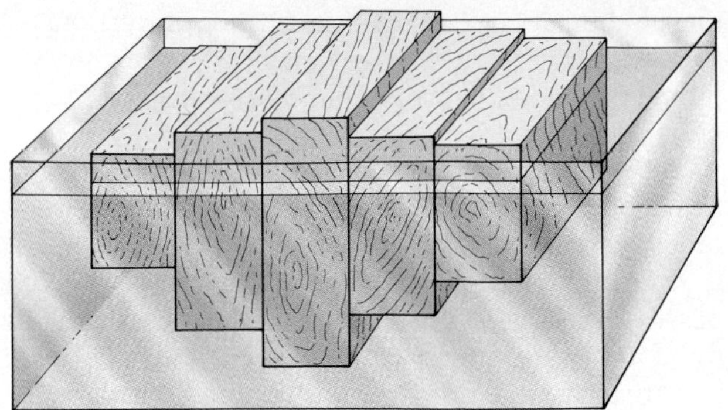

FIGURE 20.6
This drawing illustrates how wooden blocks of different thicknesses float in water. In a similar manner, thick sections of crustal material float higher than thinner crustal slabs.

what happens to a ship as cargo is being loaded and unloaded.) Evidence for this type of movement exists, strongly supporting the concept of isostatic adjustment. A classic example is provided by Ice Age glaciers. When continental ice sheets occupied portions of North America during the Pleistocene epoch, the added weight of the 3-kilometer-thick mass of ice caused downwarping of the earth's crust. In the 8000 to 10,000 years since the last ice sheet melted, uplifting of as much as 330 meters has occurred in Canada's Hudson Bay region, where the thickest ice had accumulated.

As the foregoing example illustrates, isostatic adjustment can account for considerable crustal movement. Thus, we can now understand why, as erosion lowers the summits of mountains, the crust will rise in response to the reduced load. However, each episode of isostatic uplift is somewhat less than the elevation loss due to erosion. The processes of uplifting and erosion will continue until the mountain block reaches "normal" crustal thickness (Figure 20.7). When this occurs, the mountains will be eroded to near sea level, and the deeply buried portions of the

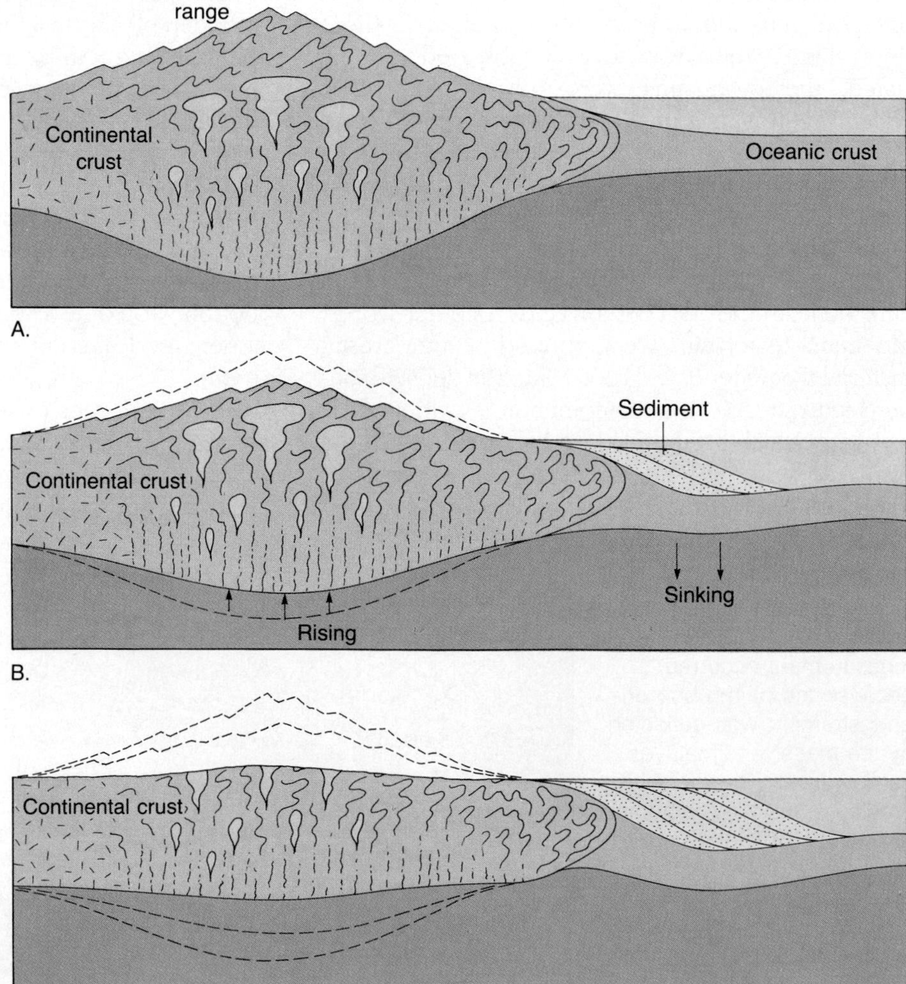

FIGURE 20.7
This sequence illustrates how the combined effect of erosion and isostatic adjustment results in a thinning of the crust in mountainous regions.

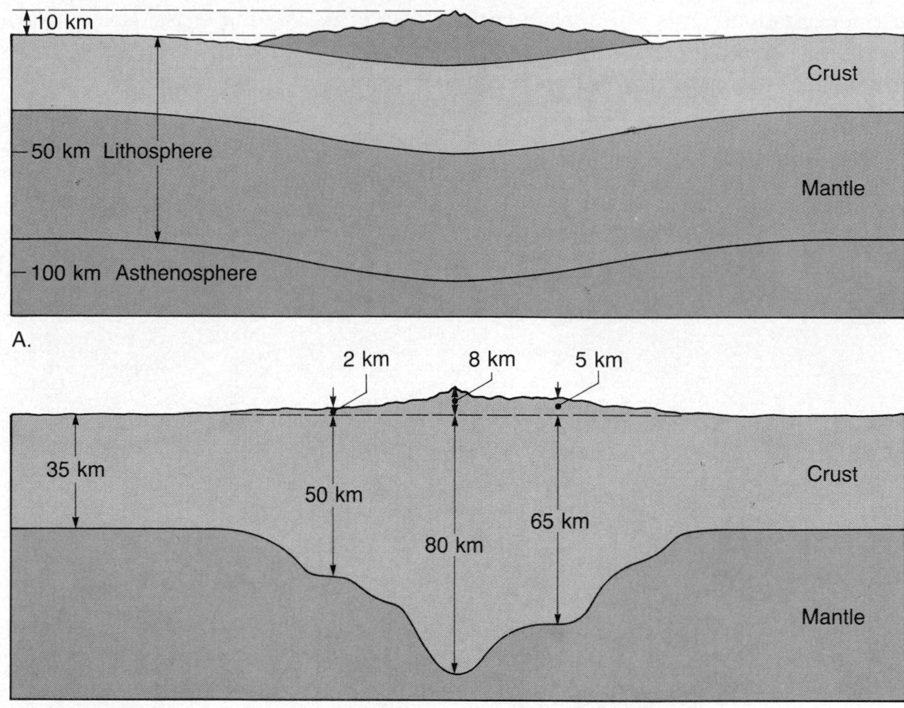

FIGURE 20.8 Two mechanisms proposed to explain how mountain ranges are supported. **A.** Mountains can be supported by flexing the lithosphere downward into the weak asthenosphere. In this situation the load is isostatically compensated over a large region. **B.** Mountains can be supported by the buoyancy of light crustal roots floating in the denser mantle (like icebergs floating in seawater). (Modified after Peter Molnar)

mountains will be exposed at the surface. In addition, as the mountains are worn down, the weight of the eroded sediment deposited on the adjacent continental margin will cause the margin to subside.

The concept that mountains are supported by light crustal roots is essentially correct, but somewhat oversimplified. In recent years it has been learned that in some mountainous areas, the earth's strong outer layer, the lithosphere, helps support these elevated masses. In such cases, isostatic balance is achieved because the weight of the load is spread over a broad region and not confined to a deep root embedded in the mantle (Figure 20.8A). Evidence that this phenomenon operates in nature is provided by the island of Hawaii. Here the extrusion of huge quantities of lava on the ocean floor has built this massive volcanic structure. The weight of the volcanic material depresses the lithosphere several hundred meters. The result is a deep moatlike structure surrounding the island.

Like the island of Hawaii, the Himalayas are believed to be partly supported by strong, thick lithosphere, consisting of relatively cold crust and mantle material. In the Himalayas the crustal unit is about 55 kilometers thick. Although this is much thicker than the crust found in the rest of India, it is far less than the 80-kilometer-thick roots that would be needed to support these mountains (Figure 20.8B). Also, like the island of Hawaii, the weight of the Himalayas has depressed the lithosphere, producing a large, linear trough south of the range. Thousands of meters of sediment derived from the highlands presently fills this depression, producing the Ganges Plain (see Figure 20.15C).

Other mountains that are believed to be at least partly supported by strong lithosphere are the Alps and the Canadian Rockies. In the case of the Alps the supporting European plate is about half as thick as the plate that supports the Himalayas. This fact helps explain why the Alps are only about one-half as high as the Himalayas.

Although major mountain chains are supported primarily by isostatic buoyancy, another important factor is also known to affect their height. With the development of the plate tectonics theory, it became clear that some form of deep circulation is occurring in the mantle. Although the exact nature of this movement is still not fully understood, it is known that some regions are underlain by zones of hot, upwelling mantle material. In these regions, the buoyancy of hot rising magma accounts for broad upwarped zones in the overlying lithosphere. This situation, in which hot spots cause doming of the surface prior to continental rifting, was presented in Chapter 19.

To summarize, we have learned that the earth's major mountain chains consist of unusually thick sections of deformed crustal material that has been elevated above the surrounding terrain. The support for these massive features comes from the buoyancy of deep crustal roots and from the strength of thick lithosphere. As erosion lowers the peaks, isostatic adjustment gradually raises the mountains in response. Eventually, the deepest portions of the mountains are brought to the shallower depths of the surrounding crust. The question still to be answered is, How do these thick sections of the earth's crust come into existence?

FIGURE 20.9
Maroon Bells, Colorado. This mountain range is just a small portion of the North American Cordillera. (Photo courtesy of Stockhouse, Inc.)

MOUNTAIN STRUCTURES

Mountain building has operated during the recent geologic past in several locations around the world (Figure 20.9). These young mountainous belts include the American Cordillera, which runs along the western margin of the Americas from Cape Horn to Alaska; the Alpine-Himalaya chain, which extends from the Mediterranean through Iran to northern India and into Indochina; and the mountainous terrains of the western Pacific, which include mature island arcs such as Japan, the Philippines, and Sumatra (see Figure 20.2). Most of these young mountain belts have come into existence within the last 100 million years. Some, including the Himalayas, began their growth as recently as 45 million years ago.

In addition to these young mountain belts, several chains of Paleozoic- and Precambrian-age mountains exist on the earth as well. Although these older structures are deeply eroded and topographically less prominent, they clearly possess the same structural features found in younger mountains. Typical of this older group are the Appalachians in the eastern United States and the Urals in the Soviet Union.

Although major mountain belts differ from one another in particular details, all possess the same basic structures. Mountain chains generally consist of roughly parallel ridges of folded and faulted sedimentary and volcanic rocks, portions of which have been strongly metamorphosed and intruded by somewhat younger igneous bodies (Figure 20.10). In most cases the sedimentary rocks formed from enormous accumulations of deep-water marine deposits that occasionally exceeded 15 kilometers in thick-

FIGURE 20.10
Highly deformed sedimentary strata in the Rocky Mountains of British Columbia. These sedimentary rocks are continental shelf deposits that were displaced toward the interior of Canada by low-angle thrust faults. (Photo by John Montagne)

ness, as well as from thinner continental shelf deposits. Moreover, most of these deformed sedimentary rocks are older than the mountain-building event. These facts indicate that an extensive period of quiescent deposition along a continental margin was followed by a dramatic episode of deformation.

Careful study of mountainous terrains has revealed that the period of mountain building is generally quite long, in some cases exceeding 100 million years. In addition, reconstruction of these events has shown that deformation generally progressed from the margin of the continent toward the interior, so that the deep-water sediments were the first to be deformed. These sediments, which consist of poorly sorted sandstones, volcanic debris, and shales, have been intensely folded, faulted, and strongly metamorphosed as if squeezed by a gigantic vise whose moving jaw migrated from the sea toward the land. In many mountain belts a period of volcanism, with the emplacement of granitic intrusions, accompanies this episode of deformation.

Next to be deformed are the shallow-water strata of the continental shelf, composed of relatively clean sandstones, limestones, and shales. These beds are usually deformed by folding and thrust faulting, which causes large slices of rock to slide up and over younger layers (see Chapter 15). The effects of metamorphism, when present, are usually of low grade.

For an extended period after the mountain-building episode had ended, the area experiences regional uplift. Generally, little additional deformation is associated with this activity. As the deformed strata are elevated to great heights, the processes of erosion accelerate, carving the deformed strata into a mountainous landscape.

Over the years, several hypotheses have been put forward regarding the formation of the earth's major mountain belts. One early proposal suggested that mountains are simply wrinkles in the earth's crust, produced as the planet cooled from its original semimolten state. As the earth lost heat, it contracted and shrank. In response to this process the crust was deformed much as the peel of an orange wrinkles as the fruit dries out. However, neither this nor any other early hypotheses were able to withstand careful scrutiny.

With the development of the plate tectonics theory, a different model for understanding orogenesis became available. According to the theory of plate tectonics, mountain building occurs at convergent plate boundaries. Here colliding plates provide the horizontal compressional stress to fold, fault, and metamorphose the thick accumulations of sediments that are deposited along the flanks of landmasses. In addition, partial melting of the subducted oceanic crust provides a source of magma that intrudes and further deforms these deposits.

MOUNTAIN BUILDING

In order to unravel the events that produce mountains, studies are conducted in areas that exhibit ancient mountain structures as well as in regions where orogenesis is thought to be in progress. Of particular interest are active subduction zones, where crustal fragments are converging. At most modern-day subduction zones, volcanic arcs are forming. This situation is typified by Alaska's Aleutian Islands and by the Andean arc of western South America. Although all volcanic arcs are similar, Aleutian-type subduction zones occur where two oceanic plates converge, whereas Andean-type subduction zones are situated where oceanic crust is being thrust beneath a continental mass. Consequently, the events that generated these particular volcanic arcs followed somewhat different evolutionary paths. Further, although the development of a volcanic arc does result in the formation of mountainous topography, this activity is viewed as just one of the phases in the development of a major mountain belt.

At sites where oceanic crust is being subducted, continental blocks are often being rafted toward one another. Recent studies indicate that the most important cause of orogenesis is the collision of two or more of these crustal fragments. Collisions can occur between a continental block and a variety of crustal masses, including island archipelagos such as the Aleutian Islands or small continental fragments such as Madagascar. We shall consider these sites of mountain building in the following sections.

Mountain Building and Island Arcs

Volcanic island arcs form where two oceanic plates converge and one subducts beneath the other. Partial melting of the subducting plate and some of the overlying mantle rocks generates magma that migrates upward to form the igneous portion of the developing arc system (Figure 20.11A). Over an extended period, numerous episodes of volcanism, coupled with the buoyancy created by intrusive igneous masses, gradually increase the size and elevation of the developing arc. The arc's greater height accelerates the erosion rate and consequently the amount of sediment added to the adjacent sea floor and to the back-arc basin.

FIGURE 20.11
The development of a mature volcanic island arc at an oceanic-oceanic convergent boundary.

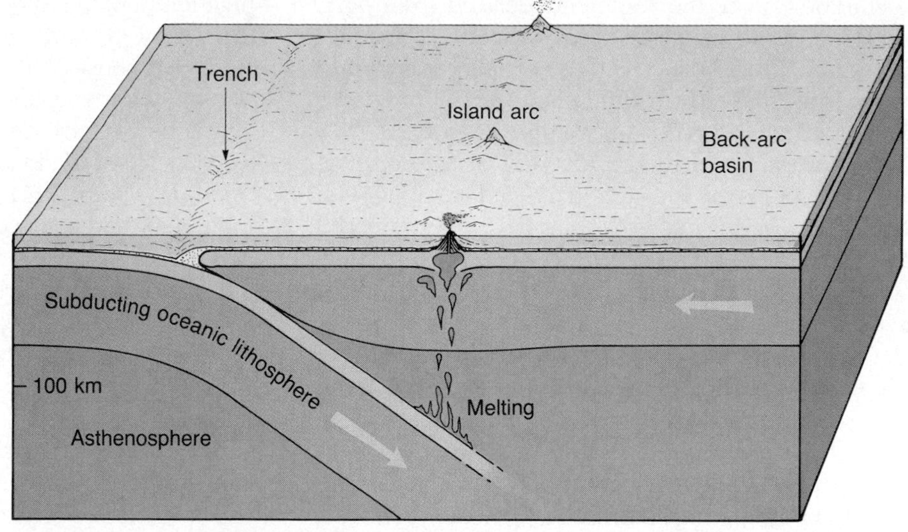

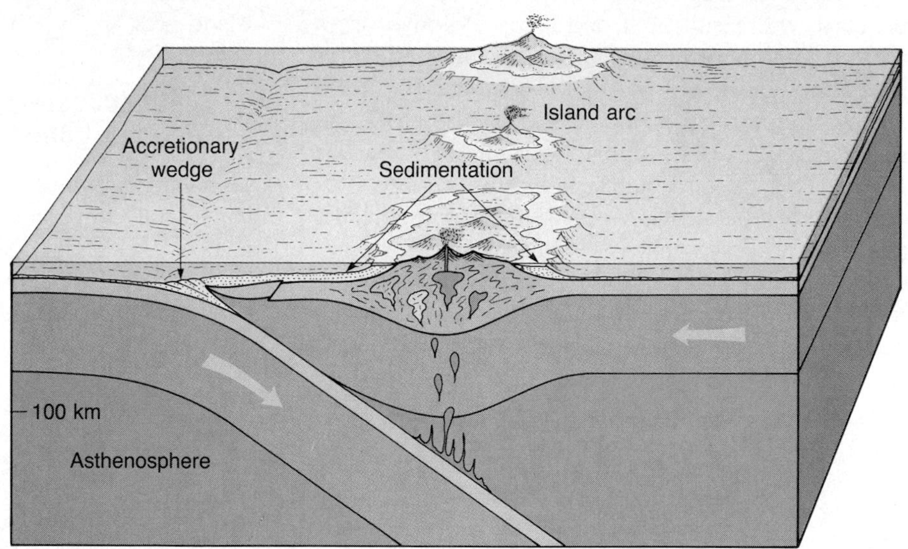

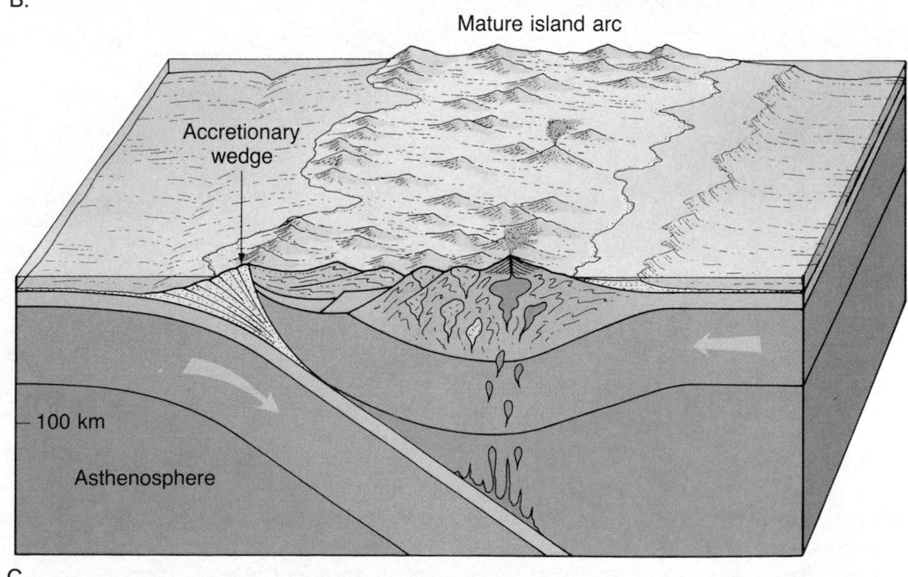

In addition to the sediments derived from land, deep-water deposits are also scraped off the descending oceanic plate. As these sediments are piled up in front of the overriding plate, they form what is known as an **accretionary wedge** (Figure 20.11B). The compressional stresses exerted by the converging plates cause the accretionary wedge, along with the slivers of oceanic crust that have been sheared from the descending plate, to become intricately folded and cut by numerous thrust faults. Prolonged subduction is believed to generate a thick wedge of deformed material lying parallel to, and seaward of, the igneous portion of the arc. Continued growth can build an accretionary wedge that is eventually large enough to stand above sea level (Figure 20.11C).

Landward of the trench, in the volcanic arc, sediments are also being deformed and metamorphosed. Whereas metamorphism in the accretionary wedge is primarily the result of strong compressional forces created by the converging plates, metamorphism in the vicinity of the volcanic arc is associated with the emplacement of large magma bodies. Therefore, the metamorphic rocks found in volcanic arcs contain minerals typical of a high-temperature regime.

These diverse activities result in the creation of a mature island arc composed of two roughly parallel orogenic belts (Figure 20.11C). The landward segment is the volcanic arc, which is made up of volcanoes and large intrusive bodies intermixed with high-temperature metamorphic rocks. The belt located seaward of the volcanic arc is the accretionary wedge. It consists of folded, faulted, and metamorphosed sediments and volcanic debris.

Geologists have only recently come to realize the significance of island arcs in the process of mountain building (Figure 20.12). There is now general agreement that the processes operating at modern island arcs represent one of the stages in the formation of the earth's major mountain belts. Because island arcs are carried by moving oceanic plates, it is possible for two arcs to collide and be *sutured* together to form a larger crustal fragment. Moreover, island arcs may also be accreted to continent-sized blocks, in which case they become incorporated into a mountain belt (see Figure 20.19). In the next section we will consider orogenesis along continental margins in light of what has been learned from the geology of island arcs.

Subduction-type Orogenesis Along Continental Margins

Mountain building along continental margins involves the convergence of an oceanic plate and a plate whose leading edge contains continental crust. Exemplified by the Andes Mountains, this type of convergence generates structures resembling those of a developing volcanic island arc.

FIGURE 20.12
Three of many volcanic islands that comprise the Aleutian arc. This narrow band of volcanism results from the subduction of the Pacific plate. In the distance is the Great Sitkin volcano (1772 meters) which the Aleuts call the "Great Emptier of Bowels," because of its frequent activity. (Photo by Bruce D. Marsh)

The first stage in the development of an Andean-type mountain belt occurs prior to the formation of the subduction zone. During this period the continental margin is **passive**; that is, it is not a plate boundary but a part of the same plate as the adjoining oceanic crust. The East Coast of the United States provides a present-day example of a passive continental margin. Here, as at other passive continental margins surrounding the Atlantic, deposition of sediment on the continental shelf is producing a thick wedge of shallow-water sandstones, limestones, and shales (Figure 20.13A). Beyond the continental shelf, turbidity currents are depositing sediments on the continental slope and rise.

At some point the continental margin becomes active; a subduction zone forms and the deformation process begins (Figure 20.13B). A good place to examine an active continental margin is the west coast of South America. Here the Nazca plate is being subducted beneath the South American plate along the Peru-Chile trench (Figure 18.15). This subduction zone probably formed in conjunction with the breakup of the supercontinent Pangaea. As South America separated from Africa and migrated west-

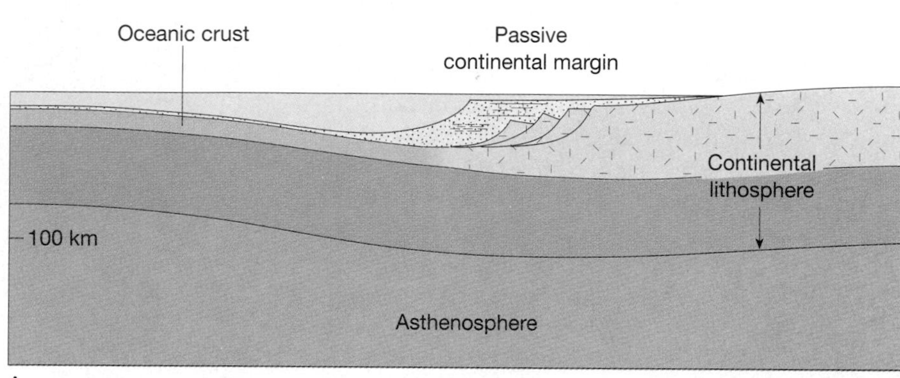

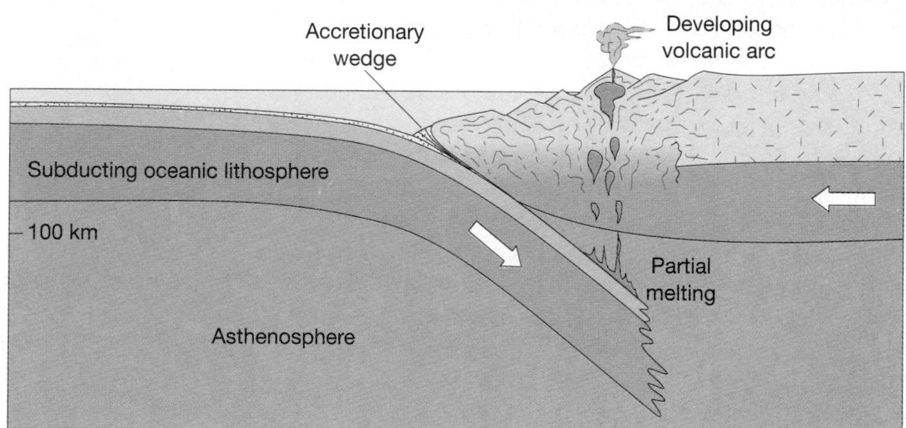

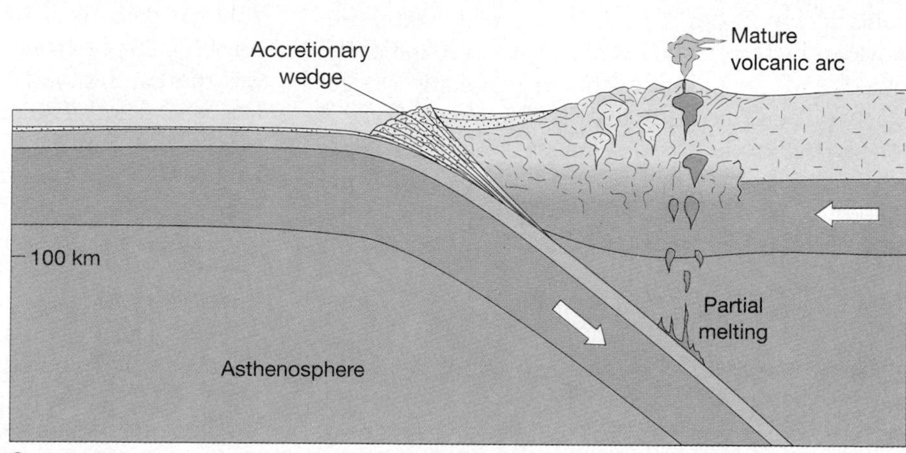

FIGURE 20.13
Generalized diagrams illustrating orogenesis along an Andean-type subduction zone. **A.** Passive continental margin with extensive wedge of sediments. **B.** Plate convergence generates a subduction zone, and partial melting produces a developing volcanic arc. **C.** Continued convergence and igneous activity further deform and thicken the crust, elevating the mountain belt, while the accretionary wedge grows in size.

ward, the oceanic crust adjacent to the west coast of South America was bent and thrust under the continent.

In an idealized Andean-type event, convergence of the continental block and the subducting oceanic plate leads to deformation and metamorphism of the continental margin. Once the oceanic plate descends to about 100 kilometers, partial melting generates magma that migrates upward, intruding and further deforming these strata (Figure 20.13B). During the development of the volcanic arc, sediment derived from the land as well as that scraped from the subducting plate is plastered against the landward side of the trench. Recall that this chaotic accumulation of sedimentary and metamorphic rocks with occasional scraps of oceanic crust is called an accretionary wedge (Figure 20.13B). Prolonged subduction can build an accretionary wedge that is large enough to stand above sea level (Figure 20.13C).

Andean-type mountain belts, like mature island arcs, are composed of two roughly parallel zones. The landward segment is the volcanic arc, which is made up of volcanoes and large intrusive bodies intermixed with high-temperature metamorphic rocks. The belt located seaward of the volcanic arc is the accretionary wedge. It consists of folded, faulted, and metamorphosed sediments and volcanic debris.

One of the best examples of an inactive Andean-type orogenic belt is found in the western United States and includes the Sierra Nevada and the Coast Ranges of California (Figure 20.14). These parallel mountain belts were produced by the subduction of a portion of the Pacific basin under the western edge of the North American plate. The Sierra Nevada batholith is a remnant of a portion of the volcanic arc that was produced by several surges of magma over tens of millions of years. Subsequent uplifting and erosion have removed most evidence of past volcanic activity and exposed a core of crystalline metamorphic and igneous rocks. In the trench region, sediments scraped from the subducting plate and those provided by the eroding volcanic arc were intensely folded and faulted into the accretionary wedge which presently constitutes the Franciscan Formation of California's Coast Ranges. Uplifting of the Coast Ranges took place only recently, as evidenced by the unconsolidated sediments still mantling portions of these highlands.

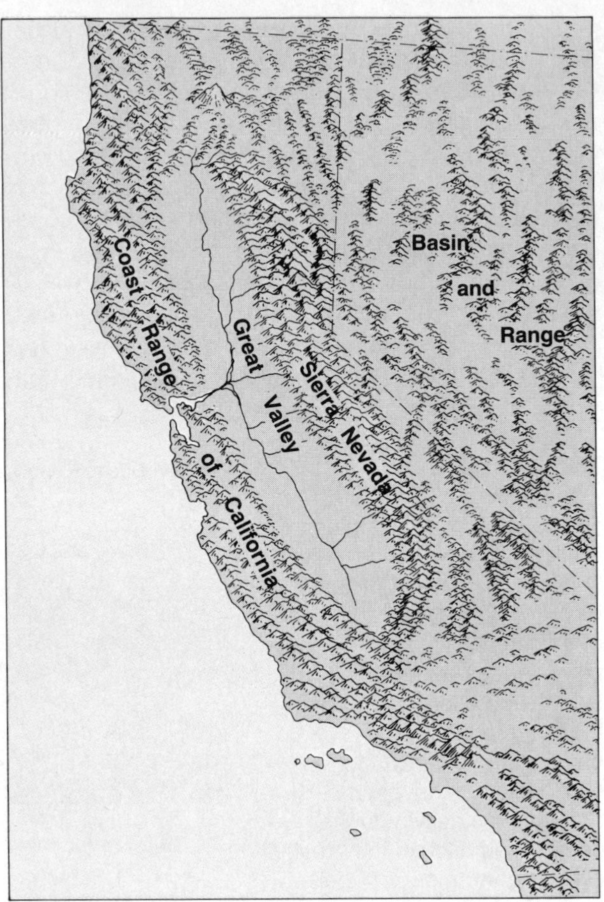

FIGURE 20.14
Map of the California Coast Ranges and the Sierra Nevada.

Continental Collisions

So far we have discussed the formation of mountain belts in which the leading edge of just one of the converging plates contained continental crust. However, it is possible that both of the colliding plates may be carrying continental crust. Because continental lithosphere is apparently too buoyant to undergo any appreciable amount of subduction, a collision between the continental fragments eventually results (Figure 20.15). An example of such a collision began about 45 million years ago when India collided with Asia. India, which was once part of Antarctica, split from that continent and moved a few thousand kilometers due north before the collision occurred. The result was the formation of the spectacular Himalaya Mountains and the Tibetan Highlands. Although most of the oceanic crust separating India from Asia prior to the collision was subducted, some was caught up in the squeeze along with sediment that lay offshore. These materials can now be found elevated high above sea level. After such a collision, the subducted oceanic plate is believed to decouple from the rigid continental plate and continue its downward path.

The spreading center that propelled India northward is still active; hence, India continues to be

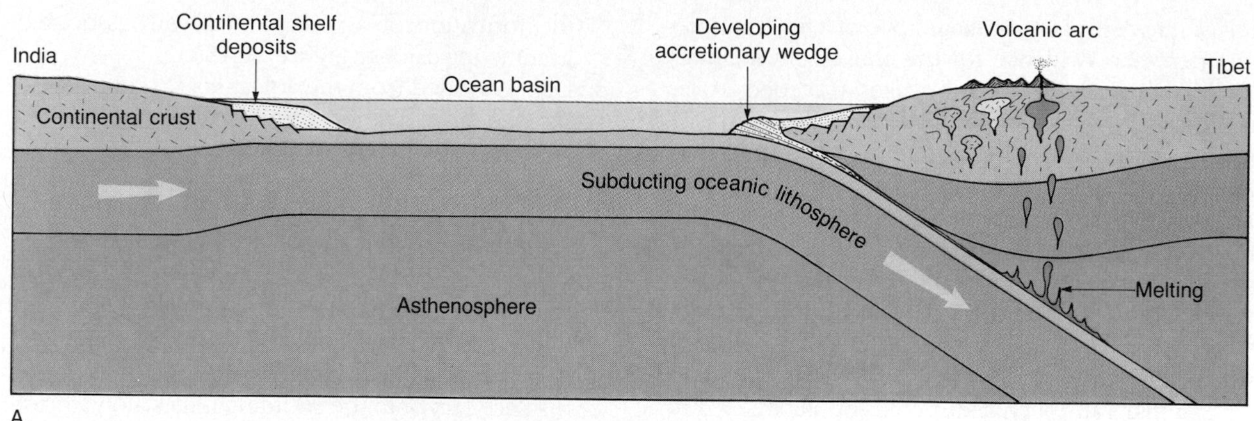

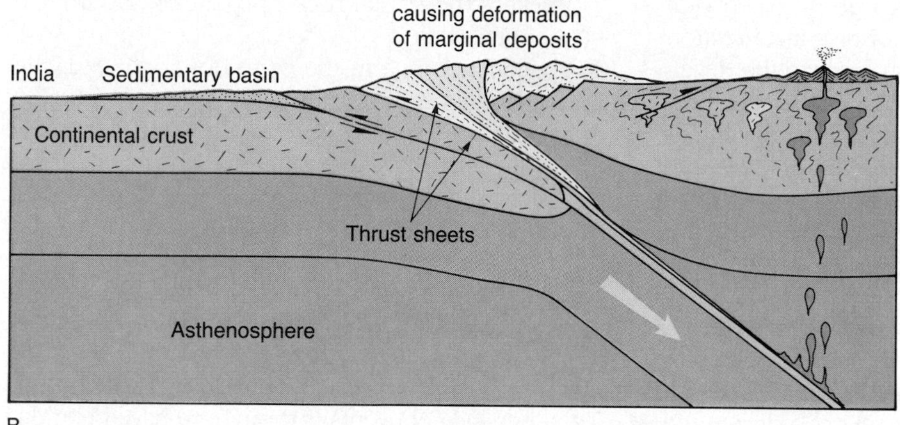

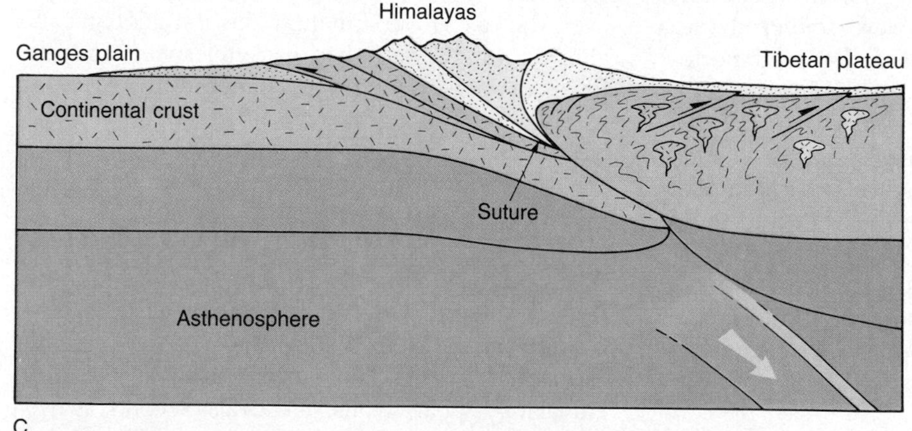

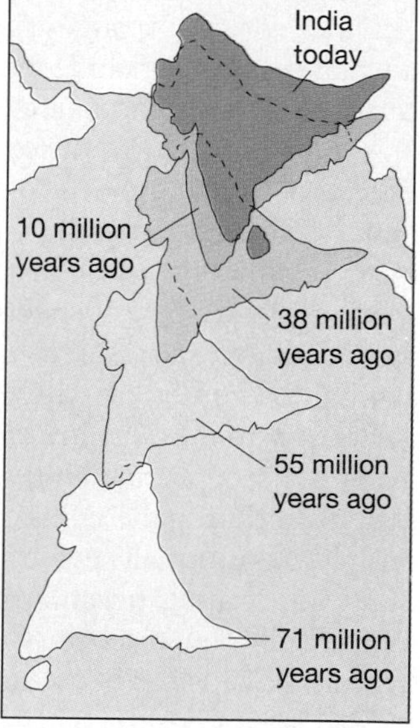

FIGURE 20.15
Simplified diagrams showing the northward migration and collision of India with the Eurasian plate. **A.** Converging plates generated a subduction zone, while partial melting of the subducting oceanic slab produced a volcanic arc. Sediments scraped from the subducting plate were added to the accretionary wedge. **B.** Eventually the two landmasses collided, deforming and elevating the accretionary wedge and continental shelf deposits. In addition, slices of the Indian crust were thrust up onto the Indian plate. **C.** Continued northward movement further displaced the crustal slices and shelf sediments onto the Indian plate to produce the bulk of the Himalayas. In addition, sediments were thrust onto the Eurasian plate capping the Tibetan Plateau. **D.** Position of India in relation to Eurasia at various times. (Modified after Peter Molnar)

thrust into Asia at an estimated rate of a few centimeters per year. Evidence for this movement includes the numerous severe earthquakes recorded as far north as China and Mongolia. However, earthquake activity occurring off the southern coast of India may indicate that a new subduction zone is in the making. If formed, it would provide a disposal site for the floor of the Indian Ocean, which is continually being produced at a spreading center located to the southwest. Should this occur, India's northward journey would come to an end and the growth of the Himalayas would cease.

A similar but much older collision is believed to have taken place when the European continent collided with the Asian continent to produce the Ural Mountains, which extend in a north-south direction through the former Soviet Union. Prior to the development of plate tectonics, geologists had difficulty explaining the existence of mountain ranges such as the Urals, which are located deep within the continental interiors. How could thousands of meters of marine sediments be deposited and then become highly deformed while situated in the middle of a large landmass?

Other mountain ranges showing evidence of continental collisions include the Alps and the Appalachians. The Alps are thought to have formed as a result of a collision between Africa and Europe during the closing of the Tethys Sea. During this closure, enormous quantities of sediment were squeezed into huge recumbent folds that were displaced northward as large thrust sheets. Incorporated within these sediments are shreds of oceanic crust trapped along the suture zone where these landmasses met.

Erosion in the Alps has already exposed the relatively large core of high-grade metamorphic rocks generated during the collision. However, large igneous plutons, normally associated with the subduction of an intervening oceanic plate, are lacking in the Alpine belt. The absence of appreciable igneous activity in the Alps has been attributed to the closing of a small ocean basin.

In summary, the development of a mountain system produced by a continental collision is believed to occur as follows:

1. After the breakup of a continental landmass, a thick wedge of sediments is deposited along the passive continental margins, thereby increasing the size of the newly formed continent.
2. For reasons not yet understood, the ocean basin begins to close and the continents begin to converge.
3. Plate convergence results in the subduction of the intervening oceanic slab and initiates an extended period of igneous activity. This activity results in the formation of a volcanic arc with associated granitic intrusions (Figure 20.15A).
4. Debris eroded from the volcanic arc and material scraped from the descending plate add to the wedges of sediment along the continental margins.
5. Eventually, the continental blocks collide. This event, which often involves igenous activity, severely deforms and metamorphoses the entrapped sediments (Figure 20.15B). Continental convergence causes these deformed materials, and occasionally slabs of crustal material, to be displaced up onto the colliding plates along thrust faults (Figure 20.15C). This activity shortens and thickens the crustal rocks, producing an elevated mountain belt.
6. Finally, a change in the plate boundary ends the growth of the mountains. Only at this point do the processes of erosion become the dominant forces in altering the landscape. Large quantities of coarse sediments are deposited in basins found within and adjacent to the mountains. Prolonged erosion coupled with isostatic adjustments eventually reduce this mountainous landscape to the average thickness of the continents (see Figure 20.7).

This sequence of events is thought to have been duplicated many times throughout geologic time. However, the rate of deformation and the geologic and climatic settings varied in each instance. Thus, the formation of each mountain chain must be regarded as a unique event.

Orogenesis and Continental Accretion

When originally formulated, the plate tectonics theory suggested two mechanisms for orogenesis. First, continental collisions were proposed to explain the formation of such mountain systems as the Alps, Himalayas, Appalachians, and Urals. Second, as typified by the Andes, orogenesis associated with the subduction of oceanic lithosphere was thought to be the underlying tectonic process for many circum-Pacific mountain chains. Recent investigations now indicate that yet another mechanism of mountain building exists. This new proposal suggests that relatively small crustal fragments collide and merge with continental margins and that through this process of collision and accretion, many of the mountainous regions rimming the Pacific have been generated.

What is the nature of the small crustal fragments and where did they come from? Researchers believe that prior to their accretion to a continental block, some of the fragments may have been microcontinents similar in nature to the present-day island of Madagascar. Many others were island arcs, such as Japan, the Philippines, and the Aleutian Islands, which presently rim the Pacific. In addition, others may have been located below sea level and are represented today by submerged platforms rising high above the floor of the western Pacific (see Figure 19.12). Over one hundred of these so-called oceanic plateaus originated as submerged continental fragments, as extinct volcanic island arcs, or as submerged volcanic chains associated with hot spot activity.

The widely accepted view today is that as oceanic plates move, they carry the embedded island arcs and microcontinents to a subduction zone. Here the upper portions of these thickened zones are peeled from the descending plate and thrust in relatively thin sheets upon the adjacent continental block. This newly added material increases the width of the continent and may later be overridden and displaced farther inland when collisions occur with other fragments. Furthermore, segments of continental crust are continually being displaced along transform faults where they may collide and merge with other crustal fragments. This process is exemplified in western North America where a portion of California and the Baja Peninsula are being displaced toward the northwest along the San Andreas fault. Given enough time, this sliver of continental crust may be accreted to Alaska.

Geologists refer to these accreted crustal blocks as **terranes**. Simply, the term *terrane* designates any crustal fragment whose geologic history is distinct from the adjoining terranes. (Do not confuse the term *terrane* with the word *terrain,* which refers to the lay of the land.) Terranes come in a variety of shapes and sizes. Some are just small volcanic islands. Others, such as the one composing the entire Indian subcontinent, are much larger.

The idea that orogenesis occurs in association with the accretion of small crustal fragments to a continental mass arose principally from studies conducted in the northern portion of the North American Cordillera (Figure 20.16). Here it was learned that some mountainous areas, principally those in the orogenic belts of Alaska and British Columbia, contain fossil and magnetic evidence to indicate that the strata originated nearer the equator.

It is now believed that many other terranes found in the North American Cordillera were once scattered throughout the eastern Pacific much as we find island arcs and oceanic plateaus distributed in the western Pacific today. Over the last 200 million years, these fragments migrated toward and collided with the western margin of North America (Figure 20.17). Apparently, this activity resulted in the piecemeal addition of fragments to the entire Pacific coast from the Baja Peninsula to northern Alaska. In a like manner, many of the modern microcontinents will eventually be accreted to active continental margins, thus resulting in the formation of new orogenic belts.

Although it is now widely accepted that some of the orogenic belts rimming the Pacific were gener-

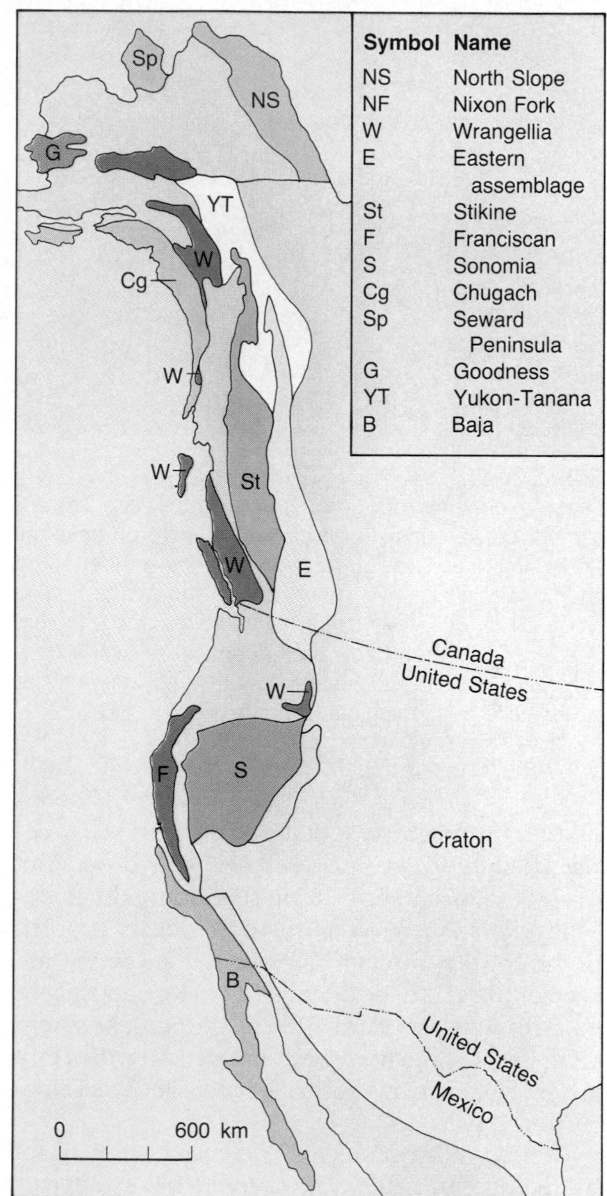

FIGURE 20.16
Map showing terranes thought to have been added to North America during the past 200 million years. Data from paleomagnetic and fossil evidence indicate that some of these terranes originated thousands of kilometers to the south of their present location. Others, such as the Yukon-Tanana terrane, are probably displaced parts of the North American continent. (Redrawn after P. Coney and others)

FIGURE 20.17
These exposed strata of the so-called Chulitna Terrane located in south-central Alaska consist of a distinctive suite of rocks found nowhere else in North America. The light and dark bands on the left consist of limestones and basalts which have been folded and overturned such that they presently overlie younger red sandstones and conglomerates of late Triassic age. Presumably these beds were overturned and deformed as they were accreted to Alaska about 90 million years ago. (Photo by David L. Jones, U.S. Geological Survey)

ated by microcontinent collisions, many of the details of these events must still be worked out. The removal of thrust sheets from subducting plates evidently plays a key role in this accretionary process. However, the manner in which thin sheets are peeled from the descending oceanic lithosphere remains uncertain. Further, in many locations where accretion is thought to have occurred, evidence of the volcanic activity normally associated with subduction is lacking.

Despite these problems and many other unanswered questions, the plate tectonics theory appears to hold the greatest promise for understanding the origin and evolution of the earth's major mountain belts. The geologic history of each mountain system will surely be reevaluated in terms of this model. Such work will shed new light on their evolutionary histories and will also be useful in evaluating the theory itself. In this way new insights into the workings of our dynamic planet will emerge.

THE APPALACHIANS: AN EXAMPLE OF MOUNTAIN BUILDING

The Appalachian Mountains provide great scenic beauty along the eastern margin of North America from Alabama to Newfoundland. In addition, mountains that formed contemporaneously with the Appalachians are found in the British Isles, Scandinavia, Western Europe, and Greenland. The orogeny that generated this extensive mountain system lasted nearly 300 million years and intensely metamorphosed and deformed the rocks of the Appalachian central core (Figure 20.18).

Detailed studies over the last 100 years indicate that the Appalachians were generated by three distinct mountain-building phases over nearly the entire Paleozoic era (Figure 20.19). Each orogeny was followed by a long period of tectonic quiescence and extensive erosion. Although little evidence for the first two episodes of mountain building is visible

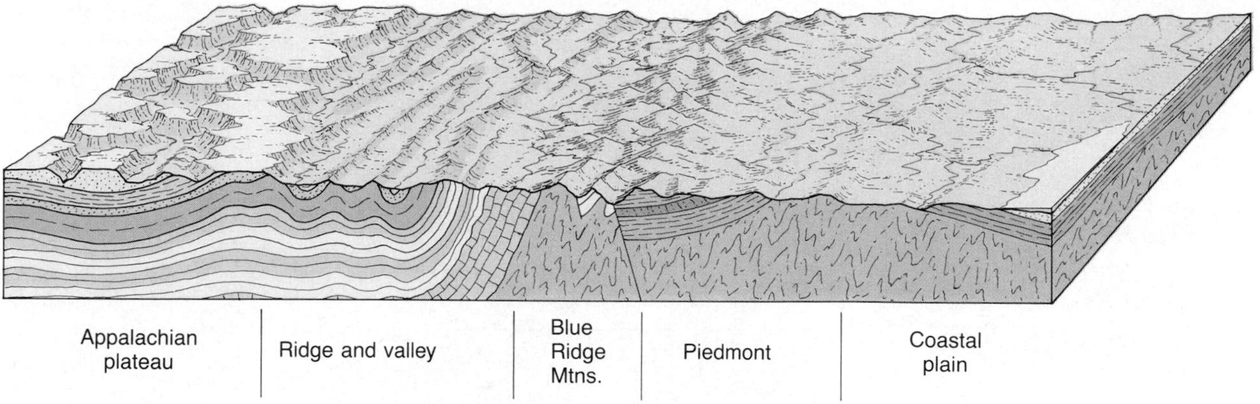

FIGURE 20.18
Major provinces of the Appalachian Mountains. (After Erwin Raisz, from "Stream Sculpture on the Atlantic Slope," by Douglas Johnson, © 1931, used with permission of Columbia University Press)

today on the landscape of the eastern United States, evidence for the third and most extensive phase is relatively abundant. This evidence shows that rocks deformed by earlier activity experienced a westward displacement of up to 250 kilometers. In addition, previously undeformed strata lying as far inland as western Pennsylvania and central Tennessee were folded and displaced along low-angle thrust faults. The remnants of this latter activity still dominate the topography of extensive portions of eastern North America. However, these present-day landscapes are just faint reminders of the lofty mountains that once flanked our East Coast.

Until recently no model regarding mountain building could adequately explain why three distinct orogenic events should occur in roughly the same tectonic setting. However, with the advent of the plate tectonics theory, a scenario has unfolded for the formation of the Appalachians that appears to comply with geologic evidence. According to this view, the Appalachian Mountains evolved through the collisions of several crustal blocks, including North America (and Greenland), Europe, and Northern Africa. Although they have since separated, these landmasses were juxtaposed as part of the supercontinent Pangaea less than 200 million years ago. Geologists believe that the joining of these continents, as well as one or more microcontinents, provided the impetus to generate this extensive mountain chain.

This oversimplified scenario begins about 600 million years ago as the proto-Atlantic begins to close, causing widely scattered continental blocks to converge (Figure 20.19A). Several subduction zones probably formed, with one being some distance from the ancient coastline of proto-North America. The igneous activity associated with subduction gave rise to a volcanic island arc, not unlike those which presently rim the western Pacific.

Studies in the southern Appalachians indicate that at least one microcontinent was situated between the developing island arc and the North American plate. Between 500 and 450 million years ago the basin between this crustal fragment and proto-North America began to close (Figure 20.19B). The ensuing collision first deformed the thick wedge of sediments which flanked the east coast of North America. This orogeny also deformed and metamorphosed the microcontinent as it was thrust onto the continent. The metamorphosed remnants of this event are recognized today as the crystalline rocks of the Blue Ridge and Piedmont regions of the Appalachians. In addition to the pervasive regional metamorphism, igneous activity emplaced numerous plutonic bodies along the entire continental margin, particularly in New England.

The second episode of mountain building occurred between 400 and 350 million years ago (Figure 20.19C). Like its predecessor, this orogeny was characterized by extensive metamorphism, faulting, and intrusive igneous activity, as well as the accretion of at least one terrane to North America. In the southern Appalachians the proto-Atlantic continued to close, resulting in the collision of a volcanic island arc and the ancient North American plate. The remnant of this volcanic arc is the Carolina Slate Belt, which runs parallel to and extends eastward from the Piedmont region. In the northern Appalachians, a collision between ancestral Europe and North America closed the northernmost portion of the proto-

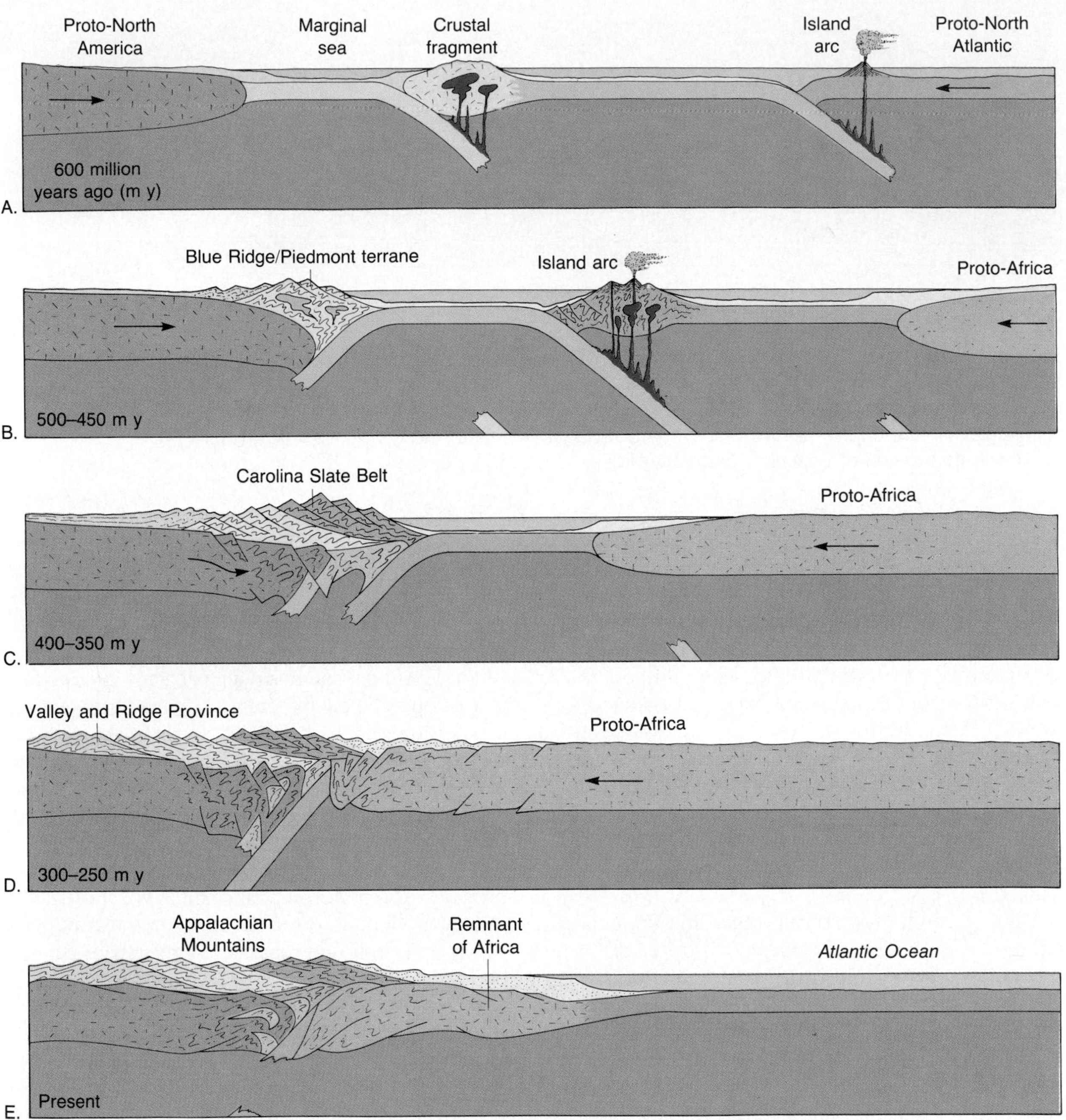

FIGURE 20.19
These simplified diagrams depict the development of the southern Appalachians as the ancient North Atlantic was closed during the formation of Pangaea. Three separate stages of mountain-building activity spanned more than 300 million years. (Adapted from Zvi Ben-Arraham, Jack Oliver, Larry Brown, and Frederick Cook)

Atlantic. This activity further deformed the region from New England to Newfoundland, and it also produced the Caledonian Mountains of the British Isles and Scandinavia. Recall that the excellent matchup between the northern Appalachians and the mountainous belts of northwestern Europe was cited by Alfred Wegener as evidence for the existence of Pangaea.

The final orogeny occurred about 300 to 250 million years ago when Africa collided with North America (Figure 20.19D). This last phase is thought to have displaced the earlier accreted terranes farther inland

FIGURE 20.20
The Valley and Ridge province. This portion of the Appalachian Mountains consists of folded and faulted sedimentary strata that were displaced landward with the closing of the proto-Atlantic. (LANDSAT image courtesy of Phillips Petroleum Company, Exploration Projects Section)

Coast of the United States (Figure 20.19E). If the Atlantic closes again, these sediments will surely be deformed and metamorphosed into a spectacular mountain system located seaward of the present Appalachians. In this manner, another sliver of land will be added to the continent.

The model proposed for the formation of the Appalachians has benefited greatly from seismic-reflection profiling. This technique, which was developed for petroleum exploration, studies the reflection of sound waves from boundaries that separate rocks of different properties. Continuing refinements in this procedure have allowed geologists to map geologic structures located deep within the crust where drilling is currently too costly or simply not feasible.

Several seismic profiles have been made across the southern Appalachians by the Consortium for Continental Reflection Profiling (COCORP). This group of geologists and geophysicists, organized by scientists from Cornell University, has provided evidence for the existence of low-angle thrust faults at depths between 4 and 18 kilometers. One zone of displacement extends from the Valley and Ridge province under the Blue Ridge and Piedmont provinces. Recall that one tenet of the model just presented is that crustal fragments were displaced landward over younger sedimentary beds that once flanked the east coast of North America. Thus, these newly discovered thrust faults support the hypothesis that one or more microcontinents were accreted to North America during the development of the Appalachian Mountains.

POST-OROGENIC UPLIFTING

Recall that mountains stand high above the surrounding lowlands in part because they are buoyed up by light crustal roots. Then, as erosion lowers the mountains, isostatic adjustment gradually raises them again. Such uplift usually occurs in short, jerky motions that elevate the region a few centimeters, or rarely a few meters, at a time. Separating these events are long spans of relative inactivity.

The Sierra Nevada of California provides an excellent example of uplift that occurred long after the original mountain-building episodes ceased. This spectacular mountain range, which includes Mount Whitney, the highest point in the United States outside of Alaska, is bounded by faults on its eastern flanks (Figure 20.21). Seismological studies indicate that this 4-kilometer-high mountain range contains a substantial crustal root that extends to a depth of nearly 60 kilometers. For comparison, the crustal thickness of the adjacent Great Valley is about 27 kilometers. This deep crustal root was most likely formed between 120 and 80 million years ago by the emplacement of large amounts of magma that generated the Sierra Nevada batholiths (Figure 20.22A).

During the period of igneous activity and crustal thickening, the buoyant crustal roots beneath the Si-

FIGURE 20.21
The east face of the Sierra Nevada. Mount Whitney, the highest point in the United States outside of Alaska, is shown in the center of the photo. (Photo by James E. Patterson)

FIGURE 20.22
Recent uplift in the Sierra Nevada of California. **A.** Between 120 and 80 million years ago the emplacement of batholiths generated the thick crustal root of the Sierra Nevada. **B.** The buoyancy of the deep root caused broad regional upwarping in the area of the Sierra Nevada. **C.** Between 5 and 10 million years ago tensional forces broke the lithosphere on the eastern flanks of the Sierra. This resulted in uplift and westward tilting of the mountain block.

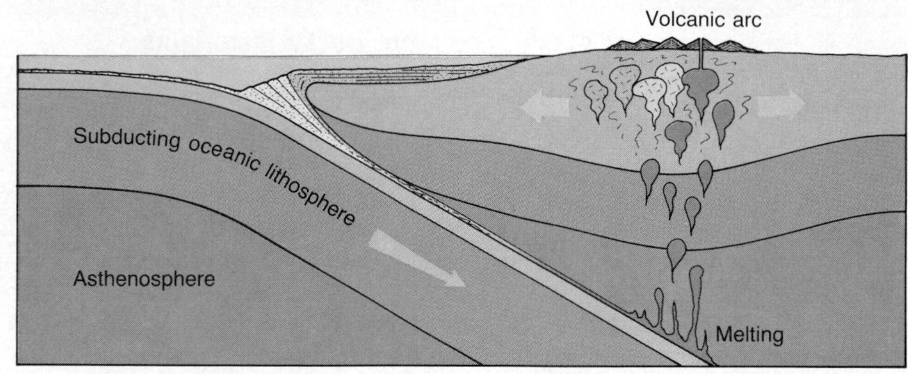

A. Emplacement of Sierra Nevada batholith

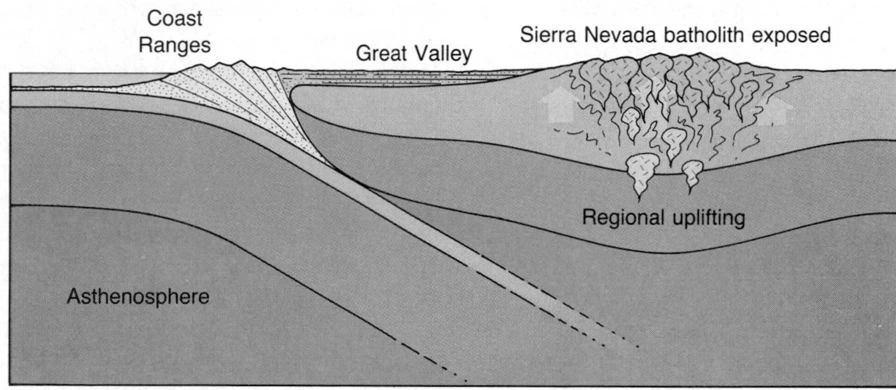

B. Erosion accompanied by broad regional upwarping

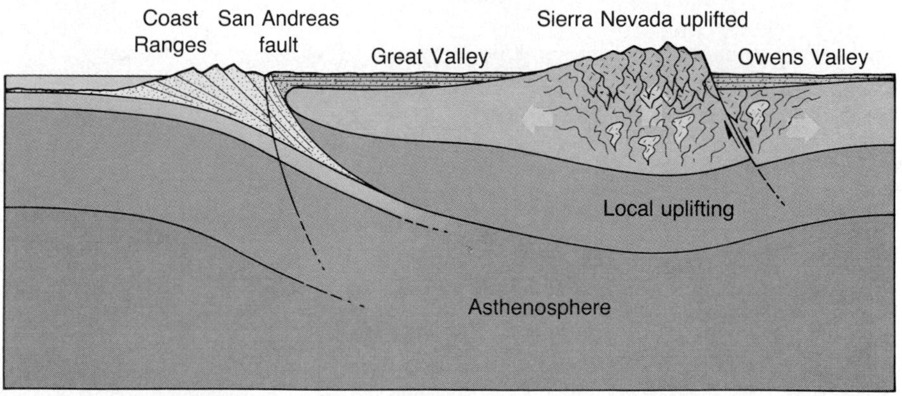

C. Tensional forces cause faulting and isostatic uplifting of Sierra Nevada

erra Nevada caused broad regional upwarping (Figure 20.22B). Accompanying this event were accelerated rates of erosion which dissected and lowered this uplifted terrain. Then, about 10 million years ago, tensional forces became dominant and tore apart the crust to produce the faults that border the eastern flanks of the Sierras. Once the buoyant crust was detached from the adjacent lithosphere, the Sierra Nevada was isostatically uplifted to its present height in a relatively short time (Figure 20.22C).

Today, the Sierra Nevada appears to be in isostatic balance, floating at its proper level. As erosion continues to remove material from these elevated structures, isostatic adjustments will produce additional uplift. However, each episode of isostatic uplift is somewhat less than the elevation loss due to erosion. These processes will operate until the mountain block reaches "normal" crustal thickness (see Figure 20.7). When that occurs, the mountains will be eroded to near sea level, and the igneous and metamorphic rocks which formed at great depths will be exposed at the surface.

BOX 20.2

The Rocky Mountains

The portion of the Rocky Mountains that extends from southern Montana to New Mexico was produced by a period of deformation known as the *Laramide Orogeny.* This event, which created some of the most picturesque scenery in the United States, began in the late Cretaceous and peaked about 60 million years ago in early Tertiary time. The mountain ranges generated during the Laramide Orogeny include the Front Range of Colorado, the Sangre de Cristo of New Mexico and Colorado, and the Bighorns of Wyoming (see Figure 20.9). These mountains are structurally much different from the northern Rockies, which include the Canadian Rockies and those portions of the Rockies found in Idaho, western Wyoming, and western Montana. Whereas the latter ranges are composed of thick sequences of sedimentary rocks that were deformed by folding and low-angle thrust faulting, the middle and southern Rockies were pushed almost vertically upward as part of broad upwarping of the crust. In general, these mountains consist of Precambrian basement rocks overlain by relatively thin layers of younger strata. Since the time of deformation, much of the sedimentary cover has been eroded from the highest portions of the uplifted blocks, exposing the igneous and metamorphic cores. Examples include a number of granitic outcrops that project as steep summits, such as Pikes Peak and Longs Peak in Colorado's Front Range. In many areas, remnants of the sedimentary strata that once covered this region are visible as prominent angular ridges called *hogbacks* flanking the crystalline cores of the mountain ranges (Figure 20.C).

FIGURE 20.C
Hogback ridges in the Rocky Mountains of Colorado. Shown is a view looking south along the east flank of the Front Range in the Roxborough Park area. These upturned sedimentary rocks are remnants of strata that once covered the Precambrian igneous and metamorphic core of the mountains to the west (right). (Photo by James M. Soule, Colorado Geological Survey)

Although the Rockies have been extensively studied for over a century, there is still a good deal of debate regarding the mechanisms that led to uplift. According to the plate tectonics model, most mountain belts are produced along continental margins in association with convergent plate boundaries. Here, crustal buckling forces generate thick sequences of folded, faulted, and metamorphosed strata which are often intruded by massive igneous bodies. However, this model does not fit the middle and southern Rocky Mountains which consist of elongated, uplifted blocks of Precambrian basement rocks separated by sediment-filled basins.

One hypothesis for the formation of the middle and southern Rocky Mountains has been gaining widespread support. According to this proposal, a subducted plate of oceanic lithosphere remained nearly horizontal as it pushed eastward under North America as far inland as the Black Hills of South Dakota (see Figure 20.B in Box 20.1). As the subducted slab scraped beneath North America, compressional stresses shortened and thickened the rocks in the lower crust. Furthermore, this event locally uplifted blocks of Precambrian basement rocks along high-angle faults to produce the Rockies and intermountain basins. Thus, the Laramide Orogeny may be associated with a special type of convergent plate boundary. Whereas plate subduction along a typical Andean-type plate boundary occurs at a steep angle, the middle and southern Rocky Mountains may have been produced by the nearly horizontal subduction of an oceanic plate.

THE ORIGIN AND EVOLUTION OF CONTINENTAL CRUST

Earlier in this chapter, we learned that the theory of plate tectonics provides a model from which to examine the formation of the earth's major mountainous belts. But what roles have plate tectonics and mountain building played in the events that led to the origin and evolution of the continents? At this time no single answer to this question has met with overwhelming acceptance. The lack of agreement among geologists can in part be attributed to the complex nature and antiquity of most continental material, which makes deciphering its history very difficult. Whereas oceanic crust has a relatively simple layered structure and a rather uniform composition, the much older continental crust consists of a collage of highly deformed and metamorphosed igneous and sedimentary rocks. Moreover, in many places, such as the Plains states, the basement rocks are mantled by thick accumulations of much younger sedimentary rocks—a fact that inhibits detailed study. Nevertheless, during the last two decades, great strides have been made in unraveling the secrets held by the rocks composing the stable continental interiors.

At one extreme is a proposal that most, if not all, continental crust formed early in the earth's history. According to this view, the continental crust originated during a primeval molten stage and coincided with the segregation of material that produced the earth's core and mantle (see Chapter 22). During this period of chemical segregation, the lighter silica-rich mineral constituents rose to the surface to produce a scum of continental-type rocks. The heavier material, silica-poor but enriched in iron and magnesium, sank to form mantle rock. Shortly after this segregation, the heat liberated by radioactive decay set the crust in motion. Thereafter a process which may have resembled plate tectonics continually reworked and recycled the continental crust. Through such activity, the primitive continental crust was deformed, metamorphosed, and even remelted. The essence of this hypothesis is that the total amount of continental crust has not changed appreciably since its origin; only its distribution and shape have been modified by tectonic activity.

An opposing view, which has gained support in recent years, contends that the continents have grown larger through geologic time by the gradual accretion of material derived from the upper mantle. A main tenet of this hypothesis is that the primitive crust of an oceanic type and the continents were small or possibly nonexistent. Then, through the chemical differentiation of mantle material, the continents slowly grew.

This view proposes that the formation of continental material takes place in multiple stages as shown in Figure 20.23. The first step occurs in the upper mantle directly beneath the oceanic ridges. Here partial melting of the rock peridotite yields basaltic magma which rises to form oceanic crust. The

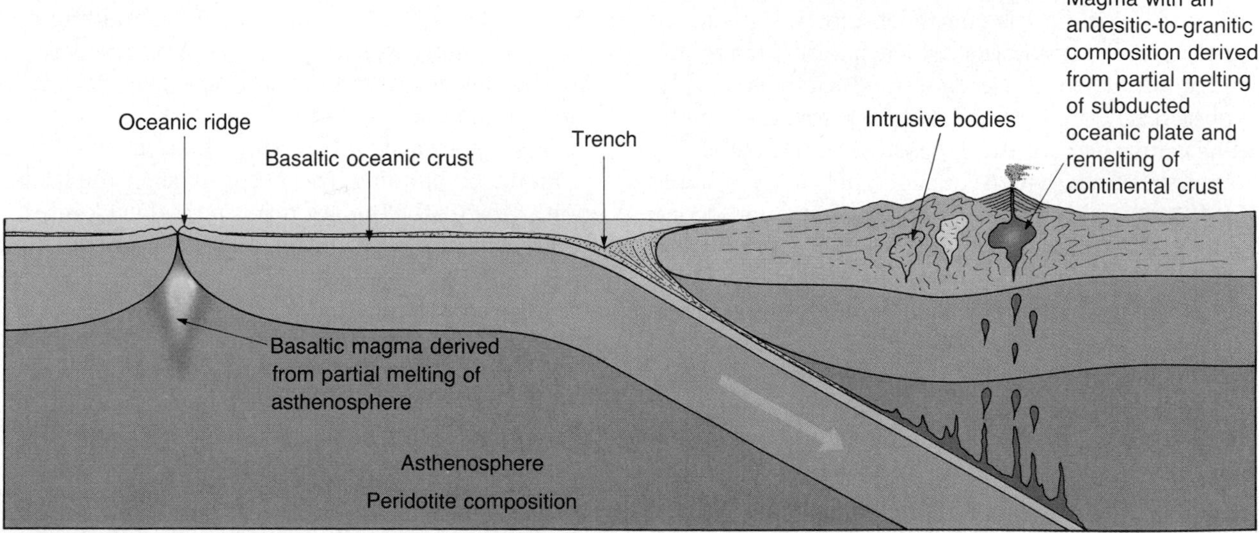

FIGURE 20.23
The multistage process for transforming material from the asthenosphere into continental crust. Once continental crust is generated, its low density apparently keeps it afloat indefinitely.

rocks of the ocean floor are higher in silica, potassium, and sodium and lower in iron and magnesium than the rocks of the upper mantle from which they were derived. As new ocean floor is generated at the ridge crests, older oceanic crust is being destroyed at the oceanic trenches. In trench regions, the subducted oceanic crust is heated sufficiently to cause partial melting. This gives rise to some relatively light, silica-rich rocks which are then emplaced in volcanic arcs. This episode of chemical differentiation produces a magma that is further enriched in silica, potassium, and sodium as compared to its parent material. The subducted oceanic crust, depleted of its lighter constituents, continues to sink and is no longer involved in the process of generating crustal rocks.

According to one view, the earliest continental rocks came into existence at a few isolated island arcs. Once formed, these island arcs coalesced to form larger continental masses, while deforming the volcanic and sedimentary rocks that were deposited in the intervening oceans. Eventually this process generated masses of continental crust having the size and thickness of modern continents.

Evidence supporting the view of continental growth comes from research in regions of plate subduction, such as Japan and the western flanks of the Americas. Equally important, however, has been the research conducted in the stable interiors of the continents, particularly in the shield areas (see Figure 20.2). Recall that all continents have these vast, flat-lying expanses of highly deformed and metamorphosed igneous and sedimentary rocks. The most common shield areas consist of immense bodies of granite and granodiorite, which have been strongly metamorphosed into gneisses and later intruded by younger igneous bodies. In addition to these granite-gneiss terranes are the greenstone belts, so named because of the green tinge common to volcanic rocks of basaltic composition that have undergone low-grade regional metamorphism. The greenstone belts also contain sedimentary rocks that were strongly deformed and metamorphosed during what appears to be a mountain-building episode.

Recent indications are that the rocks of the shield areas are mineralogically and structurally similar to the rocks found at active continental margins where oceanic crust is being consumed. More specifically, the granite-gneiss terranes are chemically similar to the intrusive igenous bodies, such as the Sierra Nevada batholith, which extend along the west coast of the Americas. The ancient rock masses and these younger Mesozoic batholiths are not identical. However, the differences can be explained by the fact that the rocks of the granite-gneiss terranes located in shield areas formed at great depth and were exposed only after long periods of uplift and erosion. On the other hand, the exposed portions of the younger Mesozoic batholiths formed much nearer the surface and thus lack the signs of deep-seated metamorphic alteration. Further, the rocks of the greenstone belts appear to have a chemical composition similar to that of the lavas and sediments found near modern volcanic arcs, such as the islands of Japan. The rocks of the greenstone belts are thought to have been deformed along with slivers of the ocean floor and firmly plastered to the continental margin during closings of ancient ocean basins. These rocks would therefore be equivalent to the rocks of an accretionary wedge, such as the one found in the Coast Range of California.

Radiometric dating of rocks from shield areas, including those in Minnesota and Greenland, has revealed that the oldest terranes formed some 3.8 billion years ago. This date is believed to represent one of the earliest periods of mountain building. At that time, possibly only 10 percent of the present continental crust existed. The next major period of continental evolution may have taken place between 3 and 2.5 billion years ago as indicated by radiometric dates of similar terranes found in the shield areas of Canada, Africa, and western Australia. This period of mountain building is believed to have generated the Superior and Churchill Cratons shown in Figure 20.24. The locations of these ancient continental blocks during their formation is not known. However, about 1.9 billion years ago these two continental fragments collided to produce the Trans-Hudson orogen (Figure 20.24). This mountain-building episode was not restricted to North America because ancient deformed strata of similar age are found on other continents.

It is not known with certainty how many periods of mountain building have occurred since the formation of the earth. The last major period evidently coincided with the closing of the proto-Atlantic and other ancient ocean basins during the formation of the supercontinent Pangaea.

If the continents do in fact grow by accretion of material to their flanks, then the continents have grown larger at the expense of oceanic crust. This view assumes the buoyancy and indestructibility of continental crust. Even the sediment derived from the erosion of continental material that is subducted along with oceanic plates melts and returns to the continents. Although continental crust apparently remains afloat indefinitely, some continents are occasionally fragmented and carried along in a conveyor-belt fashion until they collide with other landmasses. Presently, Australia, which separated from Antarctica,

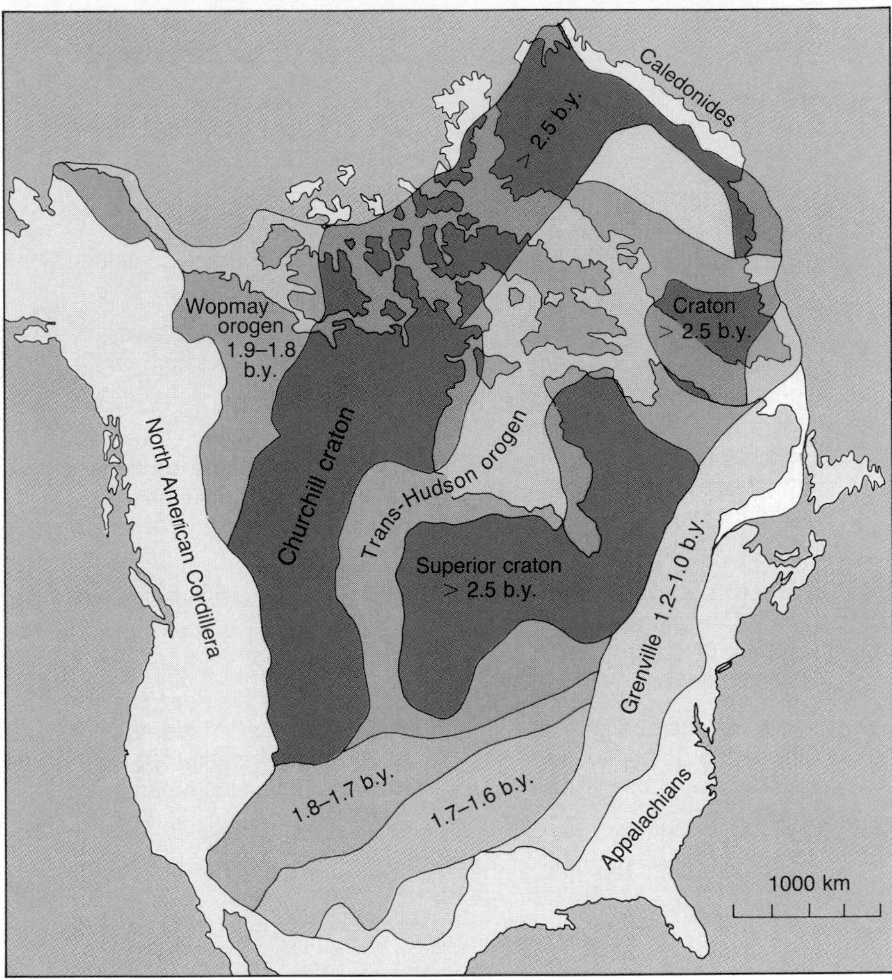

FIGURE 20.24
This map shows the major Precambrian mountain belts (orogens) and cores of ancient continental blocks (cratons) and their ages in billions of years (b.y.). It appears that North America was assembled from crustal blocks that were joined by processes very similar to modern plate tectonics. These ancient collisions produced mountainous belts that include remnant volcanic island arcs which were trapped by the colliding continental fragments.

is being rafted northward and will probably join Asia in much the same manner as India did about 45 million years ago. Thus, according to this view, fragmentation and the formation of new crustal rocks that accompanied the reshuffling of these fragments are responsible for the present volume, structure, and configuration of continents.

A word of caution is in order. The views set forth in this section to explain the origin and evolution of the continents are still somewhat speculative. Plate tectonics appears to be the major force in crustal evolution over the last 2 billion years. However, during the early history of the earth, the heat released by the decay of uranium, thorium, and potassium must have been more than twice as great as it is today. Was plate tectonics active early in the earth's history, only at a different rate, or were there much different processes in operation? Was the primitive crust composed primarily of continental rocks, or was it of the oceanic type? These are some of the questions that still require answers.

REVIEW QUESTIONS

1. List three lines of evidence in support of the crustal uplift concept.
2. What evidence initially led geologists to conclude that mountains have deep crustal roots?
3. What happens to a floating object when weight is added? Subtracted? How does this principle apply to changes in the elevation of mountains? What term is applied to the adjustment that causes crustal uplift of this type?
4. List one line of evidence in support of the idea that the lithosphere tries to remain in isostatic balance.
5. Name a mountain range that is believed to be partly supported by strong, thick lithosphere.
6. What is an accretionary wedge? Briefly explain its formation.
7. What is a passive margin? Give an example. Give an example of an active continental margin.
8. In what way are the Sierra Nevada and the western Andes similar?
9. Would the discovery of a sliver of oceanic crust in a continental interior tend to support or refute the theory of plate tectonics? Why?
10. Why do geologists believe that the northward journey of India is near an end?
11. How does the plate tectonics theory help explain the existence of fossil marine life in rocks atop the Ural Mountains?
12. In your own words, briefly enumerate the steps involved in the formation of a major mountain system according to the plate tectonics model.
13. Define the term *terrane.* How is it different from the term *terrain?*
14. On the basis of current knowledge, describe the major difference between the evolution of the Appalachian Mountains and the North American Cordillera.
15. Briefly describe the formation of fault-block mountains.
16. Compare the forces of deformation associated with fault-block mountains to those of most other major mountain belts.
17. Contrast the opposing views on the origin of the continental crust.

KEY TERMS

accretionary wedge (p. 526)
isostasy (p. 519)
isostatic adjustment (p. 519)
orogenesis (p. 517)
passive margin (p. 527)
terrane (p. 531)

21

Energy and Mineral Resources

Opposite: Hundreds of wind generators at Altamont Pass, California, producing electricity. (Photo by Schafer and Hill/Peter Arnold, Inc.) (Top photo by E. J. Tarbuck)

Materials from the earth are the basis of modern civilization. The mineral and energy resources that are extracted from the crust are the raw materials from which the products used by society are made. Few people who live in highly industrialized nations realize the quantity of resources needed to maintain their present standard of living. Figure 21.1 shows the annual per capita consumption of several important metallic and nonmetallic mineral resources for the United States. This is each person's prorated share of the materials required by industry to provide the vast array of products modern society demands. Figures for other highly industrialized countries such as Canada, Australia, and several nations in western Europe are comparable.

The number of different mineral resources required by modern industries is large. Although some countries, including the United States, have substantial deposits of many important minerals, no nation is entirely self-sufficient. This reflects the fact that important deposits are limited in number and localized in occurrence. All countries must rely on international trade to fulfill at least some of their needs. Figure 21.2 shows that this is certainly true for the United States. Sometimes materials are imported because outside sources are less costly. However, in most instances, imports make up for the lack of adequate domestic deposits.

Resources are commonly divided into two broad categories. Some are classified as **renewable**, which

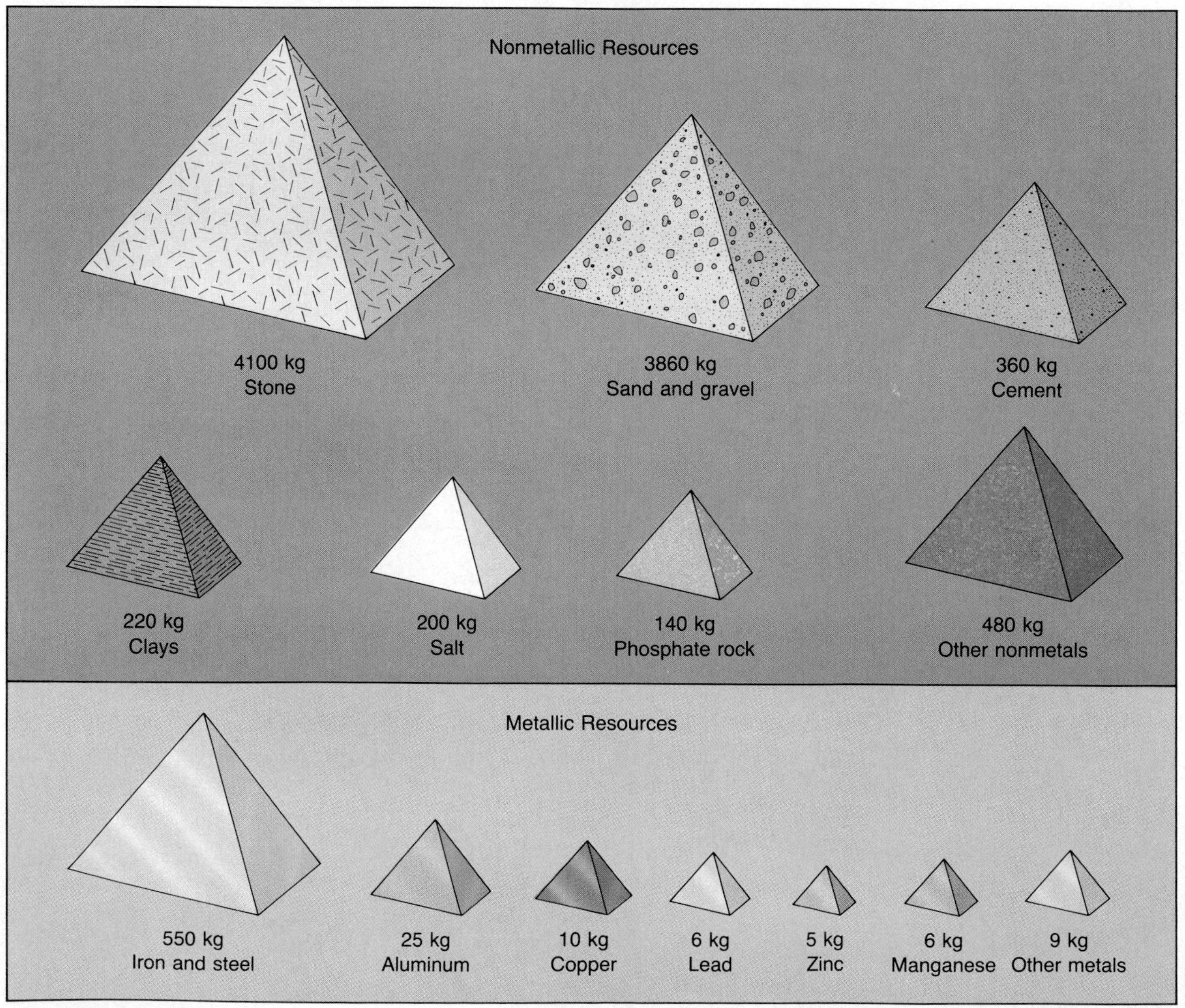

FIGURE 21.1
The annual per capita consumption of nonmetallic and metallic mineral resources for the United States is nearly 10,000 kilograms. About 94 percent of the materials used are nonmetallic. (After U.S. Bureau of Mines)

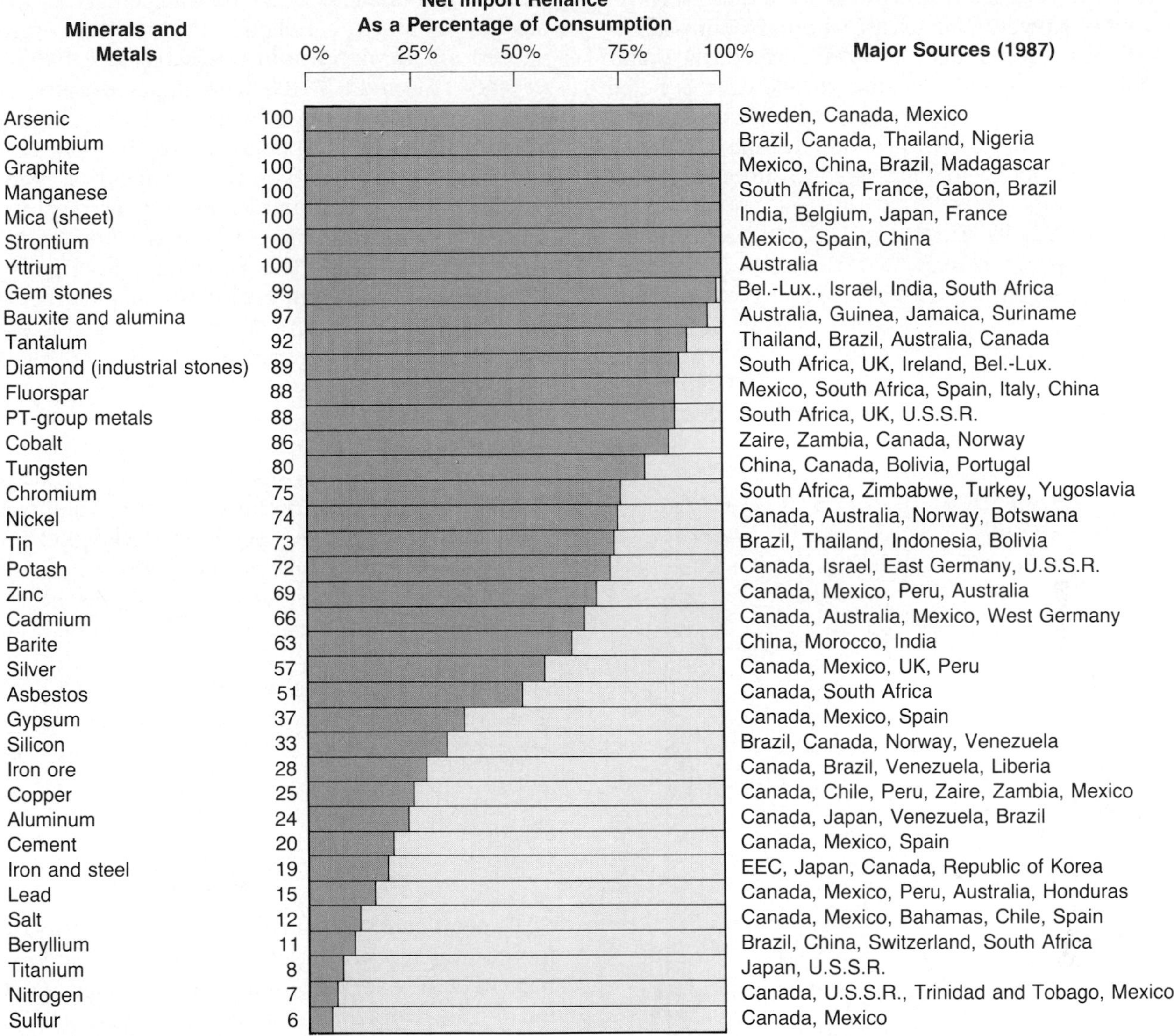

FIGURE 21.2
Net import reliance of the United States for selected nonfuel mineral materials as a percentage of consumption, 1987. All countries must rely on international trade to fulfill at least some of their needs. Nearly 50 different countries are represented on this list of sources. (From U.S. Bureau of Mines)

means that they can be replenished over relatively short time spans. Plants and animals for food, natural fibers for clothing, and forest products for lumber and paper are all common examples. Energy from flowing water, wind, and the sun are also considered renewable. By contrast, many other basic resources are classified as **nonrenewable**. Important metals such as iron, aluminum, and copper fall into this category, as do our most important fuels: oil, natural gas, and coal. Although these and other resources continue to be formed in the earth, the processes that create them are so slow that significant deposits take millions of years to accumulate. In essence, the earth contains fixed quantities of these substances. When the present supplies are mined or pumped from the ground, there will be no more. Although some nonrenewable resources, such as aluminum, can be used over and over again, others, such as oil, cannot be recycled.

Occasionally, some resources can be placed in either category depending on how they are used. Groundwater is one such example. Where it is pumped from the ground at a rate that is no greater than the rate at which the aquifer is recharged, groundwater can be classified as a renewable resource. However, in places where groundwater is withdrawn faster than it is replenished, the water table drops steadily and the groundwater is being "mined" just like other nonrenewable resources.*

A glance at Figure 21.3 shows that the population of our planet is growing rapidly. Although it took until the beginning of the nineteenth century for the number to reach 1 billion, just 130 years were needed for the population to double to 2 billion. Between 1930 and 1975 the figure doubled again to 4 billion, and by the year 2000 an estimated 7 billion people will inhabit our planet. Clearly, as population grows, the demand for resources expands as well. However, the rate of mineral and energy resource usage has climbed far more rapidly than the overall growth of population. This fact results from an increasing standard of living. This is certainly true in the United States, where about 6 percent of the world's population uses approximately 30 percent of the annual production of mineral and energy resources.

*The problem of declining water table levels is discussed in Chapter 11.

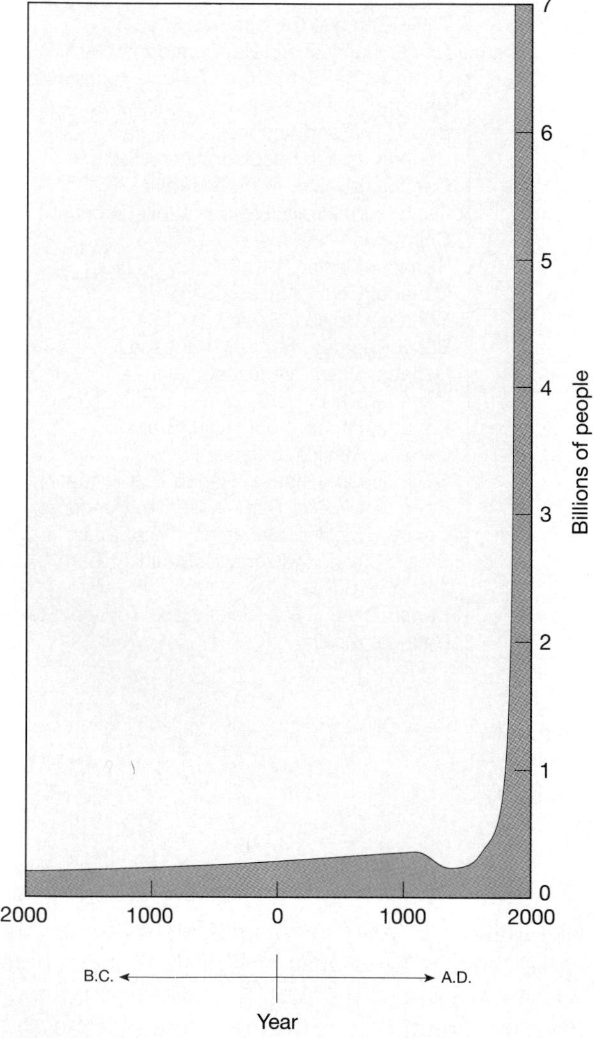

FIGURE 21.3
Growth of world population, It took until 1800 for the number to reach 1 billion. By the year 2000 an estimated 7 billion people will inhabit the planet. The demand for basic resources is growing faster than the rate of population increase. (Data from the Population Reference Bureau)

How long will the remaining supplies of basic resources be adequate to sustain the rising standard of living in today's industrialized countries and still provide for the growing needs of developing regions? How much environmental deterioration are we willing to accept in pursuit of basic resources? Can alternatives be found? If we are to cope with an increasing per capita demand and a growing world population, it is important that we have some understanding of our present and potential resources.

ENERGY RESOURCES

Coal, petroleum, and natural gas are the primary fuels of our modern industrial economy. Nearly 90 percent of the energy consumed in the United States today comes from these basic fossil fuels. In the 1970s, when the "oil crisis" became headline news, so too did interest in these vital energy resources. Although major shortages of oil and gas may not occur for many years, proven reserves are declining. Despite exploration in very remote regions and severe environments, new sources of oil are not keeping pace with consumption. Unless large, new petroleum reserves are discovered, which is possible, but not likely, a greater share of our future needs will have to come from coal and the development of alternative energy sources. Two alternatives, tar sands and oil shale, are sometimes mentioned as promising new sources of liquid fuels. Other alternative energy sources will also become more important in the years to come. In the following sections, we will briefly examine the fuels that have traditionally supplied our energy needs as well as sources that will provide an increasing share of our future requirements.

COAL

Coal is quite different from other sedimentary rocks. Nevertheless, it is often grouped with biochemical sedimentary rocks. However, unlike other rocks in this category, which are calcite- or silica-rich, coal is made of organic matter. Close examination of a piece of coal under a microscope or magnifying glass often reveals the presence of various plant structures such as leaves, bark, and wood that have been chemically altered but are nevertheless still identifiable. This supports the conclusion that coal is the end product of the burial of large amounts of plant material over extended periods.

Along with oil and natural gas, coal is commonly called a **fossil fuel**. Such a designation is certainly appropriate since each time we burn coal we are using energy from the sun that was stored by plants many millions of years ago. We are indeed burning a "fossil" (Figure 21.4).

The initial stage in coal formation is the accumulation of large quantities of plant remains. However, special conditions are required for such accumulations, because dead plants readily decompose when exposed to the atmosphere or other oxygen-rich environments. One important environment that allows for the buildup of plant material is a swamp. Since stagnant swamp water is oxygen deficient, complete decay (oxidation) of the plant material is not possible. Rather, the plants are attacked by certain bacteria that partly decompose the organic material and liberate oxygen and hydrogen. As these elements escape, the percentage of carbon gradually increases. The bacteria are not able to finish the job of decomposition because they are themselves destroyed by acids liberated from the plants.

The partial decomposition of plant remains in an oxygen-poor swamp creates a layer of *peat,* a soft, brown material in which plant structures are still easily recognized. With shallow burial, peat is changed to *lignite,* a soft, brown coal. Burial increases the temperature of sediments as well as the pressure on them. The higher temperatures bring about chemical reactions within the plant materials and yield water and organic gases (volatiles). As the load increases, the water and volatiles are pressed out and the proportion of fixed carbon (solid combustible material) increases. The greater the carbon content, the higher the coal will rank as a fuel. During burial the coal also becomes increasingly compact. For example,

FIGURE 21.4
Restoration of a Pennsylvanian coal swamp. Luxuriant plant growth coupled with an oxygen-poor, swampy environment allowed for the accumulation of great quantities of plant material, which eventually were transformed into coal. (Photo courtesy of the Field Museum of Natural History, Chicago)

deeper burial transforms lignite into a harder, more compacted black coal called *bituminous.* Compared to the peat from which it formed, a bed of bituminous coal may be only one-tenth as thick.

Lignite and bituminous coals are sedimentary rocks, but a later product, called *anthracite* (a very hard, black coal) is a metamorphic rock (Figure 21.5). Anthracite forms when sedimentary layers are subjected to the folding and deformation associated with mountain building. The heat and pressure of mountain building cause a further loss of volatiles and water, thus increasing the concentration of fixed carbon. Although anthracite is a clean-burning fuel, only a relatively small amount is mined. This stems from the fact that anthracite is not widespread and is more difficult and expensive to extract than the flat-lying layers of bituminous coal.

Coal has been an important fuel for centuries. In the nineteenth and early twentieth centuries, cheap and plentiful coal powered the industrial revolution. By 1900 coal was providing 90 percent of the energy produced in the United States. Although still important, coal currently provides less than 20 percent of the United States' energy needs and ranks third as a source of fuel behind oil and natural gas (Figure 21.6). Until the mid-twentieth century coal was an important domestic heating fuel as well as a power source for industry. However, its direct use in the home was largely replaced by oil, natural gas, and electricity. These fuels were preferred because they were more readily available and cleaner to use. Nevertheless, coal remains the major fuel used in power plants to generate electricity, and it is therefore indirectly an important source of energy used in the home as well. More than 70 percent of present-day coal usage is for the generation of electricity. As oil reserves gradually diminish in the years to come, the use of coal will probably increase. Expanded coal production is possible because the world has enormous reserves. In the United States, coal fields are widespread and contain supplies that should last for hundreds of years (Figure 21.7).

FIGURE 21.5
Anthracite coal often displays a conchoidal fracture and is glassy appearing. Because anthracite cannot be easily broken apart as bituminous can, it is also called *hard coal.* Anthracite is the result of low-to-intermediate grade metamorphism of bituminous.

Although coal enjoys the advantage of being plentiful, its recovery and use present a number of problems. Surface mining can turn the countryside into a scarred wasteland if careful (and costly) reclamation is not carried out to restore the land. Although underground mining does not scar the landscape to the same degree, it is more costly in terms of human life and health. Underground mining is no longer a pick and shovel operation, but rather is highly mechanized. Yet, despite modern equipment and safety precautions, it is still inherently dangerous. Each year many miners are injured and lose their lives because of roof falls and gas explosions.

Air pollution is a major problem associated with the burning of coal. Much coal contains significant quantities of sulfur. In spite of efforts to remove sulfur before the coal is burned, some remains; when the coal is burned, the sulfur is converted into noxious sulfur oxide gases. Through a series of complex chemical reactions in the atmosphere, the sulfur oxides are converted to sulfuric acid, which then falls to the earth's surface as rain or snow. This acid precipitation can have severe ecological effects over widespread areas. Although steps have been taken and will continue to be, the air pollution produced by burning coal is still a significant threat to our environment.

Since none of the problems just mentioned is likely to prevent the increased use of this important and abundant fuel, stronger efforts must be made to correct the problems associated with the mining and use of coal.

OIL AND NATURAL GAS

Petroleum and natural gas are found in similar environments and typically occur together. Both are mixtures of hydrocarbon compounds, that is, compounds consisting of hydrogen and carbon, and may also contain small quantities of other elements such as sulfur, nitrogen, and oxygen. Like coal, petroleum

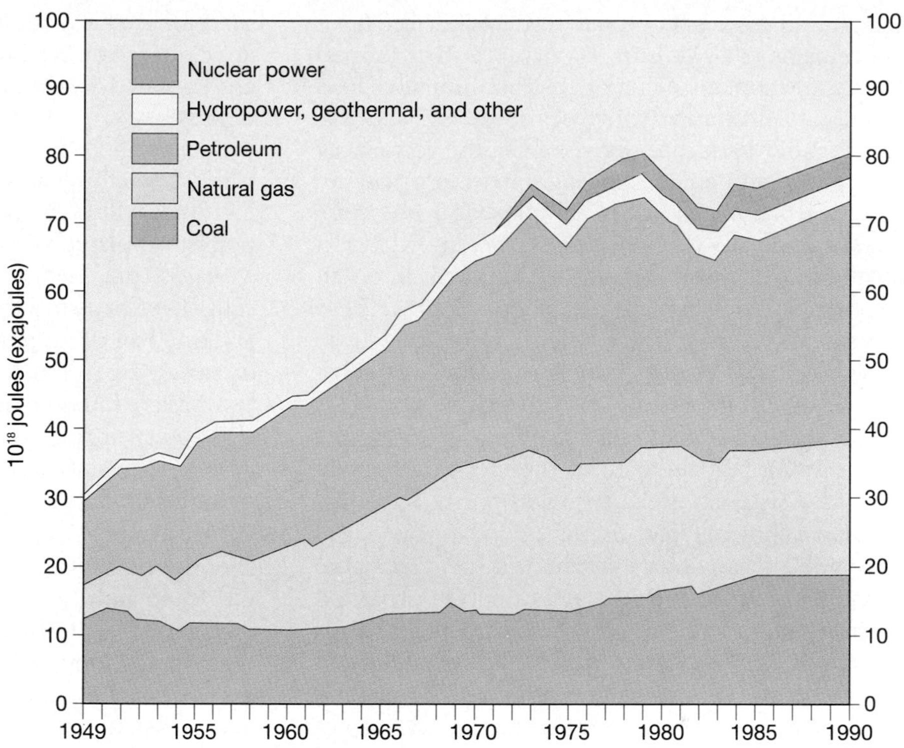

FIGURE 21.6
Consumption of energy in the United States, 1949–1990. (From Energy Information Administration)

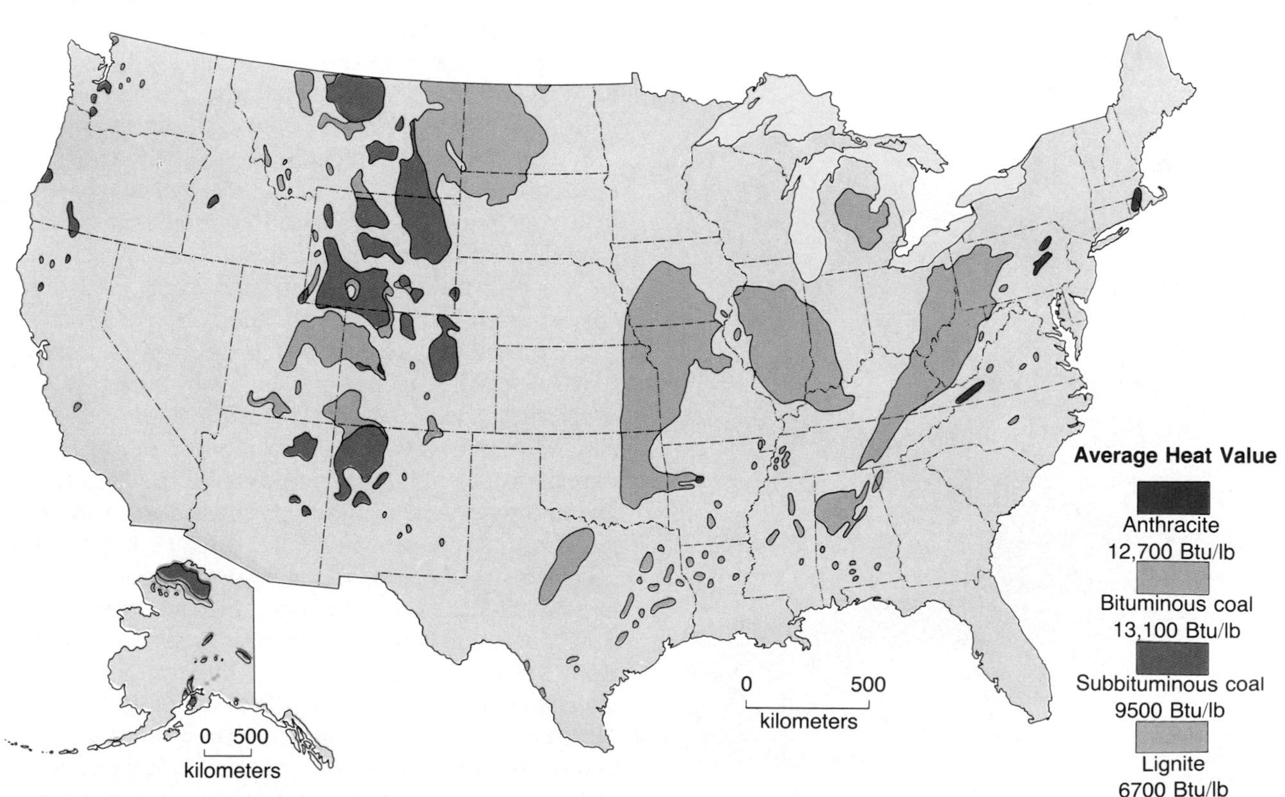

FIGURE 21.7
Coal fields of the United States. (Courtesy of the Bureau of Mines, U.S. Department of the Interior)

and natural gas are biological products derived from the remains of organisms. However, coal is formed mostly from plant material that accumulated in a swampy environment above sea level. Oil and gas, on the other hand, are derived from the remains of both plants and animals having a marine origin.

Petroleum formation is complex and not completely understood. Nonetheless, we know that it begins with the accumulation of sediment in ocean areas that are rich in plant and animal remains. These accumulations must occur where biological activity is high, such as in nearshore areas. However, most marine environments are oxygen-rich, which leads to the decay of organic remains before they can be buried by other sediments. Therefore, accumulations of oil and gas are not as widespread as the marine environments that support abundant biological activity. This limiting factor notwithstanding, large quantities of organic matter are buried and protected from oxidation in many offshore sedimentary basins (Figure 21.8). With increasing burial over millions of years, chemical reactions gradually transform some of the original organic matter into the liquid and gaseous hydrocarbons we call petroleum and natural gas.

Unlike the organic matter from which they formed, the newly created petroleum and natural gas are mobile. These fluids are gradually squeezed from the compacting, mud-rich layers where they originated into adjacent permeable beds such as sandstone, where openings between sediment grains are larger. The rock layers containing the oil and gas are saturated with water, and, since they are not as dense as water, oil and gas migrate upward through the water-filled pore spaces of the enclosing rocks. Unless something acts to halt this upward migration, the fluids will eventually reach the surface, at which point the volatile components will evaporate.

FIGURE 21.8
Offshore exploration in the relatively shallow waters of the continental shelf along the coast of Louisiana. (Photo by Richard J. Erickson)

A geologic environment that allows for economically significant amounts of oil and gas to accumulate is termed an **oil trap.** Although, as we shall see, several geologic structures may act as oil traps, all have two basic conditions in common: a porous, permeable **reservoir rock** that will yield petroleum and natural gas in sufficient quantities to make drilling worthwhile; and a **cap rock**, such as shale, that is virtually impermeable to oil and gas. The cap rock keeps the upwardly mobile oil and gas from escaping at the surface.

Figure 21.9 illustrates some common oil and natural gas traps. One of the simplest traps is an anticline, an uparched series of sedimentary strata (Figure 21.9A). As the strata are folded, the rising oil and gas collect at the apex of the fold. Because of its lower density, the natural gas collects above the oil. Both rest upon the denser water that saturates the reservoir rock. One of the world's largest oil fields, El Nala in Saudi Arabia, is the result of an anticlinal trap, as is the famous Teapot Dome in Wyoming. *Fault traps* form when strata are displaced in such a manner as to bring a dipping reservoir rock opposite an impermeable bed as shown in Figure 21.9B. In this case the upward migration of the oil and gas is halted at the fault zone. In the Gulf coastal plain region of the United States important accumulations of oil occur in association with *salt domes.* In such areas, which are characterized by thick accumulations of sedimentary strata, layers of rock salt occurring at great depth have been forced to rise in columns by the pressure of overlying beds. These rising salt columns gradually deform the overlying strata. Since oil and gas migrate to the highest level possible, they accumulate in the upturned sandstone beds adjacent to the salt column (Figure 21.9C). Yet another important geologic circumstance that may lead to significant accumulations of oil and gas is termed a *stratigraphic trap.* These oil-bearing structures result primarily from the original pattern of sedimentation rather than structural deformation. The stratigraphic trap illustrated in Figure 21.9D exists because a sloping bed of sandstone thins to the point of disappearance.

When the lid created by the cap rock is punctured by drilling, the oil and natural gas, which are under pressure, migrate from the pore spaces of the reservoir rock to the drill hole. On rare occasions the

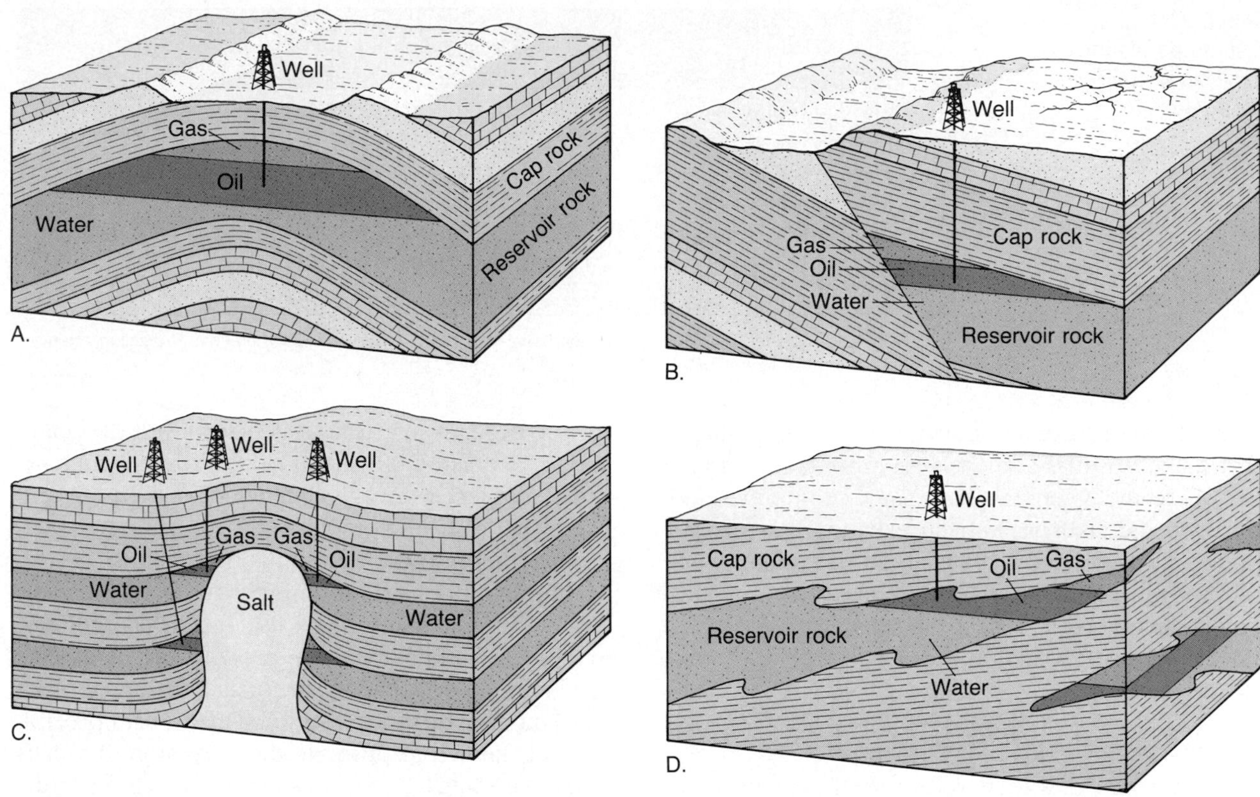

FIGURE 21.9
Common oil traps. **A.** Anticline. **B.** Fault trap. **C.** Salt dome. **D.** Stratigraphic (pinch-out) trap.

fluid pressure may force oil up the drill hole to the surface. Usually, however, a pump is required to lift the oil out.

A drill hole is not the only means by which oil and gas can escape from a trap. Traps can be broken by natural forces as well. For example, earth movements may create fractures that allow the hydrocarbon fluids to escape, or surface erosion may breach a trap with similar results. The older the rock strata, the greater the chance that deformation or erosion has affected a trap. Indeed, not all ages of rock yield oil and gas in the same proportions. The greatest production comes from the youngest rocks, those of Cenozoic age. Older Mesozoic rocks produce considerably less, followed by even smaller yields from the still older Paleozoic strata. There is virtually no oil produced from the most ancient rocks, those of Precambrian age.

SOME ENVIRONMENTAL CONSEQUENCES OF BURNING FOSSIL FUELS

Most of the energy used to run our modern world is obtained by burning fossil fuels. Coal, oil, and natural gas provide nearly 90 percent of the energy demands in the United States. Humanity faces a broad array of human-caused environmental problems. Among the most serious are the impacts on the atmosphere that occur because of fossil fuel combustion. Urban air pollution, acid rain, and global (greenhouse) warming are all closely linked to the use of these basic energy resources.

Urban Air Pollution

To people living in cities, poor air quality is a serious matter that frequently affects their health and well-being. The urban atmosphere has been accurately described as a giant chemical reactor that can produce a remarkable variety of undesirable products. Table 21.1 lists major primary pollutants and the sources that produce them. The significance of the transportation category is obvious. Fuel combustion for transportation accounts for nearly half of our pollution (by weight). The tens of millions of cars and trucks on the roads are the greatest contributors in this category. The table also shows that the second major source of primary pollutants is fuel combustion by stationary sources such as electrical-generating plants. When chemical reactions take place among primary pollutants, secondary pollutants are

TABLE 21.1
Estimated Nationwide Emissions (millions of metric tons/year).

Source	Particulates	Sulfur Oxides	Nitrogen Oxides	Volatile Organics	Carbon Monoxide	Total
Transportation	1.3	1.0	8.1	6.2	41.2	57.8
Fuel combustion from stationary sources	1.6	16.4	10.8	0.8	7.6	37.2
Industrial process	2.6	3.4	0.6	8.5	4.7	19.8
Solid waste disposal	0.3	0.0	0.1	0.6	1.7	2.7
Miscellaneous	0.9	0.0	0.2	2.4	6.0	9.5
Total	6.7	20.8	19.8	18.5	61.2	127.0

SOURCE: National Air Pollutant Emission Estimates, 1940–1988, U.S. Environmental Protection Agency Publication No. EPA-450/4-90-001, 1990.

formed. The noxious mixture of gases and particles that make up urban smog is an important example that is created when volatile organic compounds and nitrogen oxides from vehicle exhausts react in the presence of sunlight. In 1990 about 120 million people in the United States lived in areas in which air-quality levels did not meet the standards of the U.S. Environmental Protection Agency.

Acid Rain

Environmental problems related to the atmosphere are not confined to local urban areas. Because some emissions remain airborne for extended periods, the atmosphere's circulation can spread their influence far beyond the localities that produce them. Consequently, some problems affect broad regions, and others are truly worldwide in scope. Acid rain represents a serious environmental problem that occurs on a regional scale.

Each year in the United States, nearly 40 million tons of sulfur dioxide and nitrogen oxides are emitted into the atmosphere (Table 21.1). Through a series of complex chemical reactions, some of these pollutants are converted into acids, which then fall to the earth's surface as rain or snow. Another portion is deposited directly on surfaces and converted into acid after coming into contact with rain, dew, or fog.

Unlike atmospheric pollutants, acid precipitation has not yet been directly linked to any adverse effects on human health. Even so, its damaging effects on the environment are believed to be already considerable in some areas and imminent in others. The best-known effect is in thousands of lakes in Scandinavia, the northeastern United States, and eastern Canada. Accompanying this have been substantial increases in dissolved aluminum, which is leached from the soil by the acidic waters, and which in turn is toxic to fish and wildlife. As a consequence, some lakes are virtually devoid of fish while others are approaching this condition. Further, since ecosystems are characterized by many interactions at many levels of organization, evaluating the effects of acid precipitation on these complex systems is difficult and expensive, and therefore far from complete.

In addition to the thousands of lakes that can no longer support fish, preliminary research indicates that acid precipitation may also reduce agricultural crop yields and impair the productivity of forests. Acid rain not only harms the foliage but also leaches nutrient minerals from the soil. Finally, acid precipitation is known to promote the corrosion of metals and contributes to the destruction of stone structures.

Carbon Dioxide and Global Warming

The third of the environmental impacts, warming of the lower atmosphere, is global in scale. Unlike the two problems discussed so far, this issue is not associated with any of the primary pollutants listed in Table 21.1. Rather, the connection between global warming and burning fossil fuels relates to a basic product of combustion—carbon dioxide. Carbon dioxide (CO_2) is a gas that is found naturally in the atmosphere and that is being augmented by fuel combustion. Although CO_2 represents just 0.035 percent of clean, dry air, it is nevertheless meteorologically significant. The importance of carbon dioxide lies in the fact that it is transparent to incoming, short-wavelength solar radiation, but it is not transparent to some of the longer-wavelength, outgoing terrestrial radiation. A portion of the energy leaving the ground is absorbed by carbon dioxide and subsequently reemitted, part of it toward the surface, thereby keeping the air near the ground warmer than it would be without carbon dioxide. Thus, carbon dioxide is one of the gases responsible for warming the lower atmosphere. The process is called the *greenhouse effect.* Because carbon dioxide is an important heat absorber, any change in the air's carbon dioxide content could alter temperatures in the lower atmosphere.

Although the proportion of carbon dioxide in the air is relatively uniform at any given time, its percentage has been rising steadily for more than a century. Much of this rise is the result of burning ever-increasing quantities of fossil fuels.* From the mid-nineteenth century until the early 1990s, there has been an increase of more than 25 percent in the carbon dioxide content of the air.

If we assume that the use of fossil fuels will continue to increase at projected rates, current estimates indicate that the atmosphere's present carbon dioxide content of about 355 parts per million (ppm) will approach 400 ppm by the year 2000 and will reach 600 ppm by sometime in the second half of the next century. With such an increase in carbon dioxide, the enhancement of the greenhouse effect would be much more dramatic and measurable than in the past. When it is assumed that the atmosphere's carbon dioxide content will reach projected levels, the most realistic models predict a global surface temperature increase of between 1.5°C and 4.5°C. A change of this magnitude would be unprecedented in human history. Such an increase would come close to equaling the warming that has taken place since the peak of the most recent glacial stage 18,000 years ago, except that it would occur *much* more rapidly. The effects of a rapid temperature change are a matter of great concern and much uncertainty. Consequently, global warming is and will continue to be a major scientific and political issue for years to come.

Coal, oil, and natural gas are vital energy sources that power the modern world. However, the benefits that these basic and relatively low cost fuels provide do not come without environmental costs. Among the consequences are three serious atmospheric impacts: urban air pollution, acid rain, and global warming. People are clearly altering the composition of the air. Not only is this influence felt locally and regionally, but it extends worldwide and to many tens of kilometers above the earth's surface.

TAR SANDS AND OIL SHALE

At present the world's developed countries appear to have adequate, perhaps even abundant, supplies of relatively low-cost light oil. However, in the years to come, these supplies will dwindle. When this occurs, a lower grade of hydrocarbon will have to be substituted in order to meet the continuing requirements for liquid fuels. To satisfy some of the demand, fuels derived from tar sands and oil shales may be used.

Tar sands, sometimes called *oil sands,* are usually mixtures of clay and sand combined with water and varying amounts of a black, highly viscous tar known as bitumen. The use of the term *sand* is somewhat misleading because not all deposits are associated with sands and sandstones. Some occur in other materials, including shales and limestones. The oil in these deposits is very similar to heavy crude oils pumped from wells. The major difference between conventional oil reservoirs and tar sand deposits is in the viscosity of the oil they contain. In tar sands, the oil is much more viscous and cannot be recovered using conventional techniques.

Substantial tar sand deposits occur in several locations around the world, including Venezuela, the former Soviet Union, and the states of Utah and California. However, it is the Canadian province of Alberta that is believed to have the largest tar sand deposits, and it is here that synthetic crude oil has been produced since the opening of a small plant in 1967 (Figure 21.10). Subsequent development has led to production that exceeded 180,000 barrels per day in 1987, a figure that represented more than 11 percent of Canada's oil production. The Canadian government estimates that the amount of synthetic oil produced from tar sands will grow to between 230,000 and 350,000 barrels per day by the year 2000.

Most of Canada's tar sand processing has involved surface mining of shallow deposits. Once the material is dug from the ground, the bitumen is separated from the sand with hot water and steam and then chemically upgraded into synthetic crude oil that can be refined into various products. In addition to the scars that surface mining leaves on the landscape, this technique has two additional drawbacks. First, the amount of bitumen recovered is relatively small, and second, the process produces large volumes of waste consisting of oil, water, and clay that must be stored indefinitely in ever-growing ponds. This sludge is toxic to wildlife and could potentially contaminate nearby streams with organic chemicals and metal compounds if it were to escape from these reservoirs.

Only about 10 percent of Alberta's tar sands can be processed by surface mining. The remaining 90 percent is buried too deeply and must therefore be processed in place using wells. To accomplish this task, hot fluids are injected into the deep deposits to reduce the viscosity of the bitumen, which is then pumped to the surface by the same well or a second nearby well. An increasing percentage of Canada's

*Although the use of fossil fuels is the primary means by which humans add CO_2 to the atmosphere, the clearing of forests, especially in the tropics, also contributes substantially. Carbon dioxide is released as vegetation is burned or decays.

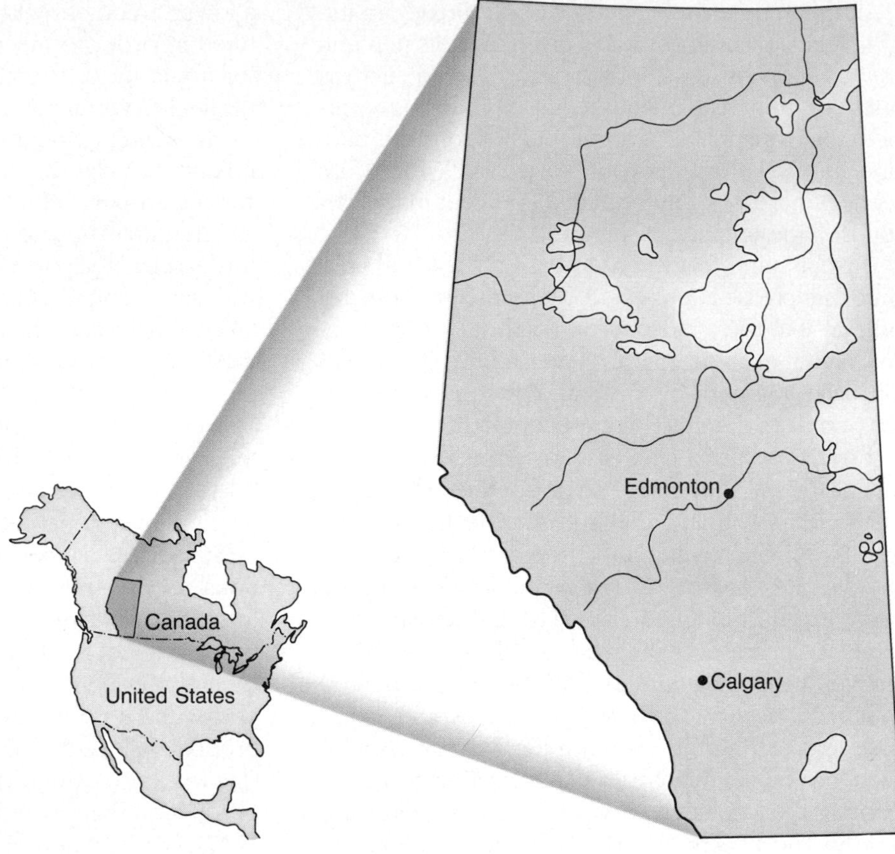

FIGURE 21.10
In North America the largest tar sand deposits occur in the Canadian province of Alberta. Production of synthetic oil began in 1967. By 1987, production exceeded 180,000 barrels per day. Production is expected to climb to between 230,000 and 350,000 barrels per day by the year 2000. (Data from Alberta Oil Sands Technology and Research Authority)

production is being accomplished by this method. During the 1990s the Canadian government plans to invest hundreds of millions of dollars to improve the technology of in-place extraction. The aims are to increase the percentage of bitumen that is recovered and to reduce environmental impacts.

In recent years the development of our oil shale resources has been suggested as a partial solution to the problem of dwindling fuel supplies. Indeed, an enormous amount of untapped oil is locked in the rock called oil shale. Worldwide, the U.S. Geological Survey estimates that there are more than 3000 billion barrels of oil contained in shales that would yield more than 38 liters (10 gallons) of oil per ton of shale. This figure, however, is misleading because less than 200 billion barrels are known to be recoverable with present technology. Roughly half of the worldwide supply is in the Green River Formation that encompasses portions of Colorado, Utah, and Wyoming (Figure 21.11). Within this region the oil shales are part of sedimentary layers that accumulated at the bottoms of two vast, shallow lakes 40 to 50 million years ago.

Oil shale is a very fine-grained sedimentary rock that contains enough organic matter to yield at least 38 liters of oil per ton. Although many other organic-rich shales contain some oil, these shales are such a low grade that they usually are not called oil shale. Oil shale does not actually contain oil but rather a waxy hydrocarbon material called *kerogen*. When heated to about 480°C (900°F), this complex solid mixture of carbon compounds decomposes into hydrocarbons and a carbonaceous residue. When cooled, the hydrocarbons condense into a liquid called shale oil. The oil from shale produces a fuel that can be burned in boilers or with proper (but expensive) conditioning can be upgraded for other uses.

Oil shales are a low-grade energy resource in comparison with other fuels. In terms of heat output per ton, the amount of heat energy in oil shale is about one-eighth that in crude oil. However, estimated resources of shale oil in the United States are about 14 times as great as those of conventionally recoverable oil, and estimates will probably increase as more geological information is gathered on less well-known deposits.* The relatively low heat energy in oil shale results from the dilution of kerogen by

*This fact as well as other information in this section is based upon a discussion of oil shale appearing in *National Energy Resource Issues*, U.S. Geological Survey Bulletin 1850, 1988, pp. 57–62.

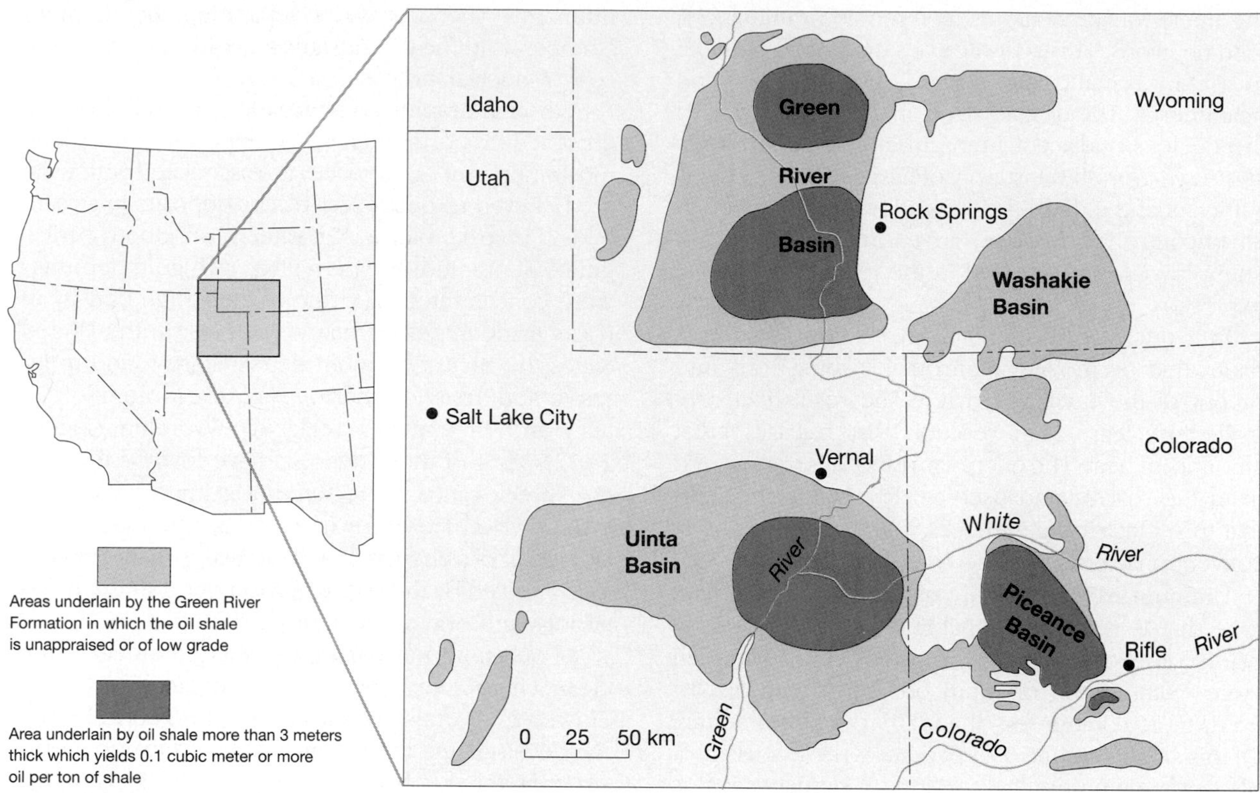

FIGURE 21.11
Distribution of oil shale in the Green River Formation of Colorado, Utah, and Wyoming. Government and industry have invested large sums to make these oil shales an economic resource, but costs have always been higher than the price of oil. However, as prices of competing fuels rise, these vast deposits will become economically more attractive. (After D. C. Duncan and V. E. Swanson, U.S. Geological Survey Circular 523, 1965)

the large proportion of mineral matter in the shales. This mineral matter contributes to mining, processing, and waste-disposal problems and to costs. Unlike crude oil, which is sent directly to the refinery, or coal, which commonly requires only crushing, screening, and perhaps washing before it is burned, oil shales require significant handling and processing before they are refined. In addition, attention must be given to environmental concerns in disposing of waste materials. At the present time, oil is plentiful and relatively inexpensive on world markets, and oil shale development efforts and the research that underlies these efforts have been almost completely abandoned by industry. Nevertheless, the U.S. Geological Survey suggests that ". . .the large amount of oil that potentially can be extracted from oil shales in the United States probably ensures its eventual inclusion in the national energy mix."*

*Ibid., p. 57.

ALTERNATE ENERGY SOURCES

An examination of Figure 21.6 clearly shows that we live in the era of fossil fuels. Nearly 95 percent of the world's energy needs are derived from these nonrenewable resources. Although there is at present an adequate supply, it is clear that a heavy reliance on coal, oil, and natural gas cannot continue indefinitely. Thus, we must look to alternate sources of energy to fuel our ever-increasing needs. In this section we will examine several possible sources, including nuclear, solar, wind, hydroelectric, geothermal, and tidal energy.

Nuclear Energy

Roughly five percent of the energy demands of the United States is being met by nuclear power plants. The fuel for these facilities comes from radioactive materials that release energy by the process of **nuclear fission**. Fission is accomplished by bombarding the nuclei of heavy atoms, commonly uranium-235, with neutrons. This action causes the uranium nuclei to split into smaller nuclei and to emit neutrons and heat energy. The ejected neutrons, in turn, bombard the nuclei of adjacent uranium atoms, producing a *chain reaction.* If the supply of fissionable material is sufficient and if the reaction is allowed to proceed in an uncontrolled manner, an enormous amount of energy would be released in the form of an atomic explosion.

In a nuclear power plant, the fission reaction is controlled by moving neutron-absorbing rods into or out of the nuclear reactor. The result is a controlled nuclear chain reaction that releases great amounts of heat. The energy produced is transported from the reactor and used to drive steam turbines that turn electrical generators, just as occurs in most conventional power plants.

Uranium-235 is the only naturally occurring isotope that is readily fissionable and is therefore the primary fuel used in nuclear power plants.* Although large quantities of uranium ore have been discovered, most contain less than 0.05 percent uranium. Of this small amount, 99.3 percent is the nonfissionable isotope uranium-238 and just 0.7 percent consists of the fissionable isotope uranium-235. Since most nuclear reactors operate with fuels that are at least 3 percent uranium-235, the two isotopes must be separated in order to concentrate the fissionable uranium-235. The process of separating the uranium isotopes is difficult and substantially increases the cost of nuclear power.

Although uranium is a rare element in the earth's crust, it does occur in enriched deposits. Some of the most important occurrences are associated with what are believed to be ancient placer deposits in stream beds. For example, in Witwatersrand, South Africa, grains of uranium ore (as well as rich gold deposits) were concentrated by virtue of their high density in rocks made largely of quartz pebbles. In the United States the richest uranium deposits are found in Jurassic and Triassic sandstones in the Colorado Plateau and in younger rocks in Wyoming (Figure 21.12). Most of these deposits have formed through the precipitation of uranium compounds from groundwater. Here precipitation of uranium occurs as a result of a chemical reaction with organic matter, as evidenced by the concentration of uranium in fossil logs and organic-rich, black shales.

At one time nuclear power was heralded as the clean, cheap source of energy to replace fossil fuels. However, several obstacles have emerged to hinder the development of nuclear power as a major energy source. Not the least of these is the skyrocketing costs of building nuclear facilities. More important perhaps is the concern over the possibility of a serious accident at one of the nearly 200 nuclear plants in existence worldwide. The 1979 accident at Three Mile Island near Harrisburg, Pennsylvania, helped bring this point home. Here, a malfunction led the plant operators to believe there was too much water in the primary system instead of too little. This con-

*Thorium, although not capable by itself of sustaining a chain reaction, can be used with uranium-235 as a nuclear fuel.

FIGURE 21.12
In the United States the richest uranium deposits occur in the Colorado Plateau and in Wyoming. This mine is producing ore in southern Utah. (Photo by Michael Collier)

fusion allowed the reactor core to lie uncovered for several hours. Although there was little danger to the public, substantial damage to the reactor resulted. Unfortunately, the 1986 accident at Chernobyl in the Soviet Union was far more serious. In this incident, the reactor ran out of control and two small explosions lifted the roof of the structure, allowing pieces of uranium to be thrown over the immediate area. During the 10 days that it took the Soviets to quench the fire that ensued, high levels of radioactive material were carried by the atmosphere and deposited as far away as Norway. In addition to the 18 persons who died within six weeks of the accident, many thousands more face an increased risk of death from cancers associated with the fallout.

It should be emphasized that the concentrations of fissionable uranium-235 and the design of reactors are such that nuclear power plants cannot explode like an atomic bomb. The dangers arise from the possible escape of radioactive debris during a meltdown of the core or other malfunction. In addition, hazards such as the disposal of nuclear waste and the relationship that exists between nuclear energy programs and the proliferation of nuclear weapons must be considered as we evaluate the pros and cons of employing nuclear power.

Solar Energy

The term *solar energy* generally refers to the direct use of the sun's rays to supply energy for the needs of people. The simplest, and perhaps most widely used, solar collectors are south-facing windows. As sunlight passes through the glass, energy is absorbed by objects in the room. These objects, in turn, radiate heat that warms the air in the room. In the United States the use of south-facing windows, along with better-insulated and more airtight construction, is widely practiced, substantially reducing heating costs.

More-elaborate systems used for home heating involve an active solar collector. These roof-mounted devices are usually large, blackened boxes that are covered with glass. The heat collected in the system can be transferred by the circulation of air or by fluids that are circulated through a network of tubes. Solar collectors are also used successfully to heat water for domestic and commercial needs. Unlike space heating, this application of solar energy benefits from the fact that hot water is required throughout the year, even in the tropics. In Israel, for example, about 20 percent of all homes are equipped with some type of solar device. Although solar energy is free, the necessary equipment and its installation are not. The initial costs of setting up a system, including a supplemental heating unit, can be substantial. Nevertheless, over the long term, solar energy is economical in many parts of the United States and will become even more cost effective as the prices of other fuels increase.

Research is currently underway to improve the technologies for concentrating solar energy so as to produce high-temperature heat. One method being examined uses an array of mirrors that track the sun and keep its rays focused on a receiving tower. A prototype facility, with 2000 mirrors, has been constructed near Barstow, California. Solar energy focused on the tower heats water in pressurized panels to over 500°C. The superheated water is then transferred to turbines, which turn electrical generators. This pilot project has demonstrated that solar collectors have the potential of becoming commercially viable.

Another type of collector uses solar cells that convert the sun's energy directly into electricity. Currently, this technology is used only for special applications such as solar-powered calculators and to provide electrical power for space vehicles. A large experimental facility that uses solar cells is located near Hesperia, California (Figure 21.13). However,

FIGURE 21.13
Panels of solar cells at the ARCO solar installation located near Hesperia, California. Electricity provided by the facility is used by customers of Southern California Edison. (Photo by E. J. Tarbuck)

the future of this technology is uncertain. In addition to being relatively inefficient, solar cells are presently very expensive to produce and are easily damaged by the elements.

Wind Energy

Wind has been used for centuries as an almost free and nonpolluting source of energy. Sailing ships and wind-powered grist mills represent two of the early ways that this renewable resource was harnessed. Further, as rural America was settled, there was a strong reliance on wind power to pump water and later to generate electricity. The world's largest wind generator was a two-bladed, 52-meter-diameter turbine developed and tested at Grandpa's Knob, Vermont, between 1941 and 1945. However, during much of the post–World War II period, abundant supplies of fossil fuels and the promise of cheap nuclear power caused a decline in the funding needed to support the development of large-scale wind turbines in the United States.

Following the "energy crisis" that was precipitated by the oil embargo in the 1970s, interest in wind power once again increased. In 1980 the federal government initiated a program to develop wind-power systems. A project sponsored by the United States Department of Energy involved setting up experimental wind farms in mountain passes known to have strong persistent winds (Figure 21.14). One of these facilities, located at Altamont Pass near San Francisco, employs over 2000 wind turbines (see chapter-opening photo). Presently, California is planning to develop wind farms capable of supplying 8 percent of the state's electricity by the turn of the century.

Although the future for wind power appears promising, it is not without difficulties. There are many technical problems to overcome in building large, efficient turbines. In addition, noise pollution and the costs of large tracts of land in populated areas present other significant obstacles to development. One proposed solution is to erect very large wind turbines on offshore platforms. The wind energy would then be used to produce hydrogen by the electrolysis of water. This combustible gas would then be piped to land and used as a fuel.

Hydroelectric Power

People have used falling water as an energy source for centuries. Through most of history, the mechanical energy produced by water wheels was used to power mills and other machinery. Today the power generated by falling water is used to drive turbines and produce electricity, hence the term **hydroelectric power.** In the United States, hydroelectric power plants contribute about 4 percent of the country's needs. Most of this energy is produced at large dams, which allow for a controlled flow of water (Figure 21.15). The water impounded in a reservoir is a form of stored energy that can be released at any time to produce electricity.

Although water power is considered a renewable resource, the dams built to provide hydroelectricity have finite lifetimes. All rivers carry suspended sediment that is deposited as soon as the dam is built. Eventually the sediment will completely fill the reservoir. The length of time for this to happen depends on the quantity of suspended material transported by the river. It ranges between about 50 and 300 years. An example is provided by Egypt's huge Aswan High

FIGURE 21.14
This wind farm near Palm Springs, California, consists of several hundred wind turbines. (Photo by Michael Collier)

FIGURE 21.15
Lake Powell is the reservoir that was created when Glen Canyon Dam was built across the Colorado River. As water in the reservoir is released, it drives turbines and produces electricity. Eventually the reservoir will be filled by sediment deposited by the Colorado River. (Photo by Michael Collier)

Dam, which was completed in the 1960s. It is now estimated that half the volume of the reservoir will be filled with sediment from the Nile River by the year 2025.

The availability of appropriate sites is an important limiting factor in the development of large-scale hydroelectric power plants. Good sites must be able to provide a significant height for the water to fall and a high rate of flow. Hydroelectric dams are found in many parts of the United States, with the greatest concentrations occurring in the Southeast and the Pacific Northwest. Although most of the best sites in the United States have already been developed, the total amount of power produced by hydroelectric sources may still increase in the years to come. However, the relative share provided by this source may decline because other alternate energy sources may increase at a faster rate.

In recent years a different type of hydroelectric power production has come into use. Called a *pumped water storage system,* it is actually a type of energy management. During times when demand for electricity is low, unneeded power produced by nonhydroelectric sources is used to pump water from a lower reservoir to a storage area at a higher elevation. Then, when demand for electricity is great, the water stored in the higher reservoir is available to drive turbines and produce electricity to supplement the power supply.

Geothermal Energy

Geothermal energy is produced by tapping naturally occurring steam and hot water located beneath the surface in regions where subsurface temperatures are high due to relatively recent volcanic activity. Although the development of geothermal power plants has grown quite rapidly in recent years, the idea of using natural steam to generate electricity is not new. As early as 1904, natural steam vents at Larderello, Italy, were used to make power. By 1990, the U.S. Geological Survey reported that nearly 200 separate power units in 17 countries were operating, with a combined capacity of almost 4800 megawatts (million watts). In addition to the United States, countries that lead in the use of geothermal energy to produce electricity are the Philippines, Indonesia, Mexico, Italy, and New Zealand.

The first commercial geothermal power plant in the United States was built in 1960 at The Geysers, north of San Francisco. By 1986, development at this location had grown to almost 1800 megawatts, enough to satisfy the electricity-generating needs of San Francisco and Oakland. However, this peak soon passed, and the production of electricity began to decline. (See Box 21.1.) In addition to The Geysers, geothermal development is occurring elsewhere in the western United States, including Nevada, Utah, and the Imperial Valley in southern California.

Geothermal energy is not used exclusively for generating electricity. In Iceland's capital, Reykjavik, steam and hot water are pumped into buildings throughout the city for space heating and are used to warm greenhouses where fruits and vegetables are grown all year. In the United States, localities in several western states use hot water from geothermal sources for space heating.

The most favorable geological factors for a geothermal reservoir of commercial value include:

1. A potent source of heat, such as a large magma chamber. The chamber should be deep enough to ensure adequate pressure and a slow rate of cooling and yet not be so deep that the natural circulation of water is inhibited. Magma chambers of this type are most likely to occur in regions of recent volcanic activity.

BOX 21.1

The Geysers: Running Out of Steam

In 1847, when the man who discovered The Geysers saw the plumes of steam rising from the ground, he thought he had come upon "the gates of hell" (Figure 21.A). A little more than a century later, developers viewed this area as an energy bonanza.

For twenty years (1960–1980), the pace of development of The Geysers was gradual. By 1981, there were 14 generating units dotting the area with a combined output of 943 megawatts. From this point onward, development accelerated, spurred by government incentives, rising oil prices, and the promise of cheap, smog-free energy. By 1985, the outlook for energy production at The Geysers was bright. The area contained 75 percent of the country's installed geothermal electrical-generating capacity, the largest complex of its kind in the world.

Just six years later, in 1991, the promising expansion of The Geysers was over. Although the geothermal field was supplying a full 6 percent of California's electrical power, it was not producing at expected levels. Installed generating capacity had grown to more than 2000 megawatts, but production was only 1500 megawatts. Moreover, steam pressure in the wells was falling rapidly. The problem was straightforward: The Geysers' geothermal field was running out of steam. By the mid-to-late 1990s, electrical output was projected to fall to half of the 1987 level.

When the facts were examined, it was clear that there had never been enough water in the crevices and fractures of the rocks to sustain the expansion of electrical-generating capacity. In most geothermal fields, drilling penetrates zones of superheated water, the volume of which can be measured. However, at The Geysers, wells tap only steam, and steam provides no measure of the volume of water giving rise to it. The reservoir was likened to a boiling teakettle that steams vigorously until it suddenly runs out of water. The amount of water present is not known until it is all gone. One writer observed, "The underlying problem at The Geysers—and a danger for geothermal development everywhere—is overdevelopment of a poorly understood resource. . . . "*

FIGURE 21.A
The Geysers, a field of steaming fumaroles 115 kilometers north of San Francisco, California. The natural steam beneath these hills was first tapped to power electrical-generating plants in 1960. Today it appears that the field is running out of steam. (Photo courtesy of Pacific Gas and Electric)

*Richard A. Kerr, "Geothermal Tragedy of the Commons," *Science* 253 (12 July 1991): 134. Much of the information in this box was based on this article.

2. Large, porous reservoirs with channels connected to the heat source, near which water can circulate and be stored; and
3. Capping rocks of low permeability that inhibit the flow of water and heat to the surface. A deep, well-insulated reservoir is likely to contain much more stored energy than an uninsulated, but otherwise similar, reservoir.

In some locations, both hot water and steam are present in the reservoir. When a well taps such a source, the steam and hot water rise, and the water changes to superheated steam as it approaches the surface. The steam is then piped directly into turbines to produce electric power. More commonly, however, a geothermal reservoir contains only hot water, which when tapped yields a mixture of steam and water. In such situations, the water must be removed before the steam can be used to generate power.

It should be emphasized that geothermal power is not an inexhaustible source of energy. When hot fluids are pumped from volcanically heated reservoirs, water cannot be replaced and then heated sufficiently to recharge the reservoir. Experience has shown that the output of steam and hot water from individual wells usually does not last for more than 10 to 15 years. Therefore, more wells must be drilled to maintain power production. Eventually, of course, the field is depleted.

Electrical power generation from geothermal sources is a relatively clean and nonpolluting alternative to other power plants that use steam. Geothermal plants do not produce the air pollution associated with burning fossil fuels; nor do they produce the radioactive wastes associated with nuclear power plants. Although geothermal power production has several attributes, there can also be some adverse environmental effects. The superheated water brought to the surface at some locations contains significant quantities of dissolved minerals. In addition to corroding generating equipment, this water can have a detrimental effect on fish and plant life if it is discharged into streams. Moreover, when fluids are withdrawn, the material in the reservoir may compact, which in turn could lead to surface subsidence. Since the wells remove heat from the reservoir, surface subsidence could also result as the cooling rocks below contract.

As with other alternate methods of power production, geothermal sources are not expected to provide a high percentage of the world's growing energy needs. Nevertheless, in regions where its potential can be developed, its use will no doubt continue to grow.

Tidal Power

With increased interest in the rising costs and eventual depletion of petroleum, greater attention is being focused upon alternate energy sources. Although several methods of generating electrical energy from the oceans have been proposed, the ocean's energy potential remains largely untapped. The development of tidal power is the principal example of energy production from the ocean.

Tides have been used as a source of power for centuries. Beginning in the twelfth century, water wheels driven by the tides were used to power gristmills and sawmills. During the seventeenth and eighteenth centuries, much of Boston's flour was produced at a tidal mill. Today, far greater energy demands must be met, and more sophisticated ways of using the force created by the perpetual rise and fall of the ocean must be employed.

Tidal power is harnessed by constructing a dam across the mouth of a bay or an estuary in a coastal area having a large tidal range (Figure 21.16). The narrow opening between the bay and the open ocean magnifies the variations in water level that occur as the tides rise and fall. The strong in-and-out flow that results at such a site is then used to drive turbines and electrical generators.

Tidal energy utilization is exemplified by the tidal power plant at the mouth of the Rance River in France (Figure 21.17). By far the largest yet constructed, this plant went into operation in 1966 and produces enough power to satisfy the needs of Brittany and also contribute to the demands of other regions. Much smaller experimental facilities near Murmansk in the former Soviet Union and near Taliang in China are also being used to generate electricity. The United States has not yet tapped its tidal power potential, although a site at Passamaquoddy Bay in Maine, where the tidal range approaches 15 meters (50 feet), has been under review for more than 50 years.

Along most of the world's coasts it is not possible to harness tidal energy. If the tidal range is less than 8 meters (25 feet) or if narrow, enclosed bays are absent, tidal power development is uneconomical. For this reason, the tides will never provide a very high portion of our ever-increasing electrical energy requirements. Nevertheless, the development of tidal power may be worth pursuing as Paul R. Ryan points out:

> Although total tidal power potential represents only a relatively small proportion of world energy requirements, its realization would nevertheless save a significant amount of fossil fuels. Tidal projects

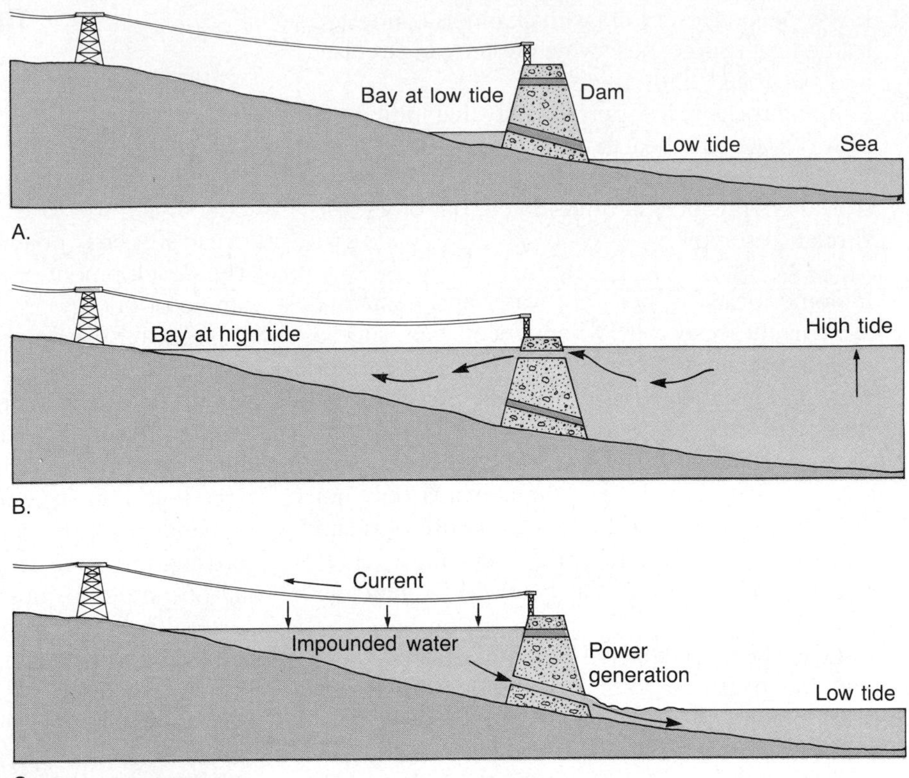

FIGURE 21.16
Simplified diagram showing the principle of the tidal dam.

FIGURE 21.17
The world's first tidal power station to produce electricity was built across the Rance River estuary in 1966. The tide that rushes up this estuary on the northern coast of Brittany is one of the highest in the world, reaching 13.5 meters (44 feet). (Courtesy of Phototeque/Electricite de France)

> worldwide have been estimated to have a potential energy output of 635,000 gigawatts, the equivalent of more than a billion barrels of oil, a year.*

In addition, electricity produced by the tides consumes no exhaustible fuels and hence creates no noxious wastes.

MINERAL RESOURCES

The outer layer of the earth, which we call the crust, is only as thick when compared to the remainder of the earth as a peach skin is to a peach, yet it is of supreme importance to us. We depend on it for fossil fuels and as a source of such diverse minerals as the talc for baby powder, salt to flavor food, and gold for world trade. In fact, on occasion, the availability or absence of certain earth materials has altered the course of history. As the material requirements of modern society grow, the need to locate additional supplies of useful minerals also grows and becomes more challenging as well.

Mineral resources are the endowment of useful minerals ultimately available commercially. Resources include already identified deposits from which minerals can be extracted profitably, called **reserves**, as well as known deposits that are not yet

*"Harnessing Power from the Tides: State of the Art," *Oceanus,* Vol. 22 (1980): p. 64.

economically or technologically recoverable. Deposits inferred to exist, but not yet discovered, are also considered as mineral resources. In addition the term **ore** is used to denote those useful metallic minerals that can be mined at a profit. In common usage, the term *ore* is also applied to some nonmetallic minerals such as fluorite and sulfur. However, materials used for such purposes as building stone, road aggregate, abrasives, ceramics, and fertilizers are not usually called ores; rather, they are classified as industrial rocks and minerals.

Recall that more than 98 percent of the earth's crust is composed of only eight elements, and, except for oxygen and silicon, all other elements make up a relatively small fraction of common crustal rocks (see Table 2.2). Indeed, the natural concentrations of many elements are exceedingly small. A deposit containing the average percentage of a valuable element is worthless if the cost of extracting it exceeds the value of the material recovered. To be considered of value, an element must be concentrated above the level of its average crustal abundance. Generally, the lower the crustal abundance, the greater the concentration must be. For example, copper makes up about 0.0135 percent of the crust. However, for a material to be considered a copper ore, it must contain a concentration that is about 50 times this amount. Aluminum, on the other hand, represents 8.13 percent of the crust and must be concentrated to only about four times its average crustal percentage before it can be extracted profitably.

It is important to realize that a deposit may become profitable to extract or lose its profitability because of economic changes. If demand for a metal increases and prices rise, the status of a previously unprofitable deposit changes, and it becomes an ore. The status of unprofitable deposits may also change if a technological advance allows the useful element to be extracted at a lower cost than before. This situation was illustrated recently at the copper mining operation located at Bingham Canyon, Utah, the largest open-pit mine on earth. (See Box 21.2.) Mining was halted here in 1985 because outmoded equipment had driven the cost of extracting the copper beyond the current selling price. The owners responded by replacing an antiquated 1000-car railroad with conveyor belts and pipelines for transporting the ore and waste. These devices achieved a cost reduction of nearly 30 percent and returned this mining operation to profitability.

Over the years, geologists have been keenly interested in learning how natural processes produce localized concentrations of essential metallic minerals. One well-established fact is that occurrences of valu-

TABLE 21.2
Occurrences of metallic minerals.

Metal	Principal Ores	Geological Occurrences
Aluminum	Bauxite	Residual product of weathering
Chromium	Chromite	Magmatic segregation
Copper	Chalcopyrite Bornite Chalcocite	Hydrothermal deposits; contact metamorphism; enrichment by weathering processes
Gold	Native gold	Hydrothermal deposits; placers
Iron	Hematite Magnetite Limonite	Banded sedimentary formations; magmatic segregation
Lead	Galena	Hydrothermal deposits
Magnesium	Magnesite Dolomite	Hydrothermal deposits
Manganese	Pyrolusite	Residual product of weathering
Mercury	Cinnabar	Hydrothermal deposits
Molybdenum	Molybdenite	Hydrothermal deposits
Nickel	Pentlandite	Magmatic segregation
Platinum	Native platinum	Magmatic segregation; placers
Silver	Native silver Argentite	Hydrothermal deposits; enrichment by weathering processes
Tin	Cassiterite	Hydrothermal deposits; placers
Titanium	Ilmenite Rutile	Magmatic segregation; placers
Tungsten	Wolframite Scheelite	Pegmatites; contact metamorphic deposits; placers
Uranium	Uraninite (pitchblende)	Pegmatites; sedimentary deposits
Zinc	Sphalerite	Hydrothermal deposits

BOX 21.2

Bingham Canyon, Utah: The Largest Open-Pit Mine

The method used to extract resources from the earth depends upon the nature and location of the material being sought. Many resources are acquired through mining. For large numbers of people, the image of a mine is that of an underground shaft or tunnel. Nevertheless, surface mining accounts for about two-thirds of the world's production of solid mineral resources. *Quarries,* in which gravel or building stone is extracted, are one relatively common type of surface mine. Others are called *strip mines* and *open-pit mines.* Strip mining is employed when a resource, such as coal, occurs near the surface in flat-lying layers. The overlying vegetation, soil, and rock are stripped off, and then the coal or other desired material is removed. When mining is completed, the land can be reclaimed by filling the void with the overburden and then restoring the topsoil and replanting vegetation.

In contrast to strip mines, open-pit mines are the preferred method of extraction where there is a large, three-dimensional body of ore near the surface. As the removal of great tonnages progresses, the excavation enlarges.

FIGURE 21.B
Aerial view of Bingham Canyon copper mine near Salt Lake City, Utah. This huge open-pit mine is about 4 kilometers across and 900 meters deep. Although the amount of copper in the rock is less than 1 percent, the huge volumes of material removed and processed each day (about 200,000 tons) yield significant quantities of metal. (Photo by Michael Collier)

able mineral resources are closely related to the rock cycle. That is, the mechanisms that generate igneous, sedimentary, and metamorphic rocks, including the processes of weathering and erosion, play a major role in producing concentrated accumulations of useful elements. Moreover, with the development of the theory of plate tectonics, geologists have added another tool for understanding the processes by which one rock is transformed into another.

MINERAL RESOURCES AND IGNEOUS PROCESSES

Some of the most important accumulations of metals, such as gold, silver, copper, mercury, lead, platinum, and nickel, are produced by igneous processes (Table 21.2). These mineral resources, like most others, result from processes that concentrate desirable materials to the extent that extraction is economically feasible.

The igneous processes that generate some of these metal deposits are quite straightforward. For example, as a large magma body cools, the heavy minerals that crystallize early tend to settle to the lower portion of the magma chamber. This type of magmatic differentiation is particularly active in large basaltic magmas where chromite (ore of chromium), magnetite, and platinum are occasionally generated. Layers of chromite, interbedded with other heavy minerals, are mined from such deposits in the Stillwater Complex of Montana. Another example is the Bushveld Complex in South Africa, which contains

With time a conical pit is formed with terraced benches spiraling down to the bottom.

The world's largest open-pit mine is the mammoth Bingham Canyon copper mine located about 40 kilometers (25 miles) southwest of Salt Lake City, Utah (Figure 21.B). A mountain once stood where this huge open-pit mine is now. At the top, the rim is nearly 4 kilometers (2.5 miles) across and covers almost 8 square kilometers (3 square miles). The excavation reaches a depth of 900 meters (3000 feet). If a steel tower were erected at the bottom, it would have to be five times taller than the Eiffel Tower to reach the top of the pit.

Underground mining operations that were seeking veins of silver and lead first began at Bingham Canyon in the late nineteenth century. Later, copper was discovered. The ore is known as *porphyry copper.* Similar deposits occur at several locations in the American Southwest and at a number of other sites in a belt that stretches from southern Alaska to northern Chile. As in other places in this belt, the ore at Bingham Canyon is disseminated throughout a porphyritic plutonic rock. The deposit formed after magma was intruded to shallow depths. Following this, shattering created an extensive system of fractures that were penetrated by hydrothermal solutions from which the ore minerals precipitated. Although the percentage of copper in the rock is small, the total volume of copper is huge. Since open-pit operations started in 1906, 5 billion tons of material have been removed, yielding more than 12 million tons of copper. Significant amounts of gold, silver, and molybdenum have also been recovered. The ore body is far from exhausted. Over the next 25 years plans call for an additional 3 billion tons of material to be removed and processed. This largest of artificial excavations has generated most of Utah's mineral production for more than 80 years and has been called the "richest hole on earth."

Like many mines that have been operating for an extended period, the Bingham pit was unregulated during most of its history. Development occurred prior to the present-day appreciation and awareness of the environmental impacts of mining and prior to effective environmental legislation. Today, problems and issues related to groundwater and surface water contamination, air pollution, solid and hazardous wastes, and land reclamation are receiving more serious and long overdue attention.

over 70 percent of the world's known reserves of platinum.

Magmatic segregation is also important in the late stages of the magmatic process. This is particularly true of granitic magmas in which the residual melt can become enriched in rare elements and heavy metals. Further, because water and other volatile substances do not crystallize along with the bulk of the magma body, these fluids make up a high percentage of the melt during the final phase of solidification. Crystallization in a fluid-rich environment, where ion migration is enhanced, is believed to result in the formation of crystals several centimeters, or even a few meters, in length. The resulting rocks, called **pegmatites**, are composed of these unusually large crystals.

Feldspar masses the size of houses have been quarried from a pegmatite located in North Carolina. Gigantic crystals of muscovite measuring a few meters across have been found in Ontario, Canada. In the Black Hills, crystals as large as telephone poles of the lithium-bearing mineral spodumene have been mined. The largest of these was more than 12 meters (40 feet) long. Not all pegmatites contain such large crystals, but these examples emphasize the special conditions that must exist during their formation.

Most pegmatites are granitic in composition and consist of unusually large crystals of quartz, feldspar, and muscovite. Feldspar is used in the production of ceramics and muscovite is used for electrical insulation and glitter. Further, pegmatites often contain some of the least abundant elements. Thus, in addition to the common silicates, some pegmatites include semiprecious gems such as beryl, topaz, and tourmaline. Moreover, minerals containing the elements lithium, cesium, uranium, and the rare earths* are occasionally found. Most pegmatites are located within large igneous masses or as dikes or veins which cut into the country rock that surrounds the magma chamber (Figure 21.18).

Not all late-stage magmas produce pegmatites; nor do all have a granitic composition. Rather, some magmas become enriched in iron or occasionally copper. For example, at Kirava, Sweden, magma composed of over 60 percent magnetite solidified to

*The rare earths are a group of 15 elements (atomic numbers 57 through 71) that posses similar properties. They are useful catalysts in petroleum refining and are used to improve color retention in television picture tubes.

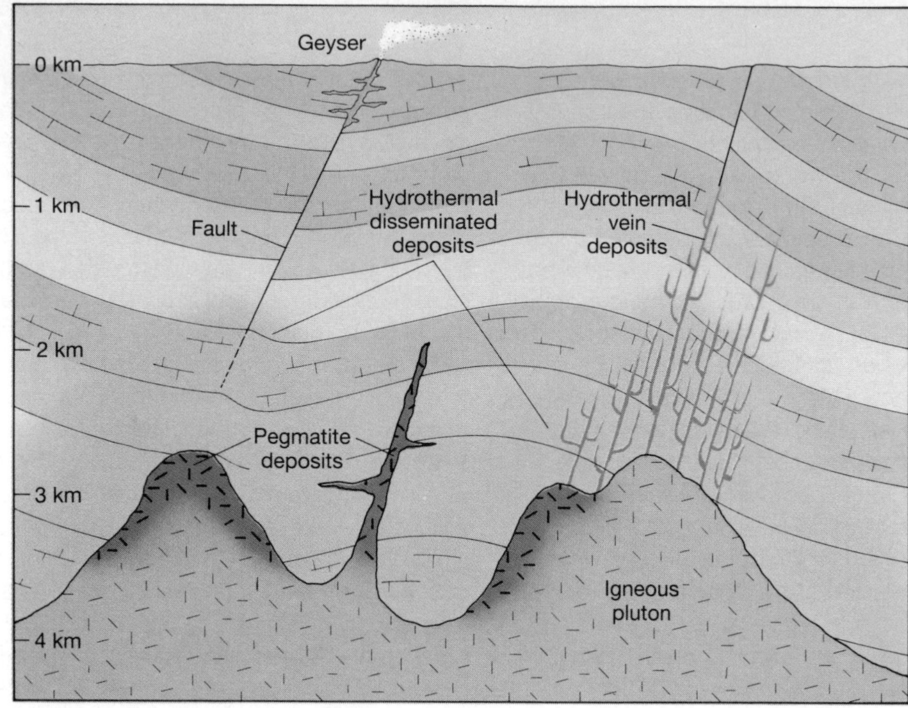

FIGURE 21.18
Illustration of the relationship between a parent igneous body and the associated pegmatite and hydrothermal deposits.

produce one of the largest iron deposits in the world.

Another economically important mineral with an igneous origin is diamond. Although best known as gems, diamonds are used extensively as abrasives. Diamonds are thought to originate at depths of nearly 200 kilometers, where the confining pressure is great enough to generate this high-pressure form of carbon. Once crystallized, the diamonds are carried upward through pipe-shaped conduits that increase in diameter toward the surface. In diamond-bearing pipes, nearly the entire pipe contains diamond crystals that are disseminated throughout an ultramafic rock called *kimberlite.* The most productive kimberlite pipes are those found in South Africa. The only equivalent source of diamonds in the United States is located near Murfreesboro, Arkansas, but this deposit is exhausted and serves today merely as a tourist attraction.

Among the best-known and most important ore deposits are those generated from **hydrothermal** (hot-water) **solutions.** Included in this group are the gold deposits of the Homestake mine in South Dakota; the lead, zinc, and silver ores near Coeur d'Alene, Idaho; the silver deposits of the Comstock Lode in Nevada; and the copper ores of the Keweenaw Peninsula in Michigan (Figure 21.19).

The majority of hydrothermal deposits are thought to originate from hot, metal-rich fluids that are remnants of late-stage magmatic processes. During solidification, liquids plus various metallic ions accumulate near the top of the magma chamber. Because of their mobility, these ion-rich solutions can migrate great distances through the surrounding rock before they are eventually deposited, usually as sulfides of various metals (Figure 21.18). Some of this fluid moves along openings such as fractures or bedding planes, where it cools and precipitates the metallic ions to produce **vein deposits** (Figure

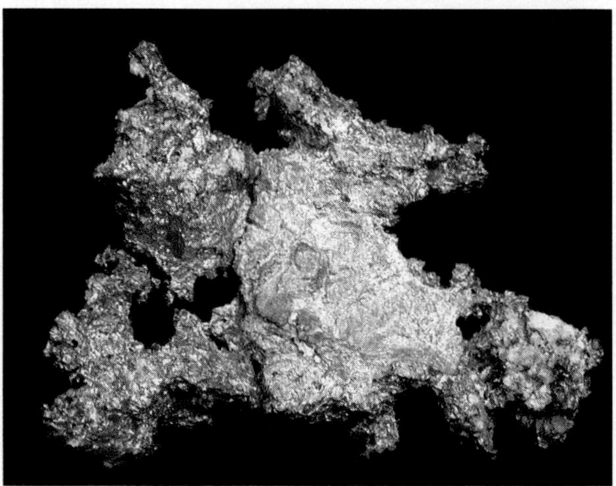

FIGURE 21.19
Native copper from northern Michigan's Keweenaw Peninsula is an excellent example of a hydrothermal deposit. At one time this area was an important source of copper, but it is now largely depleted. (Photo by E. J. Tarbuck)

21.20). Many of the most productive deposits of gold, silver, and mercury occur as hydrothermal vein deposits.

Another important type of accumulation generated by hydrothermal activity is called a **disseminated deposit**. Rather than being concentrated in narrow veins and dikes, these ores are distributed as minute masses throughout the entire rock mass. Much of the world's copper is extracted from disseminated deposits, including those at Chuquicamata, Chile, and the huge Bingham Canyon copper mine in Utah. Because these accumulations contain only 0.4 to 0.8 percent copper, between 125 and 250 kilograms of ore must be mined for every kilogram of metal recovered. The environmental impact of these large excavations, including the problems of waste disposal, is significant.

Some hydrothermal deposits have been generated by the circulation of ordinary groundwater in regions where magma was emplaced near the surface. The Yellowstone National Park area is a modern example of such a situation. When groundwater invades a zone of recent igneous activity, its temperature rises, greatly enhancing its ability to dissolve minerals. These migrating hot waters remove metallic ions from the intrusive igneous rocks and carry them upward, where they may be deposited as an ore body. Depending on the conditions, the resulting accumulations may occur as vein deposits, as disseminated deposits, or, where hydrothermal solutions reach the surface in the form of hot springs or geysers, as surface deposits.

With the advent of the plate tectonics theory it became clear that some hydrothermal deposits had their origins along ancient oceanic ridges. A well-known example is found on the island of Cyprus, where copper has been mined for more than 4000 years. Apparently, these deposits represent ores that formed on the sea floor at an ancient oceanic spreading center. Since the mid-1970s, active hot springs and metal-rich sulfide deposits have been detected at several sites, including study areas along the East Pacific Rise and the Juan de Fuca Ridge. The deposits are forming where heated seawater, rich in dissolved metals and sulfur, gushes from the sea floor as particle-filled clouds called *black smokers* (Figure 21.21). Among the most spectacular examples to date are those that were photographed along the East Pacific Rise by scientists aboard the *Alvin* (see Figure 19.C). As shown in Figure 21.22, seawater is believed to infiltrate the hot oceanic crust along the flanks of the ridge. As the water moves through the newly formed material, it is heated and chemically interacts with the basalt, extracting and transporting sulfur, iron, copper, and other metals. Near the ridge axis, the hot, metal-rich fluid rises along faults. Upon reaching the sea floor, the spewing liquid mixes with the cold seawater, and the sulfides precipitate to form massive sulfide deposits.

MINERAL RESOURCES AND METAMORPHIC PROCESSES

The role of metamorphism in producing mineral deposits is frequently tied to igneous processes. For example, many of the most important metamorphic ore deposits are produced by contact metamorphism. Here the country rock is recrystallized and chemically altered by heat, pressure, and hydrothermal solutions emanating from an intruding igneous body. The extent to which the country rock is altered

FIGURE 21.20
Light-colored vein deposits emplaced along a series of fractures in dark-colored igneous rock. (Photo by James E. Patterson)

FIGURE 21.21
This is one of two dozen vents called "black smokers" that were found by the *Alvin* at 21°N latitude on the East Pacific Rise in May, 1979. The "smoke" is actually hot, mineral-rich water that has circulated through the ocean crust and picked up iron, copper, and zinc. When the hot solution hits the cold seawater, it precipitates sulfide ores that now coat the vents. (Courtesy of Woods Hole Oceanographic Institution)

depends on the nature of the intruding igneous mass as well as the nature of the host rock. Some resistant materials, such as quartz sandstone, may show very little alteration, whereas others, including limestone, may exhibit the effects of metamorphism for several kilometers from the igneous pluton. As hot, ion-rich fluids move through limestone, chemical reactions take place which produce useful minerals such as garnet and corundum. Further, these reactions release carbon dioxide, which greatly facilitates the outward migration of metallic ions. Thus, extensive aureoles of metal-rich deposits commonly surround igneous plutons that have invaded limestone strata. The most common metallic minerals associated with contact metamorphism are sphalerite (zinc), galena (lead), chalcopyrite (copper), magnetite (iron), and bornite (copper). The hydrothermal ore deposits may be disseminated throughout the altered zone or exist as concentrated masses that are located either next to the intrusive body or at the periphery of the metamorphic zone.

Regional metamorphism can also generate useful mineral deposits. Recall that at convergent plate boundaries the oceanic crust, along with sediments that have accumulated at the continental margins, are carried to great depths. In these high-temperature, high-pressure environments the mineralogy and texture of the subducted materials are altered, producing deposits of nonmetallic minerals such as talc and graphite.

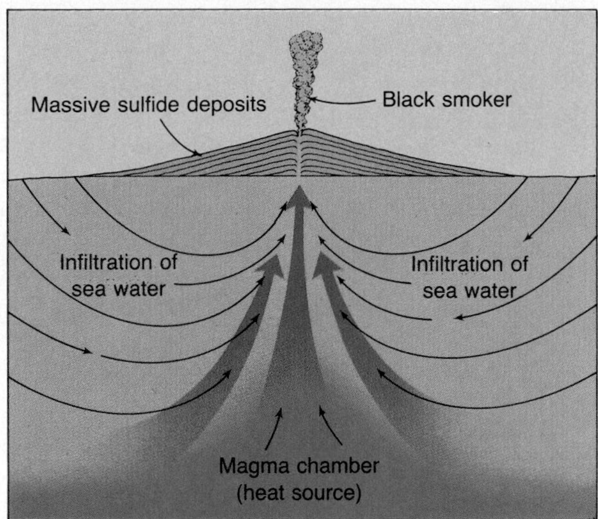

FIGURE 21.22
Massive sulfide deposits can result from the circulation of seawater through the oceanic crust along active spreading centers. As seawater infiltrates the hot basaltic crust, it leaches sulfur, iron, copper, and other metals. The hot, enriched fluid returns to the sea floor near the ridge axis along faults and fractures. Some metal sulfides may be precipitated in these channels as the rising fluid begins to cool. When the hot liquid emerges from the sea floor and mixes with cold seawater, the sulfides precipitate to form massive deposits.

WEATHERING AND ORE DEPOSITS

Weathering creates many important mineral deposits by concentrating minor amounts of metals that are scattered through unweathered rock into economically valuable concentrations. Such a transformation is often termed **secondary enrichment** and takes place in one of two ways. In one situation chemical weathering coupled with downward-percolating water removes undesirable materials from decomposing rock, leaving the desirable elements enriched in the upper zones of the soil. The second way is basically the reverse of the first. That is, the desirable elements that are found in low concentrations near the surface are removed and carried to lower zones, where they are redeposited and become more concentrated.

The formation of *bauxite,* the principal ore of aluminum, is one important example of an ore created as a result of enrichment by weathering processes (Figure 21.23). Although aluminum is the third most abundant element in the earth's crust, economically valuable concentrations of this important metal are not common, due to the fact that most aluminum is tied up in silicate minerals from which it is extremely difficult to extract.

Bauxite forms in rainy tropical climates in association with laterites. In fact, bauxite is sometimes referred to as aluminum laterite. When aluminum-rich source rocks are subjected to the intense and prolonged chemical weathering of the tropics, most of the common elements, including calcium, sodium, and silicon, are removed by leaching. Because aluminum is extremely insoluble in natural solutions, hydrated aluminum oxide (bauxite) becomes concentrated in the soil. Thus, the formation of bauxite depends upon climatic conditions in which chemical weathering and leaching are pronounced as well as on the presence of an aluminum-rich source rock. Important deposits of nickel and cobalt are also found in laterite soils that develop from igneous rocks rich in ferromagnesian minerals.

Many copper and silver deposits result when weathering processes concentrate metals that are dispersed through a low-grade primary ore. Usually such enrichment occurs in deposits containing pyrite (FeS_2), the most common and widespread sulfide mineral. Pyrite is important because when it chemically weathers, sulfuric acid forms, which enables percolating waters to dissolve the ore metals. Once dissolved, the metals gradually migrate downward through the primary ore body until they are precipitated. Deposition takes place because of changes that occur in the chemistry of the solution when it

FIGURE 21.23
Bauxite is the ore of aluminum and forms as a result of weathering processes under tropical conditions. Its color varies from red or brown to nearly white. (Photo by E. J. Tarbuck)

reaches the groundwater zone (the zone beneath the surface where all pore spaces are filled with water). In this manner the small percentage of dispersed metal can be removed from a large volume of rock and redeposited as a higher-grade ore in a smaller volume of rock. This enrichment process is responsible for the economic success of many copper deposits, including the one at Miami, Arizona. Here the ore was upgraded from less than 1 percent copper in the primary deposit to as much as 5 percent copper in some localized zones of enrichment. When pyrite weathers (oxidizes) near the surface, residues of iron oxide remain. The presence of these rusty-colored caps at the surface indicates the possibility of an enriched ore below, and this represents a visible sign for prospectors.

PLACER DEPOSITS

Sorting typically results in like-sized grains being deposited together. However, sorting according to the specific gravity of particles also occurs. This latter type of sorting is responsible for the creation of **placers**, which are deposits formed when heavy minerals are mechanically concentrated by currents. Placers associated with streams are among the most common and best known, but the sorting action of waves can also create placers along the shore. Placer deposits usually involve minerals that are not only heavy but also tough and chemically resistant enough to withstand destruction by weathering processes and transporting currents. Placers form because heavy minerals settle quickly from a current, whereas less dense particles remain suspended and are carried onward. Common sites of accumulation include point bars on the insides of meanders as well as cracks, depressions, and other irregularities on stream beds.

Many economically important placer deposits exist, with accumulations of gold the best known. Indeed, it was the placer deposits discovered in 1848 that led to the famous California gold rush. Years later, similar deposits created a gold rush to Alaska as well. Panning for gold by washing sand and gravel from a flat pan to concentrate the fine "dust" at the bottom was a common method used by early prospectors to recover the precious metal, and it is a process similar to that which created the placer in the first place.

In addition to gold, other heavy and durable minerals form placers. These include platinum, diamonds, and tin. The Ural Mountains contain placers rich in platinum, and placers are important sources of diamonds in southern Africa. Significant portions of the world's supply of cassiterite, the principal ore of tin, have come from placer deposits in Malaysia and Indonesia. Cassiterite is often widely disseminated through granitic igneous rocks. In this state, the mineral is not sufficiently concentrated to be extracted profitably. However, as the enclosing rock dissolves and disintegrates, the heavy and durable cassiterite grains are set free. Eventually the freed particles are washed to a stream where they are deposited in placers that are significantly more concentrated than the original deposit. Similar circumstances and events are common to many minerals mined from placers.

In some cases, if the source rock for a placer deposit can be located, it too may become an important ore body. By following placer deposits upstream, one can sometimes locate the original deposit. This is how the gold-bearing veins of the Mother Lode in California's Sierra Nevada batholith were found, as well as the famous Kimberly diamond mines of South Africa. The placers were discovered first, their source at a later time.

Placers have undoubtedly formed throughout geologic time, but most productive deposits are of Cenozoic age. Placer deposits are typically small in volume and accumulate at the earth's surface above local base level. Consequently, most placer deposits are eroded before they can be buried and preserved. However, buried placers do exist and are mined in some areas.

NONMETALLIC MINERAL RESOURCES

Mineral resources that are not used as fuels or processed for the metals they contain are referred to as **nonmetallic mineral resources**. These materials are extracted and processed either to make use of the nonmetallic elements they contain or for the physical and chemical properties they possess. People often do not realize the importance of nonmetallic minerals because they see only the products that resulted from their use and not the minerals themselves. That is, many nonmetallics are used up in the process of creating other products. Examples include the fluorite and limestone that are part of the steelmaking process, the abrasives required to make a piece of machinery, and the fertilizers needed to grow a food crop. (Table 21.3).

The quantities of nonmetallic minerals used each year are enormous. A glance at Figure 21.1 reminds us that the per capita consumption of nonfuel resources in the United States totals more than 11 metric tons, of which about 94 percent are nonmetallics. Nonmetallic mineral resources are commonly di-

TABLE 21.3
Occurrences and uses of nonmetallic minerals.

Mineral	Uses	Geological Occurrences
Apatite	Phosphorus fertilizers	Sedimentary deposits
Asbestos (chrysotile)	Incombustible fibers	Metamorphic alteration
Calcite	Aggregate; steelmaking; soil conditioning; chemicals; cement; building stone	Sedimentary deposits
Clay minerals (kaolinite)	Ceramics; china	Residual product of weathering
Corundum	Gemstones; abrasives	Metamorphic deposits
Diamond	Gemstones; abrasives	Kimberlite pipes; placers
Fluorite	Steelmaking; aluminum refining; glass; chemicals	Hydrothermal deposits
Garnet	Abrasives; gemstones	Metamorphic deposits
Graphite	Pencil lead; lubricant; refractories	Metamorphic deposits
Gypsum	Plaster of Paris	Evaporite deposits
Halite	Table salt; chemicals; ice control	Evaporite deposits; salt domes
Muscovite	Insulator in electrical applications	Pegmatites
Quartz	Primary ingredient in glass	Igneous intrusions; sedimentary deposits
Sulfur	Chemicals; fertilizer manufacture	Sedimentary deposits; hydrothermal deposits
Sylvite	Potassium fertilizers	Evaporite deposits
Talc	Powder used in paints, cosmetics, etc.	Metamorphic deposits

vided into two broad groups—building materials and industrial minerals. Since some substances have many different uses, they are found in both categories. Limestone, perhaps the most versatile and widely used rock of all, is the best example. As a building material, it is used not only as crushed rock and building stone, but in the making of cement as well. Moreover, as an industrial mineral, it has many uses in manufacturing and agriculture.

Besides aggregate (sand, gravel, and crushed rock) and cut stone, the other important building materials include gypsum for plaster and wallboard, clay for tile and bricks, and cement, which is made from limestone and shale. Cement and aggregate go into the making of concrete, a material that is essential to practically all construction. Aggregate gives concrete its strength and volume, and cement binds the mixture into a rock-hard substance. Just two kilometers of four-lane highway require more than 85 metric tons of aggregate. On a smaller scale, 50 tons of aggregate are needed just to build an average house.

Because most building materials are widely distributed and present in almost unlimited quantities, they have little intrinsic value. Their economic worth comes only after the materials are removed from the ground and processed. Since their per-ton value as compared to metals and industrial minerals is low, mining and quarrying operations are usually undertaken to satisfy local needs. Except for special types of cut stone used for buildings and monuments, transportation costs greatly limit the distances that most building materials can be moved.

Many nonmetallic resources are classified as industrial minerals. In some instances these materials are important because they are sources of specific chemical elements or compounds. Such minerals are used in the manufacture of chemicals and the production of fertilizers. In other cases their importance is related to the physical properties they exhibit. Examples would include minerals such as corundum and garnet that are used as abrasives. Although supplies are generally plentiful, most industrial minerals are not nearly as abundant as building materials. Moreover, deposits are far more restricted in distribution and extent. As a result, many of these nonmetallic resources must be transported considerable distances which, of course, adds to their cost. Unlike most building materials, which need a minimum of processing before they are ready to use, many indus-

FIGURE 21.24
Common salt is an important and versatile nonmetallic resource. This thick deposit at Grand Saline, Texas, is being exploited using conventional underground mining techniques. (Photo by Morton Salt Company, provided by John S. Shelton)

trial minerals require considerable processing to extract the desired substance at the proper degree of purity for its ultimate use.

The growth in world population toward 7 billion by the year 2000 requires that the production of basic food crops continue to expand. Therefore fertilizers, primarily nitrate, phosphate, and potassium compounds, are extremely important to agriculture. The synthetic nitrate industry, which derives nitrogen from the atmosphere, is the source of practically all the world's nitrogen fertilizers. The primary source of phosphorus and potassium however, remains the earth's crust. The mineral apatite is the primary source of phosphate. In the United States most production comes from marine sedimentary deposits in Florida and North Carolina. Although potassium is an abundant element in many minerals, the primary commercial sources are evaporite deposits containing the mineral sylvite. In the United States, deposits near Carlsbad, New Mexico, have been especially important.

Because of its diverse uses, sulfur is an important nonmetallic resource. In fact, the quantity of sulfur used is considered one index of a country's level of industrialization. More than 80 percent is used to produce sulfuric acid. Although its principal use is in the manufacture of phosphate fertilizer, sulfuric acid has a large number of other applications as well. Sources include deposits of native sulfur associated with salt domes and volcanic areas, as well as common iron sulfides such as pyrite. In recent years an increasingly important source has been the sulfur removed from coal, oil, and natural gas in order to make these fuels less polluting.

Common salt, known by the mineral name *halite,* is another important and versatile resource. It is among the more prominent nonmetallic minerals used as a raw material in the chemical industry. In addition, large quantities are used to "soften" water and to keep streets and highways free of ice. Of course most people are aware that it is also a basic nutrient and a part of many food products.

Salt is a common evaporite, and thick deposits are exploited using conventional underground mining techniques (Figure 21.24). Subsurface deposits are also tapped using brine wells in which a pipe is introduced into a salt deposit and water is pumped down the pipe. The salt dissolved by the water is brought to the surface through a second pipe. In addition, seawater continues to serve as a source of salt as it has for centuries. The salt is harvested after the sun evaporates the water. Box 6.1 looks at this process more closely.

REVIEW QUESTIONS

1. Contrast renewable and nonrenewable resources. Give one or more examples of each.
2. How is bituminous coal different from lignite? How is anthracite different from bituminous?
3. More than 70 percent of present-day coal usage is for what purpose?
4. List three impacts on the atmospheric environment of burning fossil fuels.
5. What is an oil trap? List two conditions common to all oil traps.
6. List two drawbacks associated with the processing of tar sands recovered by surface mining.
7. Although the United States has huge oil shale deposits, shale oil is not produced commercially. Explain.
8. What is the primary fuel for nuclear fission reactors?
9. List two obstacles that have hindered the development of nuclear power as a major energy source.
10. Briefly describe two methods by which solar energy might be used to produce electricity.
11. Explain why dams built to provide hydroelectricity do not last indefinitely.
12. Is geothermal power considered an inexhaustible energy source? Explain.
13. What advantages does tidal power production offer? Is it likely that tides will ever provide a significant proportion of the world's electrical energy requirements?
14. Contrast *resource* and *reserve.*
15. What might cause a mineral deposit that had not been considered an ore to be reclassified as an ore?
16. List two general types of hydrothermal deposits.
17. Metamorphic ore deposits are often related to igneous processes. Provide an example.
18. Name the primary ore of aluminum and describe its formation.
19. A rusty-colored zone of iron oxide at the surface may indicate the presence of a copper deposit at depth. Briefly explain.
20. Briefly describe the way in which minerals accumulate in placers. List four minerals that are mined from such deposits.
21. Which is greater, the per capita consumption of metallic or nonmetallic mineral resources?
22. Nonmetallic resources are commonly divided into two broad groups. List the two groups and some examples of materials that belong to each. Which group is most widely distributed?

KEY TERMS

cap rock (p. 554)
disseminated deposit (p. 571)
fossil fuel (p. 551)
geothermal energy (p. 563)
hydroelectric power (p. 562)
hydrothermal solution (p. 570)
mineral resource (p. 566)
nonmetallic mineral resource (p. 574)
nonrenewable resource (p. 549)
nuclear fission (p. 559)
oil trap (p. 554)
ore (p. 567)
pegmatite (p. 569)
placer (p. 574)
renewable resource (p. 548)
reserve (p. 566)
reservoir rock (p. 554)
secondary enrichment (p. 573)
vein deposit (p. 570)

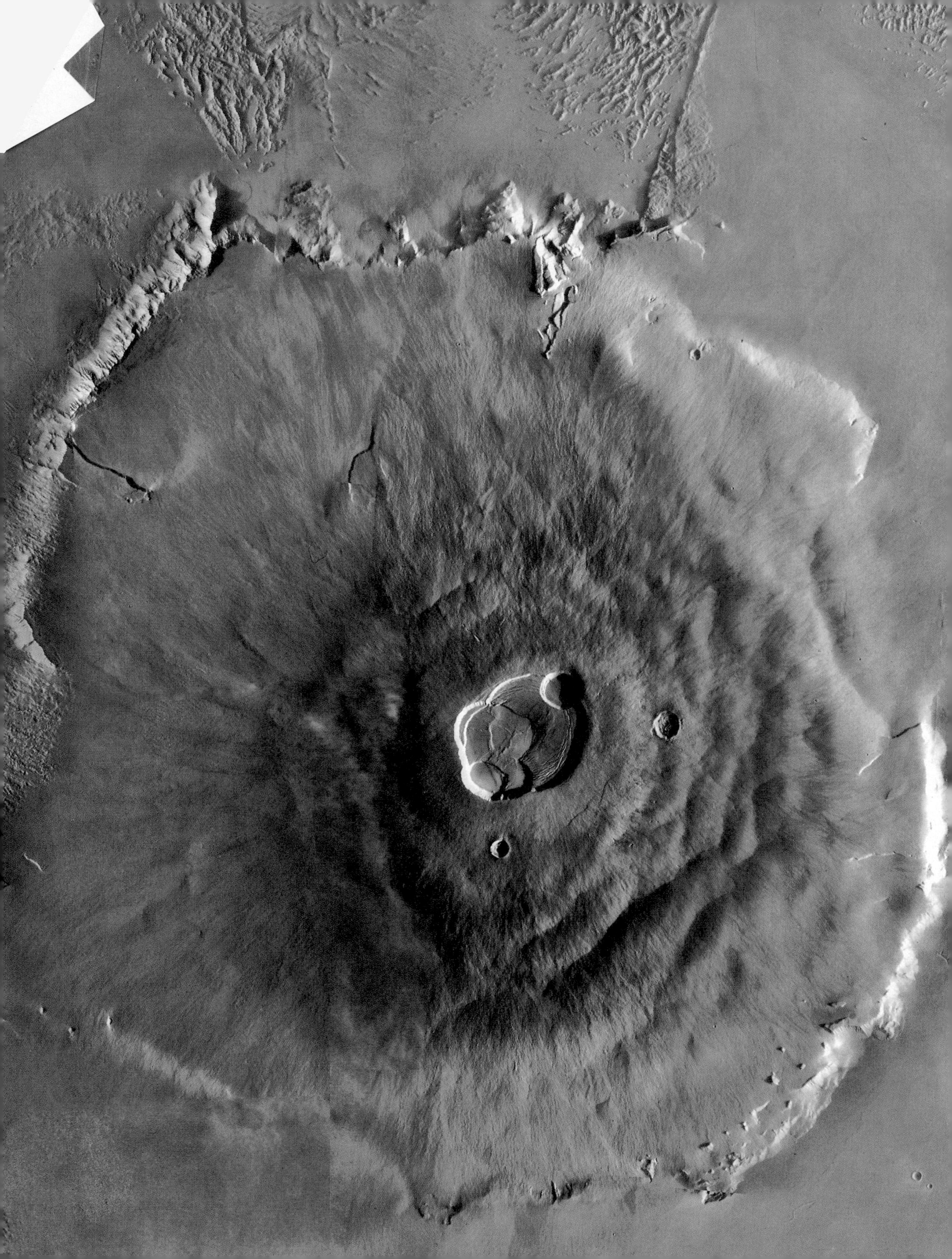

22
Planetary Geology

Opposite: Image of Mons Olympus, an inactive shield volcano on Mars that covers an area about the size of the state of Ohio. (Photo courtesy of the U.S. Geological Survey) (Top photo courtesy of NASA)

The sun is the hub of a huge rotating system consisting of nine planets, their satellites, and numerous small but nevertheless interesting bodies, including asteroids, comets, and meteoroids. An estimated 99.85 percent of the mass of the solar system is contained within the sun, while the planets collectively make up most of the remaining 0.15 percent. The planets, in order from the sun, are Mercury, Venus, Earth, Mars, Jupiter, Saturn, Uranus, Neptune, and Pluto. Under the control of the sun's gravitational force, each planet maintains an almost circular orbit and travels in a counterclockwise direction around the sun. Further, the orbits of all the planets lie within 3 degrees of the plane of the sun's equator, except for those of Mercury and Pluto, which are inclined 7 and 17 degrees, respectively.

When people first came to recognize that the planets are "worlds" much like Earth, a great deal of interest was generated. A primary concern has always been the possibility of intelligent life existing elsewhere in the universe. This expectation has not as yet come to pass. Nevertheless, since all of the planets most probably formed from the same primordial cloud of dust and gases, they should provide valuable information concerning Earth's history. Recent space explorations have been organized with this goal in mind. To date, Mercury, Venus, Mars, Jupiter, Saturn, Uranus, Neptune, and the moon have been explored by space probes (Figure 22.1).

THE PLANETS: AN OVERVIEW

Careful examination of Table 22.1 will show that the planets fall quite nicely into two groups: the **terrestrial** (Earthlike) **planets** of Mercury, Venus, Earth, and Mars; and the **Jovian** (Jupiterlike) **planets** of Jupiter, Saturn, Uranus, and Neptune. Pluto is not included in either category. Pluto's position at the far edge of the solar system and its small size make this planet's true nature a mystery. The most obvious difference between these groups is their size. The largest terres-

FIGURE 22.1
Painting of *Voyager 2* as it might have appeared when it encountered Uranus on January 24, 1986. (By Don Davis, courtesy of NASA)

TABLE 22.1
Planetary data.

Planet	Symbol	Mean Distance from Sun			Period of Revolution	Inclination to Ecliptic	Orbital Velocity	
		AU	Millions of miles	Millions of kilometers			mi/s	km/s
Mercury	☿	0.387	36	58	88^d	7° 00′	29.5	47.9
Venus	♀	0.723	67	108	225^d	3° 24′	21.8	35.0
Earth	⊕	1.000	93	150	365.25^d	0° 00′	18.5	29.8
Mars	♂	1.524	142	228	687^d	1° 51′	14.9	24.1
Jupiter	♃	5.203	483	778	12^{yr}	1° 19′	8.1	13.1
Saturn	♄	9.539	886	1427	29.5^{yr}	2° 30′	6.0	9.6
Uranus	⛢	19.180	1780	2869	84^{yr}	0° 46′	4.2	6.8
Neptune	♆	30.060	2790	4498	165^{yr}	1° 46′	3.3	5.4
Pluto	♇	39.440	3670	5900	248^{yr}	17° 12′	2.9	4.7

Planet	Period of Rotation	Diameter		Relative Mass (Earth = 1)	Average Density (g/cm^3)	Polar Flattening (%)	Eccentricity	Number of Known Satellites
		miles	kilometers					
Mercury	59^d	3015	4878	0.056	5.1	0.0	0.206	0
Venus	243^d	7526	12,112	0.82	5.3	0.0	0.007	0
Earth	$23^h56^m04^s$	7920	12,742	1.00	5.52	0.3	0.017	1
Mars	$24^h37^m23^s$	4216	6800	0.108	3.94	0.5	0.093	2
Jupiter	$\sim 9^h50^m$	88,700	143,000	318.000	1.34	6.5	0.048	16
Saturn	$\sim 10^h25^m$	75,000	121,000	95.200	0.70	10.5	0.056	18
Uranus	$\sim 17^h20^m$	29,000	47,000	14.600	1.55	7.0	0.047	15
Neptune	$\sim 16^h$	28,900	45,000	17.300	1.64	2.5	0.008	8
Pluto	6.4^d	~1500	~2400	~0.01(?)	~1.5(?)	?	0.250	1

trial planet (Earth) has a diameter only one-quarter as great as the diameter of the smallest Jovian planet (Neptune), and its mass is only one-seventeenth as great. Hence, the Jovian planets are often called *giants.* Also, because of their relative locations, the four Jovian planets are referred to as the *outer planets,* while the terrestrial planets are called the *inner planets.* As we shall see, there appears to be a correlation between the locations of these planets and their sizes.

Other dimensions along which the two groups markedly differ include density, composition, and rate of rotation. The densities of the terrestrial planets average about 5 times the density of water, whereas the Jovian planets have densities that average only 1.5 times that of water. One of the outer planets, Saturn, has a density of only 0.7 gram per cubic centimeter. The variations in the compositions of the planets are largely responsible for these differences.

The substances of which both groups of planets are composed can be divided into three groups based upon their melting points. They are called *gases, rocks,* and *ices.* The gases are those materials with melting points near absolute zero,* −273°C, and consist of hydrogen and helium. The rocky materials are made principally of silicate minerals and metallic iron, which have melting points exceeding 700°C. The ices have intermediate melting points and include ammonia (NH_3), methane (CH_4), carbon dioxide (CO_2), and water (H_2O).

The terrestrial planets are composed mostly of dense rocky and metallic material with minor amounts of gases. The Jovian planets, on the other hand, contain a large percentage of hydrogen and helium, with varying amounts of ices (mostly water, ammonia, and methane), which accounts for their low densities. The outer planets are also thought to contain as much rocky and metallic material as the terrestrial planets, and this material may be concentrated in a small central core.

The Jovian planets have very thick atmospheres

*Absolute zero is the lowest possible temperature.

consisting of varying amounts of hydrogen, helium, methane, and ammonia. By comparison, the terrestrial planets have meager atmospheres at best. Planets lose gases from their atmospheres by the process of *evaporation.* Evaporation occurs more rapidly for planets with higher atmospheric temperatures and lower surface gravities, and more slowly for those with lower atmospheric temperatures and higher surface gravities. Thus, a planet's ability to retain an atmosphere depends on its atmospheric temperature and surface gravity. Simply stated, a gas molecule can evaporate from a planet if it reaches a speed known as the *escape velocity.* For Earth, this velocity is 11 kilometers (7 miles) per second. Any material, including a rocket, must reach this speed before it can leave Earth and go into space. The Jovian planets, because of their greater surface gravities, have higher escape velocities than the terrestrial planets. Consequently, it is more difficult for gases to evaporate from them. Also, because the molecular motion of a gas is temperature dependent, at the low temperatures of the Jovian planets even the lightest gases are unlikely to acquire the speed needed to escape. On the other hand, a comparatively warm body with a small surface gravity, like our moon, is unable to hold even the heaviest gas and thus lacks an atmosphere. The slightly larger terrestrial planets of Earth, Venus, and Mars retain some heavy gases (as compared to hydrogen), but even their atmospheres make up only an infinitesimally small portion of their total mass.

It is hypothesized that the primordial cloud of dust and gas from which all the planets are thought to have condensed had a composition somewhat similar to that of Jupiter. However, unlike Jupiter, the terrestrial planets are nearly void of light gases and ices. Were the terrestrial planets once much larger? Did they contain these materials but lose them because of their close proximity to the sun? In the following section we will consider the evolutionary histories of these two diverse groups of planets in an attempt to answer these questions.

ORIGIN AND EVOLUTION OF THE PLANETS

The orderly revolution of all nine planets along the sun's equatorial plane leads most astronomers to conclude that the planets formed at essentially the same time and from the same primordial material as the sun. This **nebular hypothesis** suggests that all bodies of the solar system formed from an enormous nebular cloud consisting of approximately 80 percent hydrogen, 15 percent helium, and a few percent of all the other heavier elements known to exist (Figure 22.2). The heavier substances in this frigid cloud of dust and gases consisted mostly of elements such as silicon, aluminum, iron, and calcium—the substances of the common rocky materials. Also prevalent were the other familiar elements, including oxygen, carbon, and nitrogen. Astronomical studies indicate that these latter substances existed in a variety of organic compounds. Thus, the stuff out of which life has formed was present long before the solar system came to be.

About 5 billion years ago, and for reasons not yet fully understood, this huge cloud of minute rocky fragments and gases began to contract under its own gravitational influence. The contracting mass of material is assumed to have had some component of rotational motion. As this rotating cloud gravitationally contracted, it rotated faster and faster for the same reason an ice skater does when the arms are drawn toward the body. This rotation caused the nebular cloud to assume a disk-like shape. Within this rotating disk, relatively small, eddy-like contractions formed the nuclei from which the planets would eventually develop. However, the greatest concentration of material was gravitationally pulled toward the center, forming the *protosun.*

As more and more of this gas plunged inward, the temperature of the central mass continued to increase. The nebular material located near the protosun reached temperatures of several thousand degrees and was completely vaporized. However, at distances beyond the orbit of Mars, the temperatures probably always remained very low. Here, at −200°C, the dust fragments were most likely covered with a thick layer of water ice, and ices of carbon dioxide, ammonia, and methane. The disk-shaped cloud also contained appreciable amounts of the lighter gases, namely, hydrogen and helium, which had not been consumed by the protosun.

In a relatively short time after the protosun formed, the temperature in the inner portion of the nebula dropped significantly. This temperature de-

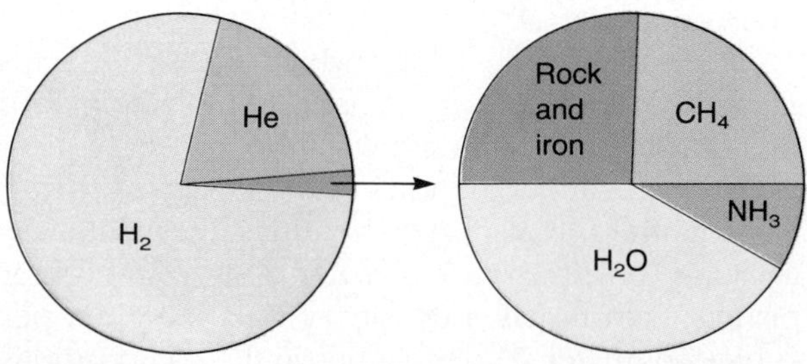

FIGURE 22.2
Composition of the primordial cloud of dust and gases from which the solar system is thought to have evolved. (Data from W.B. Hubbard)

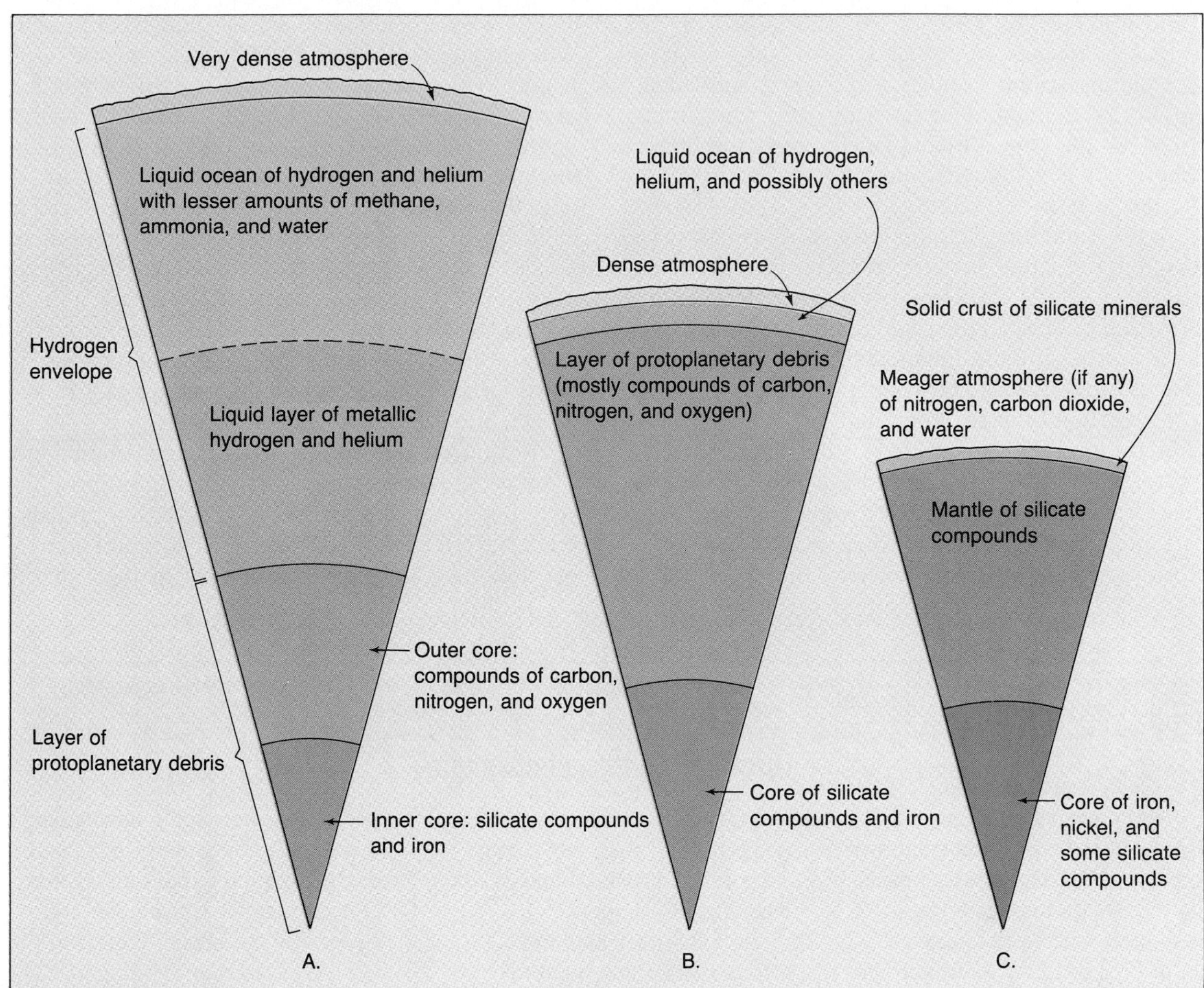

FIGURE 22.3
Idealized models for the internal structure of the Jovian and terrestrial planets. **A.** Jupiter and Saturn. **B.** Uranus and Neptune. **C.** Terrestrial planets. (Data from W.B. Hubbard et al.)

crease caused those substances with high melting points to condense into perhaps sand-sized particles. Materials such as iron and nickel solidified first. Next to condense were the elements of which the rock-forming minerals are composed. As these fragments collided, they joined into larger, asteroid-sized objects which in a few tens of millions of years accreted into the four inner planets we call Mercury, Venus, Earth, and Mars. As more and more of the nebular debris was swept up by these *protoplanets,* the inner solar system began to clear, allowing sunlight to heat the planets' surfaces. Due to their relatively high temperatures and weak gravitational fields, the inner planets were unable to accumulate an appreciable amount of the lighter components of this nebular cloud. These materials, namely, hydrogen, ammonia, methane, and water, were eventually whisked from the inner solar system by the solar winds.

Shortly after the four terrestrial planets formed, the decay of radioactive isotopes within them plus the heat from the colliding particles produced at least some melting of the planets' interiors. Melting, in turn, caused the heavier elements, principally iron and nickel, to sink, while the lighter silicate minerals floated upward. During this period of chemical differentiation, gaseous materials were allowed to escape from the planets' interiors, much like what happens during a volcanic event on Earth. The hottest and second smallest planet, Mercury, was unable to retain even the heaviest of these gases. Mars, on the other hand, being slightly larger and cooler than Mercury, retained a thin layer of carbon dioxide and some water in the form of ice. The largest of the terrestrial planets, Venus and Earth, have surface gravitations strong enough to retain a substantial amount of the heavier gases. However, when compared to the four Jovian planets, even the atmospheres of these planets must be looked upon as meager at best.

At the same time that the terrestrial planets were forming, the larger Jovian planets, along with their extensive satellite systems, were also developing. However, because of the frigid temperatures existing far from the sun, the fragments out of which these planets formed contained a high percentage of ices—water, carbon dioxide, ammonia, and methane. Perhaps by random chance, two of the outer planets, Jupiter and Saturn, grew many times larger (by mass) than Uranus and Neptune. For comparison, Jupiter is 318 and Saturn 95 times more massive than the earth. However, Uranus and Neptune have masses about 14 and 17 times greater, respectively, than that of Earth. When Jupiter and Saturn reached a certain size, estimated to be about 10 Earth masses, their surface gravitation was sufficient to attract and hold even the lightest materials—hydrogen and helium. It is thought that these gases gravitationally collapsed onto these large protoplanets as they swept through their region of the solar system. Thus, much of their size is attributable to the large envelope of light elements, which exists as a dense liquid below a thick hydrogen-rich atmosphere. Jupiter and Saturn therefore consist of a central core of ices and rock, and a much larger outer envelope containing mostly hydrogen and helium (Figure 22.3A).

By contrast, the smaller Jovian planets, Uranus and Neptune, grew more slowly and contain proportionately much smaller amounts of hydrogen and helium. Nevertheless, hydrogen, methane, and ammonia are still the major constituents of their dense

BOX 22.1

Origin of Earth's Atmosphere

Earth's atmosphere is unlike that of any other body in the solar system. No other planet is as hospitable or exhibits the same life-sustaining mixture of gases as Earth (Figure 22.A). However, the atmosphere did not always consist of the same relatively stable mixture of gases that we breathe today. On the contrary, the present mixture of gases that makes up our atmosphere is the result of very gradual change, a slow evolutionary process that began soon after Earth came into being 4.5 to 5 billion years ago.

Scientists believe that Earth's earliest atmosphere was swept away by solar winds, vast streams of particles emitted by the sun. Earth cooled, a solid crust formed and the gases that had been dissolved in the molten rock were gradually released, a process called *degassing.* Thus, an atmosphere believed to be made up of gases similar to those released during volcanic eruptions came into being. The principal components of this "new" atmosphere were probably water vapor, carbon dioxide, and nitrogen.

As Earth continued to cool, clouds formed and great rains commenced. At first, the water evaporated before reaching the surface or was quickly boiled away. This step helped to speed the cooling of Earth's surface. When Earth had cooled sufficiently, the torrential rains continued, filling the ocean basins. Not only did this event diminish the amount of water vapor in the air, but it also carried away much carbon dioxide.

We are now faced with an interesting paradox. If Earth's primitive atmosphere resulted from volcanic degassing, it could not have contained free oxygen because free oxygen is not emitted during this process. How, then, did our present oxygen-rich atmosphere come into existence? Scientists have proposed two probable sources of the free oxygen in the atmosphere. It is known that water vapor that is carried into the upper atmosphere is dissociated into hydrogen and oxygen by the action of the sun's ultraviolet radiation. Hydrogen, being a very light gas, escapes the atmosphere, whereas the heavier oxygen atoms remain and combine to form molecular oxygen (O_2). Although there is little doubt that some of the atmosphere's free oxygen was created in this manner, this very slow process is not adequate to account for the present percentage of oxygen in our atmosphere.

A second, more important source of oxygen is believed to have been (and, indeed, continues to be) green plants. By the process of *photosynthesis,* plant life uses sunlight to generate oxygen by changing water and atmospheric carbon dioxide into organic matter. This method of oxygen production obviously implies the need for life on Earth prior to the

atmospheres, and perhaps a thin outer ocean of hydrogen exists on these planets as well. Thus, Uranus and Neptune are proposed to have a small rocky-iron core and a large mantle of water, ammonia, and methane surrounded by a thin ocean of liquid hydrogen (Figure 22.3B). Consequently, these planets structurally resemble Jupiter and Saturn without their large hydrogen-helium envelopes.

In many respects the development of the outer planets with their large satellite systems roughly parallels the events that formed the solar system as a whole. Like their parents, the satellites of the outer planets are composed primarily of icy materials with lesser amounts of rocky substances. However, because of their small size, they could not contain appreciable amounts of hydrogen and helium.

In the remainder of this chapter, we will consider each planet in more detail, as well as some minor members of the solar system. First, however, a discussion of our moon, Earth's companion in space, is appropriate.

THE MOON

Only one natural satellite, the moon, accompanies Earth on its annual flight around the sun. This planet-satellite system is unique in the solar system, because the moon is unusually large compared to its parent planet. The diameter of our moon is 3475 kilometers, and from the calculation of its mass, its density is 3.3 times that of water. This density is comparable to that of crustal rocks on Earth but a fair amount less than Earth's average density. Geologists have suggested that this difference can be accounted for if the moon's iron core is rather small. The gravitational attraction at the lunar surface is one-sixth of that experienced on Earth's surface. This difference allows

time that free oxygen was present in the atmosphere. Scientists believe that the first life forms, probably bacteria, carried out their metabolic processes without oxygen. Even today many of these anaerobic bacteria still exist. Next, primitive green plants evolved, which, in turn, supplied most of the free oxygen to support higher forms of life. Slowly the amount of oxygen in the atmosphere increased. Evidence from the geologic record of this ancient time suggests that the first free oxygen combined with substances dissolved in water, especially iron. Then, once these mineral oxidation needs were met, substantial quantities of free oxygen began to accumulate in the atmosphere. The fossil record reveals that by the beginning of the Paleozoic era, some 570 million years ago, organisms that require oxygen were abundant in the sea. Two hundred million years later, during the Devonian period, land plants became widespread. Thus, the makeup of the atmosphere is directly related to the life forms on Earth, and its composition evolved through time from an oxygen-free environment to one that contained significant amounts of free oxygen.

FIGURE 22.A
Like the atmospheres of other planets in our solar system, the atmosphere of Mars is very different from that of Earth. The Martian atmosphere is only 1 percent as dense as that of Earth and is composed primarily of carbon dioxide with very small amounts of water vapor. The red color of the Martian surface and the salmon color of the sky are believed to be accurately reproduced. The color of the sky results from dust particles suspended in the atmosphere. (Photo courtesy of NASA)

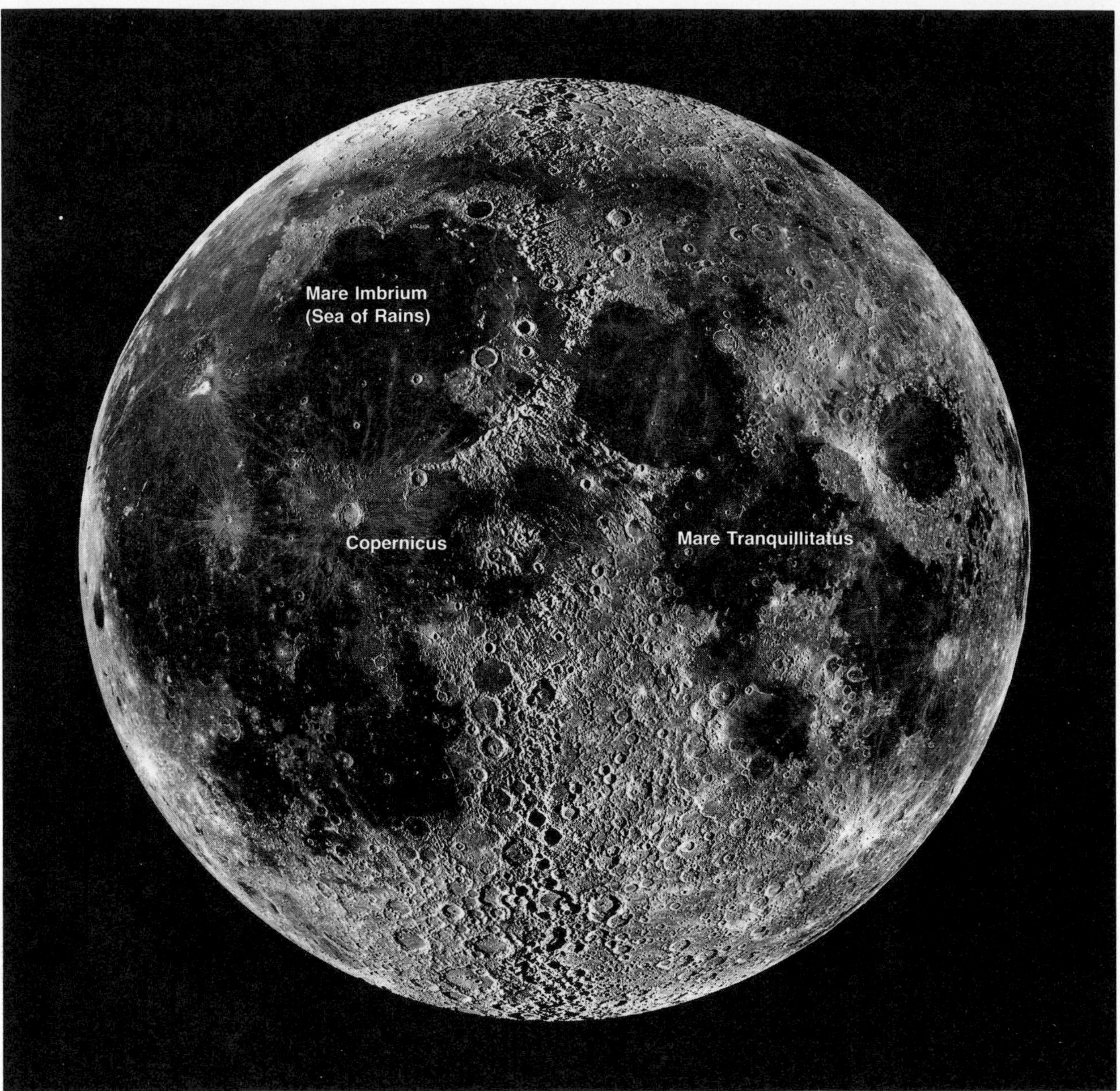

FIGURE 22.4
Telescopic view of the lunar surface. (Courtesy of Lick Observatory)

an astronaut to lift a "heavy" life-support system with relative ease. If it were not necessary to carry such a load, an astronaut could jump six times higher than on Earth.

The Lunar Surface

When Galileo first pointed his telescope toward the moon, he saw two different types of terrain (Figure 22.4). The dark areas he observed are now known to be fairly smooth lowlands, while the bright regions are densely cratered highlands. Because the dark regions resembled seas on Earth, they were later named **maria** (singular, *mare,* Latin for "sea"). This name is unfortunate because the moon's surface is totally void of water.

Today we know that the moon has no atmosphere and lacks water as well. Therefore, the processes of weathering and erosion which continually modify Earth are virtually lacking. In addition, tectonic events such as earthquakes and volcanic eruptions do not occur on the moon. However, since the moon is unprotected by an atmosphere, tiny particles (micrometeorites) continually bombard its surface and ever so gradually smooth the landscape. Rocks, for example, can become slightly rounded on top if ex-

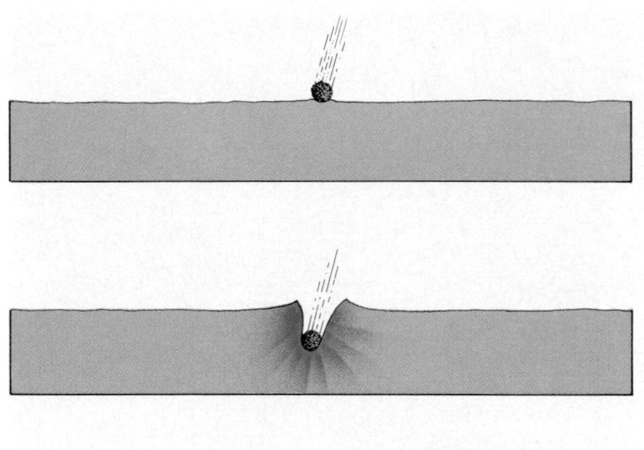

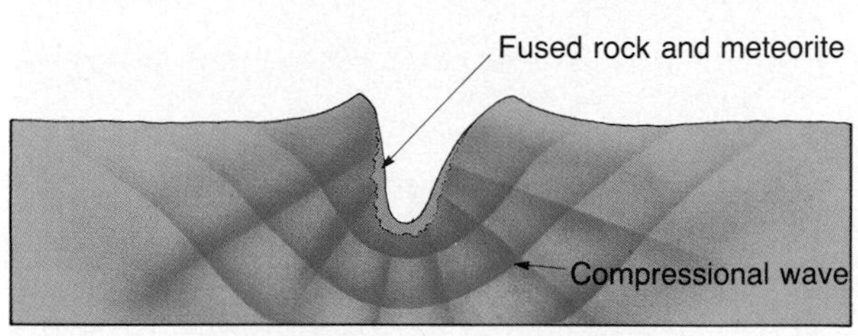

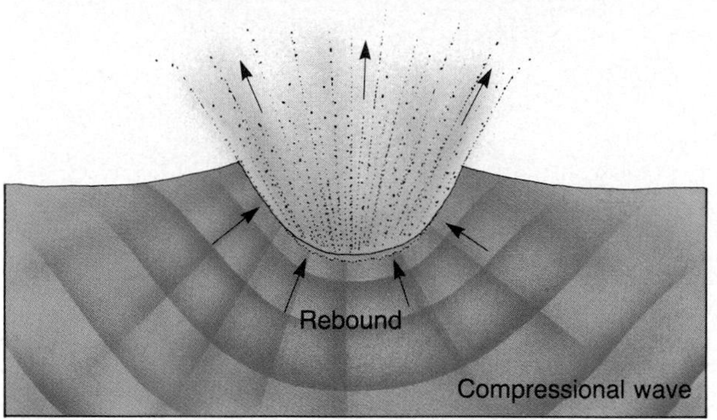

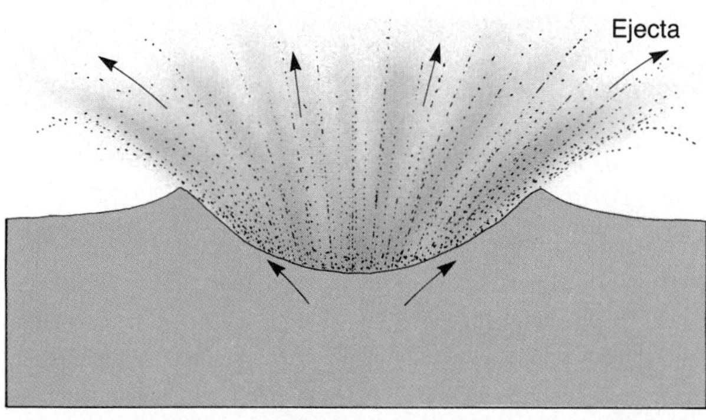

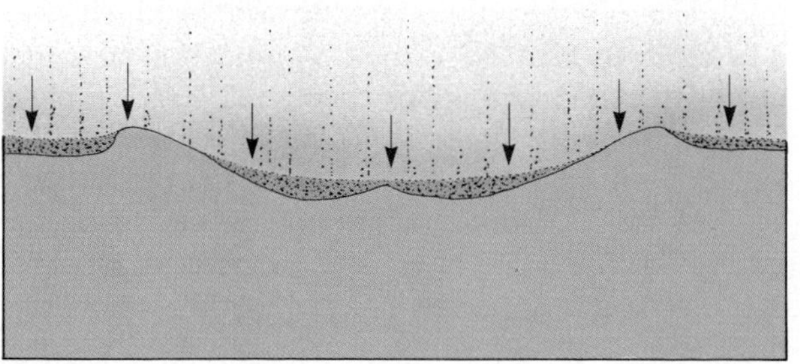

FIGURE 22.5
Formation of an impact crater. The energy of the rapidly moving meteoroid is transferred into heat energy and compressional waves. The rebound of the compressed rock causes debris to be ejected from the crater, and the heat melts some material, producing glass beads. Small secondary craters are formed by the material "splashed" from the impact crater. (After E. M. Shoemaker)

posed at the lunar surface for a long enough period. Nevertheless, it is unlikely, except for the addition of a few large craters, that the moon has changed appreciably in the last 3 billion years.

The most obvious features of the lunar surface are craters. They are so profuse that craters within craters within craters are the rule. The larger ones seen in the lower portion of Figure 22.4 are about 250 kilometers (150 miles) in diameter, and these often overlap. Most craters were produced by the impact of rapidly moving debris (meteoroids), which was considerably more abundant in the early history of the solar system than it is today. By contrast, Earth has only about a dozen easily recognized impact craters. This difference can be attributed to Earth's atmosphere, which burns up small debris before it reaches the ground. In addition, evidence for many of the craters that did form early in Earth's history has since been destroyed by erosion or tectonic processes.

The formation of an impact crater is illustrated in Figure 22.5. Upon impact, the high-speed particle compresses the material it strikes; then almost instantaneously the compressed rock rebounds, ejecting material from the crater. This process is analogous to the splash that occurs when a rock is dropped into water, and it often results in the formation of a central peak as seen in the large crater in Figure 22.6. Most of the ejected material *(ejecta)* lands near the crater, building a rim around it. The heat generated by the impact is sufficient to melt some of the impacted rock. Astronauts have brought back samples of glass beads produced in this manner, as well as rock formed when angular fragments and dust were welded together by the impact. The latter material is called **lunar breccia**.

A meteoroid only 3 meters wide can blast out a 150-meter (500-foot) crater. A few of the large craters such as Copernicus, shown in Figure 22.4, formed from the impact of bodies a kilometer, or more, in diameter. This large crater is thought to be relatively young because of the bright **rays** ("splash" marks) that radiate outward for hundreds of kilometers. These bright rays consist of fine debris ejected from

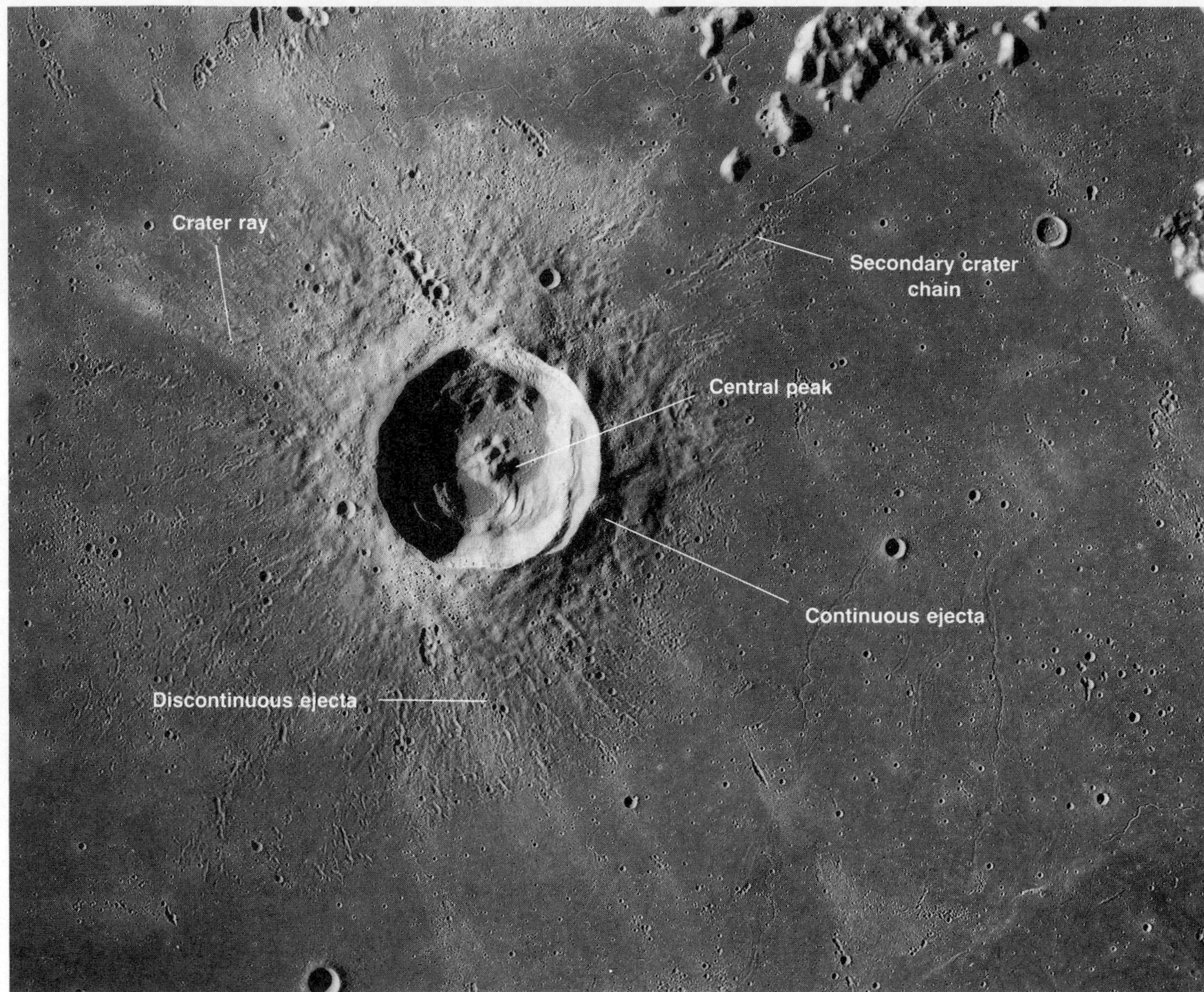

FIGURE 22.6
The 20-kilometer-wide lunar crater Euler located in the southwestern Mare Imbrium. Clearly visible are the bright rays, central peak, and secondary craters, and the large accumulation of ejecta near the crater rim. (Photo courtesy of NASA)

the primary crater, including impact-generated glass beads, as well as material displaced during the formation of smaller, secondary craters.

The densely pock-marked highland areas make up most of the lunar surface. In fact, all of the "back" side of the moon is characterized by such topography. Within the highland regions are mountain ranges that have been named for mountainous terrains on Earth. The highest lunar peaks reach elevations approaching 8 kilometers, only 1 kilometer less than Mount Everest.

Although highlands predominate, the less rugged maria have attracted most of the interest. The origin of maria basins as enormous impact craters produced by the violent impact of at least a dozen asteroid-sized bodies was hypothesized before the turn of the century by the noted American geologist G. K. Gilbert (Figure 22.7A). However, it remained for the *Apollo* missions to determine what filled these depressions to produce the relatively flat topography. Apparently, the craters were flooded, layer upon layer, by very fluid basaltic lava, in which case they somewhat resemble the Columbia Plateau in the northwestern United States (Figure 22.7B). The photograph in Figure 22.8 reveals the forward edge of a lava flow that is "frozen" in place. Astronauts have also viewed and photographed the layered nature of maria. The layers are often over 30 meters thick, and the total thickness of the material that fills the maria must approach thousands of meters.

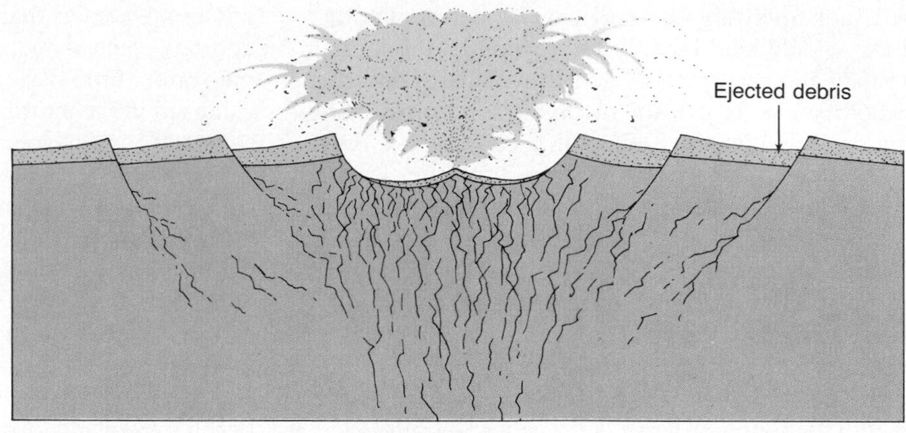

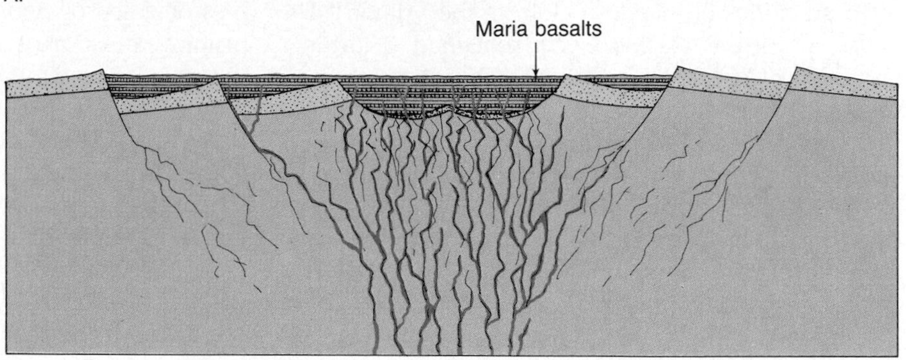

FIGURE 22.7
Formation of lunar maria. **A.** Impact of an asteroid-sized mass produced a huge crater hundreds of kilometers in diameter and disturbed the lunar crust beyond the crater. **B.** Filling of the impact area with fluid basalts, perhaps derived from partial melting deep within the lunar mantle.

On several occasions the lava flowed beyond the impact crater, engulfing the surrounding lowlands. If the rim of a remnant crater can be seen above the lava, an estimate of the flow's thickness can be made. Many geologists believe that the impacts that produced the maria basins were great enough to fracture the lunar crust some distance away. Examples of basins in which all of the disturbed regions were filled to overflowing include Mare Tranquillitatis (Sea of Tranquility), the site where astronauts first placed footprints on lunar soil, and Mare Imbrium (Sea of Rains). Some maria basalts fill only the central crater; these appear as dark, smooth crater floors in Figure 22.4.

All lunar terrains are mantled with a layer of gray, unconsolidated debris derived from a few billion years of meteoric bombardment (Figure 22.9). This soil-like layer, properly called **lunar regolith**, is composed of igneous rocks, breccia, glass beads, and fine particles commonly called *lunar dust.* As meteoroid after meteoroid collided with the lunar surface, the thickness of the regolith increased, while the size of the bombarded debris diminished. In the maria that have been explored by *Apollo* astronauts, the lunar regolith is apparently just over 3 meters thick, but it is believed to form a thicker mantle upon the older highlands.

Lunar History

Although the moon is our nearest planetary neighbor and astronauts have sampled its surface, much is

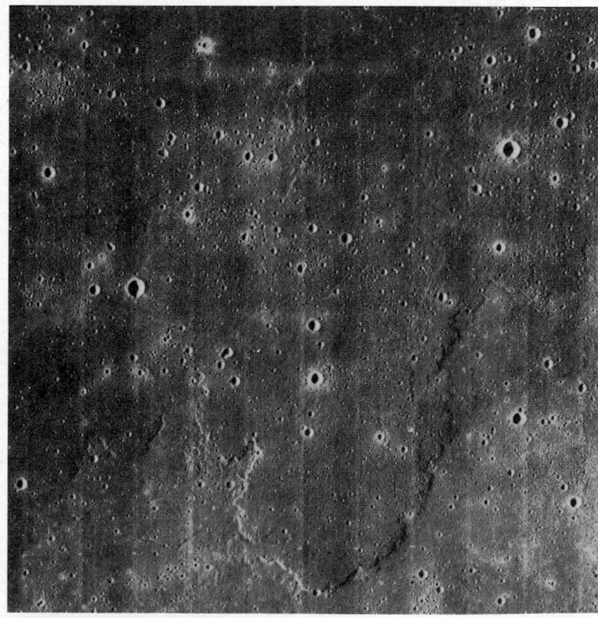

FIGURE 22.8
Margin of a lava flow on the surface of Mare Imbrium. (Photo courtesy of National Space Data Center)

still unknown about its origin. Until recently, the most widely held hypothesis argued that the formation of the moon paralleled that of Earth and the other planets. That is, the moon formed from minute rock fragments and gases which composed a disk-shaped structure that orbited the protosun. Debris from this disk collided and accumulated into larger masses which, in turn, accreted into planetary-sized bodies.

A new hypothesis, which has recently gained support from many scientists, suggests that a giant body collided with Earth to produce the moon. The resulting explosion due to the impact of a Mars-sized body with a semimolten Earth is thought to have ejected huge quantities of mantle rock from the primordial Earth. A portion of this ejecta remained in orbit around Earth, while the remainder either escaped or impacted upon Earth's surface. In a manner similar to that proposed in the earlier hypothesis, the material orbiting Earth then began to accumulate, eventually producing the moon. Though the giant-impact hypothesis provides a plausible mechanism for the moon's formation, many questions must be answered before this proposal can be considered viable.

Despite the fact that the origin of the moon is still debated, planetary geologists have been able to work out some of the basic details of the moon's history, using among other things variations in crater density (quantity per unit area). Simply stated, the higher the crater density, the longer the topographic feature has existed. During its early history the moon was continually impacted as it swept up debris from the solar nebula. This continuous bombardment and perhaps radioactive decay generated enough heat to melt the moon's outer shell, and quite possibly the rest of the moon as well. When a large percentage of the debris had been gathered, the outer layer of the moon began to cool and form a crystalline crust. From samples obtained by *Apollo* astronauts, the rocks of the primitive lunar crust are thought to be composed of a high percentage of a calcium-rich feldspar (anorthite). This feldspar mineral crystallized early and, because it was less dense than the remaining melt, floated to the top and formed a surface scum. While this process was taking place, iron and other heavy metals probably sank to form a small central core. Even after the crust had solidified, its surface was continually bombarded. Remnants of the original crust occupy the densely cratered highlands, which

FIGURE 22.9
Astronaut Harrison Schmitt sampling the lunar surface. Notice the footprints left in the lunar regolith. (Photo courtesy of NASA)

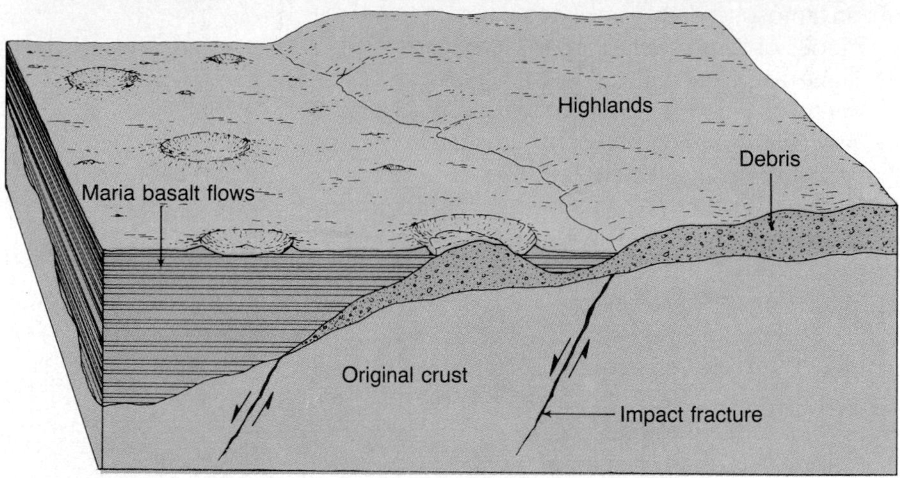

FIGURE 22.10
Mare Imbrium and associated mountainous region. A few isolated peaks are standing above the lava plain, and numerous craters dot the highlands and adjacent mare. (Photo courtesy of NASA)

have been estimated to be as much as 4.5 billion years old.

The last period of heavy bombardment recorded in the lunar highlands occurred almost 500 million years after the crust had formed. It is not known with certainty whether this final episode of bombardment was simply a clean-up phase where the remaining large particles in the Earth-moon orbit were swept up or whether it was an influx of bodies from farther out in the solar system.

The next major event in the moon's evolution was the formation of maria basins (Figure 22.10). The meteoroids that produced these huge pits ejected mountainous quantities of lunar rock into piles rising 5 kilometers or more. The Apennine mountain range, which typifies such an accumulation, was produced in conjunction with the formation of the Imbrium basin, the site of the exploration conducted by the *Apollo 15* astronauts. The crater density of the ejected material is greater than that of the surface of the associated mare, confirming that an appreciable time elapsed between the formation and filling of these basins. Radiometric dating of the maria basalts puts their age between 3.2 and 3.8 billion years, somewhat younger than the initial crust. In places, the lava flows overlap the highlands, another testimonial to the lesser age of the maria deposits.

The last prominent features to form on the lunar surface were the rayed craters, as exemplified by the crater Copernicus (see Figure 22.4). Material ejected from these "young" depressions is clearly seen blanketing the surface of the maria and many older rayless craters. By contrast, the older craters have rounded rims, and their rays have been erased by the impact of small debris. However, even a relatively young crater like Copernicus must be millions of years old. Had it formed on Earth, erosional forces would have long since removed it.

Evidence carried back from the lunar landings indicates that most, if not all, of the moon's tectonic activity ceased about 3 billion years ago. The youngest maria lava flows are almost as old as the oldest rocks so far discovered on Earth. If photos of the moon taken several hundreds of millions of years ago were available, they would reveal that the moon has changed little in the intervening years. By all standards of measure the moon is a dead body wandering lifelessly through space and time.

MERCURY: THE INNERMOST PLANET

Mercury, the innermost and swiftest planet, has a diameter of 4878 kilometers, not much larger than the moon. Also, like the moon, it absorbs most of the sunlight that strikes it, reflecting only 6 percent into space. This is characteristic of a terrestrial body without an atmosphere. By way of comparison one of the icy moons of Saturn reflects more than 60 percent of the light that it encounters.

Mercury's close proximity to the sun makes viewing from earthbound telescopes difficult at best. The first good glimpse of this planet came in the spring of 1974, when *Mariner 10* passed within 800 kilometers of its surface (Table 22.2). Its striking resemblance to the moon was immediately evident from the high-resolution images that were radioed back. In fact, the similarity was so great that a project scientist remarked that these photos could be substituted for ones of the back side of the moon and most laymen would not know the difference.

Mercury has densely cratered highlands much like the moon and vast smooth terrains which resemble maria. However, unlike the moon, Mercury is a very dense planet, which implies that this planet contains an iron core perhaps larger than Earth's. In addition, Mercury has very long scarps that cut across the plains and craters alike. One proposal is that these scarps resulted from crustal shortening as the planet cooled and shrank.

A typical period of daylight and darkness on Mercury lasts 88 days. This causes the night temperatures to drop as low as −173°C (−280°F) and the noontime temperatures to exceed 427°C (800°F), hot enough to melt tin and lead. Mercury has the greatest temperature extremes of any planet. The odds of life as we know it existing on Mercury are nil.

TABLE 22.2
Summary of significant space probes.

Name	Date	Mission
Mariner 2	1962	Fly-by of Venus (first to any planet)
Mariner 4, 6, 7	1965 1969 1969	Fly-by missions to Mars
Mariner 9	1971	Orbiter of Mars
Apollo 8	1968	Astronauts circled the moon and returned to Earth.
Apollo 11	1969	First astronaut landed on the moon.
Apollo 17	1972	Last of six *Apollo* missions to carry astronauts to the moon
Mariner 10	1974	Orbited the sun, allowing it to pass Mercury several times; fly-by mission to Venus
Pioneer 10, 11	1973 1974	First close-up views of Jupiter
Venera 8, 9, 10	1972 1975	Soviet landers on Venus (operative about one hour each)
Venera 13, 14	1982	First color images of Venus
Viking 1, 2	1976	Orbiters and landers of Mars
Voyager 1	1979 1980	Fly-by of Jupiter Fly-by of Saturn
Voyager 2	1979 1981 1986 1989	Fly-by of Jupiter Fly-by of Saturn Fly-by of Uranus Fly-by of Neptune
Galileo	1990 1992 1995	Fly-by of Earth and Venus Fly-by of Earth Orbiter of Jupiter
Magellan	1990	Radar-imaging orbiter of Venus

VENUS: THE VEILED PLANET

Venus, second only to the moon in brilliance in the night sky, is named for the goddess of love and beauty. It orbits the sun in a nearly perfect circle once every 225 days. Because Venus is similar to Earth in size, density, and mass, and because it is located in the same part of the solar system, it has been referred to as "Earth's twin." Because of these similarities, it is hoped a detailed study of Venus will provide geologists with a better understanding of Earth's evolutionary history.

Venus is shrouded in thick clouds impenetrable to visible light. Nevertheless, radar mapping by unmanned spacecraft, as well as by earthbound instruments, have revealed a varied topography with fea-

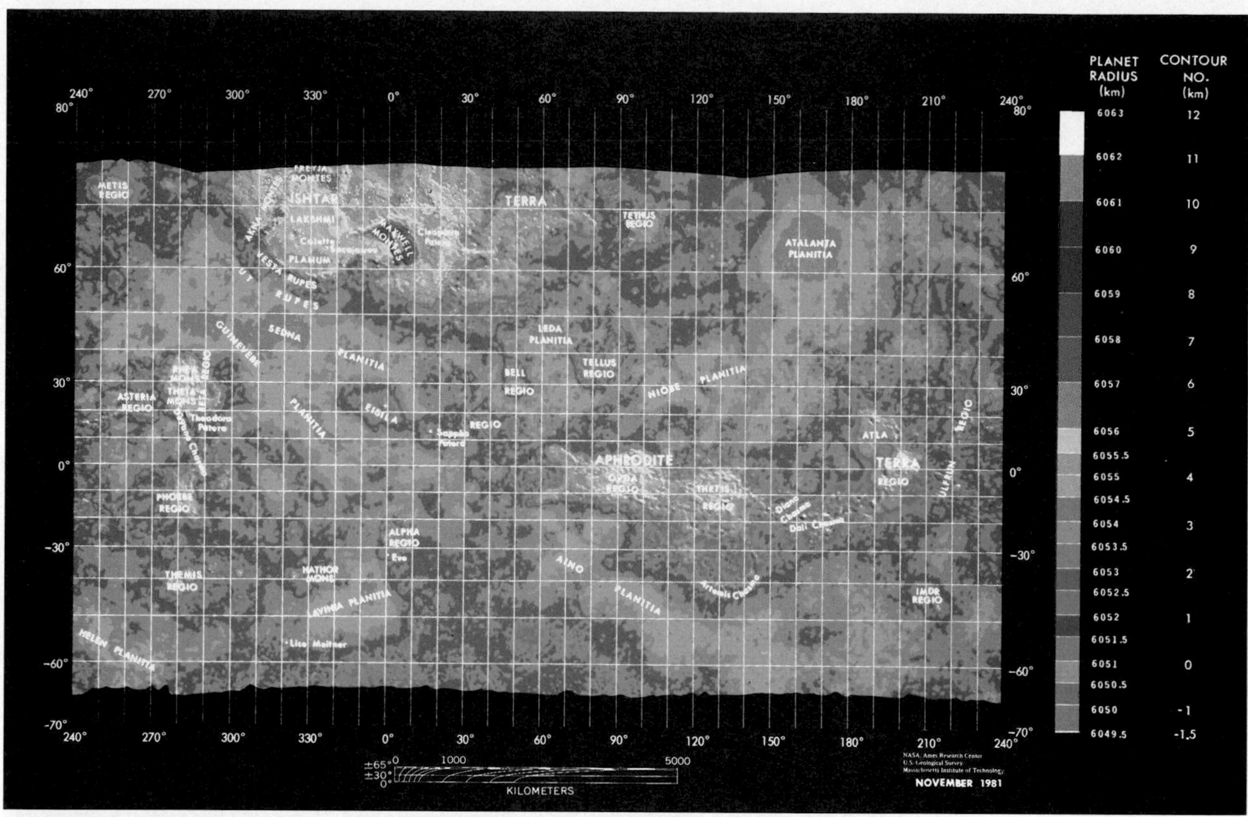

FIGURE 22.11
Map of the topography of Venus generated by computer and radar altimeter data from the *Pioneer/Venus* spacecraft. (Courtesy of U.S. Geological Survey)

tures somewhat between those of Earth and Mars (Figure 22.11). Simply, radar pulses in the microwave range are sent toward the Venusian surface, and the heights of features such as plateaus and mountains are measured by timing the return of the radar echo. Recently, *Magellan* scientists began receiving crisp radar images with a resolution that permits structures the size of Mount St. Helens to be clearly defined. These new data have confirmed earlier proposals that basaltic volcanism and tectonic deformation are the dominant processes operating on Venus. Further, based on the low density of impact craters, researchers concluded that volcanism and tectonic deformation were also very active during the recent geologic past (Figure 22.12).

Over 80 percent of the Venusian surface consists of relatively subdued plains that are mantled by volcanic flows. Some lava channels extend for tens to hundreds of kilometers and have narrow widths of from 1 to 2 kilometers. One of these lava channels meanders 6800 kilometers across the planet. Thousands of volcanic structures have already been identified. Most are small shield volcanoes a few hundred meters high with diameters of 2 to 8 kilometers. In addition, over 50 volcanoes with diameters in excess of 100 kilometers have also been mapped. One of these, Sapas Mons, is approximately 400 kilometers (249 miles) across and 1.5 kilometers (0.9 mile) high. Many of the flows from this volcano erupted from the flanks rather than the summit. This eruptive style is shared by large shield volcanoes on Earth, such as those found in Hawaii. Also like the Hawaiian volcanoes, these large structures appear to be associated with the upwelling of hot plumes within the planet's interior. Other volcanic structures discovered on Venus are circular, pancake-shaped domes about 25 kilometers in diameter and nearly 1 kilometer high (Figure 22.13). These domes are thought to be the result of outpouring of very viscous lava, much like volcanic domes on Earth.

Only 8 percent of the Venusian surface consists of highlands that may be likened to continental areas on Earth. The highlands of Venus are dominated by large plateau-like structures; however, discrete mountain ranges, including Maxwell Montes, which rises 11 kilometers above the lowlands, also exist.

Tectonic activity on Venus seems to be driven by upwelling and downwelling of material in the plan-

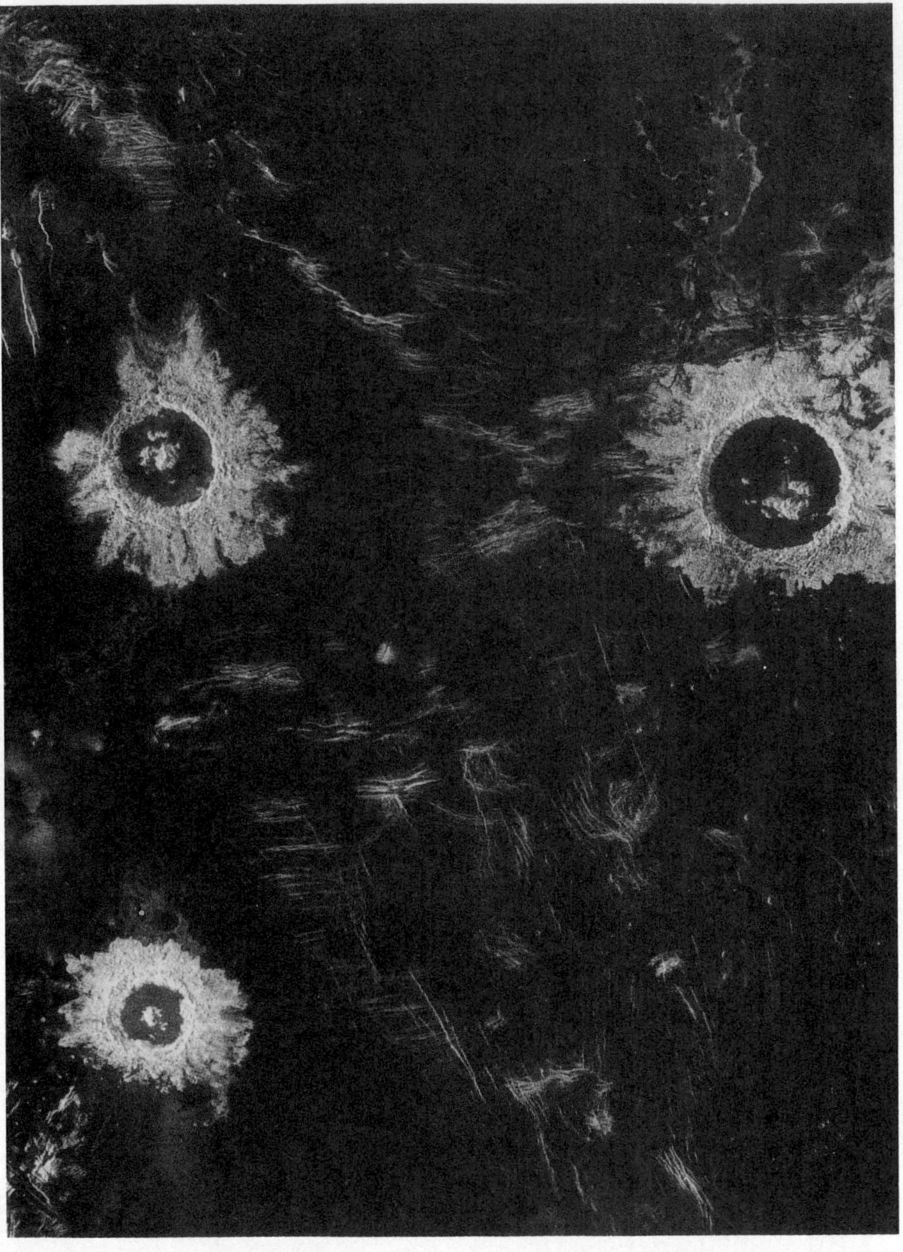

FIGURE 22.12
Three large meteorite impact craters, with diameters from 37 to 50 kilometers, located in the Lavinia region of Venus. (Photo courtesy of Jet Propulsion Laboratory)

et's interior. Parallel sets of faults or fractures, as shown in Figure 22.14, are widespread on the plains. However, no evidence for appreciable horizontal displacement, as occurs along the San Andreas fault, has yet to be detected. Some fracture zones are associated with numerous volcanic structures, suggesting that the region may have been uplifted and stretched as a result of convective upwelling. The greatest deformation on Venus is seen in the highlands which reach elevations in excess of 10 kilometers. One hypothesis proposes that the mountain areas formed where convective currents crumpled and thickened the crust. Although mantle convection still operates on Venus, the process of plate tectonics which recycles rigid lithosphere does not appear to have contributed to the present Venusian topography.

Before the advent of space vehicles, Venus and Mars were considered the most hospitable sites in the solar system (Earth excluded) and the logical places to expect living organisms. Unfortunately, evidence from *Mariner* fly-by space probes and Soviet unmanned landings on Venus indicate differently. The surface of Venus reaches temperatures of 475°C (900°F), and the Venusian atmosphere is composed mostly of carbon dioxide (97 percent). Only minor amounts of water vapor and nitrogen have been detected. Its atmosphere contains an opaque cloud deck about 25 kilometers thick which begins approx-

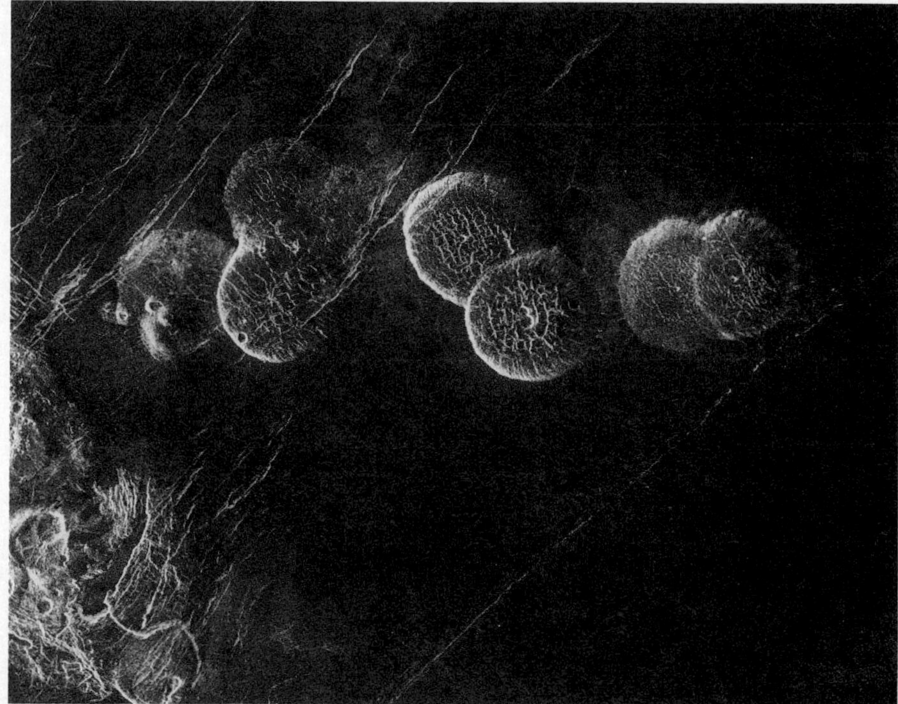

FIGURE 22.13
These domelike volcanic structures on Venus average 25 kilometers in diameter and are less than 1 kilometer high. They are interpreted as very thick lava flows. (Photo courtesy of Jet Propulsion Laboratory)

imately 70 kilometers from the surface. Although the unmanned Soviet *Venera 8* survived less than an hour on the Venusian surface, it determined that the atmospheric pressure on this planet is 90 times greater than that found on Earth's surface. This hostile environment makes it unlikely that life as we know it exists on Venus, and it makes manned space flights to Venus improbable in the foreseeable future.

FIGURE 22.14
Two sets of parallel fractures indicating tectonic activity on Venus unlike any seen before in the solar system. (Photo courtesy of Jet Propulsion Laboratory)

MARS: THE RED PLANET

Mars has evoked greater interest than any other planet, for astronomers as well as for laymen. When we think of intelligent life on other worlds, "little green Martians" come to mind. Interest in Mars stems mainly from this planet's accessibility to observation. All other planets within telescopic range have their surfaces hidden by clouds, except for Mercury, whose nearness to the sun makes viewing difficult. Through the telescope, Mars appears as a reddish ball interrupted by some permanent dark regions that change in intensity during the Martian year. The most prominent telescopic features of Mars are its brilliant white polar caps.

The dramatic photographs of Mars radioed back from the first successful fly-by probe, *Mariner 4,* revealed a crater-pocked planet more akin to the moon than to Earth. As it turned out, these images were taken of the highly cratered southern hemisphere which exhibits a topography sharply different from that of the "younger" northern hemisphere. Follow-

BOX 22.2

Venus and the Runaway Greenhouse Effect: A Lesson for Planet Earth?

Carl Sagan, one of the foremost experts on extraterrestrial life, in a description of planetary environments, called Earth "the Heaven of the solar system" and Venus "the Hell." Why should the environments of two planets that are nearly the same size and are located in close proximity to one another be so dramatically different? The primary reason is the blistering temperature of the Venusian atmosphere, which is caused by a runaway greenhouse effect. The greenhouse effect occurs when a planet's atmosphere acts like the glass in a greenhouse and allows visible solar energy to penetrate to the surface. Also, like the glass in the greenhouse, certain gases in the planet's atmosphere are quite opaque to outgoing radiation. Thus, as in a greenhouse, the surface temperatures of these planets are warmer than expected because the heat is temporarily trapped.

Since both Earth and Venus experience a greenhouse effect, why do the surface temperatures of Venus reach 475°C (900°F) when Earth is very hospitable? The answer is that carbon dioxide, which is a main contributor to the greenhouse effect, makes up 97 percent of the Venusian atmosphere (Figure 22.B). The primary source of carbon dioxide is outgassing during volcanic eruptions. Why does Earth, which has numerous active volcanoes, have an atmosphere that consists of only a small fraction of 1 percent carbon dioxide? The answer can be found in Earth's abundant plant life. Plants use carbon dioxide and water in photosynthesis to generate organic matter, while oxygen is released as a byproduct. Thus, over millions of years plant life has altered our atmosphere, making it carbon dioxide-poor and oxygen-rich. In addition, carbon dioxide dissolved in seawater is used by a vast number of organisms for the production of carbonate shells. These shells are eventually deposited as sediment on the ocean floor. Consequently, huge quantities of carbon dioxide from Earth's atmosphere are continually being converted into organic matter and carbonate sediments. Apparently, Venus does not possess living organisms or a chemical mechanism capable of removing atmospheric carbon dioxide. Hence, our neighboring planet has a carbon dioxide concentration that has reached extreme proportions.

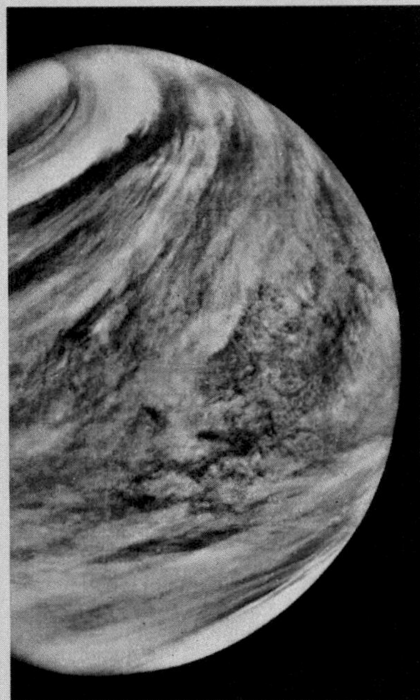

FIGURE 22.B
Venus is shrouded in a hot, cloud-filled atmosphere composed mainly of carbon dioxide. (Photo courtesy of NASA)

The discovery of a hostile Venusian atmosphere provides a good example of the practical benefits that humans might derive from space exploration. These data clearly reveal what can happen to a planet's environment when the greenhouse effect runs rampant. Some climatologists believe that a similar situation may be in the making on Earth. Through the burning of fossil fuels, we are converting increasingly larger amounts of oxygen and organic materials into carbon dioxide and water. For that reason, we are altering Earth's atmosphere by rapidly undoing what has taken plant life millions of years to accomplish.

We should expect that further studies of the Venusian atmosphere will provide insights into how Earth's atmosphere evolved and how it will respond to changes induced by human activity. Such insights could keep our planet from becoming, as some have suggested, "hothouse Earth."

up missions revealed a dynamic planet which at some time in its history experienced volcanic activity, wind erosion, and crustal movements with associated "marsquakes." Furthermore, *Viking* orbital images provided unmistakable evidence for erosion caused by flowing water.

The Martian atmosphere is only 1 percent as dense as that of Earth, much less than that found at the top of Mount Everest, and is composed primarily of carbon dioxide with only very small amounts of water vapor. Data radioed back to Earth confirmed speculations that the polar caps of Mars are made of

FIGURE 22.15
This spectacular picture of the Martian landscape by the *Viking 1* lander shows a dune field with features remarkably similar to many seen in the deserts of Earth. The dune crests indicate that recent wind storms were capable of moving sand over the dunes in the direction from lower right to upper left. The large boulder at the left is about 10 meters from the spacecraft and measures 1 by 3 meters. (Photos courtesy of NASA)

water ice covered by a relatively thin layer of frozen carbon dioxide. As the winter nears in a given hemisphere, we see the equatorward growth of that ice cap as additional carbon dioxide is deposited. This finding is compatible with the temperatures observed at the polar caps. Here temperatures reach a chilly −125°C, cold enough to soldify carbon dioxide.

Although the atmosphere of Mars is very thin, extensive dust storms do occur and may be responsible for the color changes observed from Earth-based telescopes. Winds up to 270 kilometers (170 miles) per hour can persist for weeks. Images radioed back by *Viking 1* revealed a Martian landscape remarkably similar to a rocky desert on Earth (Figure 22.15). Sand dunes are abundant, and many Martian impact craters have flat bottoms because they are partially filled with dust. Thus, unlike the craters of the moon, the oldest craters on Mars might be completely obscured by deposits of windblown material.

When *Mariner 9,* the first artificial satellite to orbit another planet, reached Mars in 1971, a large dust storm was raging, and only the ice caps were initially observable. It was spring in the southern hemisphere, and that polar cap was receding rapidly. About midsummer, when the rate of evaporation should have been greatest, the size of the polar cap showed little change. A residual cap remained through the summer, providing strong evidence that the residual polar cap is made of water ice. (Water ice has a much lower rate of evaporation than frozen carbon dioxide.)

When the dust cleared, images of the northern hemisphere revealed numerous large volcanoes. The largest, called Mons Olympus, covers an area the size of the state of Ohio and is no less than 23 kilometers (75,000 feet) in elevation, over twice as high as any mountain on Earth. This gigantic volcano and others that were observed closely resemble the shield volcanoes found on Earth, being similar to those of Hawaii (see chapter-opening photo). Their extreme size is thought to result from the absence of plate movements on Mars. Therefore, rather than a chain of volcanoes forming as we find in Hawaii, single, larger cones developed.

Impact craters are notably less abundant in the region where the volcanic activity appears most prevalent. This indicates that at least some of the volcanic topography formed more recently in that planet's history, somewhat after the early period of heavy bombardment. Nevertheless, age determinations based on crater densities obtained from *Viking* pho-

FIGURE 22.16
Valles Marineris. Shown here is a small portion of this vast canyon of Mars that extends for about 5000 kilometers, nearly the distance across the conterminous United States. Notice the huge landslides along the canyon wall that have acted to enlarge the structure. (Photo courtesy of U.S. Geological Survey)

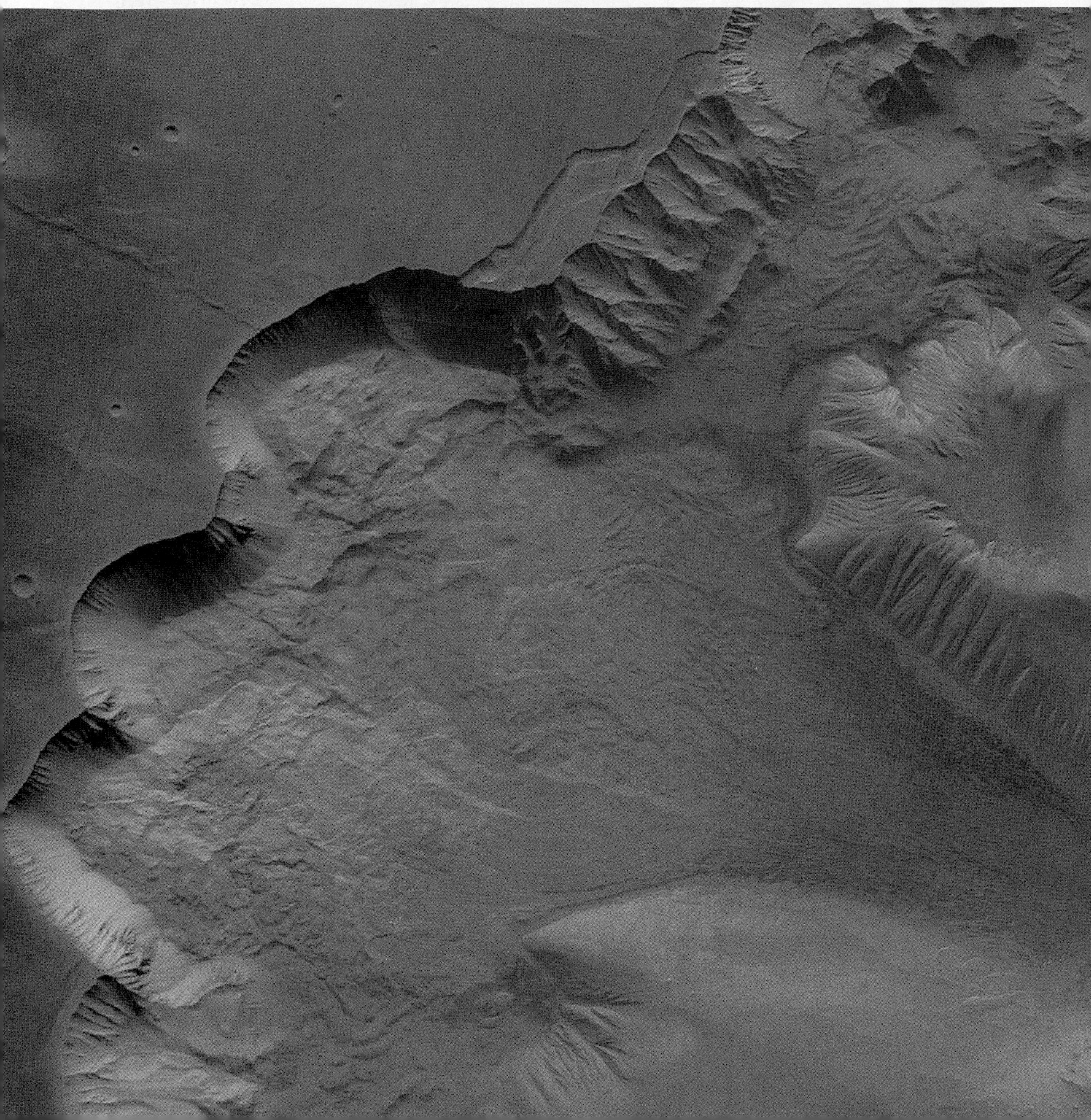

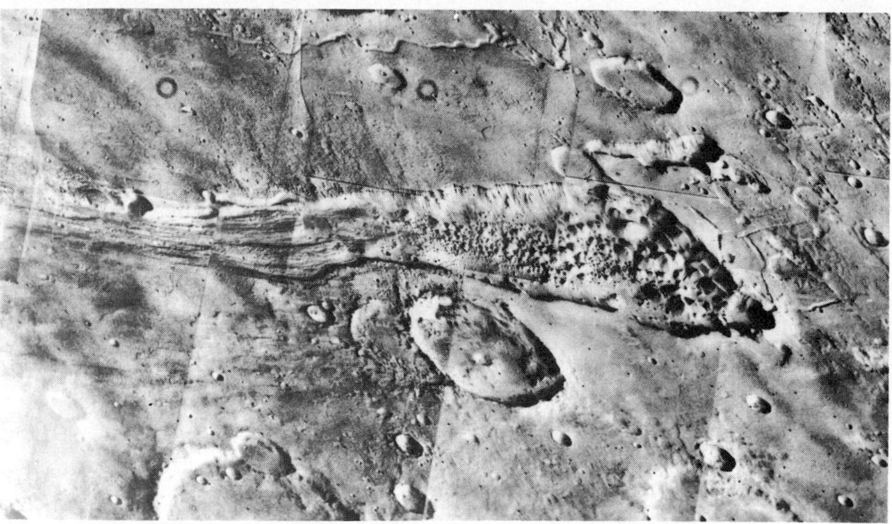

FIGURE 22.17
Extending across the middle of this photograph is a "stream" channel that most probably formed as a result of the melting of subsurface ice. As the ice melted, the unsupported land above collapsed to produce the hummocky topography to the right. The flood of meltwater carved the channel to the left. (Photo courtesy of NASA)

tographs indicate that most Martian surface features are old by Earth standards. The highly cratered Martian southern hemisphere is probably similar in age to the comparable lunar highlands (3.5–4.5 billion years old). The discovery of several highly cratered and weathered volcanoes on Mars further indicates that volcanic activity began early and had a long history. However, even the relatively fresh-appearing volcanic features of the northern hemisphere may be older than 1 billion years. This fact, coupled with the absence of "marsquake" recordings by *Viking* seismographs, points to a tectonically dead planet.

Another surprising find made by *Mariner 9* was the existence of several canyons, which dwarf even the Grand Canyon of the Colorado. One of the largest, Valles Marineris, is roughly 6 kilometers deep, up to 160 kilometers wide, and extends for almost 5000 kilometers along the Martian equator (Figure 22.16). This vast chasm is thought to have formed by slippage of crustal material along huge faults in the crustal layer. In this respect, it would be comparable to the rift valleys of Africa.

Not all valleys on Mars have a tectonic origin. Many Martian valleys have tributaries exhibiting drainage patterns similar to those of stream valleys found on Earth. In addition, *Viking* orbiter photographs have also revealed streamlined features that are unmistakably ancient islands located in what is now a dry stream bed. When these streamlike channels were first discovered, some observers speculated that a thick, water-laden atmosphere capable of generating torrential downpours once existed on Mars. But what happened to this water? The present Martian atmosphere contains only trace amounts of water. Moreover, the environment of Mars is far too harsh to allow water to exist as a liquid. Despite these difficulties, the work of flowing water still remains the most acceptable explanation for many of the Martian channels.

Many planetary geologists do not accept the premise that Mars once had an active water cycle similar to that on Earth. Rather, they believe that many of the large streamlike valleys were created by the collapse of surface material caused by the slow melting of subsurface ice (Figure 22.17). If this is the case, these large valleys would be more akin to features formed by mass-wasting processes on Earth. Some of the smaller channels, they contend, have been cut by the gradual release of subsurface water, which flowed out of the ground like springs. Further, the heat from volcanic activity or meteoroid impact may have rapidly melted subsurface ice which, in turn, caused local flooding. These floods may have carved some of the channels. An occasional flood would also account for the streamlined features found in what are presently dry river beds. The question of whether rainfall, gradual seepage of groundwater, or the collapse of surface material created the branching valleys of Mars will be debated for some time. The answer might not come until we obtain images of Mars that reveal greater detail than those presently available.

Because of their small size, the two satellites of Mars, Phobos and Deimos, were not discovered until 1877. Phobos is closer to its parent than any other natural satellite. Only 5500 kilometers from the Martian surface, Phobos requires just 7 hours and 39 minutes for one revolution. Deimos, which is smaller and 20,000 kilometers away, revolves in 30 hours and 18 minutes. *Mariner 9* revealed that both satellites are irregularly shaped and have numerous impact craters, much like their parent. The maximum diameter of Phobos is 24 kilometers, and the maximum diameter of Deimos is only about 15 kilometers. It is believed that these moons were asteroids that were captured by Mars. One of the most interesting coincidences in astronomy is the close resemblance between Phobos and Deimos and the two fictional satellites of Mars described by Jonathan Swift in *Gulliver's Travels,* written about 150 years before they were actually discovered.

JUPITER: THE LORD OF THE HEAVENS

Jupiter, the largest planet in our solar system, has a mass $2\frac{1}{2}$ times greater than the combined mass of all the remaining planets, satellites, and asteroids. It is truly a giant among planets. In fact, had Jupiter been about ten times larger, it would have evolved into a small star. Despite its great size, however, it is only one one-thousandth as massive as the sun. Jupiter also rotates more rapidly than any other planet, completing one rotation in slightly less than ten hours. The effect of Jupiter's rapid rotation causes the equatorial region to be slightly bulged and the polar dimension to be flattened.

When viewed through a telescope or binoculars, Jupiter appears to be covered with alternating bands of multicolored clouds aligned parallel to its equator. The most striking feature on the disk of Jupiter is the great red spot (Figure 22.18). Although its color varies greatly in intensity, it has been an ever-present feature since it was first discovered more than three centuries ago. When *Voyager 2* swept by Jupiter in July, 1979, the size of the red spot was 11,000 kilometers by 22,000 kilometers, which equals two Earth-sized circles placed side by side. On occasion it has grown even larger. Although the great red spot varies in size, it does remain the same distance from the Jovian equator. The cause of the spot has been attributed to everything from volcanic activity to a large cyclonic storm. Images obtained by *Pioneer 11* as it moved to within 42,000 kilometers of Jupiter's cloud tops in December, 1974, support the latter view. The great red spot apparently is a counterclockwise-rotating storm caught between two jetstream-like bands of atmosphere flowing in opposite directions. This huge hurricane-like storm rotates once every 12 Earth days. Although several smaller storms have

FIGURE 22.18
A mosaic of Jupiter and its four Galilean satellites as seen from *Voyager 1*. Ganymede is located at lower left, Callisto at lower right. Europa is adjacent to Jupiter, and Io appears to the left in the distance. (Photo courtesy of NASA)

been observed in other regions of Jupiter's atmosphere, none has survived for more than a few days.

Structure of Jupiter

Jupiter's atmosphere is composed primarily of hydrogen and helium, with methane, ammonia, water, and sulfur compounds as minor constituents. The wind systems generate the light- and dark-colored bands that encircle this giant. Unlike the winds on Earth, which are driven by solar energy, Jupiter gives off nearly twice as much heat as it receives from the sun. Thus, it is the heat emanating from Jupiter's interior that produces huge convection currents in the atmosphere. The light-colored zones are regions where gases are ascending and cooling (Figure 22.18). The light color is thought to be produced by ammonia crystallizing into "snowflakes" near the cloud tops where temperatures of −120°C have been measured. The cold material from the light zones spills over onto the lower dark belts, where air is descending and heating. The reason for the rusty-brown color of the dark belts is not known for certain; however, they may be colored by sulfur compounds.

Atmospheric pressure at the top of the clouds is equal to sea-level pressure on Earth, but because of Jupiter's immense gravity, the pressure increases rapidly toward its surface. At 1000 kilometers below the clouds, the pressure is great enough to liquefy hydrogen. Consequently, the surface of Jupiter is thought to be a gigantic ocean of liquid hydrogen. Less than halfway into Jupiter's interior, pressures of unimaginable magnitude cause liquid hydrogen to turn into liquid metallic hydrogen. Although scientists have never seen this unusual material, its properties are thought to be predictable. Jupiter is also believed to contain as much rocky and metallic material as is found in the terrestrial planets, such as Earth. Whether this "Earthy" material is scattered as small bits or makes up a central core is not known with certainty, but the latter idea is generally accepted.

Jupiter's Moons

Jupiter's satellite system, consisting of sixteen moons, resembles a miniature solar system. The four largest satellites were discovered by Galileo and travel in nearly circular orbits around the parent with periods of from two to seventeen days (Figure 22.18). The two largest of the Galilean satellites, Calisto and Ganymede, surpass Mercury in size, while the two smaller ones, Europa and Io, are about the size of the earth's moon. These Galilean moons can be observed with a small telescope and are interesting in their own right. Because their orbits are along Jupiter's equatorial plane and since they all have the same orbital direction, these moons most probably formed from "leftover" debris in much the same way as the planets did. By contrast, the four outermost satellites are very small (20 kilometers in diameter), revolve in a direction that is opposite that of the outer moons, and have orbits that are steeply inclined to the Jovian equator. These satellites appear to have been asteroids that passed near enough to be captured gravitationally by Jupiter.

The images obtained by *Voyager 1* and *2* in 1979 revealed to the surprise of almost everyone that each of the four Galilean satellites has a character all its own. The entire surface of Callisto, the outermost of the Galilean satellites, is densely cratered, much like the surfaces of Mercury and the moon. However, in this instance the impacts occurred in a crust that appears to be a dirty, frozen ocean of water ice. The largest features discovered were sets of bright, concentric rings surrounding impact craters. These bright rings are thought to resemble the ripples produced when a pebble is dropped into calm water. They are probably composed of blocks of ice, which were deformed and uplifted by the impact that generated the central crater.

Ganymede, the largest Jovian satellite, contains the most diverse terrain. Like our moon, it has densely cratered regions and very smooth areas, where a younger icy layer covers the older cratered surface. In addition, Ganymede has numerous parallel grooves. This suggests some type of tectonic activity has occurred in the distant past. In some places structural features show evidence of lateral displacement along strike-slip faults, a phenomenon previously found only on Earth.

Europa, smallest of the Galilean satellites, has an icy surface that is crisscrossed by many linear features. Although some of these linear markings are thousands of kilometers long, they have very little surface expression. Further, this satellite is notably void of large impact craters. Europa is as smooth as a billiard ball on the same scale. Therefore, the present surface of Europa must have formed sometime after the early period of bombardment, when rocky chunks were far more numerous in the solar system. The crust of Europa may be a thick, frozen ice layer that caps a slushy ocean.

The innermost of the Galilean moons, Io, is the only volcanically active body other than Earth and Triton so far discovered in our solar system. To date, eight active sulfurous volcanic centers have been discovered (Figure 22.19). Umbrella-shaped plumes have been seen rising from the surface of Io to heights approaching 200 kilometers. The surface of

FIGURE 22.19
Io, the geologically most active satellite known. The bizarre appearance of this satellite is the result of frequent "volcanic" activity. (Photo courtesy of NASA)

Io is very colorful, a result of its sulfurous "rocks," which change colors from red to yellow and from black to white depending on temperature (Figure 22.20). The source of heat for this volcanic activity is thought to be tidal energy generated by a "tug-of-war" between Jupiter and the Galilean satellites. Since Io is gravitationally locked to Jupiter, the same side always faces this giant. The gravitational influence of Jupiter and the other nearby satellites pulls and pushes on Io's tidal bulge as its slightly eccentric orbit takes it alternately closer and farther from Jupiter. This gravitational flexing of Io is transformed into heat energy.

One of the most interesting discoveries made by *Voyager 1* was the ring system of Jupiter. Thought to be less than 30 kilometers thick, the apparently continuous ring may extend outward from the surface of Jupiter to a distance equal to twice the diameter of the planet. This ring system is believed to be unlike that found on Saturn. Rather than particles held in planetary-type orbits, the particles composing Jupiter's ring appear to be temporarily entrapped by the planet's intense magnetic field. The source of the ring material may be sulfur from the volcanoes of Io, which is ionized and trapped by Jupiter's magnetic field.

Years of investigation will be needed to unravel all of the mysteries uncovered by the *Voyager 1* and *2* missions. Even when this task is accomplished, our knowledge of Jupiter and its satellites will be far

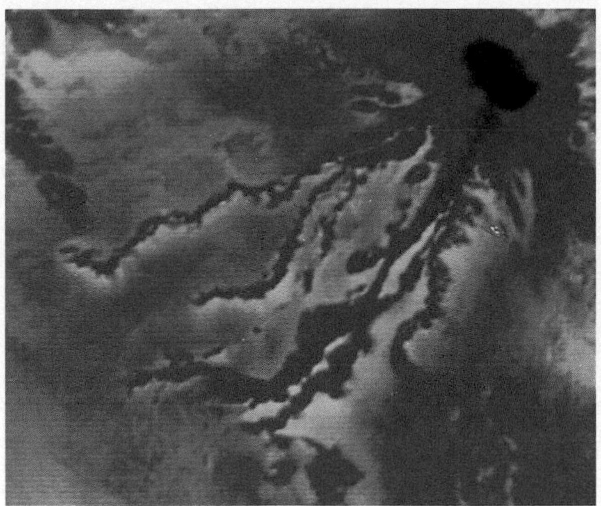

FIGURE 22.20
One of Io's volcanoes, Ra Patera, has a central caldera that is nearly 30 kilometers across. The "lava flows" extending from the summit area are believed to be liquid sulfur. (Photo courtesy of NASA)

from complete. Unfortunately, many of the remaining secrets and questions will have to go undiscovered and unanswered for many years because only a single planetbound probe is currently being developed in the United States.

SATURN: THE ELEGANT PLANET

Requiring 29½ years to make one revolution, Saturn is almost twice as far from the sun as Jupiter, yet its atmosphere, composition, and internal structure are thought to be remarkably similar. The most prominent feature of Saturn is its system of rings. These rings were discovered by Galileo and appeared to him as two smaller bodies adjacent to the planet because he could not resolve them with his primitive telescope. Their ring nature was revealed 50 years later by the Dutch astronomer Christian Huygens. Until the recent discovery that Jupiter, Uranus, and Neptune have very faint ring systems, this spectacular phenomenon was thought to be unique to Saturn.

In 1980 and 1981, fly-by missions of the nuclear-powered *Voyager 1* and *2* unmanned space vehicles came within 100,000 kilometers of the surface of Saturn. More information on Saturn and its satellite system was gained in a few days than had been acquired since Galileo first viewed this elegant planet telescopically in 1610. Some of the information acquired by these space probes is as follows:

1. Saturn's atmosphere is very dynamic with winds roaring at speeds approaching 1500 kilometers per hour.
2. Large cyclonic "storms" similar to, although much smaller than, Jupiter's great red spot occur in Saturn's atmosphere.
3. Verification was made of the existence of the sixth and innermost ring system, and evidence was gathered for a seventh ring system.
4. The icy rings of Saturn were discovered to be more complex than expected. Each of the seven rings is made of numerous ringlets resembling the grooves on a phonograph record, while the rings in the faint *F* ring are intertwined in stable kinked and braidlike configurations. Further, the *B* ring develops perplexing outwardly radiating spokes which survive for hours at a time.
5. Three additional moons were discovered, bringing the total so far detected to seventeen. The larger two of these icy worlds, although similar in size, display a surprising range of geologic evolution.

The Rings of Saturn

When viewed from Earth, Saturn's rings appear to consist of three rather distinct, concentric bands, which have classically been called the *A, B,* and *C* rings (Figure 22.21). The *A* ring is the outermost of the bright rings and is separated from the brightest ring (*B* ring) by a large gap. This space, called the **Cassini gap**, is easily seen in a photo of Saturn. Having a width of 5000 kilometers, the Cassini gap would be large enough to accommodate Earth's moon. During this century evidence has been gathered for the existence of four other very faint bands; the *D* ring is located inside and the *F, G,* and *E* rings are located outside the classically known rings. In 1980, *Voyager 1* revealed that the seven rings are actually composed of hundreds or perhaps thousands of smaller ringlets. Even the Cassini gap is not empty but rather contains a number of very faint bands. From Earth, Saturn's rings can be viewed on edge once every 15 years and appear as an extremely fine line. Satellite images reveal the thickness of the ring system to be no more than a few hundred meters, while its lateral extent exceeds 200,000 kilometers.

Although none of the images obtained so far has the resolution needed to "see" the fine structures of the rings, the rings are undoubtedly composed of relatively small particles (moonlets) which orbit the planet much like any other satellite. Radar observations indicate that most of the particles of the *A, B,* and *C* rings are no larger than 10 meters and the more abundant particles are perhaps as small as 10

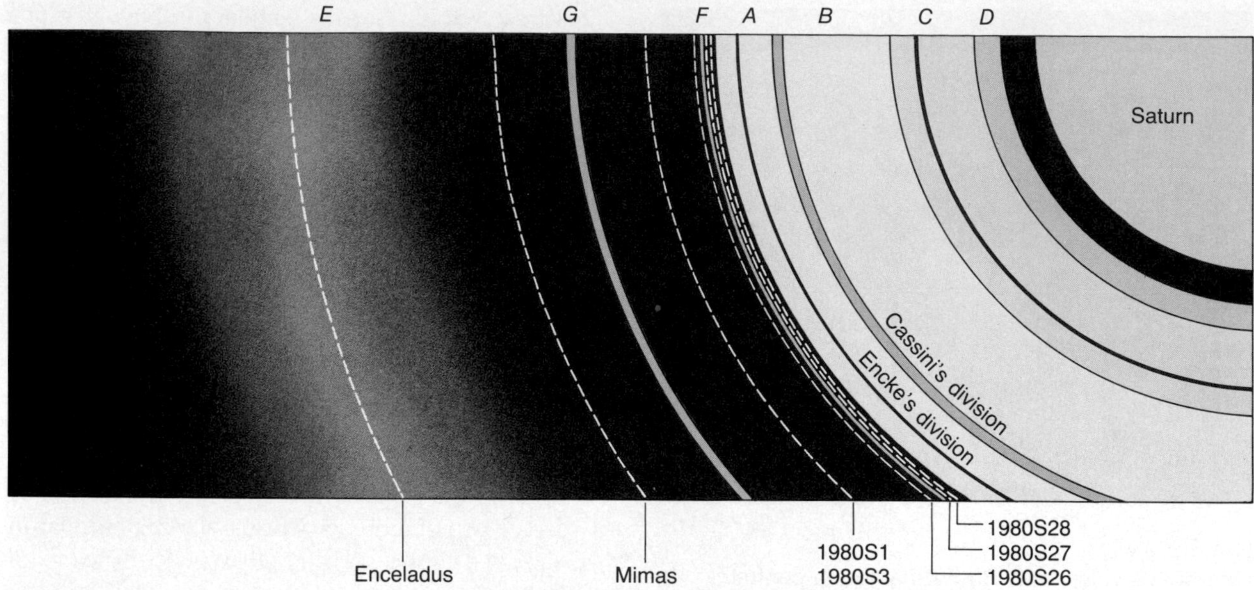

FIGURE 22.21
A polar view of the ring system of Saturn is shown diagrammatically to scale. (From "Rings in the Solar System," by James B. Pollack and Jeffrey N. Cuzzi.)

centimeters. Further studies of reflected light obtained from *Voyager* data revealed that a large fraction of the particles in the *F* ring and certain portions of the *B* ring are extremely minute. Information on the rings' composition has also been obtained from studies of how these particles reflect or absorb light of varying wavelengths. It has been determined that these moonlets are good reflectors of visible light and poor reflectors of near-infrared wavelengths, which is characteristic of water ice. Hence, it is thought that at least the outer surfaces of these particles are coated with a highly reflective layer of water ice. Saturn's rings therefore are actually swarms of small ice-covered debris that orbit the planet in nearly circular paths that are aligned with the equatorial plane of the planet. Most probably they are composed of material which failed to accrete into moons at the time the Saturnian system was forming.

The origin of the rings is believed to be related to their distance from the surface of Saturn. Since they are very close, the disruptive gravitational force of the parent prevented the individual particles from accreting into a larger satellite. Stated another way, objects cannot be held together by self-gravitation when they are within the influence of a stronger gravitational field. In fact, should one of Saturn's large icy satellites approach closer than the outer edge of the bright rings, it would be destroyed and its remains distributed among the rings. The gravitational (tidal) force of Saturn pulling on the near side of such a satellite would be enough greater than the force pulling on its far side that it would tear the satellite apart. Although the rings could be the remains of a satellite destroyed in this fashion, these moonlets are more likely part of the primordial material from which Saturn formed.

Beyond the outermost bright ring (*A* ring), some moonlets have accreted to form very small satellites having diameters on the order of 100 kilometers. Five of these asteroid-sized moons have been discovered orbiting within the faint outer rings, and others probably exist. The gravitational influence of these so-called shepherding satellites is believed to be responsible for keeping the moonlets confined within the rings, thereby producing the sharp edges that are observed. Planetary geologists are very interested in the gravitational interaction of the objects that comprise Saturn's ring system. It is hoped that this information will reveal how material from the primordial cloud of dust and gases condensed to produce the planets. This information would be valuable in reconstructing Earth's early history.

The Moons of Saturn

The Saturnian satellite system consists of seventeen known bodies, all but two of which have nearly circular, counterclockwise orbits along Saturn's equatorial plane (Figure 22.22). The other two, Iapetus and Phoebe, have orbits inclined 15 degrees and 150

FIGURE 22.22
Montage of the Saturnian satellite system. Dione is in foreground; Tethys, Mimas distant right; Enceladus, Rhea off ring's left; and Titan upper right. (Photo courtesy of NASA)

degrees, respectively, to the plane of the system, with the latter exhibiting clockwise motion. Based on estimates of their densities, which range from 1.1 to 1.5 gram per cubic centimeter (except for Titan), and their high reflectivity, these satellites are probably composed mostly of water ice as well as ices of ammonia and methane, with lesser amounts of rocky material. Thus, located over 1.4 billion kilometers from the sun, these moons must be frozen, lifeless worlds. Despite their frigid character and similar compositions, these bodies display surprising diversity.

The eight smallest satellites are irregularly shaped and have diameters that range from roughly 200 kilometers to less than 30 kilometers. The outer three are trapped in stable regions located 60 degrees either ahead of or behind the orbit of a larger satellite. Two others, 1980S1 and 1980S3, are locked in the same orbit (co-orbital) but travel at slightly different velocities. As the faster innermost satellite begins to overtake its companion, it gravitationally acquires orbital momentum from its companion, which thrusts it into a larger, slower orbit. At the same time, its companion drops into a lower, faster orbit. Thus,

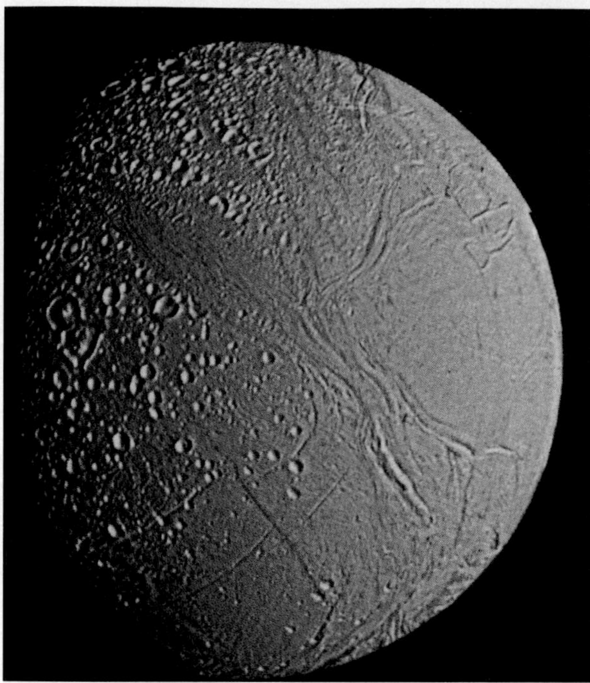

FIGURE 22.23
View of Enceladus, one of Saturn's moons, showing its cratered regions as well as areas that exhibit riftlike valleys. (Photo courtesy of NASA)

the faster inner moon becomes the slower outer moon, and vice versa. This "celestial dance" is repeated about once every four years. The remaining three unnamed moons are called ring shepherds because of their apparent role of "herding" the particles within the rings. Two are found on each side of the narrow *F* ring, and a third is located on the outer edge of the *A* ring. These satellites may be responsible for the sharpness of these ring edges.

Except for Titan, which has a diameter of 5000 kilometers, the largest satellites of Saturn are all smaller than Earth's moon. Phoebe, the smallest, is only about 200 kilometers across, while the two largest, Rhea and Iapetus, are about 1500 kilometers in diameter. Because of their small size, it seems unlikely that an internal heat engine such as that found on Earth could have operated within these bodies. However, several of these moons exhibit structural features such as fractures, rift-like valleys, and even folded terrain, indicating that some type of ancient tectonic activity occurred (Figure 22.23). It has been suggested that early in these moons' histories, ammonia and methane ices, which have a much lower melting point than water ice, could have been set in motion by a very small internal heat source.

The largest Saturnian moon is the only satellite in the solar system known to have a substantial atmosphere. Because of its dense gaseous cover, the atmospheric pressure at the surface of Titan is about one and one-half times that experienced at Earth's surface. Although Titan's atmosphere was predicted to be composed largely of methane, data from *Voyager 1* revealed that this was not the case. Rather, scientists discovered that as much as 80 percent of Titan's atmosphere is nitrogen, with methane probably accounting for less than 6 percent. The orange color of Titan's atmosphere may be the result of photochemical "smog" composed of hydrocarbon molecules, including ethylene and hydrogen cyanide. Further, this planet-sized moon appears to have polar ice caps which show seasonal variations in size. Its surface, if unfrozen, would be an ocean of liquid nitrogen.

URANUS AND NEPTUNE: THE TWINS

If any two planets in the solar system can be considered twins, Uranus and Neptune can. Besides being similar in size, they appear a pale greenish blue color, attributable to the methane in their atmospheres. Their structure and composition are believed to be similar as well, but because of its greater orbital distance, Neptune experiences somewhat lower temperatures.

The unique feature of Uranus is that its axis of rotation lies only 8 degrees from the plane of its orbit (see Figure 22.1). Its rotational motion, therefore, has the appearance of rolling, rather than spinning like a top as the other planets. Because the axis

FIGURE 22.24
This computer-assembled mosaic of the Uranian satellite Miranda was produced using data obtained by *Voyager 2* in 1986. In addition to its crater-marked areas, Miranda contains regions where folded ridges produce curvilinear patterns unique to this satellite. (Photo courtesy of NASA)

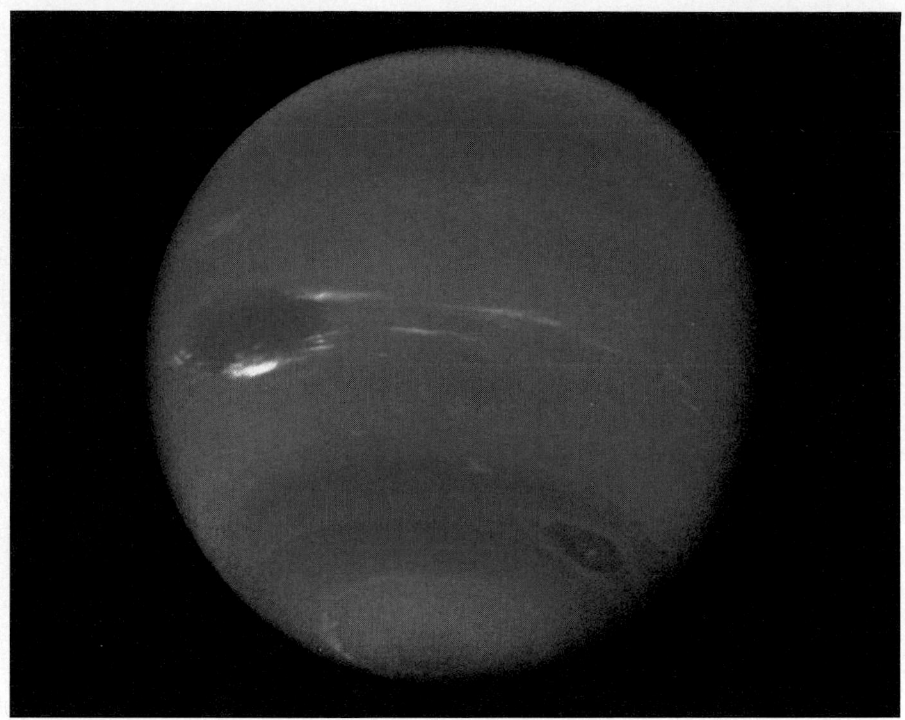

FIGURE 22.25
This image of Neptune shows the great dark spot. Also visible are bright cirruslike clouds that travel at high rates of speed around the planet. A second oval spot can be seen at 54 degrees south latitude on the east limb of the planet. (Photo courtesy of the Jet Propulsion Laboratory)

of Uranus is inclined almost 90 degrees, the sun is nearly overhead at one of the poles once each revolution, and then half a revolution later, it is nearly overhead at the other.

A surprise discovery in 1977 revealed that Uranus is surrounded by rings, much like those encircling Jupiter. This find occurred as Uranus passed in front of a distant star and blocked its view, a process called **occultation.** Observers saw the star "wink" briefly five times before the primary occultation and again five times afterward. Later studies have indicated that Uranus has at least nine distinct belts of debris orbiting its equatorial region.

The first close-up views of Uranus and its satellites were transmitted to Earth by *Voyager 2* in January, 1986. Studies of these images revealed ten previously unknown satellites that are much smaller than the five moons known from Earth-based observations. Two additional rings were also discovered, along with evidence that others may encircle the planet.

Spectacular views of the five largest moons of Uranus showed quite varied terrains. Some contain long, deep canyons and linear scars, whereas others possess large, smooth areas on otherwise crater-ridden surfaces. A spokesperson for the Jet Propulsion Laboratory described Miranda, the innermost of the five largest moons, as having a greater variety of landforms than any other body yet examined in the solar system (Figure 22.24). In particular, Miranda contains three vast areas that exhibit families of grooves that encircle one another much like the lanes of a gigantic racetrack.

Even when the most powerful telescope is focused on Neptune, it appears as a bluish, fuzzy disk. Until *Voyager 2*'s 1989 encounter with Neptune, astronomers knew very little about this planet except that it had an atmosphere consisting mostly of hydrogen, helium, and methane, that it had arcs or incomplete rings, and that it had two moons. However, *Voyager 2*'s 12-year and nearly 3-billion-mile journey provided investigators with so much new information about Neptune and its satellites that it will take years to analyze.

Images from *Voyager 2* show that Neptune has a dynamic atmosphere much like those of Jupiter and Saturn (Figure 22.25). Winds that exceed 1000 kilometers per hour (600 miles per hour) encircle the planet, making it one of the windiest places in the solar system. In addition, Neptune's dynamic atmosphere contains an Earth-sized blemish called the great dark spot that is reminiscent of Jupiter's great red spot. Like the great red spot, this atmospheric disturbance is assumed to be a large rotating storm. Perhaps the most surprising features of the Neptunian atmosphere are the white, cirruslike clouds that occupy a layer about 50 kilometers above the main cloud deck. Although these clouds appear to change configurations, one group seems permanently attached to the great dark spot. At the extremely low temperatures found on Neptune, frozen methane is the most likely constituent of these upper-level clouds.

Six new satellites ranging in size from 50 to 400 kilometers in diameter were discovered in the *Voy-*

ager images. This brings the total number of known satellites for Neptune to eight. All of the newly discovered moons orbit the planet in prograde, circular orbits. *Voyager* images also confirmed the existence of a ring system around Neptune's center. At least two narrow and two broad rings were discovered circling the planet. The outermost of these rings has three thicker arclike segments. These brighter regions should be spread relatively evenly around the ring unless they are restrained by the gravitational influence of ring shepherds (satellites). Although no such satellites were discovered in the *Voyager 2* images, objects smaller than 12 kilometers would not be detected by the cameras aboard that spacecraft.

Triton, Neptune's largest moon, proved to be a most interesting object. With a diameter of roughly 2700 kilometers (nearly as large as Earth's moon), Triton is the one large moon in the solar system that has a highly inclined retrograde orbit. Because Triton orbits Neptune in the direction opposite to the direction in which all the planets travel, Triton undoubtedly formed independently of Neptune. Eventually it wandered close to Neptune and was gravitationally captured by the parent planet.

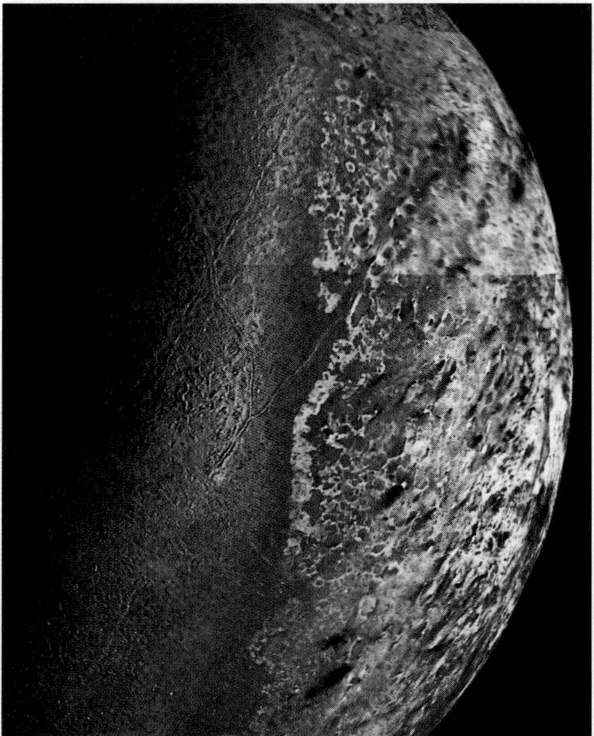

FIGURE 22.26
A photomosaic of the surface of Neptune's moon Triton. The large south polar cap is shown on the left half of the image. A covering of seasonal ice (probably nitrogen) covers the region. Since spring in Triton's southern hemisphere extends from 1960 to the year 2000, some of the polar cap has evaporated. (Photo courtesy of the Jet Propulsion Laboratory)

Much has already been learned about Triton. It has the lowest surface temperature of any body in the solar system, 38 K (−391°F). Further, its very thin atmosphere is composed mostly of nitrogen, with methane present in small amounts. The surface of Triton apparently consists largely of water ice, covered with layers of solid nitrogen and methane. Despite the low surface temperatures, Triton displays volcaniclike activity. Two active plumes were discovered that extended to an altitude of 8 kilometers and downwind for more than 100 kilometers. Presumably, solar energy is absorbed more readily by the relatively dark surface layers of methane ice. Such surface warming vaporizes some of the underlying nitrogen ice. As subsurface pressures increase, explosive eruptions eventually result.

Triton's surface is apparently geologically very young because it lacks heavily cratered areas. As can be seen in Figure 22.26, parts of its surface consist of pits and dimples that are crisscrossed by ridges. Other areas seem to have frozen lakes that resemble lunar maria. Many of these lakes contain terraces, suggesting a succession of meltings and refreezings of water ice.

As *Voyager 2* advances toward interstellar space, our understanding of Neptune and its rings and satellites will continue to improve. In the meantime, both *Voyager* spacecraft will continue to explore the outer reaches of the solar system, sending back measurements of the intensity of the solar wind. If these spacecraft continue to operate as well in the future as they have in the past, they should be able to collect and transmit data for another 25 years. By that time both vehicles will be over 100 astronomical units from the sun and may cross the ***heliopause,*** where true interstellar space begins and where there is no longer any influence from the sun.

PLUTO: PLANET X

Pluto lies on the fringe of the solar system almost 40 times farther from the sun than Earth. It is 10,000 times too dim to be visible with the unaided eye. Because of its great distance and slow orbital speed, it takes Pluto 248 years to orbit the sun. Since its discovery in 1930, it has completed less than one-fifth of a revolution. Pluto's orbit is noticeably elongated (highly eccentric), causing it to occasionally travel inside the orbit of Neptune where it currently resides. There is little likelihood that Pluto and Neptune will ever collide, because their orbits are inclined to each other and do not actually cross.

In June, 1978, a moon was discovered orbiting Pluto. Although this satellite is too small to be observed visually, it has been resolved by the Hubble Space Telescope. The satellite is about 20,000 kilometers from the planet, or 20 times closer than our moon. This discovery greatly altered earlier estimations of Pluto's size. Current data suggest that Pluto has a diameter of less than 3000 kilometers, making it the smallest planet in the solar system.

The average temperature of Pluto is estimated at −210°C, cold enough to solidify any gas that might be present. Pluto could not have an atmosphere. Consequently, Pluto might best be described as a large, dirty iceball made up of a mixture of frozen gases with lesser amounts of rocky substances.

A recent proposal suggests that Pluto was once a satellite of Neptune and was displaced from its original orbit when it collided with a large foreign object. The discovery of a satellite around Pluto is considered evidence that this event broke the Neptunian satellite into two pieces and sent them into an elongated orbit around the sun.

MINOR MEMBERS OF THE SOLAR SYSTEM

Asteroids

Asteroids are relatively small bodies that have been likened to "flying mountains." The largest, Ceres, is 800 kilometers (500 miles) in diameter, but most of the 50,000 that have been observed are only about one kilometer across. The smallest asteroids are assumed to be no larger than grains of sand. Most asteroids lie between the orbits of Mars and Jupiter and have periods ranging from three to six years. Some asteroids have very eccentric orbits and travel very close to Earth and our moon. Many of the most recent impact craters on the moon were probably caused by collisions with asteroids.

Because many asteroids have irregular shapes, planetary geologists first speculated that they may have formed from the breakup of a planet that once occupied an orbit between Mars and Jupiter. However, the total mass of the asteroids is estimated to be only one one-thousandth that of Earth, which itself is not a large planet. What, then, happened to the remainder of the original planet? Others have hypothesized that several larger bodies once coexisted in close proximity and that their collisions produced numerous smaller ones. The existence of several "families" of asteroids has been used to support this latter explanation. However, no conclusive evidence has been found for either hypothesis.

Comets

Comets are among the most spectacular and unpredictable bodies in the solar system. They have been compared to large, dirty snowballs, since they are made of frozen gases (water, ammonia, methane, carbon dioxide, and carbon monoxide) which hold together small pieces of rocky and metallic materials. Many comets travel along very elongated orbits that carry them beyond Pluto. On their return, these comets are visible only after they are within the orbit of Saturn.

When first observed, comets appear very small, but as they approach the sun, solar energy begins to vaporize the frozen gases, producing a glowing head called the **coma** (Figure 22.27). The size of the coma varies greatly from one comet to another. Some exceed the size of the sun, but most approximate the size of Jupiter. Within the coma, a small glowing nucleus with a diameter of only a few kilometers can sometimes be detected. As they approach the sun, some comets develop a tail that extends for millions of kilometers. Despite the enormous size of their tails and comas, comets are thought to have insignificant masses.

The tail of a comet points away from the sun in a slightly curved manner (Figure 22.27). This fact led early astronomers to propose that the sun had a repulsive force that pushed the particles of the coma to form the tail. Today, two solar forces are known to contribute to the formation of the tail. One, radiation pressure, pushes dust particles away from the coma, and the second, solar wind, is responsible for moving the ionized gases, particularly carbon monoxide. Usually, a single tail composed of both types of materials is produced, but two somewhat separate tails like those of the comet Halley can form (Figure 22.28).

As a comet moves away from the sun, the gases begin to condense, the tail disappears, and the comet once again returns to "cold storage." The material that was blown from the coma to form the tail is lost from the comet forever. Consequently, it is believed that most comets cannot survive more than a few hundred close encounters with the sun. Once all the gases are expended, the remaining material—a swarm of disconnected metallic and stony particles—continues to orbit the sun, but without a coma or a tail.

Little is known about the origin of comets. The most widely accepted hypothesis considers them to be members of the solar system that formed at great distances from the sun. Accordingly, millions of comets are believed to form a spherical cloud located beyond the orbit of Pluto. It is proposed that the

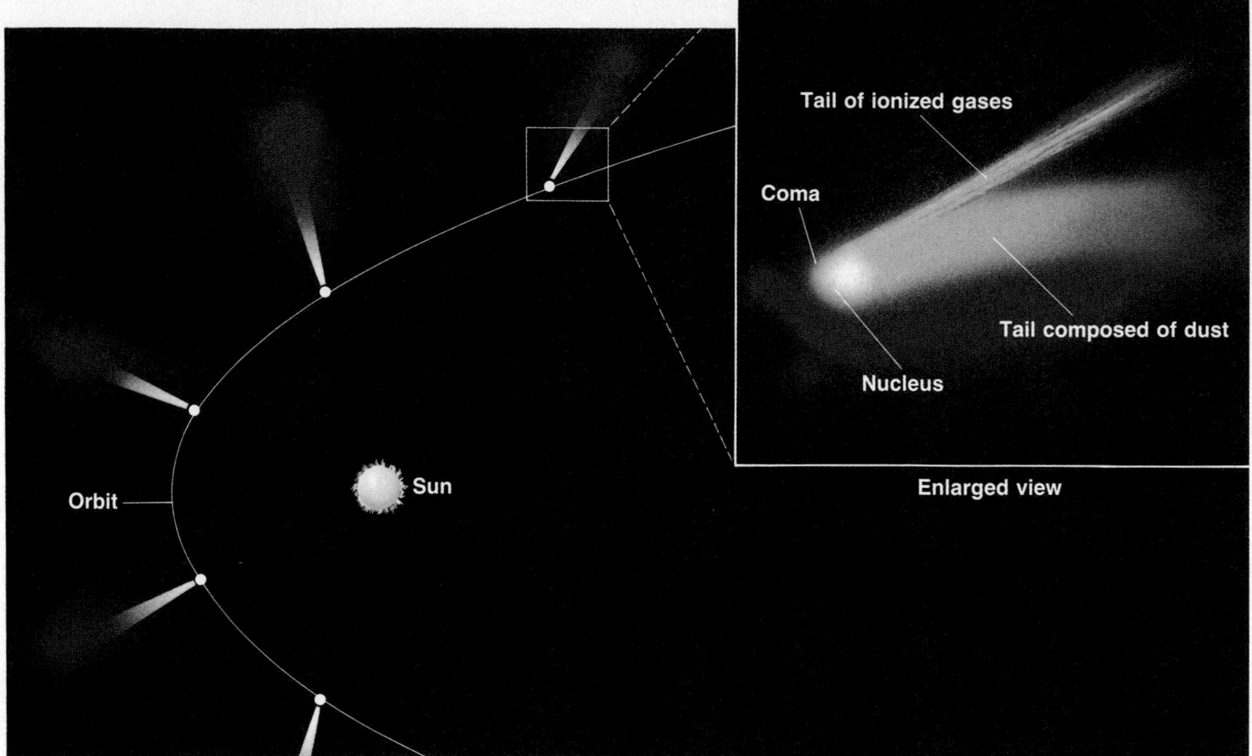

FIGURE 22.27
Orientation of a comet's tail as it orbits the sun.

gravitational effect of stars passing nearby sends some of them into highly eccentric orbits which carry them toward the center of our solar system. Here, the gravitation of the larger planets, particularly Jupiter, alters their orbit and reduces their period of revolution. Many short-period comets of this type have been discovered. However, since they have a short life expectancy, we can be reasonably certain that they are always being replaced by other long-period comets that are gravitationally deflected toward the sun.

One of the most spectacular events of modern times has been attributed to the collision of our planet with a comet. In 1908, in a remote region of Siberia, a brilliant fireball exploded with such a violent force as to create a shock wave that rattled windows and was heard hundreds of kilometers away. Trees were scorched and flattened for a distance of 30 kilometers from the impact zone. Although numerous craters resulted, some 50 meters wide, no large metallic fragments were found. The notable lack of impact fragments points strongly to the explosion of the nucleus of a comet that was vaporized as it penetrated our atmopshere.

Meteoroids

Nearly everyone has seen a meteor, popularly called a "shooting star." This streak of light, which lasts for a few seconds at most, occurs when a small solid particle, a **meteoroid**, enters Earth's atmosphere from interplanetary space. The friction between the meteoroid and the air heats both and produces visible light. Although an occasional meteoroid is as large as an asteroid, most are about the size of sand grains and weigh less than one one-thousandth of a gram. Consequently, they vaporize before reaching Earth's surface. Some, called **micrometeorites**, are so tiny and their rate of fall so slow that they drift down like dust. Each day, the total number of meteoroids that enter Earth's atmosphere must reach into the thousands. After sunset, a half dozen or more are bright enough to be seen with the naked eye each hour from a single spot on Earth.

Occasionally, the number of meteor sightings increases dramatically to 60 or more per hour. These spectacular displays, called **meteor showers**, result when Earth encounters a swarm of meteoroids traveling in the same direction and at nearly the same speed. The close association of these swarms to the orbits of some short-term comets strongly suggests that they represent material lost by the comets. Some swarms not associated with the orbits of known comets are probably the remains of the nucleus of a defunct comet. The meteor showers that occur regularly each year around August 12 are believed to be the remains of the tail of Comet 1862 III, which has a period of 110 years. Meteoroids associated with comets are small and not known to reach the ground. Most meteoroids large enough to survive the fall are thought to originate in the belt of asteroids, where a chance collision sends them toward Earth. Earth's gravitational force does the rest.

FIGURE 22.28
Photograph of Halley's comet showing two somewhat separate tails. The straight gas tail has a bluish cast, whereas the slightly curved dust tail exhibits a yellow coloration. (Photo by William Liller, NASA International Halley Watch, Easter Island)

The remains of meteoroids, when found on Earth, are referred to as **meteorites.** A few very large meteoroids have blasted out craters on Earth's surface that are not much different in appearance from those found on the lunar surface. Perhaps the most famous of these is Meteor Crater in Arizona (Figure 22.29). This huge crater is about 1.2 kilometers across and 170 meters deep, and it has an upturned rim that rises 50 meters above the surrounding countryside. Over 30 tons of iron fragments have been found in the immediate area, but attempts to locate the main metallic body have been unsuccessful. Judging from the amount of erosion, it appears that the impact occurred within the last 20,000 years.

Prior to lunar exploration, meteorites were the only samples of extraterrestrial material that could be directly examined (Figure 22.30). Depending upon their composition, meteorites can generally be put into one of three categories: (1) **irons**—mostly iron with 5–20 percent nickel; (2) **stony**—silicate minerals with inclusions of other minerals; and (3) **stony-irons**—mixtures. Although stony meteorites are probably more common, most meteorite finds are irons. This is understandable, since irons tend to withstand impact and weather more slowly, and are much easier for a lay person to distinguish from terrestrial rocks than are stony meteorites. One rare kind of meteorite, called a *carbonaceous chondrite,* was found to contain some simple amino acids and other organic compounds, which are the basic building blocks to life. This discovery confirms similar findings in observational astronomy which indicate that numerous organic compounds exist in the frigid realm of outer space.

If the composition of meteorites is representative of the material that makes up Earthlike planets, as some planetary geologists believe, then Earth must contain a much larger percentage of iron than is indicated by the surface rock. This is one of the reasons why geologists suggest that the core of Earth may be mostly iron and nickel. In addition, the dating of meteorites has indicated that our solar system has an age that certainly exceeds 4.5 billion years. This "old age" has also been confirmed by data obtained from lunar samples.

FIGURE 22.29
Meteor Crater, about 32 kilometers (20 miles) west of Winslow, Arizona. (Photo by Michael Collier)

FIGURE 22.30
Iron meteorite found near Meteor Crater, Arizona. (Courtesy of Meteor Crater, Northern Arizona, USA)

REVIEW QUESTIONS

1. By what criteria are the planets categorized as either Jovian or terrestrial?
2. What three types of materials are thought to make up the planets? How are they different? How does their distribution account for the density differences between the terrestrial and Jovian planetary groups?
3. Briefly describe the events that are thought to have led to the formation of the solar system.
4. Why are large-rayed craters considered to be relatively young features on the lunar surface?
5. Outline the steps in the evolutionary history of the moon.
6. How are the maria of the moon similar to the Columbia Plateau?
7. Why are meteorite craters more common on the moon than on Earth, even though the moon is much smaller?
8. Although some Martian valleys appear to be the products of stream erosion, what fact makes it unlikely that Mars ever had a water cycle like that found on Earth?
9. What surface features does Mars have that are also common on Earth?
10. What is the speed of a point on the Jovian equator? See Table 22.1 for the circumference of Jupiter and the length of a Jovian day.
11. Briefly describe the four Galilean satellites of Jupiter.
12. Why are the four outer satellites of Jupiter thought to have been captured?
13. What evidence indicates that Saturn's rings are composed of individual moonlets rather than solid disks?
14. What is unique about Saturn's satellite Titan?
15. Which of the five largest moons of Uranus has the most varied terrain?
16. What three bodies in the solar system exhibit volcaniclike activity?
17. What do you think would happen if Earth passed through the tail of a comet?
18. How do meteoroids and meteorites differ? How are they alike?
19. It has been estimated that Halley's comet has a mass of 100 billion tons. Further, this comet is thought to lose 100 million tons of material during the few months that its orbit brings it close to the sun. With an orbital period of 76 years, what is the maximum remaining life-span of Halley's comet?

KEY TERMS

asteroid (p. 609)
Cassini gap (p. 603)
coma (p. 609)
comet (p. 609)
iron meteorite (p. 611)
Jovian planet (p. 580)
lunar breccia (p. 587)
lunar regolith (p. 589)
maria (p. 586)
meteorite (p. 611)
meteoroid (p. 610)
meteor shower (p. 610)
micrometeorite (p. 610)
nebular hypothesis (p. 582)
occultation (p. 607)
ray (p. 587)
stony meteorite (p. 611)
stony-iron meteorite (p. 611)
terrestrial planet (p. 580)

APPENDIX A
Metric and English Units Compared

UNITS

1 kilometer (km)	=	1000 meters (m)
1 meter (m)	=	100 centimeters (cm)
1 centimeter (cm)	=	0.39 inch (in.)
1 mile (mi)	=	5280 feet (ft)
1 foot (ft)	=	12 inches (in.)
1 inch (in.)	=	2.54 centimeters (cm)
1 square mile (mi^2)	=	640 acres (a)
1 kilogram (kg)	=	1000 grams (g)
1 pound (lb)	=	16 ounces (oz)
1 fathom	=	6 feet (ft)

CONVERSIONS

Length

When you want to convert:	Multiply by:	To find:
inches	2.54	centimeters
centimeters	0.39	inches
feet	0.30	meters
meters	3.28	feet
yards	0.91	meters
meters	1.09	yards
miles	1.61	kilometers
kilometers	0.62	miles

Area

When you want to convert:	Multiply by:	To find:
square inches	6.45	square centimeters
square centimeters	0.15	square inches
square feet	0.09	square meters
square meters	10.76	square feet
square miles	2.59	square kilometers
square kilometers	0.39	square miles

Volume

When you want to convert:	Multiply by:	To find:
cubic inches	16.38	cubic centimeters
cubic centimeters	0.06	cubic inches
cubic feet	0.028	cubic meters
cubic meters	35.3	cubic feet
cubic miles	4.17	cubic kilometers
cubic kilometers	0.24	cubic miles
liters	1.06	quarts
liters	0.26	gallons
gallons	3.78	liters

Masses and Weights

When you want to convert:	Multiply by:	To find:
ounces	28.35	grams
grams	0.035	ounces
pounds	0.45	kilograms
kilograms	2.205	pounds

Temperature

When you want to convert degrees Fahrenheit (°F) to degrees Celsius (°C), subtract 32 degrees and divide by 1.8.

When you want to convert degrees Celsius (°C) to degrees Fahrenheit (°F), multiply by 1.8 and add 32 degrees.

When you want to convert degrees Celsius (°C) to kelvins (K), delete the degree symbol and add 273.

When you want to convert kelvins (K) to degrees Celsius (°C), add the degree symbol and subtract 273.

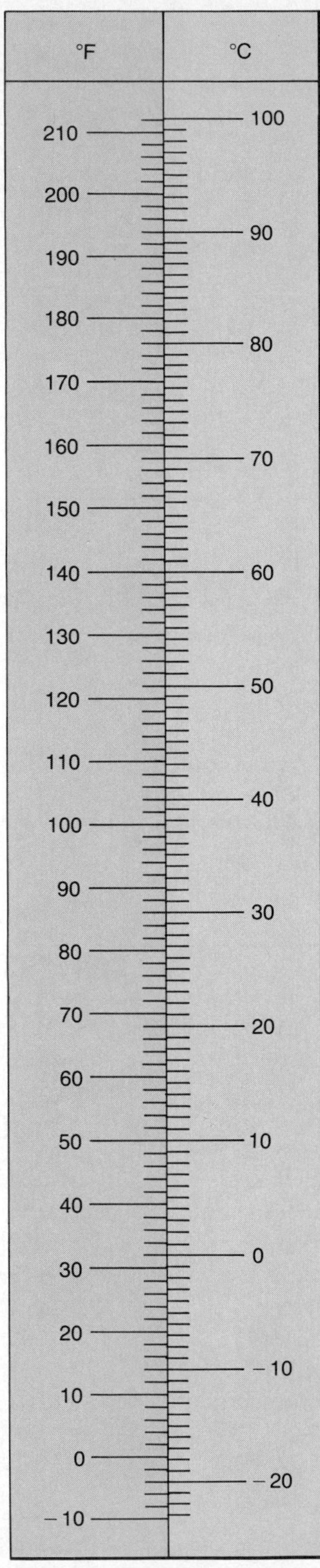

APPENDIX B

Periodic Table of the Elements

Key	
1 **H** 1.0080 Hydrogen	Atomic number Symbol of element Atomic weight Name of element

■ Inert gas
□ Gas
□ Liquid

Solid — all others

	Light Metals		Transitional Elements — Heavy Metals										Nonmetals					
	I A	II A	III B	IV B	V B	VI B	VII B	VIII B			I B	II B	III A	IV A	V A	VI A	VII A	VIII A
1	1 **H** 1.0080 Hydrogen																	2 **He** 4.003 Helium
2	3 **Li** 6.939 Lithium	4 **Be** 9.012 Beryllium											5 **B** 10.81 Boron	6 **C** 12.011 Carbon	7 **N** 14.007 Nitrogen	8 **O** 15.9994 Oxygen	9 **F** 18.998 Fluorine	10 **Ne** 20.183 Neon
3	11 **Na** 22.990 Sodium	12 **Mg** 24.31 Magnesium											13 **Al** 26.98 Aluminum	14 **Si** 28.09 Silicon	15 **P** 30.974 Phosphorus	16 **S** 32.064 Sulfur	17 **Cl** 35.453 Chlorine	18 **Ar** 39.948 Argon
4	19 **K** 39.102 Potassium	20 **Ca** 40.08 Calcium	21 **Sc** 44.96 Scandium	22 **Ti** 47.90 Titanium	23 **V** 50.94 Vanadium	24 **Cr** 52.00 Chromium	25 **Mn** 54.94 Manganese	26 **Fe** 55.85 Iron	27 **Co** 58.93 Cobalt	28 **Ni** 58.71 Nickel	29 **Cu** 63.54 Copper	30 **Zn** 65.37 Zinc	31 **Ga** 69.72 Gallium	32 **Ge** 72.59 Germanium	33 **As** 74.92 Arsenic	34 **Se** 78.96 Selenium	35 **Br** 79.909 Bromine	36 **Kr** 83.80 Krypton
5	37 **Rb** 85.47 Rubidium	38 **Sr** 87.62 Strontium	39 **Y** 88.91 Yttrium	40 **Zr** 91.22 Zirconium	41 **Nb** 92.91 Niobium	42 **Mo** 95.94 Molybdenum	43 **Tc** (99) Technetium	44 **Ru** 101.1 Ruthenium	45 **Rh** 102.90 Rhodium	46 **Pd** 106.4 Palladium	47 **Ag** 107.870 Silver	48 **Cd** 112.40 Cadmium	49 **In** 114.82 Indium	50 **Sn** 118.69 Tin	51 **Sb** 121.75 Antimony	52 **Te** 127.60 Tellurium	53 **I** 126.90 Iodine	54 **Xe** 131.30 Xenon
6	55 **Cs** 132.91 Cesium	56 **Ba** 137.34 Barium	57 TO 71	72 **Hf** 178.49 Hafnium	73 **Ta** 180.95 Tantalum	74 **W** 183.85 Tungsten	75 **Re** 186.2 Rhenium	76 **Os** 190.2 Osmium	77 **Ir** 192.2 Iridium	78 **Pt** 195.09 Platinum	79 **Au** 197.0 Gold	80 **Hg** 200.59 Mercury	81 **Tl** 204.37 Thallium	82 **Pb** 207.19 Lead	83 **Bi** 208.98 Bismuth	84 **Po** (210) Polonium	85 **At** (210) Astatine	86 **Rn** (222) Radon
7	87 **Fr** (223) Francium	88 **Ra** 226.05 Radium	89 TO 103															

Lanthanide series	57 **LA** 138.91 Lanthanum	58 **Ce** 140.12 Cerium	59 **Pr** 140.91 Praseodymium	60 **Nd** 144.24 Neodymium	61 **Pm** (147) Promethium	62 **Sm** 150.35 Samarium	63 **Eu** 151.96 Europium	64 **Gd** 157.25 Gadolinium	65 **Tb** 158.92 Terbium	66 **Dy** 162.50 Dysprosium	67 **Ho** 164.93 Holmium	68 **Er** 167.26 Erbium	69 **Tm** 168.93 Thullium	70 **Yb** 173.04 Ytterbium	71 **Lu** 174.97 Lutetium
Actinide series	89 **Ac** (227) Actinium	90 **Th** 232.04 Thorium	91 **Pa** (231) Protactinium	92 **U** 238.03 Uranium	93 **Np** (237) Neptunium	94 **Pu** (242) Plutonium	95 **Am** (243) Americium	96 **Cm** (247) Curium	97 **Bk** (249) Berkelium	98 **Cf** (251) Californium	99 **Es** (254) Einsteinium	100 **Fm** (253) Fermium	101 **Md** (256) Mendelevium	102 **No** (254) Nobelium	103 **Lw** (257) Lawrencium

APPENDIX C
Common Minerals of the Earth's Crust

Mineral or Group Name	Composition	Cleavage/ Fracture	Color	Hardness	Other Properties/Comments
Albite See *Plagioclase feldspar*					
Amphibole (common member: hornblende)	Complex family of hydrous, Ca, Na, Mg, Fe, Al silicates	Two at 60 and 120 degrees	Deep green to black	5–6	Forms elongated crystals. Commonly found in igenous and metamorphic rocks.
Anorthite See *Plagioclase feldspar*					
Augite See *Pyroxene*					
Bauxite	Mixture of weathered clay minerals	Irregular fracture	Varied, reddish brown common	Variable	Earthy luster commonly contains small spheres. Ore of aluminum.
Biotite	$K(Mg,Fe)_3(AlSi_3O_{10})(OH)_2$	Perfect cleavage in one direction	Black to dark brown	2–2.5	Splits into thin, flexible sheets. Common mica found in igneous and metamorphic rocks.
Bornite	Cu_5FeS_4	Uneven fracture	Brownish bronze on a fresh surface	3	Tarnishes to a variegated purple blue; hence, called peacock ore. High specific gravity (5). Ore of copper.
Calcite	$CaCO_3$ Calcium carbonate	Three perfect cleavages at 75 degrees	White or colorless	2.5–3	Common in sedimentary rocks. When transparent exhibits double refraction. Reacts with weak acid.
Chalcedony	SiO_2 Silicon dioxide	Conchoidal fracture	White when pure. Often multicolored	5–6.5	Microcrystalline form of quartz. Multicolored. Called agates when banded. Opal is an amorphous variety.
Chalcopyrite	$CuFeS_2$	Irregular fracture	Brass yellow	3.5–4	Usually massive. Specific gravity 4–4.5. Ore of copper.
Chlorite	$(Mg,Fe)_5(Al,Fe)_2Si_3O_{10}(OH)_8$	One direction of cleavage	Light to dark green	2–2.5	Occurs as mass of flaky scales. Common in metamorphic rocks.

Mineral or Group Name	Composition	Cleavage/ Fracture	Color	Hardness	Other Properties/Comments
Cinnabar	HgS	One direction, but not generally observed	Scarlet red	2.5	Occurs in masses mixed with other materials. Often dull earthy luster. Important ore of mercury.
Clay minerals (common member: kaolinite)	Complex group of hydrous aluminum silicates	Irregular fracture	Buff to brownish gray	1–2.5	Found in earthy masses as a main constituent of soil. Also abundant in shales and other sedimentary rocks.
Corundum	Al_2O_3	Two good cleavages with striations	Variable; red, blue, yellow and green	9	Important gemstone. Red variety called ruby; blue variety is sapphire. Also used as an abrasive.
Dolomite	$CaMg(CO_3)_2$	Three good cleavages at 75 degrees	Variable; white when pure	3.5–4	Similar to calcite, but will effervesce with acid only when powdered. Common in sedimentary rocks.
Epidote	Complex Ca, Fe, and Al silicate	One good cleavage, one poor	Yellow-green to dark green	6–7	Commonly occurs as small elongated crystals in metamorphic rocks.
Feldspar See *Orthoclase feldspar* and *Plagioclase feldspar*					
Fluorite	CaF_2	Perfect cleavage in 4 directions	Colorless; violet, green, or yellow	4	Commonly found with ores of metals.
Galena	PbS	Three cleavages at right angles	Silver gray	2.5	Shiny metallic mineral with high specific gravity (7.6). Ore of lead.
Garnet	Complex family of silicate minerals containing Ca, Mg, Fe, Mn, Al, Ti, Cr	Uneven to conchoidal fracture	Various colors; commonly deep red to brown	6.5–7.5	Forms 12- or 24-sided crystals commonly found in metamorphic rocks.
Graphite	C	One direction of cleavage	Steel gray	1–2	Occurs in scaly, foliated masses. Used as a lubricant. Greasy feel.
Gypsum	$CaSO_4 \cdot 2H_2O$	Cleavage good in one direction poor in two others	Colorless to white	2	Occurs as tabular crystals, or fibrous or finely crystalline masses. Common in sedimentary layers. Used for plaster.
Halite	NaCl	Three cleavages at right angles	Colorless to white	2.5	Common table salt. Occurs as granular masses. Common sedimentary mineral.
Hematite	Fe_2O_3	Uneven fracture	Reddish brown to steel gray	5.5–6.5	Occurs as earthy masses. High specific gravity (4.8–5.5). Important ore of iron.

Mineral or Group Name	Composition	Cleavage/ Fracture	Color	Hardness	Other Properties/Comments
Hornblende See *Amphibole*					
Kaolinite See *Clay minerals*					
Kyanite	Al_2SiO_5	One good direction of cleavage	White to light blue	5–7	Forms long, bladed or tabular crystals. Common mineral in metamorphic rocks.
Labradorite See *Plagioclase feldspar*					
Limonite (goethite)	Mixture of hydrous iron oxides	Uneven fracture	Yellowish to brown	1–5.5	Earthy masses. Forms from the alteration of other iron-rich minerals. Gives rock surfaces and soils a yellow color.
Magnetite	Fe_3O_4	Uneven fracture	Black	5.5–6.5	Submetallic to metallic luster. Magnetic. High specific gravity (5). Generally occurs in granular masses. Ore of iron.
Malachite	$Cu_2CO_3(OH)_2$	Uneven fracture	Bright green	3.5–4	Effervesces in acid. Ore of copper.
Mica See *Biotite* and *Muscovite*					
Muscovite	$KAl_3Si_3O_{10}(OH)_2$	Perfect cleavage in one direction	Colorless to light gray	2–2.5	Splits into thin elastic sheets. Transparent in thin sheets. Common in all rock types.
Olivine	$(Mg,Fe)_2SiO_4$	Conchoidal fracture	Olive to dark green	6.5–7	Occurs as granular masses or grains in dark-colored igneous rocks.
Orthoclase feldspar (K feldspar)	$KAlSi_3O_8$	Two cleavages at right angles	White to gray. Frequently salmon pink	6	Forms elongated crystals in igneous rocks. Also, commonly found in sedimentary and metamorphic rocks.
Plagioclase feldspar	$NaAlSi_3O_8$ (albite) $CaAl_2Si_2O_8$ (anorthite)	Two cleavages at nearly right angles	White to gray	6	Forms elongated crystals in igneous rocks. Also commonly found in sedimentary and metamorphic rocks. Striations on some cleavage planes.
Pyrite	FeS_2	Uneven fracture	Brass yellow	6–6.5	Occurs as granular masses or well-formed cubic crystals. High specific gravity (4.8–5.2). Often called "fool's gold."
Pyroxene (common member: augite)	Complex family of Mg, Fe, Ca, Na, and Al silicates	Good cleavage in two directions at nearly right angles	Green to black	5–6	Occurs as individual grains in igneous and metamorphic rocks.

Mineral or Group Name	Composition	Cleavage/ Fracture	Color	Hardness	Other Properties/Comments
Quartz	SiO_2	Conchoidal fracture	Colorless when pure	7	Common in all rock types. Often lightly colored, including gray, pink, yellow, and violet.
Serpentine	$Mg_3Si_2O_5(OH)_4$	Uneven fracture	Light to dark green	2.5–5	Fibrous variety is asbestos. Occurs most often in metamorphic rocks.
Sillimanite	Al_2SiO_5	One direction of cleavage	White to gray	6–7	High-grade metamorphic mineral.
Sphalerite	ZnS	Six directions of cleavage	Yellow to brown	3.5–4	Moderate specific gravity (4.1–4.3). Smell of sulfur when powdered. Ore of zinc.
Staurolite	$FeAl_4(SiO_4)_2(OH)_2$	Cleavage not prominent	Brown to reddish brown	7	Elongated crystals, occasionally twinned to form a cross-shaped crystal. Commonly found in metamorphic rocks.
Sulfur	S	Irregular fracture	Yellow	1.5–2.5	Bright yellow mineral most often associated with sedimentary deposits, in coal, and near volcanoes.
Talc	$Mg_3(Si_4O_{10})(OH)_2$	Good cleavage in one direction	White to light green	1–1.5	Soapy feel. Found in foliated masses consisting of thin flakes or scales. Most often associated with metamorphic rocks.
Wollastonite	$CaSiO_3$	Two perfect cleavages	Colorless to white	4.5–5	Forms fibrous or bladed crystals. Common in contact metamorphic rocks.

APPENDIX D

Topographic Maps

A map is a representation on a flat surface of all or a part of the earth's surface drawn to a specific scale. Maps are often the most effective means for showing the locations of both natural and manmade features, their sizes, and their relationships to one another. Like photographs, maps readily display information that would be impractical to express in words.

While most maps show only the two horizontal dimensions, geologists, as well as other map users, often require that the third dimension, elevation, be shown on maps. Maps that show the shape of the land are called **topographic maps**. Although various techniques may be used to depict elevations, the most accurate method involves the use of contour lines.

Contour Lines

A **contour line** is a line on a map representing a corresponding imaginary line on the ground that has the same elevation above sea level along its entire length. While many map symbols are pictographs, resembling the objects they represent, a contour line is an abstraction that has no counterpart in nature. It is, however, an accurate and effective device for representing the third dimension on paper.

Some useful facts and rules concerning contour lines are listed as follows. This information should be studied in conjunction with Figure D.1.

1. Contour lines bend upstream or upvalley. The contours form Vs that point upstream, and in the upstream direction the successive contours represent higher elevations. For example, if you were standing on a stream bank and wished to get to the point at the same elevation directly opposite you on the other bank, without stepping up or down, you would need to walk upstream along the contour at that elevation to where it crosses the stream bed, cross the stream, and then walk back downstream along the same contour.
2. Contours near the upper parts of hills form closures. The top of a hill is higher than the highest closed contour.
3. Hollows (depressions) without outlets are shown by closed, hatched contours. Hatched contours are contours with short lines on the inside pointing downslope.
4. Contours are widely spaced on gentle slopes.
5. Contours are closely spaced on steep slopes.
6. Evenly spaced contours indicate a uniform slope.
7. Contours usually do not cross or intersect each other, except in the rare case of an overhanging cliff.
8. All contours eventually close, either on a map or beyond its margins.
9. A single high contour never occurs between two lower ones, and vice versa. In other words, a change in slope direction is always determined by the repetition of the same elevation either as two different contours of the same value or as the same contour crossed twice.
10. Spot elevations between contours are given at many places, such as road intersections, hill summits, and lake surfaces. Spot elevations differ from control elevation stations, such as bench marks, in not being permanently established by permanent markers.

Relief

Relief refers to the difference in elevation between any two points. *Maximum relief* refers to the difference in elevation between the highest and lowest points in the area being considered. Relief determines the **contour interval**, which is the difference in elevation between succeeding contour lines that is

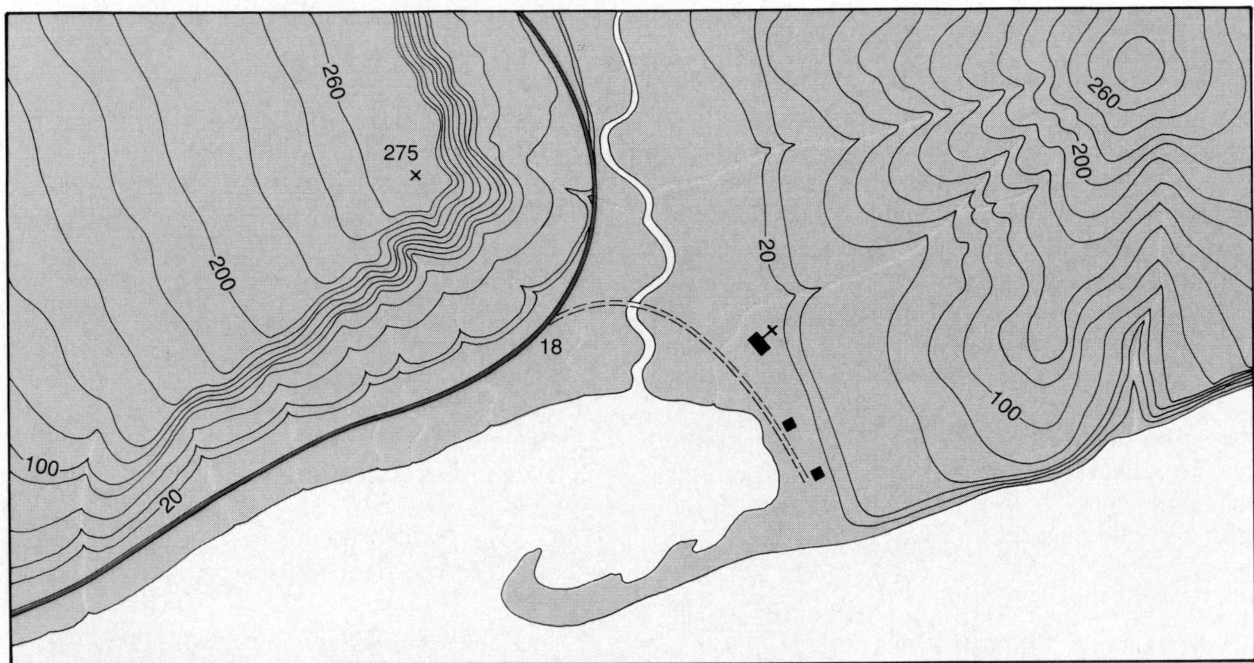

FIGURE D.1
Perspective view of an area and a contour map of the same area. These illustrations show how features are depicted on a topographic map. The upper illustration is a perspective view of a river valley and the adjoining hills. The river flows into a bay, which is partly enclosed by a hooked sandbar. On either side of the valley are terraces through which streams have cut gullies. The hill on the right has a smoothly eroded form and gradual slopes, whereas the one on the left rises abruptly in a sharp precipice, from which it slopes gently, and forms an inclined plateau traversed by a few shallow gullies. A road provides access to a church and the two houses situated across the river from a highway that follows the seacoast and curves up the river valley. The lower illustration shows the same features represented by symbols on a topographic map. The contour interval (vertical distance between adjacent contours) is 20 feet. (After U.S. Geological Survey)

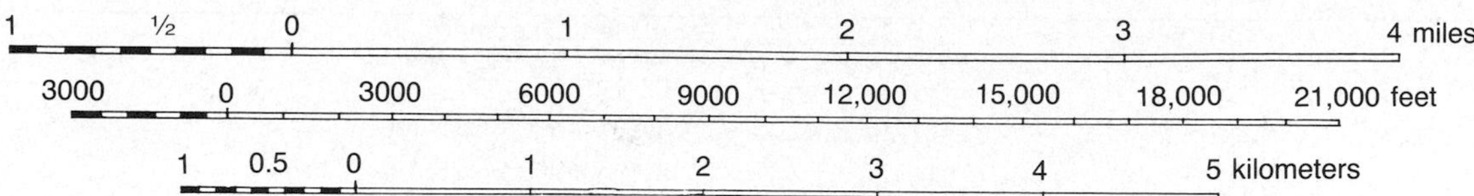

FIGURE D.2
Graphic scale.

used on topographic maps. Where relief is low, a small contour interval, such as 10 or 20 feet, may be used. In flat areas, such as wide river valleys or broad, flat uplands, a contour interval of 5 feet is often used. In rugged mountainous terrain, where relief is many hundreds of feet, contour intervals as large as 50 or 100 feet are used.

Scale

Map **scale** expresses the relationship between distance or area on the map to the true distance or area on the earth's surface. This is generally expressed as a ratio or fraction, such as 1:24,000 or 1/24,000. The numerator, usually 1, represents map distance, and the denominator, a large number, represents ground distance. Thus, 1:24,000 means that a distance of 1 unit on the map represents a distance of 24,000 such units on the surface of the earth. It does not matter what the units are.

Often, the graphic or bar scale is more useful than the fractional scale, because it is easier to use for measuring distances between points. The graphic scale (Figure D.2) consists of a bar divided into equal segments, which represent equal distances on the map. One segment on the left side of the bar is usually divided into smaller units to permit more accurate estimates of fractional units.

Topographic maps, which are also referred to as *quadrangles,* are generally classified according to publication scale. Each series is intended to fulfill a specific type of map need. To select a map with the proper scale for a particular use, remember that large-scale maps show more detail and small-scale maps show less detail. The sizes and scales of topographic maps published by the U.S. Geological Survey are shown in Table D.1.

Color and Symbol

Each color and symbol used on U.S. Geological Survey topographic maps has significance. Common topographic map symbols are shown in Figure D.3. The meaning of each color is as follows:

Blue—water features
Black—works of man, such as homes, schools, churches, roads, and so forth
Brown—contour lines
Green—woodlands, orchards, and so forth
Red—urban areas, important roads, public land subdivision lines

TABLE D.1
National topographic maps.

Series	Scale	1 Inch Represents	Standard Quadrangle Size (latitude-longitude)	Quadrangle Area (square miles)	Paper Size E-W N-S Width Length (inches)
$7\frac{1}{2}$-minute	1:24,000	2000 feet	$7\frac{1}{2}' \times 7\frac{1}{2}'$	49–70	22 × 27*
Puerto Rico $7\frac{1}{2}$-minute	1:20,000	about 1667 feet	$7\frac{1}{2}' \times 7\frac{1}{2}'$	71	$29\frac{1}{2} \times 32\frac{1}{2}$
15-minute	1:62,500	nearly 1 mile	15′ × 15′	197–282	17 × 21*
Alaska 1:63,360	1:63,360	1 mile	15′ × 20′–36′	207–281	18 × 21**
U.S. 1:250,000	1:250,000	nearly 4 miles	1° × 2°†	4580–8669	34 × 22‡
U.S. 1:1,000,000	1:1,000,000	nearly 16 miles	4° × 6°†	73,734–102,759	27 × 27

SOURCE: U.S. Geological Survey.
*South of latitude 31 degrees, $7\frac{1}{2}$-minute sheets are 23 × 27 inches; 15-minute sheets are 18 × 21 inches.
**South of latitude 62 degrees, sheets are 17 × 21 inches.
†Maps of Alaska and Hawaii vary from these standards.
‡North of latitude 42 degrees, sheets are 29 × 22 inches; Alaska sheets are 30 × 23 inches.

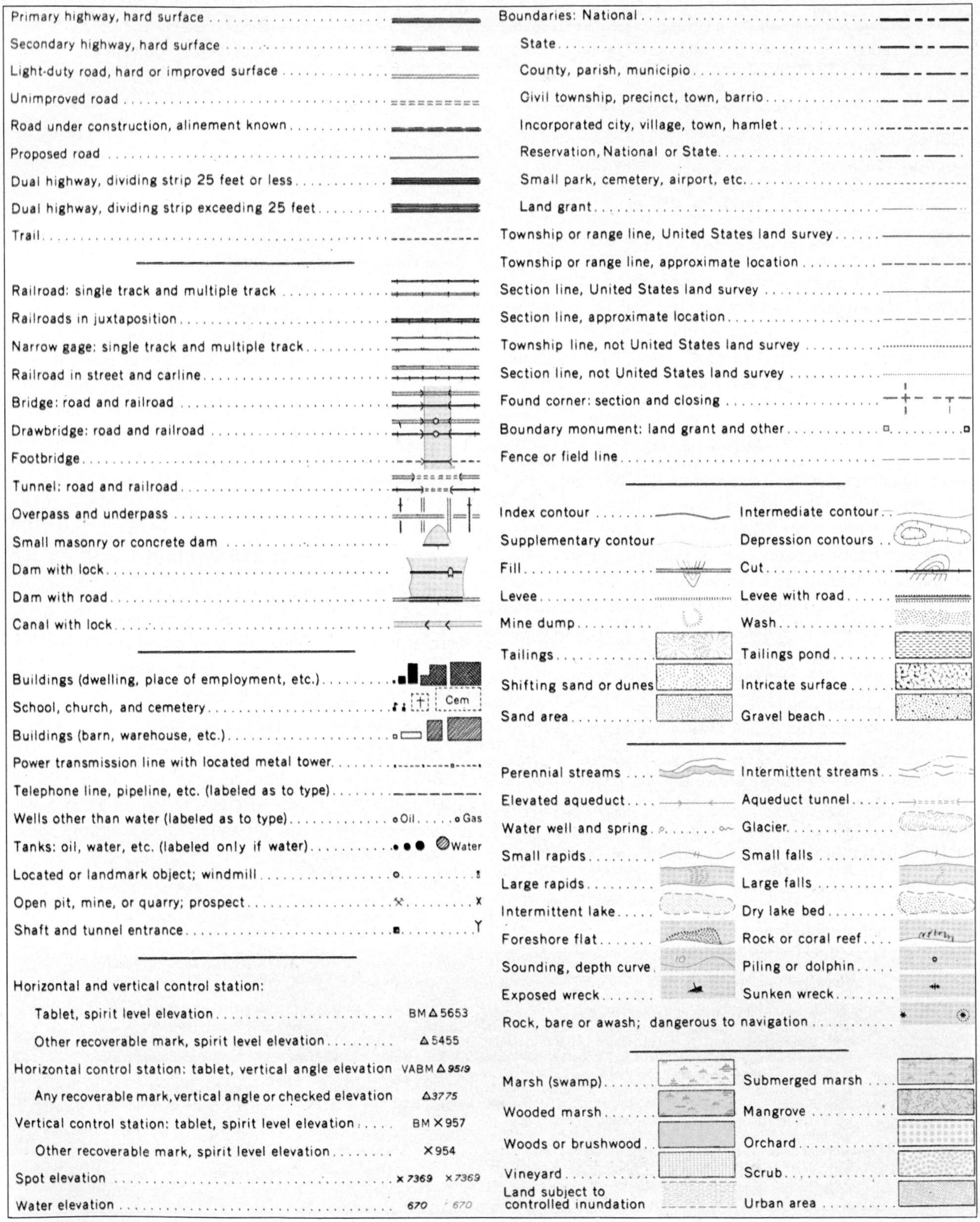

Primary highway, hard surface
Secondary highway, hard surface
Light-duty road, hard or improved surface
Unimproved road
Road under construction, alinement known
Proposed road
Dual highway, dividing strip 25 feet or less
Dual highway, dividing strip exceeding 25 feet
Trail

Railroad: single track and multiple track
Railroads in juxtaposition
Narrow gage: single track and multiple track
Railroad in street and carline
Bridge: road and railroad
Drawbridge: road and railroad
Footbridge
Tunnel: road and railroad
Overpass and underpass
Small masonry or concrete dam
Dam with lock
Dam with road
Canal with lock

Buildings (dwelling, place of employment, etc.)
School, church, and cemetery — Cem
Buildings (barn, warehouse, etc.)
Power transmission line with located metal tower
Telephone line, pipeline, etc. (labeled as to type)
Wells other than water (labeled as to type) — Oil ... Gas
Tanks: oil, water, etc. (labeled only if water) — Water
Located or landmark object; windmill
Open pit, mine, or quarry; prospect
Shaft and tunnel entrance

Horizontal and vertical control station:

- Tablet, spirit level elevation — BM△5653
- Other recoverable mark, spirit level elevation — △5455

Horizontal control station: tablet, vertical angle elevation — VABM△*9519*

- Any recoverable mark, vertical angle or checked elevation — △*3775*

Vertical control station: tablet, spirit level elevation — BM×957

- Other recoverable mark, spirit level elevation — ×954

Spot elevation — ×*7369* ×*7369*
Water elevation — *670* *670*

Boundaries: National
- State
- County, parish, municipio
- Civil township, precinct, town, barrio
- Incorporated city, village, town, hamlet
- Reservation, National or State
- Small park, cemetery, airport, etc.
- Land grant

Township or range line, United States land survey
Township or range line, approximate location
Section line, United States land survey
Section line, approximate location
Township line, not United States land survey
Section line, not United States land survey
Found corner: section and closing
Boundary monument: land grant and other
Fence or field line

Index contour — Intermediate contour
Supplementary contour — Depression contours
Fill — Cut
Levee — Levee with road
Mine dump — Wash
Tailings — Tailings pond
Shifting sand or dunes — Intricate surface
Sand area — Gravel beach

Perennial streams — Intermittent streams
Elevated aqueduct — Aqueduct tunnel
Water well and spring — Glacier
Small rapids — Small falls
Large rapids — Large falls
Intermittent lake — Dry lake bed
Foreshore flat — Rock or coral reef
Sounding, depth curve — Piling or dolphin
Exposed wreck — Sunken wreck
Rock, bare or awash; dangerous to navigation

Marsh (swamp) — Submerged marsh
Wooded marsh — Mangrove
Woods or brushwood — Orchard
Vineyard — Scrub
Land subject to controlled inundation — Urban area

FIGURE D.3
U.S. Geological Survey topographic map symbols. (Variations will be found on older maps.)

APPENDIX E

Landforms of the Conterminous United States

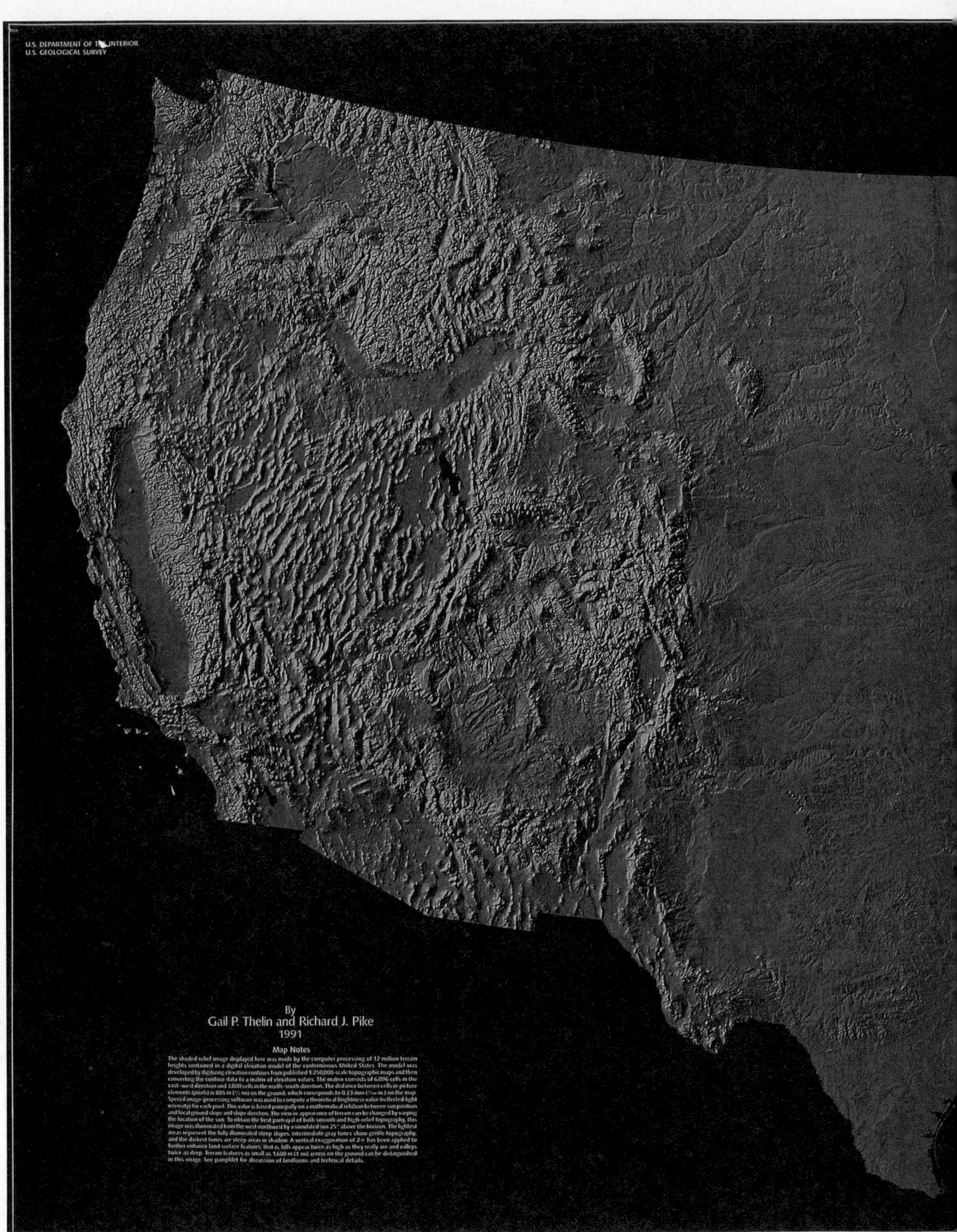
U.S. DEPARTMENT OF T INTERIOR
U.S. GEOLOGICAL SURVEY
By
Gail P. Thelin and Richard J. Pike
1991
Map Notes
The shaded-relief image displayed here was made by the computer processing of 12 million terrain heights contained in a digital elevation model of the conterminous United States. The model was developed by digitizing elevation contours from published 1:250,000-scale topographic maps and then converting the contour data to a matrix of elevation values. The matrix consists of 6,096 cells in the east–west direction and 3,800 cells in the north–south direction. The distance between cells or picture elements (pixels) is 805 m (½ mi) on the ground, which corresponds to 0.23 mm (1/100 in.) on the map. Special image-processing software was used to compute a theoretical brightness value (reflected-light intensity) for each pixel. This value is based principally on a mathematical relation between sun position and local ground slope and slope direction. The view or appearance of terrain can be changed by varying the location of the sun. To obtain the best portrayal of both smooth and high-relief topography, this image was illuminated from the west-northwest by a simulated sun 25° above the horizon. The lightest areas represent the fully illuminated steep slopes, intermediate gray tones show gentle topography, and the darkest tones are steep areas in shadow. A vertical exaggeration of 2× has been applied to further enhance land-surface features; that is, hills appear twice as high as they really are and valleys twice as deep. Terrain features as small as 1,600 m (1 mi) across on the ground can be distinguished in this image. See pamphlet for discussion of landforms and technical details.

MISCELLANEOUS INVESTIGATIONS SERIES
MAP I-2206
Explanatory pamphlet accompanies map
Landforms of the Conterminous United States—
A Digital Shaded-Relief Portrayal

Glossary

Aa A type of lava flow that has a jagged, blocky surface.

Ablation A general term for the loss of ice and snow from a glacier.

Abrasion The grinding and scraping of a rock surface by the friction and impact of rock particles carried by water, wind, and ice.

Absolute dating Determination of the number of years since the occurrence of a given geologic event.

Abyssal plain Very level area of the deep-ocean floor, usually lying at the foot of the continental rise.

Accretionary wedge A large wedge-shaped mass of sediment that accumulates in subduction zones. Here sediment is scraped from the subducting oceanic plate and accreted to the overriding crustal block.

Active layer The zone above the permafrost that thaws in summer and refreezes in winter.

Aftershock A smaller earthquake that follows the main earthquake.

Alluvial fan A fan-shaped deposit of sediment formed when a stream's slope is abruptly reduced.

Alluvium Unconsolidated sediment deposited by a stream.

Alpine glacier A glacier confined to a mountain valley, which in most instances had previously been a stream valley.

Angle of repose The steepest angle at which loose material remains stationary without sliding downslope.

Angular unconformity An unconformity in which the older strata dip at an angle different from that of the younger beds.

Antecedent stream A stream that continued to downcut and maintain its original course as an area along its course was uplifted by faulting or folding.

Anthracite A hard, metamorphic form of coal that burns cleanly and hot.

Anticline A fold in sedimentary strata that resembles an arch.

Aphanitic A texture of igneous rocks in which the crystals are too small for individual minerals to be distinguished with the unaided eye.

Aquiclude An impermeable bed that hinders or prevents groundwater movement.

Aquifer Rock or sediment through which groundwater moves easily.

Archeon eon The second eon of Precambrian time. The eon following the Hadean and preceding the Proterozoic. It extends between 3.8 and 2.5 billion years.

Arête A narrow, knifelike ridge separating two adjacent glaciated valleys.

Arkose A feldspar-rich sandstone.

Artesian well A well in which the water rises above the level where it was initially encountered.

Assimilation In igneous activity, the process of incorporating country rock into a magma body.

Asteroid One of thousands of small planetlike bodies, ranging in size from a few hundred kilometers to less than one kilometer across. Most asteroids' orbits lie between those of Mars and Jupiter.

Asthenosphere A subdivision of the mantle situated below the lithosphere. This zone of weak material exists below a depth of about 100 kilometers and in some regions extends as deep as 700 kilometers. The rock within this zone is easily deformed.

Astronomical theory A theory of climatic change first developed by the Yugoslavian astronomer Milankovitch. It is based upon changes in the shape of the earth's orbit, variations in the obliquity of the earth's axis, and the wobbling of the earth's axis.

Atmosphere The gaseous portion of a planet; the planet's envelope of air. One of the traditional subdivisions of the earth's physical environment.

Atoll A continuous or broken ring of coral reef surrounding a central lagoon.

Atom The smallest particle that exists as an element.

Atomic mass unit A mass unit equal to exactly one-twelfth the mass of a carbon-12 atom.

Atomic number The number of protons in the nucleus of an atom.

Atomic weight The average of the atomic masses of isotopes for a given element.

Aureole A zone or halo of contact metamorphism found in the country rock surrounding an igneous intrusion.

Back swamp A poorly drained area on a floodplain resulting when natural levees are present.

Bajada An apron of sediment along a mountain front created by the coalescence of alluvial fans.

Barchan dune A solitary sand dune shaped like a crescent with its tips pointing downwind.

Barchanoid dune Dunes forming scalloped rows of sand oriented at right angles to the wind. This form is intermediate between isolated barchans and extensive waves of transverse dunes.

Barrier island A low, elongate ridge of sand that parallels the coast.

Basal slip A mechanism of glacial movement in which the ice mass slides over the surface below.

Basalt A fine-grained igneous rock of mafic composition.

Base level The level below which a stream cannot erode.

Basin A circular downfolded structure.

Batholith A large mass of igneous rock that formed when magma was emplaced at depth, crystallized, and was subsequently exposed by erosion.

Baymouth bar A sandbar that completely crosses a bay, sealing it off from the main body of water.

Beach drift The transport of sediment in a zigzag pattern along a beach caused by the uprush of water from obliquely breaking waves.

Beach nourishment Process in which large quantities of sand are added to the beach system to offset losses caused by wave erosion. Building beaches seaward improves beach quality and storm protection.

Bed See *strata.*

Bedding plane A nearly flat surface separating two beds of sedimentary rock. Each bedding plane marks the end of one deposit and the beginning of another having different characteristics.

Bed load Sediment moved along the bottom of a stream by moving water, or particles moved along the ground surface by wind.

Belt of soil moisture A zone in which water is held as a film on the surface of soil particles and may be used by plants or withdrawn by evaporation. The uppermost subdivision of the zone of aeration.

Benioff zone The zone of inclined seismic activity that extends from a trench downward into the asthenosphere.

Biochemical Describing a type of chemical sediment that forms when material dissolved in water is precipitated by water-dwelling organisms. Shells are common examples.

Biogenous sediment Sea-floor sediments consisting of material of marine-organic origin.

Biosphere The totality of life-forms on Earth. The parts of the lithosphere, hydrosphere, and atmosphere in which living organisms can be found.

Bituminous coal The most common form of coal, often called soft, black coal.

Blowout (deflation hollow) A depression excavated by wind in easily eroded materials.

Body wave A seismic wave that travels through the earth's interior.

Bottomset bed A layer of fine sediment deposited beyond the advancing edge of a delta and then buried by continued delta growth.

Bowen's reaction series A concept proposed by N. L. Bowen that illustrates the relationships between magma and the minerals crystallizing from it during the formation of igneous rocks.

Braided stream A stream consisting of numerous intertwining channels.

Breakwater A structure protecting a nearshore area from breaking waves.

Breccia A sedimentary rock composed of angular fragments that were lithified.

Cactolith A quasi-horizontal chonolith composed of anastomosing ductoliths, whose distal ends curl like a harpolith, thin like a sphenolith, or bulge discordantly like an akmolith or ethmolith.

Caldera A large depression typically caused by collapse or ejection of the summit area of a volcano.

Caliche A hard layer, rich in calcium carbonate, that forms beneath the *B* horizon in soils of arid regions.

Calving Wastage of a glacier that occurs when large pieces of ice break into the water.

Capacity The total amount of sediment a stream is able to transport.

Capillary fringe A relatively narrow zone at the base of the zone of aeration. Here water rises from the water table in tiny threadlike openings between grains of soil or sediment.

Cap rock A necessary part of an oil trap. The cap rock is impermeable and hence keeps upwardly mobile oil and gas from escaping at the surface.

Cassini gap A wide gap in the ring system of Saturn between the *A* ring and the *B* ring.

Catastrophism The concept that the earth was shaped by catastrophic events of a short-term nature.

Cavern A naturally formed underground chamber or series of chambers most commonly produced by solution activity in limestone.

Cementation One way in which sedimentary rocks are lithified. As material precipitates from water that percolates through the sediment, open spaces are filled and particles are joined into a solid mass.

Cenozoic era A time span on the geologic time scale beginning about 65 million years ago following the Mesozoic era.

Chemical sedimentary rock Sedimentary rock consisting of material that was precipitated from water by either inorganic or organic means.

Chemical weathering The processes by which the internal structure of a mineral is altered by the removal and/or addition of elements.

Cinder cone A rather small volcano built primarily of pyroclastics ejected from a single vent.

Cirque An amphitheater-shaped basin at the head of a glaciated valley produced by frost wedging and plucking.

Clastic A sedimentary rock texture consisting of broken fragments of pre-existing rock.

Cleavage The tendency of a mineral to break along planes of weak bonding.

Col A pass between mountain valleys where the headwalls of two cirques intersect.

Color A phenomenon of light by which otherwise identical objects may be differentiated.

Column A feature found in caves that is formed when a stalactite and stalagmite join.

Columnar joints A pattern of cracks that forms during cooling of molten rock to generate columns.

Coma The fuzzy, gaseous component of a comet's head.

Comet A small body that generally revolves about the sun in an elongated orbit.

Compaction A type of lithification in which the weight of overlying material compresses more deeply buried sediment. It is most important in fine-grained sedimentary rocks such as shale.

Competence A measure of the largest particle a stream can transport; a factor dependent on velocity.

Composite cone A volcano composed of both lava flows and pyroclastic material.

Compound A substance formed by the chemical combination of two or more elements in definite proportions and usually having properties different from those of its constituent elements.

Compressional stress A stress that pushes together material on either side of a real or imaginary plane.

Concordant A term used to describe intrusive igneous masses that form parallel to the bedding of the surrounding rock.

Cone of depression A cone-shaped depression in the water table immediately surrounding a well.

Confining pressure An equal, all-sided pressure.

Conformable layers Rock layers that were deposited without interruption.

Conglomerate A sedimentary rock composed of rounded gravel-sized particles.

Contact metamorphism Changes in rock caused by the heat from a nearby magma body.

Continental drift A hypothesis, credited largely to Alfred Wegener, that suggested all present continents once existed as a single supercontinent. Further, beginning about 200 million years ago, the supercontinent began breaking into smaller continents which then "drifted" to their present positions.

Continental margin That portion of the sea floor adjacent to the continents. It may include the continental shelf, continental slope, and continental rise.

Continental rise The gently sloping surface at the base of the continental slope.

Continental shelf The gently sloping submerged portion of the continental margin extending from the shoreline to the continental slope.

Continental slope The steep gradient that leads to the deep-ocean floor and marks the seaward edge of the continental shelf.

Convergent boundary A boundary in which two plates move together, causing one of the slabs of lithosphere to be consumed into the mantle as it descends beneath an overriding plate.

Correlation Establishing the equivalence of rocks of similar age in different areas.

Covalent bond A chemical bond produced by the sharing of electrons.

Crater The depression at the summit of a volcano, or that which is produced by a meteorite impact.

Creep The slow downhill movement of soil and regolith.

Crevasse A deep crack in the brittle surface of a glacier.

Cross-bedding Structure in which relatively thin layers are inclined at an angle to the main bedding. Formed by currents of wind or water.

Cross-cutting A principle of relative dating. A rock or fault is younger than any rock (or fault) through which it cuts.

Crust The very thin outermost layer of the earth.

Crystal An orderly arrangement of atoms.

Crystal form The external appearance of a mineral as determined by its internal arrangement of atoms.

Crystal settling During the crystallization of magma, the earlier-formed minerals are denser than the liquid portion and settle to the bottom of the magma chamber.

Crystallization The formation and growth of a crystalline solid from a liquid or gas.

Curie point The temperature above which a material loses its magnetization.

Cut bank The area of active erosion on the outside of a meander.

Cutoff A short channel segment created when a river erodes through the narrow neck of land between meanders.

Darcy's law When permeability is uniform, the velocity of groundwater increases as the slope of the water table increases. It is expressed by the formula: $V = K(h/l)$, where V is velocity, h the head, l the length of flow, and K the coefficient of permeability.

Daughter product An isotope resulting from radioactive decay.

Debris slide A slide involving downslope movement of relatively dry, unconsolidated regolith and rock debris. The mass does not exhibit backward rotation as in a slump but slides or rolls forward.

Deep-ocean basin The portion of sea floor that lies between the continental margin and the oceanic ridge system. This region comprises almost 30 percent of the earth's surface.

Deep-ocean trench A narrow, elongated depression of the sea floor.

Deflation The lifting and removal of loose material by wind.

Deformation General term for the processes of folding, faulting, shearing, compression, or extension of rocks as the result of various natural forces.

Delta An accumulation of sediment formed where a stream enters a lake or an ocean.

Dendritic pattern A stream system that resembles the pattern of a branching tree.

Density The weight per unit volume of a particular material.

Desalination The removal of salts and other chemicals from seawater.

Desert One of the two types of dry climate; the driest of the dry climates.

Desert pavement A layer of coarse pebbles and gravel created when wind removed the finer material.

Detrital sedimentary rocks Rocks that form from the accumulation of materials that originate and are transported as solid particles derived from both mechanical and chemical weathering.

Differential weathering The variation in the rate and degree of weathering caused by such factors as mineral makeup, degree of jointing, and climate.

Dike A tabular-shaped intrusive igneous feature that cuts through the surrounding rock.

Dip The angle at which a rock layer or fault is inclined from the horizontal. The direction of dip is at a right angle to the strike.

Dip-slip fault A fault in which the movement is parallel to the dip of the fault.

Discharge The quantity of water in a stream that passes a given point in a period of time.

Disconformity A type of unconformity in which the beds above and below are parallel.

Discontinuity A sudden change with depth in one or more of the physical properties of the material making up the earth's interior. The boundary between two dissimilar materials in the earth's interior as determined by the behavior of seismic waves.

Discordant A term used to describe plutons that cut across existing rock structures, such as bedding planes.

Disseminated deposit Any economic mineral deposit in which the desired mineral occurs as scattered particles in the rock but in sufficient quantity to make the deposit an ore.

Dissolved load That portion of a stream's load carried in solution.

Distributary A section of a stream that leaves the main flow.

Diurnal tide A tide characterized by a single high and low water height each tidal day.

Divergent boundary A boundary in which two plates move apart, resulting in upwelling of material from the mantle to create new sea floor.

Divide An imaginary line that separates the drainage of two streams; often found along a ridge.

Dome A roughly circular upfolded structure.

Drainage basin The land area that contributes water to a stream.

Drawdown The difference in height between the bottom of a cone of depression and the original height of the water table.

Drift The general term for any glacial deposit.

Drumlin A streamlined symmetrical hill composed of glacial till. The steep side of the hill faces the direction from which the ice advanced.

Dry climate A climate in which yearly precipitation is less than the potential loss of water by evaporation.

Dune A hill or ridge of wind-deposited sand.

Earthflow The downslope movement of water-saturated, clay-rich sediment. Most characteristic of humid regions.

Earthquake Vibration of the earth produced by the rapid release of energy.

Ebb current The movement of tidal current away from the shore.

Echo sounder An instrument used to determine the depth of water by measuring the time interval between emission of a sound signal and the return of its echo from the bottom.

Effluent stream A stream channel that intersects the water table. Consequently, groundwater feeds into the stream.

Elastic deformation Nonpermanent deformation in which rock returns to its original shape when the stress is released.

Elastic rebound The sudden release of stored strain in rocks that results in movement along a fault.

Electron A negatively charged subatomic particle that has a negligible mass and is found outside an atom's nucleus.

Element A substance that cannot be decomposed into simpler substances by ordinary chemical or physical means.

Eluviation The washing out of fine soil components from the *A* horizon by downward-percolating water.

Emergent coast A coast where land formerly below sea level has been exposed by crustal uplift or a drop in sea level or both.

End moraine A ridge of till marking a former position of the front of a glacier.

Energy-level shell The region occupied by electrons with a specific energy level.

Entrenched meander A meander cut into bedrock when uplifting rejuvenated a meandering stream.

Eon The largest time unit on the geologic time scale, next in order of magnitude above era.

Ephemeral stream A stream that is usually dry because it carries water only in response to specific episodes of rainfall. Most desert streams are of this type.

Epicenter The location on the Earth's surface that lies directly above the focus of an earthquake.

Epoch A unit of the geologic time scale that is a subdivision of a period.

Era A major division on the geologic time scale; eras are divided into shorter units called periods.

Erosion The incorporation and transportation of material by a mobile agent, such as water, wind, or ice.

Esker Sinuous ridge composed largely of sand and gravel deposited by a stream flowing in a tunnel beneath a glacier near its terminus.

Estuary A funnel-shaped inlet of the sea that formed when a rise in sea level or subsidence of land caused the mouth of a river to be flooded.

Eugeosyncline The portion of a geosyncline seaward of the miogeosyncline in which predominantly deep-water deposits accumulate, including graywackes, lava flows, volcanic debris, and shales.

Evaporite A sedimentary rock formed of material deposited from solution by evaporation of the water.

Evapotranspiration The combined effect of evaporation and transpiration.

Exfoliation dome Large, dome-shaped structure, usually composed of granite, formed by sheeting.

Exotic stream A permanent stream that traverses a desert and has its source in well-watered areas outside the desert.

Extrusive Igneous activity that occurs at the earth's surface.

Facies A portion of a rock unit that possesses a distinctive set of characteristics that distinguishes it from other parts of the same unit.

Fall A type of movement common to mass-wasting processes that refers to the free falling of detached individual pieces of any size.

Fault A break in a rock mass along which movement has occurred.

Fault-block mountain A mountain formed by the displacement of rock along a fault.

Fault scarp A cliff created by movement along a fault. It represents the exposed surface of the fault prior to modification by weathering and erosion.

Faunal succession Fossil organisms succeed one another in a definite and determinable order, and any time period can be recognized by its fossil content.

Fetch The distance that the wind has traveled across the open water.

Fiord A steep-sided inlet of the sea formed when a glacial trough was partially submerged.

Firn Granular recrystallized snow. A transitional stage between snow and glacial ice.

Fissility The property of splitting easily into thin layers along closely spaced, parallel surfaces, such as bedding planes in shale.

Fission (nuclear) The splitting of a heavy nucleus into two or more lighter nuclei caused by the collision with a neutron. During this process a large amount of energy is released.

Fissure eruption An eruption in which lava is extruded from narrow fractures or cracks in the crust.

Flood basalts Flows of basaltic lava that issue from numerous cracks or fissures and commonly cover extensive areas to thicknesses of hundreds of meters.

Floodplain The flat, low-lying portion of a stream valley subject to periodic inundation.

Flood current The tidal current associated with the increase in the height of the tide.

Flow A type of movement common to mass-wasting processes in which water-saturated material moves downslope as a viscous fluid.

Flowing artesian well An artesian well in which water flows freely at the earth's surface because the pressure surface is above ground level.

Fluorescence The absorption of ultraviolet light, which is re-emitted as visible light.

Focus (earthquake) The zone within the earth where rock displacement produces an earthquake.

Fold A bent layer or series of layers that were originally horizontal and subsequently deformed.

Foliated A texture of metamorphic rocks that gives the rock a layered appearance.

Foliation A term for a linear arrangement of textural features often exhibited by metamorphic rocks.

Foreset bed An inclined bed deposited along the front of a delta.

Foreshocks Small earthquakes that often precede a major earthquake.

Fossil The remains or traces of organisms preserved from the geologic past.

Fossil fuel General term for any hydrocarbon that may be used as a fuel, including coal, oil, natural gas, bitumen from tar sands, and shale oil.

Fracture Any break or rupture in rock along which no appreciable movement has taken place.

Frost wedging The mechanical breakup of rock caused by the expansion of freezing water in cracks and crevices.

Fumarole A vent in a volcanic area from which fumes or gases escape.

Geology The science that examines the earth, its form and composition, and the changes that it has undergone and is undergoing.

Geosyncline A large linear downwarp in the earth's crust in which thousands of meters of sediment have accumulated.

Geothermal energy Natural steam used for power generation.

Geothermal gradient The gradual increase in temperature with depth in the crust. The average is 30°C per kilometer in the upper crust.

Geyser A fountain of hot water ejected periodically from the ground.

Glacial erratic An ice-transported boulder that was not derived from the bedrock near its present site.

Glacial striations Scratches and grooves on bedrock caused by glacial abrasion.

Glacial trough A mountain valley that has been widened, deepened, and straightened by a glacier.

Glacier A thick mass of ice originating on land from the compaction and recrystallization of snow that shows evidence of past or present flow.

Glass (volcanic) Natural glass produced when molten lava cools too rapidly to permit recrystallization. Volcanic glass is a solid composed of unordered atoms.

Glassy A term used to describe the texture of certain igneous rocks, such as obsidian, that contain no crystals.

Gondwanaland The southern portion of Pangaea consisting of South America, Africa, Australia, India, and Antarctica.

Graben A valley formed by the downward displacement of a fault-bounded block.

Graded bed A sediment layer characterized by a decrease in sediment size from bottom to top.

Graded stream A stream that has the correct channel characteristics to maintain exactly the velocity required to transport the material supplied to it.

Gradient The slope of a stream; generally expressed as the vertical drop over a fixed distance.

Greenhouse effect Carbon dioxide and water vapor in a planet's atmosphere absorb and re-radiate infrared wavelengths, effectively trapping solar energy and raising the temperature.

Groin A short wall built at a right angle to the seashore to trap moving sand.

Groundmass The matrix of smaller crystals within an igneous rock that has porphyritic texture.

Ground moraine An undulating layer of till deposited as the ice front retreats.

Groundwater Water in the zone of saturation.

Guyot A submerged flat-topped seamount.

Hadean eon The first eon on the geologic time scale. The eon ending 3.8 billion years ago that preceded the Archean eon.

Half-life The time required for one-half of the atoms of a radioactive substance to decay.

Hanging valley A tributary valley that enters a glacial trough at a considerable height above the floor of the trough.

Hardness A mineral's resistance to scratching and abrasion.

Head The vertical distance between the recharge and discharge points of a water table. Also the source area or beginning of a valley.

Headward erosion The extension upslope of the head of a valley due to erosion.

Historical geology A major division of geology that deals with the origin of the earth and its development through time. Usually involves the study of fossils and their sequence in rock beds.

Hogback A narrow, sharp-crested ridge formed by the upturned edge of a steeply dipping bed of resistant rock.

Horizon A layer in a soil profile.

Horn A pyramid-like peak formed by glacial action in three or more cirques surrounding a mountain summit.

Horst An elongate, uplifted block of crust bounded by faults.

Hot spot A concentration of heat in the mantle capable of producing magma which, in turn, extrudes onto the earth's surface. The intraplate volcanism that produced the Hawaiian Islands is one example.

Hot spring A spring in which the water is 6–9°C (10–15°F) warmer than the mean annual air temperature of its locality.

Humus Organic matter in soil produced by the decomposition of plants and animals.

Hydraulic gradient The slope of the water table. Expressed as h/l, where h is the head and l the length of flow.

Hydroelectric power Electricity generated by falling water that is used to drive turbines.

Hydrogenous sediment Sea-floor sediment consisting of minerals that crystallize from seawater. An important example is manganese nodules.

Hydrologic cycle The unending circulation of the earth's water supply. The cycle is powered by energy from the sun and is characterized by continuous exchanges of water among the oceans, the atmosphere, and the continents.

Hydrolysis A chemical weathering process in which minerals are altered by chemically reacting with water and acids.

Hydrosphere The water portion of our planet; one of the traditional subdivisions of the earth's physical environment.

Hydrothermal solution The hot, watery solution that escapes from a mass of magma during the latter stages of crystallization. Such solutions may alter the surrounding country rock and are frequently the source of significant ore deposits.

Hypothesis A tentative explanation that is then tested to determine if it is valid.

Ice cap A mass of glacial ice covering a high upland or plateau and spreading out radially.

Ice-contact deposit An accumulation of stratified drift deposited in contact with a supporting mass of ice.

Ice sheet A very large, thick mass of glacial ice flowing outward in all directions from one or more accumulation centers.

Ice shelf Forming where glacial ice flows into bays, it is a large, relatively flat mass of floating ice that extends seaward from the coast but remains attached to the land along one or more sides.

Igneous rock Rock formed from the crystallization of magma.

Immature soil A soil lacking horizons.

Inclusion A piece of one rock unit contained within another. Inclusions are used in relative dating. The rock mass adjacent to the one containing the inclusion must have been there first in order to provide the fragment.

Index fossil A fossil that is associated with a particular span of geologic time.

Index mineral A mineral that is a good indicator of the metamorphic environment in which it formed. Used to distinguish different zones of regional metamorphism.

Inertia Objects at rest tend to remain at rest and objects in motion tend to stay in motion unless either is acted upon by an outside force.

Infiltration The movement of surface water into rock or soil through cracks and pore spaces.

Infiltration capacity The maximum rate at which soil can absorb water.

Influent stream A stream channel that is above the water table level. Water seeps downward from the channel to the zone of saturation to produce an upward bulge in the water table.

Inner core The solid innermost layer of the earth, about 1216 kilometers (754 miles) in radius.

Inselberg An isolated mountain remnant characteristic of the late stage of erosion in a mountainous arid region.

Intensity (earthquake) An indication of the destructive effects of an earthquake at a particular place. Intensity is affected by such factors as distance to the epicenter and the nature of the surface materials.

Interior drainage A discontinuous pattern of intermittent streams that do not flow to the ocean.

Intrusive rock Igneous rock that formed below the earth's surface.

Ion An atom or molecule that possesses an electrical charge.

Ionic bond A chemical bond between two oppositely charged ions formed by the transfer of valence electrons from one atom to the other.

Irons One of the three main categories of meteorites. This group is composed largely of iron with varying amounts of nickel (5–20 percent). Most meteorite finds are irons.

Island arc A chain of volcanic islands generally located a few hundred kilometers from a trench where active subduction of one oceanic slab beneath another is occurring.

Isostasy The concept that the earth's crust is "floating" in gravitational balance upon the material of the mantle.

Isostatic adjustment Compensation of the lithosphere when weight is added or removed. When weight is added, the lithosphere will respond by subsiding, and when weight is removed there will be uplift.

Isotopes Varieties of the same element that have different mass numbers; their nuclei contain the same number of protons but different numbers of neutrons.

Jetties A pair of structures extending into the ocean at the entrance to a harbor or river that are built for the purpose of protecting against storm waves and sediment deposition.

Joint A fracture in rock along which there has been no movement.

Jovian planet One of the Jupiter-like planets: Jupiter, Saturn, Uranus, and Neptune. These planets have relatively low densities.

Kame A steep-sided hill composed of sand and gravel originating when sediment collected in openings in stagnant glacial ice.

Kame terrace A narrow, terrace-like mass of stratified drift deposited between a glacier and an adjacent valley wall.

Karst A topography consisting of numerous depressions called sinkholes.

Kettle holes Depressions created when blocks of ice become lodged in glacial deposits and subsequently melt.

Klippe A remnant or outlier of a thrust sheet that was isolated by erosion.

Laccolith A massive igneous body intruded between preexisting strata.

Lag time The amount of time between a rainstorm and the occurrence of flooding.

Lahar Mudflows on the slopes of volcanoes that result when unstable layers of ash and debris become saturated and flow downslope, usually following stream channels.

Laminar flow The movement of water particles in straight-line paths that are parallel to the channel. The water particles move downstream without mixing.

Lateral moraine. A ridge of till along the sides of a valley glacier composed primarily of debris that fell to the glacier from the valley walls.

Laterite A red, highly leached soil type found in the tropics that is rich in oxides of iron and aluminum.

Laurasia The northern portion of Pangaea consisting of North America and Eurasia.

Lava Magma that reaches the earth's surface.

Lava dome A bulbous mass associated with an old-age volcano, produced when thick lava is slowly squeezed from the vent. Lava domes may act as plugs to deflect subsequent gaseous eruptions.

Law A formal statement of the regular manner in which a natural phenomenon occurs under given conditions; e.g., the "law of superposition."

Law of superposition In any undeformed sequence of sedimentary rocks, each bed is older than the one above and younger than the one below.

Leaching The depletion of soluble materials from the upper soil by downward-percolating water.

Liquefaction The transformation of a stable soil into a fluid that is often unable to support buildings or other structures.

Lithification The process, generally cementation and/or compaction, of converting sediments to solid rock.

Lithosphere The rigid outer layer of the earth, including the crust and upper mantle.

Local base level See *Temporary base level.*

Loess Deposits of windblown silt, lacking visible layers, generally buff-colored, and capable of maintaining a nearly vertical cliff.

Longitudinal dunes Long ridges of sand oriented parallel to the prevailing wind; these dunes form where sand supplies are limited.

Longitudinal profile A cross section of a stream channel along its descending course from the head to the mouth.

Longshore current A nearshore current that flows parallel to the shore.

Long (L) waves These earthquake-generated waves travel along the outer layer of the earth and are responsible for most of the surface damage. L waves have longer periods than other seismic waves.

Low-velocity zone A subdivision of the mantle located between 100 and 250 kilometers and discernible by a marked decrease in the velocity of seismic waves. This zone does not encircle the earth.

Lunar breccia A lunar rock formed when angular fragments and dust are welded together by the heat generated by the impact of a meteoroid.

Lunar regolith A thin, gray layer on the surface of the moon, consisting of loosely compacted, fragmented material believed to have been formed by repeated meteoritic impacts.

Luster The appearance or quality of light reflected from the surface of a mineral.

Magma A body of molten rock found at depth, including any dissolved gases and crystals.

Magma mixing The process of altering the composition of a magma through the mixing of material from another magma body.

Magmatic differentiation The process of generating more than one rock type from a single magma.

Magnetometer A sensitive instrument used to measure the intensity of the earth's magnetic field at various points.

Magnitude (earthquake) The total amount of energy released during an earthquake.

Manganese nodules A type of hydrogenous sediment scattered on the ocean floor, consisting mainly of manganese and iron and usually containing small amounts of copper, nickel, and cobalt.

Mantle The 2885-kilometer (1789-mile) thick layer of the earth located below the crust.

Maria The smooth areas on our moon's surface that were incorrectly thought to be seas.

Massive An igneous pluton that is not tabular in shape.

Mass number The sum of the number of neutrons and protons in the nucleus of an atom.

Mass wasting The downslope movement of rock, regolith, and soil under the direct influence of gravity.

Meander A looplike bend in the course of a stream.

Meander scar A floodplain feature created when an oxbow lake becomes filled with sediment.

Mechanical weathering The physical disintegration of rock, resulting in smaller fragments.

Medial moraine A ridge of till formed when lateral moraines from two coalescing alpine glaciers join.

Melt The liquid portion of magma excluding the solid crystals.

Mercalli intensity scale A 12-point scale originally developed to evaluate earthquake intensity based upon the amount of damage to various structures.

Mesozoic era A time span on the geologic time scale between the Paleozoic and Cenozoic eras—from about 225 to 65 million years ago.

Metallic bond A chemical bond present in all metals that may be characterized as an extreme type of electron sharing in which the electrons move freely from atom to atom.

Metamorphic rock Rock formed by the alteration of preexisting rock deep within the earth (but still in the solid state) by heat, pressure, and/or chemically active fluids.

Metamorphism The changes in mineral composition and texture of a rock subjected to high temperatures and pressures within the earth.

Meteorite Any portion of a meteoroid that survives its traverse through the earth's atmosphere and strikes the surface.

Meteoroid Any small solid particle that has an orbit in the solar system.

Meteor shower Numerous meteoroids traveling in the same direction and at nearly the same speed. They are thought to be material lost by comets.

Micrometeorite A very small meteorite that does not create sufficient friction to burn up in the atmosphere, but slowly drifts down to the earth.

Mid-ocean ridge A continuous mountainous ridge on the floor of all the major ocean basins and varying in width from 500 to 5000 kilometers (300 to 3000 miles). The rifts at the crests of these ridges represent divergent plate boundaries.

Migmatite A rock exhibiting both igneous and metamorphic rock characteristics. Such rocks may form when light-colored silicate minerals melt and then crystallize, while the dark silicate minerals remain solid.

Mineral A naturally occurring, inorganic crystalline material with a unique chemical structure.

Mineral resource All discovered and undiscovered deposits of a useful mineral that can be extracted now or at some time in the future.

Miogeosyncline The portion of a geosyncline landward of the eugeosyncline and characterized by deposits of clean sandstones, limestones, and shales.

Mohorovičić discontinuity (Moho) The boundary separating the crust and the mantle, discernible by an increase in seismic velocity.

Mohs scale A series of ten minerals used as a standard in determining hardness.

Monocline A one-limbed flexure in strata. The strata are usually flat lying or very gently dipping on both sides of the monocline.

Mouth The point downstream where a river empties into another stream or water body.

Mud crack A feature in some sedimentary rocks that forms when wet mud dries out, shrinks, and cracks.

Mudflow The flowage of debris containing a large amount of water; most characteristic of canyons and gullies in dry, mountainous regions.

Natural levees The elevated landforms composed of alluvium that parallel some streams and act to confine their waters, except during floodstage.

Neap tide The lowest tidal range, occurring near the times of the first and third quarters of the moon.

Nebular hypothesis A model for the origin of the solar system that supposes a rotating nebula of dust and gases that contracted to form the sun and planets.

Neutron A subatomic particle found in the nucleus of an atom. The neutron is electrically neutral with a mass approximately equal to that of a proton.

Nonclastic A term for the texture of sedimentary rocks in which the minerals form a pattern of interlocking crystals.

Nonconformity An unconformity in which older metamorphic or intrusive igneous rocks are overlain by younger sedimentary strata.

Nonflowing artesian well An artesian well in which water does not rise to the surface because the pressure surface is below ground level.

Nonfoliated Metamorphic rocks that do not exhibit foliation.

Nonmetallic mineral resource Mineral resource that is not a fuel or processed for the metals it contains.

Nonrenewable resource Resource that forms or accumulates over such long time spans that it must be considered as fixed in total quantity.

Normal fault A fault in which the rock above the fault plane has moved down relative to the rock below.

Normal polarity A magnetic field the same as that which presently exists.

Nuclear fission The splitting of atomic nuclei into smaller nuclei, causing neutrons to be emitted and heat energy to be released.

Nucleus The small, heavy core of an atom that contains all of its positive charge and most of its mass.

Nuée ardente Incandescent volcanic debris buoyed up by hot gases that moves downslope in an avalanche fashion.

Oblique-slip fault A fault having both vertical and horizontal movement.

Occultation The disappearance of light resulting when one object passes behind an apparently larger one. For example, the passage of Uranus in front of a distant star.

Oceanic ridge system See *Mid-ocean ridge.*

Octet rule Atoms combine in order that each may have the electron arrangement of a noble gas; that is, the outer energy level contains eight neutrons.

Oil trap A geologic structure that allows for significant amounts of oil and gas to accumulate.

Ophiolite complex The sequence of rocks that make up the oceanic crust. The three-layer sequence includes an upper layer of pillow basalts, a middle zone of sheeted dikes, and a lower layer of gabbro.

Ore Usually a useful metallic mineral that can be mined at a profit. The term is also applied to certain nonmetallic minerals such as fluorite and sulfur.

Original horizontality Layers of sediment are generally deposited in a horizontal or nearly horizontal position.

Orogenesis The processes that collectively result in the formation of mountains.

Outer core A layer beneath the mantle about 2270 kilometers (1410 miles) thick which has the properties of a liquid.

Outlet glacier A tongue of ice normally flowing rapidly outward from an ice cap or ice sheet, usually through mountainous terrain to the sea.

Outwash plain A relatively flat, gently sloping plain consisting of materials deposited by meltwater streams in front of the margin of an ice sheet.

Oxbow lake A curved lake produced when a stream cuts off a meander.

Oxidation The removal of one or more electrons from an atom or ion. So named because elements commonly combine with oxygen.

Pahoehoe A lava flow with a smooth-to-ropy surface.

Paleomagnetism The natural remnant magnetism in rock bodies. The permanent magnetization acquired by rock which can be used to determine the location of the magnetic poles and the latitude of the rock at the time it became magnetized.

Paleontology The systematic study of fossils and the history of life on the earth.

Paleozoic era A time span on the geologic time scale between the Precambrian and Mesozoic eras—from about 600 million to 225 million years ago.

Pangaea The proposed supercontinent which 200 million years ago began to break apart and form the present landmasses.

Parabolic dune A sand dune similar in shape to a barchan dune except that its tips point into the wind. These dunes often form along coasts that have strong onshore winds, abundant sand, and vegetation that partly covers the sand.

Parasitic cone A volcanic cone that forms on the flank of a larger volcano.

Parent material The material upon which a soil develops.

Partial melting The process by which most igneous rocks melt. Since individual minerals have different melting points, most igneous rocks melt over a temperature range of a few hundred degrees. If the liquid is squeezed out after some melting has occurred, a melt with a higher silica content results.

Passive margin An inactive continental margin that is characterized by a thick accumulation of undeformed sediments and sedimentary rocks.

Pater noster lakes A chain of small lakes in a glacial trough that occupies basins created by glacial erosion.

Pedalfer Soil of humid regions characterized by the accumulation of iron oxides and aluminum-rich clays in the *B* horizon.

Pediment A sloping bedrock surface fringing a mountain base in an arid region, formed when erosion causes the mountain front to retreat.

Pedocal Soil associated with drier regions and characterized by an accumulation of calcium carbonate in the upper horizons.

Pegmatite A very coarse-grained igneous rock (typically granite) commonly found as a dike associated with a large mass of plutonic rock that has smaller crystals. Crystallization in a water-rich environment is believed to be responsible for the very large crystals.

Peneplain In the idealized cycle of landscape evolution in a humid region, an undulating plain near base level associated with old age.

Perched water table A localized zone of saturation above the main water table created by an impermeable layer (aquiclude).

Peridotite An igneous rock of ultramafic composition thought to be abundant in the upper mantle.

Period A basic unit of the geologic time scale that is a subdivision of an era. Periods may be divided into smaller units called epochs.

Permafrost Any permanently frozen subsoil. Usually found in the subarctic and arctic regions.

Permeability A measure of a material's ability to transmit water.

Phaneritic An igneous rock texture in which the crystals are roughly equal in size and large enough so the individual minerals can be identified with the unaided eye.

Phanerozoic eon That part of geologic time represented by rocks containing abundant fossil evidence. The eon extending from the end of the Proterozoic eon (570 million years ago) to the present.

Phenocryst Conspicuously large crystal embedded in a matrix of finer-grained crystals.

Physical geology A major division of geology that examines the materials of the earth and seeks to understand the processes and forces acting beneath and upon the earth's surface.

Piedmont glacier A glacier that forms when one or more alpine glaciers emerge from the confining walls of mountain valleys and spread out to create a broad sheet in the lowlands at the base of the mountains.

Pillow lava Basaltic lava that solidifies in an underwater environment and develops a structure that resembles a pile of pillows.

Pipe A vertical conduit through which magmatic materials have passed.

Placer Deposit formed when heavy minerals are mechanically concentrated by currents, most commonly streams and waves. Placers are sources of gold, tin, platinum, diamonds, and other valuable minerals.

Plastic deformation Permanent deformation that results in a change in size and shape through folding or flowing.

Plastic flow A type of glacial movement that occurs within the glacier, below a depth of approximately 50 meters, in which the ice is not fractured.

Plate One of numerous rigid sections of the lithosphere that moves as a unit over the material of the asthenosphere.

Plate tectonics The theory which proposes that the earth's outer shell consists of individual plates which interact in various ways and thereby produce earthquakes, volcanoes, mountains, and the crust itself.

Playa The flat central area of an undrained desert basin.

Playa lake A temporary lake in a playa.

Playfair's law A well-known and oft-quoted statement by John Playfair which states that a valley is the result of the work of the stream that flows in it.

Pleistocene epoch An epoch of the Quaternary period beginning about 2.5 million years ago and ending about 10,000 years ago. Best known as a time of extensive continental glaciation.

Plucking (quarrying) The process by which pieces of bedrock are lifted out of place by a glacier.

Pluton A structure that results from the emplacement and crystallization of magma beneath the surface of the earth.

Pluvial lake A lake formed during a period of increased rainfall. For example, this occurred in many non-glaciated areas during periods of ice advance elsewhere.

Point bar A crescent-shaped accumulation of sand and gravel deposited on the inside of a meander.

Polymorphs Two or more minerals having the same chemical composition but different crystalline structures. Exemplified by the diamond and graphite forms of carbon.

Porosity The volume of open spaces in rock or soil.

Porphyritic An igneous rock texture characterized by two distinctively different crystal sizes. The larger crystals are called phenocrysts, whereas the matrix of smaller crystals is termed the groundmass.

Porphyry An igneous rock with a porphyritic texture.

Pothole A depression formed in a stream channel by the abrasive action of the water's sediment load.

Precambrian All geologic time prior to the Paleozoic era.

Primary (P) wave A type of seismic wave that involves alternating compression and expansion of the material through which it passes.

Principle of faunal succession Fossil organisms succeed one another in a definite and determinable order, and any time period can be recognized by its fossil content.

Principle of original horizontality Layers of sediment are generally deposited in a horizontal or nearly horizontal position.

Proterozoic eon The eon following the Archean and preceding the Phanerozoic. It extends between 2500 and 570 million years ago.

Proton A positively charged subatomic particle found in the nucleus of an atom.

P wave The fastest earthquake wave, which travels by compression and expansion of the medium.

Pyroclastic An igneous rock texture resulting from the consolidation of individual rock fragments that are ejected during a violent eruption.

Pyroclastic flow A highly heated mixture, largely of ash and pumice fragments, traveling down the flanks of a volcano or along the surface of the ground.

Pyroclastic material The volcanic rock ejected during an eruption. Pyroclastics include ash, bombs, and blocks.

Radial drainage A system of streams running in all directions away from a central elevated structure, such as a volcano.

Radioactivity The spontaneous decay of certain unstable atomic nuclei.

Radiocarbon (carbon-14) The radioactive isotope of carbon, which is produced continuously in the atmosphere and used in dating events as far back as 75,000 years.

Radiometric dating The procedure of calculating the absolute ages of rocks and minerals that contain certain radioactive isotopes.

Rainshadow desert A dry area on the lee side of a mountain range. Many middle-latitude deserts are of this type.

Rapids A part of a stream channel in which the water suddenly begins flowing more swiftly and turbulently because of an abrupt steepening of the gradient.

Rays Bright streaks that appear to radiate from certain craters on the lunar surface. The rays consist of fine debris ejected from the primary crater.

Recessional moraine An end moraine formed as the ice front stagnated during glacial retreat.

Rectangular pattern A drainage pattern characterized by numerous right angle bends that develops on jointed or fractured bedrock.

Refraction A change in direction of waves as they enter shallow water. The portion of the wave in shallow water is slowed, which causes the wave to bend and align with the underwater contours.

Regional metamorphism Metamorphism associated with large-scale mountain building.

Regolith The layer of rock and mineral fragments that nearly everywhere covers the earth's land surface.

Rejuvenation A change in relation to base level, often caused by regional uplift, that causes the forces of erosion to intensify.

Relative dating Rocks and structures are placed in their proper sequence or order. Only the chronological order of events is determined.

Renewable resource A resource that is virtually inexhaustible or that can be replenished over relatively short time spans.

Reserve Already identified deposits from which minerals can be extracted profitably.

Reservoir rock The porous, permeable portion of an oil trap that yields oil and gas.

Residual soil Soil developed directly from the weathering of the bedrock below.

Reverse fault A fault in which the material above the fault plane moves up in relation to the material below.

Reverse polarity A magnetic field opposite to that which presently exists.

Richter scale A scale of earthquake magnitude based on the motion of a seismograph.

Rift A region of the earth's crust along which divergence is taking place.

Rift zone See *Rift.*

Rills Tiny channels that develop as unconfined flow begins producing threads of current.

Ripple marks Small waves of sand that develop on the surface of a sediment layer by the action of moving water or air.

Roche moutonnée An asymmetrical knob of bedrock formed when glacial abrasion smoothes the gentle slope facing the advancing ice sheet and plucking steepens the opposite side as the ice overrides the knob.

Rock A consolidated mixture of minerals.

Rock avalanche The very rapid downslope movement of rock and debris. These rapid movements may be aided by a layer of air trapped beneath the debris, and they have been known to reach speeds in excess of 200 kilometers per hour.

Rock cleavage The tendency of rock to split along parallel, closely spaced surfaces. These surfaces are often highly inclined to the bedding planes in the rock.

Rock cycle A model that illustrates the origin of the three basic rock types and the interrelatedness of earth materials and processes.

Rock flour Ground-up rock produced by the grinding effect of a glacier.

Rockslide The rapid slide of a mass of rock downslope along planes of weakness.

Runoff Water that flows over the land rather than infiltrating into the ground.

Salinity The proportion of dissolved salts to pure water, usually expressed in parts per thousand (‰).

Saltation Transportation of sediment through a series of leaps or bounces.

Salt flat A white crust on the ground produced when water evaporates and leaves its dissolved materials behind.

Schistosity A type of foliation characteristic of coarser-grained metamorphic rocks. Such rocks have a parallel arrangement of platy minerals such as the micas.

Scoria Hardened lava that has retained the vesicles produced by the escaping gases.

Sea arch An arch formed by wave erosion when caves on opposite sides of a headland unite.

Sea-floor spreading The hypothesis first proposed in the 1960s by Harry Hess which suggested that new oceanic crust is produced at the crests of mid-ocean ridges, which are the sites of divergence.

Seamount An isolated volcanic peak that rises at least 1000 meters (3300 feet) above the deep-ocean floor.

Sea stack An isolated mass of rock standing just offshore, produced by wave erosion of a headland.

Seawall A barrier constructed to prevent waves from reaching the area behind the wall. Its purpose is to defend property from the force of breaking waves.

Secondary enrichment The concentration of minor amounts of metals that are scattered through unweathered rock into economically valuable concentrations by weathering processes.

Secondary (S) wave A seismic wave that involves oscillation perpendicular to the direction of propagation.

Sediment Unconsolidated particles created by the weathering and erosion of rock, by chemical precipitation from solution in water, or from the secretions of organisms, and transported by water, wind, or glaciers.

Sedimentary rock Rock formed from the weathered products of pre-existing rocks that have been transported, deposited, and lithified.

Seismic sea wave A rapidly moving ocean wave generated by earthquake activity, which is capable of inflicting heavy damage in coastal regions.

Seismogram The record made by a seismograph.

Seismograph An instrument that records earthquake waves.

Seismology The study of earthquakes and seismic waves.

Settling velocity The speed at which a particle falls through a still fluid. The size, shape, and specific gravity of particles influence settling velocity.

Shadow zone The zone between 105 and 140 degrees distance from an earthquake epicenter which direct waves do not penetrate because of refraction by the earth's core.

Shear Stress that causes two adjacent parts of a body to slide past one another.

Sheeted dikes A large group of nearly parallel dikes.

Sheet flow Runoff moving in unconfined thin sheets.

Sheeting A mechanical weathering process characterized by the splitting off of slablike sheets of rock.

Shelf break The point at which a rapid steepening of the gradient occurs, marking the outer edge of the continental shelf and the beginning of the continental slope.

Shield A large, relatively flat expanse of ancient metamorphic rock within the stable continental interior.

Shield volcano A broad, gently sloping volcano built from fluid basaltic lavas.

Silicate Any one of numerous minerals that have the silicon-oxygen tetrahedron as their basic structure.

Silicon-oxygen tetrahedron A structure composed of four oxygen atoms surrounding a silicon atom that constitutes the basic building block of silicate minerals.

Sill A tabular igneous body that was intruded parallel to the layering of pre-existing rock.

Sinkhole A depression produced in a region where soluble rock has been removed by groundwater.

Slaty cleavage The type of foliation characteristic of slates in which there is a parallel arrangement of fine-grained metamorphic minerals.

Slide A movement common to mass-wasting processes in which the material moving downslope remains fairly coherent and moves along a well-defined surface.

Slip face The steep, leeward surface of a sand dune which maintains a slope of about 34 degrees.

Slump The downward slipping of a mass of rock or unconsolidated material moving as a unit along a curved surface.

Snowfield An area where snow persists throughout the year.

Snowline The lower limit of perennial snow.

Soil A combination of mineral and organic matter, water, and air; that portion of the regolith that supports plant growth.

Soil horizon A layer of soil that has identifiable characteristics produced by chemical weathering and other soil-forming processes.

Soil profile A vertical section through a soil showing its succession of horizons and the underlying parent material.

Solifluction Slow, downslope flow of water-saturated materials common to permafrost areas.

Solum The *O*, *A*, and *B* horizons in a soil profile. Living roots and other plant and animal life are largely confined to this zone.

Solution The change of matter from the solid or gaseous state into the liquid state by its combination with a liquid.

Sorting The degree of similarity in particle size in sediment or sedimentary rock.

Specific gravity The ratio of a substance's weight to the weight of an equal volume of water.

Speleothem A collective term for the dripstone features found in caverns.

Spheroidal weathering Any weathering process that tends to produce a spherical shape from an initially blocky shape.

Spit An elongate ridge of sand that projects from the land into the mouth of an adjacent bay.

Spring A flow of groundwater that emerges naturally at the ground surface.

Spring tide The highest tidal range. Occurs near the times of the new and full moons.

Stalactite The iciclelike structure that hangs from the ceiling of a cavern.

Stalagmite The columnlike form that grows upward from the floor of a cavern.

Star dune An isolated hill of sand that exhibits a complex form and develops where wind directions are variable.

Steppe One of the two types of dry climate. A marginal and more humid variant of the desert that separates it from bordering humid climates.

Stock A pluton similar to but smaller than a batholith.

Stony-irons One of the three main categories of meteorites. This group, as the name implies, is a mixture of iron and silicate minerals.

Stony meteorite One of the three main categories of meteorites. Such meteorites are composed largely of silicate minerals with inclusions of other minerals.

Strata Parallel layers of sedimentary rock.

Stratified drift Sediments deposited by glacial meltwater.

Stratovolcano See *Composite cone.*

Streak The color of a mineral in powdered form.

Stream A general term to denote the flow of water within any natural channel. Thus, a small creek and a large river are both streams.

Stream piracy The diversion of the drainage of one stream resulting from the headward erosion of another stream.

Stress The force per unit area acting on any surface within a solid. Also known as *directed pressure.*

Striations (glacial) Scratches or grooves in a bedrock surface caused by the grinding action of a glacier and its load of sediment.

Strike The compass direction of the line of intersection created by a dipping bed or fault and a horizontal surface. Strike is always perpendicular to the direction of dip.

Strike-slip fault A fault along which the movement is horizontal.

Subduction The process by which oceanic lithosphere plunges into the mantle along a convergent zone.

Subduction zone A long, narrow zone where one lithospheric plate descends beneath another.

Submarine canyon A seaward extension of a valley that was cut on the continental shelf during a time when sea level was lower, or a canyon carved into the outer continental shelf, slope, and rise by turbidity currents.

Submergent coast A coast whose form is largely the result of the partial drowning of a former land surface due to a rise of sea level or subsidence of the crust, or both.

Subsoil A term applied to the *B* horizon of a soil profile.

Superposed stream A stream that cuts through a ridge lying across its path. The stream established its course on uniform layers at a higher level without regard to underlying structures and subsequently downcut.

Superposition, law of In any undeformed sequence of sedimentary rocks, each bed is lower than the one above and younger than the one below.

Surf A collective term for breakers; also the wave activity in the area between the shoreline and the outer limit of breakers.

Surface waves Seismic waves that travel along the outer layer of the earth.

Surge A period of rapid glacial advance. Surges are typically sporadic and short-lived.

Suspended load The fine sediment carried within the body of flowing water or air.

S wave An earthquake wave, slower than a P wave, that travels only in solids.

Swells Wind-generated waves that have moved into an area of weaker winds or calm.

Syncline A linear downfold in sedimentary strata; the opposite of anticline.

Tabular Describing a feature such as an igneous pluton having two dimensions that are much longer than the third.

Talus An accumulation of rock debris at the base of a cliff.

Tarn A small lake in a cirque.

Tectonics The study of the large-scale processes that collectively deform the earth's crust.

Temporary (local) base level The level of a lake, resistant rock layer, or any other base level that stands above sea level.

Tensional stress The type of stress that tends to pull a body apart.

Terminal moraine The end moraine marking the farthest advance of a glacier.

Terrace A flat, benchlike structure produced by a stream, which was left elevated as the stream cut downward.

Terrane A crustal block bounded by faults, whose geologic history is distinct from the histories of adjoining crustal blocks.

Terrestrial planet One of the Earthlike planets: Mercury, Venus, Earth, and Mars. These planets have similar densities.

Terrigenous sediment Sea-floor sediments derived from terrestrial weathering and erosion.

Texture The size, shape, and distribution of the particles that collectively constitute a rock.

Theory A well-tested and widely accepted view that explains certain observable facts.

Thrust fault A low-angle reverse fault.

Tidal current The alternating horizontal movement of water associated with the rise and fall of the tide.

Tidal delta A deltalike feature created when a rapidly moving tidal current emerges from a narrow inlet and slows, depositing its load of sediment.

Tidal flat A marshy or muddy area that is alternately covered and uncovered by the rise and fall of the tide.

Tide Periodic change in the elevation of the ocean surface.

Till Unsorted sediment deposited directly by a glacier.

Tillite A rock formed when glacial till is lithified.

Tombolo A ridge of sand that connects an island to the mainland or to another island.

Topset bed An essentially horizontal sedimentary layer deposited on top of a delta during floodstage.

Transform boundary A boundary in which two plates slide past one another without creating or destroying lithosphere.

Transpiration The release of water vapor to the atmosphere by plants.

Transported soil Soils that form on unconsolidated deposits.

Transverse dunes A series of long ridges oriented at right angles to the prevailing wind; these dunes form where vegetation is sparse and sand is very plentiful.

Travertine A form of limestone ($CaCO_3$) that is deposited by hot springs or as a cave deposit.

Trellis drainage A system of streams in which nearly parallel tributaries occupy valleys cut in folded strata.

Trench An elongate depression in the sea floor produced by bending of oceanic crust during subduction.

Truncated spurs Triangular-shaped cliffs produced when spurs of land that extend into a valley are removed by the great erosional force of a valley glacier.

Tsunami The Japanese word for a seismic sea wave.

Turbidite Turbidity current deposit characterized by graded bedding.

Turbidity current A downslope movement of dense, sediment-laden water created when sand and mud on the continental shelf and slope are dislodged and thrown into suspension.

Turbulent flow The movement of water in an erratic fashion often characterized by swirling, whirlpool-like eddies. Most streamflow is of this type.

Ultimate base level Sea level; the lowest level to which stream erosion could lower the land.

Unconformity A surface that represents a break in the rock record, caused by erosion and nondeposition.

Uniformitarianism The concept that the processes that have shaped the earth in the geologic past are essentially the same as those operating today.

Valence electron The electrons involved in the bonding process; the electrons occupying the highest principal energy level of an atom.

Valley glacier See *Alpine glacier.*

Valley train A relatively narrow body of stratified drift deposited on a valley floor by meltwater streams that issue from the terminus of an alpine glacier.

Vein deposit A mineral filling a fracture or fault in a host rock. Such deposits have a sheetlike, or tabular, form.

Vent A pipelike conduit that connects a magma chamber to a volcanic crater.

Ventifact A cobble or pebble polished and shaped by the sandblasting effect of wind.

Vesicles Spherical or elongated openings on the outer portion of a lava flow that were created by escaping gases.

Vesicular A term applied to igneous rocks that contain small cavities called vesicles, which are formed when gases escape from lava.

Viscosity A measure of a fluid's resistance to flow.

Volcanic Pertaining to the activities, structures, or rock types of a volcano.

Volcanic arc Mountains formed in part by igneous activity associated with the subduction of oceanic lithosphere beneath a continent. Examples include the Andes and the Cascades.

Volcanic bomb A streamlined pyroclastic fragment ejected from a volcano while still semimolten.

Volcanic neck An isolated, steep-sided, erosional remnant consisting of lava that once occupied the vent of a volcano.

Volcano A mountain formed from lava and/or pyroclastics.

Waterfall A precipitous drop in a stream channel that causes water to fall to a lower level.

Water gap A pass through a ridge or mountain in which a stream flows.

Water table The upper level of the saturated zone of groundwater.

Wave-cut cliff A seaward-facing cliff along a steep shoreline formed by wave erosion at its base and mass wasting.

Wave-cut platform A bench or shelf along a shore at sea level, cut by wave erosion.

Wave height The vertical distance between the trough and crest of a wave.

Wave length The horizontal distance separating successive crests or troughs.

Wave of oscillation A water wave in which the wave form advances as the water particles move in circular orbits.

Wave of translation The turbulent advance of water created by breaking waves.

Wave period The time interval between the passage of successive crests at a stationary point.

Weathering The disintegration and decomposition of rock at or near the surface of the earth.

Welded tuff A pyroclastic deposit composed of particles fused together by the combination of heat still contained in the deposit after it has come to rest and the weight of overlying material.

Well An opening bored into the zone of saturation.

Wilson cycle The complex cycle of ocean basin openings and closings named in honor of J. Tuzo Wilson, the Canadian geologist who proposed their existence.

Wind gap An abandoned water gap. These gorges typically result from stream piracy.

Xenolith An inclusion of unmelted country rock in an igneous pluton.

Xerophyte A plant highly tolerant of drought.

Yazoo tributary A tributary that flows parallel to the main stream because a natural levee is present.

Zone of accumulation The part of a glacier characterized by snow accumulation and ice formation. The outer limit of this zone is the snowline.

Zone of aeration The area above the water table where openings in soil, sediment, and rock are not saturated but filled mainly with air.

Zone of fracture The upper portion of a glacier consisting of brittle ice.

Zone of saturation The zone where all open spaces in sediment and rock are completely filled with water.

Index